Methods in Enzymology

Volume 224
MOLECULAR EVOLUTION: PRODUCING THE
BIOCHEMICAL DATA

METHODS IN ENZYMOLOGY

EDITORS-IN-CHIEF

John N. Abelson   Melvin I. Simon

DIVISION OF BIOLOGY
CALIFORNIA INSTITUTE OF TECHNOLOGY
PASADENA, CALIFORNIA

FOUNDING EDITORS

Sidney P. Colowick and Nathan O. Kaplan

Methods in Enzymology

Volume 224

Molecular Evolution: Producing the Biochemical Data

EDITED BY

Elizabeth A. Zimmer

LABORATORY OF MOLECULAR SYSTEMATICS
NATIONAL MUSEUM OF NATURAL HISTORY
SMITHSONIAN INSTITUTION
WASHINGTON, D.C.

Thomas J. White

ROCHE DIAGNOSTIC RESEARCH
ALAMEDA, CALIFORNIA

Rebecca L. Cann

DEPARTMENT OF GENETICS AND MOLECULAR BIOLOGY
UNIVERSITY OF HAWAII AT MANOA
HONOLULU, HAWAII

Allan C. Wilson

DIVISION OF BIOCHEMISTRY AND MOLECULAR BIOLOGY
UNIVERSITY OF CALIFORNIA
BERKELEY, CALIFORNIA

ACADEMIC PRESS, INC.

A Division of Harcourt Brace & Company

San Diego   New York   Boston   London   Sydney   Tokyo   Toronto

Copyright © 1993 by ACADEMIC PRESS, INC.

All Rights Reserved.

No part of this publication may be reproduced or transmitted in any form or by any means, electronic or mechanical, including photocopy, recording, or any information storage and retrieval system, without permission in writing from the publisher.

Academic Press, Inc.
1250 Sixth Avenue, San Diego, California 92101-4311

United Kingdom Edition published by
Academic Press Limited
24–28 Oval Road, London NW1 7DX

International Standard Serial Number: 0076-6879

International Standard Book Number: 0-12-182125-0

PRINTED IN THE UNITED STATES OF AMERICA
93 94 95 96 97 98   EB   9 8 7 6 5 4 3 2 1

Table of Contents

Section III. Comparing Macromolecules: Exploring Evolutionary Pattern and Process

Contributors to Volume 224

Article numbers are in parentheses following the names of contributors.
Affiliations listed are current.

DARRILYN G. ALBRIGHT (21), *Laboratory of Molecular Systematics, National Museum of Natural History, Smithsonian Institution, Washington, D.C. 20560*

MARC W. ALLARD (34), *Human Genome Center, University of Michigan, Ann Arbor, Michigan 48109*

JEAN-PIERRE BACHELLERIE (25), *Laboratoire de Biologie Moléculaire Eucaryote, Institut de Biologie Cellulaire du CNRS, 31062 Toulouse, France*

MARK A. BATZER (16), *Human Genome Center, Biology and Biotechnology Research Program, Lawrence Livermore National Laboratory, Livermore, California 94551*

GREGORY C. BEAULIEU (18), *Department of Biology, Western Washington University, Bellingham, Washington 98225*

ARNOLD J. BENDICH (18), *Department of Botany, University of Washington, Seattle, Washington 98195*

MEREDITH BLACKWELL (5), *Department of Botany, Louisiana State University, Baton Rouge, Louisiana 70803*

JUDITH A. BLAKE (1), *Laboratory of Molecular Systematics, National Museum of Natural History, Smithsonian Institution, Washington, D.C. 20560*

BRUNELLA MARTIRE BOWDITCH (21), *Laboratory of Molecular Systematics, National Museum of Natural History, Smithsonian Institution, Washington, D.C. 20560*

TIMOTHY BOWEN (38), *Department of Genetics, University of Leicester, Leicester LE1 7RH, United Kingdom*

BARBARA H. BOWMAN (29), *Roche Molecular Systems, Inc., Alameda, California 94501*

MICHAEL J. BRAUN (1, 21), *Laboratory of Molecular Systematics, National Museum of Natural History, Smithsonian Institution, Washington, D.C. 20560*

DIANE M. BRIDGE (4), *Department of Biology, Yale University, New Haven, Connecticut 06511*

ROY J. BRITTEN (17), *Kerckhoff Marine Laboratory, California Institute of Technology, Corona del Mar, California 92625*

JANICE BRITTON-DAVIDIAN (7), *Laboratoire de Genetique et Environnement, Institut des Sciences de L'Evolution, Université Montpellier II, 34095 Montpellier, France*

CAROL J. BULT (6), *Laboratory of Molecular Systematics, National Museum of Natural History, Smithsonian Institution, Washington, D.C. 20560*

REBECCA L. CANN (3), *Department of Genetics and Molecular Biology, University of Hawaii at Manoa, Honolulu, Hawaii 96822*

ROSE ANN CATTOLICO (13), *Department of Botany, University of Washington, Seattle, Washington, 98195*

SHENG-YUNG CHANG (32), *Roche Molecular Systems, Inc., Alameda, California 94501*

RUSSELL L. CHAPMAN (5), *Department of Botany, Louisiana State University, Baton Rouge, Louisiana 70803*

JOBY MARIE CHESNICK (13), *Department of Biology, Lafayette College, Easton, Pennsylvania 18042*

ROB DESALLE (4, 14), *Department of Entomology, American Museum of Natural History, New York, New York 10024*

PRESCOTT L. DEININGER (16), *Department of Biochemistry and Molecular Biology, Louisiana State University Medical Center, New Orleans, Louisiana 70112, and Laboratory of Molecular Genetics, Alton Ochsner Medical Foundation, New Orleans, Louisiana 70121*

MATHEW DICK (4), *Department of Biology, Yale University, New Haven, Connecticut 06511*

GABRIEL A. DOVER (38), *Department of Genetics, University of Leicester, Leicester LE1 7RH, United Kingdom*

DANIEL E. DYKHUIZEN (45), *Department of Ecology and Evolution, State University of New York at Stony Brook, Stony Brook, New York 11794*

ANDREW D. ELLINGTON (47), *Department of Chemistry, Indiana University, Bloomington, Indiana 47405*

ROBERT A. FELDMAN (3), *Department of Genetics and Molecular Biology, University of Hawaii at Manoa, Honolulu, Hawaii 96822*

INGRID FELGER (23), *Papua New Guinea Institute of Medical Research, Madang, Papua New Guinea*

LEONARD A. FREED (3), *Department of Zoology, University of Hawaii at Manoa, Honolulu, Hawaii 96822*

MATTHEW GEORGE (14), *Department of Biochemistry, Howard University, Washington, D.C. 20059*

THOMAS J. GIVNISH (2), *Department of Botany, University of Wisconsin, Madison, Wisconsin 53706*

DAVID GOLDMAN (8), *Laboratory of Neurogenetics, National Institute on Alcohol Abuse and Alcoholism, National Institutes of Health, Bethesda, Maryland 20892*

WILLIAM J. HAHN (2), *Department of Botany, University of Wisconsin, Madison, Wisconsin 53706*

BARRY G. HALL (44), *Department of Biology, University of Rochester, Rochester, New York 14627*

JOHN M. HANCOCK (38), *Molecular Neurobiology Group, Research School of Biological Sciences, Australian National University, Canberra City, ACT 2601, Australia*

BERNHARD HAUER (44), *BASF, Aktiengesellschaft, ZHB-Biotechnologie, D-6700 Ludwigshafen, Germany*

DAVID M. HILLIS (34), *Department of Zoology, The University of Texas at Austin, Austin, Texas 78712*

KENT E. HOLSINGER (33), *Department of Ecology and Evolutionary Biology, University of Connecticut, Storrs, Connecticut 06269*

JOHN A. HUNT (23), *Department of Genetics and Molecular Biology, John. A. Burns School of Medicine, University of Hawaii at Manoa, Honolulu, Hawaii 96822*

DAVID M. IRWIN (40), *Department of Clinical Biochemistry, and Banting and Best Diabetes Centre, University of Toronto, Toronto, Ontario, Canada M5G 1L5*

ROBERT K. JANSEN (33), *Department of Botany, The University of Texas at Austin, Austin, Texas 78712*

SUE JINKS-ROBERTSON (46), *Department of Biology, Emory University, Atlanta, Georgia 30322*

CAROLE JOISSON (10), *Laboratoire d'Immunochimie, Institut de Biologie Moleculaire et Cellulaire, CNRS, 67084 Strasbourg, France*

ELDON R. JUPE (39), *Department of Molecular Genetics, University of Cincinnati College of Medicine, Cincinnati, Ohio 45267*

LAUREN N. W. KAM-MORGAN (36), *Division of Biochemistry and Molecular Biology, University of California, Berkeley, California 94720*

YUN-TZU KIANG (6), *Department of Plant Biology, University of New Hampshire, Durham, New Hampshire 03824*

JACK F. KIRSCH (36, 42), *Division of Biochemistry and Molecular Biology, University of California, Berkeley, California 94720*

THOMAS D. KOCHER (28), *Department of Zoology, University of New Hampshire, Durham, New Hampshire 03824*

CLETUS P. KURTZMAN (24), *Microbial Properties Research, National Center for Agricultural Utilization Research, Agricultural Research Service, United States Department of Agriculture, Peoria, Illinois 61604*

SHIRLEY KWOK (32), *Roche Molecular Systems, Inc., Alameda, California 94501*

THOMAS B. LAVOIE (36), *Bristol-Myers Squibb Pharmaceutical Research Institute, Princeton, New Jersey 08544*

ENRIQUE P. LESSA (31), *Laboratorio de Evolución, Instituto de Biologia, Montevideo 11200, Uruguay*

ANDRÉS RUIZ LINARES (38), *Department of Genetics, Stanford University Medical School, Stanford, California 94305*

J. KOJI LUM (3), *Department of Genetics and Molecular Biology, University of Hawaii at Manoa, Honolulu, Hawaii 96822*

BARBARA LUNDRIGAN (37), *Museum of Zoology and Department of Biology, University of Michigan, Ann Arbor, Michigan 48109*

BRUCE A. MALCOLM (42), *Division of Drug Discovery and Development, Chiron Corporation, Emeryville, California 94608*

SANDRA L. MARTIN (22), *Department of Cellular and Structural Biology, University of Colorado School of Medicine, Denver, Colorado 80262*

RICHARD B. MEAGHER (26), *Department of Genetics, University of Georgia, Athens, Georgia 30602*

MICHAEL M. MIYAMOTO (34), *Department of Zoology, University of Florida, Gainesville, Florida 32611*

STEPHEN J. O'BRIEN (8), *Laboratory of Viral Carcinogenesis, National Cancer Institute, National Institutes of Health, Frederick, Maryland 21701*

SVANTE PÄÄBO (30), *Department of Zoology, University of Munich, D-8000 Munich 2, Germany*

STEPHEN R. PALUMBI (29), *Department of Zoology, and Kewalo Marine Laboratory, University of Hawaii at Manoa, Honolulu, Hawaii 96822*

THOMAS D. PETES (46), *Department of Biology, University of North Carolina, Chapel Hill, North Carolina 27599*

ELLEN M. PRAGER (11), *Division of Biochemistry and Molecular Biology, University of California, Berkeley, California 94720*

LIANG-HU QU (25), *Biotechnology Research Center, Zhongshan University, Guangzhou 510 275, People's Republic of China*

A. LANE RAYBURN (15), *Department of Agronomy, University of Illinois, Urbana, Illinois 61801*

CAROL A. REEB (3), *Department of Genetics and Molecular Biology, University of Hawaii at Manoa, Honolulu, Hawaii 96822*

BARBARA REINHOLD-HUREK (35), *Department of Biological Sciences, Center for Molecular Genetics, State University of New York at Albany, Albany, New York 12222*

SCOTT O. ROGERS (18), *Environmental Science and Forestry, Syracuse University, Syracuse, New York 13210*

STEVEN H. ROGSTAD (20), *Department of Biological Sciences, University of Cincinnati, Cincinnati, Ohio 45221*

STEVEN ROSENBERG (42), *Division of Drug Discovery and Development, Chiron Corporation, Emeryville, California 94608*

CARL W. SCHMID (16), *Departments of Chemistry and Genetics, University of California, Davis, California 95616*

JULIE F. SENECOFF (26), *Department of Genetics, University of Georgia, Athens, Georgia 30602*

ANDY SHIH (32), *Innovir Laboratories, New York, New York 10021*

PHOEBE SHIH (42), *Division of Biochemistry and Molecular Biology, University of California, Berkeley, California 94720*

DAVID A. SHUB (35), *Department of Biological Sciences, Center for Molecular Genetics, State University of New York at Albany, Albany, New York 12222*

JERRY L. SLIGHTOM (19), *Molecular Biology Unit, The Upjohn Company, Kalamazoo, Michigan 49007*

JAMES F. SMITH (2), *Department of Biology, Boise State University, Boise, Idaho 83725*

SANDRA J. SMITH-GILL (36), *Laboratory of Genetics, National Cancer Institute, National Institutes of Health, Bethesda, Maryland 20892*

MARK S. SPRINGER (17), *Department of Biology, University of California, Riverside, California 92521*

DAVID STAHL (27), *Departments of Veterinary Pathobiology and Microbiology, University of Illinois, Urbana, Illinois 61801*

LINDA STATHOPLOS (9), *Conservation Analytical Laboratory, Smithsonian Institution, Washington, D.C. 20560*

DIANA B. STEIN (12), *Department of Biological Sciences, Mount Holyoke College, South Hadley, Massachusetts 01075*

CARO-BETH STEWART (43), *Department of Biological Sciences, State University of New York at Albany, Albany, New York 12222*

YOUNGBAE SUH (1), *Natural Products Research Institute, Seoul National University, Seoul 110-460, South Korea*

KENNETH J. SYTSMA (2), *Department of Botany, University of Wisconsin, Madison, Wisconsin 53706*

W. KELLEY THOMAS (28, 30), *Division of Biochemistry and Molecular Biology, University of California, Berkeley, California 94720*

PRISCILLA K. TUCKER (37), *Museum of Zo-ology and Department of Biology, University of Michigan, Ann Arbor, Michigan 48109*

NOREEN TUROSS (9), *Conservation Analytical Laboratory, Smithsonian Institution, Washington, D.C. 20560*

MARC H. V. VAN REGENMORTEL (10), *Laboratoire d'Immunochimie, Institute de Biologie Moléculaire et Cellulaire, CNRS, 67084 Strasbourg, France*

CARL WETTER (10), *Department of Botany, University of Saarbrücken, Saarbrücken, Germany*

WARD C. WHEELER (4), *Department of Invertebrates, American Museum of Natural History, New York, New York 10024*

HOLLY A. WICHMAN (22), *Department of Biological Sciences, University of Idaho, Moscow, Idaho 83843*

ANNIE K. WILLIAMS (14), *Department of Entomology, American Museum of Natural History, New York, New York 10024*

JOHN G. K. WILLIAMS (21), *Pioneer Hi-bred International, Inc., Johnston, Iowa 50131*

ALLAN C. WILSON† (11, 42), *Division of Biochemistry and Molecular Biology, University of California, Berkeley, California 94720*

GRAEME WISTOW (41), *Section on Molecular Structure and Function, National Eye Institute, National Institutes of Health, Bethesda, Maryland 20892*

ELIZABETH A. ZIMMER (39), *Laboratory of Molecular Systematics, National Museum of Natural History, Smithsonian Institution, Washington, D.C. 20560*

METHODS IN ENZYMOLOGY

VOLUME 74. Immunochemical Techniques (Part C)
Edited by JOHN J. LANGONE AND HELEN VAN VUNAKIS

VOLUME 75. Cumulative Subject Index Volumes XXXI, XXXII, and XXXIV–LX
Edited by EDWARD A. DENNIS AND MARTHA G. DENNIS

VOLUME 76. Hemoglobins
Edited by ERALDO ANTONINI, LUIGI ROSSI-BERNARDI, AND EMILIA CHIANCONE

VOLUME 77. Detoxication and Drug Metabolism
Edited by WILLIAM B. JAKOBY

VOLUME 78. Interferons (Part A)
Edited by SIDNEY PESTKA

VOLUME 79. Interferons (Part B)
Edited by SIDNEY PESTKA

VOLUME 80. Proteolytic Enzymes (Part C)
Edited by LASZLO LORAND

VOLUME 81. Biomembranes (Part H: Visual Pigments and Purple Membranes, I)
Edited by LESTER PACKER

VOLUME 82. Structural and Contractile Proteins (Part A: Extracellular Matrix)
Edited by LEON W. CUNNINGHAM AND DIXIE W. FREDERIKSEN

VOLUME 83. Complex Carbohydrates (Part D)
Edited by VICTOR GINSBURG

VOLUME 84. Immunochemical Techniques (Part D: Selected Immunoassays)
Edited by JOHN J. LANGONE AND HELEN VAN VUNAKIS

VOLUME 85. Structural and Contractile Proteins (Part B: The Contractile Apparatus and the Cytoskeleton)
Edited by DIXIE W. FREDERIKSEN AND LEON W. CUNNINGHAM

VOLUME 86. Prostaglandins and Arachidonate Metabolites
Edited by WILLIAM E. M. LANDS AND WILLIAM L. SMITH

VOLUME 87. Enzyme Kinetics and Mechanism (Part C: Intermediates, Stereochemistry, and Rate Studies)
Edited by DANIEL L. PURICH

VOLUME 88. Biomembranes (Part I: Visual Pigments and Purple Membranes, II)
Edited by LESTER PACKER

VOLUME 89. Carbohydrate Metabolism (Part D)
Edited by WILLIS A. WOOD

VOLUME 90. Carbohydrate Metabolism (Part E)
Edited by WILLIS A. WOOD

VOLUME 211. DNA Structures (Part A: Synthesis and Physical Analysis of DNA)
Edited by DAVID M. J. LILLEY AND JAMES E. DAHLBERG

VOLUME 212. DNA Structures (Part B: Chemical and Electrophoretic Analysis of DNA)
Edited by DAVID M. J. LILLEY AND JAMES E. DAHLBERG

VOLUME 213. Carotenoids (Part A: Chemistry, Separation, Quantitation, and Antioxidation)
Edited by LESTER PACKER

VOLUME 214. Carotenoids (Part B: Metabolism, Genetics, and Biosynthesis)
Edited by LESTER PACKER

VOLUME 215. Platelets: Receptors, Adhesion, Secretion (Part B)
Edited by JACEK J. HAWIGER

VOLUME 216. Recombinant DNA (Part G)
Edited by RAY WU

VOLUME 217. Recombinant DNA (Part H)
Edited by RAY WU

VOLUME 218. Recombinant DNA (Part I)
Edited by RAY WU

VOLUME 219. Reconstitution of Intracellular Transport
Edited by JAMES E. ROTHMAN

VOLUME 220. Membrane Fusion Techniques (Part A)
Edited by NEJAT DÜZGÜNEŞ

VOLUME 221. Membrane Fusion Techniques (Part B)
Edited by NEJAT DÜZGÜNEŞ

VOLUME 222. Proteolytic Enzymes in Coagulation, Fibrinolysis, and Complement Activation (Part A: Mammalian Blood Coagulation Factors and Inhibitors)
Edited by LASZLO LORAND AND KENNETH G. MANN

VOLUME 223. Proteolytic Enzymes in Coagulation, Fibrinolysis, and Complement Activation (Part B: Complement Activation, Fibrinolysis, and Nonmammalian Blood Coagulation Factors)
Edited by LASZLO LORAND AND KENNETH G. MANN

VOLUME 224. Molecular Evolution: Producing the Biochemical Data
Edited by ELIZABETH A. ZIMMER, THOMAS J. WHITE, REBECCA L. CANN, AND ALLAN C. WILSON

VOLUME 225. Guide to Techniques in Mouse Development
Edited by PAUL M. WASSARMAN AND MELVIN L. DEPAMPHILIS

VOLUME 226. Metallobiochemistry (Part C: Spectroscopic and Physical Methods for Probing Metal Ion Environments in Metalloenzymes and Metalloproteins
Edited by JAMES F. RIORDAN AND BERT L. VALLEE

Section I

Practical Issues in Performing Comparative Molecular Studies

Section I. Practical Issues in Performing Comparative Molecular Studies

Section I deals with the immediate practical issues that arise when a broad comparative molecular study is planned. Because the targeted audience includes both molecular and organismal specialists, issues critical both to proper identification of materials and to proper preservation of biochemical structure and function are addressed here. For organismal biologists used to equipping their research program with microscopes and field gear, the variety of equipment and protocols that can be used in comparative molecular genetic studies may at first seem overwhelming. However, rational strategies can be developed in order to include a molecular component in a comparative study, ones which maximize equipment sharing and minimize equipment and reagent costs. In [1] Suh *et al.* summarize the utility of common methodologies for examining various levels of taxonomic diversity, discuss effective strategies for their implementation, and provide listings of the equipment likely to be needed for comparative molecular studies. Several other publications[1-3] are also of particular interest. Chapters [2]–[5] outline strategies for obtaining, shipping, and storing samples suitable for macromolecular comparisons of the entire spectrum of living organisms. Although strategies for collection and documentation of samples may be similar across phyla, unique biological features of each group may dictate the need for special protocols for preservation, sampling, and obtaining permits. We therefore include chapters specific for higher plants [2], vertebrates [3], invertebrates [4], and fungi and algae [5]. Chapter [3] by Cann and colleagues includes a discussion of population level sampling issues. Although no chapter is included that deals specifically with bacterial samples, a catalog listing available bacteria, phages, and recombinant DNA vectors can be obtained from the American Type Culture Collection (ATCC, Rockville, MD). Further reference to the ATCC can be found in [3] and [5].

[1] D. M. Hillis and C. Moritz (eds.), "Molecular Systematics." Sinauer, Sunderland, Massachusetts, 1990.

[2] D. D. Blumberg, this series, Vol. 152, p. 3.

[3] C. Orrego, *in* "PCR Protocols: A Guide to Methods and Applications" (M. A. Innes, D. H. Gelfand, J. J. Sninsky, and T. J. White, eds.), p. 447. Academic Press, New York, 1990.

[1] Equipping and Organizing Comparative Molecular Genetics Laboratories

By Youngbae Suh, Judith A. Blake, and Michael J. Braun

Introduction

Rapid advances in biochemical and molecular genetic technology are revolutionizing many fields of biological inquiry. Nowhere is the effect of these new tools more keenly felt than in evolutionary biology.[1] For here, the new technology is not merely changing the course of research within a discipline; it is merging what were previously unrelated fields. Thus, modern systematists must now concern themselves with such unlikely topics as protein structure and function, while molecular geneticists must consider their data in the light of evolutionary theory. This melding process can be especially difficult for biologists without extensive training in biochemistry or molecular genetics who nonetheless find molecular data sufficiently compelling to try to use the techniques themselves.

Our purpose in this chapter is to give guidance to the newcomer on how to organize and equip a laboratory effectively for comparative molecular research. General articles on laboratory setup can be found in molecular methodology publications.[2,3] In molecular evolutionary studies, however, the focus on comparative analysis dictates special emphasis on streamlining repetitive procedures, adapting techniques for a variety of organisms, and managing many samples efficiently. We shall endeavor to orient researchers to the general considerations involved in planning a laboratory for molecular evolutionary studies. We briefly describe the techniques available and criteria for choosing among them. We suggest rationales for selecting equipment and strategies for the physical organization of the laboratory. Throughout, we highlight possible problems and solutions.

Setting Goals

Starting a molecular laboratory requires both consideration of the ultimate questions being addressed by the research and evaluation of the

---

[1] M. Clegg, J. Felsenstein, W. Fitch, M. Goodman, D. Hillis, M. Riley, F. Ruddle, D. Sankoff, P. Arzberger, M. Courtney, P. Harriman, C. Lynch, J. Plesset, M. Weiss, and T. Yates, *Mol. Phylogenet. Evol.* **1,** 84 (1992).

[2] D. D. Blumberg, this series, Vol. 152, p. 3.

[3] C. Orrego, *in* "PCR Protocols" (M. A. Innis, D. H. Gelfand, J. J. Sninsky, and T. J. White, eds.), p. 447. Academic Press, New York, 1990.

current research environment available to the researcher. The rapid prolif-
eration and simplification of molecular techniques presents investigators
with what at times may seem a bewildering array of possible approaches to
problems. Yet, from the viewpoint of many evolutionary biologists, these
procedures are still expensive, laborious, and time-consuming. Thus, it is
important at the outset to assess the resources and support available for the
contemplated research, and to pay attention to choosing techniques that
are well suited to the problems at hand.

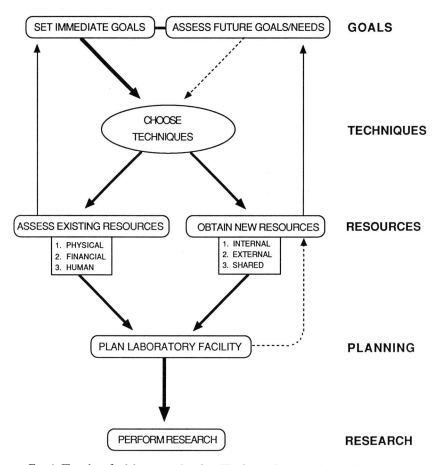

FIG. 1. Flowchart for laboratory planning. The five major stages in planning a laboratory
facility are listed to the right. Arrows in the chart indicate the direction of progress through
stages and how information gleaned at one stage can cause reconsideration of decisions made
at a previous stage.

The following issues will merit consideration: (1) What frequency and intensity of laboratory work is expected? Some scientists need to obtain molecular data on one or a few specific questions; for others it will be an ongoing focus of research. (2) What technique or techniques will be used? Success often hinges on choosing a technique well suited to the research problem. The scope of the laboratory operation, in turn, will increase with the number and complexity of techniques to be implemented. (3) What resources already exist? What resources are obtainable? Given the cost and complexity of molecular research, it is crucial to assess the physical, technical, and human resources of the home institution and the local environment, and to be realistic about opportunities for obtaining new resources. (4) Can resources be shared? Virtually all molecular investigators share some laboratory resources. The extent of sharing may range from a few pieces of large equipment in a traditional department comprising many single-investigator laboratories to sharing of a single, centralized molecular facility among several investigators. Collaboration among research groups can provide the means to address scientific questions without setting up multiple independent research units.

Careful consideration of these basic issues at the outset will lead to more effective planning and a cost-efficient facility better designed to meet the needs of the individual and the institution. A flowchart (Fig. 1) helps illustrate the interrelationships of these issues and the feedback effects they may have on one another.

Choosing Techniques

The physical organization of a laboratory will be dictated by the techniques to be implemented there. Thus, it is essential to decide what techniques will be used immediately and what techniques might be attempted at a later date. Six major comparative molecular techniques are in common use today in evolutionary research. These are isozyme electrophoresis, comparative immunological methods [especially microcomplement fixation (MC'F)], DNA–DNA hybridization, restriction enzyme analysis, random amplification of polymorphic DNA (RAPD), and DNA sequencing. Two other methods, molecular cloning and the polymerase chain reaction (PCR), are not inherently comparative, but they provide the foundation on which a number of DNA techniques are based. We first consider the major factors involved in choosing among techniques, then discuss the advantages and disadvantages of each. Table I summarizes information presented below that will be helpful in selecting techniques.

TABLE I
CONSIDERATIONS FOR CHOOSING TECHNIQUES[a]

Technique	Optimal divergence range	Complexity	Cost	Time and labor	Nature of samples	Comparability
Isozyme electro-phoresis	Lo/Med	Lo	Lo	Lo	Multiple tissues	Good
Microcomplement fixation	Med	Hi	Med	Med/Hi	Purified protein	Fair
DNA–DNA hy-bridization	Med	Hi	Med	Med/Hi	Large quantities of DNA	Fair
Restriction enzyme analysis	Lo	Med	Hi	Med/Hi	HMW DNA	Good
DNA sequencing	Lo/Med/Hi	Hi	Hi	Hi	Any purified DNA	Excellent
Random amplifica-tion of polymor-phic DNA	Lo	Med	Med	Lo	HMW DNA	Fair
Polymerase chain reaction	N.A.	Med	Med	Lo	Any purified DNA	N.A.
Molecular cloning	N.A.	Hi	Hi	Hi	Any purified DNA	N.A.

[a] N.A., Not applicable, since technique is not a comparative method, but forms the basis for others; HMW, high molecular weight; Lo, low; Med, medium; Hi, high.

Matching Resolving Power of Techniques to Genetic Divergence of Organisms

Most research problems in molecular evolutionary genetics ultimately involve questions of relationship. However, the degree of relationship may vary dramatically, from comparisons of related individuals within a population to comparisons of taxa that diverged early in the history of life on Earth. There is a general correlation between degree of relationship among organisms and the level of genetic similarity among their genomes. Again, the level of similarity can vary widely: the genomes of clonally reproducing organisms may be nearly 100% identical, whereas those of distantly divergent taxa may retain sequence similarity only in a few, highly conserved genes. The level of genetic similarity among the genes or genomes to be compared is the primary determinant of the applicability of the various comparative molecular techniques. Each technique is best suited to a particular range of divergences among the organisms being compared. Each technique probes genetic questions at different levels of resolution, from direct DNA sequence comparisons to inferred amino acid sequence variation. Some techniques, such as DNA sequencing, can be adapted to a broad range of divergences; others are not so versatile. Failure to appreciate the relationship between gene or genome divergence and resolving power of a technique can lead to a data set that is largely irrelevant to the original research focus.

Complexity

To use some methods successfully requires more background and experience in biochemistry than others. Most comparative techniques involve repetitive procedures that seem deceptively simple to learn when all is working well. When a technique fails to work properly, however, considerable expertise may be required to discover the problem and get the research back on track.

Cost

It is convenient to think of costs as falling into three categories, namely, overhead costs of setting up and maintaining the laboratory, consumables cost of the supplies actually used up in a particular project, and personnel costs. At present, initial overhead costs might range from $20,000 or less for a laboratory concentrating on isozyme surveys to several hundred thousand dollars for a laboratory fully equipped to implement a wide range of techniques. Setup costs will vary dramatically depending on preexisting resources and physical location, so it is important to explore the local environment of the proposed laboratory (see section on assessing environment and resources below). The annual cost of consumable supplies will range from a few thousand dollars per active researcher for isozyme work to $10,000–$15,000 per person for the more expensive techniques. At many university laboratories, personnel costs may not be a major factor if salaries and stipends come from sources other than research funds. At other institutions, however, where all expenditures come from a common pool, personnel costs are likely to be an important budget item.

Time and Labor

The major comparative techniques differ substantially in the amount of time and effort they require. Some labor-intensive techniques are more appropriate when relatively small numbers of comparisons are to be made; other techniques can be readily applied to large numbers of samples.

Nature of Organism and/or Samples

The techniques differ in the amount and type of sample required. The physical size of the organism may limit the amount of tissue that can be obtained from each individual, and may make some techniques difficult or impossible to use. Some techniques work best with certain kinds of tissue. Other techniques (e.g., isozyme electrophoresis) benefit from the availability of several tissue types. Most techniques work best with fresh or frozen material, but some can be used with samples that are thousands of years old.

Comparability across Studies

The ease with which a data set can be compared to other studies is an important consideration. DNA sequencing provides the greatest advantages in this regard. Because the method involves the absolute determination of nucleotide sequence, the data can be compared to other sequences determined completely independently. Some data derived from other techniques can be compared in a general way (e.g., isozyme genetic distance, sequence divergence estimated from restriction enzyme analysis), whereas other comparisons are best made side by side in the same laboratory (e.g., establishing homology of isozyme alleles or restriction fragment bands).

Techniques: Pros and Cons

Isozyme Electrophoresis

The examination of proteins by electrophoresis has been the single most popular technique in molecular evolutionary genetics. It can provide useful information for studies at the population and species level.[4-6] For some taxa where genetic distances tend to be low, such as birds, the useful range of isozyme electrophoresis extends to the genus and even family level. It is the least expensive technique, both in terms of overhead and cost of consumables. Large sample sizes can be handled readily, and the technical demands are relatively minimal. Electrophoresis and staining conditions are often similar throughout major groups of organisms, although some optimization is required for most studies.[4-7] In addition, gene expression varies by tissue, so having a variety of tissue types available for study increases the number of genes that can be assayed. Although data from separate studies cannot be directly combined, the large body of comparative data available in the literature often provides a helpful background for interpretation of isozyme data.

[4] C. J. Bult and Y.-T. Kiang, this volume [6].
[5] J. Britton-Davidian, this volume [7].
[6] R. W. Murphy, J. W. Sites, Jr., D. G. Buth, and C. H. Haufler, in "Molecular Systematics" (D. M. Hillis and C. Moritz, eds.), p. 45. Sinauer, Sunderland, Massachusetts, 1990.
[7] D. Goldman and S. J. O'Brien, this volume [8].

An important aspect of isozyme work is that provision must be made for the careful handling of tissues to keep them solidly frozen at all times. Proteins are particularly sensitive to thawing events. This means that dry ice or liquid nitrogen procedures must be used during the collection and transportation of the tissues; in addition, at least a $-70°$ freezer is needed for the storage of tissues.

Comparative Immunological Methods

A number of immunological techniques have been used in comparative studies.[8,9] The most important of these is microcomplement fixation (MC′F), a quantitative technique that has played a key role in many classic studies of molecular evolution and molecular systematics. By selecting proteins with different rates of evolution, a broad range of divergences can be examined. The cost of the technique is moderate, but biochemical expertise is required and the labor involved is substantial. Protein must be purified from some or all taxa for antibody production, and, for those taxa, a sizable tissue or serum sample is needed. Antibody production itself is usually done in rabbits, so an animal care facility must be available. Like isozyme electrophoresis, the large body of immunological distance data already available ensures the continued value of this technique for certain investigations.

DNA–DNA Hybridization

Solution hybridization studies of DNA have been used extensively in the analysis of genome complexity and organization, and intensively in phylogenetic studies of some groups.[10,11] The technique has the attractive feature that it ideally allows all (or some significant fraction) of two genomes to be compared at once. It is relatively simple to perform in the laboratory, and neither the cost nor the time and labor involved is prohibitive for small to medium sized numbers of samples. From a physicochemical standpoint, however, the technique is much more complicated than it appears at first, because what is really taking place in the hybridization vessel is many thousands or millions of separate hybridization reactions, each with its own kinetics and thermal stability. Careful attention to experimental design, laboratory technique, and data interpretation is essential. Also, pairwise comparisons between all samples are often desirable,

---

[8] M. H. V. Van Regenmortel, C. Joisson, and C. Wetter, this volume [10].

[9] L. R. Maxson and R. D. Maxson, in "Molecular Systematics" (D. M. Hillis and C. Moritz, eds.), p. 127. Sinauer, Sunderland, Massachusetts, 1990.

[10] M. S. Springer and R. J. Britten, this volume [17].

[11] S. D. Werman, M. S. Springer, and R. J. Britten, in "Molecular Systematics" (D. M. Hillis and C. Moritz, eds.), p. 204. Sinauer, Sunderland, Massachusetts, 1990.

so the cost and labor involved goes up rapidly with the number of samples. Relatively large quantities of DNA are needed, which may be hard to come by for some smaller organisms or those with low DNA content per unit mass. Data can be compared to other studies in a general way, but adding new taxa to a data set generally will require making pairwise comparisons with some or all previously studied taxa.

Restriction Enzyme Analysis

The comparative analysis of restriction site maps or restriction fragment patterns is a powerful tool for evolutionary studies at lower taxonomic levels. The site mapping approach can be extended to somewhat higher levels of divergence than fragment comparisons. The fragment comparison approach is much less time-consuming than site mapping, but it may lead to difficulty in determining homology for fragments in regions that have undergone structural rearrangements or show high degrees of variability. Large numbers of samples can be analyzed routinely, and the biochemical complexity of the technique is moderate. There are several methodological approaches to restriction analysis (end labeling, Southern blotting, PCR followed by digestion); all are fairly expensive to perform. Well-purified, high molecular weight DNA is a prerequisite for the end labeling and Southern blotting methods. The mitochondrial and chloroplast genomes have been popular targets for restriction analysis; for such projects the organellar genome must be extensively purified, or cloned probes must be available. In these cases, it is desirable to have tissue samples rich in the organelle. For example, vertebrate blood is a poor source of mitochondrial DNA; liver or heart samples work much better. Similarly, young leaf samples are the best source of chloroplast DNA.

Random Amplification of Polymorphic DNA

Random amplification of polymorphic DNA (RAPD) is an *in vitro* amplification technique (see section on polymerase chain reaction below) that eliminates the need for any prior characterization of the genome to be analyzed.[12-14] Based on the random amplification of DNA segments using single short primers of arbitrary sequence, RAPD readily detects genetic polymorphisms in most organisms. The method is simple and rapid to

---

[12] B. M. Bowditch, D. G. Albright, J. Williams, and M. J. Braun, this volume [21].
[13] J. G. Williams, A. R. Kubelik, K. J. Livak, J. A. Rafalski, and S. V. Tingey, *Nucleic Acids Res.* **18,** 6531 (1990).
[14] H. Hadrys, M. Balick, and B. Schierwater, *Mol. Ecol.* **1,** 55 (1992).

perform, and large numbers of individuals can be analyzed at a reasonable expense. Only small amounts of DNA are required, and the same set of primers can be used on any organism. The method does demand careful replication of reaction conditions to achieve consistent amplifications, and good quality (i.e., high molecular weight) DNA is probably desirable. RAPD markers have already been used extensively in gene mapping research, and they are likely to see regular use in population and species level studies, especially if the organisms in question are small, poorly known, or otherwise difficult to study.[15] There is some concern at present about their use in DNA typing or fingerprinting studies owing to the occasional appearance of nonparental bands in offspring of known parentage.[16] In phylogenetics, the method is likely to be useful only among very closely related organisms because of the rapid evolution of band patterns. Another limitation is that the majority of RAPD markers show dominance, so that heterozygotes cannot be distinguished from homozygotes for the dominant allele.

Nucleic Acid Sequencing

Nucleic acid sequencing reveals the discrete order of nucleotides that encode all genetic information. In principle, sequence data can provide the greatest possible resolution for most studies, yet sequencing is the most expensive and labor-intensive of the common comparative techniques. Therefore, other procedures should be considered first to determine whether they could be employed effectively to answer questions of interest. A key issue here is whether the sequence of a single gene is adequate to answer the question or whether a broader survey of the genome is needed. Sequencing should be considered especially when conservative genes need to be compared among very divergent organisms, or when the level of resolution obtained by other techniques is inadequate. Preparations for sequencing prior to the late 1980s typically involved extensive molecular cloning procedures, making it impractical for many comparative studies. PCR has short-circuited the process of producing purified template for sequencing and opened a new era in molecular evolutionary biology by making it feasible to determine and compare the exact nucleotide sequence of homologous genes for many dozens, even hundreds, of individuals or species. An especially attractive feature of sequence data is that, because it is determined absolutely, the sequence can freely be compared to any other alignable sequence determined by the same or any other investigator.

[15] R. C. Wilkerson, T. Parsons, D. G. Albright, T. A. Klein, and M. J. Braun, *Insect Mol. Biol.* in press (1993).
[16] M. F. Reidy, W. J. Hamilton III, and C. F. Aquadro, *Nucleic Acids Res.* **20**, 918 (1992).

An advance in DNA sequencing is the development of automated sequencing instruments that replace the traditional radioactive labels with laser-activated fluorescent labels and that read sequences from gels directly into computer files. The development of automatic sequencers is a promising new strategy for DNA sequencing that increases the speed of data acquisition, but they are still too expensive for most individual laboratories to be able to justify their purchase.[17]

Polymerase Chain Reaction

The polymerase chain reaction technique allows for the direct *in vitro* amplification of DNA. It is not itself a comparative technique but forms the basis for others. PCR has revolutionized molecular evolutionary studies by making it possible to amplify similar segments of DNA from many organisms quickly and simply. It is possible to accomplish in 1 day by PCR what would have taken months or years to do by standard cloning techniques. PCR methods have greatly extended the range of feasibility of molecular investigations beyond previous limits on the age, quality, and quantity of material available for study.[18,19] The technique is relatively simple in concept and practice. It requires little time, labor, and cost compared to the results achieved. Yet the method is so extraordinarily versatile that it has affected experimental strategies throughout molecular genetics. This is especially true of evolutionary research, where comparison of homologous genes is all-important.

Beside greatly facilitating techniques such as DNA sequencing and molecular cloning, PCR technology has spawned a number of new comparative techniques that are providing invaluable information for evolutionary research. One of these (RAPD) has been mentioned in some detail above. Others include methods for the analysis of polymorphism at hypervariable genetic loci such as variable number tandem repeats (minisatellites)[20] and di- or trinucleotide repeats (microsatellites).[21] The highly informative, single-locus genetic markers provided by these methods are finding wide application in studies requiring genetic mapping, individual identification, parentage analysis, or other forms of DNA typing (fingerprinting).

Despite its versatility, there are some limitations to PCR that should

[17] D. M. Hillis, A. Larson, S. K. Davis, and E. A. Zimmer, *in* "Molecular Systematics" (D. M. Hillis and C. Moritz, eds.), p. 318. Sinauer, Sunderland, Massachusetts, 1990.

[18] E. M. Golenberg, D. E. Gianasi, M. T. Clegg, C. J. Smiley, M. Durbin, D. Henderson, and G. Zirawski, *Nature (London)* **344,** 656 (1990).

[19] S. Pääbo, J. A. Gifford, and A. C. Wilson, *Nucleic Acids Res.* **16,** 9775 (1988).

[20] A. J. Jeffreys, R. Neumann, and V. Wilson, *Cell (Cambridge, Mass.)* **60,** 473 (1990).

[21] J. W. Weber and P. E. May, *Am. J. Hum. Genet.* **44,** 388 (1989).

be noted. For a specific target segment to be amplified, prior knowledge of the sequence of its termini must be available so that specific primers can be synthesized. With current methods, the enzymatic synthesis of DNA required for PCR is relatively inefficient for long molecules. Thus, amplification of segments longer than a few kilobases may be difficult. The exquisite sensitivity of the technique makes it particularly vulnerable to contamination, especially by previously amplified PCR products.[22] For this reason, a number of precautions are necessary in order to prevent contamination, and it is advisable to physically isolate pre-PCR from post-PCR protocols if at all possible (see section on space concerns below).[3,22] Finally, PCR amplifies all viable target molecules in a sample more or less indiscriminately. Thus, if the population of target molecules is heterogeneous, the PCR product is likely to be heterogeneous as well. This feature of PCR is advantageous in some situations[23] but disadvantageous in others.[24]

Molecular Cloning

Standard molecular cloning techniques provide a means for the dissection of complex genomes and the *in vivo* amplification of specific sequences.[25] In many cases, little or no prior characterization of the sequence of interest is necessary. Like PCR, cloning is not in itself a comparative technique but provides the foundation on which several of the other techniques rest. For example, restriction analysis often involves the use of cloned probes. DNA sequencing following cloning is a commonly used method that consistently results in high-quality sequencing gels. Cloning of PCR amplification products dramatically enhances the efficiency of cloning and reduces the need for extensive screening efforts to find clones containing the desired insert.

Molecular cloning involves the use of bacteria, bacteriophage, and plasmids; thus, attention to sterile technique is imperative, and a working knowledge of bacterial and phage genetics is useful. Of the techniques discussed here, it is the most demanding in terms of molecular genetic expertise. Both the cost and the labor involved vary according to the method employed, but can be substantial. Now that many genes have been well characterized from a variety of organisms, some comparative molecu-

[22] S. Kwok, *in* "PCR Protocols" (M. A. Innis, D. H. Gelfand, J. J. Sninsky, and T. J. White, eds.), p. 142. Academic Press, New York, 1990.
[23] S. Pääbo and A. C. Wilson, *Nature (London)* **334**, 387 (1988).
[24] D. A. Lawlor, C. D. Dickel, W. W. Hauswirth, and P. Parham, *Nature (London)* **349**, 785 (1991).
[25] S. L. Martin and H. A. Wichman, this volume [22].

lar laboratories will not find it necessary to attempt *de novo* cloning. Still, the many cloning strategies available today provide such exquisite control and flexibility in manipulating DNA that they are sure to remain the backbone of mainstream molecular genetics and allow the development of new comparative molecular approaches.

Assessing Environment and Resources

Once research goals have been set and comparative techniques chosen, the next critical step in planning a laboratory is assessing the local environment and the resources available for research. As discussed above, the comparative molecular techniques now available differ greatly in complexity as well as in the type and cost of equipment required. Some can be performed with minimal laboratory facilities on a modest budget, whereas others are likely to be implemented successfully only at institutions where a substantial amount of molecular genetics expertise and activity already exists. In some instances, a survey of existing resources will cause the researcher to reconsider decisions on goals or techniques (Fig. 1). It is important to assess resources in several general categories.

Physical Plant

The physical space designated to house a new laboratory may vary in its state of preparation for laboratory work. Ideally, the space will have been originally designed as a laboratory, but in some cases it will be little more than empty rooms. The space itself and the physical plant it is housed in should be viewed with the following questions in mind. Is adequate bench space and cabinetry already in place, or will it have to be installed? Are sinks and plumbing sufficient to support the contemplated work? Is purified water available? Is a fume hood present or available nearby? Does the general ventilation of the entire space meet applicable standards for laboratories? Is the electrical power supply sufficient? Will the number and location of receptacles allow flexibility in equipment location? Are appropriate high-capacity circuits available for large equipment?

Instrumentation

All molecular genetic techniques require some amount of specialized scientific equipment. Because many items of equipment are expensive, it is a good idea to determine what instrumentation exists at the department or institution level, and to what extent it can be shared. Table II provides information on equipment required for the various comparative techniques and the ease with which each instrument can be shared. Items that typically are shared include large equipment such as high-speed and ultra-

TABLE II

MAJOR EQUIPMENT NEEDED FOR COMPARATIVE MOLECULAR GENETICS LABORATORY[a]

Item	Isozyme electrophoresis	MC'F	DNA–DNA hybridization	Restriction analysis	RAPD	Sequencing	PCR	Cloning
Large Equipment								
Autoclave	n	o	o	s	o	o	o	s
Ultracentrifuge	n	o	o	s	o	o	o	s
High-speed centrifuge	n	i/s	i/s	i/s	i/s	i/s	i/s	i/s
Ultracold freezer (−70° or below)	i/s	i/s	i/s	i/s	i/s	i/s	i/s	i/s
Crushed ice machines	s	s	s	s	s	s	s	s
UV/visible spectrophotometer	n	s	s	s	s	s	s	s
Polaroid MP-4 camera or equivalent	s	n	o	s	s	s	s	s
UV transilluminator	n	n	o	s	s	s	s	s
Darkroom with film developing facilities	o	n	n	s	n	s	s	s
Scintillation counter	n	n	s	s	n	s	n	s
Shaking incubator (ambient to 65°)	n	o	o	o	o	n	n	n
Laminar flow hood	n	n	n	n	o	n	o	o
Digitizer	o	n	n	o	o	o	n	n
Oligonucleotide synthesizer	n	n	n	o	o/s	o/s	o/s	n
General Equipment								
Freezer (−20°)	i	i	i	i	i	i	i	i
Refrigerator	i	i	i	i	i	i	i	i
Low-speed centrifuge	o	i/s	i/s	o	o	o	o	i/s
Microcentrifuge	i	i/s	i/s	i	i	i	i	i
Oven (ambient to 200°)	n	i/s	i/s	i/s	i/s	i/s	i/s	i/s
Stationary incubator (ambient to 68°)	s	n	n	s	s	s	s	i
Shaking water bath (ambient to 68°)	n	i	s	o	n	n	n	n
Water bath (ambient to 95°)	s	i	i	i	i	i	i	i
Top-loading balance	i	i	i	i	i	i	i	i
Analytical balance	i/s	i/s	i/s	i/s	i/s	i/s	i/s	i/s
pH meter	i	i	i	i	i	i	i	i
Stirring hot plate	i	i	i/s	i	i	i	i	s
Electrical homogenizers, blenders	i	i/s	i/s	s	s	s	s	s
Geiger counter	n	n	i/s	i/s	n	i	n	i/s
Pipettors (20, 200, 1000 μl)	i	i	i	i	i	i	i	i
Power supply (250 V, 150 mA)	i	o	n	i	i	i	i	i
Power supply (2500–3000 V, 300 mA)	n	n	n	n	n	i	n	n
Gel boxes (several varieties)	i	i	n	i	i	i	i	i
X-Ray film cassettes	n	n	n	i	n	i	n	i
Intensifying screens	n	n	n	i	n	o	n	i
Gel dryer	o	n	n	n	n	i	n	n
Vacuum centrifuge dryer	n	n	o	o	o	o	o	o
DNA thermal cycler	n	n	n	o	i	o	i	o
Liquid nitrogen tank	s	o	s	s	s	s	s	s
Microwave oven	o	n	n	i	i	n	i	i
Heating block	o	n	o	i	n	i	n	o
Vortex	i	i	i	i	i	i	i	i
Vacuum pump	n	i/s	o	o	n	i	n	n

[a] i, Needed for individual laboratory; o, optional; s, needed on a shared access basis; n, not used for that technique.

centrifuges, scintillation counters, autoclaves, automatic dishwashers, darkrooms and associated photographic equipment, and coldrooms. Depending on the size and needs of both the new and preexisting laboratories, it may be convenient to share other items, such as spectrophotometers, thermal cyclers, and ultracold freezers.

Core Facilities and Support Services

Most research institutions offer at least some centralized support services. Common examples include animal care facilities, greenhouses, glassware washing, chemical stockrooms, radiation safety support, and hazardous waste disposal. It is important to assess not only the existence, but the quality and reliability of such services. A central computing facility, for example, exists at most institutions, but not all maintain access to genetic databases or software for analysis of DNA sequences. Licensing fees for the more comprehensive software packages may cost several thousand dollars annually. The precise services offered by the institution will often determine the extent to which the individual investigator must act independently.

A trend at larger institutions is to establish a core facility to provide access to some of the more expensive molecular genetic technologies on a "pay as you go" basis. Such facilities generally perform oligonucleotide synthesis to provide primers for PCR and sequencing. They may also offer DNA sequencing services, which, for a small project or pilot survey, may represent a cost-effective alternative to setting up an independent laboratory.

Human Resources

The wide range of molecular techniques available and the rapid pace of new developments make it difficult for an individual to stay apprised of all aspects of the field that might impinge on the research at hand. For this reason, other researchers involved in molecular genetics are a key element of the local environment. This may include other members of the same research group, scientists who share the laboratory facility, or others doing various kinds of molecular genetics research at the same or nearby institutions. The collective molecular expertise in the local environment (and how well one can hope to make use of it) should be carefully assessed, and it may influence decisions on the choice of technique and the scope of research envisioned.

In addition to the local intellectual environment, communication among researchers via electronic bulletin boards is becoming a common route to seeking answers to laboratory technical problems. Electronic mail subscription lists include such groups as a Molecular Evolution Discussion Group and a Methods and Reagents Discussion Group. Information about these bulletin boards can be obtained through standard LISTSERV subscription lists from BIOSCI@NET.BIO.NET or via USENET.[26]

---

[26] K. Hoover and D. Kristofferson, *Plant Mol. Biol. Rep.* **10,** 228 (1992).

Selecting Equipment

The techniques that are to be implemented will determine what equipment is necessary. Some equipment may already be available in or near the laboratory, but it is likely that other items will have to be purchased. Once it is clear that a certain instrument must be purchased, there are still many factors that will affect selection of the exact make and model best suited to a particular laboratory. Some key issues to keep in mind are the following. Is the model in question adequate to support the contemplated technique(s)? Will other techniques require the same or similar instruments? Will the particular model support those techniques as well? Is a more versatile model available? Does its versatility justify any additional cost? Can the instrument (and its associated costs) be shared? How reliable is the model in question? Are support and maintenance available from the vendor or other sources? How expensive is maintenance? Will changing technology or new research goals make a different model significantly more desirable in the near future?

Organizing the Laboratory

The physical organization of the laboratory will, of course, depend on many factors peculiar to a particular facility and research program. Despite the many differences among laboratories, however, some major considerations are common to all. The location and utilization of equipment need to be well thought out in terms of space and facilities available, potential technologies to be employed, safety concerns, and efficiency. The following discussion assumes that a preexisting laboratory space will be occupied with little opportunity for altering the placement of benches, sinks, hoods, etc. If the space is to be newly built or renovated for work, then a number of design issues come into play; these are briefly treated in the next section.

Space Concerns

Within the physical limitations imposed by a preexisting laboratory design, much can be done to streamline operations through careful placement of equipment and designation of dedicated areas for certain procedures. For example, the space allocated to the laboratory operation may be arranged in one or several rooms. If separate rooms are available, it may be desirable to dedicate one or more rooms to purposes that, for reasons of safety, quality control, or convenience, are logically separated from other laboratory functions. Radioisotope work is often carried out in an auxiliary room for safety reasons. Large items of equipment may be well placed in a separate room, especially if they generate noise or heat or if they need to be shared. A small room adjoining the main laboratory is a good location for a camera stand and light box to record electrophoresis results, because the room lights can be turned out without interrupting other workers. Tissue collections and associated activities, a major commitment of many molecular systematics laboratories, can be placed in a room with few laboratory accoutrements, as long as precautions are taken to guard against freezer failure.

A growing number of comparative laboratories isolate pre-PCR and post-PCR procedures in separate rooms. Contamination of samples or reagents with even minute amounts of a previously amplified PCR product can be a serious problem, especially for those working with ancient or degraded DNA. Such products are guaranteed to contain perfect primer binding sites, and they may amplify better than the target DNA in a new sample. "Pre-PCR" protocols include sample storage, genomic DNA isolation, and the initial preparation of PCR reactions. "Post-PCR" protocols include PCR itself, purification and electrophoretic analysis of PCR products, DNA sequencing, and secondary amplifications.[3]

Another aspect of spatial organization will be decisions on allocation of personal bench space versus a "workstation" arrangement of laboratory bench tops. A standard space configuration of a single-investigator laboratory involves the designation of small personal bench spaces and the establishment of common bench areas for shared equipment and setup of space-consuming operations. However, comparative laboratory work involves moderate to high degrees of repetition, and sharing of molecular facilities among many investigators is becoming more common. Under these conditions, a workstation arrangement may be desirable, in which standardized protocols are performed at workstations which are outfitted and maintained for that express purpose.

Facilities Considerations

The exact configuration of utilities and services within the laboratory will play a role in determining optimal organization. For example, the position of a fume hood may be the decisive factor in choosing where DNA extractions and other protocols involving organic solvents are carried out. This in turn may dictate placement of water baths and centrifuges used in DNA extraction. The position and size of sinks should be considered as well. Inherently messy protocols, such as Southern blotting and DNA sequencing, should take place near sinks to minimize problems with spills and cleanup. Electrical receptacles and the capacity of circuits will also play a role in the positioning of equipment. Equipment requiring voltages other than the standard line voltage will need special circuits and receptacles. Many pieces of laboratory equipment have significant amperage demands, so care must be taken not to overload circuits. This may mean that apparatuses with heating or cooling elements need to be spaced evenly around the laboratory.

Beyond simple issues of equipment placement, power and plumbing can be serious problems within the laboratory infrastructure. With the increased use of computer technology and the development of sophisticated equipment, the quality of electrical power needed for laboratory use has risen. Power surges and sags will shut down computers and computer-operated equipment, and low voltages will stress and wear relay switches and solenoids in machinery. Surge protectors can be useful for most electric spikes and sags. Some equipment, such as water baths and incubators, restart after a short power loss. Other apparatuses, such as ultracentrifuges and ultrafreezers, will restart but are vulnerable to damage from voltage fluctuations. Electronic alarms, backup cooling systems, and a predetermined strategy for responding to freezer failure are all highly desirable. It is notable that most PCR machines do not restart automatically, and interruption of the run may spell a total loss of the reactions and time involved. Long computer analysis programs also will be irreversibly interrupted. Uninterruptable power sources can be purchased that will clean the power and switch to a battery backup within nanoseconds if necessary.

Plumbing can become a significant concern in molecular laboratories, and thought must be given to designating sinks for specific purposes. A large sink 3 to 5 feet long and 12 to 18 inches deep will be wanted for glassware washing and DNA sequencing purposes. On one sink, it may be useful to have a second faucet installed so that one of the faucets may be permanently connected to an aspirator pump for pulling vacuums. In addition, purified water (distilled and/or deionized and charcoal filtered) is required for accurate quantification of sensitive reactions. House-distilled water in some older installations may not be of the quality required. A purification system can be installed at individual sinks to supplement house distillation if needed.

Occasionally, a researcher will be asked to participate in the design of new laboratory facilities or substantial renovation to existing facilities. Although a detailed discussion of laboratory design is beyond the scope of this chapter, a few crucial issues can be stressed. First, seek architectural, engineering, and construction firms with experience in laboratory design and construction. This is a highly specialized area that presents technical problems new to many firms. Second, plan on investing much time and effort in the process. Work closely with the architects, engineers, and builders to familiarize yourself with the design process from the ground up.

Make sure they understand the needs and expectations of the end users of the facility. Third, do not assume anything will be automatically "taken care of." Ask many questions. Things that seem obvious to you may not be obvious to the architect, and minor oversights, especially in the early stages, can result in major problems later. Fourth, remember that you will have to live with the result. The architects, engineers, and institutional administrators will move on to other projects, but you and your colleagues will have to work in the resulting facility, good or bad, for years to come.

Organizing around Techniques

With multiple operations often occurring simultaneously, the organization of laboratory equipment needs to accommodate both the grouping of equipment by technique and the sharing of equipment among the different technologies. It is often the case in molecular evolutionary laboratories that several researchers will carry out similar research programs using the same molecules (or genes) and similar methods on different organisms. This requires multiple users to share similar sets of experimental facilities. Laboratory areas are most efficiently organized with designated areas for major techniques such as nucleic acid isolations, Southern blotting, PCR, and sequencing. This situation lends itself to the workstation concept mentioned above. A DNA sequencing workstation, for example, would typically include bench space for pouring gels, one or two gel rigs and associated power supplies, a gel drying unit and associated vacuum pump (aspirator type), and a variety of sequencing paraphernalia, all located near a large sink for washing the large glass plates used for pouring gels. Careful organization of stations around the techniques to be performed there can result in significant streamlining of research.

Safety

Safety concerns revolve around general laboratory safety, hazardous waste management, and radiation safety. Included in any laboratory operation should be the recognition of the potential hazards associated with the laboratory work space, and training in safety procedures should be included in the orientation of new laboratory members. Work with organic solvents or other noxious materials that are potentially volatile or aerosol-forming should be carried out in a fume hood. Emergency showers and eyewashes should be available in each section of the laboratory. Working with aqueous solutions and electrical equipment presents the constant hazard of electrical shock. To minimize this hazard, all electrophoresis apparatuses should be fitted with safety shields that cover buffer compartments during use, and circuits servicing areas near sinks should have ground fault circuit interrupters installed.

Use and disposal of potentially hazardous chemicals are becoming heavily regulated. Proper storage, maintenance of Material Safety Data Sheets (MSDS) about each chemical, and the control and documentation of hazardous waste disposal are among the required responses to laboratory safety. An institution Safety Officer can provide necessary information about reporting and disposal requirements.

Reagents labeled with radioactive isotopes (^{32}P, ^{35}S, ^{3}H) are commonly used in molecular biology despite recent development of chemiluminescent labeling techniques.[26,27] Obtaining appropriate licensing, working behind shields, monitoring work areas with radiation detectors, and disposing of radioactive waste properly are aspects of this work. Institutions with researchers using radioactive materials often employ a Radiation Safety Officer who will be cognizant of the local and federal rules and regulations. Clearly, close cooperation with this person is important and will result in the most satisfactory working conditions.

Planning for the Future

In attempting to equip and organize a comparative molecular laboratory, one must keep in mind that molecular biology is an extremely dynamic field. The rapid pace of technological innovation makes change the rule rather than the exception. A well-planned laboratory facility will have the capacity for change built into it. Standard features of modern design which contribute to flexibility include gridded drop ceilings that provide access to utilities and services and modular bench units that can be easily rearranged as the need arises. Even quite simple measures, such as buying tables with lockable wheels and installing extra circuits and receptacles so that electrical power is readily available throughout, can make the laboratory space substantially more adaptable to the demands of new technology.

Change affects not only methodological aspects of laboratory work but equipment requirements as well. For example, thermal cyclers are a necessity in molecular laboratories today but were unheard of before the advent of PCR. PCR technology has also made it possible to do a good deal of comparative molecular work without the aid of ultracentrifuges or steam sterilizers, items that are expensive but irreplaceable components of a standard molecular genetics laboratory. Clearly, it behooves the researcher

[27] P. M. Gillevet, *Nature (London)* **348,** 657 (1990).

to stay as current as possible with technological developments in order to anticipate changing demands on space and equipment.

Several recent innovations seem likely to have an impact on comparative molecular laboratories. Nonradioactive DNA labeling techniques promise to reduce the need for radioisotopes and associated safety concerns.[27,28] New applications of PCR continue to appear at a dizzying rate. These methods are sure to increase the need for thermal cyclers. Perhaps the most dramatic changes will come in laboratory automation. Automated instruments for the isolation of nucleic acids from various biological samples are available from several sources. Automated laboratory workstations equipped with industrial robot systems are also available commercially. Digital scanning systems allow automated capture and interpretation of data from two-dimensional images such as autoradiographs of restriction fragment length polymorphism (RFLP) or DNA sequencing gels. Automated DNA sequencers that combine gel electrophoresis with data capture have been available for several years. It seems likely that complete automation of DNA sequencing, from template preparation to direct data acquisition into computer files, will be possible. The potential effect that such extensive automation could have on molecular evolutionary research would be dramatic. Although laboratory automation is in its infancy, it is a rapidly growing field to which close attention should be given.

Finally, it should be noted that it is possible to overreact to new technology. The constant onslaught of new techniques and apparatus can be fascinating, and the temptation to switch to the latest (and trendiest) thing can be strong. However, the ultimate goal of answering evolutionary questions must be kept in mind, and the value of older techniques should not be underestimated. The real issue is how well a particular approach addresses the problem at hand. Sometimes an older, more mundane technique actually provides a better solution than the latest innovation. Many of the currently available techniques are likely to provide reliable, cost-effective approaches to questions of evolutionary interest for years to come.

Acknowledgments

We thank Debra Blue for typing and Drs. Carol J. Bult and Elizabeth A. Zimmer for reading and commenting on the manuscript.

[28] G. M. Church and S. Kietter-Higgins, *Science* **240,** 185 (1988).

[2] Collection and Storage of Land Plant Samples for Macromolecular Comparisons

By KENNETH J. SYTSMA, THOMAS J. GIVNISH, JAMES F. SMITH, and WILLIAM J. HAHN

Introduction

The plant molecular systematist is confronted with three major practical issues when initiating a survey of land plants: collection of plant tissue (location, type of tissue, quantity, preservation, transport, etc.), method of macromolecule extraction, and molecular technique used. These three factors are closely intertwined, and each can, and often does, have an influence on the others. The practical issues discussed in this chapter must therefore be considered in the broader context of issues relating to macromolecule isolation (see [12] in this volume) and the kind of molecular information to be obtained (see Section II in this volume).

This chapter deals specifically with three practical issues encountered in large-scale macromolecular systematic studies of land plants: (1) obtaining the plant tissue (including sources, tissues, collection, preservation, permits, and vouchers); (2) transport of the plant tissue; and (3) storage of plant tissue or macromolecules. Attention will focus on those studies involving DNA, but also on proteins (namely, isozymes) when issues dealing with the latter differ from those involving DNA.[1,2]

Sources of Land Plant Samples

Sources of land plant tissue have included field collections, botanical gardens and greenhouses, spores and seeds from seed banks or field collected, herbarium specimens, and fossils. In a cursory review of published macromolecular systematic studies on land plants, the great majority of DNA-based studies utilized field-collected tissue as the primary source and botanical gardens (and sometimes seed banks) as the secondary source. The exceptions include, for the most part, studies of agronomically important species and their immediate relatives that used greenhouse- or field-grown tissue as the primary source. In sharp contrast, slightly less than one-third of the isozyme-based studies utilized field-collected tissue, depending instead on greenhouse-grown tissue from field-collected seeds.

---

[1] D. J. Crawford, "Plant Molecular Systematics." Wiley, New York, 1990.
[2] R. W. Murphy, J. W. Sites, Jr., D. G. Buth, and C. H. Haufler, *in* "Molecular Systematics" (D. M. Hillis and C. Moritz, eds.), p. 45. Sinauer, Sunderland, Massachusetts, 1990.

METHODS IN ENZYMOLOGY, VOL. 224

This discrepancy highlights the importance of the ease (or difficulty) of macromolecular extraction when considering sources of plant tissue. Isozyme activity and ease of extraction generally decrease with maturity of plant tissue, mandating the use of seedlings grown from seed collections or young fronds.[3,4] DNA extraction from most field-collected tissue, however, is not difficult. In the few cases where extraction is difficult (e.g., Cactaceae, Onagraceae), the problem is generally overcome not by using younger tissue, but by experimentation with different extraction protocols.[5,6]

Field Collections

Field collections represent the most important source for land plant tissue, and arguably the best, as there is no question where the material originated. Additional advantages include the ability to collect within and among populations, to obtain rare or poorly collected plant species (these are often not represented in botanical gardens or seed banks), and, in the long run, to reduce the time prior to DNA or isozyme extraction (if the alternative is to use seed-grown material). Disadvantages include the time and expense associated with expeditions to remote areas along with the additional (and sometimes major) problems of plant tissue transport and permits (see below). Nevertheless, many interesting and important molecular systematic studies cannot be done without these field collections, particularly those involving a biogeographic bent or rare and/or narrowly distributed species. Those interested in pursuing molecular studies requiring visitation and collection in foreign countries with even limited logistical support from local botanists are urged to see the recommendations outlined in Mori and Holm-Nielsen.[7] Detailed checklists of floras, botanical gardens, herbaria, information on threatened species, and other useful addresses for all countries and island groups in the world are available in Davis *et al.*[8] Additional information is available from the Association of Systematic Collections (Washington, D.C.).

An alternative source of field-collected material is through procurement by local botanists who are either specialists for the group under study

[3] C. R. Werth, *Virginia J. Sci.* **36**, 53 (1985).
[4] D. E. Soltis, C. H. Haufler, D. C. Darrow, and G. J. Gastony, *Am. Fern J.* **73**, 9 (1983).
[5] R. Wallace, personal communication (1992).
[6] J. F. Smith, K. J. Sytsma, J. S. Shoemaker, and R. L. Smith, *Phytochem. Bull.* **23**, 2 (1991).
[7] S. A. Mori and L. B. Holm-Nielsen, *Taxon* **30**, 87 (1981).
[8] S. D. Davis, S. J. M. Droop, P. Gregerson, L. Henson, C. J. Leon, J. L. Villa-Lobos, H. Synge, and J. Zantovska, "Plants in Danger. What Do We Know?" International Union for the Conservation of Nature and Natural Resources, Cambridge, England, 1986.

or knowledgeable about specific collection sites. The *Directory and Guide to Resource Persons* of the American Society of Plant Taxonomists is invaluable in this regard as it contains both geographical and specialty listings. Although the use of local scientists can greatly add to the breadth of the study, a number of practical issues remain. These include providing detailed directions concerning shipment of the plant material and the necessity of providing required permits if crossing international boundaries (see below). A final issue that must be considered is whether the support in obtaining plant material has been so extensive and critical to the success of the project that the collector(s) should be considered a collaborator and thus a joint author.

Botanical Gardens

An emerging important source of land plant tissue is the botanical garden (including university-supported greenhouses). The botanical garden, long viewed as of secondary importance in comparison with the research herbarium, has become the primary source of plant tissue for molecular studies at higher levels. Botanical gardens are intrinsically teaching tools (both for scientists, students, and the general public) and thus maintain a large diversity of land plant families and genera, a collection ideal for selective sampling when involved in higher level molecular studies. The botanical garden, however, has also been very useful in, if not critical to, several molecular studies at the genus level and below (e.g., *Viburnum, Populus, Ulmus,* and *Pritchardia* in the laboratory of K.J.S.). A worldwide listing of botanical gardens[9] is an important resource base.

Despite the obvious benefits of obtaining plant tissue from botanical gardens (large diversity, ease of collection and transport, and savings in both costs and time), several potential problems remain. Foremost is the lack of complete voucher information (collector, date, exact locality) attached directly (or indirectly in records) to the garden specimen. A more serious problem is the possibility of errant identifications or label switches. The latter problems can be circumvented by making appropriate vouchers for subsequent identification by experts in each group (see below). A second practical problem exists if the burden for collection and transport of the plant material rests on botanical garden personnel. Most directors of botanical gardens are more than willing to provide access to their collection for molecular studies as it enhances the role, and thus continued support, of the garden. However, many botanical gardens do not have the

---

[9] D. M. Henderson and H. T. Prentice, "International Directory of Botanical Gardens. Regnum Vegetabile," Vol. 95. International Bureau for Plant Taxonomy and Nomenclature, Utrecht, The Netherlands, 1977.

time or personnel to oversee the multitude of requests for shipment of plant material; nor should they bear the expenses associated with the shipment of the plant material. Personal contact with a knowledgeable scientist at the botanical garden can minimize these problems.

Seed and Spore Banks

Seed banks are an increasingly important, and often underutilized, source for land plant tissue. Not surprisingly, molecular studies of agronomically important groups have primarily used seed banks of the cultivated species and their wild relatives (e.g., *Brassica, Solanum, Triticum, Zea, Glycine*). The seed bank of *Solanum* (Sturgeon Bay, WI) is annually updated and planted out to maintain the collection; this collection, in turn, has been essential for the large molecular systematic research program on the cultivated potato and wild relatives.[10] A number of botanical gardens maintain seed exchange lists (e.g., *Index Seminum* by the Botanic Gardens of Indonesia). The Royal Botanical Garden at Kew has the most extensive listing of fully documented species maintained in their seed bank. Through the far-sighted efforts of specialists during the last few decades, many families of noncultivated plants are well-represented in the seed bank (e.g., Onagraceae).

Three problems with seed sources, either from seed banks or university greenhouse collections, have been encountered in past molecular systematic studies. These include contaminated seed source, errors in handling or labeling, and misidentification.[11] Needless to say, these problems can be circumvented by vouchering all plant tissue grown from seed (see below).

Spores from ferns and fern allies can be ideal sources for obtaining tissue of these plants. Although fronds and rhizomes are the most used, spores and their resulting gametophytes have also been used.[12] The American Fern Society maintains a spore exchange program, and information can be obtained by consulting the Bulletin of the American Fern Society.

Herbarium Specimens

A number of studies have indicated that DNA can be successfully extracted from herbarium specimens (personal observation).[13-15] It is very unlikely that isozymes will ever be routinely extracted in an active state

[10] D. Spooner, personal communication (1992).
[11] J. Palmer, personal communication (1992).
[12] D. Stein, personal communication (1992).
[13] O. S. Rogers and A. J. Bendich, *Plant Mol. Biol.* **5,** 69 (1985).
[14] J. J. Doyle and E. E. Dickson, *Taxon* **36,** 715 (1987).
[15] M. M. Pyle and R. P. Adams, *Taxon* **38,** 576 (1989).

from herbarium specimens, although certain dehydrated tissue can provide good activity for some systems (see below).[16] DNA can be obtained from herbarium specimens, perhaps more often than not, but its relatively poor quality precludes efficient and routine restriction site analysis (personal observation).[17] However, these DNAs have proved to be very good templates for polymerase chain reaction (PCR) amplification of specific sequences (personal observation).[18] As herbarium specimens encompass all named species and because they are fully vouchered and retrievable, this source of plant tissue undoubtedly will become more important in the future as PCR and sequencing allow rapid and extensive surveys at all taxonomic levels.

Fossils

The extraction of DNA from mummified and fossilized plant tissue is one of the most exciting developments in molecular systematics. Mummified plant tissues up to 44,600 years old have yielded analyzable DNA.[13,19] The recent publications of *rbc*L sequences from Miocene fossils (17–20 million years old) of *Magnolia*[20] and *Taxodium*[21] and the amplification of similar-aged DNA from *Platanus* and *Pseudofagus*[21] open a new source for land plant tissue and offer enormous possibilities for research in molecular systematics and evolution. The latter report clearly invalidates the objections[22] to the authenticity of the Miocene fossil plant DNA and hence of their systematic utility.

Collection of Land Plant Tissues

Permits for Collection of Land Plant Tissue

The first step in initiating the collection of plant tissue is determining which permits are required. Because many permits (particularly those needed overseas) can take 6 months or longer to obtain, it is imperative that the application for permits begin as soon as possible. Within the

---

[16] A. Liston, L. H. Rieseberg, R. P. Adams, N. Do, and G. Zhu, *Ann. Mo. Bot. Gard.* **77**, 859 (1990).
[17] M. Chase and J. Hills, *Taxon* **40**, 215 (1991).
[18] E. Conti, J. Doyle, D. Soltis, R. Price, M. Chase, J. Rodman, personal communication (1992).
[19] F. Rollo, A. La Marca, and A. Amici, *Theor. Appl. Genet.* **73**, 501 (1987).
[20] E. M. Golenberg, D. E. Giannasi, M. T. Clegg, C. J. Smiley, M. Durbin, D. Henderson, and G. Zurawski, *Nature (London)* **344**, 656 (1990).
[21] P. S. Soltis, D. E. Soltis, and C. J. Smiley, *Proc. Natl. Acad. Sci. U.S.A.* **89**, 449 (1992).
[22] S. Pääbo and A. C. Wilson, *Curr. Biol.* **1**, 45 (1991).

United States, permits are often mandatory for plant collections made within city, county, state, and national parks, as well as for those made in wilderness areas, conservancies, and other protected sites. Such permits are usually specific as to species and quantity of material allowed for collection and expire within 6 to 12 months. Many require submitting a detailed proposal, depositing a set of herbarium specimens, and presenting periodic field or progress reports (personal observation).[23]

Permits for collecting plant material in tropical regions outside the United States (e.g., Latin America, Africa, or Australasia) typically require 1 year or more of advance work. Many countries have quite different rules (Bolivia has no application process, whereas the process in Peru is lengthy and complex), and regulations and requirements are apt to change without notice. In addition, the regulatory bureaus in some countries simply do not have the time or resources (or perhaps the inclination) to respond to correspondence regarding permits from foreign nationals. It is thus suggested that contact first be made with botanists in the host countries (the listings in Holmgren et al.[24] and Davis et al.[8] are ideal for this purpose), as well as with their respective consulates. Mail is often slow and uncertain, so that establishing contact with international colleagues via telephone, fax (facsimile), telex, or express mail is frequently desirable. Additional (and more practical) sources of information include the Association of Systematic Collections (Washington, D.C.), fellow scientists who have recently traveled in the countries of concern, and institutions (e.g., Smithsonian Institution, Missouri Botanical Garden, New York Botanical Garden) that maintain staff or projects in various countries.

Permits are often the greatest hurdle (other than obtaining support) involved in collecting plants in foreign countries. For helicopter expeditions to the tepuis of Amazonas Territory in southern Venezuela, for example, seven different permits are needed from the Ministry of Environment (plant collecting permit and phytosanitary certificate), the Institute of National Parks, the Indian Affairs Bureau (including separate paperwork from both the national office in Caracas and the territorial office in Puerto Ayacucho), the National Guard, and the Governor of the Amazonas Territory. (In addition, in one case we had to be deputized as members of the Ministry of Environment in order to circumvent a blanket ban on tepui travel imposed after the filming of *Arachnophobia*.) Even with up to 1 year of lead time to obtain these permits with the help of a local botanist, 1 week or more of permit negotiations in Caracas and Puerto

---

[23] B. Baldwin, personal communication (1992).
[24] P. K. Holmgren, N. H. Holmgren, and L. C. Barnett, "Index Herbariorum, Part I: The Herbaria of the World." New York Botanical Gardens, Bronx, New York, 1990.

Ayacucho is often necessary to finalize the paperwork. A duplicate set of herbarium specimens (usually including the unicates) must be deposited in a national or local herbarium; often, a preliminary field trip report and/or a complete copy of field notes must be submitted after the expedition before the investigator is permitted to leave the country.

The U.S. Department of Agriculture (USDA) permit for entry of plant material into the United States is a general requirement for all work involving field collections from other countries. This requirement also holds for material originating in Canada, although packages clearly marked "Plant Material for Scientific Study" but without a USDA permit have not been stopped or delayed.[25] This permit can be obtained for specific taxa and investigators, although we recommend a "blanket" permit that covers a number of researchers and plant groups. International shipment of endangered plants listed in the CITES (Convention on International Trade in Endangered Species of Wild Fauna and Flora) protocol (e.g., many orchids and carnivorous plants) is strongly controlled, and a specific waiver must be sought in order to ship them to the United States.

Copies of the USDA import permit should be carried at all times and forwarded to colleagues who will be mailing plant material from other countries, so that copies may be included inside the package and outside with the shipping label. Many countries require a phytosanitary certificate to attest to the pest-free nature of the plant material being shipped before it may even be exported from that country; in our experience, a phytosanitary certificate is neither necessary nor sufficient for passing material through U.S. Customs, although it does eliminate one potential reason for rejecting material. Material of economically important groups (e.g., members of the Rutaceae, Poaceae, Orchidaceae) are subject to especially stringent inspection for insects, arthropods, viruses, and other pathogens.

Vouchers for Land Plant Collections

An essential, and sometimes overlooked, part of plant tissue collection is the herbarium voucher. Herbarium specimen vouchers are usually obtained with field-collected material, but they are often neglected when the tissue is obtained via seeds or through an intermediate source (botanical garden, colleague, etc.). Botanical garden collections are numbered by accession, and the original voucher information (collector and number) can usually be traced in records maintained by the botanical garden. This information should be obtained and recorded, but it is also prudent to have a second voucher made at the same time as the tissue is collected. Label

---

[25] D. Soltis, personal communication (1992).

switches or errors in collecting do occur, and the voucher at time of collection is extremely important to determine these kinds of mistakes.

The specimen should be properly dried, labeled, and identified by competent systematists or even specialists. Photographic records are helpful and sometimes must take the place of a plant specimen in situations when endangered or threatened species are involved and populations have already been vouchered (e.g., Hawaiian Lobeliaceae). Special arrangements must be made if the voucher for the molecular systematic study is unicate and will be carried out of the country, because most foreign countries require that the first set of plant specimens be deposited at the national or a local herbarium. The herbarium in which the vouchers are deposited (see Holmgren et al.[24]) should be cited in subsequent publications.

General Methods for Collection of Land Plant Tissue

Plant tissue collection for DNA and protein studies has primarily utilized leaf tissue, but seeds, roots, flowers, stems, pollen, spores, and gametophytes have all been used successfully. For example, DNA extraction from the parasitic *Cuscuta* required using only internode tissue to prevent DNA contamination from its host species[26]; likewise, the green stems of a leafless *Koeberlinia* provided an adequate source of DNA.[27]

The collection of plant tissue is quite different from animal tissue collection. The discussion of collection of plant and animal tissue by Dessauer et al.[28] is detailed and helpful. However, the recommendations for procedures unique to plant tissue collection are somewhat misleading and outdated, especially when tropical collections are involved. Plant tissue can now be collected and transported as either fresh tissue (leaves and/or shoot cuttings) or preserved tissue; the latter either as cryopreserved tissue (liquid nitrogen or dry ice) or as dried tissue (air-dried, herbarium-dried, lyophilized, or chemically dried). Ambient-temperature liquid chemical preservation techniques (such as those routinely done for herbarium plant specimens in the tropics) so far have been ineffective in maintaining adequate yields of high-quality DNA.[15] It should be stressed again that the manner of collecting plant tissue is dictated by several other factors: what macromolecule (DNA, RNA, or isozymes) will be examined, what type of nucleic acid extraction method will be used (or, more impor-

[26] R. Olmstead, personal communication (1992).
[27] R. Price, personal communication (1992).
[28] H. C. Dessauer, C. J. Cole, and M. S. Hafner, in "Molecular Systematics" (D. M. Hillis and C. Moritz, eds.), p. 25. Sinauer, Sunderland, Massachusetts, 1990.

tantly, what method will work), how the tissue will be preserved (if done) and transported, and how pure and intact the macromolecules must be.

Young, actively growing leaves or shoots are the best tissues to collect. Tissue from different plants should always be kept separate even if the level of analysis to be examined does not necessarily require this (e.g., chloroplast DNA restriction site analysis among populations). Future studies involving single-copy genes, introgression, hybridization, or recombination require a knowledge of the specific source(s) of the macromolecules. Once collected the plant tissue should be carefully cleaned of dirt, epiphytes, fungi, insects, etc., and then blotted dry. This is critical when transport occurs across international boundaries or between Hawaii and the mainland, as discussed above.

Fresh Tissue

The collection and transport of fresh material have been and will continue to be most important, except perhaps in tropical areas. Although not as logistically simple as chemical preservation (see below), fresh tissue routinely provides the highest yield and quality of DNA amenable to amplification and restriction digestion (personal observation).[17] If collected as whole plants or shoot cuttings and placed in large bags, the leaf tissue is maintained the longest, provides an immediate voucher when some of the leaf tissue is separated, and allows for later propagation in greenhouses.

Contrary to past ideas and practices (e.g., Dessauer *et al.*[28]), most plant tissue does not have to be immediately placed on ice or frozen. Expeditions of several days to a week's duration in Venezuelan tepuis, Panama, Philippines, Thailand, Malaysia, and Hawaii in which we have been involved indicate that a diverse array of land plant tissues can be kept in stable condition if placed in Ziploc bags or larger storage bags and then placed in a Styrofoam box away from light and fluctuating temperatures. If wet ice or chemical ice (blue ice) is available or if the trip is of short duration, a small amount of ice in the cooler to be brought into the field is recommended. Once a continuous source of refrigeration is encountered, the plant tissue should then be kept at this cooler refrigerated temperature until processed or placed in long-term ultracold storage (see below).

The most important point to remember in deciding how to collect fresh plant tissue is that fluctuating heat exposure or warm up from cold temperatures should be avoided. Because of the relatively more sensitive nature of the activity of proteins in isozyme analysis, fresh plant tissue might require more rapid exposure to cooler temperatures. However, leaf tissue collected in Panama (in a cooler with ice) resulted in very adequate activities of isozymes and as well yielded high-quality DNA, despite the elapse of 8 days from collection to ultracold storage of some samples.[29-31] Ziploc bags, of various sizes but of sturdy nature, are ideal for maintaining and subsequent transporting leaf or shoot tissue. Most tissue can be maintained in very good condition without any moisture added to the bag, as the tissue will transpire naturally. Some collectors prefer to add a damp, but not dripping, paper towel to increase the moisture content of the atmosphere, but this might not be suitable for delicate leaves as it can promote waterlogging and rotting (personal observation). The bag should be clearly marked on the outside with permanent marker ink as well as in the inside on a loose piece of paper (permanent marker ink or heavy pencil); both markers should include all information pertinent to identifying the plant tissue (date, species, locality, collector, voucher number, etc.).

Cryopreserved Tissue

Cryopreservation (using liquid nitrogen and dry ice) of plant tissue has been done effectively on some long-term expeditions in the Amazon basin[32] and to remote Venezuelan tepuis,[33] and in the molecular studies of Miocene fossil tissues.[20,21] These methods involve considerably more planning and logistical support as sources of liquid nitrogen and dry ice are needed. Medical and veterinarian clinics, universities, hospitals, welding supply companies, and mining operations are possible sources of liquid nitrogen.[28,34] Dry ice is generally available from airlines if requested by their customers.[23] Authorization for shipment of these chemicals on airlines is absolutely mandatory, and both are classified as "Restricted Articles" (see below). Most molecular studies on plants, unlike those with animals, do not require cryopreservation.

Dried Tissue

Various methods of dehydrating plant tissue prior to transport and macromolecular extraction have met with some success with a limited sampling of plant taxa. These have included lyophilization,[35,36] dehydra-

[29] K. J. Sytsma and B. A. Schaal, *Evolution* 39, 582 (1985).
[30] K. J. Sytsma and B. A. Schaal, *Evolution* 39, 594 (1985).
[31] K. J. Sytsma and B. A. Schaal, *Plant Syst. Evol.* 170, 97 (1990).
[32] B. Boom, personal communication (1992).
[33] V. Funk, personal communication (1992).
[34] H. C. Dessauer and M. S. Hafner (eds.), "Collections of Frozen Tissues: Value, Management, Field and Laboratory Procedures, and Directory of Existing Collections." Association of Systematics Collections, Univ. of Kansas Press, Lawrence, Kansas, 1984.
[35] M. A. Saghai-Maroof, K. M. Soliman, R. A. Jorgensen, and R. W. Allard, *Proc. Natl. Acad. Sci. U.S.A.* 81, 8014 (1984).
[36] J. L. Hamrick and M. D. Loveless, *Biotropica* 18, 201 (1986).

tion via a food drier,[37] air-drying,[13–15,38] and herbarium specimen (forced-air or heat) drying.[13,14] However, as pointed out by Chase and Hills[17] and confirmed by experiments in our laboratories, these techniques cannot be universally applied to all plants. Moreover, when these methods do work, the DNA is often sufficiently degraded (or degrades in time) to preclude their use for restriction site analysis (personal observation).[17] Importantly, however, these same DNA samples often amplify nicely with the PCR (personal observation).[17] The warning of Chase and Hills[17] to test various preservation techniques prior to large-scale collections should be heeded to prevent "discovering failure after two months of fieldwork"!

Methods of preserving collected plant tissue in $CaSO_4$ (Drierite) or silica gel[16,17] offer the best, and sometimes only, alternative solution to collecting plant tissue when fresh tissue collection and transport is not possible. Tissue collected by these methods has been tested for DNA extraction, digestion of DNA with restriction enzymes, or sequencing of DNA. Tested species include a diversity of angiosperm taxa from remote areas: numerous species from China[16]; numerous species from Argentina and Brazil[17]; Myrtales, Proteales, Theales, and Malvales from Australia, Madagascar, and Africa[39]; Asteraceae from Juan Fernandez Islands[40]; and Bromeliaceae from South America.[41] This form of chemical preservation is simple to use; the chemicals are readily available and easy to carry or mail to/from remote areas, and thus ideal for obtaining species that exist in remote areas or are to be collected by helpers in other countries. As with other methods of plant tissue collection, experimentation with the methods before planning long collecting trips is critical. We have found that Malagasy *Sarcolaena* (Sarcolaenaceae) leaf tissue preserved with Drierite provides high molecular weight DNA that amplified readily for *rbc*L sequencing, but the yield of total DNA precluded use for restriction site mapping analysis.[39]

Details concerning these two methods of collection and preservation are found in Liston *et al.*[16] and Chase and Hills,[17] and the following information comes from the latter. Silica gel is a more efficient desiccant than Drierite as it can be obtained in smaller mesh size (28–200, grade 12 from Fisher Scientific, Pittsburgh, PA), allowing for greater surface area coverage of the leaf tissue. Usually 4–6 g of fresh leaves are placed in small (12 × 8 cm) Ziploc plastic bags. Leaf tissue will dry faster and thus with less DNA degradation if first torn into smaller pieces. Subsequently, 50–60 g of silica gel (minimum 10:1 gram ratio of silica gel to leaf tissue) is

[37] T. H. Tai and S. D. Tanksley, *Plant Mol. Biol. Rep.* **8,** 297 (1990).
[38] E. Knox, personal communication (1992).
[39] K. Sytsma and T. Givnish, unpublished observations (1992).
[40] D. Crawford, personal communication (1992).
[41] G. Brown, personal communication (1992).

added to the bag. Drying should take place within 12 hr; this is determined by checking to see if leaf tissue snaps cleanly when bent. Longer exposures to silica gel might be required, as in many monocots, but longer periods to reach desiccation usually result in DNA degradation. If the silica gel is to be reused, about 5% (w/w) indicating silica gel (6-16 mesh, grade 42; deep violet-blue color) should be added. The hydrated silica gel (indicator silica gel will be whitish pink) can be rejuvenated (dehydrated back to original violet-blue color) at oven temperatures (175°) for 1 hr or 4–5 hr on a light bulb plant dryer. The dried plant tissue in the Ziploc bags, with trace amounts of indicator silica gel to verify that tissue is not rehydrating, should then be kept in tightly sealed plastic boxes. Extreme caution is necessary when reusing these drying agents to prevent cross-contamination of samples in case where PCR amplification will be used.

Transport of Land Plant Samples

Fresh Tissue Transport

The transport of fresh tissue has been done routinely in two ways. For short transport times (2–3 days maximum), leaf tissue or shoot cuttings in Ziploc bags can be carried at ambient temperature in Styrofoam containers (to prevent fluctuations in temperature) or mailed directly via overnight or express mail couriers, with or without Styrofoam containment. This is often the method of transport used when securing plant tissue from botanical gardens within the United States or Canada. It is important that the botanical garden be provided the account number to bill the receiver for express mail.

A second and more reliable, but generally more expensive, transport method involves placing the bags of tissue directly into an ice source in a Styrofoam container. Wet ice is most often used but tends to warm more quickly and often leads to water-soaked packages after several days. To prevent cold temperature burn with any form of ice, the Ziploc bags should be separated from the ice source by several thicknesses of newspaper. An alternative strategy for backpacking trips (up to 4 days) is to carry a small amount of dry ice wrapped up in canvas, and this is placed with the wet ice to prevent the latter from melting. As the plant tissue is added to the container, excess dry ice is discarded, and the plant tissue is kept cooled with the frozen wet ice only.

Fresh tissue transport from remote areas is more difficult and involves phytosanitary, export, and import permits (see above). We have had no problem in carrying fresh tissue on wet ice in Styrofoam containers as airline hold or carryon baggage from Central and South America, Asia,

Australasia, and Hawaii. When longer periods of travel are expected, fresh tissue can be readily shipped on wet ice via international mail couriers, such as DHL, from many countries in South America and Australasia, as well as from Africa.[38] Small packages cost minimally $60–70 and require specific documentation to pass customs and agricultural inspection points, in both the shipping and receiving countries (see above). The mail courier, however, handles all passage until reaching the destination. Using airlines to ship packages is more involved and risky. Many airlines refuse to handle a package originating overseas unless it has cleared customs, thus requiring a contact person to intercept and pass the package through the entry port. In all cases, a copy of the USDA permit, phytosanitary certificate, and letter identifying and describing the purpose of the plant tissue should be attached inside the package and outside with the shipping label.

Alternatively, and as a last resort, leaf tissue can be air-mailed in small envelopes with attached copies of all permits. The envelopes usually arrive within 1 or 2 weeks and normally do not go through customs and inspection.

Frozen Tissue Transport

Transport of frozen tissue on dry ice or in liquid nitrogen Dewar containers is relatively easy if no air shipment is required. Transport via airline is possible but difficult. Airlines permit up to 200 kg of dry ice in a single package. The "Shipper's Certification for Restricted Articles" is required, and the package must be marked as Hazard Class ORM-A.[28] Airlines also permit up to 50 kg of liquid nitrogen to be transported if the container is nonpressurized and of the metal Dewar type. The Dewar container must be well labeled with "Nonpressurized Liquid Nitrogen" and "This Side Up" to prevent careless handling and spillage. Alternatively, the liquid nitrogen can be discarded before short flights or a "dry shipper" can be used that prevents liquid nitrogen spillage. Details on cryopreservation, sources of the chemicals, and shipping regulations are provided in Dessauer et al.[28]

Dried Plant Tissue Transport

Dried plant tissue is the easiest to transport but still can require permits (as with herbarium voucher specimens) to clear customs. A small amount of the desiccant in contact with the leaf tissue in Ziploc bags or plastic bottles is recommended. Indicator desiccant is recommended to avoid possible confusion concerning the nature of the "white powder."[17] Customs and agricultural inspection officials in both exiting and entering countries are less prone to cause delays with plant tissue that is obviously dried and brittle.

Storage of Land Plant Samples

Storage of Plant Tissue

The third major issue to address in macromolecular systematic studies deals with storage of the plant tissue when it comes into the laboratory. The first approach is to immediately place the tissue into ultracold storage ($-70°$ or colder). The chest-type freezers have more efficient use of interior space and maintain colder temperatures than the upright models, although the latter require less floor space. The leaf tissue should be separated from stem tissue, rinsed with distilled water, blotted dry, weighed, torn into small pieces if necessary, and placed into clean Ziploc bags with all information recorded on the outside and inside of the bag. Permanent marker eventually deteriorates at extreme cold temperature, as do many kinds of tape, so placing a label inside the bag is necessary. Alternatively, the tissue can be powdered under liquid nitrogen with a mortar and pestle and placed in small bags, plastic tubes, or plastic bottles.

Owing to space limitations some laboratories have stored plant tissue at $-20°$ (after flash freezing in liquid nitrogen or prior storage at $-70°$) with apparently no DNA degradation after 6–8 months.[42] We have seen marked degradation of DNA when leaf tissue (*Caryota*, Palmae) was stored at $-20°$ versus $-70°$. The lower temperature storage is recommended, and this is required if the tissue will also be the source for isozyme analysis. We have noticed no difference in DNA quality (however, isozyme activity has not been tested) between tissue flash-frozen in liquid nitrogen before being placed in the ultracold freezer relative to tissue simply placed in the freezer.

The second approach to storage of plant tissue is to minimize or even eliminate storage in ultracold freezers and to store only the DNA permanently. Plant tissue is placed in the refrigerator as it arrives, and then DNA extraction is done as soon as possible. This is the preferred method (e.g., J. Palmer laboratory)[11] in studies where fast DNA extraction methods work well, tissue can be processed quickly, and no backup tissue source for additional extractions of either DNA or proteins is needed or expected. The benefits of this approach include the following: (1) the need is diminished for ultracold storage space, which is costly in terms of purchasing, electrical power, and maintenance; (2) space is then available for more

---

[42] S. Downie and J. Doyle, personal communication (1992).

important tissues where a backup tissue source is required; (3) DNA samples are quickly and efficiently obtained; and (4) the not uncommon breakdown of ultracold freezers (enhanced unfortunately by the continual opening and closing necessitated in the first method) will not be so serious because DNA samples will have already been obtained.

Inventory of Plant Samples

Keeping track of plant tissue that has arrived into the laboratory, been placed into ultracold storage, and extracted for DNA (or proteins), along with the quantity of DNA that is left, is a difficult but important task, especially when a large number of concurrent molecular studies are underway involving many different scientists. Our laboratory operates under a decentralized system where each project investigator maintains a separate record (most often on a spreadsheet) of each plant sample. Other laboratories[11] operate under a completely centralized system where all incoming plant tissue and resulting DNAs are assigned numbers on a first-come basis, with information being recorded under these numbers. With either approach it is critical to include, or have referenced, all voucher information (date of collection, locality, collector, number), quantity of plant tissue remaining, type of DNA extraction, quantity and quality of DNA, date of extraction, and a photograph of the electrophoresed DNA stained with ethidium bromide.

To facilitate finding the frozen plant tissue or DNA sample listed on the inventory sheet, both the ultracold freezer and the preferred nonfrostless ($-20°$) freezer for DNA storage should be well organized. As mentioned above, we maintain separate boxes to hold frozen plant tissue for each project. Metal (Revco) racks that fit snugly into chest freezers are ideal for DNA storage. These racks are designed to hold 7 or 11 Revco boxes (8 cm square, 2 inches high) partitioned inside with dividers to carry up to 100 1.5-ml microcentrifuge tubes. The contents of both the ultracold storage boxes and the DNA storage boxes should be clearly marked to permit rapid entry into and exit from the freezers.

Acknowledgments

This chapter would not have been possible without the detailed information supplied by B. Baldwin, M. Chase, D. Crawford, S. Downie, J. Doyle, R. Olmstead, J. Palmer, R. Price, J. Rodman, D. Soltis, and D. Stein. Experimentation on methods of collecting and shipping tropical plant material was made possible by National Science Foundation Grants BSR 8516573, 8806520, 8815173, 9007293, and 9016260, and by grants from the National Geographic Society and the Nave Fund (University of Wisconsin, Madison, WI).

[3] Collection and Storage of Vertebrate Samples

By Rebecca L. Cann, Robert A. Feldman, Leonard A. Freed,
J. Koji Lum, and Carol A. Reeb

Introduction

Biologists have an obligation to "sample" natural populations using methodologies that will ensure the long-term viability of the taxa they study. To do this, they must realize that estimates of the census size (N) of a population and the effective population size (N_e) can be orders of magnitude different, depending on the geologically recent past and on the breeding structure of present demes.[1] Most population genetic models assume equilibrium conditions when, in fact, few evolutionarily interesting populations actually fulfill the conditions of these models. Species listed as threatened and endangered manifestly violate them. Even seemingly abundant and unlisted populations can experience precipitous declines, as has been noted for migratory birds of North America.[2] Regulatory agencies may legally permit oversampling of them as bureaucratic procedures lag behind new information from the field. The biodiversity crisis of this planet should compel all biologists, whether they work with common or rare species, to examine the methodological alternatives to destructive sampling whenever possible.

The goal of this chapter is to provide guidelines for investigators who, in the course of their research, must obtain, hold, transport, process, and archive samples from vertebrates. There are legal, logistic, and technical issues that need to be considered. Timely recognition and planning can make the difference between successful research and a wasted effort.

Relationships with Regulatory Agencies

State and federal agencies are responsible for issuing various permits regulating the capture, holding, and sampling procedures that pertain to most free-ranging wild or feral animals. They dictate the numbers and circumstances under which scientists can collect blood, hair, feathers, urine, saliva, semen, milk, eggs, venom, body tissues, and whole carcasses. Road kills are usually not exempted from these regulations. Investigators

---

[1] J. C. Avise, *Annu. Rev. Genet.* **25**, 45 (1991).
[2] J. Terborgh, "Where Have All The Birds Gone?" Princeton Univ. Press, Princeton, New Jersey, 1989.

should determine whether the taxa of interest have any special restrictions or are unregulated, as in the case of designated pest species.

Researchers are further urged to contact personnel within relevant agencies before planning experiments, especially if the investigator desires unstable or easily degraded materials such as mRNAs of certain age or tissue specificity. If annual restrictions on take (such as sex, breeding versus nonbreeding season) apply to the organism, such material may only be obtainable at certain times. Such basic information potentially influences the starting dates of grants and the academic schedules of graduate students.

Most institutions receiving federal grants also have internally constituted animal care advisory committees, which screen research proposals that include use of vertebrates. Unless investigators plan experiments that require animals to be housed on-site or that utilize protocols which can be expected to produce significant but unavoidable stress, permission to take small quantities of blood, urine, feathers, feces, or fur from healthy animals is usually granted routinely.

Besides specifying numbers to be sampled and the times at which animals can be taken from the wild, regulatory agencies often dictate the preferred or permitted method of collecting. In the case of passerine birds, the use of anesthesia for bleeding and even minor abdominal surgery, such as laparotomies, may be counterproductive.[3] Dessauer *et al.*[4] cover the effect of anesthesia on proteins. The time permitted to hold animals for processing while collecting bodily fluids and making morphological measurements may be limited to only 1 hr to minimize stress. Handling of birds, bats, and any animal with chemical defenses requires special skills, of which laboratory biologists may be unaware. The best strategy is usually one of close collaboration with a field biologist skilled and certified to handle these species.

Additional considerations include vaccinations of investigators against rabies or tuberculosis, especially when working with some mammals and their parasites. These preventative measures are needed to protect the researcher. Animal necropsy should include protective clothing and the containment of possible biohazards, in order to avoid contracting diseases such as psittacosis, erysipelas, rickettsial infections, and brucellosis.[5] The

[3] L. Oring, K. P. Able, D. W. Anderson, L. F. Baptista, J. C. Barlow, A. S. Gaunt, F. B. Gill, and J. C. Wingfield, Ad Hoc Committee on the Use of Wild Birds in Research, Report of the American Ornithologists' Union, Cooper Ornithological Society, Wilson Ornithological Society, Supplement to *Auk,* **105,** (1988).

[4] H. C. Dessauer, C. J. Cole, and M. S. Hafner, *in* "Molecular Systematics" (D. M. Hillis and C. Moritz, eds.), p. 25. Sinauer, Sunderland, Massachusetts, 1990.

National Wildlife Health Center, located in Madison, WI (see Appendix), has a Resource Health Team available to examine animal specimens for possible diseases, and it may have frozen materials of interest to molecular evolutionists concerned with pathogen – host dynamics.

For endangered species, an even higher level of screening exists regarding which personnel may actually handle animals. Investigators should include on their permits all students and technicians who perform field and laboratory protocols with materials from endangered species. Changes in research protocols or personnel employed should be promptly reported. Heavy fines for violation of permit conditions are a strong incentive for careful attention to detail.[6] Preparation of DNA from endangered species carries legal responsibilities akin to the handling of feathers, bones, or skins from these animals.

The Convention on International Trade in Endangered Species of Wild Fauna and Flora (CITES) may cover additional regulations concerning importation and shipment of endangered species. It is folly to cross international boundaries with these animals, without collaborators fluent in the permitting process and in the languages in question on both sides. If the animal must be shipped alive, state and federal regulations usually require quarantine periods, as well as agricultural inspections of cage litter for noxious weeds or invertebrates (see Appendix for addresses of appropriate agencies involved). Shipping live animals between the continental United States and either Alaska or Hawaii should be avoided whenever possible, as it requires lengthy quarantine and, except for captive breeding experiments, is usually unnecessary in this age of DNA amplification.

Information regarding federal regulations can be obtained from U.S. Fish and Wildlife Service branch offices, or from the address listed in the Appendix. State regulations vary, depending on jurisdiction (fisheries versus land and natural resources offices), but statutes on collections are usually similar. Collecting in state or national parks, wildlife refuges, marine sanctuaries, or at monuments requires permits, and fewer regulations may pertain to collections on the land of a private owner, except for threatened or endangered species. We emphasize that endangered species as individual organisms are not biologically fragile, but rather are endangered at the population level because of historical mistakes made by humans.

---

[5] M. Friend (ed.), "Field Guide to Wildlife Diseases," U.S. Department of the Interior, Fish and Wildlife Service Resource Publication No. 167, Washington, D.C., 1987.

[6] C. Holden (ed.), *Science* **255**, 406 (1992).

Existing Sources and Vouchers

Various sources, such as the teaching guide to the text *Biology*,[7] contain extensive lists of national parks, botanical gardens, zoos, museums, and biological supply houses that may serve as collection sources for molecular evolutionary studies. Before beginning a new research project, we recommend investigators first consult the directory of existing collections of frozen tissues, compiled by Dessauer and Hafner,[8] which gives great detail concerning the holdings of public and private institutions both in the United States and abroad.

Finally, we wish to stress the dynamic nature of systematics and the necessary changes in nomenclature that accompany the discovery of new species or the reworking of evolutionary relationships within previously described taxa. Some molecular evolutionists may be naive about the importance of voucher specimens and the necessity of retaining information about the actual animals from which they have sampled nucleic acids or proteins. Voucher specimens "physically and permanently document data in an archival report by (1) verifying the identity of the organism(s) used in the study; and (2) by so doing, ensure that a study which otherwise could not be repeated can be accurately reviewed or reassessed."[9]

Voucher specimens both serve as a backup for future work and act as important documentation for ongoing systematic investigations. This is especially true for research across hybrid zones, geologically disturbed areas, or regions characterized by high levels of migration and dispersal. Amphibian, reptile, bird, and fish projects may be totally jeopardized if these specimens are not stored correctly, and museum curators should be consulted about which storage method (liquid, frozen, or desiccated) is preferred for each taxa. Skins are usually preserved for birds and mammals, whereas wet storage is preferred for amphibians and reptiles. General methods for preparing vertebrate specimens to be stored as vouchers are discussed by Hall[10] and Wagstaffe and Fidler.[11] Specific information re-

---

[7] P. H. Raven, G. B. Johnson, and V. E. Zenger, "Instructor's Manual to Accompany Raven and Johnson *Biology*," p. 33. Times Mirror/Mosby College Publ., St. Louis, Missouri, 1986.

[8] H. C. Dessauer and M. S. Hafner, "Collections of Frozen Tissues," p. 46. Association of Systematics Collections, Univ. of Kansas Press, Lawrence, Kansas, 1984.

[9] W. L. Lee, B. M. Bell, and J. F. Sutton, "Guidelines for Acquisition and Management of Biological Specimens." Association of Systematics Collections, Univ. of Kansas Press, Lawrence, Kansas, 1982.

[10] E. R. Hall, University of Kansas Museum of Natural History Miscellaneous Publication 30 (1962).

[11] R. Wagstaffe and J. H. Fidler, "Preservation of Natural History Specimens, Volume 2: Zoology/Vertebrates." H. F. G. Witherby Ltd., London, 1968.

garding preparation of bird specimens and shipment of samples to national institutions is included in Ref. 5.

Factors Affecting Utility of Collected Specimens

Preservation

Investigators concerned with preservation of nucleic acids and proteins in modern vertebrate collections may benefit from reviewing the discussion of researchers recorded at the Royal Society meeting in March 1991, "Molecules through time: Fossil molecules and biochemical systematics."[12] Chemists addressed the preservation of both soluble components and insoluble debris, primarily the linkages and the groups susceptible to modifications. For proteins, those include peptide bonds, side-chain functional groups (amino, carboxyl, hydroxyl, and sulfhydryl groups), and chiral centers (α-carbon atoms of amino acid units subject to racemization). In nucleic acids (DNA and RNA), the ester linkages with phosphate molecules and the carbon–nitrogen (sugar–base) linkages are the most susceptible. Heterocyclic rings and amino groups are chemically active, but rapid desiccation seems to place an upper limit on endogenous hydrolytic damage, while oxidative attack continues. More extensive discussion of these processes can be found in papers compiled from the conference[12] and elsewhere in this volume.[13]

Given present knowledge, we have collected samples for nucleic acid extraction and subsequent DNA sequencing in two ways. The first is the traditionally conservative approach employing cold temperatures and combinations of wet ice, dry ice, and liquid nitrogen. Ample discussion of these methods is contained in Ref. 4, and this strategy represents the best approach for proteins and RNA preservation. The second, which relies on rapid desiccation, is a good strategy to use in the less developed countries of the Pacific where reliability of shipping varies because of storms, fuel availability, and/or wind conditions. In the case where blood is taken, it can be spotted on paper (Whatman, Clifton, NJ, No. 1 or 3 mm) and left to dry in bright sun for 30 min, placed in envelopes or tubes, and then mailed in regular mail. Researchers have published methods for recovering DNA from surfaces of paper.[14,15] If muscle tissue is taken, approximately 1 g of tissue is sliced off and centered in a 15-ml or 45-ml conical test tube

---

[12] G. Eglinton and G. B. Curry (eds.), *Philos. Trans. R. Soc. London B,* 315 (1991).
[13] N. Tuross and L. Stathoplos, this volume [9].
[14] E. A. Ostrander, C. L. Maslen, and L. M. Hallick, *Focus* **9**(4), 9 (1987).
[15] E. R. B. McCabe, S.-Z. Huang, W. K. Seltzer, and M. L. Law, *Hum. Genet.* **75,** 213 (1987).

containing rock salt. This procedure gives acceptably high yields of total genomic DNA. Polymerase chain reaction (PCR) from these samples has resulted in mitochondrial (mt) DNA fragments up to 0.3 kilobase (kb) in size (C. Reeb, 1992, unpublished results). Various collectors may actually prefer rapid desiccation because it requires no refrigeration or flammable substances. Once back in the laboratory, the tissue can simply be rinsed under distilled water and run through standard DNA extraction protocols. We have not attempted isozyme analysis on these samples, however, and cannot evaluate this method for protein work.

Routine use of a mannitol–sucrose buffer[16] at ambient temperatures is reported by some investigators to be compatible with mtDNA surveys and allows a field biologist to collect materials for up to 2 weeks without refrigerating them. When no ice, salt, or buffer is available, saturating the tissue with ethanol or rubbing alcohol (2-propanol) with EDTA added is reported as compatible with DNA hybridization studies (C. Sibley, quoted in Ref. 4), as well as DNA sequencing. We have also collected blood from birds, wallabies, and bats using no refrigeration and a scaled-down method published previously by Quinn and White,[17] which generates high molecular weight total genomic DNA compatible with PCR and sequencing. Microhematocrit tubes of blood blown into 500 μl of 10 mM Tris, 10 mM NaCl, and 2 mM EDTA, pH 8.0, are stable over months if refrigeration is impossible and still give ample mtDNA sequences.

Other fixatives, such as 10% buffered-neutral formalin (BNF) (v/v), acetone, and OmniFix (American Histology Reagent Co., Stockton, CA) have been evaluated along with 95% ethanol for their specific ability to yield DNA fragments of moderate size on enzymatic amplification.[18] Researchers have demonstrated that paraffin-embedded tissues should be considered as potential sources for molecular evolutionary studies, but that fragments of DNA from human β-hemoglobin genes, of the order of 1 kb, were impossible to amplify from formalin-fixed tissues after 24 hr. Only 95% ethanol storage of tissues resulted in PCR products of 1 kb after 30 days. Fixatives listed by these authors as giving poor results were Zamboni's, Clarke's, paraformaldehyde, formalin–ethanol–acetic acid, and methacarn. Even worse results were obtained with highly acidic fixatives such as Carnoy's, Zenker's, or Bouin's.

Zenker's solution is a common tissue fixative (5 parts glacial acetic acid mixed with 95 parts saturated mercuric chloride and 5% potassium dichromate) and may affect the stability of DNA in tissues preserved in it, owing to high heavy metal concentrations which promote phosphodiester break-

[16] R. Lansman, R. O. Shade, J. F. Shapira, and J. C. Avise, *J. Mol. Evol.* **17**, 214 (1981).
[17] T. Quinn and B. White, *Mol. Biol. Evol.* **4**, 126 (1987).
[18] C. E. Greer, J. K. Lund, and M. M. Manos, *PCR Methods Appl.* **1**, 46 (1991).

age. It is also unfortunate that Bouin's solution (75 parts picric acid, 25 parts 100% formalin, mixed as 95 : 5 parts glacial acetic acid) was formerly the fixative of choice for many studies of infectious or parasitic diseases in animals, where, traditionally, a half-fist-sized piece of tissue was preserved with a 10-fold excess of solution. However, tissues preserved even under these conditions may be useful to molecular parasitology, if investigators use the strategy of designing PCR primers to yield small (<200 nucleotides) amplification products. After fixation with these solutions, preserved samples would usually be placed in formalin or ethanol. A simple spot test for the identification of formalin versus ethanol-preserved specimens, based on a color change in the acid/base indicators sodium sulfite and sodium metabisulfite, has been reported by Waller and McAllister.[19] This should replace the "sniff test" and lower the potential exposure of researchers to inhalation of potentially toxic or carcinogenic substances.

Ambient temperatures are not always sufficient for successful extraction of DNA from hair shafts. In the case of plucked human hairs left in envelopes in a desk drawer for over 1 year, the rate of successful mtDNA amplification was low (<10%) compared to recently plucked hairs prepared and amplified using the same protocols. We therefore recommend that, once hairs and feathers are taken, they should be frozen at −70° until extracted for nucleic acids.

Labeling and Containers

We caution investigators to label materials they collect and to be conservative in reporting the information under which vertebrate samples are taken (more information rather than less). Labeling markers should be permanent and nonrunning. Because markers come in a variety of colors, species can be color coded. If in doubt about the permanence of a marker, test it first by duplicating the collection conditions with regard to water and temperature. Labels should be written directly on the sample container, and, if affixed, they should be tied with string as well as taped. Tape has a way of freeing itself during handling, freezing, and thawing, and extra care must be taken to ensure that labels stay with appropriate samples. Permanent labels tied to the specimen itself are usually safest.

Bags and foil packages are an important safeguard against failed labeling. Many laboratories have zip-lock bags in different sizes, and as a rule twice as many bags should be taken to the field as you anticipate the team

[19] R. Waller and D. E. McAllister, *in* "Proceedings of the 1985 Workshop on Care and Maintenance of Natural History Collections" (J. Waddington and D. M. Rudkin, eds.), p. 93. Life Sciences Miscellaneous Publications of the Royal Ontario Museum, Toronto, 1986.

will need. Dissection, if necessary, can be done on the clean surface of these bags, thus avoiding soil microbial contamination which could potentially complicate infectious disease studies. Excess bags can also be used to wrap samples and to help in avoiding cold-burning tissues if electrophoresis of proteins is anticipated.

After samples are labeled and bagged, they can be placed in a larger bag or container with a summary of that collection for the site or the day included as a descriptor attached outside the individual bags or containers. This method is most appropriate where investigators work from a base camp, to which they return periodically. We have substituted knee-high nylon stockings for zip-lock bags in the case of liquid nitrogen (LN) collections as an excellent, low-cost alternative to aluminum canes. The stockings stretch to accommodate virtually any shape, can be handled easily even when cold, are washable, and can be soaked in bleach to disinfect the material and then reused many times without fear of cross-contamination in case of popped tubes. We also suggest that if any vial, tube, or container in question is of unknown composition, test it first to see if it can withstand ultracold conditions.

Documentation

If working as a team, it is best to decide ahead of time which member has the documentation duties and to assign a backup. All members of the team should understand what system will be used to identify and log samples. This includes employing unambiguous abbreviations, briefing all members about the order of information to appear on the labels, and specifying any other critical information about the provenance of the sample which may disappear when field notes are transcribed. Water-resistant pocket notebooks and pens with waterproof ink are as essential to successful collecting as are sterile tubes and bags.

Photographs and sound recordings are no replacement for actual vouchers, but they are an excellent backup in documentation. We have made extensive photographic slide collections of endangered birds released after bleeding, to record plumage variation, changes in growth, skin lesions, and missing appendages. We refer to these slides for disease studies and ecological monitoring. Slides can be easily duplicated and shared among colleagues, and they are also useful to evaluate the stability of colored leg-band combinations over time. We have found, like other researchers, that some color bands fade rapidly over time in high UV light environments, making identification of individual birds in the field problematic in low-light conditions. When behavioral and laboratory studies are combined, certainty of identity through extensive documentation of samples can allow testing of more rigorous hypotheses about behavior and the mechanisms leading to its genetic correlations.

Notebook computers represent an important new innovation in field documentation during collecting, both in the outdoors and during visits to museum collections. We now routinely use one model of notebook computer, the Everex Tempo LX, which has been highly rated due to its ruggedness.[20] Battery operation for greater than 3 hr in the field, with 1-hr recharging periods from 110 V outlets or car batteries (as direct power sources), plus the ability to link via internal or facsimile (fax) modems to remote laboratory computer facilities, has improved the maintenance of up-to-date database files. We emphasize that primary documentation still needs to be done by hand, as backup systems for field computers are primitive and subject to failure, the mildest being damage to floppy disks in transport.

Shipping

At this time, dry ice and liquid nitrogen are regulated substances that require special labeling. Currently, 200 kg (70 lb) of dry ice is the limit for any single container, and it must be marked as hazard class ORM-A (carbon dioxide, solid) on the outside of the package. Cargo facilities may require one to state pound equivalents, so use both metric and English equivalents for packages. The special black-and-white hazard sticker is a must for acceptance as air cargo. Dry ice may be available from a variety of sources, notably hospitals, meat packers, ice cream wholesalers, and gas dealers. Internationally, the best source of information about the availability of dry ice is often from the air cargo offices of airlines. Up to 50 kg of liquid nitrogen per unpressurized Dewar container can be shipped on passenger aircraft, but it too must be marked as a hazardous substance. Policies for carrying hazardous materials on passenger aircraft will vary with the carrier, so check first. The labeling must read "Nitrogen (liquid, nonpressurized)" with upright arrows and carry the notation "DO NOT DROP—HANDLE WITH CARE." Carriers also require a green label stating "NONFLAMMABLE GAS."

We have successfully used a "Dry Shipper" (such as the Arctic Express Thermolyne CY 50915/50905 available from VWR Scientific, San Francisco, CA) and prefer it for cryogenic collections because of the ease with which samples can be transported by air. The absorbent walls of the shipper are precharged with liquid nitrogen and the flask then maintains ultracold conditions for upward of 3 weeks if left upright. No special restrictions apply to transport of samples in this container, because there is no liquid nitrogen left to spill or pressurize.

[20] J. S. Finnie, *PC/Computing* **4**(11), 136 (1991).

Mark all labels with permanent ink, and cover them with a plastic sheet. A laminated list of materials included, with pertinent names, addresses, and telephone numbers, should be included inside the package and will help recipients determine if the package was tampered with. When necessary, mark the outside of the package "Biomedical samples for scientific research — Please rush."

Before shipping, ensure that holidays (civic and religious) will not impede or delay the delivery of materials. It is also wise to avoid shipping over a weekend. Become familiar with local customs, and do not assume that just because shipping is within the United States you know the rules. (For example, most businesses shut down in Louisiana during Mardi Gras, which is a mobile holiday each year, and Hawaii has a number of state holidays uncelebrated on the mainland.) Recipients should be called and/ or faxed concerning the expected arrival dates, with the airbill number noted. Recipients should know if they can expect delivery or if they must pick up their samples at another location. Finally, recipients should bring multiple copies of any permits or documentation necessary for pickup. If transporting across international boundaries, make sure that colleagues on both sides are intimately involved in the process of shipment.

Storage in the Laboratory

The best long-term storage conditions are those that minimize variation in temperatures, light, and liquid volumes. For nucleic acids and proteins, unextracted samples are best stored carefully wrapped at $-70°$ or in liquid nitrogen, after extensive archiving and efficient packaging to minimize temperature drops on retrieval. Prepared DNAs that have undergone chemical purification (extraction, precipitation, and dialysis) are best stored at $5°$,[21] or in ethanol at $-20°$ and $-70°$. High salt (> 1 M) in the presence of high Na_2EDTA at pH 8.5 (Tris) buffer is recommended for long-term storage, along with storage in light-protected CsCl at $5°$. High molecular weight DNAs are stable at $20°$ for several days in buffer [1 mM Tris, 0.1 mM EDTA] and can often be sent unrefrigerated if the need arises. RNA preparations usually require either freshly sacrificed animal tissues or $-70°$ frozen cells. Storage of tissues for subsequent RNA work almost always involves getting access to cryogenic conditions for successful nuclease inhibition.

One important safeguard against contamination is the practice of aliquoting DNA samples into multiple units, so that the loss of a given tube to sloppy bench practices can be contained. Novices in the laboratory may

[21] R. W. Davis, D. Botstein, and J. R. Roth, "Advanced Bacterial Genetics," p. 213. Cold Spring Harbor Laboratory, Cold Spring Harbor, New York, 1980.

not be aware of the extreme sensitivity of the PCR, and even experienced researchers make pipetting mistakes when distracted. Truly irreplaceable samples should be aliquoted into separate tubes with positive displacement pipettes or pipettes with plugs.

A rich literature concerning storage of tissues for protein work already exists, and an excellent summary is contained in Ref. 4. For long-term preservation, $-20°$ and $-70°$ freezers equipped with alarms or backup generators should be used. Temperatures should be checked on a regular basis, especially after storms or following building maintenance. Tissue samples in 2% 2-phenoxyethanol (v/v), with glycerol or dimethyl sulfoxide (DMSO), are normally stored at $-20°$.

Museum Collections

Museums, as well as individual laboratories, vary in their ability to protect collections from light, insects, high humidity, molds, damage to labels, and the ravages of temperature fluctuations. Curators are enlisting the aid of chemists to help them evaluate the biological changes their materials undergo over time. Investigators should expect that new advances in curation technologies will alter storage methods of solid materials in the next 10 years. These advances will have a direct impact on the way individual investigators store their own materials and how they utilize museum collections.

Researchers should understand that the job of a curator is to keep the materials intact, and these conservative dictates are intellectually opposed to the needs of an experimentalist seeking to do destructive analysis. Rather than view the curator as a block to progress, the experimentalist needs to design laboratory protocols that make maximum use of existing materials or that sample precious materials in nondestructive ways. Advances in DNA extraction technologies, such as those making use of ion-exchange resins to block inhibitors or to act as carriers,[22] will help convince scientists of the value of vertebrate samples in museums and pathology collections for molecular evolutionary studies. It is also not necessary to use exhibition quality specimens if partial but well-documented items will suffice. In some cases, as exhibitions are moved, some damage may occur which cannot be repaired by the curatorial staff (feathers falling out of feathered cloaks, for instance). Destructive work with these materials might yield the same biological information as work on more high-profile specimens, and it should be tried first.

Museums vary also in their policies regarding access to materials for biological studies at the DNA or protein level. Some museums have more

[22] J. Singer-Sam, R. L. Tanguay, and A. D. Riggs, *Amplification* 3, 11 (1989).

flexible regulations than others, depending on their experience with individual investigators and their familiarity with the techniques proposed. Almost all museums are adopting guidelines that require the return of unused nucleic acids or extracts from their specimens for use by future investigators. Researchers are urged to contact heads of research departments in these museums, in order to keep abreast of changes in policies about access to and amounts available from any given specimen. Graduate students may be required to submit supplementary materials from their thesis advisors, which include evidence that their advisor is knowledgeable about the techniques involved (i.e., peer-reviewed publications with these methods).

Sources for Nucleic Acid Extractions

We have had success amplifying mtDNA from small (<0.5 g) quantities of 500-year-old bone, single feathers, fish scales, plucked hair, muscle, skin, whole blood (liquid and dried), and serum. We have not evaluated yields of nucleic acids or proteins from these sources compared to feces, stomach contents, shed skins, or molted hairs and feathers. It seems unlikely that we are especially skilled in extracting DNA, and we encourage all investigators to start with small quantities and to titrate down their methodologies to using the least amount of material they need to complete their project successfully, allowing for repeated measurements.

Progress with methods developed for early disease detection and ancient samples will continue to affect work with more easily accessible materials, and investigators should not become complacent that they are doing the best possible work given existing conditions. The *Ancient DNA Newsletter* (address in Appendix) will continue to provide updates and late-breaking news of technological advances relevant to the issues addressed here. *DNA Amplifications,* the free newsletter of Perkin-Elmer Cetus Corporation (Norwalk, CT), as well as the free journal *BioTechniques* (Eaton Publishing Co., Natick, MA) often contain well-documented protocols.

Finally, investigators should check on the availability of commercial materials, either through standard biological supply houses such as those listed in Ref. 7, centers for mammalian genetic research, cDNA libraries which may have been custom made and now are being sold commercially (Clonetech), or clones and cell lines available through the American Type Culture Collection (ATCC) collections (addresses in Appendix). Documentation of the source of these materials should not be relied on, as the primary concerns of molecular biologists and field biologists may be different. Still, if the purpose of using these materials is simply to provide an outside reference specimen, the evolutionary distance involved may be sufficiently broad to justify the inattention to detail about subspecies that is usually lacking for many of the commercially available samples.

Conclusions

Rapid publication of DNA sequences gives many investigators the opportunity to design and use oligonucleotide primers for evolutionary comparisons that would never have been attempted 20 years ago. We applaud all researchers who continue the tradition of making this information available to a wide audience of potential users. It embodies the recognition that knowledge of a DNA sequence from an organism is just the first step in the study of phylogenetic reconstruction and evolutionary inference. It emphasizes the objective, quantitative nature of new evolutionary biology studies and delineates the growth of this field out of the problems which plagued past generations of biologists. It is ironic that just as we have learned how to procure and extract this information, the taxa we depend on are disappearing at an unparalleled rate. One basic biological problem is estimating the true number of species that exist today. It is currently known with less precision than an equally basic physical estimate, the distance from the earth to the moon. The number of scientists who are concerned with knowing how many species inhabit the earth is growing each year. We think this is one example of the progress in biology that can be attributed to molecular evolutionary studies.

Appendix:

This appendix contains a listing of addresses and telephone numbers of importance in molecular evolutionary studies.
1. To obtain current endangered and threatened listings, and federal permits:
Office of Management Authority
U.S. Fish and Wildlife Service
4401 North Fairfax Drive
Arlington, Virginia 22203
(703) 358-1732
2. To find out if the sample from a vertebrate is disease-free:
National Wildlife Health Center
6006 Schroeder Road
Madison, Wisconsin 53711
FTS 364-5411, (608) 271-4640
[NWHC Health Resource Team, FTS 364-5422, (608) 264-5422, and (608) 271-4640]
3. To find out if the importation requires special permits because of disease risks:
Office of Biosafety
U.S. Department of Health and Human Services, Public Health Service
Centers for Disease Control
1600 Clifton Road

Atlanta, Georgia 30333
(404) 639-3883
4. To find out about quarantine requirements in shipping:
Veterinary Services: Import–Export Products
Animal and Plant Health Inspection Service
U.S. Department of Agriculture
6505 Belcrest Road
Hyattsville, Maryland 20782
(301) 436-7830
5. To locate sources of cloned materials, cell lines, or cultures of live microorganisms:
American Type Culture Collection
12301 Parklawn Drive
Rockville, Maryland 20852
1-800-638-6597
FAX 1-301-231-5826
Especially useful is their *Catalogue of Recombinant DNA Materials,* 2nd Edition, 1991, edited by D. R. Maglott and W. C. Nierman, ISBN 0-930009-36-3, also available through electronic access and in PC/diskette form.
6. To find out if a cDNA library has already been constructed from the species or tissue of choice, and is available commercially:
Clonetech Laboratories, Inc.
4030 Fabian Way
Palo Alto, California 94303
1-800-662-2566
FAX 415-424-1352
7. To request information about the history of, or availability of, established strains of mice:
The Jackson Laboratory
Bar Harbor, Maine 04609
8. To be placed on the mailing list for the *Ancient DNA Newsletter* address your request to:
Dr. Robert K. Wayne, Head of Conservation Genetics
The Zoological Society of London
Institute of Zoology
Regent's Park
London NW1 4RY.

[4] Collection and Storage of Invertebrate Samples

By MATHEW DICK, DIANE M. BRIDGE, WARD C. WHEELER, and ROB DESALLE

Introduction

Increasingly, researchers whose primary training lies in molecular biology are turning to comparative studies of various sorts. Their interest may lie in the comparative molecular biology of organisms related to well-established invertebrate systems such as *Drosophila melanogaster* or *Caenorhabditis elegans,* in studying the evolution of developmental systems or gene families, or in using their own molecular data sets to investigate phylogenetic relationships among higher taxa. As the diversity of studies increases, so will the requirements for a diversity of invertebrate organisms. Furthermore, with the development of techniques that can rapidly and efficiently generate sequence information for large numbers of individuals and taxa, knowledge of how to identify, store, and document tissue source material becomes extremely important.

In this chapter we describe (1) how to obtain invertebrate organisms for use in molecular studies through field collection, stock centers, or purchase; (2) how to select the right tissue for the right purpose; (3) how to store tissue for future use; and (4) how to document the source of tissue so that information will not be lost. Although collection of invertebrate organisms is not nearly as tightly regulated as that of vertebrates, regulations that do apply are discussed. Detailed coverage of collecting methods for all invertebrates is beyond the scope of this chapter. Our goal is to direct the investigator to the most useful sources of information from which to proceed.

Newer techniques have allowed the isolation of nucleic acids from preserved specimens.[1] Because DNA isolation requires the destruction of preserved material, this aspect of obtaining tissue necessitates new procedures in the handling of museum specimens and in the protocols that investigators will follow to gain access to museum specimens. As an example, we outline the procedures that the American Museum of Natural History currently has in effect, or intends to place into effect. Other museums will undoubtedly have different procedures, so we recommend strongly that individual institutions be consulted when material is needed.

Collection of Specimens and Tissues from the Field

Regulations

Regulations that do pertain to collection of invertebrates fall into several categories: where collecting may be done, what may be collected, and how specimens may be transported. The multiplicity of state, federal, and international regulations that may apply in certain instances is beyond the scope of this chapter, but some general guidelines can be given.

---

[1] S. Paabo, *Proc. Natl. Acad. Sci. U.S.A.* **86,** 1939 (1989).

Where Collecting May Be Done. In the United States, specific permits are needed to collect any organisms on certain federal lands, such as national wildlife refuges, national parks, and national monuments; this also applies to state parks and streams in some states. Inquiries regarding permits should be addressed to the manager, superintendent, or ranger in charge. In some states, for example, California, all intertidal and nearshore marine organisms are protected, and any collecting done in the intertidal zone or by scuba close to shore requires a permit. If there is any doubt concerning regulations in a particular state, inquiries should be made to the pertinent natural resource agency of that state. Similar considerations apply to refuges, parks, and other protected lands in most countries.

What May Be Collected. Unless invertebrates occur on specifically protected lands, few restrictions apply to which ones may be taken. There are, however, exceptions. Taking of certain species of "shellfish" (e.g., edible molluscs, crustaceans, echinoderms) is controlled by either federal or provincial sport- and commercial-fishing regulations in many countries. In the United States, authority for these regulations in freshwater and nearshore marine waters resides with the state governments. In general, a collecting permit to obtain shellfish will usually eliminate the need to obtain a fishing license as well.

Endangered and threatened species comprise another special category. In the United States, the Fish and Wildlife Service (Department of Interior) maintains a "List of Endangered and Threatened Wildlife," including some invertebrates. A federal permit must be obtained for the collection, possession, or transportation of any species on the list. Furthermore, the United States is a signing nation on the Convention on International Trade in Endangered Species of Wild Fauna and Flora (CITES). Pursuant to this treaty, there exists an international endangered species list, which includes some invertebrates. Importation into the United States (or any other participant nation) of any species (alive, dead, or in part) on this list requires a permit, and the consequences of being caught doing so without a permit may be rather severe. The U.S. and international endangered species lists are published and updated in the U.S. Code of Federal Regulations, Title 50, parts 17 and 23. Copies of the lists and permit information may be obtained at Office of Management Authority, U.S. Fish and Wildlife Service, 4401 North Fairfax Drive, Arlington, Virginia 22203 (telephone 703-358-1732). Many states in the United States also have lists of endangered and threatened species for which permission to collect must be obtained from the state natural resource agency.

Transportation of Specimens. Transportation involves getting specimens out of the source country and into the destination country. Many so-called underdeveloped countries have strict regulations regarding collection of natural history specimens and their transportation out of the country. One must apply to the appropriate authorities in each country. There are no prohibitions per se against bringing specimens or tissue preserved in alcohol or formalin, or as extracted nucleic acids, into the United States, unless the tissue comes from an endangered species (see above). For some molecular purposes, however, it may be necessary to import live material if the organism is small and needs to be cultured to obtain enough nucleic acid to work with, or if mechanistic studies are to be done. Several federal offices regulate the importation of live organisms which may potentially be harmful to crops, livestock, or humans. Contact addresses of these offices, type of function, and name of permit given are as follows.

1. Plant Protection and Quarantine
   Biological Assessment and Taxonomic Support
   Animal and Plant Health Inspection Service
   U.S. Department of Agriculture
   6505 Belcrest Road
   Hyattsville, Maryland 20782
   (telephone 301-436-5055; telefax 301-436-8700)
   Function: Regulates the importation and transportation of plant pests (e.g., insects), pathogens, vectors, and articles that may harbor these organisms.
   Contact for: Application and permit to move live plant pests and noxious weeds.
2. Veterinary Services: Import–Export Products
   Animal and Plant Health Inspection Service
   U.S. Department of Agriculture
   6505 Belcrest Road
   Hyattsville, Maryland 20782
   (telephone 301-436-7830; telefax 301-436-8226)
   Function: Regulates the importation and transportation of all animal-origin materials that could represent a disease risk to U.S. livestock.
   Contact for: Application for permit to import controlled material, or import or transport organisms or vectors.
3. Office of Biosafety
   U.S. Department of Health and Human Services
   Public Health Service
   Centers for Disease Control
   1600 Clifton Road
   Atlanta, Georgia 30333
   (telephone 404-639-3883; telefax 404-639-2294)

Function: Regulates the importation of agents of human disease or vectors that potentially harbor these agents.

Contact for: Application for permit to import or transport agents or vectors of human disease.

Even though one does not technically need a permit to import material that has been inactivated by preservation or extraction, customs officials may not be able to judge whether particular specimens or tissue are safe, and it is within their power to confiscate questionable material until its safety can be verified. To avoid such problems, the researcher should contact the appropriate agency above; if a permit is not required, then a letter to that effect should be obtained from the agency.

How to Collect Invertebrates

Invertebrate organisms are everywhere: in the sea, in freshwater habitats, in most terrestrial habitats, and in or on other organisms. To obtain a particular species, one must determine from the literature or a specialist where the organism can be found. If one merely needs representatives of higher taxa, there are some general considerations. Nearly half of the metazoan phyla are restricted to the marine environment,[2] and most of the other phyla are richer in marine species than in freshwater or terrestrial species. A few phyla are better represented in freshwater habitats (e.g., Rotifera, Nematomorpha, Tardigrada), but only one phylum of free-living organisms (Onycophora) is restricted to a nonmarine environment.

In the marine environment, diversity of organisms varies with the coast (in North America, Pacific Coast intertidal communities are generally richer than those on the Atlantic), with exposure and substrate type (protected rocky shore is likely to have greater diversity than exposed rocky shore or mud flat), and with latitude (intertidal communities are likely to be more diverse at midlatitudes than at high or low latitudes; subtidal shelf communities tend to be more diverse at low latitudes). The method of collection is determined by the size, location, and behavior of the required organism(s). Table I lists general collection methods used for invertebrates both in aquatic and terrestrial habitats; several sources are specific for *Drosophila* or other insects. Specialized collection methods for particular groups can often be found in field guides and identification manuals specific for those groups.

---

[2] R. D. Barnes, "Invertebrate Zoology." Saunders, Philadelphia, 1980.

TABLE I
GENERAL METHODS FOR COLLECTING INVERTEBRATES

Location	Method	Refs.[a]
Aquatic environments		
Benthic		
Within substrate	Grabs, core samplers, anchor dredge	1–3
On substrate surface	Trawl, dredge, epibenthic sledge, bottom plane, bottom plankton net, baited traps, scuba	1–3
Water column (large and small plankton, invertebrate nekton)	Plankton net, dip net, "snatch" bottles, night-light attraction, midwater trawl, scuba	2–4
Streams	Picking by hand, dip net, baited traps, screens	5
Intertidal, shoreline	Picking by hand (knife, shovel, bucket), small dip net, baited traps, shovel and sieve	
Terrestrial environments		
Flying insects	Insect nets, light traps, bait traps	5–9
In soil and leaf litter	Pick by hand, screens, Berlese funnel, emergence cage, Baermann funnel	5–9
Parasites and other symbionts (terrestrial, aquatic)		
Ectoparasites	Pick from host, bag and chloroform host, pick from nests and burrows	5, 10
Endoparasites	Dissection of host, Baermann funnel, blood smears	10

[a] Key to references: (1) N. A. Holm and A. D. McIntyre, "Methods for the Study of Marine Benthos." International Biological Programme Handbook 16, Blackwell Scientific Publications, Oxford, 1971. (2) C. Schlieper, "Research Methods in Marine Biology." Univ. of Washington Press, Seattle, 1972. (3) J. H. Thorp and A. P. Covich, "Ecology and Classification of North American Freshwater Invertebrates." Academic Press, San Diego, 1991. (4) M. Omori and T. Ikeda, "Methods in Marine Zooplankton Ecology." Wiley, New York, 1984. (5) B. P. Beirne, "Collecting, Preparing and Preserving Insects." Publication 932, Canada Dept. of Agriculture, Ottawa, 1963. (6) J. W. Knudson, "collecting and Preserving Plants and Animals." Harper & Row, New York, 1972. (7) J. E. H. Martin, "The Insects and Arachnids of Canada, Part 1: Collecting, Preparing, and Preserving Insects, Mites, and Spiders." Canada Dept. of Agriculture, Ottawa, 1977. (8) H. L. Carson, and J. N. Thompson, eds.), p. 2. Academic Press, New York, 1983. (9) D. J. Borror, D. W. De Long and C. A. Triplehorn, "An Introduction to the Study of Insects." Saunders College Publ., New York, 1989. (10) M. H. Prichard and G. O. W. Kruse, "The Collection and Preservation of Animal Parasites." Univ. of Nebraska Press, Lincoln, 1982.

Specimen Identification and Documentation

Organisms from which molecular data are obtained should be properly identified and documented. This is important for systematic and evolutionary studies because knowledge of the species involved may be crucial to interpretation, because published sequence information may be utilized in ways other than for which it was initially obtained, and because verifiability is one of the tenets of empirical science. A sequence reported as coming from "*Genus* sp." is, from a systematic standpoint, so poorly documented that the future utility of the information is seriously impaired. Given the effort required to obtain sequence information, it behooves the molecular researcher to devote some care to proper identification and documentation.

Identification. There are a number of ways in which reliable identification can be accomplished. One way is the taxonomic equivalent of "cloning by phone," that is, to let a specialist in a particular group of organisms provide identified material for nucleotide extraction, identify specimens which one intends to use, or identify voucher specimens from the population which one has used. Alternatively, if one needs an organism which is to be treated as representative of some higher taxon, there are within most groups common, easily recognized species which may be identified by the nonspecialist with some confidence. Into this category fall many of the invertebrate species sold live by bulk supply houses (see below).

Systematists are generally willing to help with identification, but they are often backlogged with work and may not be able to do so in a timely manner. The task of identification then falls to the actual investigator. Several types of literature for the identification of invertebrates are as follows, with an example of each referenced.

Regional pictorial identification guides. Written primarily for the layman, pictorial guides (e.g., Voss[3]) are easiest to use but least inclusive and reliable. Color photographs or line drawings are accompanied by short descriptions and range of occurrence of the most common species occurring in a region. The danger in using these guides is that several closely related species may occur, only one of which is included. Furthermore, these guides avoid use of morphological characters that may be necessary for correct identification, considering the ontogenetic, phenotypic, geographical, or ecotypic variation found among organisms in the field.

Field guides. Field guides vary greatly in utility; some (e.g., Meinkoth[4])

[3] G. L. Voss, "Seashore Life of Florida and the Caribbean." E. A. Seemann Publishing, Miami, 1976.
[4] N. A. Meinkoth, "The Audubon Society Field Guide to North American Seashore Creatures." Alfred A. Knopf, New York, 1981.

are little more than pictoral identification guides, whereas others (e.g., Schultz[5]) consist of keys. Some (e.g., Bland and Jaques[6]) allow identification to the level of order or family, and identification to species must be followed up in a monograph; others are specific for a particular group (e.g., Schultz[5]) and may be fairly complete and reliable.

Identification manuals. A selected list of identification manuals is given in Table II. These are collections of detailed keys written by specialists. They are generally excellent for identifying the better known species in the region they cover, but they require effort in learning morphological terminology. Some progressive institutions publish series of monographs for the identification of selected groups; these monographs are well illustrated, contain keys, and are extremely "user friendly." Two exemplary series are the handbooks published by the British Columbia Provincial Museum (Publications, British Columbia Provincial Museum, Victoria, BC V8V 1X4, Canada; volumes are available for crabs, univalves, bivalves, sea stars, and dragonflies) and the Synopses of the British Fauna (New Series) published under the auspices of the Linnean Society of London; a list of available titles may be found in any recent volume of that series (e.g., Hayward[7]). Although regional in coverage, the monographs are useful over a wider range than their titles indicate.

Systematic literature. The basis for all of the guides and manuals is the primary systematic literature scattered in books, monographs, and journals. Sims[8] has referenced primary systematic literature for invertebrates worldwide by taxonomic group and region. Additional access to the primary literature on a particular group can be obtained from the often extensive bibliographies found in identification manuals. Finally, research museums, marine stations, and agricultural stations maintain specialized reference libraries and microscope facilities, and access is generally possible for identifying organisms.

The ease with which organisms can be identified varies with taxonomic group and region. Some taxa, for example, butterflies and nearshore marine molluscs, which have been collected by amateurs for centuries, are simply better known than others; careful use of field guides and identifica-

---

[5] G. A. Schultz, "How to Know the Marine Isopod Crustaceans." W. C. Brown, Dubuque, Iowa, 1969.

[6] R. G. Bland and H. E. Jaques, "How to Know the Insects." W. C. Brown, Dubuque, Iowa, 1978.

[7] P. J. Hayward, "Ctenostome Bryozoans." Synopses of the British Fauna (New Series), No. 33. E. J. Brill, London, 1985.

[8] R. W. Sims, "Animal Identification, A Reference Guide; Volume 1, Marine and Brackish Water Animals; Volume 2, Land and Freshwater Animals (not Insects); Volume 3, Insects." British Museum (Natural History), London, and Wiley, Chichester, 1980.

TABLE II
SELECTED IDENTIFICATION MANUALS FOR NORTH AMERICAN INVERTEBRATES

Organisms/region	Refs.[a]
Protozoa	1
Marine invertebrates	
Alaska	2, 3
Pacific Northwest	3
California	4
Atlantic Northeast	5
Atlantic Southeast	6
Caribbean	7
Freshwater invertebrates	8–11
Terrestrial invertebrates: insects	12

[a] Key to references: (1) J. J. Lee, S. H. Hutner, and E. C. Bovee, "An Illustrated Guide to the Protozoa." Society of Protozoologists, Lawrence, Kansas, 1985. (2) D. W. Kessler, "Alaska's Saltwater Fishes and Other Sea Life." Alaska Northwest Books, Bothell, 1985. (3) E. N. Kozloff, "Marine Invertebrates of the Pacific Northwest." Univ. of Washington Press, Seattle, 1987. (4) R. I. Smith and J. T. Carlton, "Light's Manual: Intertidal Invertebrates of the Central California Coast." Univ. of California Press, Berkeley, 1975. (5) K. L. Gosner, "A Field Guide to the Atlantic Seashore: Invertebrates and Seaweeds of the Atlantic Coast from the Bay of Fundy to Cape Hatteras." Houghton Mifflin, Boston, 1979. (6) E. E. Ruppert and R. S. Fox, "Seashore Animals of the Southeast: A Guide to Common Shallow-water Invertebrates of the Southeastern Atlantic Coast." Univ. of South Carolina Press, Columbia, 1988. (7) W. Sterrer and C. Schoepfer-Sterrer (eds.), "Marine Fauna and Flora of Bermuda: A Systematic Guide to the Identification of Marine Organisms." Wiley, New York, 1986. (8) R. L. Usinger, "Aquatic Insects of California, with Keys to North American Genera and California Species." Univ. of California Press, Berkeley, 1956. (9) R. W. Pennak, "Fresh-water Invertebrates of the United States: Protozoa to Mollusca." Wiley, New York, 1989. (10) J. H. Thorp and A. P. Covich, "Ecology and Classification of North American Freshwater Invertebrates." Academic Press, San Diego, 1991. (11) R. W. Merritt and K. W. Cummins (eds.), "An Introduction to the Aquatic Insects of North America." Kendall/Hunt Publ. Dubuque, Iowa, 1978. (12) D. J. Borror, D. W. De Long, and C. A. Triplehorn, "An Introduction to the Study of Insects." Saunders College Publ., New York, 1989.

tion manuals will generally allow correct identification. Other taxa, such as free-living nematodes and marine amphipods, are quite diverse, have small individuals, and are notoriously difficult to identify; only a specialist has a reasonable hope for correct identification. In general, the taxonomy of invertebrates is better known in north temperate regions than anywhere else, for historical reasons. For many parts of the world, comprehensive identification manuals simply do not exist.

Documentation. We recommend that organisms from which molecular data, especially sequences, are published be documented in such a way that their identity can be verified should a question arise. This is currently not standard procedure, but should be, especially for field-collected organisms. Documentation of sequence information should include placement of voucher specimens, from the same local population as the experimental animals, into a permanent depository (i.e., a museum). Publication of sequences should include the locality where the organisms were collected, the location of the voucher specimens, and, if possible, their catalog number. The catalog number will allow access to further information, such as when the specimens were collected, who collected them, and who identified them.

For long-term storage with retention of characters necessary for identification, invertebrates must be properly preserved, and different groups have different requirements (see references in Table I).[9-12] For example, insects are generally preserved simply by drying; some soft-bodied aquatic invertebrates severely contract or fragment when placed directly in fixative and must be relaxed first, whereas others require sectioning for identification and must be specially fixed; organisms with calcareous tests or shells dissolve in formalin and must be stored in buffered formalin or alcohol. One of the best guides to preservation of marine invertebrates is that of Mueller[13]; it can be obtained through the University of Alaska Museum (Fairbanks, AK 99701).

Ordering Specimens from Biological Supply Companies

For *Caenorhabditis, Drosophila,* and *Tribolium castanium,* stock centers exist from which particular strains or species may be ordered (Table III). Live representatives of most animal phyla and pure cultures of some protists can be obtained relatively cheaply from supply companies (Table III). Ordering specimens is convenient in that they arrive identified, although voucher specimens should still be preserved (see above). Organisms should be ordered as far in advance as possible, since their appearance in the field may be seasonal or unpredictable.

[9] R. Wagstaffe and J. H. Fidler, "The Preservation of Natural History Specimens, Volume 1, Invertebrates." H. F. & G. Witherby, London, 1955.
[10] E. Morholt, P. F. Brandwein, and A. Joseph, "A Sourcebook for the Biological Sciences." Harcourt, Brace, and World, New York, 1966.
[11] H. D. Russell, "Notes on Methods for Narcotization, Killing, Fixation and Preservation of Marine Organisms." Marine Biological Laboratory, Woods Hole, Massachusetts, 1963.
[12] S. Satyamurti, "The Preservation of Biological Specimens." Dept. of Museology (Fac. Fine Arts), Maharaja Sayajirao Univ. Baroda, India, 1965.
[13] G. J. Mueller, "Field Preparation of Marine Specimens." Univ. of Alaska Museum, Fairbanks, 1972.

TABLE III
INSTITUTIONS SUPPLYING LIVE SPECIMENS

Supplier	Address	Taxa available
Stock centers		
Caenorhabditis Genetic Center	c/o Mark Edgley 311 Tucker Hall University of Missouri Columbia, MO 65211 (314) 882-7384	Over 1300 species and strains of *Caenorhabditis*
Drosophila melanogaster Stock Center	Dept. of Biology Indiana University Bloomington, IN 47401 (812) 855-5783	
National *Drosophila* Species Resource Center	c/o Dr. Jong S. Yoon Dept. of Biological Sciences Bowling Green State University Bowling Green, OH 43403-0212 (419) 372-2742 or 372-2096	Approximately 300 species of *Drosophila*
Tribolium Stock Center	USA Grain Marketing USDA Agricultural Research Station 1515 College Avenue Manhattan, KS 66502 (913) 776-2710	Approximately 500 strains of *Tribolium castanium*
Supply companies		
Ardea Enterprises	11025 44th Street SE Snohomish, WA 98290 (206) 334-7720	Representation of 6 invertebrate phyla: marine species
Carolina Biological Supply	Burlington, NC 27215 (919) 584-0381 or Powell Laboratories Gladstone, OR 97027 (503) 656-1641	Representatives of 12 invertebrate phyla, some protists: freshwater and marine species
Fisher Scientific, Educational Materials Division	4901 W. LeMoyne Street Chicago, IL 60651 (800) 621-4769 or (312) 378-7770	Representatives of 7 invertebrate phyla, some protists: freshwater and marine species
Gulf Specimen Co.	P.O. Box 237 Panacea, FL 32346 (904) 984-5297	Representatives of 12 invertebrate phyla: marine species
Marine Biological Laboratory, Woods Hole	Dept. of Marine Resources Woods Hole, MA 02543 (508) 548-3705	Representatives of 11 invertebrate phyla: marine species
Pacific Bio-Marine	P.O. Box 536 Venice, CA 90291 (213) 677-1056	Representatives of 18 invertebrate phyla, mesozoa, some protists: marine species
Seacology	3481 West 7th Avenue Vancouver, BC V6R 1W2, Canada (604) 737-2106	Representatives of 13 invertebrate phyla: marine species
Ward's Natural Science Establishment, Inc.	Rochester, NY, or Santa Fe Springs, CA 1-800-962-2660	Representatives of 12 invertebrate phyla, some protists: freshwater and marine species

Storage of Specimens

General Guidelines

Tissue is most stable if stored frozen at a temperature of $-70°$ or lower. Frozen tissue may be stored in an ultracold freezer, on dry ice, or in a liquid nitrogen tank. Advice on organizing frozen tissue collections is given by Dessauer *et al.*[14] Material can also be stored at room temperature in 70–80% ethanol; 95% ethanol is not used for long-term storage because it is highly flammable. Sperm containing 0.01–0.02% sodium azide can be stored refrigerated for 1 to 2 years, but it must not be frozen.[15]

Shipping and Receiving Specimens

The best way to ship material is live. Insects (*Drosophila,* in particular) can be sent on small amounts of food in vials. Most marine organisms can be shipped live in sealed plastic bags. On receipt of the organisms, however, prompt treatment of the shipment is required. Specimens obtained for rearing should be placed in proper rearing media. Specimens obtained for protein or DNA extraction should be immediately and rapidly frozen at or below $-70°$. Frozen tissue should be shipped on dry ice. A much more convenient way to send tissue, however, is in 70–95% ethanol. The DNA yield and quality from specimens sent in ethanol are usually almost as good as from fresh or frozen tissue. Sperm can be shipped on wet ice if it is collected "dry" (i.e., undiluted in water) and if sodium azide is added to a final concentration of 0.01–0.02% in order to prevent bacterial growth.[15]

Museum Collections

Museum collections are a tremendous and still largely underutilized resource for molecular systematic studies. The large research collections in the world contain huge amounts of material amenable to molecular analysis that may be otherwise unavailable due to extinction or collection difficulties. This can be especially acute in higher level studies of geographically diverse taxa. Collections can make possible, in time and resources, studies for which it would take years to gather material.

The vast majority of invertebrate specimens are in ethanol. With the notable exceptions of insects, corals, some sponges, echinoids, asteroids,

[14] H. C. Dessauer, C. J. Cole, and M. S. Hafner, *in* "Molecular Systematics" (D. M. Hillis and C. Moritz, eds.), p. 25. Sinauer, Sunderland, Massachusetts, 1990.
[15] Raff, personal communication (1992).

and mollusc shells, invertebrates simply do not dry well. Even organisms with hard chitinous exoskeletons may not preserve well because of softer parts, which can be rendered useless (at least for morphological analysis) by drying. In the case of spiders, the abdomen shrivels to a useless knot.

DNA seems to be preserved fairly well in samples in ethanol, depending on collection practices, but the majority of the DNA is usually degraded to less than 2 kilobases (kb). Other chemicals may have been added to dope the ethanol in an effort to fix or better maintain the samples. Glycerin is present in some older collections of spiders, as is 2-propanol. These practices have been abandoned because 2-propanol tends to make the samples brittle, and the glycerin sometimes thickens the specimen and its storage environment. Formaldehyde has been used along with formalin to preserve other invertebrate tissues, which can make DNA extraction quite challenging, but these practices also seem to be waning.

Insects are spectacular not only in their biological diversity but also in their collection media. When possible insects should be pinned dry. This technique is convenient, requires fairly low maintenance, and is much lighter than liquid storage. They are, however, frequently interred in ethanol, as well as collected in ethanol, and later dried out and pinned. Fortunately, both pinned and ethanol-preserved samples yield usable DNA.

All is not quite this simple, however. A wide variety of more exotic additives have been used in wet insect collections, including formamide, picric acid, formalin, formaldehyde, and glycerin. Because any and all of these materials may affect DNA extraction and subsequent analysis, it is crucial to know how the samples were collected and something of their maintenance history. Because cnidarians and ctenophores are traditionally kept in solutions of 10% formalin (as are ctenostome bryozoans), we have not had much success in extracting DNA from museum collections of these taxa. Overall, most of the wet and all of the dry materials are potentially useful to molecular studies. Given the tremendous resources of some of the large institutional collections, some intractable problems may be resolved, and some groups of invertebrates whose molecular biology is poorly known may be sampled more completely.

Loan procedures and conditions vary from institution to institution. Because the practice of loaning specimens for DNA work is fairly novel to many museums, some institutions are still in the process of establishing policies for the distribution of materials in collections that will be destroyed during investigations. The best way to obtain information on getting material is through contacting the curator in charge of the groups of interest.

Special Considerations

Tissue Preferences

The best tissue to use depends on the organism to be studied and on the type of molecular work to be done. The goal is to use parts of the organism that are relatively free of compounds damaging the nucleic acid of interest or that interfere with cloning, polymerase chain reaction (PCR), or sequencing.

In general, gamete or muscle material is best for DNA preparation. Ripe specimens can often be induced to spawn (for taxon-specific methods see Strathman[16]). Alternatively, gonads or muscle tissue can be dissected out. Using specific tissues has the added benefit of reducing the risk of contamination with parasites, gut fauna, or recently eaten prey. Some caveats are that the eggs of species with large eggs should be avoided, and that PCR may not work well with DNA extracted from sperm, because of DNA viscosity.[17] For mitochondrial DNA preparation, eggs are preferable to sperm. Embryos are a good source of insect mitochondrial DNA.

If gamete or muscle tissue is difficult to obtain, body tissue with the stomach removed is an alternative. Animals too small to be dissected should be starved for at least 2 days if possible and used whole. If all the above procedures are impractical, it is possible that good results may be obtained with whole, unstarved specimens, especially if PCR primers or DNA probes are sufficiently taxon-specific to prevent contamination with foreign tissue from causing problems.

Whatever portions of the organism are selected, these should be used, frozen, or preserved immediately, before tissue begins to degrade. Fresh tissue is preferable if purified mitochondrial DNA is being prepared. This is because freezing may break mitochondrial membranes, reducing yield at the step in a protocol where mitochondria are pelleted. However, if genomic DNA or total cellular DNA with a mitochondrial DNA fraction is desired, frozen tissue can give good DNA quality and yield. Tissue should be frozen very quickly by covering with liquid nitrogen or dry ice.

Different Applications Requiring Different Precautions

High molecular weight DNA is required for several molecular biology applications such as cloning or mapping of restriction fragment length polymorphisms (RFLP). In handling tissues slated for these types of stud-

---

[16] M. F. Strathmann, "Reproduction and Development of Marine Invertebrates of the Northern Pacific Coast." Univ. of Washington Press, Seattle, 1987.
[17] Palumbi, personal communication (1992).

ies, care should be taken not to allow the tissue to warm after initial storage. Similarly, if protein analysis is the goal of the study, extreme care should be taken to maintain the specimens at temperatures that will prevent denaturation of proteins.

Although high molecular weight DNA is the preferred template for PCR, degraded DNAs will often amplify if the target fragment is small enough and the primers specific enough. DNA slated for PCR analysis can often be used from alcohol-preserved or museum specimens, which sometimes produce fairly degraded DNA.

Summary

The validity of any comparative study is dependent on the reliability of the identification of the samples in the study. Not all researchers are experts in the field of identification of samples, nor do all researchers have quick and ready access to expert systemetists who can accomplish the task of identification. The importance of verification of sample identity for comparative studies is vital. We describe several methods by which researchers can obtain and identify samples from the wild, and we suggest methods by which voucher samples can be obtained for future reference to these collected samples. We outline alternatives to collection of samples from the wild, such as purchase from stock centers and biological supply companies. Museum collections can also be extremely helpful in obtaining complete organismal samples for comparative studies.

[5] Collection and Storage of Fungal and Algal Samples

By MEREDITH BLACKWELL and RUSSELL L. CHAPMAN

Introduction

Samples to be used for taxonomic and systematic study of algae, fungi, and lichens always have been important, but the current combination of explicit phylogenetic analysis and molecular approaches in studies of the systematics and evolutionary biology of these taxa requires some special considerations and approaches. Obtaining and storing fungal and algal samples is a major component in an exciting new assault on interesting, and previously untractable, questions in the biology of fungi and algae.

The problems of applying evolutionary analysis to organisms such as fungi, lichen symbionts, and algae have been due in part to (1) the availability of relatively few morphological characters, some of which may be missing in certain closely related taxa; (2) an inability to distinguish morphological features as discrete characters or multiple states of the same character; and (3) the difficulty in establishing character polarity by outgroup rooting or other means. The emphasis of taxonomists usually has been on distinguishing among or recognizing distinct taxa of these difficult organisms; thus many groups, especially at higher taxonomic levels, are based on autapomorphic characters. Until recently, few attempts[1-4] were made to apply phylogenetic analysis methods to fungi and algae.

The increasing application of explicit phylogenetic methods has occurred concurrently with the use of molecular data. The use of molecular approaches such as nucleic acid sequencing has required greater reliance on computer-aided data analysis to deal with the vast number of new characters. The combination of new sources of characters with new methods of data analysis having the capability of establishing character polarity in the absence of an extensive fossil record has kindled an exciting new era in the taxonomy and systematics of fungi and algae. The advent of simpler molecular techniques, particularly those requiring small amounts of sample, allows a new group of biologists to exploit the methods in conjunction with morphological characters. The advances hold hope for eventual answers to several hundred years of evolutionary questions by providing for the testing of hypotheses based primarily on morphological characters with independent data sets. The revolution brought about by additional molecular characters is just beginning. The importance of obtaining and storing fungal and algal samples for study remains fundamentally important in the new era.

This chapter provides a cursory review of the practical aspects of obtaining fungal and algal samples for molecular studies. It is directed toward the molecular biologist interested in initiating studies on the organisms, but it also should generally be useful to mycologists and phycologists, particularly beginning students.

Culture Collections

The obvious place to obtain fungi and algae for molecular analysis is from culture collections where they are maintained in pure culture, usually with species determinations. Because the services of culture collections,

[1] Q. Wheeler and M. Blackwell, *in* "Fungus–Insect Relations, Perspectives in Ecology and Evolution" (Q. Wheeler and M. Blackwell, eds.), p. 5. Columbia Univ. Press, New York, 1984.
[2] R. Currah, *Am. J. Bot.* **71**, 161 (Abstr.) (1984).
[3] D. R. Reynolds, *Mycotaxon* **27**, 377 (1986).
[4] E. C. Theriot, *J. Phycol.* **25**, 407 (1989).

including maintenance, are expensive, costs must be defrayed by user fees in many cases. Researchers must remember to budget for these necessary expenses.

In addition to well-known major collections with broad general holdings, there are many smaller or specialized culture collections found throughout the world. For example, if isolates from a particular geographical locality are required, a smaller collection in the region of interest might be best for providing the culture; an insect fungal pathogen can be obtained from the specialist collection of entomopathogenic fungi maintained by the U.S. Department of Agriculture (USDA) (ARSEF) at the U.S. Plant, Soil, and Nutrition Laboratory (Ithaca, NY). Cultures of many isolates of wood-rotting basidiomycetes are available from the USDA Forest Service Center for Forest Mycology (Madison, WI); besides having large numbers of multiple isolates of species, the laboratory maintains voucher specimens for every culture on hand. Among examples of specialized algal collections are the *Chlamydomonas* Genetics Center maintained at Duke University (Durham, NC), which houses a large collection of *Chlamydomonas reinhardtii* mutants and numerous strains of other *Chlamydomonas* species; all strains are axenic, and a printout of file information on each strain is provided. A collection of prokaryotic nitrogen-fixing blue-green algae is maintained by the Soil Microbiology Division of the International Rice Research Institute (Manila, Philippines), and a collection of algae with a high potential for use in biomass energy production is available at the Solar Energy Research Institute (Golden, CO).

More detailed information on the specialized collections mentioned above and on hundreds of collections worldwide is available from the World Federation for Culture Collections (WFCC) (see address below), which maintains a data center, the World Data Center on Microorganisms (WDC), also a component of the Microbial Resources Centers (MIRCEN) network. In its major role of disseminating information on culture collections, the WDC has two important publications that are essential guides to cooperating fungal and algal collections throughout the world. The *World Catalogue of Algae*[5] was published with the cooperation of 39 algal collections in 16 countries. The *World Directory of Collections of Cultures of Microorganisms*[6] in its third edition, includes fungi from 345 culture collections worldwide. In addition, users may search the listings of all the

[5] S. Miyachi, O. Nakayama, Y. Yokohama, Y. Hara, M. Ohmori, K. Komagata, H. Sugawara, and Y. Ugawa, "World Catalogue of Algae," 2nd Ed. Japan Scientific Societies Press, Tokyo, 1989.

[6] J. E. Staines, V. F. McGowan, and V. A. D. Skerman, "World Directory of Collections of Cultures of Microorganisms," 3rd Ed. World Data Centre, Univ. of Queensland, Brisbane, Australia, 1986.

collections for individual species on the WDC electronic database that is accessible 24 hr a day. A guide to the database *(Guide to World Data Center on Microorganisms — A List of Culture Collections in the World)* is available from the WFCC, WDC, RIKEN, Wako, Saitama, Japan. Users must first register with the WDC to obtain a "user-ID" and "password" for access. Registration may be done easily by electronic mail, telex, or postal service using the following addresses:

DIALCOM: 42:CDT0007
JUNET: r35118@rkna50.riken.go.jp
Telex: 2962818 RIKEN J
Postal service: WDC/RIKEN
　　　　　　　　2-1 Hirosawa, Wako
　　　　　　　　Saitama 351-01, Japan

Cultures of some fungi are stored lyophilized and can be shipped immediately on order; however, others must be grown out on agar or in liquid culture so that shipping may be delayed several weeks. Additional delays can be encountered by the necessity of obtaining permits for shipping of certain organisms, particularly plant pathogenic fungi. This process involves both state and federal agriculture official approval in the United States that may take several months.[7,8] Some fungi or their secondary products are the cause of virulent or chronic disease in animals and require special procedures.[7] No one should attempt to work with these organisms without proper facilities and a thorough appreciation of the risks involved. Culture collection staff can advise on the correct procedures that may be necessary. Literature on worldwide importation and postal requirements is cited in the *International Mycological Directory*.[9] Algal samples are usually provided as liquid- or agar-grown cultures and are rarely subject to any requirements for special permits; however, the viability of the cultures (especially those with specific temperature requirements) is often a greater concern. Although many phototrophic algae can survive several days without light, a prolonged delay in shipping time will kill some algae, as will exposure to extreme temperatures.

Most culture collection curators take elaborate precautions to ensure that cultures are correctly identified. Some even return transferred cultures

---

[7] S. C. Jong and W. B. Atkins, *in* "Fungi Pathogenic for Humans and Animals" (D. H. Howard, ed.), p. 153. Dekker, New York, 1985.

[8] D. L. Hawksworth and K. Allner, *in* "Living Resources for Biotechnology, Filamentous Fungi" (D. L. Hawksworth and B. E. Kirsop, eds.), p. 54. University Press, New York, 1988.

[9] G. S. Hall and D. L. Hawksworth, "International Mycological Directory," 2nd Ed. International Mycological Association, CAB International Mycological Institute, Oxon, United Kingdom, 1990.

to the depositing scientist for authentication before cataloging. The vast majority of fungal and algal cultures will be identified correctly, but mistakes can be made by the depositor, collection employees, or even the researcher. Similarly, contamination with other cultures is a real possibility. To ensure against incorrect identifications we routinely grow out cultures to confirm identity if possible, and we always keep the original or verified subcultures stored. For fungi, it is sometimes useful also to keep dried culture vouchers. If there is any doubt about the purity and identity of batch cultures used for nucleic acid extraction, they should be grown on agar and examined microscopically for comparison with the stored culture. When a number of similar isolates are being handled, extreme care must be taken to avoid contamination that cannot be detected easily. Even with great care against contamination during all phases of work, problems may occur, and a final check is the inclusion of two closely related forms that occur as nearest relatives in the analysis.

Ploidy or nuclear number should be a consideration when choosing cultures. Highly variable DNA regions in single-copy or repetitive genes (e.g., ribosomal RNA genes) may vary at single base positions in the homologous chromosomes of diploid or dikaryotic isolates.[10] Haploid cultures of many fungi with gametic meiosis can be acquired easily from single spores, whereas isolates from basidiomata, ascomata, or mass spore cultures of heterothallic basidiomycetes and ascomycetes provide strains with nuclear variation. Diploid material is difficult to avoid in some organisms such as oomycetes, but it is important to recognize the possibility of variation. DNA can be extracted from single spores[11] and may eliminate some ploidy problems as well as provide the means for a variety of intraspecific level studies. For most algae in culture collections, the ploidy is known or can be inferred from what is known about the life cycles. Life cycle information can be obtained from a number of sources such as mycology and phycology textbooks.[12,13]

Larger culture collections usually offer multiple strains of species. Here again, informed choices should be made. Whenever possible one should use so-called type cultures, that is, those isolated from the type collection of a species at the time of its description. If generic level questions are

[10] M. Gardes, T. J. White, J. A. Fortin, T. D. Bruns, and J. W. Taylor, *Can. J. Bot.* **69**, 180 (1991).

[11] S. B. Lee and J. W. Taylor, *in* "PCR Protocols: A Guide to Methods and Applications" (M. A. Innis, D. H. Gelfand, J. J. Sninsky, and T. J. White, eds.), p. 282. Academic Press, San Diego, California, 1990.

[12] C. J. Alexopoulos and C. W. Mims, "Introductory Mycology." Wiley, New York, 1979.

[13] H. C. Bold and M. J. Wynne, "Introduction to the Algae." Prentice-Hall, Englewood Cliffs, New Jersey, 1985.

involved in the project, cultures of the type species of the genus always should be used. Strains may differ in metabolism and other genetic characters, and these differences may be important in choosing a culture. The American Type Culture Collection (Rockville, MD) has a helpful publication[14] that lists processes and products associated with many isolates in the ATCC collection, and some culture collection catalogs contain information including literature references to strains.[15]

Herbaria

There is renewed interest in herbaria because they provide a source of potentially useful DNA from identified organisms, including type specimens.[16-18] Herbarium specimens of fungi as old as 50 years[17] and almost 100 years[19] have provided adequate template for polymerase chain reaction (PCR) amplification of ribosomal genes. In fact, Swann and colleagues[19] were able to obtain DNA sequences from a sample of only about 1000 spores of a rust fungus specimen and a single powdery mildew ascoma. As Bruns and co-workers[17] have shown, cultures established at the time of specimen collection are sometimes available and provide for comparisons to demonstrate the integrity and usefulness of DNA from herbarium specimens. Although DNA may be partially degraded, primer-directed amplification, especially of multiple copy genes, may overcome the problem. Generally, specimens fast-dried at moderate temperatures or edges of larger specimens provide the best DNA template.[17] Herbarium curators should consider including provisions for maintenance of DNA samples as part of their guidelines (see below). Herbarium specimens are especially important for rare or nonculturable species; they will be important to type studies. Finally, in addition to providing DNA template from tissue, herbarium specimens of fungi and algae may contain viable propagules for establishing cultures.[20,21]

[14] M. J. Edwards, "Microbes and Cells at Work: An Index to ATCC Strains with Special Applications." American Type Culture Collection, 12301 Parklawn Drive, Rockville, Maryland, 1988.

[15] M. I. Krickevsky, B. O. Fabricius, and H. Sugawara, in "Living Resources for Biotechnology, Filamentous Fungi" (D. L. Hawksworth and B. E. Kirsop, eds.), p. 31. Cambridge Univ. Press, New York, 1988.

[16] F. Rollo, A. Amici, R. Sabvi, and A. Garbuglia, Nature (London) 335, 774 (1988).

[17] T. D. Bruns, R. Fogel, and J. W. Taylor, Mycologia 82, 175 (1990).

[18] E. M. Golenberg, D. E. Giannasi, M. T. Clegg, C. J. Smiley, M. Durbin, D. Henderson, and G. Zurawski, Nature (London) 344, 656 (1990).

[19] E. C. Swann, G. S. Saenz, and J. W. Taylor, MSA Newsl. 42, 36 (Abstr.) (1991).

[20] A. S. Sussman, in "The Fungi" (G. C. Ainsworth and A. S. Sussman, eds.), Vol. 3, p. 447. Academic Press, New York, 1968.

[21] D. J. Young, R. L. Gilbertson, and S. M. Alcorn, Mycologia 74, 504 (1982).

Field Collections

Field collections of fungi, lichens, and algae are useful for establishing cultures or extracting DNA directly. The *Mycology Guidebook*[22] is an important resource for information on isolation of fungi from nature, and Volume 1 of the *Handbook of Phycological Methods*[23] provides comparable information for the isolation of algae. Many species of fungi and algae have never been cultured, so fresh collections may be essential, especially with species for which inadequate herbarium material is available. Although the amount of material needed for DNA extraction has been reduced by the use of the PCR, culturing even small amounts or using herbarium specimens of some fungi and algae may require more time and resources than would a collection from nature. If extraction cannot be performed in a reasonable time, field-collected specimens should be refrigerated or even frozen. Liquid nitrogen is useful for indefinite storage of moderate-sized specimens.

An important caution to be heeded in using fresh material directly is to ensure that all contaminating organisms are removed (washed, picked, etc.) from the desired specimen. This may not always be possible, and use of taxon-specific primers in the PCR may overcome the problem. For example, the presence of bacterial contaminants on an eukaryote would be of little consequence if eukaryote-specific primers were being used; in the case of lichens, fungus- or alga-specific primers might be used.[24]

Voucher specimens must be prepared, so save enough of the specimens, being certain they contain diagnostic features for positive identification. Vouchers should be deposited in established herbaria if they warrant it. Most herbaria probably would not have space for or desire large numbers of specimens of a single species from one locality or specimens lacking diagnostic characters. We feel it is desirable to keep vouchers for a number of years in case questions should arise about published work. Voucher specimens should be kept dry and insect-free in a special herbarium cabinet, preferably in a herbarium and perhaps at the home institution or in the care of the researcher if they are not appropriate for the herbarium.

Another potential problem in using field-collected specimens is obtaining species identifications. This problem is addressed more fully below. However, it may be necessary to contact a specialist in the group of interest. If a home institution mycologist or phycologist cannot be of help, he or she can usually suggest a specialist in the group of interest. Publishing

[22] R. B. Stevens, "Mycology Guidebook." Univ. of Washington Press, Seattle, 1974.
[23] J. R. Stein, "Handbook of Phycological Methods: Culture Methods and Growth Measurements." Cambridge Univ. Press, New York, 1973.
[24] A. Gargas and J. W. Taylor, *MSA Newsl.* **42**, 14 (Abstr.) (1991).

taxonomists also can be located by looking through taxon indices of primary or reference journals such as *Biological Abstracts*. If all else fails, officers of mycological and phycological societies (usually listed in society-published journals) can make suggestions of appropriate taxonomists. Society newsletters often will publish requests for specific specimens. Fungal and algal "forays" are held periodically by both professional and amateur groups. It is especially important to check identifications of specimens that are acquired more informally, and, again, collectors are cautioned to keep vouchers in case there are questions later.

For fungi and some algae fresh field-collected specimens often are a source of insects and mites. Not only can these arthropods destroy specimens, but they can also destroy a culture collection by contamination as they move from one tightly sealed culture to the next. Mites also are suspected vectors in horizontal gene transfer between *Drosophila* spp.,[25] and perhaps this form of "contamination" may be detected in fungi and algae one day.

Storage of Cultures

Length of culture viability is dependent on species (or strain in some cases), age of culture at storage, propagules present, culture medium (affecting nutritional state), and method of storage. For these reasons storage of cultures is a research area in itself and cannot be covered adequately here. In-depth discussions of the techniques as well as primary references can be found in various reviews.[7,26-28] Some culture collections provide the service of specimen storage preparation, and the price may well be worth it if long-term storage is the goal.

Over a period of 2–4 weeks most fungi survive well at room temperature. Parafilm wrapping retards desiccation of the medium and increases length of viability; contamination is also lessened. Fungal stock cultures usually remain viable for at least 6 months on agar slants stored in a refrigerator (5°). Cultures in screw-capped bottles (15 to 20 ml capacity) remain viable for longer periods in our experience. Although there are some auxotrophic algae that may be quite similar to fungi in short-term storage requirements, most are phototrophic or photoauxotrophic and

[25] M. A. Houck, J. B. Clark, K. R. Peterson, and M. G. Kidwell, *Science* **253**, 1125 (1991).

[26] S. C. Jong, *in* "Biotic Diversity and Germplasm Preservation, Global Imperatives" (L. Knutson and A. K. Stoner, eds.), p. 241. Kluwer Academic Publ., Dordrecht, The Netherlands, 1989.

[27] D. Smith, *in* "Living Resources for Biotechnology, Filamentous Fungi" (D. L. Hawksworth and B. E. Kirsop, eds.), p. 75. Cambridge Univ. Press, New York, 1988.

[28] O. Holm-Hansen, *in* "Handbook of Phycological Methods: Culture Methods and Growth Measurements" (J. R. Stein, ed.), p. 195. Cambridge Univ. Press, New York, 1973.

require adequate lighting. Although exact requirements will vary greatly among different types of algae, temperatures slightly below room temperature ($\sim 21°$) and dim lighting are best for storage of cultures in liquid medium or on agar. Some algae will require special conditions (e.g., bubbling with air or CO_2) even for short-term storage periods of 2–4 weeks. One common practice that leads to disaster is placing algal cultures on window sills where they will be exposed to direct sunlight for part of the day. Such exposure will readily kill many algae. Detailed information on standard practices are presented in *Handbook of Phycology.*[23]

Many fungal and algal cultures are viable after lyophilization.[29-32] This type of storage is time consuming over the short term, but over a long period it may save time, space, and expense. Lyophilization of cultures has the additional advantage of preserving the genetic integrity of strains that might change during years of active growth.[7,26,27] Fungi that produce spores also have been stored, usually for shorter periods of time, by drying in soil[33-35] or silica gel.[36,37] Cultures on agar may be covered with sterile mineral oil to lower oxygen levels (but not deplete them completely) or kept in sterile distilled water at 4°.[7,26,27] Smith[27] considers freezing and storage with a cryoprotectant in liquid nitrogen[38] to be the best general method of preserving filamentous fungi. Algae can be stored as frozen samples with cryoprotectants such as glycerol, dimethyl sulfoxide, or dried milk solids. For short-term storage a frozen sample can be kept in a standard freezer at $-20°$, in a mixture of glycol and dry ice (about $-78°$), or in liquid nitrogen at $-196°$. For long-term storage a temperature of $-40°$ or colder is considered more likely to maintain viability than a normal freezer for most algae.[28] The complexity of the processes involved in freezing fungi and algae and maintaining cell viability requires that optimal conditions be determined for each culture.[7,26,27] Even when empirical results demonstrate such optimal conditions, a program of routine, periodic monitoring of viability is probably essential for very long-term (decade or longer) storage.

[29] R. H. Haskins and J. Anastasiou, *Mycologia* **45,** 523 (1953).
[30] R. H. Haskins, *Can. J. Microbiol.* **3,** 477 (1957).
[31] W. C. Haynes, L. J. Wickerham, and C. W. Hesseltine, *Appl. Microbiol.* **3,** 361 (1955).
[32] C. W. Hesseltine, B. J. Bradle, and C. R. Benjamin, *Mycologia* **52,** 762 (1960).
[33] R. G. Atkinson, *Can. J. Bot.* **32,** 673 (1954).
[34] D. I. Fennell, *Bot. Rev.* **26,** 79 (1960).
[35] C. Booth, "The Genus *Fusarium.*" Commonwealth Mycological Institute, Kew, Australia, 1971.
[36] D. D. Perkins, *Can. J. Microbiol.* **8,** 591 (1962).
[37] D. D. Perkins, *Neurospora Newsl.* **24,** 16 (1977).
[38] S. C. Jong, *in* "The Biology and Cultivation of Edible Mushrooms" (S. T. Chang and W. A. Hays, eds.), p. 119. Academic Press, London, 1978.

Inclusion of Taxonomists in Molecular Studies

Alpha-level taxonomists are an essential but dwindling resource. It is these individuals who use their data (generally morphological characters of organisms) to propose level one hypotheses (erection of species, genera, and higher taxa) essential to later molecular studies. In addition to species identification, taxonomists can provide important information on taxon selection and particularly interesting questions about unresolved relationships and good model systems for studying broad biological questions, such as speciation processes in wind-dispersed species, broad intercontinental distributions, coevolution of symbionts, or chloroplast evolution.

Molecular techniques already are bringing new evidence to provide independent data in 100-year-old morphologically based hypotheses. For example, an extremely close relationship between basidiomycetes with highly divergent basidiomes (boletes and hypogeous forms) has been supported by molecular data.[17] On the other hand, a long-standing hypothesized relationship between red algae and ascomycetes that periodically resurfaces[39] is not supported by molecular evidence.[40,41] In 1858, Pringsheim[42] suggested a controversial relationship between some oomycetes and the Vaucheriaceae (which contain plastids with chlorophylls *a* and *c*); additional morphological evidence from flagellar apparatus structure supports this view.[43,44] The topic is the exciting subject of current molecular studies by at least two groups of workers to provide additional independent data that appear to support the idea of Pringsheim.[45,46]

Long-standing nomenclatural problems in mycology are even being addressed. Some species of ascomycetes and basidiomycetes have several morphs. Asexual stages without known sexual stages long have been placed as form species in the "Deuteromycota." Reynolds and Taylor[47,48] have questioned the need for this practice now that molecular characters are

---

[39] V. Demoulin, *Biosystems* **18,** 347 (1985).

[40] S. Kwok, T. J. White, and J. W. Taylor, *Exp. Mycol.* **10,** 196 (1986).

[41] D. Bhattacharya, H. J. Elwood, L. J. Goff, and M. L. Sogin, *J. Phycol.* **26,** 181 (1990).

[42] N. Pringsheim, *Pringsh. Jahrb. Wiss. Bot.* **1,** 284 (1859).

[43] D. J. S. Barr, *in* "Zoosporic Plant Pathogens" (S. T. Buczacki, ed.), p. 43. Academic Press, New York, 1983.

[44] D. J. S. Barr, *Mycologia* **84,** 1 (1992).

[45] G. R. Klassen, G. Hausner, and A. Belkhiri, *MSA Newsl.* **42,** 22 (Abstract) (1991).

[46] S. B. Lee and M. S. Fuller, *MSA Newsl.* **42,** 23 (Abstract) (1991).

[47] D. R. Reynolds and J. W. Taylor, *Taxon* **40,** 311 (1991).

[48] D. R. Reynolds and J. W. Taylor, *in* "Improving the Stability of Names: Needs and Options" (D. L. Hawksworth, ed.), p. 171. Koeltz, Konigstein, Germany, 1991.

available; they have initiated discussion of the use of DNA as the type element in some cases and its implications for classification under the *International Code of Botanical Nomenclature.* They envision the possibility that some species even may be discovered and known only by their DNA. Ward and colleagues[49] certainly have found this to be true of bacteria.

Although an undergraduate student can be trained to perform most of the techniques used in modern phylogenetic studies, a depth of knowledge of fungal and algal taxonomy, including the ability to identify the organisms to the species level, takes many years. Given the tens of thousands of species of algae and perhaps more than 1.5 million species of fungi,[50] a large cadre of taxonomic experts is needed in both disciplines, but the modest numbers of these scientists are declining rather than increasing. It is ironic that this shortage should occur at a time when interest in biodiversity is high and molecular approaches combined with explicit phylogenetic analysis of the data are providing the means for exciting advances in systematics and evolutionary biology. The suggestion that *Pneumocystis carinii* is a fungus[51,52] has important implications for disease treatment; there is an interest in fungal phylogenetic relationships to decrease time and expense in fungicide trials. The exciting recognition that coastal nanoplankton organisms are far more important in total primary productivity than previously thought increases the practical need to be able to identify these minute organisms, for which morphology is of limited usefulness.[53] These applications underscore the fundamental importance of taxonomy. Unfortunately, the number of taxonomists and students of major groups of fungi, lichens, and algae continues to decrease. The condition is critical and warrants mention here.

Because taxonomists are essential, they should be totally involved in molecular evolution studies as principal investigators. They should not simply be consulted toward the end of a study when identifications or discussions for papers are needed, but rather should be included from the very beginning in the project design. In fact, the pairing of taxonomists and molecular biologists should be encouraged both for efficiency and for well-conceived science.

[49] D. M. Ward, R. Weller, and M. M. Bateson, *Nature (London)* **345**, 63 (1990).
[50] D. L. Hawksworth, *Mycol. Res.* **95**, 641 (1991).
[51] J. Edman, J. A. Kovacs, H. Masur, D. V. Santi, H. J. Elwood, and M. L. Sogin, *Nature (London)* **334**, 519 (1988).
[52] J. Edman, J. A. Kovacs, H. Masur, D. V. Santi, H. J. Elwood, and M. L. Sogin, *J. Protozool.* **36**, 185 (1989).
[53] L. S. Murphy and E. M. Haugen, *Limnol. Oceanogr.* **30**, 47 (1985).

Herbarium Guidelines

We have already mentioned that herbarium curators should begin to formulate guidelines concerning extraction of DNA from specimens. Plans should be made for the best preservation of incoming specimens so that they will provide minimally degraded samples. Pilot studies can be made to obtain information needed to make informed decisions of this kind. Herbaria also should make provisions for storage of excess DNA samples once they are obtained from specimens, either at their institution or at a cooperative institution. Certain zoological museums already store frozen tissues for isozyme and other molecular studies. Ideally DNA extractions from new type specimens could be performed routinely soon after they are described. We need to begin planning for curation of these new resources that provide more characters than we have ever known.

Conclusions

Given the small amounts of nucleic acid needed for many studies, one can propose that major coordinated regional collections of samples eventually could serve the needs of the scientific community, just as major herbaria and culture collections do. Perhaps nucleic acid sample collections will be established as additions to major herbaria and/or culture collections.

The potential usefulness of, and need for, major, coordinated regional collections is clear, but the feasibility is not. The concept warrants attention because the historical record has proved that individual local collections are often maintained and sustained by the commitment and effort of one or more individual researchers, rather than by a long-term commitment of the institution of the collection. Thus, orphaned collections of specimens of all types often are lost (literally lost, discarded, or allowed to deteriorate), and long-term protection and continuity are needed both for traditional fungal and algal material (e.g., herbarium specimens, cultures) and nucleic acid samples. The latest, most modern advances on the molecular front have underscored dramatically the importance of very basic, mostly traditional, aspects of obtaining and storing fungal and algal specimens. Newer aspects, such as cryostorage and nucleic acid inventories, are now part of the game plan. Our samples are literally bits of the biodiversity of the world; as such, they are useful sources of many kinds of information. Because we do not know all there is to know, we cannot appreciate how critically important some specimens will be in future research. Some of the specimens will prove to be scientific treasurers, and since we do not know which specimens will be the most valuable, we must treat all of them *a priori* as invaluable.

Acknowledgments

We thank Drs. S. C. Jong and D. L. Hawksworth for providing helpful literature. Dr. John W. Taylor was a helpful reviewer of the manuscript. Financial assistance from the National Science Foundation (BSR-8918157 to M.B. BSR-8918564 to M. A. Buchheim and R.L.C., and BSR-9107389 to R.L.C.) is gratefully acknowledged.

Section II

Comparing Macromolecules: Exploring Biological Diversity

A. Comparisons at Protein Level
Articles 6 through 11

B. Comparisons at Nucleic Acid Level
Articles 12 through 34

Section II. Comparing Macromolecules: Exploring Biological Diversity

A. Comparisons at Protein Level

For many years biologists have successfully used the indirect measures of genetic difference available by comparing proteins for both systematics and population genetic studies. Both electrophoretic and immunological techniques have been applied extensively to analyses of large numbers of species or individuals because of their relative simplicity, low cost, and practicality. Section II introduces electrophoretic methods for detecting protein variation based on differences in size and net charge. Techniques, buffers, and stains are described by Bult and Kiang [6] for plant tissues and by Britton-Davidian [7] for vertebrates. This is necessary because aspects of their overall biochemistry do not allow absolute generalization between these two major groups. Goldman and O'Brien [8] discuss two-dimensional gel electrophoresis, a method having higher resolution and the ability to examine variation at more loci simultaneously. Although most studies use proteins isolated from contemporary or artificially preserved biological materials, procedures for isolating macromolecules from ancient specimens are advancing rapidly, as reviewed by Tuross and Stathoplos [9].

Van Regenmortel et al. [10] describe a variety of immunochemical techniques that relate antigenic variability to amino acid sequence differences. These methods can detect additional variation beyond that which changes the net charge of a protein and can be useful over a wider range of sequence divergence. Prager and Wilson [11] critically examine the calibration error involved in using an indirect method (immunological difference) to estimate actual sequence difference among proteins. By applying confidence values to the distance estimates, they test the robustness of ten hypotheses about phenetic relationships for a variety of organisms. This approach can be used to develop similar confidence limits and prediction intervals for other methods of estimating genetic distance.

B. Comparisons at Nucleic Acid Level

Researchers who study the evolution of molecules must first ensure that the materials they work with in isolation reflect with high fidelity the organization of nucleic acids in their study organisms. The chapters by Stein [12], Chesnick and Cattolico [13], DeSalle et al. [14], and Rayburn [15] mirror this concern. Manipulation of purified nucleic acids using DNA–DNA reassociation kinetics, discussed in [16] and [17], reveals the diversity and complexity of genome organization at a "macro" scale. Rogers et al. [18] discuss methods to quantify copy number in gene families, and Slightom [19] provides details that gene libraries constructed under a variety of conditions will contain representative members of most classes of sequences. Rogstad [20] and Bowditch et al. [21] discuss alternative approaches to examining global genome organization, ones that probe the nature of randomly dispersed repetitive sequences. These methods may prove essential for localizing the physical basis of quantitative genetic traits in the genomes of plants and animals, where adaptive evolution has helped shape morphological diversity. Martin and Wichman [22] and Felger and Hunt [23] utilize mobile repetitive sequences to uncover patterns that link gene expression and speciation. Species delineation in bacteria may best be estimated by some of the above techniques that provide quick estimates of genetic distance; Kurtzman [24] presents one such approach.

Chapters [25]–[30] discuss specific aspects of nucleic acid sequencing technology for cloned and enzymatically amplified regions. Lessa [31] and Chang et al. [32] deal explicitly with intraspecific variation in DNA sequences. Finally, chapters [33] and [34] are included as a practical guide to phylogenetic analysis of evolutionary patterns found in nucleic acid data. These chapters are meant to complement and update the discussion of "evolutionary tree building" found in Volume 183 of this series.

[6] One-Dimensional Electrophoretic Comparisons of Plant Proteins

By CAROL J. BULT and YUN-TZU KIANG

Introduction

The combination of horizontal slab-gel electrophoresis and *in situ* assay of enzyme activity is a versatile and powerful method for detecting protein variants (isozymes and allozymes) in plants. Such variants are potentially useful as genetic markers for mapping chromosomes and in studies of breeding systems,[1] population structure,[2,3] gene flow,[4] polyploidy,[5,6] and systematics.[7–9]

The purpose of this chapter is to provide a brief introduction to preparing tissues, gels, and buffers for electrophoresis, with an emphasis on working with plant materials. We also provide enzyme activity staining protocols that we have found to work well for a variety of tissue types and plant species. The protocols presented here are meant as an introduction to the methodology; modifications will no doubt be required to optimize results for a particular study, taxon, or tissue type. A number of comprehensive reviews and books have been published on general methods in protein electrophoresis and on the interpretation and analysis of electrophoretic data.[10–12] In addition, many technical reports are available that provide detailed protocols for working with specific groups of plants, such as conifers[13] or ferns.[14]

The technical aspects of electrophoresis are straightforward. A wick, cut from filter paper, is saturated with a tissue homogenate and inserted into

[1] D. Schoen, *Evolution* **36**, 352 (1982).
[2] M. D. Loveless and J. L. Hamrick, *Annu. Rev. Ecol. Syst.* **15**, 65 (1984).
[3] J. L. Hamrick and M. J. W. Godt, *in* "Plant Population Genetics, Breeding, and Genetic Resources" (A. H. D. Brown, M. T. Clegg, A. L. Kahler, and B. S. Weir, eds.), p. 43. Sinauer, Sunderland, Massachusetts, 1990.
[4] M. Slatkin, *Science* **236**, 787 (1987).
[5] C. Werth, *Biochem. Syst. Ecol.* **17**, 117 (1989).
[6] L. Gottlieb, *Science* **216**, 373 (1982).
[7] M. F. Mickevich and M. S. Johnson, *Syst. Zool.* **25**, 260 (1976).
[8] D. Buth, *Annu. Rev. Ecol. Syst.* **15**, 501 (1984).
[9] D. L. Swofford and S. H. Berlocher, *Syst. Zool.* **36**, 293 (1987).
[10] B. J. Richardson, P. R. Baverstock, and M. Adams, "Allozyme Electrophoresis: A Handbook for Animal Systematics and Population Studies." Academic Press, New York, 1986.
[11] S. R. Kephart, *Am. J. Bot.* **77**, 693 (1990).
[12] R. W. Murphy, J. W. Sites, Jr., D. G. Buth, and C. H. Haufler, *in* "Molecular Systematics" (D. M. Hillis and C. Moritz, eds.), p. 45. Sinauer, Sunderland, Massachusetts, 1990.

METHODS IN ENZYMOLOGY, VOL. 224

an inert gel matrix. An electric potential is then applied that results in the migration of the proteins present in the homogenate through the gel. The direction and rate of protein separation are dependent on the net charge and size of the molecule which, in turn, is determined by the amino acid sequence of the protein and the pH of the electrophoresis buffer system. Following electrophoresis, the gel matrix is sliced into thin layers and assayed for enzyme activity using histochemical staining methods.[10,15] Areas of enzyme activity are visualized as discrete colored bands, sometimes referred to as zymograms.[16]

Equipment

Equipment essential for electrophoresis includes power sources, gel molds, sample grinding trays, electrode buffer trays, reusable ice packs (or a 4° cold room), a gel slicer, staining trays, assorted glassware, variable temperature incubator(s), hot/stir plates, and a pH meter. Details on equipping a laboratory for electrophoresis are given elsewhere.[12,17]

One of the advantages of protein electrophoresis is that much of the equipment can be fashioned using materials which are readily available from hardware and electronics stores at considerable savings over many of the products offered by laboratory supply companies. For example, gel molds, grinding trays, and buffer trays can be made out of polyvinyl chloride (PVC) board or acrylic materials available from local distributors. Most laboratories use a combination of custom-made and commercial equipment.

Choice of Plant Tissue

One of the first considerations for an electrophoretic analysis of plant proteins is the type of tissue(s) to be used. Virtually any living tissue is suitable (anthers, pollen, flower buds, shoots, roots, megagametophytes, embryos). Specialized protocols for working with pollen leachate (and with organelles) are described by Morden et al.[18] We work primarily with leaves

---

[13] M. T. Conkle, P. D. Hodgkiss, L. B. Nunnaly, and S. C. Hunter, "Starch Gel Electrophoresis of Conifer Seeds: A Laboratory Manual." USDA General Technical Report, PSW-64, 1982.

[14] D. E. Soltis, C. H. Haufler, D. C. Darrow, and G. J. Gastony, Am. Fern J. 73, 9 (1983).

[15] C. E. Vallejos, in "Isozymes in Plant Genetics and Breeding, Part A" (S. D. Tanksley and T. J. Orton, eds.), p. 469. Elsevier Science, Amsterdam, 1983.

[16] R. Hunter and C. Markert, Science 125, 1294 (1957).

[17] C. R. Werth, Virginia J. Sci. 36, 53 (1985).

[18] C. W. Morden, J. Doebley, and K. F. Schertz, "A Manual of Techniques for Starch Gel Electrophoresis of Sorghum Isozymes." Texas Agricultural Experiment State Publication MP-1635, The Texas A&M University System, College Station, Texas 77843, 1987.

and seeds because they are easy to collect and to store. Leaves are best used fresh, as an appreciable loss of enzyme activity can occur after one or more freeze/thaw cycles. We have stored seeds from wild and cultivated soybeans for several years at room temperature in coin envelopes (kept dry and free from insect and fungal damage) and in the freezer ($-20°$) without any noticeable change in enzyme activity.

An important factor to consider when planning an electrophoretic survey is that the patterns and levels of expression of some enzymes vary according to developmental stage, tissue type, and/or external environmental conditions.[19] It is advisable to assay enzyme activity over the life of a plant to determine whether the patterns or levels of expression change for a given enzyme in a specific tissue type. Whenever possible, use tissues that have been germinated and/or grown in a uniform environment, such as a growth chamber or greenhouse, to control for environmentally induced variation in enzyme activity.

Buffer Systems

The choice of gel and electrode buffers is critical to obtaining consistent, well-resolved zymograms. In many cases, the appropriate electrophoresis buffer system(s) will need to be determined empirically for each enzyme. Several commonly used electrophoresis buffer systems are given in Table I.

In our experience with cultivated and wild soybean tissues, a discontinuous buffer system using 5 mM L-histidine (pH 7.0) as the gel buffer and 0.13 M Tris–40 mM citrate (pH 7.0) as the electrode buffer works well for the 23 enzymes we routinely assay. We employ this single buffer system whenever possible so that we can assay for as many enzymes as possible on a single gel.

Choice of Media

Several media are available for protein electrophoresis. The most commonly employed are hydrolyzed potato starch, cellulose acetate, and acrylamide. The choice of medium depends largely on the budget for the project and the number of enzymes to be assayed from a single sample. In all cases, it is best to use electrophoretic grade reagents. Cellulose acetate has the advantage of providing high resolution of bands, but it is relatively expensive and cannot be sliced for staining multiple enzymes from a single sample. Acrylamide is a highly versatile material but poses serious health risks. In its monomeric state, acrylamide is a neurotoxin and is readily

[19] J. G. Scandalios, *Biochem. Genet.* **3**, 37 (1969).

TABLE I
ELECTROPHORESIS BUFFER SYSTEMS FOR STUDIES OF PLANT PROTEINS[a]

Buffer system	Electrode buffer		Gel buffer		Refs.
Histidine–Tris–citrate	0.13 M Tris 40 mM Citric acid monohydrate Adjust to pH 7.0 with 1 M HCl	15.73 g 7.68 g	5 mM L-Histidine monohydrochloride Adjust to pH 7.0 with 4 M NaOH	1.05 g	b
Morpholine–Citrate	40 mM Citric acid monohydrate N-3-(3-Aminopropyl) morpholine Add N-3-(3-aminopropyl) morpholine until pH is 6.1	8.4 g ~ 10 ml	1 : 19 dilution of electrode buffer		c
Phosphate	0.50 M Potassium phosphate monobasic 0.50 M Potassium phosphate dibasic pH 7.0	87.0 g 68.0 g	1 : 10 dilution of electrode buffer		d
Lithium–borate–Tris–citrate	30 mM Lithium hydroxide 0.20 M Boric acid pH 8.3	1.2 g 11.89 g	50 mM Tris 8 mM Citric acid monohydrate Adjust to pH 8.3 using 1 M HCl; use 1 part Tris–citrate buffer and 9 parts electrode buffer as gel buffer	6.2 g 1.6 g	e
Tris–citrate	0.13 M Tris 40 mM Citric acid monohydrate Adjust to pH 7.0	16.35 g 9.04 g	1 : 15 dilution of electrode buffer		f

[a] Amounts given are for 1 liter. Store buffers at 4°.
[b] Y. T. Kiang and M. B. Gorman, in "Isozymes in Plant Genetics and Breeding, Part B"(S. D. Tanksley and T. J. Orton, eds.), p. 295. Elsevier Science Publishers, Amsterdam, 1983.
[c] J. W. Clayton and D. N. Tretiak, J. Fish. Res. Board Can. 29, 1169 (1972).
[d] C. R. Shaw and R. Prasad, Biochem. Genet. 4, 297 (1970).
[e] J. G. Scandalios, Biochem. Genet. 3, 37 (1969).
[f] M. J. Siciliano and C. R. Shaw, in "Chromatographic and Electrophoretic Techniques, Volume II Zone Electrophoresis" (I. Smith, ed.), p. 185. Year Book Medical Publ., Chicago, 1976.

absorbed through the skin. Starch is nontoxic, but often the consistency and quality varies according to manufacturer and lot (for this reason, we recommend that enough starch of the same lot be purchased to complete a study).

We have found that mixing polyacrylamide and starch expands the versatility of both. In our experience, many enzymes that are not well resolved on starch or acrylamide alone are well-resolved on a mixture of the two. The five acrylamide/starch concentrations (weight/volume) we use most often are (1) 12% starch, (2) 7% acrylamide, (3) 9% acrylamide, (4) 7% acrylamide with 2% starch, and (5) 6% acrylamide with 4% starch. For gels containing acrylamide, the total amount of gelling agent is prepared as 95% acrylamide and 5% N,N'-methylenebisacrylamide. Gel polymerization catalysts, ammonium persulfate (APS) and N,N,N',N'-tetramethylethylenediamine (TEMED), are 0.1% (w/v) and 0.2% (v/v) of the total volume of the gel buffer, respectively.

Electrophoresis

Gel Preparation

The total volume of gel solution will depend on the dimensions of the gel molds used. The dimensions we use (length by width by depth) are $21 \times 18 \times 0.3$ cm, $21 \times 18 \times 0.6$ cm, and $21 \times 18 \times 0.9$ cm. By using deeper molds, we can remove multiple layers from a single gel and assay for enzymes that are separated using the same buffer system and have similar electrophoretic mobilities, thereby conserving materials, time, and space.

The length of time that gels can be stored before use varies according to the material they are made from. Starch gels should not be stored (refrigerated) for longer than 1 day as the consistency of the gel matrix changes over time due to dehydration. Acrylamide and acrylamide/starch combination gels can be stored for several days to 1 week if they are wrapped tightly in plastic wrap and refrigerated.

Below are procedures for preparing gels for protein electrophoresis. The volume of gel buffer is for a $21 \times 18 \times 0.3$ cm gel mold unless otherwise noted.

12.5% (w/v) Starch Gel

Gel buffer	240 ml
Hydrolyzed potato starch	30 g

1. Combine the starch and buffer in a 1-liter Erlenmeyer side-arm flask containing a large (3 inch) magnetic stirring bar.
2. Stopper the flask and place it in a pan of boiling water on a magnetic stirrer/hot plate.
3. Heat the starch solution with constant stirring to 80°.

4. Degas the solution via vacuum aspiration for 30 sec and pour the hot starch into a 21 × 18 × 0.6 cm gel mold.
5. Lower a plate of glass (21 × 18 × 0.6 cm) onto the gel to produce an even surface. Avoid air pockets between the glass and the gel.
6. Allow the gel to come to room temperature and then carefully slide the glass plate off of the mold. Wrap the gel tightly with plastic wrap.
7. Refrigerate the gel for 1 hr before use.

Acrylamide Gels

	7%	9%
Gel buffer	150 ml	150 ml
Acrylamide	9.9 g	12.8 g
N,N'-Methylenebisacrylamide	0.53 g	0.67 g
APS	0.15 g	0.15 g
TEMED	300 μl	300 μl

1. Add all the chemicals, except TEMED, to the gel buffer in a 1-liter glass beaker.
2. Heat the solution to 30° with constant stirring.
3. Add TEMED to the mixture and quickly pour the solution into the gel mold.
4. After the gel sets, wrap it in plastic wrap and refrigerate for 1 hr before use.

Acrylamide–Starch Combination Gels

	7%–2%	6%–4%
Gel buffer	150 ml	150 ml
Acrylamide	9.9 g	8.5 g
N,N'-Methylenebisacrylamide	0.53 g	0.45 g
APS	0.15 g	0.15 g
TEMED	300 μl	300 μl
Hydrolyzed potato starch	3.0 g	6.0 g

1. Add acrylamide and APS to one-half of the total volume of gel buffer and mix well with constant stirring at room temperature.
2. Combine the starch and remaining volume of gel buffer in a 500-ml Erlenmeyer side-arm flask; stopper the flask and heat (with constant stirring) to 80°.
3. Degas the hot starch solution via vacuum aspiration or 30 sec, then pour it into the room temperature acrylamide solution and mix.

4. Add the TEMED, mix thoroughly, and then quickly pour the solution into the gel mold.
5. After the gel is set, wrap it in plastic wrap and refrigerate for 1 hr before use.

Sample Preparation

For the tissue grinding trays, use PVC board (2 cm thick) into which several rows of shallow (0.5 cm) wells have been drilled. We chill the grinding trays in the freezer prior to using them and keep them on blocks of reusable blue ice during sample preparation. We homogenize tissues by hand using glass rods with rounded ends. When seeds or other "tough" tissues are used, we add 2–4 drops of homogenization buffer to each seed chip and set the tray set aside, covered, at 4° for 2–4 hr. Seed coats are removed from seeds prior to grinding.

If seeds are used and they are large enough, a small chip can be cut opposite the embryo and used for an electrophoretic analysis without affecting germination. This method is particularly useful for screening seed lots for electrophoretic variants to be used in subsequent genetic analyses or selective breeding programs.

The cellular debris which results from homogenization of plant tissues can adhere to the sample wick and result in streaking of enzyme activity bands. We place a 1 × 1 cm square of lens paper between the homogenate and the sample wick to prevent adhesion of cellular debris. Sample wicks are generally made from Whatmann (Clifton, NJ) 3MM chromotography paper; other types of filter paper are equally suitable. The width and length of the wicks are adjusted according to the gel thickness and sample number per gel. Generally, we assay 24 to 32 individuals on a single gel using wicks that are 4 mm wide. We always include several standard genotypes across the gel as references for comparison of band mobilities.

Two alternative approaches to sample preparation are often used for plant tissues and deserve mention here. In the first method, tissues can be ground into a fine powder using liquid nitrogen and then suspended in extraction buffer just prior to use.[11] Another commonly used approach is to grind a fresh sample in a microcentrifuge tube using a motorized pestle with a Teflon tip.[11,17] The homogenate is then centrifuged (10,000 rpm, 4°, 20 min) to pellet the cellular debris.[19]

We use 5 mM L-histidine monohydrochloride (pH 7.0) as a homogenization buffer. For plant tissues containing secondary metabolites that can inhibit enzyme activity (e.g., resins, phenolics, tannins), complex extraction buffers are usually required.[11,14,17] Examples of extraction buffers for plant tissues are given in Table II. To develop a specialized extraction buffer, start with a simple buffer formulation and determine which additives improve the resolution of the enzyme activity bands. Volatile extraction buffer additives (e.g., 2-mercaptoethanol) should be added just before the samples are homogenized. Alternatively, the extraction buffer can be prepared complete, then aliquoted and frozen until use.

Loading Gels

To load the samples for electrophoresis, insert a scalpel or knife into the gel approximately 3 cm from the edge of the gel mold. Brace the scalpel against a straightedge and draw it across and through the gel to form a slit for samples. Take extra care when cutting through a high percentage acrylamide gel as these gels tend to crumble. Alternatively, individual sample wells can be formed using a straightedge to which thin metal

TABLE II
HOMOGENIZATION BUFFERS FOR PLANT TISSUES[a]

Components	Amount	Comments
5 mM L-Histidine monohydrochloride Adjust to pH 7.0 with 4 N NaOH	0.11 g	Described by Kiang and Gorman[b] for cultivated soybean tissue
0.2 M Tris-HCl, pH 8.0	100 ml	Described by Werth[c]
10 mM MgCl$_2$	1 ml of a 1 M solution	
30 mM Sodium (meta)bisulfite	0.50 g	
1 mM EDTA (tetrasodium salt)	0.05 g	
Adjust to pH 7.5. Add 5 g PVP-40[d] and 100 μl of 2-mercaptoethanol before use		
0.1 M Tris	1.20 g	Developed by Soltis et al.[e] for fern tissue
0.1 M Maleic acid	1.16 g	
0.25 M Sodium ascorbate	4.96 g	
30 mM Sodium diethyldithiocarbamate	0.45 g	
20 mM Sodium (meta)bisulfite	0.38 g	
Adjust to pH 7.5 with concentrated HCl. Add 5 g PVP-40 and 100 μl 2-mercaptoethanol before use		

[a] Amounts given are for 100 ml.
[b] Y. T. Kiang and M. B. Gorman, in "Isozymes in Plant Genetics and Breeding, Part B" (S. Γ Tanksley and T. J. Orton, eds.), p. 295. Elsevier Science, Amsterdam, 1983.
[c] C. R. Werth, Virginia J. Sci. 36, 53 (1985).
[d] PVP-40, Polyvinylpyrrolidone with average molecular weight of 40,000.
[e] D. E. Soltis, C. H. Haufler, D. C. Darrow, and G. J. Gastony, Am. Fern J. 73, 9 (1983).

fragments of identical size are attached and arranged evenly, like teeth on a comb. The wells are formed by inserting the metal teeth into a set gel. The teeth should not be so long as to touch the bottom panel of the gel mold.

Using forceps, remove the sample-saturated wicks from the wells of the grinding tray one at a time, blot the wick lightly on an absorbent tissue to remove excess moisture, and insert it perpendicularly into the opening in the gel. Place the wicks evenly along the sample well using the markings on the ruler as guides. Include one wick that has been soaked in a solution of tracking dye [e.g., 0.1% (w/v) bromphenol blue] at the end of the line of sample wicks. Use the mobility of the tracking dye relative to the mobility of the enzyme activity bands to calculate relative mobility (R_f) values for each band; R_f is the distance migrated by protein/distance migrated by dye.

Leave sufficient space between wicks to avoid cross-contamination of samples caused by diffusion of the homogenate. Allow at least 0.5 cm on either end of the line of wicks to eliminate band distortion arising from "edge effects." The tops of the wicks should be barely visible above the gel surface.

When all the wicks are inserted, gently press the two sections of the gel together along the suture line to remove air gaps and to ensure good contact between the wicks and the gel. Starch gels tend to shrink slightly during electrophoresis, so insert a glass or plastic rod between the gel and gel mold at the cathodal end to prevent the two sections of gel from separating during electrophoresis. Gel shrinkage is not usually a problem for acrylamide or acrylamide–starch combination gels.

Conducting Electrophoresis

If a commercial horizontal slab-gel electrophoresis apparatus is not available, Rubbermaid plastic drawer organizers ($23 \times 8 \times 5$ cm) or similar containers can serve as electrode buffer reservoirs. To modify the trays, fit the female portion of a banana plug into a small hole which has been drilled 1 cm from the top of the tray. Wind one end of a 22 cm length of thin (30-gauge) platinum wire around the inside part of the plug once or twice and then place the remaining length of wire along the bottom of the tray. Use several dollops of silicone gel to secure the wire to the tray and to seal any small gaps around the female banana plug fitting.

Place a gel, loaded with samples, on a block of reusable ice between two reservoirs containing electrode buffer. Connect the reservoirs to the power supply using electrical leads fitted with male banana plugs. Be sure that the entire bottom of the gel mold is in contact with the ice so that cooling is even. Orient the gel so that the wicks are closest to the negative pole (black or cathode). Use cellulose sponges or Handiwipes (available in most grocery stores) to wick the electrode buffer to the gel. (Kimtex wipes also can be used and are available through science supply companies.) For starch gels, a "wickless" method is described by Conkle.[13] Allow one end of a cellulose sponge (moistened with electrode buffer) to overlap the gel 0.5 cm above the line of sample wicks and place the other end of the sponge in the electrode buffer reservoir. Be careful that the sponge does not come in contact with the platinum electrode wire. Repeat on the opposite end of the gel, allowing one end of the sponge to overlap the gel by 2 cm. This completes the circuit across the gel.

Place a piece of plastic wrap over the gel and sponges. Then place a piece of thin (5 mm) PVC board or acetate sheet on top of the gel to ensure good contact between the sponges and the gels. Place a second, smaller block of reusable blue ice on top of the 5 mm PVC board to help maintain a constant temperature of 4° throughout electrophoresis. Replace the top ice as necessary. This method can be used on a bench top without refrigeration. If the gels are run in a refrigerated unit, the bottom ice can be eliminated; simply place the gels on a support such that the surface of the gel mold is even with the top of the electrode trays.

Slab gels can be electrophoresed at constant voltage, power, or amperage. We run gels at 150 or 200 V (constant voltage). Running times range from 4 to 16 hr depending on the gel thickness. Although many researchers remove the sample wicks about 15 min into electrophoresis, we have found that this step is generally not necessary. Monitor the voltage, amperage, and wattage settings at regular intervals during electrophoresis.

Enzyme Activity Staining

Following electrophoresis slice the gel horizontally into several thin layers according to gel thickness. The slicing surface we use is a piece of PVC board with 3 mm raised edges. A coping saw with a 0.009 mm wire (guitar E-string) stretched tightly across one end makes an excellent gel slicer. Carefully remove the gel from the gel mold and place it, top down, on a slicing surface. Cut a notch in the upper corner of the gel to indicate sample orientation. Next, place a plate of glass on the gel to help hold it in place while it is being sliced. Discard the first (top) layer. Place the gel layer(s) to be stained into a suitable staining tray containing the enzyme activity stain solution. Although most enzymes run toward the anode, it is a good idea to check the cathodal end for enzyme activity when trying an assay for the first time.

The Appendix at the end of this chapter lists buffer formulations for stains and 20 enzyme activity staining protocols that have been adapted from recipes found in the literature.[15,20,21] Many additional protocols can be found in the references cited herein.

Documenting Results

Because we may assay several hundred individuals for many enzymes, we do not photograph every gel. For routine analyses, we sketch the banding patterns by hand on graph paper and record the relative mobility of the bands. We use 35 mm photography (a camera mounted on a copy stand) to document gels when we test new enzymes, find new electrophoretic variants, and prepare for publications. For black-and-white high-contrast prints we use Kodak (Rochester, NY) Tri-X Pan film; for color slides we use Kodachrome or Ektachrome (ASA 64) film. If the bands of enzyme activity are intense, gels can be photocopied. Faint bands, however, are difficult to resolve with this method.

Methods for preserving intact gels include drying the gels onto filter paper using a slab gel dryer[17] or wrapping the gels in plastic wrap and freezing them at $-20°$.[10] Gels that are stained using MTT [3-(4,5-dimethylthiazol-2-yl)-2,5-diphenyl-2H-tetrazolium bromide] can be soaked in 50% glycerol, wrapped in plastic, and stored in the refrigerator.[17] Gels stained with NBT (nitro blue tetrazolium) can be fixed in a 5:5:1 mixture of methanol–water–glacial acetic acid, rinsed with water, wrapped in plastic, and stored in the refrigerator.[13,17] Some deterioration of band quality will occur with any of these methods.

Interpretation and Analysis of Electrophoretic Data

A discussion of the interpretation of enzyme activity banding patterns and the analysis of electrophoretic data is beyond the scope of this chapter. Details on the interpretation of zymograms can be found in several of the references cited herein.[10,11,22] Richardson et $al.$[10] provide a thorough discussion of data analysis and present several detailed examples of analyses taken from the literature. Among the major computer software packages currently available for data analysis are BIOSYS-1,[23] PAUP,[24] PHYLIP,[25]

[20] J. L. Brewbaker, M. D. Upadhya, Y. Makinen, and T. Macdonald, *Phys. Plant* **21**, 930 (1968).
[21] C. R. Shaw and R. Prasad, *Biochem. Genet.* **4**, 297 (1970).
[22] D. W. Moss. "Isoenzymes." Chapman & Hall, London, 1982.
[23] D. L. Swofford and R. B. Selander, *J. Hered.* **72**, 281 (1981).
[24] D. L. Swofford, Illinois Natural History Survey, 607 E. Peabody Dr., Champaign, Illinois 61820.
[25] J. Felsenstein, Dept. Genetics SK-50, Univ. of Washington, Seattle, Washington 98195.

and HENNIG86.[26] Information on these programs can be obtained by writing directly to the respective authors.

Appendix: Stain Buffer Formulations and Enzyme Activity Stain Protocols

Stain Buffer Formulations

Amounts given are for 1 liter. Buffers should be prepared with distilled water.

0.2 M Tris-HCl (pH 8.5)
   0.2 M Tris   24.22 g
   Adjust to pH 8.5 with concentrated HCl

0.1 M Acetate buffer (pH 5.0)
   A. 0.2 M Acetic acid (11.5 ml in 1 liter of distilled water)
   B. 0.2 M Sodium acetate (16.4 g $C_2H_3O_2$ Na in 1 liter of distilled water)
   Add 148 ml A + 352 ml B and bring up to 1 liter

0.1 M Phosphate buffer (pH 6.4)
   90 mM Sodium monophosphate   13.9 g
   20 mM Sodium diphosphate   5.3 g
   Adjust to pH 6.4 with 1 M HCl.

0.1 M Tris–maleate buffer (pH 5.2)
   0.10 M Tris   12.1 g
   0.12 M Maleic anhydride   11.6 g
   40 mM Sodium hydroxide   1.6 g
   Adjust to pH 5.2 with 1 M HCl

0.2 M Tris–maleate buffer (pH 3.7)
   0.2 M Tris   24.2 g
   0.2 M Maleic acid   23.2 g
   Adjust to pH 3.7 with 1 M HCl; to use as a stain buffer, dilute the stock as follows: 5 (stock) : 3 (distilled water) : 2 (0.2 M NaOH) with final pH of 5.55

10 mM Malate buffer (pH 7.0)
   10 mM L-Malic acid   1.3 g
   20 mM Tris   3.0 g
   Adjust to pH 7 with 4 M NaOH

[26] J. S. Farris, Molekylärsystematiska laboratoriet, Naturhistoriska riksmuseet, Box 50007, S 104 05 Stockholm, Sweden.

Potassium iodide solution
    Iodine   1 g
    Potassium iodide   5 g
It may take several hours for the iodine to dissolve; store solution in an amber bottle at room temperature

Enzyme Activity Stain Protocols

The enzyme activity stains presented here are organized according to enzyme function (oxidoreductase, transferase, etc.). Many of the chemicals used in the staining process are toxic and must be handled and disposed of with caution.

Gently rock the staining trays on occasion to ensure that the stain buffer is evenly distributed over the gel. Gels should be incubated in the dark at 37° unless otherwise noted. Monitor the gels closely and stop the reaction (by pouring off the stain solution and adding water or fixative) when the resolution of the bands is optimal.

The following abbreviations are used in the stain recipes presented below:

NAD,   nicotinamide adenine dinucleotide;
NADH,   nicotinamide adenine dinucleotide (reduced);
NADP,   nicotinamide adenine dinucleotide phosphate;
MTT,   3-(4,5-dimethylthiazol-2-yl)-2,5-diphenyltetrazolium bromide;
PMS,   phenazine methosulfate; and
Tris,   tris(hydroxymethyl)aminomethane.

A. Oxidoreductases

  1. Alcohol dehydrogenase (ADH),   EC 1.1.1.1
      0.2 *M* Tris-HCl   50 ml
      MTT   10 mg
      NAD   10 mg
      95% (v/v) Ethanol   2 ml
      PMS   2 mg
    Stain the gel in the dark at room temperature; keep the staining tray tightly covered

  2. Diaphorase (DIA),   EC 1.6.99.2
      0.2 *M* Tris-HCl   50 ml
      MTT   10 mg
      NADH   10 mg
      2,6-Dichlorophenol indophenol   2 mg

3. Glucose-6-phosphate dehydrogenase (GPD),   EC 1.1.1.49
   0.2 M Tris-HCl   50 ml
   MTT   10 mg
   NADP   10 mg
   Glucose 6-phosphate, disodium salt   100 mg
   PMS   1 mg

4. Isocitrate dehydrogenase (IDH), NADP active,   EC 1.1.1.42
   0.2 M Tris-HCl   50 ml
   MTT   10 mg
   NADP   10 mg
   $MgCl_2$ 120 mg
   DL-Isocitric acid   200 mg
   PMS   1 mg

5. Malate dehydrogenase (MDH),   EC 1.1.1.37
   10 mM Malate buffer   50 ml
   MTT   15 mg
   NAD   30 mg
   PMS   1 mg

6. Peroxidase (PER),   EC 1.11.1.7
   3-Amino-9-ethylcarbazole (dissolved in   25 mg
      1.25 ml dimethylformamide)
   10% $CaCl_2$ (w/v)   1 ml
   H_2O_2 (3% solution)   1 ml
   Sodium acetate   3 ml
   Distilled water (chilled)   45 ml

   Add chemicals slowly to prevent reprecipitation of 3-amino-9-
      ethylcarbazole. Stain the gel in the dark at room temperature.

7. 6-Phosphogluconate dehydrogenase (PGD),   EC 1.1.1.43
   0.2 M Tris-HCl   50 ml
   MTT   10 mg
   NADP   10 mg
   $MgCl_2$ 20 mg
   6-Phosphogluconate   15 mg
   PMS   1 mg

8. Shikimate dehydrogenase (SKD),   EC 1.1.1.25
   0.2 M Tris-HCl   50 ml
   MTT   10 mg
   NADP   10 mg
   Shikimic acid   15 mg
   PMS   1 mg

B. *Transferases*

9. Glutamate oxaloacetic transaminase (GOT);
   aspartate transaminase,   EC 2.6.1.1
     0.2 *M* Tris-HCl   50 ml
     Pyridoxal 5'-phosphate   25 mg
     L-Aspartic acid   272 mg
     Ketoglutaric acid   36 mg
     Fast Blue BB salt   112 mg

C. *Hydrolases*

10. Acid phosphatase (ACP),   EC 3.1.3.2
     0.2 *M* Acetate buffer   50 ml
     Black K salt   40 mg
     Naphthyl acid phosphate   40 mg
     Stain the gel in the dark at room temperature

11. β-Amylase (AM),   EC 3.2.1.2
     A. 0.2 *M* Acetate buffer   100 ml
        Soluble potato starch   1 g
     Heat the potato starch in buffer A until dissolved. Cool the
     solution to 30° and pour over the gel. Incubate the gel at 37°
     in the dark for 15 to 30 min. Pour off A and rinse the gel with
     distilled water

     B. Potassium iodide solution (see above)
     Pour 20 ml of B over the gel; score bands immediately

12. Endopeptidase (ENP),   EC 3.4.?.?
     0.2 *M* Tris–maleate, pH 5.5   50 ml
     Black K Salt   20 mg
     $MgCl_2$ 10 mg
     n-α-Benzoyl-DL-arginine   20 mg
        β-Naphthylamide hydrochloride (BANA)
     Stir stain solution vigorously for 10 min (protect the solution
     from exposure to light); stain the gel in the dark at room
     temperature

13. Carboxylesterase (EST),   EC 3.1.1.1
     0.1 *M* Phosphate buffer   50 ml
     Fast Blue RR salt   100 mg
     α-Naphthyl butyrate   1 ml
     100% Acetone   3 drops
     1% α-Naphthyl acetate   1 ml

Add α-naphthyl butyrate and three drops of acetone to the Fast Blue RR salt. Then add 50 ml of phosphate buffer and stir the solution vigorously over low heat. Add 1 ml of 1% α-naphthyl acetate to the warm solution. Pour the solution immediately over the gel through a single layer of cheesecloth

14. Fluorescent arylesterase (FLE),   EC 3.1.1.2
    0.2 M Acetate buffer   50 ml
    100% Acetone   10 ml
    4-Methylumbelleferyl acetate   15 mg

Dissolve 15 mg of 4-methylumbelliferyl acetate in 100% acetone. To this solution add 30 ml of acetate buffer. Saturate several Kimwipes with the stain solution and place in direct contact with the sliced gel surface. Examine the gel with a UV light source (366 nm) soon after staining as the bands fade quickly

15. Leucyl aminopeptidase (LAP),   EC 3.4.11.1
    A. 0.1 M Tris – maleate buffer   100 ml
       (25 ml stock + 75 ml distilled water)
       L-leucine-β-naphthylamide   20 mg
       (dissolved in 3 drops 100% acetone)

Incubate the gel in A at 37° for 1 hr, then pour A into a beaker for use in step B

    B. Black K salt   50 mg

Add B to A and pour back over gel; place the gel at room temperature in the dark

16. Urease (EU),   EC 3.5.1.5
    A. 0.2 M Acetate buffer   100 ml
       0.1% (w/v) Cresol red   0.1 gm

Pour A over the gel and place gel at room temperature in the dark for 10 – 15 min; pour off A

    B. 333 mM Urea   2 g
       0.1% (w/v) Cresol red   0.1 g
       0.1% (w/v) Na$_2$EDTA   0.1 g
       Distilled water   100 ml

Pour B over the gel. Place the gel at room temperature in the dark. Bands appear within 15 min. Record results immediately as the bands blur within a few hours

D. Lyase

17. Aconitase (ACO); aconitate hydratase,   EC 4.2.1.3
        0.2 *M* Tris-HCl   50 ml
        MTT   10 mg
        NADP   10 mg
        $MgCl_2$ 30 mg
        cis-Aconitic acid   8 ml
          (1% solution (w/v), pH 7.5)
        Isocitrate dehydrogenase   10 units
        PMS   1 mg

E. Isomerases

18. Mannose-6-phosphate isomerase (MPI),   EC 5.3.1.8
        0.2 *M* Tris-HCl   50 ml
        MTT   10 mg
        NAD   10 mg
        Mannose 6-phosphate   20 mg
        Glucose-6-phosphate dehydrogenase   10 units
          (NAD active)
        PMS   1 mg

19. Phosphoglucose isomerase (PGI); glucose-6-
    phosphate isomerase,   EC 5.3.1.9
        0.2 *M* Tris-HCl   50 ml
        MTT   10 mg
        NAD   10 mg
        $MgCl_2$ 20 mg
        Fructose 6-phosphate   20 mg
        Glucose-6-phosphate dehydrogenase   10 units
        PMS   1 mg

20. Phosphoglucomutase (PGM),   EC 5.4.2.1
        0.02 *M* Tris-HCl   50 ml
        MTT   10 mg
        NAD   10 mg
        $MgCl_2$ 20 mg
        D-Glucose 1-phosphate disodium salt   125 mg
        Glucose-6-phosphate dehydrogenase   10 units
          (NAD active)
        PMS   1 mg

[7] Starch Gel Electrophoresis in Vertebrates

By JANICE BRITTON-DAVIDIAN

Introduction

Starch gel electrophoresis has been used successfully in population genetics, systematics, phylogeny, and population structure studies. The purpose of this chapter is to introduce zoologists to this technique, providing guidelines for sample and gel preparation as well as a list of protein stains and electrophoresis buffers commonly used for a variety of vertebrate species. Many of the latter were originally adapted from the paper by Selander *et al.*[1] and the handbook of Harris and Hopkinson[2] with new recipes and modifications added throughout the years. It is recommended that a researcher addressing a specific taxon of animals consult the published literature for particular technical adaptations which should be available through the widespread use of this method. In addition, several excellent technical manuals are available describing electrophoretic techniques for different gel media (cellulose acetate,[3] starch, and acrylamide gels[4]) and providing background information on gel interpretation, basic genetic statistical methods, and guidelines for data sampling.

Equipment

The standard equipment needed for protein electrophoresis consists of a refrigerated centrifuge, a deep freezer ($-70°$) or a liquid nitrogen container, a crushed ice machine, a pH meter, an incubator, an electric barrel homogenizer, a power pack, UV light source, and storage boxes for 1.5-ml Eppendorf microtubes. Storing samples at temperatures lower than $-60°$ and keeping thawed samples on crushed ice allows them to be thawed and frozen many times with a minimum loss of activity or degradation of the proteins. Most of the specific electrophoretic equipment (gel molds, elec-

[1] R. K. Selander, M. H. Smith, S. Y. Yang, W. E. Johnson, and J. B. Gentry. *Univ. Texas Publ.* No. 7103, p. 49 (1971).
[2] H. Harris and D. A. Hopkinson, "Handbook of Enzyme Electrophoresis in Human Genetics." North-Holland Publ., Oxford, 1976.
[3] B. J. Richardson, P. R. Baverstock, and M. Adams, "Allozyme Electrophoresis: A Handbook for Animal Systematics and Population Studies." Academic Press, New York, 1986.
[4] N. Pasteur, G. Pasteur, F. Bonhomme, J. Catalan, and J. Britton-Davidian, "Practical Isozyme Genetics." Ellis, Horwood; Wiley, New York, 1988. Also available in French: "Manuel Technique de Génétique par Électrophorèse des Protéines." Technique et Documentation, Lavoisier, Paris, 1987.

trode trays, cutting plates, gel slicer) can be made in the laboratory from plexiglass sheets and commercially available material. Stain boxes with lids are required. As most reagents used in electrophoresis are highly toxic, all necessary precautions must be taken, and rubber gloves should be worn when handling acrylamide and stains.

Preparation of Samples

The smallest vertebrates are homogenized whole and most larger animals can be dissected and different tissues removed for storage in 1.5-ml Eppendorf microtubes at $-70°$ until homogenized. Heart, kidneys, or liver are sufficient to score more than 20 proteins; testes, spleen, brain, and muscle can be sampled for tissue-specific enzymes. Additionally, blood, muscle, and/or saliva from most vertebrates can be conveniently sampled without killing the animal. During all sample preparation steps, keep tubes on crushed ice.

Blood Samples

Blood samples are prepared by several methods: (1) cutting a suitable vein with a sharp blade and letting the blood drop into a centrifuge tube containing one drop of a heparin solution (1% w/v in physiological saline); (2) inserting a heparinized syringe in any large vein; or (3) puncturing the infraorbital sinus using a long heparinized pipette. If sufficient blood is available (>0.5 ml), add a volume of physiological saline (0.85% w/v NaCl in deionized water) slightly more than half that of the blood sample, mix gently, and centrifuge at 1000 g for 10 min. Pipette the plasma and store in a 1.5-ml Eppendorf microtube at $-70°$. The remaining red blood cells are then washed twice by filling the centrifuge tube with saline, gently mixing the tube twice, and centrifuging at 1000 g for 5 min each. Discard the supernatant and add a volume of deionized water equal to that of the red blood cell pellet and centrifuge at 20,000 g for 20 min. Pipette the supernatant into a 1.5-ml Eppendorf microtube and store at $-70°$. When dealing with very small quantities of blood (<0.5 ml), no saline should be added to the blood sample prior to the first centrifugation; after rinsing the red blood cell pellet in saline, cap the tube and place it in a deep freeze for 1 hr or more to disrupt the cells. When the sample is thawed, add a drop a deionized water, mix on a vortex, and centrifuge at 20,000 g for 20 min.

Saliva Samples

A saliva sample can be obtained from small vertebrates by rinsing the mouth of the animal several times with 20 μl of saline solution using a micropipette, collecting all the liquid, transferring it to a 1.5-ml Eppendorf microtube, and storing at −70°.

Homogenizing Tissues

Place the tissue in a centrifuge tube with an equal volume of homogenizing solution (see below). Homogenize the sample with an electric barrel homogenizer for several seconds and centrifuge at 20,000 g for 20 min. The supernatant is pipetted into 1.5-ml Eppendorf microtube and stored at −70°. However, if the supernatant appears cloudy (indicating the presence of fats, which often occurs with liver), then transfer it to a clean centrifuge tube to which 0.5 ml of toluene is added, thoroughly mix it with a vortex, then centrifuge as above. The supernatant is pipetted by inserting the pipette through the solvent phase (the best way is to blow bubbles while crossing this phase to avoid picking up toluene residues) and then transferred to a 1.5-ml Eppendorf microtube. Store at −70°.

Homogenizing Solution. Homogenizing solution contains Tris (1.2 g) and EDTA (0.37 g); bring to 1 liter with deionized water and adjust the pH to 6.8 with concentrated HCl. Last, add 4 ml of a 1% (w/v) nicotinamide adenine dinucleotide phosphate (NADP) solution.

Preparation of Gels

Starch gels are used for screening all proteins except amylases, which are run in horizontal acrylamide gels.

12% Starch Gels

Forty-eight grams of hydrolyzed starch is mixed with 400 ml of gel buffer in a 1-liter Erlenmeyer flask and heated over a Bunsen burner, using asbestos gloves to stir it constantly. When the solution thickens and becomes translucid (about 3–5 min), continue cooking the gel for 5–70 sec while vigorously shaking the flask. Degas the gel using vacuum aspiration for 20 sec. Quickly pour the gel into a 19 × 21 × 1 cm (internal dimensions) plexiglass mold. Once the gel has cooled, cover the gel mold with plastic wrap. Gels are routinely prepared the day before use and are kept at room temperature, but they are ready to use once cooled. The crucial step in preparing starch gels is the time the gel is cooked after it changes in consistency; this time may vary according to the starch brand and batch and needs to be adjusted experimentally. Undercooking will lead to fragile gel slabs, and overcooking results in brittle gels.

6% Acrylamide Gels

Six grams of acrylamide and 0.16 g of bisacrylamide are added to 100 ml of gel buffer in a 200-ml Erlenmeyer flask. Mix thoroughly over a magnetic stirrer plate, then rapidly add 70 mg of ammonium persulfate and 100 μl of TEMED (N,N,N',N'-tetramethylethylenediamine). Degas the gel for 20 sec and pour it into a 19 × 21 × 0.2. cm (internal dimensions) plexiglass mold. Carefully lower a plexiglass lid from one end of the gel mold to the other to remove all air bubbles. Place the mold under a fluorescent light or daylight. The gel polymerizes in about 30 min. Remove the lid and cover with plastic wrap.

Loading Samples

The samples are ordered in a rack, thawed, and placed on crushed ice. Uncover the larger end of the mold, against which a 4 cm wide ruler is applied. Form the sample well by slitting the gel across its width with a scalpel held vertically against the ruler. Spread a drop of a bromphenol blue solution (0.5% in deionized water) throughout the slit to use as a migration indicator. Using forceps, dip a 0.9 × 0.4 cm rectangle of What-man (Clifton, NJ) 3MM filter paper in the microtube containing the sample and place it in loading order on a large 10 × 8 cm rectangle of Whatman 1MM filter paper to remove excess liquid. Be careful to wipe the forceps after handling each sample. When all the samples are ready, insert them (starting at the left-hand side and 5 mm from the edge of the mold) into the slit by gently separating the two lips of the gel. Leave 2–3 mm between the paper wicks to avoid cross-contamination. The dimension of this mold holds 24 paper wicks. When all the samples have been loaded into the gel, use the scalpel to detach the gel on all four sides of the mold and gently push the gel to ensure that the two lips of the slit do not separate during electrophoresis. If needed, a rubber polystyrene tube can be inserted at the anodal end.

Cover the gel with plastic wrap, leaving 3 cm uncovered at each end. Place the mold over the electrode trays so that the sample origin is closest to the cathode. The contact between the gel and the electrode buffer is ensured by dipping one end of a flat sponge in the electrode tray while the other overlaps the uncovered end of the gel. Overheating during electrophoresis is avoided by covering the gel and sponges with a glass plate over which a metal tray containing a reusable frozen ice pack and/or crushed ice is set. Connect the electrodes to a power pack. The amount of voltage applied is calculated so that the product of the voltage (V in volts) by the intensity (I in milliamps) is no higher than 10,000 (= 10 W) per gel. If the power pack allows it, two gels made with the same buffer can be run in parallel; in this case, the voltage will be constant but the amperage and the product $V \times I$ will be doubled. Check the voltage and replace the ice pack and/or the crushed ice as needed during the run. The electrophoresis is stopped when the bromphenol blue is within 2–3 cm of the other side of the gel, which takes 4–7 hr. An improvement has been achieved by filling an airtight plastic bag with crushed ice, sealing it shut, and placing it directly over the sponge wicks and the gel; this method produces a better cooling system which allows one to increase the voltage, resulting in a shorter run and sometimes better resolution of the bands.

Staining Procedures

Disconnect the power pack, remove the sponge wicks (which can be stored in the electrode trays until further use) and gently separate the two lips of the slit. Remove the paper samples, cut away all unnecessary edges of the gel, make a nick in the top and bottom right-hand side of the gel (to indicate the side the last sample is at), lift the gel, and place it on a plexiglass cutting plate with 2 mm high ridges. Be sure to include the cathodal end of the gel as some isozymes are negatively charged. Remove all air bubbles between the gel and the glass plate. Use a string cutter (made with a guitar string) to slice the gel horizontally, keeping one hand on the gel to prevent it from slipping off the plate. Remove the top of the gel and place it on a second cutting plate. Gently lift the anodal end of the gel slab and transfer it to a staining box. Four slabs can be cut from the gel and stained for four different isozymes.

Four methods are used to stain the gel slice: (1) for the box stain, the staining solution is poured directly on the slab, and the box is gently rocked every 5 min; (2) for the agar overlay method, the staining solution is mixed just prior to staining with warm agar and poured onto the gel; (3) for paper overlay staining, a rectangle of Whatman 3MM filter paper precut to the size of the gel slab is soaked in the staining solution and applied to the gel slab, with care to avoid air bubbles, and the paper is removed when the gel is sufficiently stained; and (4) for the paintbrush stain, the staining solution is painted on the gel with a soft paintbrush in the approximate area of migration known for the enzyme. The overlay methods are sometimes required to reduce diffusion of the bands; however, the last three methods can be used efficiently to reduce the volume and hence the cost of the stain. A list of protocols for staining enzyme and nonenzymatic proteins is provided below. Bands start appearing from 10 min to 1 hr.

Once the stain has reached the desired intensity, stop the reaction by squirting 50 ml of a fixative solution containing 10% (v/v) (for agar overlays) or 5% (v/v) acetic acid (for the other types of stains). Pour off the stain solution for box stains before soaking in fixative.

Scoring

The proteins can be scored before the gel slab is fixed, while it soaks in the fixative, or under UV light. Staining of the gel slabs reveals one or several colored bands or spots. The quaternary structure of the protein must be known to determine the number of bands to expect in heterozygotes. Most loci used in electrophoresis have codominant alleles, which ensures that the heterozygotes can be accurately scored since they will express both allelic products. Care must be taken to distinguish allelic products from isozymes that correspond to different loci coding for proteins with the same enzyme function. This is not too difficult in vertebrates since isozymes are often localized in different tissues or, if expressed in the same tissue, they usually migrate in different areas of the gel. In some cases, the subunits of the isozymes may form one or several heteropolymer molecules; this is common for lactate dehydrogenase (LDH) in mammals and glucose phosphate isomerase (GPI) in some fish species. In homozygous individuals, the homopolymer products of the loci usually correspond to the extreme migrating bands. Care must be taken to include a control sample in all gels.

At each locus, electromorphs are designated numerically according to their mobility relative to an arbitrarily chosen reference electromorph designated as 100, which usually corresponds to the most common allele observed in the species under study. An increase in electromorph detection can be achieved by sequential electrophoresis where each locus is successively screened in different gel buffers.[5,6] However, one-buffer electrophoresis is sufficient to detect 97% of the genetic variability at the intrageneric level (data for rodents[5]). For data processing, the BIOSYS program[7] is highly recommended and can be ordered on request from D. L. Swofford (Center for Biodiversity, Illinois Natural History Survey, 607 E. Peabody Dr., Champaign, Illinois 61820).

Preserving Gel Slabs

For fluorescent stains, the position of the bands can be marked either on the gel by making a hole with a pencil or on a drawing of the gel. For the other types of stains, two preservation methods are currently used.

[5] D. Iskandar and F. Bonhomme, *Can. J. Genet. Cytol.* **26,** 622 (1984).
[6] C. F. Aquadro and J. C. Avise, *Genetics* **102,** 269 (1982).
[7] D. L. Swofford and R. B. Selander, *J. Hered.* **72,** 281 (1981).

Method for Agar Overlays

Let the gel slab soak in the 10% acetic acid fixative for at least 15 min. Pour off the fixative. Turn the stain box over a glass plate and let the gel slab slip off onto it. Remove the gel slab from the agar overlay. Cover the agar with a rectangle of Whatman 1MM filter paper that is slightly larger than the agar overlay. Pick up the paper with the agar stuck onto it, turn it over, place it on a flat surface, and tape the edges to avoid crinkling while the gel dries. Let dry at room temperature for 1 day or more.

Method for Box and Paintbrush Stains

Let the gel slab soak in the 5% acetic acid fixative for 1–2 hr. Pour off the fixative and soak the gel in 50 ml of 10% glycerol for at least 24 hr. The gel becomes translucid. Soak a sheet of pure cellophane paper in the 10% glycerol and spread it on a glass plate; place the gel slab on this sheet and cover it with a second sheet of glycerol-soaked cellophane paper. Make sure to remove all air bubbles between the gel slice and the two sheets of cellophane paper. Fold the cellophane tightly under the glass plate. Let dry for at least 48 hr. Overdrying will cause the gel to crumble. Remove the assemblage from the glass plate and trim the cellophane 2 cm from the gel slice.

The paper-mounted and cellophane-wrapped preserved gels are stored inside a double plastic sheet with the protocol data. These methods have been in use for several years in our laboratory with excellent results, and no deterioration of band resolution or gel quality has been detected.

Running Gels in the Field

An excellent alternative to the starch gel medium is cellulose acetate if gels are to be run in the field for species identification, population structure studies, or selection of individuals for particular marker alleles. As buffer systems are often specifically adapted to the gel medium used, we recommend consulting the manual by Richardson *et al.*[3] for technical information on cellulose acetate electrophoresis.

Protocols

Electrophoresis Buffers

In continuous buffer systems (i.e., when the same chemicals are used in the gel and electrode buffers), the electrode buffers can be reused up to 3 times by mixing the contents of the two trays and checking the pH.

TC 6.7: *Gel buffer:* Tris (8 mM), citric acid (3 mM); adjust to pH 6.7 with 1 M Tris or 1 M citric acid. *Electrode buffer:* Tris (0.223 M), citric acid (86 mM); adjust to pH 6.3 with 1 M Tris or 1 M citric acid

TC 6.4: *Gel buffer:* Same as above, but adjust pH to 6.4. *Electrode buffer:* Same as above but adjust pH to 6.0

TC 8.0: *Gel buffer:* 1 : 29 dilution of electrode buffer; adjust to pH 8.0 with 1 M Tris or 1 M citric acid. *Electrode buffer:* Tris (0.62 M), citric acid (0.14 M); adjust to pH 8.0 with 1 M Tris or 1 M citric acid

TG 8.5: *Gel buffer:* Tris (20 mM), glycine (60 mM); adjust to pH 8.5 with 1 M Tris or 1 N HCl. *Electrode buffer:* Tris (50 mM), glycine (0.39 M); adjust to pH 8.1 with 1 M Tris or 1 N HCl

TCB 8.7: *Gel buffer:* Tris (70 mM), citric acid (5 mM); adjust to pH 8.7 with 1 M Tris or 1 M citric acid. *Electrode buffer:* NaOH (60 mM), boric acid (0.3 M); adjust to pH 8.2 with 1 M NaOH or 1 M boric acid

TEB 8.6: *Stock solution:* Tris (0.9 M), EDTA (20 mM), boric acid (0.5 M). *Gel buffer:* 1:39 dilution of stock solution. *Electrode buffer:* 1:14 dilution of stock solution; in both cases, adjust to pH 8.6 with 1 M Tris or 1 M boric acid

THCl 8.5: *Gel buffer:* Tris (10 mM); adjust pH to 8.5 with 1 M Tris or concentrated HCl. *Electrode buffer:* same as TCB 8.7

TM 6.9: *Gel buffer:* 1:9 dilution of electrode buffer; adjust to pH 6.9 with 1 M Tris or 1 M DL-malic acid. *Electrode buffer:* Tris (0.1 M), DL-malic acid (0.1 M), EDTA (10 mM), MgCl$_2$ (10 mM); adjust to pH 6.9 with NaOH. If further adjustments are necessary, use 1 M Tris or 1 M malic acid

TMgP 8.2: *Gel buffer:* 1:19 dilution of electrode buffer; adjust to pH 8.2 with 1 M KOH. *Electrode buffer:* Tris (0.1 M), KH$_2$PO$_4$ (0.1 M), MgSO$_4$ (1 mM); adjust to pH 8.2 with 1 M KOH

LiOH 8.3: *Gel buffer:* Tris (46 mM), citric acid (7 mM), 100 ml of electrode buffer, and complete to 1 liter with water; adjust to pH 8.3 with 1 M Tris or 1 M citric acid. *Electrode buffer:* Lithium hydroxide (50 mM), boric acid (0.19 M); adjust to pH 8.1 with 1 M LiOH or 1 M boric acid

Protein Stains

The left-hand column describes the staining protocol with the name of the protein abbreviated in parentheses, followed by a letter giving the quaternary structure (M, monomer; D, dimer, T, trimer; TT, tetramer).

Nearly all the stains can be prepared in 50-ml flasks while the gel is running and stored at 4°, in some cases adding only a few chemicals just prior to pouring it on the gel slab. In such cases, write the quantity of the reagents to be added on a tape over the top of the flask. The right-hand column lists the number of loci, the tissue in which the protein is scored (P, plasma; Ha, hemolysate; K, kidney; L, liver; H, heart; S, saliva; M, muscle), and the electrophoresis buffers used in our laboratory for rodents/fish. Where specified, cystamine or NADP is added to the gel buffer and mixed prior to pouring into the mold. The agar overlay is prepared with 1.5% (w/v) agarose: boil the solution and keep it at 37° on a magnetic hot plate stirrer. All stock staining solutions (see below) are stored in capped bottles at 4°. β-NAD, β-NADP, MTT [3-(4,5-dimethylthiazol-2-yl)-2,5-diphenyl-2H-tetrazolium bromide], NBT (nitro blue tetrazolium), and PMS (phenazine methosulfate) are all 1% solutions in deionized water.

Alcohol dehydrogenase (ADH), D: *Box stain*

Tris buffer 1	40 ml	**1 locus**
MgCl$_2$ (0.5 M)	0.2 ml	**L**
Ethanol (95%, v/v)	3 ml	**TC 8.0**
NAD	2 ml	
	Add before staining:	
NBT	1 ml	
MTT	0.3 ml	
PMS	0.5 ml	

Amylase (AMY), M: *Box stain*

Phosphate buffer 6	70 ml	**1 locus**
Soluble starch	700 mg	**S, P**
		TG 8.5

Incubate for 30 min at 37°; rock the box often to spread the stain. Pour off stain, rinse several times with tap water, then add 70 ml of iodine solution (0.5 g of I$_2$ is diluted in 20 ml ethanol; take 5 ml of this solution, add 5 g of KI and bring to 1 liter with deionized water). When the white bands appear, pour off the iodine solution, which can be reused several times

Carbonic anhydrase (CAR, carbonate dehydratase); M: *Box stain*

Phosphate buffer 2	50 ml	**1 locus**
Fluorescein diacetate (1% in ace-		**H**
tone)	1 ml	**TEB 8.6**

After 30 min. observe under UV light

Esterase—4MU (EST), M (D): *Box stain*

Acetate buffer	50 ml	**1 locus**

4-Methylumbelliferyl acetate   **Ha**
  (0.5% in acetone)   1 ml   **TM 6.9**
The bands appear when exposed to UV light

Esterase — Naphthol esters (EST), M: *Box stain*
Method 1:
  Phosphate buffer 1   50 ml   **5 loci/3 loci**
  α-Naphthyl propionate   1 ml   **P, H, K/L, K**
  β-Naphthyl propionate   1 ml   **LiOH 8.3/**
  Fast Blue RR   20 mg   **TC 6.0, TC 8.0**
α- and β-naphthyl propionate are 1% solutions in acetone

Method 2:
  Phosphate buffer 3   50 ml   **1 locus**
  α-Naphthyl butyrate (1% in ace-   **Ha**
    tone)   1 ml   **TM 6.9**
  Fast Garnet GBC   20 mg
α- and β-naphthyl acetate (1 or 2% in acetone) can also be used as substrates

Glucose dehydrogenase (GDH or HPD), D: *Box stain*
  Phosphate buffer 5   40 ml   **1 locus**
  D-Glucose   9 g   **K/L**
  NAD   2 ml   **LiOH 8.3/**
                                 Add before staining: **TC 6.7**
  MTT and PMS   0.5 ml each

Glucose-phosphate isomerase (GPI), D: *Agar overlay*
  Tris buffer 1   5 ml   **1 locus/2 loci**
  MgCl$_2$ (0.5 *M*)   1 ml   **Ha, K, L/M, L**
  Fructose 6-phosphate   10 mg   **TM 6.9/**
  NAD   1 ml   **TCB 8.7**
                                 Add before staining:
  Glucose-6-phosphate dehydro-
    genase   6 μl
                                       (=17 units)
  NBT   1 ml
  MTT and PMS   0.5 ml each
  Agar   10 ml

Glutamate oxaloacetate transaminase (GOT; aspartate transaminase),
  D: *Box or agar overlay*
  Tris buffer 1   40 ml   **2 loci**
  L-Aspartic acid   40 mg   **K, L/L, M**
  Ketoglutaric acid   80 mg   **TM 6.9/TC 6.0,**
  Pyridoxal 5'-phosphate   1 mg   **TC 8.0**

Incubate the gel slab for 15 min, then add:

| Fast Blue BB | 80 mg | |

α-Glycerophosphate dehydrogenase (GPD or GDC), D: *Box stain*

Tris buffer 1	40 ml	**1 locus/2 loci**
α-DL-glycerophosphate	250 mg	**K/M, L**
$MgCl_2$ (0.5 M)	0.2 ml	**TC 8.0/TC 8.0,**
NAD	2 ml	**TC 6.7**
	Add before staining:	
NBT	1 ml	
MTT	0.3 ml	
PMS	0.5 ml	

Glyoxalase (GLO), T?: *Paper and agar overlay*

Glutathione (reduced)	15 mg	**1 locus**
	Add before staining:	**K, Ha/L**
Phosphate buffer 4	15 ml	**TEB 8.6/ TC 6.7**
Methylglyoxal	0.45 ml	

Use this solution for a paper overlay and incubate at 37° for 30 min. Remove the paper and prepare an agar overlay:

Tris buffer 1	15 ml	
MTT	1.8 ml	
2,6-Dichlorophenol indophenol		
DCIP; 1% in water	20 drops	
	(freshly	
	filtered)	
Agar	15 ml	

Isocitrate dehydrogenase (IDH), D: *Agar overlay*

Tris buffer 1	5 ml	**2 loci**
Trisodium DL-isocitrate (0.1 M)	1 ml	**K/M, L**
$MnCl_2$ (0.25 M)	0.3 ml	**TC 6.7/TC 6.7**
$MgCl_2$ (0.5 M)	1 ml	**+ NADP**
NADP	0.3 ml	
	Add before staining:	
NBT, MTT, and PMS	0.3 ml each	
Agar	10 ml	

Lactate dehydrogenase (LDH), TT: *Box stain*

Tris buffer 1	35 ml	**2 loci/ 3 loci**
Lithium DL-lactate (0.5 M)	6 ml	**K, L, H/M, H**
NAD	1 ml	**TC 6.7/TCB 8.7**
	Add before staining:	
NBT	0.3 ml	
PMS	0.5 ml	

Malate dehydrogenase (MOR or MDH), D: *Box stain*

Tris buffer 1	35 ml	**2 loci**
Malic acid (2 M)	5 ml	**K, L/M**
MgCl$_2$ (0.5 M)	0.3 ml	**TC 6.7/**
NAD	2 ml	**TC 8.0**
	Add before staining:	
NBT and MTT	1 ml each	
PMS	0.5 ml	

See next stain for 2 M malic acid solution

Malic enzyme (MOD or ME), TT: *Paintbrush or agar overlay*

Tris buffer 1	5 ml	**2 loci**
MgCl$_2$ (0.5 M)	1.5 ml	**K, L/M**
Malic acid (2 M)	1 ml	**TC 8.0/TC 8.0**
NADP	0.1 ml	**+ NADP**
	Add before staining:	
PMS	0.1 ml	
NBT and MTT	0.2 ml each	
Agar (optional)	10 ml	

To prepare malic acid (2 M), place 536.4 g of malic acid in a glass beaker and put the beaker in a basin with crushed ice over a magnetic stirrer. Pour 500 ml of water onto the malic acid, then very slowly add 320 g of NaOH. The solution must not become cloudy nor heat up. When the NaOH is completely dissolved, bring the volume to 2 liters with water. Adjust to pH 7.0 with NaOH flakes

Mannose-phosphate dehydrogenase (MPI), M: *Agar overlay*

Tris buffer 1	10 ml	**1 locus**
D-Mannose 6-phosphate	20 mg	**Ha K**
Pyruvate	20 mg	**TC 6.4,**
NAD	1 ml	**LiOH 8.3,**
NADP	0.5 ml	**TC 8.0**
	Add before staining:	
Phosphoglucose isomerase	15 μl	
	(= 10 units)	
Glucose-6-phosphate dehydro-genase	6 μl	
	(= 17 units)	
MTT	1 ml	
PMS	0.25 ml	
Agar	10 ml	

Phosphoglucomutase (PGM), M: *Agar overlay*

Tris buffer 1	5 ml	**2 loci/1 locus**
$MgCl_2$ (0.5 M)	1 ml	**Ha, L/M**
Glucose 1-phosphate	300 mg	**TC 6.4, TM 6.9/**
NAD	1 ml	**TCB 8.7**

Add before staining:

Glucose-6-phosphate dehydrogenase	6 μl	
	(= 17 units)	
NBT	1 ml	
MTT and PMS	0.5 ml each	
Agar	10 ml	

Phosphogluconate dehydrogenase (PGD), D: *Paintbrush or agar overlay*

Tris buffer 1	5 ml	**1 locus**
$MgCl_2$ (0.5 M)	1.5 ml	**K, Ha/L**
6-Phosphogluconic acid	20 mg	**TC 8.0/TC 6.7**
NADP	0.1 ml	**+ NADP**

Add before staining:

PMS	0.1 ml	
NBT and MTT	0.2 ml each	
Agar (optional)	10 ml	

Purine-nucleoside phosphorylase (NP), T: *Agar overlay*

Inosine	15 mg	**1 locus**

Add before staining:

Phosphate buffer 5	5 ml	**Ha, L**
Xanthine oxidase	10 μl	
	(= 0.3 units)	**TEB 8.6**
PMS	0.25 ml	
MTT	1 ml	
Agar	10 ml	

Pyruvate kinase (PK), TT: *Paper overlay*

Phosphoenol pyruvate	30 mg	**2 loci**
Fructose 1,6-diphosphate	30 mg	**L, H**
$MgSO_4 \cdot 7H_2O$	80 mg	**TMgP 8.2,**
KCl	80 mg	**LiOH 8.3**
NADH	20 mg	
ADP	60 mg	

Add before staining:

Tris buffer 2	16 ml	
Lactate dehydrogenase	50 μl	
	(= 140 units)	

Observe under UV light

Sorbitol dehydrogenase (SDH), TT: *Box stain*

Tris buffer 1	40 ml	**1 locus**
Sorbitol	250 mg	**K, L**
$MgCl_2$ (0.5 M)	0.2 ml	**TC 8.0**
NAD	2 ml	
	Add before staining:	
NBT	1 ml	
MTT	0.3 ml	
PMS	0.5 ml	

Superoxide dismutase (SOD), D: *Box stain*

Tris buffer 1	40 ml	**1 or 2 loci**
$MgCl_2$ (0.5 M)	0.2 ml	**K, L**
	Add before staining:	
MTT	1 ml	**TC 8.0,**
PMS	0.5 ml	**TC 6.7**

General proteins (PT): *Box stain*

Incubate gel slab for 30 min in 50 ml of a Coomassie blue solution (10% Coomassie blue in 450 ml methanol, 450 ml water, and 100 ml acetic acid). Remove the stain, which can be reused many times. The slab is colored a dark blue. Rinse off the slab several times in tap water and then soak in 50 ml of 5% acetic acid fixative. The fixative is removed and replaced by fresh fixative until dark blue bands appear on a light blue background **3 loci/4 loci P, LiOH 8.3 Ha, THCl + 1% cystamine/M, TCB 8.7**

Stains nonenzymatic proteins: **Albumin (ALB), M; hemoglobin (HBB), TT; transferrin (TRF), M**

Stain Buffers

Acetate (50 mM, pH 6.0)		
Sodium acetate trihydrate	6.8 g	
Water	To 1 liter	
Phosphate 1 (40 mM, pH 7.0)		
$NaH_2PO_4 \cdot 2H_2O$	3.12 g	
Na_2HPO_4 (anhydrous)	2.84 g	
Water	To 1 liter	
Phosphate 2 (0.1 M, pH 6.5)		
$NaH_2PO_4 \cdot 2H_2O$	9.98 g	
Na_2HPO_4 (anhydrous)	5.11 g	
Water	To 1 liter	

Phosphate 3 (0.1 M, pH 6.0)
$NaH_2PO_4 \cdot 2H_2O$ 12.5 g
Na_2HPO_4 (anhydrous)   2.8 g
Water   To 1 liter

Phosphate 4 (0.2 M, pH 6.5)
$NaH_2PO_4 \cdot 2H_2O$ 20.8 g
Na_2HPO_4 (anhydrous)   9.47 g
Water   To 1 liter

Phosphate 5 (0.2 M, pH 7.5)
Na_2HPO_4 (anhydrous)   28.4 g
Water   To 1 liter
Adjust to pH 7.5 with HCl (1 N)

Phosphate 6 (50 mM, pH 6.9)
KH_2PO_4 3.0 g
Na_2HPO_4 (anhydrous)   3.2 g
NaCl   0.4 g
Water   To 1 liter

Tris buffer 1 (0.2 M, pH 8.0)
EDTA   0.4 g
Tris   54.2 g
Water   To 1 liter
Adjust to pH 8.0 with about 11 ml of
    concentrated HCl

Tris buffer 2 (0.5 M, pH 7.1 or 8.0)
Tris   60.55 g
Water   To 1 liter
Adjust to pH 7.1 or 8.0 with concentrated HCl

Tris buffer 3 (50 mM, pH 8.7)
Tris   6.05 g
HCl (1 N)   3.3 ml
NaCl   6.7 g
Water   To 1 liter
Adjust to pH 8.7 with 1 M Tris or 1 N HCl

Acknowledgments

The author wishes to acknowledge the contributions of N. Pasteur and R. K. Selander to the development of this technique in our laboratory. Publication number 93-023 of the Institut des Sciences de l'Evolution (URA 327 CNRS).

[8] Two-Dimensional Protein Electrophoresis in Phylogenetic Studies

By David Goldman and Stephen J. O'Brien

Introduction

The capability of two-dimensional (2D) protein electrophoresis to detect protein modifications and primary sequence differences causing alteration in isoelectric point was identified by O'Farrell,[1] who introduced the method, and exploited for this purpose by Steinberg et al.[2] For molecular systematics, the 10-fold increase in number of typed protein loci as compared to allozyme methods normalizes the variance arising from differing evolutionary rates of proteins, and the large number of sites screened [e.g., relative to screens for nuclear and mitochondrial DNA restriction fragment length polymorphisms (RFLPs)] reduces the variance in detected mutational divergence that has been stochastically accumulated over evolutionary time scales. If we assume that a structural gene has on average 1000 coding nucleotides and, further, that 25–30% of all nucleotide substitutions are resolved by aqueous electrophoresis or isoelectric focusing,[3] then a screen of 400 2D electrophoresis protein spots will evaluate 400 × 1000 × 0.25 = 100,000 nucleotides. No other procedure permits large-scale screening for protein genetic differences.

Two-dimensional protein electrophoresis was first used for molecular systematics by Aquadro and Avise[4] and by Ohnishi et al.,[5] who determined that the rate of 2D electrophoresis protein divergence was less than allozyme divergence rates. In phylogenetic studies on the hominoid primates[6] and on the Ursidae,[7] the authors optimized methods for interpreting the genetic differences observed between taxa, evaluating them as single or multistep electrophoretic shifts. These approaches lead to a refined estimate of the average rate of amino acid substitution for the cellular proteins visualized by 2D electrophoresis, and that rate is approximately one-third the rate for allozymes. In addition to the methods described here, which have been directly used by the authors, several other options and refine-

---

[1] P. H. O'Farrell, J. Biol. Chem. **250**, 4007 (1975).

[2] R. A. Steinberg, P. H. O'Farrell, U. Friedrich, and P. Coffino, Cell (Cambridge, Mass.) **10**, 381 (1977).

[3] M. Nei and R. Chakraborty, J. Mol. Evol. **2**, 323 (1973).

[4] C. F. Aquadro and J. C. Avise, Proc. Natl. Acad. Sci. U.S.A. **76**, 6500 (1981).

[5] S. Ohnishi, M. Kawanishi, and T. K. Watanabe, Genetica **61**, 55 (1983).

[6] D. Goldman, P. Rathnagiri, and S. J. O'Brien, Proc. Natl. Acad. Sci. U.S.A. **84**, 3307 (1987).

[7] D. Goldman, P. Rathnagiri, and S. J. O'Brien, Evolution **43**, 282 (1989).

ments of the 2D electrophoresis methodology are mentioned that are described in greater depth elsewhere.

Methods

Preparation and Labeling of Cellular Proteins

Fibroblasts were selected because they are readily cultivated and radiolabeled and are available from skin explants of a variety of species. Autoradiograms of radiolabeled cellular proteins are more diverse and easily analyzable for molecular systematics than alternatives such as silver-stained serum or erythrocyte protein patterns. We have achieved nearly 100% success rates at minimal discomfort and risk to human subjects by establishing cultures from 3 mm punch biopsies from the upper buttock. Local lidocaine anesthesia is used. Samples are collected following informed consent and under an approved human research protocol. Other species are sampled by dart gun[8] or while sedated. Skin samples can be collected from various body sites without compromising the 2D electrophoresis metric, which does not rely on quantitative differences in protein expression. To comply with the Convention on International Trade in Endangered Species of Wild Fauna and Flora (CITES), tissues collected from wild and captive exotic animals must be obtained under specific permits issued by the U.S. Fish and Wildlife Service.

Skin punches are viably frozen by gradual freezing in the vapor phase of liquid N_2[9] or are transferred directly to T-25 Falcon tissue culture flasks gassed with CO_2 and containing Eagle's minimum essential medium (EMEM) with 15% fetal bovine serum (FBS), 1 mg% gentamicin, and 4 mM L-glutamine. To minimize the volume of medium to 0.3–0.4 ml and to oppose closely to the tissue a substrate for fibroblast growth, punches are cut with a scalpel into four equal pieces, and the pieces are cultured between two 10×50 mm glass coverslips in a 16×93 mm screw-top Leighton tube (both from Fisher Scientific, Pittsburgh, PA). Cells are fed twice a week by pipetting excess medium from the bottom of the tube and adding 0.5 ml of fresh medium. Progress of cell outgrowth is monitored microscopically for 4–5 weeks. The coverslips are then removed, placed in a 60×15 mm petri dish, gently washed with 10 ml phosphate-buffered saline (PBS), and layered with 0.02% trypsin and 0.02% EDTA in PBS for 2 min, after which all trypsin is removed except for a thin film. Cells are resuspended in growth medium (total volume

[8] W. B. Karesh, F. Smith, and H. Frazier-Taylor, *Conserv. Biol.* **3**, 261 (1987).

[9] W. S. Modi, W. G. Nash, A. C. Ferrari, and S. J. O'Brien, *Gene Anal. Tech.* **4**, 75 (1987).

~ 10 ml) and grown in T-25 flasks. They are fed twice a week until sheeted and propagated in splits of 1 : 3.

Prior to radiolabeling, cells are grown to confluence, trypsinized (0.02% trypsin and 0.02% EDTA in PBS), and resuspended in growth medium. Fibroblasts are seeded in 96-well plates at 10^5 cells/well and incubated for 24 hr in 5% CO_2/95% (v/v) air at 37°. Six wells per cell line are sufficient. The confluent cells are radiolabeled by replacing the cell growth medium with 100 μl EMEM without methionine and with 15% charcoal-filtered or dialyzed FBS, 4 mM L-glutamine, and 1 μCi/μl [^{35}S]methionine (New England Nuclear, Boston, MA, > 1100 Ci/mmol). After 3 hr of incubation, radiolabeling is terminated by shaking off the medium, washing twice with PBS, and blotting dry.

Sample Preparation

Fifteen microliters of a solution containing 2% sodium dodecyl sulfate (SDS), 5% mercaptoethanol, 20% glycerol, 2% Nonidet P-40 (NP-40), and 2% ampholytes [a 4 : 1 mixture of 5/7 and 3/10 ampholyte (LKB, Piscataway, NJ)] is added to each well and repipetted. Lysates are pooled and briefly digested with 10 μl DNase/RNase (final concentration 10 μg/ml). The stock of deoxyribonuclease I from bovine pancreas (Worthington, Freehold, NJ, 100 μg/ml) and RNase A (Worthington, 50 μg/ml) is made up in 0.5 mM Tris-HCl and 50 mM MgCl$_2$, pH 7.0, and is kept frozen at −70°. Samples are mixed by repipetting and are then quickly frozen in liquid nitrogen and lyophilized.

Two-Dimensional Protein Electrophoresis

Detailed methods for 2D protein electrophoresis have previously been provided in this series.[10] To analyze 300–400 proteins at an acceptable level of resolution, we have used the original methods of O'Farrell.[1] Fifty micrograms of labeled protein [1 × 10^6 counts/min (cpm)] is loaded onto each isoelectric focusing (IEF) gel in 16 μl of the lysate solution with 8 M urea. IEF is for 20 hr at 300–500 V in 5% polyacrylamide tube gels that are prefocused with 2% ampholytes [a 4 : 1 mixture of 5/7 and 3/10 ampholyte (LKB)]. The pH 5–7 range shows the greatest abundance of proteins in a wide variety of species and tissues. Voltage is increased to 1000 V for the last hour. SDS–polyacrylamide gel electrophoresis (SDS-PAGE) is performed at 15–20 mA/gel using 10% acrylamide slab gels that are 0.8 mm thick. Fibroblast proteins are visualized by autoradiography using Kodak (Rochester, NY) XAR-2 film exposed for 2 weeks.

---

[10] J. I. Garrels, this series, Vol. 100, p. 411.

Patton *et al.*[11] have suggested three significant improvements in 2D protein electrophoresis analytic procedures: (1) incorporation of a 0.08 mm thread into the IEF gel to minimize stretching and breakage; (2) the use of the free base of glycine instead of glycine hydrochloride for buffers to reduce the presence of salts, which slow electrophoretic separations and increase the size of protein spots through diffusion; and (3) tight control of temperature and avoidance of gel heating during SDS-PAGE for improved pattern reproducibility and optimal resolution. A convenient target temperature, with Peltier cooling (Millipore, Bedford, MA), is 20–23°.

Alternative Procedures for Larger Format Gels. More proteins (> 500) may be resolved for phylogenetic analysis using larger format gels.[10] Samples are resuspended in a solution containing 9.95 M urea, 4% NP-40, 2% ampholytes [containing a 4:1 ratio of 5/7 and 3/10 (LKB)], and 100 mM dithiothreitol (DTT). For IEF, glass tubes 21 cm long and 0.047 inches in internal diameter are used, and the pH gradient is established with pH 5–7 ampholytes (LKB).

Semiautomated Gel Analysis

For computerized densitometry and precision measurement of protein spot positions, there are several commercially available systems.[11] We use an inexpensive, user-interactive system that is implemented on a general purpose scientific workstation. Images are scanned using a Sierra Scientific (Sunnyvale, CA) High Resolution CCD camera (Model MS 4030) and the Data Translation 8-bit image capture board on a Macintosh II Fx computer with 8 MB of memory. IMAGE (a scientific, public domain, image analysis package developed by Wayne S. Rasband, NIMH, Bethesda, MD) is used to display and analyze autoradiograms.

Relative Densities. To define gene dosage effects, relative densities are measured as follows.[12] (1) A sampling window (usually an ellipse) is drawn to enclose only the protein spot of interest and a larger surrounding background region. (2) The background is the modal density within this window. (3) Protein spot density is calculated as (average density within window minus background) times area within window. (4) Densities are normalized between gels by measuring 20–30 proteins that by visual inspection are invariant and nonsaturating and then calculating by linear

---

[11] W. F. Patton, M. G. Pluskal, W. M. Skea, J. L. Buecker, M. F. Lopez, R. Zimmermann, L. M. Belanger, and P. D. Hatch, *BioTechniques* **8**, 518 (1990).

[12] D. Goldman, and C. R. Merril, R. J. Polinsky, and M. H. Ebert, *Clin. Chem.* **28**, 1021 (1982).

regression the slope of the relationship of these densities from one gel compared to another.

Protein Spot Positions. Protein spot positions can be measured to provide quantitative, statistical evidence of positional identity or difference for proteins in different gels. To accomplish this, the coordinates of the protein spot and of landmark (invariant) protein spots are recorded. Among six great apes 60% of the proteins were positionally invariant,[6] and in a study of bears and procyonids 63% of proteins showed no positional variation.[7] The mean protein spot position is determined from several gels, and outliers are defined statistically.

Detection of Charge Shifts and Other Differences

A set of taxa are compared protein spot by protein spot. Missing spots and alternate isoelectric point forms of a protein are recorded.[6] It is important to note that shifts in isoelectric point are frequently accompanied by minor alterations in apparent molecular mass.[2] Identification of novel charge variants of a protein is aided by the simultaneous comparison of members of a group of related taxa. When several taxa are "missing" a protein, careful inspection usually reveals that all or most have an alternate isoelectric point form.[6,7] It must also be realized that charge substitutions alter the isoelectric point of lower molecular weight proteins more than higher molecular weight proteins.[2]

Two related drawbacks of the 2D protein electrophoresis method are the difficulty of establishing sequence similarities of proteins between different species and our lack of knowledge as to the identity or function of most of the cellular proteins visualized by 2D electrophoresis. Efforts are underway to characterize proteins separated by 2D electrophoresis via functional studies and direct sequencing. Meanwhile, some progress has been made in the linkage mapping and identification of panels of polymorphic human proteins identified by 2D electrophoresis.[13] Polymorphic proteins also exhibit the highest levels of interspecific genetic variation. If it is thought necessary to provide additional confirmation of protein homology, this may be accomplished by several approaches, including differential radiolabeling with various amino acid precursors to detect compositional differences and by silver staining[14,15] using conditions in which particular proteins produce a defined color.

[13] D. Goldman, S. J. O'Brien, S. Lucas, and M. Dean, *Genomics* **11,** 875 (1993).
[14] C. R. Merril, D. Goldman, and M. L. Van Keuren, this series, Vol. 96, p. 230.
[15] C. R. Merril, D. Goldman, and M. L. Van Keuren, this series, Vol. 104, p. 441.

Calculation of Two-Dimensional Protein Electrophoresis Molecular Metric

Double charge shifts are due to two mutations approximately 99% of the time.[6,16] The following weightings are therefore used: single charge difference, 1; double charge difference, 2; heterozygous, 0.5; missing or extra polypeptide, 0.5; shift in molecular weight, 0.5. The Nei genetic distance[17] is computed using a modification for distance measures based on low numbers of individuals. Intraspecies genetic variation incorporated into the Nei genetic distance cannot be assumed to be equivalent between species. Precise measurement of intraspecies average heterozygosity *(H)* will greatly improve the accuracy of distances between closely related species. Early 2D protein electrophoresis studies revealed scant amounts of within-species genetic variation; however, later studies detected substantial amounts of polymorphism, validating the use of 2D electrophoresis as a method of detecting protein genetic variation. For the human, average heterozygosity for 2D electrophoresis cellular proteins was found to range from 2 to 3%.[18-20] However, the human has a lower level of 2D electrophoresis genetic variability as compared to several other hominoid primates (range 1.9 – 3.9%).[21]

Calibration of Two-Dimensional Protein Electrophoresis Molecular Metric. The top half of Table I displays Nei genetic distances generated from a comparison of various great ape species and the crab-eating macaque *(Macaca fascicularis)* across 383 loci.[6] These 2D electrophoresis molecular distances were calibrated based on the estimated time of divergence of the orangutan to yield distances in millions of years before present (myr b.p., lower half of Table I). The average rate of amino acid substitution resulting in 2D electrophoresis protein charge alteration is 0.86%/myr.[6] Rates for isozymes range from 2.1 to 2.6%/myr.[6]

Limits to Usefulness of Two-Dimensional Protein Electrophoresis Molecular Metric

Sampling Error. In a reversal of the relative rate test, we can calculate a conservative estimate of the error inherent in the metric by assuming that rates of substitution are constant in different lineages and comparing the distance of several more closely related taxa to outgroups (Table II). The

---

[16] M. Nei, *Am. Nat.* **106**, 283 (1972).

[17] M. Nei, *Genetics* **89**, 583 (1978).

[18] D. Goldman and C. R. Merril, *Am. J. Hum. Genet.* **28**, 1021 (1983).

[19] D. Goldman, L. R. Goldman, P. Rathnagiri, S. J. O'Brien, J. A. Egeland, and C. R. Merril, *Am. J. Hum. Genet.* **37**, 898 (1985).

[20] S. M. Hanash, L. J. Baier, D. Welch, and M. Galteau, *Am. J. Hum. Genet.* **39**, 317 (1986).

[21] D. N. Janczewski, D. Goldman, and S. J. O'Brien, *J. Hered.* **81**, 375 (1990).

TABLE I

GENETIC DISTANCE AND DIVERGENCE TIMES BASED ON 383 FIBROBLAST PROTEINS[a]

	Human	Chimpanzee	Gorilla	Orangutan	Siamang	Crested gibbon	Macaque
Human	—	0.070	0.097	0.118	0.173	0.158	0.330
Chimpanzee	8.1	—	0.107	0.103	0.166	0.147	0.303
Gorilla	11.3	12.4	—	0.115	0.184	0.143	0.309
Orangutan	13.7	12.0	13.4	—	0.175	0.153	0.327
Siamang	20.1	19.3	21.4	20.3	—	0.100	0.420
Crested gibbon	18.4	17.1	16.6	17.8	11.6	—	0.333
Macaque	38.4	35.2	35.9	38.0	48.8	38.7	—

[a] Above the diagonal are Nei genetic distances (D') corrected for small sample size and back mutation. Below the diagonal are divergence times in millions of year (myr) b.p., calibrated using an estimated divergence date for the orangutan from the other great apes of 13 myr b.p. From left to right, species were *Homo sapiens* ($n = 5$), *Pan troglodytes* ($n = 2$), *Gorilla gorilla* ($n = 1$), *Pongo pygmaeus abelii* ($n = 2$), *Symphalangus syndactylus* ($n = 1$), *Hylobates concolor* ($n = 1$), and *Macaca fascicularis* ($n = 1$). From D. Goldman, P. Rathnagiri, and S. J. O'Brien, *Proc. Natl. Acad. Sci. U.S.A.* **84,** 3307 (1987).

TABLE II

COMPARISON OF MULTIPLE TAXA TO OUTGROUPS TO ESTIMATE
ERROR INHERENT IN TWO-DIMENSIONAL PROTEIN
ELECTROPHORESIS METRIC[a]

Outgroup	Comparisons	D' (mean)	SD ($n = 1$)	SD (%)
Number of loci = 383; great apes compared to outgroups[b]				
Macaque	6	0.337	0.042	12.6
Gibbon	4	0.150	0.007	4.4
Siamang	4	0.175	0.007	4.2
Orangutan	3	0.112	0.008	7.1
Number of loci = 289; bears compared to outgroups[c]				
Lesser panda	7	0.198	0.009	4.8
Raccoon	7	0.187	0.007	3.8
Giant panda	7	0.156	0.006	3.9
Spectacled bear	6	0.090	0.003	2.9

[a] D', Nei distances, see text. SD, standard deviation.

[b] D. Goldman, P. Rathnagiri, and S. J. O'Brien, *Proc. Natl. Acad. Sci. U.S.A.* **84,** 3307 (1987).

[c] D. Goldman, P. Rathnagiri, and S. J. O'Brien, *Evolution* **43,** 282 (1989).

coefficients of variation range from 2.9 to 12.6% but are generally less than 5%. Therefore, it is just within the power of the 2D electrophoresis technique to resolve the human/chimpanzee/gorilla trichotomy.[6] The data favoring an earlier gorilla divergence are human/chimpanzee, 0.070; human/gorilla, 0.97; and chimpanzee/gorilla, 0.107.

Brief Evolutionary Time Scales. Because the 2D electrophoresis divergence rate is 0.86%/myr, an average of 3.44 differences will accumulate per million years in a data set of 400 loci. The method is therefore too insensitive at time scales of less than several million years.

Longer Evolutionary Time Scales. The 2D electrophoresis method has not been shown to be effective for comparing taxa that are more divergent than 0.35 (~40 myr b.p.). Interspecies 2D electrophoresis comparisons become taxing with highly divergent species because the number of invariant landmark proteins is reduced. In addition there is evidence for a ceiling effect in which conservation of protein charge owing to functional constraints causes compression of 2D electrophoresis molecular distances between highly divergent taxa.[7]

Cladistic Analysis

An advantage of 2D protein electrophoresis analysis for phylogenetic inference is that the data can be analyzed either by using phenetic methods and data from a distance matrix (e.g., using Nei's genetic distance or by unit coding the specific protein characters and using cladistic methods for constructing phylogenetic trees based on maximum parsimony. Both approaches are available in widely used computer packages (e.g., PHYLIP, NJTREE, PAUP, UPGAM, BIOSYS) and have been employed on 2D protein electrophoresis data sets.[6,7,21]

Concluding Remarks

Two-dimensional protein electrophoresis molecular distance measures are based on a large sample of randomly sampled proteins that are constrained only in that they fall within a certain range of isoelectric point, mass, and abundance. Improvements in 2D electrophoresis methodology, such as threaded isoelectric focusing gels, ease the difficulty of obtaining high-quality gels. The major methodological difficulty remains the analysis of autoradiograms. For this application, analysis is a time-consuming process that has not been greatly eased by computerization. However, when the amount of time spent for gel analysis is weighed against the number of locus tests performed, 2D protein electrophoresis emerges as an efficient method of screening for genetic changes in protein isoelectric point.

Acknowledgments

We are grateful to Longina Akhtar and Mary Eichelberger for assistance with tissue culture. We thank P. Rathnagiri and D. N. Janczewski, our collaborators in 2D protein electrophoresis phylogenetic studies. We are obliged to the directors and veterinarians of cooperating zoological institutions. In compliance with the Convention on International Trade in Endangered Species of Wild Fauna and Flora (CITES), tissues from exotic animals were collected under specific U.S. Fish and Wildlife Permits: Endangered and Threatened Species, Captive Bred, issued to S. O'Brien. The content of this publication does not necessarily reflect the views or policies of the Department of Health and Human Services, nor does the mention of trade names, commercial products, or organizations imply endorsement by the U.S. Government.

[9] Ancient Proteins in Fossil Bones

By NOREEN TUROSS and LINDA STATHOPLOS

Introduction

Proteins from ancient bones are of interest to a wide range of disciplines, for applications as diverse as the diagnosis of disease in archeological remains[1] to the reconstruction of phylogenetic relationships of extinct taxa.[2,3] Protein from vertebrate hard tissue is also the preferred substrate for dating animal remains within "radiocarbon time," roughly the last 50,000 years.[4] Collagen is the major protein in vertebrate hard tissue, and it is used for both radiometric and stable isotopic analyses of ancient bones.[4] Collagen has been reported in a wide range of ancient bones including Jurassic dinosaurs,[5,6] emus,[7] and Pleistocene whales,[8] by a variety of techniques including amino acid analysis, radioimmunoassay (RIA), and Western blot.

In contemporary bone from a mature animal, native type I collagen makes up approximately 90% (by weight) of the protein and can only be brought into solution by partial enzymatic digestion or acid hydrolysis.[9] In

[1] D. J. Ortner, N. Tuross, and A. I. Stix, *Hum. Biol.* **64,** 337 (1992).

[2] J. M. Lowenstein, V. M. Sarich, and B. J. Richardson, *Nature (London)* **291,** 409 (1981).

[3] J. M. Lowenstein and O. A. Ryder, *Experientia* **41,** 1192 (1985).

[4] R. E. M. Hedges and C. J. A. Wallace, *J. Archaeol. Sci.* **5,** 377 (1978).

[5] T.-Y. Ho, *Proc. Natl. Acad. Sci. U.S.A.* **54,** 26 (1965).

[6] R. W. G. Wyckoff and F. D. Davidson, *Comp. Biochem. Physiol.* **55B,** 95 (1976).

[7] M. J. Rowley, P. V. Rich, T. H. Rich, and I. R. Mackay, *Naturwissenschaften* **73,** 620 (1986).

[8] N. Tuross, M. L. Fogel, and P. E. Hare *Geochim. Cosmochim. Acta* **52,** 929 (1988).

[9] G. M. Herring, *in* "The Biochemistry and Physiology of Bone" (G. H. Bourne, ed.), p. 127. Academic Press, New York, 1972.

ancient bone, collagen becomes partially degraded and soluble, and it can be lost from the skeletal matrix. Some of the *in situ* breakdown products come into solution with traditional protein extraction methods, including decalcification with EDTA. Collagen degradation that yields increasingly soluble breakdown products can occur quite rapidly postmortem, as seen in a series of wildebeest bones that weathered on the ground in Kenya over the course of a decade. Eight years postmortem, 40% more collagen was solubilized during decalcifying protein extraction from the aged, weathered bones than from modern bones.[10] Collagen degradation can also occur slowly, however, and in such a way that the isotopic integrity of the remaining biomolecules is intact. A fossil whale bone from Pleistocene deposits estimated to be greater than 70,000 years old had collagen present at 70% of modern whale bone levels.[8]

Bone Selection and Preparation

By measuring the elemental carbon and nitrogen composition of whole bone, large numbers of samples can be assayed for general organic preservation. Small samples (1–3 mg) of whole bone are weighed and put in lightweight tin sample boats. Elemental determinations of carbon and nitrogen are done in a Carlo-Erba CHN analyzer, which flash combusts samples in an atmosphere temporarily enriched in oxygen and separates the resulting products, N_2, CO_2, and H_2O, over a Porapak QS column. Products are quantified by a thermal conductivity detector.

Most of the organic carbon and nitrogen in bone resides in the collagen molecule. The nitrogen content of skeletal material serves as a proxy of collagen preservation; the percent nitrogen declines postmortem as collagen breakdown products are solubilized away from the mineral matrix. The carbon content as well as the observed atomic carbon to nitrogen ratio (C/N) often increase postmortem, owing to included calcium carbonate[11] or humic acid contamination.[12] Evidence from human skeletal remains also suggests that the time required for complete degradation and loss of collagen or other bone biomolecules is controlled largely by the burial conditions and depositional environment, and not primarily by sample age.[13,14]

[10] N. Tuross, A. K. Behrensmeyer, E. D. Eanes, L. W. Fisher, and P. E. Hare, *Appl. Geochem.* **4**, 261 (1989).

[11] A. A. Hassan and D. J. Ortner, *Archeomaterials* **19**, 131 (1977).

[12] T. W. Stafford, Jr., A. J. T. Jull, K. Brendel, R. C. Duhamel, and D. Donahue, *Radiocarbon* **29**, 24 (1987).

[13] M. B. Lynch and R. W. Jeffries, *J. Archeol. Sci.* **9**, 381 (1982).

[14] S. Ambrose, *Quat. Res.*

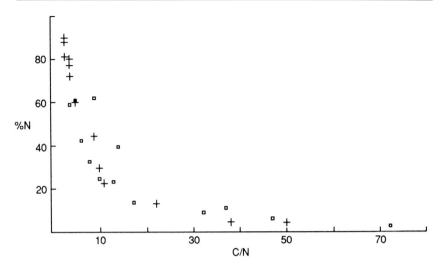

FIG. 1. Atomic carbon and nitrogen values versus percent nitrogen in whole bone from two human cemetery samples. Pieces of human femurs from Archaic [7000 years before present (b.p.)] (squares) and historic (crosses) skeletons are shown. The nitrogen content is referenced to a contemporary compact bovine bone specimen (100%).

The C/N values and nitrogen content for human whole bone samples from two cemetery sites (Fig. 1) demonstrate the variability in macromolecular preservation in buried bone. The more ancient site, Windover, is a neutral anoxic peat bog in east central Florida that was used as a cemetery approximately 7000 years ago for perhaps as long as 1000 years.[15] The more recent site, Cypress Grove, is a below-ground historic cemetery in New Orleans, a city prone to flooding, that was actively used from 1840 to 1900. The amounts of organic matter (indexed by the nitrogen content) in the two populations overlap, even though these intentionally interred human bone samples differ in age by more than 7000 years. The C/N values and nitrogen content data from Windover and Cypress Grove cemeteries varied widely between samples from the same site and illustrate that assessing biomolecular preservation of skeletal material at a site requires analysis of a large sample of bones (Fig. 1).

Once bone samples with acceptable levels of organic preservation are identified, the surface of the bone is brushed clean, and the bone is ground in an impact freezer mill (Spex, New Jersey). The mill grinds bone in a cylinder that is immersed in liquid N_2, thus preventing heat damage to macromolecules. Alternating grinding and cooling pulses of 15 sec each

---

[15] G. H. Doran and D. N. Dickel, *Fl. Anthropol.* **41,** 365 (1988).

will usually reduce the bone to a fine powder. The use of bone powder rather than chunks decreases the time required for the subsequent extractions and demineralization steps.

Protein Extraction and Concentration

All extracting solutions are prepared with molecular biology grade chemicals or the equivalent, and they are stirred with approximately 1 g of activated charcoal per 4 liters and filtered through a 0.45-μm filter prior to use in bone extractions. The powdered bone is constantly stirred at a concentration of 30 g/4 liters at 4° for 3 days in 4 M guanidine hydrochloride, 50 mM Tris, pH 7.4. After the insoluble, mineralized reside settles to the bottom of the flask, the extract is poured off, and the residue is sequentially extracted and decalcified with constant stirring at 4° for another 3 days in 0.5 M EDTA, 4 M guanidine hydrochloride, 0.5 M Tris, pH 7.4. Insoluble, cross-linked collagen will remain after these two extraction steps.

The dissociative demineralizing extraction protocol used to prepare ancient bone samples is based on techniques developed since the late 1970s and widely adopted in vertebrate calcified tissue biochemistry.[16,17] For contemporary bone samples, it is important to maintain the denaturing conditions afforded by 4 M guanidine hydrochloride (and add a series of protease inhibitors) because of endogenous protease activity and because of the tendency of extracellular matrix proteins to reaggregate in complex mixtures in nondenaturing solvents.[18,19] Although there is no evidence that proteases can be activated in ancient bone, the denaturing effect of 4 M guanidine hydrochloride minimizes macromolecular aggregation. The dissociative properties of guanidine hydrochloride are even more important in working with ancient samples than they are in contemporary bone.

Most contemporary bone will completely decalcify at a concentration of 1 g/135 ml of 0.5 M EDTA. Fossil bone, however, often has more calcium per dry weight of bone owing to protein loss and calcite inclusion, and it will saturate EDTA with calcium at this concentration, often forming a heavy white precipitate. Increasing the volume of the extract with fresh decalcifying solution will usually dissolve the precipitate.

Soluble extracts from the first guanidine and second guanidine/EDTA

[16] L. W. Fisher, P. G. Robey, M. F. Young, and J. D. Termine, this series, Vol. 145, p. 269.
[17] L. W. Fisher, G. W. Hawkins, N. Tuross, and J. D. Termine, *J. Biol. Chem.* **262,** 9702 (1987).
[18] L. W. Fisher and J. D. Termine, *in* "Current Advances in Skeletogenesis" (A. Ornoy, A. Harell, and J. Sela, eds.), p. 467. Elsevier, New York, 1985.
[19] E. J. Miller, G. R. Martin, K. A. Piez, and M. J. Powers, *J. Biol. Chem.* **242,** 5481 (1967).

solutions are filtered through Whatman (Clifton, NJ) No. 1 paper. The solutions are then separately concentrated at 4° in large volume stirred-cell units (such as those made by Amicon Danvers, MA) using proper filters. Most of the proteins of interest in bone [e.g., osteonectin, osteopontin, albumin, immunoglobulin G (IgG)] are retained on 30,000 molecular weight cutoff filters (e.g., YM30, Amicon). A notable exception is osteocalcin, a 14,000 molecular weight osteoblastic product that is retained with the use of YM10 (Amicon) filters. The bone extracts are reduced to a minimal volume, preferably less than 10 ml, although the concentration of collagen degradation products in the solute sometimes determines the final concentration volume that can be obtained with ancient bone extracts.

Desalting Bone Extracts

Aliquots (2 ml) of the concentrated extracts are desalted via FPLC (fast protein liquid chromatography) over Sepharose 6 (Pharmacia, Piscataway, NJ) or in gravity-run 10-ml pipettes over BioGel P6DG resin (Bio-Rad, Richmond, CA) in 50 mM ammonium acetate. Protein extracts are immediately frozen, freeze-dried, and stored with a desiccant.

Gel Electrophoresis and Immunodetection

Commercially available minigels, such as those made by NOVEX (San Diego, CA), can be useful in screening ancient extracts if sufficient sample is available. For optimal separation of ancient bone extracts, individually made gradient (4–20%) acrylamide slab sodium dodecyl sulfate (SDS) gels (1.0–1.5 mm) (37.5:1, acrylamide:bisacrylamide) are useful in resolving the wide range of molecular weight products that are often extracted from fossil bone. With ancient samples, identification of a single product by Western blot often requires that a large amount of protein, up to 1 mg, be applied to the gel. Traditional Laemmli[20] gel electrophoresis protocols and Western blotting to nitrocellulose as described by Towbin et al.[21] can be applied to ancient protein extracts. Because it is often necessary to load large amounts of protein, running gels overnight at 8 mA/gel at 4°–10° will often lead to greater resolution of the protein products. Transfer of the products to nitrocellulose can be accomplished in a buffer containing 0.19 M glycine, 25 mM Tris (pH 8.3), and 20% methanol. Often a 1- to 2-hr transfer time at 4°–10° is necessary to visualize the small amounts of product preserved in ancient samples. This requires that short-term

[20] U. K. Laemmli, *Nature (London)* **227**, 680 (1970).
[21] H. Towbin, T. Staehelin, and J. Gordon, *Proc. Natl. Acad. Sci. U.S.A.* **76**, 4350 (1979).

transfers (5–15 min) be done for contemporary controls that might be run on the same gel.

After blocking the nitrocellulose transfer with 3% (w/v) bovine serum albumin (BSA) in Tris-buffered saline (TBS), first antibody should be added at a concentration no greater than a 1:1000 dilution. Incubation with first antibody should proceed for a minimum of 1 hr at room temperature, or it can be allowed to continue overnight at 4°. After washing 3 times with TBS, horseradish peroxidase-conjugated second antibody is added (1:1000 in 3% (w/v) BSA–TBS) for 1 hr at room temperature. After 3 more washes with TBS (the final wash with 10% methanol added), 4-chloro-1-naphthol is used to visualize the bound antibodies.

Noncollagenous proteins also show variable preservation between samples from a single archaeological site. Gel electrophoresis (Fig. 2) of human bone extracts from the 7000-year-old Windover site exhibited a

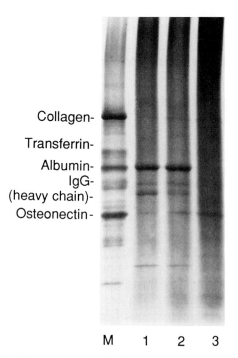

FIG. 2. Gradient (4–20%) acrylamide gel of bone extracts. Two hundred micrograms of modern bone extract (M) and 400 μg of three fossil bone extracts from the Archaic (7000 years b.p.) Windover site were stained with Coomassie Brilliant Blue. Products in the molecular weight range of albumin, IgG (heavy chain), and osteonectin are annotated, and the three proteins were identified in Western blots of this gel in both the modern and fossil bone extracts.

wide range in preservation of noncollagenous proteins comparable to the variability in collagen preservation inferred from the C/N ratios and nitrogen content. Three human bone specimens extracted with the dissociative, demineralizing protocols described above differed widely in electrophoretic patterns, suggesting small-scale variability in the burial environment.

One or more of the noncollagenous proteins may be preferentially preserved even after all collagen has been lost from bone.[22] This concept dates to the early observations of Wyckoff,[23] where amino acid patterns devoid of hydroxyproline (diagnostic for collagen) and enriched in aspartic acid were seen in many older fossil bones. A fossil mastodon bone from a putative pre-Clovis site in Venezuela[24] had an overall amino acid pattern devoid of hydroxyproline and enriched in aspartic acid. This massive bone sample, thought to be approximately 13,000 years old, had retained none of its original collagen. Although hundreds of grams of the mastodon bone were decalcified under denaturing conditions and tested for reactivity to several antisera by gel electrophoresis and Western blot, only albumin could be identified at its original molecular weight.[25] This serum-derived protein had been preferentially preserved compared to collagen and osteoblastic noncollagenous proteins.

Collagen Digestion with Bacterial Collagenase

The degradation and solubilization of collagen in ancient material has important consequences for the analysis of any other protein (or nucleic acids) in bone. Protein extraction protocols of fossil bones that involve demineralization with halide acids or chelating agents (e.g., EDTA) will yield extracts dominated (by weight) by collagen degradation products. Gel electrophoresis of these extracts often exhibits a continuous range of molecular weight fragments (Fig. 3A), in contrast to the discrete protein products observed in contemporary bone or the unusual Windover samples shown in Fig. 2. These collagen degradation products can interfere with attempts to identify and quantify noncollagenous proteins.

Bacterial collagenase, however, can be used to digest the interfering collagen degradation products to tripeptides,[26] thus revealing the pattern of

[22] P. E. Hare, in "Fossils in the Making" (A. K. Behrensmeyer and A. P. Hill, eds.), p. 208. University of Chicago Press, Chicago, 1980.
[23] R. W. G. Wyckoff, "The Biochemistry of Animal Fossils." Scientechnica, Bristol, England, 1972.
[24] A. L. Bryan, R. M. Casamiquela, J. M. Cruxent, R. Gruhn, and C. Ochsenius, Science 200, 1275 (1978).
[25] N. Tuross, Appl. Geochem. 4, 255 (1989).
[26] N. Tuross, in "Human Paleopathology: Current Synthesis and Future Options" (D. J. Ortner and A. C. Aufderheide, eds.), p. 51. Smithsonian, Washington, D.C., 1991.

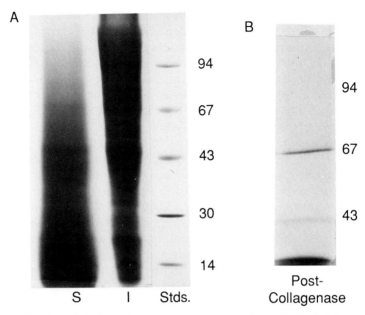

FIG. 3. Gradient (4–20%) acrylamide gel of bone extracts from a fossil whale bone (10,000 years b.p.). Soluble extract (A) shows a range of molecular weights that are stained with Coomassie Brilliant Blue. The insoluble (in guanidine/EDTA) extract was heated in gel sample buffer at 100° for 30 min, and the buffer was removed. This insoluble extract also has a range of molecular weights that tend to be higher on average than the EDTA-soluble component. The soluble extract was digested with bacterial collagenase (B), and two products with molecular weights similar to albumin and osteonectin were revealed.

noncollagenous proteins still in bone (Fig. 3B). Bacterial collagenase (Advanced Biofractures) has been used to digest up to 1 mg of fossil bone extract. The digest is then concentrated by single-use filtration cells (Centricon). The bacterial collagenase used must be totally free of other proteases, a criterion often not met by commercially available collagenase preparations, in order to retain the small amount of noncollagenous protein that many exist in ancient bone. After collagenase digestion, the extract can be further concentrated in individual units such as the Centricon 30 (Amicon) prior to gel electrophoresis. After bacterial collagenase digestion, the extremely good noncollagenous molecular preservation in a 10,000-year-old whale bone (Fig. 3B) could be seen directly on the gel. The protein products were confirmed by Western blot to albumin and osteonectin.

Conclusion

Artificial age barriers that have been erected for the persistence of organic molecules in ancient materials[27,28] continue to fall as sites of exceptional preservation are identified. Radioimmunoassay (RIA) and enzyme-linked immunosorbent assay (ELISA) have also been used to measure noncollagenous proteins in the extracts of a variety of ancient bones; osteocalcin, albumin, and hemoglobin have been reported.[29-33] The aggregate evidence indicates that, in fossil bones which contain collagen, there may well be other proteins preserved. By the time that bone has lost all of its collagen, however, the amount of hydrolyzable amino acids remaining is quite small, and the volume of bone necessary for most analyses is prohibitive. The overall preservation of the bone sample in terms of nitrogen content, total amino acid composition, and mineral integrity are far better predictive indicators of noncollagenous protein preservation than is sample age.

Ancient bone has become recognized as a source of nucleic acids as well as proteins. DNA has been extracted and sequenced from human bone 300–5500 years old,[34] although technical problems with both damaged and contaminated ancient DNA persist.[35] DNA is coextracted in the dissociative, demineralizing extraction procedure described above for protein. Often, the volume of ancient bone required for protein analysis has produced large amounts of DNA having apparent molecular weights much higher than that seen from ancient soft tissue.[36,37]

[27] A. Sillen, J. C. Sealy, and N. J. van der Merwe, *Am. Antiq.* **54,** 504 (1989).

[28] P. H. Abelson, *Year Book—Carnegie Inst. Washington* **54,** 107 (1955).

[29] M. M. W. Ulrich, W. R. K. Perizonius, C. F. Spoor, P. Sandberg, and C. Vermeer, *Biochem. Biophys. Res. Commun.* **149,** 712 (1987).

[30] N. L. Huq, *et al., Biochem. Biophys. Res. Commun.* **129,** 714 (1985).

[31] C. Cattaneo, K. Gelsthorpe, R. Phillips, and R. J. Sokol, *Nature (London)* **347,** 339 (1990).

[32] A. Ascenzi, M. Brunori, G. Citro, and R. Zito, *Proc. Natl. Acad. Sci. U.S.A.* **82,** 7170 (1985).

[33] P. R. Smith and M. T. Wilson, *J. Archeol. Sci.* **17,** 225 (1990).

[34] E. Hagelberg, B. Sykes, and R. Hedges, *Nature (London)* **342,** 485 (1989).

[35] S. Paabo, *Proc. Natl. Acad. Sci. U.S.A.* **86,** 1939 (1989).

[36] N. Tuross and M. L. Fogel, *in* "Archaeometry Proceedings" (David Scott, ed.). Getty Institute, Los Angeles, CA. In press.

[37] H. C. Lee, E. M. Pagliaro, K. M. Berka, N. L. Folk, D. T. Anderson, G. Ruano, T. P. Keith, P. Phipps, G. L. Herrin, D. D. Garner, and R. E. Gaensslen, *J. Forensic Sci.* **36,** 320 (1991).

[10] Comparative Immunological Methods

By MARC H. V. VAN REGENMORTEL, CAROLE JOISSON,
and CARL WETTER

Introduction

Studies with several protein families have shown that the extent of antigenic cross-reactivity between homologous proteins is correlated with the degree of sequence similarity between them.[1] The microcomplement fixation technique has been used extensively for measuring antigenic cross-reactivity mainly because it provides very precise quantitative data.[2] However, because the microcomplement fixation technique is very sensitive to amino acid substitutions in the antigen, it is usually not possible to demonstrate cross-reactivity in proteins that differ in amino acid sequence by more than 30%. In many families of homologous proteins, individual members differ in sequence by more than 30%, and in such cases other immunochemical techniques are needed to determine the degree of antigenic cross-reactivity between two proteins. It has been shown, for instance, that enzyme immunoassays are able to detect weak cross-reactivities between two proteins that share only 30–40% sequence homology.[3,4]

In the case of plant viruses the degree of antigenic relationship between individual members of a viral genus or family has often been measured by precipitation tests, either in liquid or in gel.[5,6] It is customary to express the degree of antigenic relatedness between two viruses by a serological differentiation index (SDI), which corresponds to the average number of 2-fold dilution steps separating homologous from heterologous precipitin titers.[7] The homologous titer refers to the titer of an antiserum (highest antiserum dilution giving a visible precipitation reaction) with respect to the antigen used for immunizing the animal, whereas a heterologous titer refers to the titer with respect to another related antigen.

The SDI is reliable as a measure of antigenic cross-reactivity only if it represents the average value calculated from a large number of bleedings

[1] A. B. Champion, E. M. Prager, D. Wachter, and A. C. Wilson, *in* "Biochemical and Immunological Taxonomy of Animals" (C. A. Wright, ed.), p. 397. Academic Press, New York, 1974.
[2] E. M. Prager and A. C. Wilson, this volume [11].
[3] D. Hoffmann and M. H. V. Van Regenmortel, *J. Mol. Evol.* **82,** 14 (1984).
[4] M. Jaegle and M. H. V. Van Regenmortel, *J. Virol. Methods* **11,** 189 (1985).
[5] J. Brandes, and C. Wetter, *Virology* **8,** 99 (1959).
[6] R. Koenig, *Virology* **72,** 1 (1976).
[7] M. H. V. Van Regenmortel and M. B. Von Wechmar, *Virology* **41,** 330 (1970).

METHODS IN ENZYMOLOGY, VOL. 224

from different immunized animals. This is due to the fact that individual antisera obtained from different animals immunized with the same antigen, or from the same animal at different times, often show considerable variation in the amount of cross-reactive antibodies they contain.[8] However, when about a dozen different antisera are analyzed, meaningful immunochemical comparisons become possible. In particular, average SDI values between two antigens obtained in reciprocal tests, when antisera are used against each of the two strains, have been shown to agree closely.[9] In the tobamoviruses, the degree of immunological similarity among individual virus species measured by SDIs was found to correlate closely with the degree of sequence similarity of their coat proteins. A 40% sequence difference between tobamovirus proteins was found to correspond to an SDI of about 4.0.[9] In the case of the tymoviruses, the grouping of individual virus species constructed on the basis of reciprocal SDIs[6] was found to agree better with sequence relationships of the individual coat proteins[10,11] than with coat protein composition[12] or cDNA–RNA hybridization tests.[13] As discussed by Rybicki,[11] tables of reciprocal SDIs are matrices of pairwise distance relationships and they can be used to calculate the evolutionary distance between proteins.

Precipitin Tests in Liquid

Quantitative precipitin tests in fluid medium are usually performed in small tubes about 7 mm in diameter. The tests are performed in tubes by mixing 0.5-ml volumes of suitable dilutions of antiserum and purified antigen and incubating them in a water bath at 25–40°. The highest dilution of antiserum that gives a visible precipitate (i.e., the antiserum titer) is recorded after 1–2 h. The visibility of precipitates is enhanced by holding the tubes over a light box in front of a black background. The most commonly used diluents for tube precipitin tests are 0.85% NaCl or phosphate-buffered saline (PBS).

Various factors that influence the results of precipitin tests have been discussed by Matthews.[14] In particular, the size of the reacting antigen

---

[8] M. H. V. Van Regenmortel, "Serology and Immunochemistry of Plant Viruses," p. 175. Academic Press, New York, 1982.

[9] M. H. V. Van Regenmortel, *Virology* **64**, 415 (1975).

[10] S. W. Ding, P. Keese and A. Gibbs, *J. Gen. Virol.* **71**, 925 (1990).

[11] E. P. Rybicki, *Arch. Virol.* **119**, 83 (1991).

[12] H. L. Paul, A. Gibbs, and B. Wittmann-Liebold, *Intervirology* **13**, 99 (1980).

[13] J. Block, A. Gibbs, and A. Mackenzie, *Arch. Virol.* **96**, 225 (1987).

[14] R. E. F. Matthews, *in* "Methods in Virology" (K. Maramorosch and H. Koprowski, eds.), Vol. 3, p. 199. Academic Press, New York, 1967.

influences to a considerable degree the antiserum titer. For instance, virus particles consisting of about 2000 identical protein subunits have been shown to be a 25 times better detector of a given quantity of homologous antibody than the dissociated protein subunits.[15] Antigens that become aggregated after purification may therefore give spuriously high antiserum titers. In general, precipitin titers are not suitable for comparing antibody levels against antigens that vary considerably in size.

Microprecipitin Tests in Droplets

Microprecipitin tests are performed in single drops (50 μl) of the mixed antigen and antiserum reactants deposited on a slide or on the bottom of a petri dish.[16,17] If glass surfaces are used, they should be rendered hydrophobic by a coat of silicone or 0.1% (w/v) Formvar in chloroform. When dishes are used, drops can be covered with a layer of mineral oil to prevent evaporation. Slides should be kept in a moist chamber. The precipitates are observed after 30–60 min at room temperature under a dark-field microscope at a magnification of 10–100$\times$. If the drops are kept at 4°, a longer incubation time (6 hr) is necessary. A complete grid titration can be performed in a single petri dish. The main advantage of the microprecipitin test is that it is economical in its use of antiserum.

Immunodiffusion Tests

Immunodiffusion tests are precipitin tests that are carried out in gel instead of free liquid. The most commonly used format is the double-diffusion technique in which antigen and antibody diffuse toward each other into a gel which initially contains neither reagent. As diffusion progresses, the two reactants meet, and precipitation occurs along a line where so-called optimal proportions are reached.[18,19] At optimal proportions of the reactants, all available antigen molecules bind all specific antibody molecules. Optimal proportions of antigen and antibody can be determined in immunodiffusion tests by examining dilution series of antibody and antigen preparations and noting at which ratio of the reactant concentrations the position and width of the band does not change with time. If one of the

[15] J. F. Shepard and T. A. Shalla, *Virology* **42**, 825 (1970).
[16] C. Wetter, *Annu. Rev. Phytopathol.* **3**, 19 (1965).
[17] D. Noordam, Center Agric. Publ. Document, Wageningen, The Netherlands, 1973.
[18] D. Ouchterlony, *Acta Pathol. Microbiol. Scand.* **25**, 186 (1948).
[19] M. H. V. Van Regenmortel, "Serology and Immunochemistry of Plant Viruses," p. 82. Academic Press, New York, 1982.

reactants is initially present in excess, the precipitin band will broaden and move with time toward the reservoir containing the less concentrated reactant. Under these conditions, the visibility of the precipitin line will be less than that at optimal proportions.

The tests can be performed either in petri dishes or on microscope slides. The gel is usually 0.6–1.5% (w/w) agar or agarose in a buffer suitable for the particular antigen being tested. In some cases, the electrolyte concentration in the gel can influence the diffusion process and the formation of precipitin lines.[20] In general, the buffer concentration should not be higher than 0.1 M.

Wells can be formed in the gel by positioning templates on the plate before pouring the agar or by using gel cutters after the agar has set. Agar plugs are then removed by suction. Different well patterns can be used.[21] After the reactants have been introduced in the wells, diffusion is allowed to proceed over periods of 24–72 hr, depending on the size and concomitant diffusion rate of the antigen.[19] Evaporation of the reactant preparations should be prevented by keeping the petri dish in a high humidity incubation chamber or by pouring a layer of light mineral oil over the gel surface. Precipitin lines can be clearly seen by examining the dish against a dark background over a box with a circular light source. Records of precipitin lines can be obtained by simple contact printing onto photographic paper or by using a Polaroid camera.

Gel precipitin titers correspond to the last antiserum dilution giving a visible precipitin line. Since precipitin lines are sharpest under conditions of optimal proportions, it is essential to work under conditions of so-called equivalence. Once the optimal equivalence ratio of concentration has been determined, both reactants should be serially diluted while keeping the equivalence ratio unchanged. Under these conditions it was possible, for instance, to obtain sharp precipitin lines with virus particles using as little as 10 μg/ml antigen. In contrast, when the concentration of the antigen was kept constant and only the antiserum was serially diluted, about 50 times more antigen was needed to detect a visible precipitin line in the same system.[22] Immunodiffusion tests have been used successfully to measure the extent of antigenic cross-reactivity among mutants, strains, and species of plant viruses.[6,23,24]

[20] C. Wetter, *Virology* **31**, 498 (1967).
[21] A. J. Crowle, "Immunodiffusion," 2nd Ed. Academic Press, New York, 1973.
[22] M. H. V. Van Regenmortel, "Serology and Immunochemistry of Plant Viruses," p. 173. Academic Press, New York, 1982.
[23] M. H. V. Van Regenmortel, *Virology* **31**, 467 (1967).
[24] J. C. Devergne and L. Cardin, *Ann. Phytopathol.* **7**, 255 (1975).

Enzyme Immunoassay

Because of its greater sensitivity and economical use of reagents, the enzyme-linked immunosorbent assay (ELISA) is replacing precipitin and immunodiffusion tests for many investigations of protein antigenicity. A great variety of ELISA procedures have been described,[25,26] but not all are equally suited for measuring the extent of antigenic cross-reactivity between proteins.

ELISA methods can be divided into direct and indirect procedures. In direct procedures, the specific antibody is itself labeled with an enzyme, whereas in indirect procedures the enzyme conjugate is an antiimmunoglobulin reagent. The conjugation of an enzyme to antibody molecules tends to reduce their affinity as well as their capacity to cross-react with distantly related antigens.[27,28] As a result, direct ELISA procedures have a narrow specificity that makes them unsuitable for measuring distant immunological relationships. Another reason is that in most cases the conjugation procedure leads to only a proportion of the enzyme and antibody molecules becoming conjugated. Because the conjugated molecules are usually not separated from the free enzyme and free antibody molecules, the unconjugated antibody that has retained its initial affinity will preferentially bind to heterologous antigens, and this may mask any cross-reaction with the labeled antibody.

In contrast, in indirect procedures the activity of the cross-reacting antibody is kept intact. When the direct and indirect ELISA procedures were compared for their ability to reveal cross-reactions between related plant viruses, it was found that viruses which differed by an SDI greater than 2–3 (measured by precipitin tests) could no longer be detected in the direct assay format.[28,29] In contrast, viruses that differed by an SDI of 7.0 could still be detected in the indirect procedure. These comparisons were done using the so-called double-antibody sandwich (DAS) ELISA format in which the wells of the microtiter plate are first coated with antibody. The antigen is then trapped by the immobilized antibody and is subsequently revealed by a labeled antibody (direct method) or by an unlabeled antibody followed by an antiimmunoglobulin conjugate. To prevent the antiimmunoglobulin conjugate from reacting with the coating antibody on the solid phase, the first and second antibodies used in the indirect procedure must

[25] R. Koenig and H. L. Paul, *J. Virol. Methods* **5,** 113 (1982).
[26] P. Tÿssen, "Practice and Theory of Enzyme Immunoassay." Elsevier, Amsterdam, 1985.
[27] R. Koenig, *J. Gen. Virol.* **55,** 53 (1981).
[28] M. H. V. Van Regenmortel and J. Burckard, *Virology* **106,** 327 (1980).
[29] J. C. Devergne, L. Cardin, J. Burckard, and M. H. V. Van Regenmortel, *J. Virol. Methods* **3,** 193 (1981).

be prepared in two different animal species, for instance, rabbits and chickens. Chicken immunoglobulins do not cross-react with mammalian immunoglobulin,[30] and they can be obtained easily from the eggs of immunized hens.[31] The need for antibodies from two animal species can be circumvented by coating the microtiter plates with F(ab')$_2$ fragments of rabbit immunoglobulins. After trapping the antigen, it is then possible to incubate with the intact rabbit immunoglobulins and to reveal the bound immunoglobulin G(IgG) by a conjugated anti-Fc reagent.[32]

Another advantage of using an ELISA procedure in which the antigen is trapped on an antibody-coated plate and not directly adsorbed to the plate is that the antigen is not denatured or disrupted by the adsorption step. It is well known that proteins become at least partly denatured when they are adsorbed to a layer of plastic during a solid-phase assay.[33-35] If the antigens that are compared are partly denatured in the assay, the observed cross-reactivity will tend to be greater than with native protein molecules. This is due to the fact that the unfolding of polypeptide chains renders the internal, more conserved regions of the molecule more accessible to the immune system and to specific antibodies.

When virus particles are directly adsorbed to the solid phase, usually in the presence of a carbonate buffer of pH 9.6, it is in fact the dissociated viral protein that becomes preferentially adsorbed. This degradation of the particles can be readily visualized by performing the first stage of the ELISA on electron microscope grids.[36] Because subunits from different viruses are usually more closely related than the corresponding intact virions (a larger portion of the polypeptide chain is antigenically expressed in the subunit, and the newly exposed surface tends to be functionally more conserved for permitting self-aggregation), the extent of cross-reactivity measured with antigen-coated plates is greater than when antibody-coated plates are used.[4]

Indirect ELISA Procedure Using Antigen-Coated Plates

1. Microtiter plates are coated by incubation of 200 μl of the antigen preparation (1-50 μg/ml) in carbonate buffer, pH 9.6 (15 mM

[30] G. A. Leslie and L. W. Clem, *J. Exp. Med.* **130**, 1337 (1969).
[31] A. Polson, M. B. Von Wechmar, and M. H. V. Van Regenmortel, *Immunol. Commun.* **9**, 475 (1980).
[32] D. J. Barbara and M. F. Clark, *J. Gen. Virol.* **58**, 315 (1982).
[33] M. E. Soderquist and A. G. Walton, *J. Colloid Interface Sci.* **75**, 386 (1980).
[34] B. Friguet, L. Djavadi-Ohaniance, and M. E. Goldberg, *Mol. Immunol.* **21**, 673 (1984).
[35] S. A. Darst, C. R. Robertson, and J. A. Berzofsky, *Biophys. J.* **53**, 533 (1988).
[36] I. Dore, E. Weiss, D. Altschuh, and M. H. V. Van Regenmortel, *Virology* **162**, 279 (1988).

Na_2CO_3, 35 mM $NaHCO_3$, 0.2 g/liter NaN_3).[28] Incubation time is 2 hr at 37° or 18 hr at 4°.

2. Three washing steps are performed with phosphate buffered saline, pH 7.4, containing 50 mM Tween 20 (PBS–T).

3. Blocking of plastic surface is accomplished by incubation with 200 μl of 1–2% bovine serum albumin (BSA) in PBS–T.

4. Plates are incubated with 200 μl of 2-fold dilutions of antiserum in PBS–T or PBS–T containing 1% BSA for 2 hr at 37°. The usual dilution range is 10^{-3} to 10^{-6}.

5. Three washing steps are performed as above.

6. Incubate for 2 hr with antiimmunoglobulin enzyme conjugate (e.g., goat anti-rabbit IgG conjugated with alkaline phosphate) diluted 1 : 1000 to 1 : 5000 in 1% BSA–PBS–T.

7. Three washing steps are performed as above.

8. Incubate for 20–60 min with appropriate enzyme substrate (e.g., 1 mg/ml p-nitrophenyl phosphate in 0.1 M diethanolamine buffer, pH 9.8, in the case of alkaline phosphatase) and read the absorbance at 405 nm.

Indirect ELISA Using Antibody-Coated Plates

1. Coat microtiter plates with immunoglobulins (1–20 μg/ml) purified from antiserum as described[37] and diluted in carbonate buffer (see ELISA procedure above).[28]

2. Perform three washing steps with PBS–T.

3. Block plastic surfaces with 2% BSA in PBS–T.

4. Incubate with antigen (0.1–2 μg/ml) in PBS–T for 2 hr at 37°.

5. Perform three washing steps with PBS–T. Subsequent incubation steps are conducted as in the ELISA procedure above. The antiimmunoglobulin conjugate must be specific for the second antibody (e.g., rabbit antibody) used to react with the trapped antigen and should not react with the first antibody (e.g., chicken antibody) used for coating the plates.

Several variations of indirect ELISA procedures have been described.[25,38]

[37] M. H. V. Van Regenmortel, "Serology and Immunochemistry of Plant Viruses," p. 50. Academic Press, New York, 1982.

[38] M. H. V. Van Regenmortel, in "Control of Virus Diseases" (E. Kurstak and R. G. Marusyk, eds.), p. 405. Dekker, New York, 1984.

Direct Biotin – Avidin ELISA

The biotin – avidin system utilizes the very high affinity of avidin for biotin,[39] corresponding to an affinity constant of 10^{15} M^{-1}. Biotin can be covalently linked to antibody without affecting its antigen-binding capacity,[40] and biotin-labeled antibodies are thus a superior immunochemical reagent for revealing distant antigenic cross-reactions compared to the enzyme-labeled antibodies used in direct ELISA.[41]

The preparation of biotinylated antibodies is very simple in view of the commercial availability of activated biotin (*N*-hydroxysuccinimidobiotin).[41] Biotin conjugates can be stored for several months at $-20°$ without loss of activity.

Procedure

1. Coat plates with 2 μg/ml rabbit immunoglobulins (3 hr at 37°).
2. Wash and block with 1% BSA in PBS – T.
3. Incubate with antigen as in ELISA procedures above.
4. Perform washing and a 2-hr incubation with biotin-labeled antibodies (usually about 1 μg/ml) prepared as described.[41]
5. Incubate for 2 hr at 37° with avidin labeled with alkaline phosphatase (diluted 1/4000 in 1% BSA – PBS – T).
6. Incubate with enzyme substrate as above.

Measurement of Serological Differentiation Index by ELISA

When the ELISA is used to measure distant antigenic relationships, the concentration of the reactants should be adjusted to allow the most sensitive detection of antibody. A grid titration of antiserum versus homologous and heterologous antigens is used to define suitable conditions for the assay. The concentration of enzyme conjugate and the substrate hydrolysis time must be chosen so as to give a rapidly increasing optical density (OD) curve with no trace of a plateau.[4]

To calculate SDI values from ELISA results, it is necessary to compare the antiserum dilutions that lead to the same absorbance for the homologous and heterologous antigens. This is illustrated in Fig. 1, where the SDI values were calculated from the points in the curves that correspond to an OD measurement of 1.0. In the case of nonparallel curves, it is preferable to calculate SDI values at OD levels of 0.5.

[39] N. M. Green, *Adv. Protein Chem.* **29**, 85 (1975).
[40] C. I. Kendall, J. Ionescu-Matiu, and G. R. Dreesman, *J. Immunol. Methods* **56**, 329 (1983).
[41] M. Zrein, J. Burckard, and M. H. V. Van Regenmortel, *J. Virol. Methods* **13**, 121 (1986).

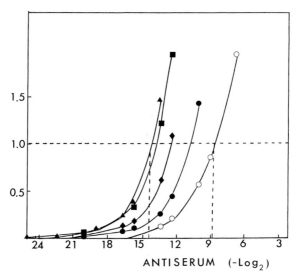

ANTISERUM $(-\log_2)$

FIG. 1. Quantitative measurement of antigenic cross-reactivity between several tobamo-viruses assessed by indirect ELISA on antigen-coated plates. ▲, Homologous reactions: ■, ◆, ●, and ○, heterologous reactions with increasingly distantly related viruses. The SDI value separating the homologous reaction (▲) from the most distant heterologous reaction (○) was $14.5 - 8.8 = 5.7$. [From M. Jaegle and M. H. V. Van Regenmortel, *J. Virol. Methods* **11**, 189 (1985).]

In a comparative study of antigenically related tobamoviruses, it was found that the SDI values from two series of reciprocal ELISA experiments were in reasonable agreement with the average SDI values calculated from precipitin tests in liquid.[4] In a subsequent study, it was shown that the number of amino acid exchanges (as an index of coat protein similarity) among different tobamoviruses was proportional to the immunological distance between the viruses expressed as the SDI.[42] In a study of geminiviruses, it was shown that the phylogeny derived from SDI values obtained in ELISA[43] agreed with that derived from sequence data.[10]

Because homologous antiserum titers in ELISA are much higher than precipitin titers (usually of the order of 10^{-6} instead of 10^{-3}), ELISA is able to detect more distant immunological relationships than precipitin tests. As a result, antigens that differ in sequence by as much as 60–70% can be shown by ELISA to be immunologically related.[3,4] When the antigens are partly denatured by direct adsorption to the plastic surface of microtiter plates, the degree of antigenic relationship between related proteins tends to be higher than when the antigen is trapped on antibody-coated plates

[42] C. Wetter, I. Dore, and M. Bernard, *J. Phytopathol* **119**, 333 (1987).
[43] E. L. Dekker, M. S. Pinner, P. G. Markham, and M. H. V. Van Regenmortel, *J. Gen. Virol.* **69**, 983 (1988).

and the intact protein conformation is retained. In the case of viruses, adsorption of viral proteins subunits directly on the plastic also increases the level of observed cross-reactivity compared to assays in which intact virus particles are trapped on antibody-coated plates.

Immunoblotting

In immunoblotting procedures, a mixture of antigens is first separated by gel electrophoresis and subsequently transferred from the gel to a membrane, where it is detected by antibody. Comparative levels of reactivity with antibody can be assessed by incubation with serially diluted antibody preparations. A comprehensive review of the technique is available.[44] Because the test is done with proteins that are at least partially degraded, relatively distant relationships can be detected.[45,46] The method has been used to calculate SDI values for proteins of potyviruses.[47]

Following separation on one- or two-dimensional gels, the antigens can be transferred electrophoretically to nitrocellulose and the resulting blots incubated with diluted antiserum (1 : 1000). After incubation with ^{125}I-labeled protein A, the amount of homologous or heterologous reaction can be quantitated by counting the radioactivity.[48] Pieces of nitrocellulose corresponding to cross-reacting antigens seen on stained autoradiograms are cut out and weighed using a Cahn microbalance (Cahn Instruments, Cerritos, CA), and γ counts are measured. Counts per unit area are also determined for a background piece for each blot, and these are used to correct the raw counts.[49] The method has been successfully used to measure the immunological distance between various prokaryotic ribosomal proteins.[49]

Monoclonal Antibodies

Although monoclonal antibodies (MAbs) are preferable to polyclonal antiserum for many types of immunological investigations,[50,51] they are not suitable for analyzing quantitatively the degree of antigenic similarity

[44] D. I. Stott, *J. Immunol. Methods* **119**, 153 (1989).
[45] E. P. Rybicki and M. B. Von Wechmar, *J. Virol. Methods* **5**, 267 (1982).
[46] W. Burgermeister and R. Koenig, *Phytopathol. Z.* **111**, 15 (1984).
[47] N. Susuki, S. Shirako, and Y. Ehara, *Intervirology* **31**, 43 (1990).
[48] R. J. Schmidt, A. M. Myers, N. W. Gillham, and J. E. Boynton, *Mol. Biol. Evol.* **1**, 317 (1984).
[49] B. L. Randolph-Anderson, N. W. Gillham, and J. E. Boynton, *J. Mol. Evol.* **29**, 68 (1989).
[50] M. H. V. Van Regenmortel, *in* "Hybridoma Technology in Agricultural and Veterinary Research" (N. J. Stern and H. R. Gamble, eds.), p. 43. Rowman & Allanheld, Totowa, New Jersey, 1984.
[51] A. M. Campbell, "Monoclonal Antibody and Immunosensor Technology." Elsevier, Amsterdam, 1991.

between proteins. Since MAbs are specific for a single epitope of the protein, it is possible, for instance, to select a MAb that does not recognize a related antigen at all (because the particular epitope is not shared between the two antigens) or, on the contrary, to select a MAb that does not differentiate between the antigens because it recognizes the same epitope present in both proteins. In some cases, a single amino acid substitution may suffice to abolish the reactivity of a MAb that recognizes the modified epitope.[52] In other cases, proteins that differ in their sequence by as much as 18% may not be distinguishable by some MAbs that recognize a conserved epitope present in both proteins.[53]

When MAbs are used to investigate the presence of antigenic cross-reactivity between antigens, it is imperative not to use excessively high reagent concentrations that may give rise to nonspecific reactions. In ELISA, antigen concentrations of $1-10$ μg/ml and MAb concentrations of $0.1-10$ μg/ml tend to produce spurious nonspecific reactions. Such conditions have led to unwarranted claims of cross-reactivity between unrelated antigens.[54] However, when defatted milk was used as blocking agent in ELISA instead of the usual BSA, the spurious cross-reactions were abolished.[55,56]

[52] D. Altschuh, Z. Al Moudallal, J. P. Briand, and M. H. V. Van Regenmortel, *Mol. Immunol.* **22,** 329 (1985).
[53] J. P. Briand, Z. Al Moudallal, and M. H. V. Van Regenmortel, *J. Virol. Methods* **5,** 293 (1982).
[54] R. G. Dietzgen, *Arch. Virol.* **91,** 163 (1986).
[55] D. Zimmermann and M. H. V. Van Regenmortel, *Arch. Virol.* **106,** 15 (1989).
[56] R. G. Dietzgen and M. Zaitlin, *Virology* **184,** 397 (1991).

[11] Information Content of Immunological Distances

By ELLEN M. PRAGER and ALLAN C. WILSON*

Introduction

For over 25 years, the microcomplement fixation method[1] has allowed rapid estimation of the approximate degree of sequence difference between monomeric, globular proteins from many different species.[2,3] With this immunological method and two other ways of comparing proteins, elec-

[1] A. B. Champion, E. M. Prager, D. Wachter, and A. C. Wilson, *in* "Biochemical and Immunological Taxonomy of Animals" (C. A. Wright, ed.), p. 397. Academic Press, London, 1974.
[2] A. C. Wilson, S. S. Carlson, and T. J. White, *Annu. Rev. Biochem.* **46,** 573 (1977).
* Deceased.

trophoresis and amino acid sequencing, it became evident that (1) proteins accumulate mutations at rather steady rates which seem unrelated to the rates of evolutionary change at higher levels of biological organization and (2) protein divergence is proportional to time measured in years rather than in generations. In addition, immunological distances were used extensively to infer phylogenetic relationships among species. For various kinds of indirect comparisons (e.g., DNA hybridization, as well as immunological comparisons), attention has been given to the mathematics of building trees from distances and the effects of measurement errors on making evolutionary inferences. However, there has not been an explicit attempt previously to attach confidence values to the distance estimates themselves, that is, to evaluate a bigger source of error, the calibration error involved in using an indirect method to estimate the "true" distance rather than a more direct method.

To assess the certainty with which the extent of difference in amino acid sequence can be inferred from immunological distance, we have estimated prediction intervals for five kinds of monomeric proteins. These estimates are made possible by the extensive calibration work done in our laboratory on proteins of known amino acid sequence. This chapter then uses prediction intervals to test the robustness of hypotheses about phenetic relationships among monomeric proteins of mammals, frogs, and birds. The probabilities of arriving at various conclusions by chance range from 1 in 20 to 1 in 10 billion.

The method offered here provides a phenetic criterion that should be met in order for a particular phylogenetic hypothesis to be taken seriously. Our approach, which evaluates the degree of molecular resemblance at the protein level, is phenetic, not phylogenetic. However, because protein evolution is fairly clocklike,[2,4] phenograms in practice resemble phylogenetic trees. In this article the use of terms like phylogenetic, outgroup, and relationship is thus with the explicit recognition that, strictly speaking, phenetic resemblance is being tested.

Methods

Immunological and Sequence Data

The left-hand side of Table I summarizes the numbers of amino acid sequences and pairwise comparisons considered for each of the five pro-

[3] D. C. Benjamin, J. A. Berzofsky, I. J. East, F. R. N. Gurd, C. Hannum, S. J. Leach, E. Margoliash, J. G. Michael, A. Miller, E. M. Prager, M. Reichlin, E. E. Sercarz, S. J. Smith-Gill, P. E. Todd, and A. C. Wilson, *Annu. Rev. Immunol.* **2**, 67 (1984).
[4] A. C. Wilson, H. Ochman, and E. M. Prager, *Trends Genet.* **3**, 241 (1987).

TABLE I
IMMUNOLOGICAL COMPARISONS OF FIVE PROTEINS[a]

Protein (number of sequences)	Number of pairwise comparisons (n)	Slope	r	95% Confidence band, width (%)	90% Prediction interval, width (%)				
					5	10	15	20	25
Myoglobin (13)	78	10.5	0.87	6.1	70	35	—	—	—
Lysozyme c (13)	75	5.1	0.90	6.1	138	69	46	35	28
Azurin (8)	21	4.6	0.79	11.6	193	97	65	49	40
Ribonuclease (13)	80	6.9	0.92	4.2	123	61	41	31	25
Albumin (8)	10	6.3	0.96	8.8	102	51	35	26	22

[a] Slopes, confidence bands, and prediction intervals are shown graphically in Fig. 1. The slope is the coefficient k for the regression line through the origin given by $y = kx$, where x is percent amino acid sequence difference and y is immunological distance. r is the correlation coefficient for the least-squares line relating immunological distance to percent sequence difference. The width of the band for the regression line determined with 95% confidence is expressed (in percent) as $100a/k$, where k is the slope of the regression line through the origin and a is the measure of uncertainty in that slope; the same result obtains if $100b/y$ is calculated for any immunological distance y, where b comes from the uncertainty in y expressed as $y \pm b$. The width of the interval for one prediction made at the 90% level of confidence is computed (in percent) as $100c/y$, from the limits $y \pm c$ defined by the upper and lower boundaries of the prediction intervals shown in Fig. 1; values of $100c/y$ are tabulated for amino acid sequence differences ranging from 5 to 25%. The number of pairwise comparisons that are the averages of reciprocal measurements of immunological distances are 66, 45, 9, 0, and 3, respectively, for myoglobin, lysozyme, azurin, ribonuclease, and albumin. The sources of the immunological distances are as follows: cetacean myoglobins [E. M. Prager, *J. Mol. Evol.* **37**, in press (1993)], bird lysozymes [I. M. Ibrahimi, E. M. Prager, T. J. White, and A. C. Wilson, *Biochemistry* **18**, 2736 (1979), and references therein; E. M. Prager, unpublished (1978, 1982)], bacterial azurins [A. B. Champion, K. L. Soderberg, A. C. Wilson, and R. P. Ambler, *J. Mol. Evol.* **5**, 291 (1975); A. B. Champion, E. L. Barrett, N. J. Palleroni, K. L. Soderberg, R. Kunisawa, R. Contopoulou, A. C. Wilson, and M. Doudoroff, *J. Gen. Microbiol.* **120**, 485 (1980); A. B. Champion, K. L. Soderberg, and A. C. Wilson, unpublished (1973)], mammalian ribonucleases [E. M. Prager, G. W. Welling, and A. C. Wilson, *J. Mol. Evol.* **10**, 293 (1978)], and mammalian serum albumins (references given in Benjamin *et al.*[3]). Sequence revisions relative to some of the information used before have been incorporated. For 5 of the 8 albumins, partial sequences were considered.[3]

teins used to derive confidence limits and prediction intervals. The sources of the immunological data are given in the footnote to Table I.

Calculations

Slopes *(k)* of regression lines through the origin were calculated according to Eq. (1):

$$k = \Sigma \, x_i y_i / \Sigma \, x_i^2 \tag{1}$$

where x_i and y_i are, respectively, percent amino acid sequence difference and immunological distance for the ith pairwise comparison.

Standard procedures described by Neter and Wasserman[5] were used to compute confidence limits for the regression lines (those going through the origin and also those derived by the method of least squares) and to compute prediction intervals. Specifically, for parameters related by a line passing through the origin, we followed the procedure outlined (in their section 5.5) and, along with the t distribution, used their equations (Eqs. 5.35 and 5.37, respectively) to place confidence intervals about the slope of the line and about a new value.

The prediction interval indicates, with the chosen degree of confidence, the region within which a newly measured immunological distance is expected to fall for a given sequence difference or, vice versa, the sequence difference which one would predict for a given immunological distance. If, for example, a 90% confidence level is chosen, there is a 5% chance that for a given sequence difference the immunological distance will be less than or equal to the value defined by the lower boundary of the prediction interval and a 5% chance that it will exceed or match that defined by the upper boundary. The half-widths of confidence limits for regression lines and prediction intervals were also calculated as a percentage of the immunological distance implied by a given regression line (see Table I).

Two methods based on the intervals computed for making one prediction were used to assess the probability that two measured immunological distances, $d_1 < d_2$, actually represent the same degree of amino acid sequence difference. Method 1 asks what the probability is of measuring a value greater than or equal to d_2 if d_1 represents the correct value as defined by the regression line (or, analogously and resulting in the same answer, the probability of measuring a value less than or equal to d_1 if d_2 is the correct value). Method 2 first makes the hypothetical assumption that the correct immunological distance is $(d_1 + d_2)/2$, that is, that the amino acid sequence differences being evaluated are identical. This method then asks what the probabilities p_1 and p_2 are of measuring, respectively, one value less than or equal to d_1 and one value greater than or equal to d_2 and takes the final probability of recording $(d_1 + d_2)/2$ as d_1 one time and d_2 the next as the product of p_1 and p_2 (in practice $p_1 = p_2$). The results from Methods 1 and 2 were averaged (except where stated otherwise). In cases where two or more proteins were compared immunologically for the same taxa, the combined probability, Π, that one is unable to make a distinction between pairs of immunological distances is computed as the product of

[5] J. Neter and W. Wasserman, "Applied Linear Statistical Models." R. D. Irwin, Homewood, Illinois, 1974.

$P_i \times P_j \times \ldots \times P_n$, where each P_i represents the probability calculated for a single protein. For calculations involving proteins for which there exist insufficient sequence data to derive a sequence–immunology correlation, the line and interval parameters determined for mammalian serum albumin were used.

Examples and Discussion

Prediction Intervals

Figure 1 shows for five proteins the 90% prediction intervals, along with regression lines and their 95% confidence limits. These parameters are

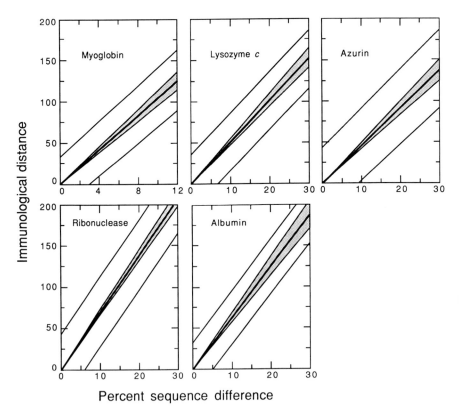

FIG. 1. Confidence limits and prediction intervals for immunological comparisons of five proteins. For each protein the heavy central line is the regression line through the origin relating immunological distance in the microcomplement fixation test to percent difference in amino acid sequence, and the shaded region portrays the 95% confidence limits for that line. The outer two lines in each graph define the boundaries of the intervals for one prediction made at the 90% level of confidence.

TABLE II
PROBABILITY THAT TWO IMMUNOLOGICAL DISTANCES REPRESENT SAME EXTENT OF
SEQUENCE DIFFERENCE: AN ALBUMIN EXAMPLE[a]

Parameters[a]: $d_1 = 6$, $d_2 = 22$, $n - 1 = 9$, s (estimated standard deviation) $= 17.4$
Selected percentiles of t distribution for $n - 1 = 9$

$1 - \alpha/2$	0.65	0.70	0.75	0.80	0.85	0.90	0.95
t	0.398	0.543	0.703	0.883	1.100	1.383	1.833

Method 1
$d_2 - d_1 = 16 = ts$, $t = 16/17.4 = 0.920$
Interpolate between $t(0.80) = 0.883$ and $t(0.85) = 1.100$ above to compute that $1 - \alpha/2 = 0.81$ for $t = 0.920$
$\alpha/2 = 19.0\%$, the probability that a distance ≥ 22 could be measured if 6 is the correct distance

Method 2
$(d_1 + d_2)/2 = 14$ is assumed as the correct value
$(d_1 + d_2)/2 - d_1 = 8 = ts$, $t = 8/17.4 = 0.460$
Interpolate between $t(0.65) = 0.398$ and $t(0.70) = 0.543$ above to compute that $1 - \alpha/2 = 0.67$ for $t = 0.460$
$\alpha/2 = 33.0\%$, the probability p_1 that a distance ≤ 6 could be measured if 14 is the correct distance
Analogously, $p_2 = 33.0\%$, the probability that a distance ≥ 22 could be measured if 14 is the correct distance
$p_1 \times p_2 = (0.33)^2 = 10.9\%$, the probability of measuring 6 one time and 22 the next time if 14 is correct

Average of Methods 1 and 2
$(19.0 + 10.9)/2 = 15.0\%$, the probability that albumin immunological distances of 6 and 22 represent the same amount of amino acid sequence difference

---

[a] This example uses the parameters determined for albumin in the course of generating Table I and Fig. 1. d_1 and d_2 are, respectively, distances between bear and giant panda and between bear and lesser panda. n is the number of pairwise comparisons used to correlate immunological distance with sequence difference (Table I). s is the estimated standard deviation of a newly predicted immunological distance (derived with Eq. 5.37 of Neter and Wasserman[5]). s can be retrieved from the graphs in Fig. 1 by taking, at any given sequence difference, the difference between the upper or lower limit of the prediction interval and the regression line (defined as $\pm c$ in the footnote to Table I) and dividing by t, since $c = ts$. (Thus, t is a measure of the difference between two immunological distances expressed in units of standard deviation; see below for more about t.) Equivalently, one can use the information at the right-hand side of Table I to calculate c and then s. For example, at a 10% sequence difference, the regression line with slope of 6.3 leads to a prediction of 63 units of immunological distance, and c is 51% of that 63, or 32.1 units. For albumin, s ranges from 17.4 at 0% sequence difference to 18.8 at 30%; for all five proteins considered, s ranged from 17.4 to 26.7. $\alpha/2$ is the probability that an immunological distance will lie on or beyond the upper or lower boundary of a given prediction interval; thus for a 90% confidence level for the prediction interval, $\alpha/2 = 0.05$. t is the value of the t distribution as a function of $1 - \alpha/2$ and $n - 1$ and is $t(1 - \alpha/2; 9)$ in this example; for the 90% prediction interval depicted for albumin in Fig. 1 and described in Table I, a t value of 1.833 was therefore used. t is relatively insensitive to n, with the drop in t for $n = 61$ versus $n = 10$ being under 9% as $1 - \alpha/2$ ranges from 0.55 to 0.95. See the text (Methods, *Calculations*) for further explanations of Methods 1 and 2.

TABLE III

IMMUNOLOGICAL TESTING OF PHYLOGENETIC HYPOTHESES[a]

A ("outgroup")	B	C	Protein	Immunological distance			Probability (%)	
				AB	AC	BC	P_i	Π
Carnivores								
1. Cat	Dog	Seal	Abumin	98	89	50	1.9	0.055
			Transferrin	148	149	110	2.9	
2. Mink	Seal	Sea lion	Albumin	43	38	24	14.6	2.0
			Transferrin	101	82	71	13.5	
3. Bear	Sea lion	Seal	Albumin	33	29	24	26.4	2.1
			Transferrin	92	101	71	8.1	
4. Lesser panda	Giant panda	Bear	Albumin	21	22	6	15.6	0.61
			Transferrin	54	52	19	3.9	
Primates								
5. Orangutan	Human	Chimpanzee	Albumin	12	10	6	30.5	4.6
			Transferrin	25	27	10	15.1	
Rodents								
6. Spiny mouse *(Acomys)*	True mouse *(Mus)*	Rat *(Rattus)*	Albumin	*140*	*131*	61	0.18	1.5×10^{-6}
			Transferrin	*126*	—	88	2.9	
			Lysozyme c	—	*173*	66	0.028	
Frogs								
7. *Gastrotheca*	*Anotheca*	*Hyla regilla*	Albumin	136	148	69	0.20	0.20
8. *Hymenochirus*	*Xenopus tropicalis*	*Xenopus laevis*	Albumin	188	180	61	0.026	0.026

	Protein					
Birds						
9. Duck Penguin Loon	Albumin	32	*30*	*17*	17.2	3.3 × 10⁻⁹
	Transferrin	160	138	31	0.027	
	Ovalbumin	*104*	95	22	0.13	
	Penalbumin	—	*122*	*38*	0.054	
10. Chicken Tinamou Rhea	Albumin	58	55	*43*	18.1	0.051
	Transferrin	136	*133*	*67*	0.28	

a For each of the 10 examples, the possibility that taxon A is closer to taxon B and taxon C is the outgroup was evaluated by calculating the probability that the AB and AC immunological distances represent the same amount of amino acid sequence difference as does the BC distance (see also Fig. 2). AB and AC were separately compared to BC and the results averaged. The immunological distances tabulated are the averages of reciprocal measurements except those in italics, which are unidirectional measurements. The sources of the immunological data are as follows: carnivores [Sarich[6,10,11]; V. M. Sarich, *Syst. Zool.* **18**, 416 (1969); V. M. Sarich, unpublished (1971–1972)], primates [J. E. Cronin, *Kroeber Anthrop. Soc. Pap.* **50**, 75 (1977); II is <5% also when gorilla replaces chimpanzee], rodents [V. M. Sarich, *in* "Evolutionary Relationships among Rodents: A Multidisciplinary Analysis" (W. P. Luckett and J.-L. Hartenberger, eds.), p. 423. Plenum, New York, 1985; M. F. Hammer, J. W. Schilling, E. M. Prager, and A. C. Wilson, *J. Mol. Evol.* **24**, 272 (1987); E. M. Prager, unpublished (1982, 1984); the transferrin values came from use of an antiserum pool to the *Praomys* protein, with *Praomys* being phenetically about as close to the house mouse *Mus domesticus* as are members of other subgenera of *Mus*], frogs [Maxson[19]; Bisbee *et al.*[21]], and birds [Ho *et al.*[23]; Prager *et al.*[24]; E. M. Prager, A. H. Brush, R. A. Nolan, M. Nakanishi, and A. C. Wilson, *J. Mol. Evol.* **3**, 243 (1974); E. M. Prager and A. C. Wilson, *J. Mol. Evol.* **9**, 45 (1976); E. M. Prager, unpublished (1973–1975)]. Differences in the rate of evolution of a given protein can occur among lineages,[4] and some antisera can yield immunological distances that are systematically high or low [J. E. Cronin and V. M. Sarich, *J. Hum. Evol.* **4**, 357 (1975)]. This chapter employs only measured immunological distances and does not describe rate tests relevant to the phylogenetic hypotheses examined. The rationale for this procedure is 2-fold. First, deviations from ideal behavior will not generate false conclusions in cases where the probability of accepting an alternative hypothesis is very low. Second, when two or more proteins are tested and found to support the same hypothesis, as is true for 8 of the 10 examples, any anomalous behavior exhibited in immunological comparisons of one protein shoulde be compensated for.

comparable for the five proteins, as summarized quantitatively in Table I. (Because the slope is steeper for myoglobin than for the other proteins, the width in percent of the prediction interval is narrower.) When least-squares lines are used (not shown), the prediction intervals are very similar to those for regression lines through the origin.

Table II illustrates how the information in Fig. 1 and Table I can be used to assess the statistical significance of immunological data bearing on phenetic relationships, with attention centered on the position of the giant panda with respect to the bear and lesser panda (see also example 4 in Table III). In essence, we estimate the probability that the bear is molecularly as similar to the lesser panda as to the giant panda.

Consider first the immunological distances of 6 and 22 for albumin. As explained in Table II, whichever method of assessing probability (1 or 2) is used, these two albumin distances are not statistically significantly different. There is an 11–19% (average, 15%) probability that 6 and 22 could be obtained by chance even if bear albumin were equally distinct in amino acid sequence from the two panda albumins.

The probability of equal distinctness in sequence is lower, 4–5%, for the transferrin distances of 19 and 52 (see Table III). Accordingly, when both pairs of results are considered together, the probability of supporting a bear–giant panda association purely by chance drops to below 1% (i.e., Π is at most 0.95%, the product of 19 and 5%). This conclusion of a statistically significant difference emerges whether one uses Method 1, which gives larger probabilities, or Method 2, and whether one uses the prediction intervals for the protein producing the most scatter (azurin, Fig. 1 and Table I) or the least (albumin), with Π being less than or equal to 3% in all cases. Finally, the same conclusions emerge, with very similar probabilities, from the use of least-squares lines and prediction intervals.

Immunological Distances and Phylogenetic Testing

Table III gives 10 examples in which immunological comparisons pointed to the overthrow of an earlier hypothesis or allowed a choice to be made between two competing hypotheses. Figure 2 depicts our test schematically, with each side of the triangles representing an immunological distance. The proportions shown on the left-hand side reflect the measured distances (i.e., AB and AC large, BC smaller). In contrast, the triangle on the right-hand side asks us to force the distances onto an alternative framework, in which A and B are the most similar of the three pairs. It asks us whether the observed distances, which seem to support the left-hand triangle, are also compatible with the right-hand triangle. In all 10 cases discussed below, the confidence with which one can predict degree of

Supported   Rejected

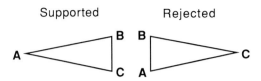

FIG. 2. Schematic representation of phenetic or phylogenetic hypotheses tested immunologically. The immunological evidence indicated that the proteins from taxa B and C resemble one another more than they do the protein from taxon A (left-hand side of figure), whereas earlier hypotheses, based on nonmolecular evidence, had suggested the organismal relationships shown on the right-hand side, with C being the outgroup taxon. See Table III and text for examples.

sequence difference from immunological distance is sufficient to reject the right-hand triangle.

Example 1. As recently as about 25 years ago morphologists thought that pinnipeds—seals, sea lions, and walruses—should be in a separate order, outside the order Carnivora, which includes the catlike carnivores (feloids) and the doglike carnivores (canoids).[6,7] The immunological comparisons, by contrast, indicated with a high degree of certainty ($\Pi <$ 0.0006, Table III) that feloids are the outgroup to canoids and pinnipeds, a view later strongly endorsed by sequences of several proteins.[8]

Examples 2 and 3. Demonstration of pinniped monophyly relative to the other nonfeloid carnivores required a finer degree of discrimination. Consequently, the conclusions, although statistically significant in opposing origin of the seals from an otterlike stock and of the sea lions and walrus from a bearlike stock, are less robust than in example 1. Amino acid sequences[8] and DNA hybridization data[9] are now available in support of pinniped monophyly.

Example 4. The conclusion by Sarich[10,11] in the early 1970s that the giant panda is allied with bears rather than with the lesser panda can be regarded as statistically robust. This view was consistent with a comprehensive anatomical study[11-13] and with fossil evidence[13] and was afterward reinforced by more molecular and cytogenetic data.[14]

[6] V. M. Sarich, *Syst. Zool.* **18,** 286 (1969).

[7] D. Morris, "The Mammals." Harper & Row, New York, 1965.

[8] D. A. Tagle, M. M. Miyamoto, M. Goodman, O. Hofmann, G. Braunitzer, R. Göltenboth, and H. Jalanka, *Naturwissenschaften* **73,** 512 (1986).

[9] Ú. Árnason and B. Widegren, *Mol. Biol. Evol.* **3,** 356 (1986).

[10] V. M. Sarich, *Nature (London)* **245,** 218 (1973).

[11] V. M. Sarich, *Trans. Zool. Soc. London* **33,** 165 (1976).

[12] D. D. Davis, *Fieldiana, Zool. Mem.* **3,** Chicago Nat. Hist. Museum, Chicago (1964).

[13] E. Mayr, *Nature (London)* **323,** 769 (1986).

[14] J. O'Brien, W. G. Nash, D. E. Wildt, M. E. Bush, and R. E. Benveniste, *Nature (London)* **317,** 140 (1985).

However, the report[8] of the α- and β-hemoglobin sequences of the giant and lesser pandas resurrected the idea of associating the giant panda with the lesser panda and argued against the possibility of grouping the giant panda with the bears. Nevertheless, the globin evidence was not statistically significant in the winning-sites test.[15,16] Convergence and parallelism or sequencing errors could account for the anomalous similarity of the globin sequences of the two pandas.[15] In any case, the best estimates suggest that the bear and giant panda differ at only 38 of 1550 amino acid positions (represented collectively by albumin, transferrin, α-globin, and β-globin), whereas the lesser panda is inferred to differ from these taxa at 85–92 positions.

Example 5. Morphologists could not rule out the possibilities that orangutans are closer to humans or to African apes than the latter two groups are to each other.[17] By contrast, the immunological evidence for two proteins considered together reached significance in placing orangutans outside a group encompassing humans and African apes. This inference was later supported at an extremely high level of significance by sequences of mitochondrial and nuclear DNA.[17]

Example 6. The example from rodents offers the biggest discrepancy known for living mammals between a morphological classification and a molecular one.[4] Rodent morphologists had considered spiny mice *(Acomys)* to be the closest living relatives of true mice *(Mus)*. Yet the immunological evidence indicates that the probability of *Acomys* proteins being phenetically closer to *Mus* than are *Rattus* proteins is less than 1 in 10 million. Recently obtained mitochondrial DNA sequences agree that *Acomys* is genetically remote from *Mus* and *Rattus*.[18]

Example 7. In example 7, which is one of many that could be listed for frogs, the molecular evidence made possible detection of a case of convergent morphological evolution. The monotypic genus *Anotheca* had been assigned morphologically to the marsupial tree-frog subfamily (Amphignathodontinae, which has since been made part of subfamily Hemiphractinae) with *Gastrotheca,*[19] whereas the albumin data grouped it with *Hyla* in the subfamily Hylinae. The chances of the morphological assignment being molecularly correct are only 1 in 500.

Example 8. From morphological evidence that did not reach statistical

[15] E. M. Prager, unpublished (1987).
[16] E. M. Prager and A. C. Wilson, *J. Mol. Evol.* **27,** 326 (1988).
[17] A. C. Wilson, E. A. Zimmer, E. M. Prager, and T. D. Kocher, *in* "The Hierarchy of Life" (B. Fernholm, K. Bremer, and H. Jörnvall, eds.), p. 407. Elsevier, Amsterdam, 1989.
[18] M. W. Allard, Ph.D. Thesis, Harvard Univ., Cambridge, Massachusetts (1990).
[19] L. R. Maxson, *Syst. Zool.* **26,** 72 (1977).

significance, Cannatella and Trueb[20] in 1988 proposed that *Xenopus laevis* lies outside a group that includes *Xenopus tropicalis* and *Hymenochirus*. This proposal was made rather than one invoking, for example, convergent morphological evolution even though the immunological evidence[21] presented a decade earlier implied that the chances of this arrangement being molecularly correct are less than 1 in 3000. Recently obtained nuclear ribosomal DNA sequences[22] strongly endorse the inference from albumin immunological distances that *X. tropicalis* groups with *X. laevis* to the exclusion of *Hymenochirus*.

Example 9. Only about 30 years ago some authorities considered penguins so distinctive in anatomy and way of life as to classify them in a superorder separate from all other bird orders.[23] If this were the case, the proteins of ducks and loons should be more similar to each other than those from either group are to penguin proteins. The immunological evidence obtained with four different proteins indicates that the chances are about 1 in 10 billion that this expectation will be fulfilled. Instead, penguin proteins are very similar to those of loons.

Example 10. Tinamous, unlike ratites, have a keel on the sternum, but they do share with ratites the paleognathous palate and unique bill structure.[24] This final example in Table III thus reflects the past uncertainty as to whether tinamous are phylogenetically allied with ratites or possibly with gallinaceous birds exemplified by the chicken. The immunological work firmly supports a tinamou–ratite grouping.

General Applicability of Criterion

There has been a tendency for evolutionary biologists to ignore immunological distances despite the extensive calibration work done correlating these distances with sequence difference and also the extensive rate tests done for albumin and transferrin distances demonstrating clocklike evolution. This chapter provides a way to integrate immunological information with other kinds of molecular information. This contribution should stimulate and facilitate a reexamination of the vast body of published immunological distances, to see whether the immunological results bearing on particular questions survive the test described here and can be viewed with statistical confidence.

[20] D. C. Cannatella and L. Trueb, *Zool. J. Linn. Soc.* **94,** 1 (1988).
[21] C. A. Bisbee, M. A. Baker, A. C. Wilson, I. Hadji-Azimi, and M. Fischberg, *Science* **195,** 785 (1977).
[22] R. O. de Sá and D. M. Hillis, *Mol. Biol. Evol.* **7,** 365 (1990).
[23] C. Y.-K. Ho, E. M. Prager, A. C. Wilson, D. T. Osuga, and R. E. Feeney, *J. Mol. Evol.* **8,** 271 (1976).
[24] E. M. Prager, A. C. Wilson, D. T. Osuga, and R. E. Feeney, *J. Mol. Evol.* **8,** 283 (1976).

Many workers have previously devoted attention to the contribution of errors in measurements to the problem of building trees from distances, as summarized in the contribution by Marshall.[25] By contrast, we have not been concerned with this relatively minor source of error. Instead, our concern has been with a bigger source of error, the calibration error, which reflects the uncertainty in the relationship of distance measured with an indirect method to that measured with a more direct method. This aspect has not been addressed by previous workers. To illustrate the magnitude that this problem can assume, we note that DNA hybridization led to an estimate of 3.3% sequence divergence between the mitochondrial DNAs of two flies (*Drosophila yakuba* and *D. teissieri*).[26] Restriction analysis done on the whole mitochondrial genome, in contrast, led to an estimate of 0.22%.[27] Sequencing of one-seventh of these fly mitochondrial DNAs produced an estimate of 0.3%,[27] similar to the latter indirect estimate but in dramatic contrast to the estimate from hybridization.

The approach presented in this chapter is, in principle, applicable to any indirect method of estimating genetic distance (e.g., protein electrophoresis, DNA hybridization, restriction fragment analysis, restriction mapping, and even amino acid sequencing as an estimator of the degree of divergence in base sequence). All that is needed to apply the criterion described here is a calibration between the indirect and the more direct measure. At present, immunological distance stands almost alone in having been extensively calibrated, with 264 pairs of proteins represented in Table I.

Acknowledgments

We thank R. P. Ambler, A. B. Champion, S. V. Edwards, R. L. Honeycutt, B. N. Jones, J. L. Meuth, M. F. Perutz, V. M. Sarich, M. Stoneking, and T. J. White for materials, unpublished data, and helpful discussion, and we thank the National Science Foundation and the National Institutes of Health for financial support.

[25] C. R. Marshall, *Mol. Biol. Evol.* **8**, 386 (1991).
[26] A. Caccone, G. D. Amato, and J. R. Powell, *Genetics* **118**, 671 (1988).
[27] M. Monnerot, M. Solignac, and D. R. Wolstenholme, *J. Mol. Evol.* **30**, 500 (1990).

[12] Isolation and Comparison of Nucleic Acids from Land Plants: Nuclear and Organellar Genes

By Diana B. Stein

Introduction

Isolation and comparisons of nucleic acids are being carried out on an ever increasing number of plant species. For many of these taxa, the procedures described previously for nuclear DNA,[1,2] chloroplast DNA,[3] and mitochondrial DNA[4-6] have been invaluable. An additional, new source detailing many aspects of nucleic acid comparisons is highly recommended.[7] This chapter describes protocols that we have found useful with plants containing many secondary metabolites. In addition, it summarizes procedures from other laboratories that may be helpful in future studies involving nucleic acid comparisons.

Isolation of DNA

Because plants produce a wide variety of substances such as tannins, mucilages, and numerous secondary compounds, a number of procedures have been designed to simplify the isolation process, to increase DNA yield, or to make possible DNA isolation from a number of recalcitrant species. No one isolation method is optimal for all taxa. Smith *et al.* compared five different extraction procedures for total DNA by using the resulting DNA in a Southern hybridization assay.[8] The first protocol involves cell breakage, collection of organelles, and cesium chloride centrifugation.[9] The remainder are modifications of procedures[10,11] using hex-

---

[1] J. C. Watson and W. F. Thompson, this series, Vol. 118, p. 57.

[2] C. Rivin, this series, Vol. 118, p. 75.

[3] J. D. Palmer, this series, Vol. 118, p. 167.

[4] M. R. Hanson, M. L. Boeshore, P. E. McClean, M. A. O'Connell, and H. T. Nivison, this series, Vol. 118, p. 437.

[5] D. M. Lonsdale, T. P. Hodge, and P. J. Stoehr, this series, Vol. 118, p. 453.

[6] A. J. Dawson, V. P. Jones, and C. J. Leaver, this series, Vol. 118, p. 470.

[7] D. M. Hillis and C. Moritz (eds.), "Molecular Systematics." Sinauer, Sunderland, Massachusetts, 1990.

[8] J. F. Smith, K. J. Sytsma, J. S. Shoemaker, and R. L. Smith, *Phytochem. Bull.* **23**, 2 (1991).

[9] E. A. Zimmer, C. J. Rivin, and V. Walbot, *Plant Mol. Biol. Newsl.* **2**, 93 (1981).

[10] M. A. Saghai-Maroof, M. Solimank, R. A. Jorgensen, and R. W. Allard, *Proc. Natl. Acad. Sci. U.S.A.* **81**, 8014 (1984).

[11] J. J. Doyle and J. L. Doyle, *Phytochem. Bull.* **19**, 11 (1987).

METHODS IN ENZYMOLOGY, VOL. 224

adecyltrimethylammonium bromide (CTAB). Hybridization of blots representing eight species isolated by each of the five methods to probes of chloroplast DNA (cpDNA) or nuclear ribosomal DNA (nrDNA) revealed that no single method was best for all species. In two cases (*Bambusa* and *Columnea*) DNA isolated by cesium chloride centrifugation from collected organelles did not give any signal when hybridized to the cpDNA probe, but in two other instances (*Fuchsia* and *Impatiens*) collecting organelles prior to lysis was required to produce DNA that would successfully hybridize with cpDNA. In our work on ferns, we have often found that isolation methods which were successful for many angiosperms were unsatisfactory when applied to the fern taxa we examined. Therefore, we first describe the isolation procedures for total and chloroplast DNA that have worked well in our laboratory on a large number of ferns and other taxa. We then summarize a number of the methods that have been published.

Chloroplast DNA Isolation

The following methodology is based on Palmer and Stein.[12]

1. Use fresh dark green leaf tissue and place leaves in the dark for 1 to 3 days. If desired, check for optimal dark treatment length by doing a time course of starch grain loss as follows. Boil a piece of leaf in 95% (v/v) ethanol, rehydrate, and stain with iodine–potassium iodide (IKI) for starch content. Examine a thin section under the microscope.

2. Rinse and dry leaves. Place each lot of 20 g tissue (use 80 to 100 g total) in 400 ml ice-cold buffer [10% (w/v) polyethylene glycol (PEG 3350, Sigma, St. Louis, MO), 0.35 M sorbitol, 50 mM Tris-HCl, pH 8.0, 5 mM EDTA, 0.1% (w/v) bovine serum albumin (BSA) (Sigma, A-4503), and 0.125% (v/v) 2-mercaptoethanol; the last component is added just before use] and hold at 4° on ice for 15 min. Work at 4° for Steps 1–4 and keep the preparation on ice or at 4° until Step 12.

3. Homogenize tissue for 7 sec in a blender and then with a Polytron (Brinkman, Westbury, NY) at high speed twice for 12 sec each.

4. Filter through two layers cheesecloth and one layer Miracloth (Calbiochem, La Jolla, CA) supported on a large powder funnel resting on a 1-liter beaker. Squeeze gently or not at all.

5. Collect organelles from the filtrate by centrifugation at 1000 g for 15 min.

6. Resuspend pellets with camel's hair artist's brush in approximately 8 ml in PEG wash buffer [10% (w/v) PEG 3350, 0.35 M sorbitol, 50 mM Tris-HCl, pH 8.0, and 25 mM EDTA]. Half fill the centrifuge bottle with wash buffer, swirl gently, and pellet organelles at 1000 g for 8 min.

---

[12] J. D. Palmer and D. B. Stein, *Curr. Genet.* **5**, 165 (1982).

7. Resuspend the pellets by brushing in 8 ml of PEG wash buffer and apply to sucrose step gradients. Gradient solutions are 10% (w/v) PEG 8000 (Sigma), 50 mM Tris-HCl, pH 8.0, and 25 mM EDTA containing 30 or 50% (w/v) sucrose. Each gradient is composed of 18 ml of 50% sucrose overlaid with 7 ml of 30% sucrose solution. The gradients are prepared 1 to 2 days earlier to allow diffusion, or one can use them directly by forming a third step of 40% between the 30 and 50% layers by mixing together 1 ml of each of the 30 and 50% sucrose solutions. Six gradients are used for approximately 80 to 100 g tissue.

8. Centrifuge for 1 hr at 25,000 rpm using a Beckman (Palo Alto, CA) SW-27 rotor.

9. Use a serological pipette to remove the chloroplasts from the 30 to 50% sucrose interface. Do not include any aggregates adhering to the wall of the centrifuge tube. The nuclei pellet through the 50% sucrose layer and can be separately recovered from the bottom of the tube.

10. Dilute the combined recovered chloroplasts slowly (over 10 to 15 min) with PEG wash buffer until the bottle is two-thirds full, then centrifuge at 1000 g for 15 min.

11. Resuspend the pellet by brushing in 8 ml of PEG wash buffer, half filling the centrifuge bottle, and centrifuge at 1000 g for 10 min. Repeat Step 9, if possible (depends on size of pellet) two more times. Each wash removes some of the contaminating nuclear DNA that may be present.

12. Prepare cesium chloride gradients by draining the pellet well. Use a Kimwipe tissue to blot excess buffer. Add 5 ml of a suspension buffer (prepare buffer as in Step 6 but omit PEG) and suspend chloroplasts at 4°. Transfer the chloroplasts to a 50-ml conical tube (with gradations) that has a cap. Use additional suspension buffer for washing and transferring to give a final volume of 16 ml. Add 4.8 ml of 10 mg/ml self-digested pronase (Calbiochem) and let sit at room temperature for 5 min. Add 4.8 ml of lysis buffer [5% (w/v) sodium sarcosinate 50 mM Tris-HCl (pH 8.0), and 25 mM EDTA] and incubate at 37° for 15 to 30 min. Add 28.1 g of cesium chloride (technical grade, Cabot Corporation, Revere, PA) ground to a powder with a mortar and pestle. Adjust the volume to 37.8 ml. When the CsCl is dissolved, add 1.2 ml of ethidium bromide (10 mg/ml), transfer to a 39-ml heat seal tube (Beckman), and seal.

13. Centrifuge at 44,000 rpm in a VTi 50 rotor (Beckman) at 20° for at least 12 hr.

14. Visualize the DNA band with a long-wavelength UV lamp. Remove the DNA to 5-ml heat seal tubes, fill the tubes with CsCl–ethidium bromide solution (1.55 g/ml) and reband at 58,000 rpm for 6 hr or 44,000 rpm for 12 hr at 20° in a VTi 65 rotor (Beckman).

15. Remove the DNA band. Extract 3 times with 2-propanol saturated

with a saturated solution of NaCl. Centrifuge at 4000 g for 5 min at room temperature to separate the phases. Transfer the DNA (bottom phase) to dialysis tubing and dialyze against 10 mM Tris-HCl (pH 8.0), 10 mM NaCl, and 0.1 mM EDTA with four changes over 2 days.

Total DNA Extraction: Large Amounts of Plant Tissue

For extracting total DNA from large amounts of plant tissue, use the above procedure, omitting the sucrose gradients (i.e., do Steps 1–6 and 10–15). After the 2-propanol extraction and dialysis for 2 hr or longer, use an equal volume of Tris-buffered phenol [prepared just before use by adding 1/10 volume of unbuffered Tris to 1 volume of water-saturated phenol (IBI, New Haven, CT)] to extract DNA twice, followed by an ether extraction. The DNA is then dialyzed as above.

The amount of tissue can be increased to 25–35 g in 400 ml buffer. The time of homogenization may be lengthened to 2 times 20 sec or longer since some nuclear breakage is acceptable and the additional grinding may increase the number of organelles liberated. To collect mitochondria, a high-speed spin (16,500 g for 20–45 min) should be added after Steps 5 and 6, and the pellet obtained should be added to the collected nuclei and chloroplasts.

Yield and DNA Quality. We generally have available large amounts of leaf tissue for our extractions, and find that if breakage is adequate we typically get good yields of total DNA (84–1920 μg from ~30 g fresh weight leaf tissue) that give strong hybridization signals with nrDNA or cpDNA probes. Yields of chloroplast DNA range from approximately 18 to 60 μg from 25 g fresh weight of tissue; these are naturally approximate values as some residual nuclear DNA is generally present.

Many of the plant compounds that might create difficulty in obtaining intact organelles are diluted out in the large volume of extraction buffer used. Moreover, the presence of high concentrations of sorbitol (or any sugar) in the extraction medium appears to inhibit polyphenol oxidase (catechol oxidase) and prevent browning. Although, cesium chloride centrifugations and phenol extractions take time, they do permit us to process shipments of ferns from the topics by a single method and to be reasonably certain that the method will work with most members of a study group. For the same reasons, we prefer not to precipitate our high molecular weight DNA as we have found that this step sometimes renders the DNA unrestrictable. Therefore, we dialyze our larger DNA preparations. Despite these precautions, we have encountered a few species in which the DNA may not be digestible (1–2%) or in which the leaves contain a material with detergentlike qualities that causes lysis of the organelles during homogenization (one species). In addition to the few species that present special problems, this method of DNA isolation is not suitable when only small amounts of tissue are available.

Total DNA Extraction: Small Amounts of Plant Tissue

For small amounts of tissue such as individual fern gametophytes, we have used a method employing CTAB. An early method[13] used dilution of sodium chloride to precipitate the CTAB-solubilized DNA; since then, 2-propanol[10,11] and ethanol[14] have been utilized. We use the latter procedure (which the authors show has been successful with a wide variety of crop plants as well as the deodar, *Cedrus deodara*), but we find that grinding the frozen tissue in liquid nitrogen is easier than grinding in dry ice. We also band the DNA isolated by this method in a 5 ml CsCl–ethidium bromide gradient (1.55 g/ml) by spinning at 58,000 rpm for 6 hr using a VTi 65 rotor (Beckman) to remove any trace impurities that might persist.

Summary of Other Extraction Methods. As noted earlier, many species contain compounds that may interfere with certain protocols. Table I shows some methods[15-34] for isolating nucleic acids that may prove useful for difficult species or provide other advantages. In general, if DNA from a

[13] M. G. Murray and W. F. Thompson, *Nucleic Acids Res.* **8**, 4321 (1980).

[14] S. O. Rogers and A. J. Bendich, *in* "Plant Molecular Biology Manual" (S. B. Gelvin and R. A. Schilperoort, eds.), A6: p. 1. Kluwer Academic Publishers, Boston, 1988.

[15] G. S. Varadarajan and C. S. Prakash, *Plant Mol. Biol. Rep.* **9**, 6 (1991).

[16] A. de Kochko and S. Hamon, *Plant Mol. Biol. Rep.* **8**, 3 (1990).

[17] J. A. Couch and P. J. Fritz, *Plant Mol. Biol. Rep.* **8**, 8 (1990).

[18] T. H. Tai and S. D. Tanksley, *Plant Mol. Biol. Rep.* **8**, 297 (1990).

[19] I. J. Mettler, *Plant Mol. Biol. Rep.* **5**, 346 (1987).

[20] D. M. Webb and S. J. Knapp, *Plant Mol. Biol. Rep.* **8**, 180 (1990).

[21] S. S. Baker, C. L. Rugh, and J. C. Kamalay, *BioTechniques* **9**, 268 (1990).

[22] J. C. Kamalay, R. Tejwani, and G. K. Rufener II, *Crop Sci.* **30**, 1079 (1990).

[23] C. Perez, J.-F. Bonavent, and A. Berville, *Plant Mol. Biol. Rep.* **8**, 104 (1990).

[24] C. M. Bowman and T. A. Dyer, *Anal. Biochem.* **122**, 108 (1982).

[25] A. M. Dally and G. Second, *Plant Mol. Biol. Rep.* **7**, 135 (1989).

[26] P. J. Calie and K. W. Hughes, *Plant Mol. Biol. Rep.* **4**, 206 (1987).

[27] J. R. Marienfeld, R. Raski, C. Friese, and W. O. Abel, *Plant Sci.* **61**, 235 (1989).

[28] L. Charbonnier, C. Primard, P. Leroy, and Y. Chupeau, *Plant Mol. Biol. Rep.* **4**, 213 (1987).

[29] E. E. White, *Plant Mol. Biol. Rep.* **4**, 98 (1986).

[30] G. Mourad and M. L. Polacco, *Plant Mol. Biol. Rep.* **6**, 193 (1988).

[31] A. Pay and M. A. Smith, *Plant Cell Rep.* **7**, 96 (1988).

[32] B. G. Milligan, *Plant Mol. Biol. Rep.* **7**, 144 (1989).

[33] C. L. Hsu and B. C. Mullin, *Plant Cell Rep.* **7**, 356 (1988).

[34] T. Kiss and F. Solymosy, *Acta Biochim. Biophys. Hung.* **22**, 1 (1987).

TABLE I
METHODS FOR ISOLATING DNA FROM PLANT TAXA

Plant group[a]	Extraction method	Advantage	Ref.
Total DNA			
Ipomea batatas (L.) Lank (sweet potato)	Modification of Dellaporta[b]; uses leaves	DNA prepared from tissue with high levels secondary metabolites, mucilaginous materials, proteins, polysaccharides	15
Abelmoschus (okra)	Modification of Dellaporta[b]; adds RNase plus organic extraction and uses cotyledons	Bypasses sticky polysaccharides in green leaves, low in cpDNA	16
Theobroma cacao (cocoa)	Extraction medium contains polyvinylpyrrolidone and diethyldithiocarbamic acid; uses leaves	DNA isolated from tissue high in polyphenolics	17
Rice, tomato	Modification of Dellaporta[b]; uses dehydrated tissue	Easier tissue breakage for handling large number of samples	18
Nicotiana tabacum, N. plumbaginifolium, Helianthus annuus, Zea mays	Minipreparation for protoplasts or cultured cells; phenol extraction of sarkosyl-lysed cells	Rapid; uses small amount of tissue (25–1000 mg)	19
Cuphea lanceolata Ait., *Cuphea viscosissima* Jacq.	CTAB extraction with chloroform and phenol washes	Separates DNA from sticky, resinous substance	20
Gossypium hirsutum	Leaves	Works with wide variety of recalcitrant tissues; yields DNA and RNA	21
Pinus strobus	Seedlings, needles, ovules, seeds, callus		
Pinus nigra	Seeds		
Pinus palustris	Seeds		
Picea abies	Callus		
Juniperus virginica	Needles		
Cycas sp.	Fronds		
Ginkgo biloba	Leaves		
Glycine max	Seeds		
	Uses tissues shown, extraction buffer for RNA, and CsCl centrifugations		
Glycine max Merr. (soybean), *Phaseolus vulgaris* L. (French bean)	Single seeds, ground, defatted with chloroform and extracted with CTAB	High yields from single seeds	2

(continue

TABLE I *(continued)*

Plant group[a]	Extraction method	Advantage	Ref.
Organelle DNA			
Beta maritima, Helianthus annuus, Helianthus petiolaris	Etiolated and mature leaves; uses high ionic strength buffer, high concentration of mercaptoethanol, PEG 6000, and chloroform	Isolates both mtDNA and cpDNA; yields restrictable DNA from leaves	23
Oryza spp. (rice)	Nonaqueous method, modified from Bowman and Dyer[24]; uses leaves	Reduces nuclear DNA background of cpDNA	25
Physcomitrella patens (moss)	Cultures gametophyte tissue; sucrose and CsCl gradients	Pure cpDNA; other methods failed	26
Physcomitrella patens (moss)	Cultures gametophyte tissue; Percoll gradients	Methods for nuclear DNA, cpDNA, and mtDNA	27
Tobacco, French bean, rapeseed	Prepares protoplasts; differential centrifugation; phenol extraction	mtDNA and cpDNA from small amounts tissue	28
Pinus manticola, conifers	PEG 6000 in all buffers, sucrose gradients; uses needles	cpDNA from difficult species	29
Zea mays and mutant	3 days etiolation and DNase treatment; uses leaves	Purer cpDNA	30
Nicotiana tabacum, Carthamus tinctorius L., *Nicotiana–Salpiglossus* cybrids, *Salpiglossus sinuata, Daucus carota* L.	Differential centrifugation, self-generated Percoll gradients; uses leaves	Yields cpDNA and mtDNA	31
Psilotum nudum, Selaginella sp., *Equisetum arvense, Aquilegia* sp., *Menthaspicata, Trifolium* spp.	Uses high salt buffer plus CTAB	Efficient for cpDNA from broad taxonomic range	32
Gossypium barbadense, G. hirsutum (cotton)	Alterations in ionic strength of media; permitted isolation in presence of high levels of phenols	mtDNA purity greatly increased	33
Vicia faba (broad bean)	Uses citric acid isolation medium, plus Percoll cushion to isolate nuclei	Yields high molecular weight nuclear DNA	34

[a] Plant taxa are listed as cited in the published paper.
[b] S. J. Dellaporta, *Plant Mol. Biol. Rep.* **1,** 19 (1983).

given species has been difficult to purify, a combination of procedures is more likely to be successful. For example, organic extractions with or without protease treatments or banding of DNA in cesium chloride are often useful additions to a rapid isolation protocol. In one instance,[35] DNAs from a wide variety of mucilaginous angiosperms were successfully isolated by a modification of the method of Dally and Second.[25] Lyophilized, powdered plant tissue was passed through density step gradients of inert organic solvents using centrifugation to partially separate organelles from the contaminating polysaccharides. In a second case, residual amounts of acidic polysaccharides that inhibit the activity of restriction enzymes were removed[36] using Elutip-d (RPC-5 type resin, Schleicher and Schuell, Keene, NH). The methods described in Table I yield total cellular DNA either because detergent is used prior to any cellular fractionation[37] or because organelles are collected and then lysed. Methods that collect organelles at 1000 g are likely to yield primarily mixed chloroplast and nuclear DNAs. An additional, higher speed centrifugation at 16,500 g for 20 min or longer is clearly necessary to collect most of the mitochondria.

Comparisons of Nucleic Acids

Southern Hybridization for Restriction Fragment Comparisons

Blot Preparation. DNA (for purified cpDNA, 0.2–0.75 μg; for cpDNA studies that use total DNA, ~1.5 μg; and for single-copy gene analysis, 10 μg or more) is digested with restriction enzymes that recognize 4 to 6 bases, according to the manufacturer's instructions. See Palmer[3] and Dowling *et al.*[38] for the number of fragments produced by digestion of organellar DNAs with specific enzymes. For cpDNA studies that use total DNA (this is the most efficient way to make restriction site comparisons of large numbers of taxa) a minimum of 6 units (U) of enzyme and 5 hr of digestion are used for enzymes that generate few fragments and up to 12 U for those that cut more frequently. The DNA fragments are separated on agarose gels in 0.1 M Tris, 12.5 mM sodium acetate, 10 mM EDTA, pH 8.1, using 0.7% gels for less frequent cutting enzymes and increasing percentages of agarose (up to 1.5%) for enzymes cutting more frequently. Electrophoresis is carried out overnight until the bromphenol blue marker dye has migrated 18 cm. The gel is stained for 10 min with 0.5 μg/ml ethidium bromide, destained for 30 min or longer, and photographed. For complete details, see Palmer[3] and Dowling *et al.*[38]

[35] R. S. Wallace, personal communication (1991).
[36] N. Do and R. P. Adams, *BioTechniques* **10**, 162 (1991).
[37] S. J. Dellaporta, *Plant Mol. Biol. Rep.* **1**, 19 (1983).
[38] T. E. Dowling, C. Moritz, and J. D. Palmer, *in* "Molecular Systematics" (D. M. Hillis and C. Moritz, eds.), p. 250. Sinauer, Sunderland, Massachusetts, 1990.

The gel is then marked with India ink to designate any lanes that will be removed for separate hybridization, orientation, etc., trimmed, and prepared for blotting. We use a modified procedure for alkaline blotting[39] developed by the manufacturer of Zetabind (Cuno Inc., a unit of Commercial Intertech Corp., Meriden, CT). The gel, on a plastic plate in a tray, is covered with 0.25 M HCl and placed on a rotary shaker for 10 min. The gel is rinsed with distilled water and transferred to 0.4 N NaOH/0.6 M NaCl for 30 min of gentle shaking. The blotting membrane is cut to size, wetted in water, and then soaked in 20× SSC (3 M NaCl, 0.3 M sodium citrate, pH 7.4) for 20 min. For single-sided blots, the gel is eased from the plate, smooth side up, onto a flat surface covered with plastic wrap. The membrane is lowered onto the gel, and any bubbles are rolled out with a pipette, taking care not to move the membrane. Three layers blotting paper moistened with 0.4 N NaOH, a 2-inch stack of paper towels, a flat plate, and a light weight are added sequentially. The transfer is allowed to proceed overnight. Marks are transferred from the gel to the membrane while it is still in place on the gel using an indelible ballpoint pen. The membrane is then washed 3 times with 2× SSC for 10 min each time, air-dried, and baked for 1 hr at 80° in a vacuum oven. Reference marks on the blot are then darkened with an indelible, felt-tip pen. We have used double-sided blots[40] when multiple replicas were required.[12] We[41] also found the procedures of Klessig and Berry[42] for blotting and hybridization were valuable for detecting the napin gene in fern genomic DNA.

Southern Hybridization. Our hybridization procedures are essentially those of Palmer.[3] New blots are washed for 1 hr at 65° in 0.1× SSC, 0.5% (w/v) sodium dodecyl sulfate (SDS); blots that have been used previously are treated with 0.4 N NaOH at 40° for 30 min and rinsed 3 times with 2× SSC for 10 min each time. If a large number of blots are to be hybridized simultaneously, 10 can be sealed into one bag and 50 ml of hybridization buffer added. Hybridization buffer is 4× SSC, 10 M EDTA, 5× Denhardt's [50× Denhardt's is 1.0% (w/v) BSA, 1.0% (w/v) Ficoll 40 (Sigma), and 1% (w/v) polyvinylpyrrolidone (PVP type 400, Sigma)], 0.5% (w/v) SDS, and 25 μg/ml of calf thymus DNA sonicated and freshly boiled. All except the last component are Millipore (Bedford, MA) filtered before the DNA is added. For heterologous probes we prehybridize and hybridize at 55°; for homologous or highly conserved DNA we use 65°. We prehybridize for a minimum of 2 hr, then replace the buffer with fresh solution. The denatured probe DNA (see below) is diluted in 5 ml buffer and added. Probe concentrations should not exceed 50 to 100 ng/ml for buffers that do

[39] K. C. Reed and D. A. Mann, *Nucleic Acids Res.* **13**, 7207 (1985).
[40] G. E. Smith and M. D. Summers, *Anal. Biochem.* **109**, 123 (1980).
[41] T. S. Templeman, D. B. Stein, and A. E. DeMaggio, *Biochem. Genet.* **26**, 595 (1988).
[42] D. Klessig and J. Berry, *Plant Mol. Biol. Rep.* **1**, 12 (1983).

not contain dextran sulfate and should be only 10 to 20 ng/ml when dextran sulfate is used. Homologous probes are hybridized overnight, but heterologous probes are hybridized for 40 hr. Blots are then washed free of unhybridized probe DNA with $2\times$ SSC, 0.5% (w/v) SDS for 10 min at room temperature, followed by 4 washes of 30 min each at the temperature used for hybridization and a final 20-min wash at room temperature in $2\times$ SSC.

The blots are drained well and wrapped while damp in plastic wrap, and an estimate is made of the strength of the signal using a Geiger counter and a label attached. We have found it convenient for the labeling of large numbers of blots to make reusable labels; the same specific symbol is used for each blot every time it is sequentially hybridized. Fluorescent paint is applied to a piece of plastic, allowed to dry, and covered with clear tape. The label is then taped to the blot, which is exposed to bright room light prior to being placed in a labeled cassette (containing an intensifying screen if needed). In a darkroom, film is placed between the screen and the blot. The paint fluoresces sufficiently to expose the film, thereby labeling each autoradiograph. Screens should be kept in the dark prior to addition of film as they can also fluoresce and cause the formation of a blurred image on the film. Cassettes are placed at $-70°$ for up to 2 weeks.

Probe Preparation. We label probe DNA by nick translation for any probe that is larger than about 2000 base pairs (bp); since normally we do not purify cloned cpDNA or nrDNA from the vector, most of our probes are nick translated. Nick translation reactions are assembled as follows: 2.5 μl buffer [0.5 M Tris, pH 7.2, 0.1 M MgSO$_4$, 10 mM dithiothreitol (DTT), 0.5 mg/ml BSA], 1.3 μl of a mixture of 1 mM dGTP, dTTP, and dCTP, 1 μl DNase I (0.1 μg/ml), and sample DNA and distilled water such that the final volume minus 1 μl DNA polymerase I (10 U/μl) and 1–10 μl [α-^{32}P]dATP will equal 25 μl. Prior to the addition of the latter two components, we incubate the DNA at 37° for 15 min[43]; this reduces background caused by high molecular weight DNA.[44] Depending on the size of the probe and the activity of the DNase, this step may not be necessary. The reaction mixture is then cooled, the DNA polymerase and [α-^{32}P]dATP are added, and the reaction is incubated at 15.5° for 90 min. Unincorporated deoxynucleoside triphosphates are removed by spun column chromatography exactly as described[43] except that our STE buffer contains 0.2% (w/v) SDS. For labeling polymerase chain reaction (PCR) products and cloned fragments isolated from vector plasmid (fragments

[43] J. Sambrook, E. F. Fritsch, and T. Maniatis, "Molecular Cloning: A Laboratory Manual," 2nd Ed. Cold Spring Harbor Laboratory, Cold Spring Harbor, New York, 1989.
[44] J. Meinkoth and G. Wahl, *Anal. Biochem.* **138,** 267 (1984).

smaller than 2000 bp), we use random priming kits such as Prime Time C Biosystem (IBI) or Random Prime DNA Labeling Kit (Boerhringer-Mannheim, Indianapolis, IN). All probe DNAs are denatured in microcentrifuge tubes for 5 min at 100° using a heat block and held on ice until diluted and used (see above).

Choice of Probes. The sequence chosen for analysis will be determined by the nature of the evolutionary comparison to be carried out. Examples of probe DNAs used for examining populational and species level comparisons, as well as for higher level systematics, are described by Dowling *et al.*[38] Table II provides some of the cloned sequences (or techniques to obtain them)[45-75] that have the potential to be useful in evolutionary studies. The list is not meant to be comprehensive but representative; there

[45] S. H. Rogstad, J. C. Patton, and B. A. Schaal, *Proc. Natl. Acad. Sci. U.S.A.* **85,** 9176 (1988).
[46] B. S. Landry and R. W. Michelmore, *Plant Mol. Biol. Rep.* **3,** 174 (1985).
[47] J. G. K. Williams, A. R. Kubelik, K. J. Livak, J. A. Rafalski, and S. V. Tingey, *Nucleic Acids Res.* **18,** 6531 (1991).
[48] S. V. Evola, F. A. Burr, and B. Burr, *Theor. Appl. Genet.* **71,** 765 (1986).
[49] B. S. Landry, R. Kesseli, H. Leung, and R. W. Michelmore, *Theor. Appl. Genet.* **74,** 646 (1987).
[50] R. Bernatzky and S. D. Tanksley, *Genetics* **112,** 887 (1986).
[51] D. M. Shah, R. C. Hightower, and R. B. Meagher, *Proc. Natl. Acad. Sci. U.S.A.* **79,** 1022 (1982).
[52] W. L. Gerlach, A. J. Pryor, E. S. Dennis, R. J. Ferl, M. M. Sachs, and W. J. Peacock, *Proc. Natl. Acad. Sci. U.S.A.* **79,** 2981 (1982).
[53] R. Broglie, G. Bellemare, S. G. Bartlett, N.-H. Chua, and A. R. Cashmore, *Proc. Natl. Acad. Sci. U.S.A.* **78,** 7304 (1981).
[54] W. F. Thompson and M. J. White, *Annu. Rev. Plant Physiol. Plant Mol. Biol.* **42,** 423 (1991).
[55] A. J. Delauney and D. P. S. Verma, *Plant Mol. Biol. Rep.* **6,** 279 (1988).
[56] J. N. M. Mol, T. R. Stuitje, A. G. M. Gerats, and R. E. Koes, *Plant Mol. Biol. Rep.* **6,** 274 (1988).
[57] R. A. Jorgensen, R. E. Cuellar, W. F. Thompson, and T. A. Kavanagh, *Plant Mol. Biol.* **8,** 3 (1987).
[58] E. A. Zimmer, E. R. Jupe, and V. Walbot, *Genetics* **120,** 1125 (1988).
[59] C. Dean, E. Pichersky, and P. Dunsmuir, *Annu. Rev. Plant Physiol. Plant Mol. Biol.* **40,** 415 (1989).
[60] K. O. Elliston, S. Imran, and J. Messing, *Plant Mol. Biol. Rep.* **6,** 22 (1988).
[61] J. H. Weil, *Plant Mol. Biol. Rep.* **6,** 30 (1988).
[62] M. J. Guiltinan, D. P. Ma, R. F. Barker, M. M. Bustos, R. J. Cyr, R. Yadegari, and D. E. Fosket, *Plant Mol. Biol.* **10,** 171 (1987).
[63] L. Montoliu, J. Rigau, and P. Puigdomenech, *Plant Mol. Biol.* **14,** 1 (1989).
[64] D. M. Lonsdale, *Plant Mol. Biol. Rep.* **6,** 266 (1988).
[65] B. M. Winning, B. Bathgate, P. E. Purdue, and C. J. Leaver, *Nucleic Acids Res.* **18,** 5885 (1990).
[66] E. K. Kaleikau, C. P. Andre, B. Doshi, and V. Walbot, *Nucleic Acids Res.* **18,** 372 (1990).
[67] E. K. Kaleikau, C. P. Andre, and V. Walbot, *Nucleic Acids Res.* **18,** 371 (1990).
[68] E. K. Kaleikau, C. P. Andre, and V. Walbot, *Nucleic Acids Res.* **18,** 370 (1990).

TABLE II
CLONED SEQUENCES OF POTENTIAL USE FOR STUDIES OF PLANT MOLECULAR EVOLUTION

Organelle	Sequence type	Comment or species sequence from	Ref.
Nuclear	Hypervariable regions	Bacteriophage M13	45
	Sequences of unknown function	How to clone them	46
	Sequences of unknown function	How to amplify them by PCR	47
	Random genomic fragments; cDNA clones	*Zea mays* (maize)	48
	Random genomic fragments; cDNA clones	*Lactuca sativa* (lettuce)	49
	Random cDNA clones	*Lycopersicon esculentum, L. pennellii* (tomato)	50
	Actin	Soybean	51
	*Adh*1, the alcohol dehydrogenase gene	Maize	52
	Chlorophyll *a/b*-binding protein	Pea	53
	Light-regulated genes	Many	54
	Nodulin genes	Many	55
	Phenylpropanoid metabolism genes	Many	56
	rDNA	Pea	57
	rDNA	Soybean	58
	Small subunit of ribulose-bisphosphate carboxylase	Several	59
	Storage proteins	Many	60
	tRNA	*Phaseolus vulgaris*	61
	Tubulin	Soybean	62
	Tubulin	Maize	63
Mitochondrial	Summary of many genes	Many	64
	ATP synthase	Maize	65
	Apocytochrome *b*	Rice	66
	Cytochrome oxidase subunit 3	Rice	67
	F_o-ATPase (*atp* 9)	Rice	68
	tRNA genes	*Triticum aestivum*	61
Chloroplast	Summary of complete and partial clone banks	Many	3, 6
	Clone bank	Tobacco, small subclones (dicot), completely sequenced	70
	Clone bank	*Oncidium excavatum* (monocot)	71
	Clone bank	*Oryza sativa* (rice, monocot) completely sequenced	72
	Clone bank	*Pseudotsuga menziesii* (Douglas fir)	73
	Clone bank	*Adiantum capillus-veneris* (fern)	74
	Clone bank	*Marchantia polymorpha* (bryophyte) completely sequenced	75
	tRNA genes	*Vicia faba, Triticum aestivum*	61

is a large number of other newly cloned and/or sequenced genes. Also, although several complete mitochondrial genomes have been cloned, they have been omitted from Table II since the extensive rearrangements undergone by most plant mitochondrial genomes[76,77] makes these clone banks of more limited use.

Data Analysis. The sizes of the fragments are determined by reference to the molecular weight markers. The analysis of restriction site data is described by Dowling *et al.*,[38] and the methods used to infer phylogenies from these data are detailed by Swofford and Olsen,[78] Holsinger and Jansen,[79] Lander,[80] and Hillis.[81]

Comments. Several factors affect the strength of signals obtained from Southern hybridization.[44] The amount of DNA left in the gel can be estimated by neutralizing the gel and restaining with ethidium bromide. A more definitive measure of transfer is the sensitivity of a blot to hybridization. For example, in our hands, dry blotting overnight transferred significantly more DNA than a Bios blotting unit (Bios Corp., New Haven, CT) of otherwise identically prepared blots as judged by Southern hybridization.

Single-sided blots should transfer more DNA more evenly to the membrane and are preferred in work when small heterologous probes are to be used and hence low signal is predicted. An example of this is use of the small, subcloned fragments of tobacco cpDNA as probes to map rearranged fern chloroplast DNA.[70,82] For blots that are intended to be

---

[69] J. D. Palmer, R. K. Jansen, H. J. Michaels, M. W. Chase, and J. R. Manhart, *Ann. Mo. Bot. Gard.* **75,** 1180 (1988).

[70] R. G. Olmstead and J. D. Palmer, *Ann Mo. Bot. Gard.* **79,** 249 (1992).

[71] M. W. Chase and J. D. Palmer, *Am. J. Bot.* **76,** 1720 (1989).

[72] H. Shimada, R. F. Whittier, J. Hiratsuka, Y. Maeda, A. Hirai, and M. Sugiura, *Plant Mol. Biol. Rep.* **7,** 284 (1989).

[73] C.-H. Tsai and S. H. Strauss, *Curr. Genet.* **16,** 211 (1989).

[74] M. Hasebe and K. Iwatsuki, *Curr. Genet.* **17,** 359 (1990).

[75] K. Ohyama, H. Fukuzawa, T. Kohchi, H. Shirai, T. Sano, S. Sano, K. Umesono, Y. Shiki, M. Takeuchi, Z. Chang, S. Aota, H. Inokuchi, and H. Ozeki, *Nature (London)* **322,** 572 (1986).

[76] M. W. Gray, *Annu. Rev. Cell Biol.* **5,** 25 (1989).

[77] J. D. Palmer, C. A. Makaroff, I. J. Apel, and M. Shirzadegan, *in* "Molecular Evolution" (M. T. Clegg and S. J. O'Brien, eds.), p. 85. Alan R. Liss, New York, 1990.

[78] D. L. Swofford and G. J. Olsen, *in* "Molecular Systematics" (D. M. Hillis and C. Moritz, eds.), p. 411. Sinauer, Sunderland, Massachusetts, 1990.

[79] K. E. Holsinger and R. K. Jansen, this volume [33].

[80] M. J. Donoghue and M. J. Sanderson, *in* "Molecular Systematics of Plants" (P. S. Soltis, D. E. Soltis, and J. J. Doyle, eds.), p. 340. Chapman and Hall, New York, New York, 1992.

[81] D. M. Hillis, M. W. Allard, and M. M. Miyamoto, this volume [34].

[82] D. B. Stein, D. S. Conant, M. E. Ahearn, E. T. Jordan, S. A. Kirch, M. Hasebe, K. Iwatsuki, M. K. Tan, and J. A. Thomson, *Proc. Natl. Acad. Sci. U.S.A.* **89,** 1856 (1992).

serially hybridized, we use the least homologous probes first and finish with the more homologous probe DNAs to maximize signal strength.

Alkaline blotting[39] offers the advantage of efficiency but has been criticized[83] as producing blots that hybridize 10 times less efficiently than those transferred in 1 M ammonium acetate, 0.02 N NaOH (after two 15-min washes in 0.25 N HCl, two 20-min washes in 0.5 N NaOH, 1.5 M NaCl, and two 30-min washes in the transfer buffer). In both studies the membrane used was Zeta probe (Bio-Rad, Richmond, CA).

We compared the modified alkaline blotting procedure we use (see above) to the earlier method suggested for Zetabind (the latter uses 0.2 N NaOH, 0.6 M NaCl to denature the DNA in the gel, followed by neutralization in 0.5 M Tris, pH 7.5, 1.5 M NaCl) by Southern hybridization and found them equally good. We also tried blotting from 0.4 N NaOH using different sides of the membrane facing the gel and found that one side of the membrane gave a stronger hybridization signal than the other. We recommend that each batch of membranes be tested for sidedness. Because the alkaline blotting procedure we use is rapid and sufficiently sensitive to permit our blots to be probed 10 or more times, we have not tested alternative membranes or buffer systems. However, it has been reported that alkaline blotting does not work equally well with all membranes.[84] For example, blotting in 0.4 N NaOH gives poor results with Hybond-N (Amersham, Arlington Heights, IL) and Nytran (Schleicher and Schuell), but these membranes are satisfactory if several washes are added after the 0.4 N NaOH step [0.25 M Tris–acetate, 0.1 M NaCl, pH 8.0 (15 min) and 25 mM Tris–acetate, 10 mM NaCl, pH 8.0 (15 min)].

Exposure of DNA in gels to UV light to fragment the DNA prior to transfer or after transfer to cross-link the DNA to the membrane appears to reduce the ability of the DNA to hybridize.[39] Moreover, nylon membranes that were UV-treated to cross-link transferred DNA lost target DNA after repeated stripping of probe DNA and rehybridization.[85] Hence, we recommend baking the blot *in vacuo*.

If the target sequence is present on the filter in a low concentration, either because it is a low copy number sequence or because a particular group of taxa yield a low concentration of chloroplast DNA in a total DNA extract, a prehybridization/hybridization buffer that contains dextran sulfate can greatly enhance the rate of hybridization (up to 100-fold[44] when nick-translated probes are used) and permit detection of the target se-

[83] G. Rigaud, T. Grange, and R. Pictet, *Nucleic Acids Res.* **15,** 857 (1987).
[84] B. Kempter, P. Luppa, and D. Neumeier, *Trends Genet.* **7,** 109 (1991).
[85] S. A. Nierzwicki-Bauer, J. S. Gebhardt, L. Linkkila, and K. Walsh, *BioTechniques* **9,** 472 (1990).

quences. We have used the following buffer with blots that had very low concentrations of cpDNA: $4 \times$ SSC, $5 \times$ Denhardt's, 10 mM EDTA, pH 8.0, 0.5% (w/v) SDS, 500 μg/ml calf thymus DNA (sonicated and boiled), and 10% (w/v) dextran sulfate.

Because we use total DNA from many species on one filter and hybridize at a reduced stringency (55°) with cloned cpDNA which is not removed from the vector, we find the technique of Clark and Hanson[86] provides a useful control for occasional spurious bands. A small amount of very pure chloroplast DNA from one member of our study group is radioactively labeled and hybridized to every blot at a higher stringency (65°). A comparison of the films obtained from cloned cpDNA and from pure cpDNA determines if any bands are anomalous.

Other Methods of Nucleic Acid Comparisons

Other methods of comparing nucleic acids include DNA–DNA hybridization and sequencing. Several excellent descriptions of these methodologies are available. These include chapters on DNA–DNA hybridization by Werman et al.,[87] Deininger and co-workers,[88] and Springer and Britten.[89] Hillis et al.[90] have written an excellent overview of all aspects of sequencing. Chapters by Bachellerie and Qu on direct sequencing from RNA[91] and by Thomas and Kocher on sequencing of PCR-amplified DNA[92] are provided in this volume. Techniques that facilitate cloning and/or amplification of plant DNAs specifically are described by Rogstad in this volume.[93]

Acknowledgments

I am grateful to A. E. C. Valinski and O. L. Stein for suggestions concerning the manuscript. The work was supported by a grant from the National Science Foundation to D. B. Stein and D. S. Conant (BSR-8818459).

[86] E. M. Clark and M. R. Hanson, *Plant Mol. Biol. Rep.* **1**, 77 (1983).
[87] S. D. Werman, M. S. Springer, and R. J. Britten, *in* "Molecular Systematics" (D. M. Hillis and C. Moritz, eds.), p. 204. Sinauer, Sunderland, Massachusetts, 1990.
[88] M. A. Batzer, C. W. Schmid, and P. L. Deininger, this volume [16].
[89] M. S. Springer and R. J. Britten, this volume [17].
[90] D. M. Hillis, A. Larson, S. K. Davis, and E. A. Zimmer, *in* "Molecular Systematics" (D. M. Hillis and C. Morits, eds), p. 318. Sinauer, Sunderland, Massachusetts, 1990.
[91] J. P. Bachellerie and L.-H. Qu, this volume [25].
[92] W. K. Thomas and T. D. Kocher, this volume [28].
[93] S. H. Rogstad, this volume [20].

[13] Isolation of DNA from Eukaryotic Algae

By Joby Marie Chesnick and Rose Ann Cattolico

Introduction

Autotrophic eukaryotes that inhabit aquatic environments have existed since the late Precambrian (900 million years ago). This long evolutionary history has culminated in a highly specialized and complex assemblage of photosynthetic organisms that display great diversity in morphological, biochemical, and genetic profiles. The algae are critical to the maintenance of a balanced global ecosystem. These plants are the anchor of the oceanic food chain, fix half of the carbon that is processed on earth, and form both symbiotic and commensal associations with other biota.

Historically, the algae are categorized into three major taxa based on pigmentation. The Rhodophyta (chlorophyll *a*, phycobilin-containing), Chromophyta (chlorophyll *a c*-containing), and Chlorophyta (chlorophyll *a b*-containing) represent major superdivisions whose status has been confirmed by chloroplast 16 S rDNA nucleotide sequence analysis.[1] Each superdivision contains a wide array of morphologically diverse representatives that include unicellular, colonial, filamentous, and pseudoparenchymatous forms.[2]

Until relatively recently, information concerning the isolation, characterization, and sequence analysis of algal nucleic acids remained minimal. Diverse algal types including those that display non-histone-associated DNA,[3] highly methylated DNA,[4] plasmids,[5,6] more than one morphologically distinct nucleus,[7,8] or synchronous reproduction[9-11] have been identified. This range of experimental organisms represents a new, untapped, exploitable resource. Application of nucleic acid methodologies allows these systems to be used in the analysis of chromosome packaging, gene

[1] T. P. Delaney, Ph.D. Thesis, Univ. of Washington, Seattle (1989).
[2] H. C. Bold and M. J. Wynne, "Introduction to the Algae." Prentice Hall, Englewood Cliffs, New Jersey, 1985.
[3] P. J. Rizzo, *Biosystems* **18**, 249 (1985).
[4] B.A. Boczar, J. Liston, and R. A. Cattolico, *Plant Physiol.* **97**, 613 (1991).
[5] L. J. Goff and A. W. Coleman, *J. Phycol.* **24**, 357 (1988).
[6] J. Woolford and R. A. Cattolico, unpublished (1989).
[7] R. W. Tomas and E. R. Cox, *J. Phycol.* **9**, 304 (1973).
[8] A. D. Greenwood, *Aust. Acad. Sci., Canberra* **32** (1974).
[9] R. A. Cattolico, J. C. Boothroyd, and S. P. Gibbs, *Plant Physiol.* **57**, 497 (1976).
[10] R. F. Jones, *Ann. N.Y. Acad. Sci.* **175**, 648 (1970).
[11] B. M. Sweeney, "Dinoflagellates," p. 343. Academic Press, New York, 1984.

expression, intracellular communication, and/or organelle function. Diverse methods in nucleic acid analysis will also be needed to discriminate among algae that have similar genotype but are either morphologically plastic or represent different life history phases (i.e., haploid bladed form versus diploid filamentous form) from those of different genotypes that lack distinguishable character traits. Sequence analysis of specific algal genes will give insight into the evolutionary origins of the autotrophic cell as well as offer a new data set on the structure–function aspects of gene products critical to the maintenance of algal survival. Last, the successful isolation of DNA from algal cells will allow investigators to address questions concerning the mechanisms by which aquatic organisms respond to a spectrum of environmental cues that dictate the utilization of specified biochemical pathways necessary for survival and reproductive success.

The complexity in morphological form and internal cellular organization among algal representatives (Table I) requires that separate protocols, unique from those methodologies devised for terrestrial plant nucleic acid isolation, be developed. This chapter presents several protocols for the extraction and isolation of algal nucleic acids.

DNA Isolation

The methodologies used for algal DNA extraction differ depending on the morphology and biochemistry of the source organism. Complex cell wall architecture, scale ornamentation, pellicles, a spectrum of storage materials that range from lipids to mucopolysaccharides, as well as varying architectures of the nucleus, chloroplast, and mitochondria among algal taxa, serve to challenge the investigator. Fractionation of algae has been successfully accomplished in our laboratory by French press, glass beads, liquid nitrogen grinding, and Dounce homogenization. Dependent on the system, we have recovered DNA from whole cells[4,12,13] or from isolated organelles.[14,15] Nuclear, chloroplast, and mitochondrial DNA yields, however, vary between species, reflecting organellar differences in DNA content among taxa.

High-quality algal DNA has been routinely recovered (Fig. 1) in our laboratory using a direct, lysis–fluorochrome dye–CsCl density gradient protocol. This method contrasts with those in which a phenol or isoamyl

[12] J. M. Chesnick, P. Kugrens, and R. A. Cattolico, *Mol. Mar. Biol. Biotech.* **1**, 18 (1991).
[13] T. Delaney and R. A. Cattolico, *Curr. Genet.* **15**, 221 (1989).
[14] N. Li and R. A. Cattolico, *Mol. Gen. Genet.* **209**, 343 (1987).
[15] J. K. Aldrich and R. A. Cattolico, *Plant Physiol.* **68**, 641 (1981).

TABLE I

REPRESENTATIVE ALGAL TAXA FROM WHICH DNA HAS BEEN RETRIEVED USING
FLUOROCHROME–CESIUM CHLORIDE DENSITY GRADIENT METHODOLOGIES

Superdivision	Morphology	Cell covering
Chlorophyta		
Division Chlorophyta		
Tetraselmis sp.	Unicellular	Scales
Oedogonium cardiacum	Multicellular filamentous	Cell wall
Cladophora glomerata	Multicellular filamentous	Cell wall
Chromophyta		
Division Cryptophyta		
Cryptomonas ovata	Unicellular	Proteinaceous cell plates and mucilage
Cryptomonas erosa	Unicellular	Proteinaceous cell plates and mucilage
Chroomonas coerulea	Unicellular	Proteinaceous cell plates and mucilage
Rhodomonas lacustris	Unicellular	Proteinaceous cell plates and mucilage
Division Chrysophyta		
Olisthodiscus luteus	Unicellular	Wall-less
Ochromonas danica	Unicellular	Wall-less
Chattonella subsalsa	Unicellular	Wall-less
Synura petersonii	Colonial	Scales
Mallomonas sp.	Colonial	Scales
Division Pyrrhophyta		
Peridinium balticum	Unicellular	Cell plates and pellicle
Peridinium foliaceum	Unicellular	Cell plates and pellicle
Peridinium inconspicuum	Unicellular	Cell plates and pellicle
Woloszynskia sp.	Unicellular	Cell plates and pellicle
Rhodophyta		
Division Rhodophyta		
Griffithsia pacifica	Coenocytic filamentous	Cell wall
Porphyra yezoensis	Multicellular filamentous (conchocelis stage)	Cell wall and mucilage

alcohol extraction step is used to purify DNA before specific DNA species are retrieved on fluorochrome dye–CsCl gradients.[16–20] The incorporation of ethidium bromide and Hoechst fluorochrome dyes in the isolation procedure is critical to the resolution of individual DNA species. The

[16] R. A. Cattolico, *Plant Physiol.* **62**, 558 (1978).
[17] L. J. Moore and A. W. Coleman, *Plant Mol. Biol.* **13**, 459 (1989).
[18] J. R. Manhart, R. W. Hoshaw, and J. D. Palmer, *J. Phycol.* **26**, 490 (1990).
[19] M. Maerz and P. Sitte, *Plant Mol. Biol.* **16**, 593 (1991).
[20] S. E. Douglas, *Curr. Genet.* **14**, 591 (1988).

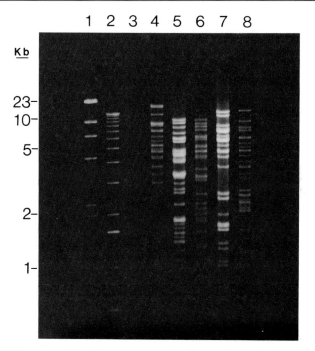

FIG. 1. Ethidium bromide-stained agarose gel (0.7%) showing *Eco*RI-digested chloroplast DNA isolated from five morphologically variant algal species using lysis–fluorochrome dye–CsCl density gradient protocols. Lane 1, λ *Hind*III DNA molecular weight markers; lane 2, 1-kb ladder (BRL, Gaithersburg, MD) molecular weight markers; lane 3, *Griffithsia pacifica* cpDNA (filamentous red alga); lane 4, *Olisthodiscus luteus* cpDNA (unicellular wall-less chromophyte); lane 5, *Cryptomonas ovata* cpDNA (unicellular thin-walled chromophyte); lane 6, *Cryptomonas erosa* cpDNA (unicellular thin-walled chromophyte); lane 7, *Peridinium foliaceum* cpDNA (thick-plated dinoflagellate).

planar ethidium ion intercalates between adjacent base pairs within the DNA double helix with significantly different binding constants dependent on base sequence.[21] Hoechst 33258 is a bibenzimidazole dye that forms an intensively fluorescent and highly specific compound when associated with DNA. Selective fluorescent enhancement by Hoechst dye occurs for AT versus GC base pairs.[22] Retrieved DNA is of sufficient purity to allow restriction enzyme digestion,[4,12–14] cloning,[13,23] reassociation kinetic analysis,[24] and polymerase chain reaction (PCR) studies.[25]

[21] T. R. Krugh and C. G. Reinhardt, *J. Mol. Biol.* **97**, 133 (1975).
[22] R. F. Steiner and H. Sternberg, *Arch. Biochem. Biophys.* **197**, 580 (1979).
[23] M. Reith and R. A. Cattolico, *Proc. Natl. Acad. Sci. U.S.A.* **83**, 8599 (1986).
[24] D. R. Ersland and R. A. Cattolico, *Biochemistry* **20**, 6886 (1981).
[25] T. Nadeau, J. M. Chesnick and R. A. Cattolico, *ISPMB Abstr.* (1991).

The fluorochrome dye–CsCl procedures described above allow for the isolation of a specific DNA species from whole cells or the retrieval of DNA from isolated organelles. However, it may be desirable and/or sufficient to have total cellular DNA in a single preparation. The application of PCR methodologies precludes the need to recover highly purified DNA from a specific organelle for the analysis of a chosen gene. Additionally, the presence of plasmids or nongenomic DNA satellite species has been described for every algal superdivision. It is not known if high-resolution centrifugation hinders the recovery of some of these novel DNAs. A whole-cell DNA recovery method involving cell lysis followed by extraction with buffered phenol (pH 7.5), without a CsCl purification step, has been previously published and is still recommended.[16]

The following protocols concentrate on the use of fluorochrome dye–CsCl density gradients for the retrieval of intact organellar genomic DNA. Three different algal morphotypes have been chosen to facilitate the discussion of DNA extraction methodologies. These morphotypes include (1) unicellular, wall-less, thinly walled, or scale-embellished forms, (2) unicellular thick-walled algae, and (3) multicellular thick-walled algae.

Unicellular Algae: Wall-less, Thinly Walled, or Scale Embellished Forms

A great number of unicellular algae have no defined cell wall. The plasma membrane is the outermost structure of the cell. In some wall-less algae, the plasma membrane may be surrounded by a calyx of mucopolysaccharide materials. Alternatively, a "thin wall" of proteinaceous plates is seen in certain algal species, or the presence of species-specific scale ornamentation may be observed (Table I). A wall-less or thin-walled morphology is advantageous because little to no mechanical methods are needed to effect cell breakage. This factor facilitates the isolation of DNA from either whole cells or isolated organelles. The following procedure for the isolation of DNA from a cell with minimal wall architecture is described as it has been applied to the wall-less alga *Olisthodiscus luteus*.[26] The same method has been successfully applied, with modification, to other wall-less algae,[27] representatives with thin walls or proteinaceous plates,[12] or scaled forms.[28] The method is also useful when recovering DNA from isolated organelles.[14,15]

Approximately 3 liters of a midexponentially growing *O. luteus* culture (i.e., 5×10^4 cells/ml) is harvested in 500-ml bottles by centrifugation at 2170 g for 10 min at 5°. Pellets that contain approximately 7.5×10^9 cells

[26] The taxonomy of this chrysophytic organism is presently under revision.

[27] N. Li, Ph.D. Thesis, Univ. of Washington, Seattle (1989).

[28] J. L. Wee, J. Chesnick, and R. A. Cattolico, *J. Phycol.* **29**, 96 (1993).

(150 μg of total DNA) are resuspended in 7 ml of lysis buffer that contains 50 mM Tris, 500 mM EDTA (pH 8.0). Following the addition of 200 μl proteinase K (10 mg/ml aqueous stock solution), the mixture is maintained at 5° for 5 min, after which an additional 200 μl proteinase K and 1 ml of 20% (w/v) sarkosyl are added. The mixture is incubated at 75° for 2 min. Solid CsCl is then added (1 g/ml) to the lysates and gently mixed until dissolved. Finally, ethidium bromide (10 mg/ml aqueous stock) is added to a final concentration of 0.15 mg/ml. The lysate density is adjusted by addition of solid CsCl to 1.05 g/ml; the mixture is placed in 12.5-ml ultracentrifuge tubes and then subjected to equilibrium density centrifugation at 190,000 g for 40 hr at 25° in a fixed-angle rotor. Whole-cell DNA, which appears as one orange fluorescent band when viewed with UV light, is centrally located in the tube. DNA retrieval is accomplished by side wall puncture of the gradients using an 18-gauge needle.

The target DNA is then pooled if necessary (i.e., for chloroplast or mitochondrial species) and the ethidium bromide removed with 2-propanol saturated with a NaCl solution. Hoechst dye 33258 (10 mg/ml aqueous stock) is added (25 μl/ml). Total retrieved solution is diluted to approximately 10 ml with lysis buffer, the density adjusted to 1.05 g/ml with solid CsCl, and the sample then transferred to 5.2-ml ultracentrifuge tubes. Three DNA species are resolved in the resulting gradient after samples are centrifuged in a vertical rotor at 190,000 g for 20 hr at 25°. These DNA species, for *O. luteus,* include nuclear (highest density), mitochondrial (middle density), and chloroplast (lowest density) DNA. Target DNAs are retrieved, pooled, and recentrifuged in a vertical rotor using the Hoechst dye–CsCl gradient conditions described above without further dye addition. This step may repeated to increase the purity of targeted DNA. After the final centrifugation step, Hoechst dye is removed using the 2-propanol method described above. Retrieved DNA is finally dialyzed against 2 liters TE buffer (10 mM Tris, 1 mM EDTA; pH 8.0) at 4°, for a total of three changes of buffer in 18 hr. After dialysis, sample volumes are reduced to 0.4 ml with 1-butanol, and the DNA is precipitated overnight at −20° using 50 μl of 3 M sodium acetate (pH 5.2) and 1 ml of 95% (v/v) ethanol. Precipitated DNA is collected by centrifugation at 14,000 g for 15–30 min; pellets are washed with cold 75% ethanol, dried, and resuspended in TE buffer. DNA can be either used immediately or stored at −20°.

Unicellular Algae: Thick Cell Covering

Thick or complex cell coverings present perhaps the greatest obstacle to the investigator in the attempt to retrieve high molecular weight DNA from algal cells. Although cellulosic walls are formed by a number of algal representatives, many other algae exhibit a cell covering composed of complex proteinaceous or cellulosic thecal plates, silica-infused cell coverings, calcium carbonate ornamentation, pellicles, loricas, or protease-resistant membrane structures. This list, though extensive, is not inclusive. Rigorous mechanical methods of cell breakage may seriously compromise the integrity of the DNA product; thus, caution is advised.

Cells with a complex cell covering may offer an advantage when retrieving DNA from nonaxenically maintained cultures. A resistant cell covering gives opportunity for extensive cell washing postharvest. It is advised that wash procedures be monitored microscopically for bacteria using either Hoechst or acridine orange dyes.[29] The protocol described below for the retrieval of DNA has been successfully applied to seven different dinoflagellate species (including several armored forms) and to the unicellular chlorophyte *Tetraselmis* sp. in our laboratory.

Approximately 8 liters of nonaxenic late log phase (1×10^5 cells/ml) cells are collected by centrifugation in 500-ml bottles at 3580 g for 20 min at 5°. Cells are resuspended in sterile growth medium, transferred to 15-ml tubes, washed, and pelleted several times using a clinical centrifuge at 950 rpm. Washed pellets are resuspended in 30 ml of cold, sterile breakage buffer (50 mM Tris, 100 mM EDTA, 0.3 M glucose; pH 8.0) which contains 500 μl of 10 mg/ml proteinase K. Cells are then disrupted in a precooled French press at 2200 psi. Immediately after cell disruption an additional 1 ml of proteinase K and 500 μl of 20% (w/v) sterile sarkosyl are added, and the mixture is incubated at 55° for 3–5 min. Following incubation, an additional 500 μl of proteinase K and 4.5 ml of 20% (w/v) sarkosyl are added, and the lysate is placed at 75° for 3–5 min. After incubation at 75°, solid CsCl (1 g/ml) is added to the approximately 35 ml of lysate and dissolved by gently agitating the tube at 75° until all salt is in solution. Finally, ethidium bromide (10 mg/ml aqueous stock) is added to a final concentration of 0.15 mg/ml. The lysate density is adjusted by addition of solid CsCl to 1.05 g/ml, and then samples are subjected to equilibrium density centrifugation in 5.2-ml ultracentrifuge tubes at 190,000 g for 40 hr at 25° in a fixed-angle rotor. The retrieval of DNA after this point follows the protocol previously described for the retrieval of DNA from unicellular wall-less algal species.

It should be noted that dinoflagellates present a difficult taxonomic assemblage for the retrieval of DNA. Not only do these organisms contain high levels of DNA (e.g., 30–200 pg/cell), but DNA species (nuclear as well as satellites) may be extensively methylated. In our experience, chloroplast

[29] R. A. Cattolico, unpublished (1989).

DNA is retrieved from these algae with minimal efficiency. Whether this recovery is due to a low chloroplast – nuclear DNA ratio within dinoflagellates or to a naturally low abundance of plastid DNA copies per dinoflagellate cell is presently unknown. Modifications of the procedure described above wherein glass beads were used to fractionate cells, and a different buffer – detergent – incubation program followed,[4] did not improve the efficiency of chloroplast DNA recovery.

Multicellular Thick-Walled Algae

Multicellular algae vary significantly in morphology and include branched or unbranched filaments, bladed forms, as well as large macroalgal kelps of pseudoparenchymatous composition. Many algae differentiate holdfast, stipe, and blade structures. Some species, such as the filamentous red alga *Griffithsia pacifica,* reach a maximum height of only a few centimeters during their growth season, whereas many kelps, such as the phaeophyte *Nereocystis* sp., maintain meristematic tissue that allows for elongation of the plant for several meters in the water column. All multicellular thick-walled algae rely heavily on the construction of their cell walls to maintain form and functionality in the aquatic environment. Frequently, walls are biochemically complex, containing compounds such as calcium carbonate or sulfated polysaccharides. It is recommended that only healthy filaments or young blades be used for extracting DNA. The following protocol describes DNA isolation from the filamentous red alga *Griffithsia pacifica.* This technique, with modification, has been successfully used for brown macrokelps[30] as well as red algal bladed[5] and conchocelis forms.[31]

Twenty to forty grams (wet weight) of nonaxenic *Griffithsia pacifica* cultures are collected by decanting all growth medium from the algal filaments. The filaments are then rigorously washed twice by shaking with sterile growth medium, removing as much liquid as possible from the filaments after each wash step. One half of the material is placed into a mortar, which has been prechilled with liquid nitrogen and is seated in ice. Liquid nitrogen is added to the mortar until all filaments are covered, and the specimen is ground to a fine powder with a prechilled pestle. The resulting powder is transferred to 30 ml isolation buffer [25 ml of 50 mM Tris, 500 mM EDTA, pH 8.0; 3.5 ml of 20% (w/v) sarkosyl; 1.5 ml of proteinase K (10 mg/ml aqueous stock)]. As the material is mixed, it will become quite viscous as the algal powder thaws. The mixture is incubated at 75° for 3 min. Solid CsCl (1 g/ml) is then added, the lysate mixed well, and incubation continued at 75° for 20 – 30 min or until a color change in

[30] S. R. Fain, L. D. Druehl, and D. L. Baillie, *J. Phycol.* **24,** 292 (1988).
[31] M. Shivji, *Curr. Genet.* **19,** 49 (1991).

the mixture occurs from red to brown to green. This procedure is then repeated for the second portion of the sample. Both samples are consequently passed through a double layer of Miracloth and the filtrate collected in sterile polypropylene tubes. Ethidium bromide (10 mg/ml aqueous stock) is added to a final concentration of 0.15 mg/ml, and the lysate density is adjusted to 1.05 g/ml by addition of solid CsCl. The resulting solution is subjected to equilibrium density centrifugation in 12.5-ml ultracentrifuge tubes at 190,000 g for 40 hr at 25° in a fixed-angle rotor. The retrieval of DNA after this point follows the protocol previously described for the retrieval of DNA from unicellular, wall-less algal species.

Acknowledgments

This research was supported by a National Science Foundation Postdoctoral Fellowship in Plant Biology DCB 8710604 to J. M. Chesnick and by Sea Grant NA 89AA-D-SG 022 and NSF Systematic Grant BSR8717563, AO#1 to R. A. Cattolico. Special thanks go to K. Suiter for manuscript preparation.

[14] Isolation and Characterization of Animal Mitochondrial DNA

By Rob DeSalle, Annie K. Williams, and Matthew George

Introduction

Since the early 1980s, the utility of mitochondrial (mt) DNA for evolutionary and population studies has become widely recognized. The characteristics of this molecule have been examined and described in detail, and data have accumulated both for the distribution and patterns of variation within the mtDNA molecule of specific organisms and for comparisons of many groups at a variety of taxonomic levels.[1-6] This information consti-

[1] A. C. Wilson, R. L. Cann, S. M. Carr, M. George, U. B. Gyllensten, K. M. Helm-By-chowski, R. G. Higuchi, S. R. Palumbi, E. M. Prager, R. D. Sage, and M. Stoneking, *Biol. J. Linn. Soc.* **26**, 375 (1985).

[2] C. Moritz, T. E. Dowling, and W. M. Brown, *Annu. Rev. Ecol. Syst.* **18**, 269 (1987).

[3] J. C. Avise, J. Arnold, R. M. Ball, E. Bermingham, T. Lamb, J. E. Neigel, C. A. Reeb and N. C. Saunders, *Annu. Rev. Ecol. Syst.* **18**, 489 (1987).

[4] R. G. Harrison, *Trends Ecol. and Evol.* **4**, 6 (1989).

[5] C. Simon, *in* "Molecular Techniques in Taxonomy" (G. M. Hewitt, A. W. B. Johnston, and J. P. W. Young, eds.), pp. 33. Springer-Verlag, Berlin 1991.

[6] M. W. Gray, *Annu. Rev. Cell Biol.* **5**, 25 (1989).

tutes an important resource for molecular biologists and molecular systematists. Decisions regarding what portion of the molecule to examine, what level of resolution is required, and which animals to include in a given study can be based on information available from previous research. Experiments can be designed from a more sophisticated vantage point, and the heartbreak of generating the wrong data for a question of import can be minimized.

Here we attempt to distill the key patterns that have emerged from the data available to date, and we present these as a practical guide for accumulating and using mtDNA data for evolutionary study. Additionally, we provide a list of protocols for isolating and comparing animal mtDNAs. Our bias in this regard will be evident. Whenever available, we favor "fast and dirty" methods so long as we have demonstrated that they work and provide consistently repeatable results. Whenever we advocate shortcuts, references are provided that direct readers to the more laborious procedures on which these are based. However, it should be pointed out that the methods provided below not only can save a great deal of time, but will, in some instances, allow the efficient production of material for analysis without the necessity for purchase of or access to very expensive pieces of equipment. In addition, although almost all manipulations of animal mtDNA can be done using the polymerase chain reaction (PCR), techniques such as Southern blotting and restriction fragment length polymorphism (RFLP) analysis and cloning of mtDNA are still useful approaches to analyzing mtDNA and are therefore also discussed.

Mitochondrial DNA Molecule: Scope of Use

Animal mtDNA has been characterized as circular, conserved in genome size, rapidly evolving, maternally inherited, and containing the genes shown in Table I. The utility of the mtDNA molecule actually lies in the fact that neither all animal mtDNAs nor all mitochondrial genes are equally well described by the five characteristics listed above. Most importantly, mitochondrial genes have been found to evolve at rates that are highly variable from gene to gene. Moreover, variations in genome size,[6-9] gene order (Table I), and inheritance have been detected. Different aspects of mtDNA sequence change have been used for a variety of evolutionary questions at many different taxonomic or hierarchical levels.

[7] M. Snyder, A. R. Fraser, J. LaRoche, K. F. Gartner-Kepkey, and F. Zouros, *Proc. Natl. Acad. Sci. U.S.A.* **84**, 7595 (1990).
[8] M. C. R. Fauron and D. R. Wolstenholme, *Nucleic Acids Res.* **11**, 2439 (1980); N. Solignac, M. Monnerot, and J. C. Mounolou, *Proc. Natl. Acad. Sci. U.S.A.* **80**, 6942 (1986).
[9] D. M. Rand and R. G. Harrison, *Genetics* **121**, 551 (1989).

TABLE I
GENE ORDERS OF ANIMAL MITOCHONDRIAL DNAs[a]

Type	Gene order	Refs.[b]
1. Insect and *Daphnia*	2-I-II-*8*-*6*-III-3-5-4-4L-6-b-1-*16*-*12*	*1–5*
2. Nonavian vertebrate	*12*-*16*-1-2-I-II-*8*-*6*-III-3-4L-4-5-6-b	*6–14*
3. Avian	*12*-*16*-1-2-I-II-*8*-*6*-III-3-4L-4-5-b-6	*15, 16*
4. Urchin	1-2-*16*-I-4L-II-*8*-*6*-III-3-4-5-6-b-*12*	*17–19*
5A. Worm-1	6-4L-*12*-1-6-2-b-III-4-I-II-*16*-3-5	*20, 21*
5B. Worm-2	1-6-2-b-III-4-I-II-*16*-3-5-6-4L-*12*	*22*
6. *Mytilus*	b-II-1-4-III-2-3-I-6-4L-5-6-*12*-*16*	*23*

[a] The basic arrangements of gene order are shown for six major arrangements of mtDNA. Gene abbreviations are as follows: 1, 2, 3, 4, 4L, 5, and 6 refer to the NADH dehydrogenase subunits 1, 2, 3, 4, 4L, 5, and 6, respectively; I, I, and III refer to the cytochrome oxidase subunits I, II, and III, respectively; *6* and *8* refer to the ATPase subunits 6 and 8, respectively; *12* and *16* refer to the 12 S and 16 S ribosomal RNAs. Note that the worms and *Mytilus* appear to lack the ATPase 8 gene (*8*). For the position of tRNAs, consult the listed reference of the animal of interest. The gene sequences are listed with the D-loop regions and AT-rich regions at the far left, followed by each gene as it appears in that organism. Consult the references for particular organisms for direction of transcription. The D-loop and AT-rich areas can vary greatly in size in some taxa such as the bark weevil [*Pissodes nemorensis;* T. M. Boyce, M. E. Zwick, and C. F. Aquadro, *Genetics* **123**, 825 (1989)], parasitic worm [*Romanomermis culicivorax;* T. O. Powers, E. G. Platzer, and B. C. Hyman, *Curr. Genet.* **11**, 71 (1989)], sea scallop [*Placopecten magellanicus;* M. Snyder, A. R. Fraser, J. La-Roche, K. F. Gartner-Kepkay, and F. Zouros, *Proc. Natl. Acad. Sci. U.S.A.* **84**, 7595 (1990)], and *Drosophila* , [M. C. R. Fauron and D. R. Wolstenholme, *Nucleic Acids Res.* **11**, 2439 (1980); M. Solignac, M. Monnerot, and J. C. Mounolou, *Proc. Natl. Acad. Sci. U.S.A.* **80**, 6942 (1986)].

[b] *Key to references: (1) Drosophila yakuba* [D. O. Clary and D. R. Wolstenholme, *J. Mol. Evol.* **22**, 252 (1985)]; *(2) Drosophila silvestris* [R. DeSalle and A. R. Templeton, *J. Hered.* **83**, 211 (1992)]; *(3) Drosophila melanogaster* [M. H. L. deBruijn, *Nature (London)* **304**, 234 (1983); R. Garesse, *Genetics* **118**, 649 (1988)]; *(4) Anopheles quadrimaculatus* [A. Cockburn, S. E. Mitchell, and J. A. Seawright, *Arch. Insect Biochem. Physiol.* **14**, 31 (1989)]; *(5) Daphnia pulex* [D. J. Stanton, J. J. Crease, and P. D. N. Hebert, *J. Mol. Evol.* **33**, 152 (1991)]; *(6) Rattus rattus* [G. Gadaleta, G. Pepe, G. De Candia, C. Quagliariello, E. Sbisa, and C. Saccone, *J. Mol. Evol.* **28**, 497 (1989)]; *(7) Gadus morhua* [S. Johansen, P. H. Guddal, and T. Johansen, *Nucleic Acids Res.* **18**, 411 (1990)]; *(8) Xenopus laevis* [B. A. Roe, D.-P. Ma, R. K. Wilson, and F. F.-H. Wong, *J. Biol. Chem.* **260**, 9759 (1985)]; *(9) Cyprinus carpio* [A. Araya, R. Amthauer, G. Leon, and M. Krauskopf, *Mol. Gen. Genet.* **196**, 43 (1984)]; *(10) Homo sapiens* [S. Anderson, A. T. Bankier, B. G. Barrell, M. H. L. de Bruijn, A. R. Coulson, J. Drouin, I. C. Eperon, D. P. Nierlich, B. A. Roe, F. Sanger, P. H. Schreirer, A. J. H. Smith, R. Staden, and I. G. Young, *Nature (London)* **290**, 457 (1981)]; *(11) Bos bos* [S. Anderson, M. H. L. de Bruijn, A. R. Coulson, I. C. Eperon, F. Sanger, and I. G. Young, *J. Mol. Biol.* **156**, 683 (1982)]; *(12) Mus musculus* [M. J. Bibb, R. A. Van Etten, C. T. Wright, M. W. Walberg, and D. A. Clayton, *Cell* (Cambridge, Mass.) **26**, 167 (1981)]; *(13) Phoca vitulina* [U. Arnason and E. Johnsson, *J. Mol. Evol.* **34**, 493 (1991)]; *(14) Balaenoptera physalus* [U. Arnason, A. Gullberg, and B. Widegren, *J. Mol. Evol.* **33**, 556 (1991)]; *(15) Gallus gallus*

Population Level Studies

The objective of any population study is to detect and quantify variation. This objective can be accomplished in many ways depending on the overall thrust of a study. Often, the aim of a study is to follow the distribution of single polymorphisms.[10,11] Assaying variation in population studies also may involve the characterization of levels of variation between individuals, within populations, between populations, or over zones of contact of populations. This objective requires methods for extensive screening of many individuals for polymorphisms.

Systematics at Low Levels of Divergence between Species, Genera, or Families

Early studies relied heavily on RFLPs to generate character state and distance data for systematic studies. However, informed choice of genes for sequencing and the use of mtDNA universal primers have made the use of RFLPs as systematic tools nearly obsolete. Table II is a summary of some of the published sequencing studies using various mtDNA genes. The percent nucleotide similarity is shown for all of the mtDNA genes except tRNAs. Table II demonstrates the wide range of patterns of sequence change for these genes, and it can be used as a rough guide detailing which genes might be useful at particular taxonomic levels.

In any systematic study the number of characters used to generate the phylogenetic inference needs to be maximized. At the same time homoplasy or convergence of characters needs to be minimized. mtDNA genes for systematic studies should take both of these factors into consideration.

[10] R. DeSalle, A. R. Templeton, I. Mori, S. Pletscher, and J. S. Johnston, *Genetics* **116**, 215 (1987).

[11] R. L. Cann, M. Stoneking, and A. C. Wilson, *Nature (London)* **325**, 31 (1987).

[P. Desjardins and R. Morals, *J. Mol. Biol.* **212**, 599 (1990)]; *(16) Coturnix japonica* [P. Desjardins and R. Morals, *J. Mol. Evol.* **32**, 153 (1990)]; *(17) Paracentrotus lividus* [P. Cantatore, M. Roberti, G. Rainaldi, M. N. Gadaleta, and C. Saccone, *J. Biol. Chem.* **264**, 10965 (1989)]; *(18) Strongylocentrotus purpuratus* [H. T. Jacobs, D. J. Elliott, V. B. Math, and A. Farquharson, *J. Mol. Biol.* **202**, 185 (1988)]; *(19) Arbacia lixula* [C. De Giori, F. De Luca, and C. Saccone, *Gene* **103**, 249 (1991)]; *(20) Caenorhabditis elegans* [D. R. Wolstenholme, J. L. Macfarlane, R. Okimoto, D. O. Clary, and J. A. Wahleithner, *Proc. Natl. Acad. Sci. U.S.A.* **84**, 1324 (1987)]; *(21) Ascaris suum* [R. Okimoto, J. L. Macfarlane, D. O. Clary, and D. R. Wolsenholme, *Genetics* **130**, 471 (1992)]; *(22) Meloidogyne javanica* [R. Okimoto, H. M. Chamberlin, J. L. Macfarlane, and D. R. Wolstenholme, *Nucleic Acids Res.* **19**, 1619 (1991)]; and *(23) Mytilus edulis* [R. J. Hoffman, J. L. Boore, and W. M. Browne, *Genetics* **131**, 397 (1992)].

TABLE II
EXAMPLES OF RANGES OF PERCENT NUCLEOTIDE SIMILARITIES OF MITOCHONDRIAL
DNA GENES[a,b]

Gene	Phylum	Class	Order	Species	Comparison (ref.)
b	60	72	73	95/93	Rodents (6)/birds (6)
I	68	79	79	93/89	Flies (7)/birds (8)
II	63	74	70	94	Flies (7)
III	65	73	79	93/92	Flies (7)/fish (9)
6	46	67	73	93/90	Flies (7)/fish (9)
8	29	62	63	93	Flies (7)
1	55	70	73	97/84	Flies (10)/birds (8)
2	44	58	66	92/96	Flies (7)/flies (10)
3	52	68	70	89	Fish (8)
4	50	61	72	90	Primates (11)
4L	48	54	75	94	Fish (9)
5	43	70	71	96/95	Flies (10)/primates (11)
6	33	62	65	89	Birds (8)
16	47	80	75	98/97	Flies (10)/bovids (12)
12	58	72	83	89/95	Birds (8)/antelopes (13)

[a] Similarities for four taxonomic levels (interphyla, interclass, interorder, and intergeneric or interspecies) are shown. Interphylum similarities, human and sea urchin comparisons[1,2]; interclass, frog to fish comparisons[3,4]; and interorder, human to cow comparisons.[1,5] Interspecies comparisons vary from gene to gene, and the specific comparison is given in the comparison column followed by the reference in parentheses. The bird comparisons from Ref. 8 are intergeneric [i.e., chicken (Gallus) to quail (Coturnix) comparisons]. All other comparisons in the species column are interspecific. For some genes more than one interspecific comparison is shown. For comparisons other than the ones presented, consult the individual publications listed for genomes that have been sequenced in their entirety. Abbreviations for genes are as in Table I. The expected number of nucleotide differences for a particular taxonomic comparison can easily be computed by taking into consideration the length of the gene of interest. It must be emphasized that no consideration of saturation of third positions is taken into account and that the similarities depicted are meant to give a raw estimate for the number of changes that occur at different taxonomic levels.

[b] Key to references: (1) S. Anderson, A. T. Bankier, B. G. Barrell, M. H. L. de Bruijn, A. R. Coulson, J. Drouin, I. C. Eperon D. P. Nierlich, B. A. Roe, F. Sanger, P. H. Schreierr, A. J. H. Smith, R. Staden, and I. G. Young, Nature (London) 290, 457 (1981); (2) P. Cantatore, M. Roberti, G. Rainaldi, M. N. Gadaleta, and C. Saccone, J. Biol. Chem. 264, 10965 (1989); (3) S. Johansen, P. H. Guddal, and T. Johansen, Nucleic Acids Res. 18, 411 (1990); (4) H. T. Jacobs, D. J. Elliott, V. B. Math, and A. Farquharson, J. Mol. Biol. 202, 185 (1988); (5) S. Anderson, M. H. L. de Bruijn, A. R. Coulson, I. C. Eperon, F. Sanger, and I. G. Young, J. Mol. Biol. 156, 683 (1982); (6) T. D. Kocher, W. K. Thomas, A. Meyer, S. V. Edwards, S. Paabo, F. X. Villablanca, and A. C. Wilson, Proc. Natl. Acad. Sci. U.S.A. 86, 6196 (1989); (7) D. R. Wolstenholme and D. O. Clary, Genetics 109, 725 (1985); (8) P. Desjardins and R. Morals, J. Mol. Evol. 32, 153 (1990); (9) W. K. Thomas and A. T. Beckenbach, J. Mol. Evol. 29, 233 (1989); (10) R. DeSalle, T. Freedman, E. M. Prager, and A. C. Wilson, J. Mol. Evol. 26, 157 (1987); (11) W. M. Brown, E. M. Prager, A. Wang, and A. C. Wilson, J. Mol. Evol. 18, 225 (1982); (12) M. M. Miyamoto, F. Kraus, and O. A. Ryder, Proc. Natl. Acad. Sci. U.S.A. 87, 6127 (1990); M. M. Miyamoto, S. M. Ranhauser, and P. J. Lapis, Syst. Zool. 38, 342 (1989); and (13) J. Gatesy, D. Yelon, R. DeSalle, and E. Vrba, Mol. Biol. Evol. 9, 433 (1992).

The first consideration (number of characters) can be maximized by picking a gene with an ample number of sequence changes for a particular taxonomic level. The second consideration (homoplasy) can be minimized by making sure that the nucleotide sequence changes have not saturated.[12-14] Table II and the references therein can be helpful in detecting genes and particular taxonomic comparisons that might be in saturation zones and hence prone to generating data with homoplasy. Even when gene sequences are saturated, measures can be taken to account for the saturation. These include character weighting nucleotide changes that are more prone to saturation. For example, it is a common practice to base phylogenetic inferences on transversions in comparisons where saturation is suspected, or to consider only nucleotide changes that implement amino acid substitutions (effectively eliminating third position changes).

Systematics at High Levels of Divergence between Orders, Classes, Phyla, or Kingdoms

As mentioned above, because there are only five possible character states for any nucleotide position in a DNA sequence (G, A, T, C, or gap), saturation of these positions occurs. DNA sequences then essentially become random and contain little or no phylogenetic information. The greater the divergence between two taxa, the higher the probability that phylogenetic information will have eroded. Consequently, DNA sequences of mtDNA genes, even of extremely slowly evolving genes such as rDNAs or tRNAs, can be troublesome for examining very deep relationships. An alternative approach using mtDNA that has shown great promise for these deep divergences is the examination of the structure (circularity or linearity[15]) or gene order of the molecule.[2,16]

Strategies for Collecting Data

Whether researchers are interested in examining mtDNA variation at the population level or at higher systematic levels, two decisions must be grappled with at the outset. First, some choices must be made regarding which portion of the genome to examine. Alternatively, surveying the whole genome might be a more appropriate strategy depending on the question being asked. Second, it must be determined what level of resolu-

[12] W. M. Brown, E. M. Prager, A. Wang, and A. C. Wilson, *J. Mol. Evol.* **18**, 225 (1982).

[13] R. DeSalle, T. Freedman, E. M. Prager, and A. C. Wilson, *J. Mol. Evol.* **26**, 157 (1987).

[14] W. Brown, M. George, and A. C. Wilson, *Proc. Natl. Acad. Sci. U.S.A.* **76**, 1967 (1979).

[15] D. M. Bridge, C. W. Cunningham, B. Schierwater, R. DeSalle, and L. W. Buss, *Proc. Natl. Acad. Sci. U.S.A.* **89**, 8750 (1992).

[16] D. Sankoff, G. Leduc, N. Antoine, B. Paquin, B. F. Lang, and R. Cedergren, *Proc. Natl. Acad. Sci. U.S.A.* **89**, 6575 (1992).

tion will be required to adequately address the questions being asked. The mosaic nature of sequence change in the mtDNA makes it ideal for spanning several taxonomic levels. Once a decision has been made as to the particular scope and depth of a study, specific protocols can be

TABLE III
STRATEGIES FOR MITOCHONDRIAL DNA ISOLATION

Level	Approach	Isolation techniques[a]
1. Single polymorphism in population studies	Allele-specific oligonucleotide (ASO)	Method 2 or 3, then PCR to amplify target sequences and assay with dot blots or Southern blots
	Gel denaturing (GD)	Method 2 or 3, then PCR to amplify target sequences and assay on alkaline gels
	RFLP assays	(A) Method 1, then cleave and end label
		(B) Method 2 or 3, then restrict and Southern blot; hybridize with labeled probe
	Sequencing	Method 1, 2, or 3, then PCR or clone to amplify target sequences; direct sequence PCR products or use M13 or modified M13 sequencing system for clones
2. Multiple polymorphisms in population studies	RFLP assays	(A) Method 1, then cleave with a battery of restriction enzymes; end label
		(B) Method 2 or 3, then cleave with a battery of restriction enzymes; Southern blot; hybridize with labeled probe
	Sequencing	Method 2 or 3, then PCR or clone to amplify target sequences; direct sequence PCR products or use M13 or modified M13 sequencing system for clones
3. Multiple characters in systematics	RFLP assays	Method 1, 2, 3, then same procedures as above
	Sequencing	Method 2 or 3, then clone or PCR and sequence as above
4. Rearrangements as characters in higher level systematics	Analysis of clones	Method 1, 2, 3, then clone and characterize by RFLP or sequencing
	PCR across gene junctions	Method 1, 2, 3, then a battery of PCR reactions to determine adjacency of genes

[a] Methods 1, 2, and 3 refer to the three main types of DNA preparations outlined in this chapter. Method 1 is the highly purified mtDNA preparation; method 2 is the crude total cellular preparation: full scale; and method 3 is the crude total cellular DNA: minipreparation.

adopted. Table III lists these levels and the techniques that are commonly used to accomplish the levels of resolution desired.

Specific DNA Isolation Strategies

DNA Isolation Techniques for Polymorphism Data

If the RFLP technique is chosen, several approaches can be taken. (1) Highly purified mtDNA can be cleaved, end labeled, and visualized with autoradiography after agarose or acrylamide gel electrophoresis. (2) Crude DNA extracts can be cleaved, separated on agarose gels, transferred to nylon membranes, and hybridized to labeled mtDNA probes appropriate for the organism under study. Clones for the entire mtDNA genome for several organisms exist (Table IV) and can be used as probes. In addition, several organisms have parts of the mtDNA cloned (Table IV), and these too can be used in RFLP hybridizations. PCR-generated fragments of the mtDNA or highly purified organismal mtDNA can also be used as probes in Southern screens. (3) Amplified DNA, either from PCR-amplified mtDNA fragments or from cloned mtDNA, can be cleaved, separated on agarose gels, and visualized by ethidium bromide staining. In this approach, cloned mtDNA or PCR-amplified DNA can originate from highly purified or crude extract total cellular DNA.

DNA Isolation Techniques for Allele-Specific Oligonucleotide Assay or Gel Denaturing and Polymerase Chain Reaction

Both allele-specific oligonucleotide (ASO) and gel denaturing (GD) assays require that considerable amounts of target DNA be available and therefore commonly use PCR-mediated DNA amplification of target sequences prior to analysis. Two types of target DNA can be used in the PCR amplification of animal mtDNA: (1) highly purified mtDNA and (2) crude total cellular DNA. Although highly purified mtDNA amplifies with great ease in the PCR reaction, often the crude total cellular DNA preparations are more than adequate for the task. In fact, if nuclear gene products are also desirable for a particular study, it is best to do a single isolation of crude total cellular DNA. Such a preparation contains sufficient amounts of nuclear and mitochondrial DNA sequences for amplification of both types of targets.

DNA Isolation Techniques for Cloning

Isolation of DNA for cloning is extremely similar to that for PCR. Highly purified mtDNA is preferred, but total cellular DNA is adequate.

TABLE IV
CLONED REGIONS OF ANIMAL MITOCHONDRIAL DNA's[a]

Type	Organism	Clones	Ref.[b]
Insect and *Daphnia*	*Drosophila yakuba*	Es	*1*
	D. silvestris	E	*2*
	D. melanogaster	Es	*3*
	Hawaiian *Drosophila*	1, *16*	*4*
	Anopheles quadrimaculatus	E	*5*
	D. melanogaster subgroup	2, I	*6*
	D. nasuta subgroup	b, 1	*7*
	Apis mellifera	*16, 12*	*8*
		b, *6, 8*	*9*
	Gryllus affirmus	*16, 12*	*10*
	Aedes albopictus	*16, 12*	*11*
	Pissodes nemorensis	I, II, *8, 6*, III, 3, 5	*12*
	Daphnia pulex	E	*13*
Nonavian vertebrate	*Rattus rattus*	Es	*14*
	Gadus morhua	Es	*15*
	Xenopus laevis	Es	*16*
	Cyprinus carpio	E	*17*
	Homo sapiens	Es	*18*
	Bos bos	Es	*19*
	Mus musculus	Es	*20*
	Odocoileus virginicanus	E	*21*
	Canis familiaris	E	*22*
	Phoca vitulina	Es	*23*
	Balaenoptera physalus	Es	*24*
	Artebius jamaisensis	E	*25*
	Microtus pennsylvanicus	I, II, *8, 6*, 6, b	*26*
	Bovidae (four species)	*12, 16*	*27*
	Hominoidea (six species)	*12*	*28*
	Hominoidea (six species)	2, I, II, *8, 6*	*29*
	Salmonid fishes	II, 8, 6, III, 3, 4L	*30*
Avian	*Gallus gallus*	Es	*31*
	Coturnix japonica	*12, 16*, 1, 2	*32*
Urchin	*Paracentrotus lividus*	Es	*33*
	Strongylocentrotus purpuratus	Es	*34*
	Arbacia lixula	E	*35*
Worm	*Caenorhabditis elegans*	Es	*36*
	Ascaris suum	Es	*37*
	Meloidogyne javanica	E	*38*
	Romanomermis culicivorax	3, *16*	*39*
Mytilus	*Mytilus edulis*	Es	*40*

[a] *Key to references: (1)* D. O. Clary and D. R. Wolstenholme, *J. Mol. Evol.* **22**, 252 (1985); *(2)* R. DeSalle and A. R. Templeton, *J. Hered.* **83**, 211 (1992); *(3)* M. H. L. de Bruijn, *Nature (London)* **304**, 234 (1983); R. Garesse, *Genetics* **118**, 649 (1988); *(4)* R. DeSalle, T. Freedman, E. M. Prager, and A. C. Wilson, *J. Mol. Evol.* **26**, 157 (1987); *(5)* A.

The choice of cloning vector will have an influence on whether highly purified mtDNA or crude cellular DNA is used. Cloning into plasmids such as pUC or pBR vectors is most easily accomplished using highly purified mtDNA. Cloning into vectors such as λgt10 is accomplished using cellular DNAs but, again, is more easily accomplished using highly purified mtDNA.

Cockburn, S. E. Mitchell, and J. A. Seawright, *Arch. Insect Biochem. Physiol.* **14**, 31 (1989); *(6)* Y. Satta and N. Takahata, *Proc. Natl. Acad. Sci. U.S.A.* **87**, 9558 (1990); L. Nigro, M. Lolignac, and P. M. Sharp, *J. Mol. Evol.* **33**, 156 (1991); *(7)* K. Tamura, *Mol. Biol. Evol.* **9**, 814 (1992); *(8)* I. Vlasak, S. Burgschwaiger, and G. Kreil, *Nucleic Acids Res.* **15**, 2388 (1987); *(9)* R. H. Crozier and Y. C. Crozier, *Mol. Biol. Evol.* **9**, 474 (1992); *(10)* D. M. Rand and R. G. Harrison, *Genetics* **121**, 551 (1989); *(11)* C.-C., HsuChen, R. M. Kotin, and D. T. Dubin, *Nucleic Acids Res.* **12**, 7771 (1984); *(12)* T. M. Boyce, M. E. Zwick, and C. F. Aquadro, *Genetics* **123**, 825 (1989); *(13)* D. J. Stanton, J. J. Crease, and P. D. N. Hebert, *J. Mol. Evol.* **33**, 152 (1991); *(14)* G. Gadaleta, G. Pepe, G. De Candia, C. Quagliariello, E. Sbisa, and C. Saccone, *J. Mol. Evol.* **28**, 497 (1989); *(15)* S. Johansen, P. H. Guddal, and T. Johansen, *Nucleic Acids Res.* **18**, 411 (1990); *(16)* B. A. Roe, D.-P. Ma, R. K. Wilson, and F. -H. Wong, *J. Biol. Chem.* **260**, 9759 (1985); *(17)* A. Araya, R. Amthauer, G. Leon, and M. Krauskopf, *Mol. Gen. Genet.* **196**, 43 (1984); *(18)* S. Anderson, A. T. Bankier, B. G. Barrell, M. H. L. de Bruijn, A. R. Coulson, J. Drouin, I. C. Eperon, D. P. Nierlich, B. A. Roe, F. Sanger, P. H. Schreierr, A. J. H. Smith, R. Staden, and I. G. Young, *Nature (London)* **290**, 457 (1981); *(19)* S. Anderson, M. H. L. de Bruijn, A. R. Coulson, I. C. Eperon, F. Sanger, and I. G. Young, *J. Mol. Biol.* **156**, 683 (1982); *(20)* M. J. Bibb, R. A. Van Etten, C. T. Wright, M. W. Walberg, and D. A. Clayton, *Cell (Cambridge, Mass.)* **26**, 167 (1981); *(21)* M. Cronin, M. E. Nelson, and D. F. Pac, *J. Hered.* **82**, 118 (1991) (1993); *(22)* N. Lehman, A. Eisenhawer, K. Hansen, L. D. Mech, R. O. Peterson, P. J. P. Gogan, and R. K. Wayne, *Evolution* **45**, 104 (1991); *(23)* U. Arnason and E. Johnsson, *J. Mol. Evol.* **34**, 493 (1992); *(24)* U. Arnason, A. Gullberg, and B. Widegren, *J. Mol. Evol.* **33**, 556 (1991); *(25)* C. J. Phillips, D. E. Pumo, H. H. Genoways, P. E. Ray, and C. A. Briskey, in "Latin American Mammology: History, Biodiversity and Conservation" (M. A. Mares and D. J. Schmidley, eds.), p. 97. Univ. of Oklahoma Press, Norman, 1991; *(26)* D. Pumo, C. J. Phillips, M. Barcia, and C. Millan, *J. Mol. Evol.* **34**, 163 (1992); *(27)* M. M. Miyamoto, S. M. Ranhauser, and P. J. Lapis, *Syst. Zool.* **38**, 342 (1989); *(28)* W. M. Brown, E. M. Prager, A. Wang, and A. C. Wilson, *J. Mol. Evol.* **18**, 225 (1982); *(29)* S. Horai, Y. Satta, K. Hayasaka, R. Kondo, T. Inoue, T. Ishida, S. Hayashi, and N. Takahata, *J. Mol. Evol.* **35**, 32 (1992); *(30)* W. K. Thomas and A. T. Beckenbach, *J. Mol. Evol.* **29**, 233 (1989); *(31)* P. Desjardins and R. Morals, *J. Mol. Biol.* **212**, 599 (1990); *(32)* P. Desjardins and R. Morals, *J. Mol. Biol.* **32**, 153 (1990); *(33)* P. Cantatore, M. Roberti, G. Rainaldi, M. N. Gadaleta, and C. Saccone, *J. Biol. Chem.* **264**, 10965 (1989); *(34)* H. T. Jacobs, D. J. Elliott, V. B. Math, and A. Farquharson, *J. Mol. Biol.* **202**, 185 (1988); *(35)* C. De Giori, F. De Luca, and C. Saccone, *Gene* **103**, 249 (1991); *(36)* R. Okimoto, J. L. Macfarlane, D. O. Clary, and D. R. Wolsenholme, *Genetics* **130**, 471 (1992); *(37)* D. R. Wolstenholme, J. L. Macfarlane, R. Okimoto, D. O. Clary, and J. A. Wahleithner, *Proc. Natl. Acad. Sci. U.S.A.* **84**, 1324 (1987); *(38)* R. Okimoto, H. M. Chamberlin, J. L. Macfarlane, and D. R. Wolstenholme, *Nucleic Acids Res.* **19**, 1619 (1991); *(39)* T. O. Powers, E. G. Platzer, and B. C. Hyman, *Curr. Genet.* **11**, 71 (1989); *(40)* R. J. Hoffman, J. L. Boore, and W. M. Brown, *Genetics* **131**, 397 (1992).

DNA Isolation Techniques for Study of Gross Rearrangements

Isolation of highly purified mtDNA, when a particular form of the mtDNA is known to exist for the organisms under study, is the most desirable approach. However, if the organisms being examined have unknown mtDNA conformations such as circular versus linear, then difficulties arise in isolation techniques. Closed circular mtDNA behaves very differently in cesium chloride gradients than linearized DNA, and the same isolation techniques cannot be used for both. In fact, linear forms will cosediment with nuclear DNA in gradients that are set up to isolate circular mtDNA forms. A novel approach to examining gene order in animal mtDNAs involves designing PCR primers that span the junction of the 30 or so mtDNA genes for a particular gene order such as the nonavian vertebrate gene order.[16] These primers are designed so that they produce expected amplification products of around 100 base pairs (bp) if the two genes represented by the primers are adjacent. Organisms of unknown gene order are then challenged with a battery of primers to determine adjacency of genes, and hence the entire gene order of the mtDNA can be determined. In addition, the products of such gene junctures can be examined at the sequence level by retaining the PCR products from the initial screens, then directly sequencing the products. Crude total cellular DNA that contains the mtDNA regardless of circularity, linearity, or broken linear conformation is appropriate for this kind of manipulation.

Specific Strategies and Protocols

As is evident from the previous section, there are basically two types of prepared mtDNA that are preferred for the various techniques used in evolutionary studies: highly purified mtDNA and crude total cellular DNA. Specific strategies and protocols for isolating animal mtDNA not only depend on whether highly purified mtDNA or crude total cellular DNA is desirable, but also on the type and the size of the organism being studied and the number of individuals targeted for study. If the detection of a single nucleotide polymorphism in many individuals is the object of the study, rapid isolation techniques are desirable, and only small amounts of target DNA are needed. If, on the other hand, the focus of the study is to collect RFLP data for many enzymes for several organisms, then the very different strategy of having to obtain large amounts of target DNA for end labeling or Southern blot analysis is needed. Table III lists the strategies most often employed for the different objectives. Table III can be used as a guide to direct one to the specific protocols discussed in this section.

We present three basic types of preparations: (1) highly purified mtDNA preparations using sucrose pad gradients and cesium chloride gradients, (2) large-scale total cellular DNA preparations, and (3) rapid minipreparations of total cellular DNA. Solignac[17] provides an alternative summary of mtDNA preparations.

Each preparation is separated into four phases: (I) The *tissue treatment phase* includes how to handle tissue so there is no cross-contamination of samples and how to grind, powder, or macerate the tissue to enhance isolation. (II) The *lysis phase* includes how to differentially lyse the cellular membrane and the mitochondrial membrane and how to isolate mito-chondria away from the rest of the cell. (III) The *purification phase* includes techniques for extracting purified and enzyme-sensitive DNA from the lysis phase. (IV) The *assay phase* shows how to assay the final product of the preparation by agarose gel electrophoresis and offers some sugges-tions as to how to assess the purity of the isolated DNA.

Method 1. Highly Purified Mitochondrial DNA Preparation

I. Tissue Preparation and Treatment

1. Depending on the size of the organism or the kind of material available, cellular DNA can be easily extracted from either tissue or organ samples or from a homogenate prepared from the whole organism.[18]

   A. For vertebrates the following tissues are recommended and are listed in order of their desirability: brain, testes or ovary, liver, kidney, heart, and skeletal muscle.

   B. For insects, such as *Drosophila,* large quantities of embryos are the best starting material, although some success can be had by using freshly collected live adults.

2. The best method for grinding or dispersing the cells for this type of preparation appears to be use of a Dounce homogenizer. One should be careful to keep all tissues on ice before and during the "Dounc-ing." Douncing or grinding, in grinding buffers appropriate for the chosen tissue (see below), is performed on ice.

3. The ground homogenate should be centrifuged at low speed (2500 rpm) for 5 min in a swinging-bucket rotor at 4°. The resulting pellet is discarded. Repetition of this low-speed spin (2500 rpm) 3 to 5 times ensures that most cellular contaminants and nuclei are pel-leted and eliminated. Because the low-speed pellets contain nuclei, it is sometimes desirable to save these pellets for nuclear DNA prepa-rations.

[17] M. Solignac, *in* "Molecular Techniques in Taxonomy" (G. M. Hewitt, A. W. B. Johnston, and J. P. W. Young, eds.), p. 295. Springer-Verlag, Berlin, 1991.
[18] R. A. Lansman, R. O. Shade, J. F. Shapira, and J. C. Avise, *J. Mol. Evol.* **17,** 214 (1981).

4. After several low-speed spins, the supernatant is pelleted by a 13,000 rpm spin for 20 min in an SS34 rotor in a Sorvall (Norwalk, CT) high-speed centrifuge (or equivalent). This pellet can then be taken to the lysis phase for further treatment.

5. Optional: Another method of ensuring purity of the mitochondrial fraction is to run a sucrose step gradient. This is done in a swinging-bucket rotor (Beckman, Palo Alto, CA, SW-28) in 30-ml tubes with an ultracentrifuge run at 25,000 rpm for 1 hr at 4°. The step gradient consists of a pad of 10 ml of 1.5 M sucrose in TE [Tris (1 mM) EDTA (0.1 mM) buffer] followed by a 10 ml layer of 1.0 M sucrose. The mitochondrial pellet from above is resuspended in 10 ml TE and layered above the 1.0 M sucrose pad. The tubes are then spun in a SW-40 rotor for 1–2 hr at 27,000 rpm (4°). Highly purified mitochondria will appear as a milky white band at the interface of the 1.0 and 1.5 M sucrose pads. This band can be collected with a Pasteur pipette and placed in a 30-ml SS34 tube. The highly purified mitochondria can then be further concentrated by spinning in an SS34 rotor for 15 min at 13,000 rpm (4°).

II. Lysis

1. The pelleted mitochondria are then resuspended in 1.0–1.5 ml TE in a 15-ml Corex tube. The mitochondrial pellet should be dispersed gently by swirling the tubes.

2. Add 0.3 ml of 10% sodium dodecyl sulfate (SDS). The tubes are left at room temperature for 10 min. At this point the cloudy mitochondrial suspension should "clear" to a translucent solution.

3. Add 0.25 ml of saturated cesium chloride to the "cleared" mitochondrial lysate. The tubes are placed on ice for 20 min.

4. Spin the tubes at 12,000 rpm for 10 min in an SS34 rotor and save the supernatant. At this point if multiple tubes for the same organism or sample have been prepared, the supernatants can be combined. Do not exceed the volume of the ultracentrifuge tubes to be used for the cesium chloride gradient, which is usually 8 ml. When all supernatants have been combined, if the volume is not 8 ml then adjust to 8 ml by adding TE.

III. Purification

1. Add 8 g of cesium chloride to the 8 ml of supernatant and 0.6 ml of 2 mg/ml propidium iodide (PI). Some researchers have used ethidium bromide, but we recommend PI for maximum separation. Mix by inverting the tube, with Parafilm over top, until the CsCl has gone into solution.

2. Adjust the density of the solution to 1.56 g/ml. This can be done by placing the tube in a rack on a top-loading digital balance of appropriate sensitivity (milligram range). The balance is then set to zero, and, using a Pipetman, 1 ml of the solution is removed. The negative reading on the balance will give the density of the solution. The density can be adjusted up by adding CsCl and adjusted down by adding TE.

3. Transfer the solution to SW50.1 ultracentrifuge tubes and counterbalance the tubes. SW50.1 tubes have a total capacity of 5 ml. Overlay the tops of the tubes with mineral oil.

4. Spin the tubes at 36,000 rpm for 42 hr at room temperature.

5. Carefully clamp the opaque centrifuge tube on a ring stand over a UV light box. The gradient can be visualized under UV light, where two distinct bands should be visible in the middle of the tube. Throughout this procedure, care should be taken to protect oneself against UV exposure.

   A. The upper band will contain nuclear DNA. The amount of nuclear DNA in the preparation will depend on the amount of nuclear contamination eliminated in earlier steps. This band is, in general, relatively broad and sometimes diffuse but will, in some cases, be barely discernible.

   B. The lower band will contain highly purified circular mtDNA. It is generally a very tight band and easily discernible from the nuclear band because of its tightness and position in the gradient. Both the mtDNA and nuclear DNA bands should be collected.

6. The mtDNA band can be collected in several ways.

   A. While observing the tube under UV light, a hole can be punched in the bottom of the tube using a syringe needle or pin. Drops from the tube are collected when the mtDNA band begins to exit the tube. The volume collected usually does not exceed 0.25 to 0.5 ml.

   B. A syringe and needle can be used to puncture the side of the tube just below the mtDNA band, and the band can be collected by drawing on the syringe under observation with UV light. The volume collected should not exceed 0.25 to 0.5 ml. If both bands are being collected, insert the syringe under one band and leave it in place while collecting the other.

7. Optional: A CsCl step gradient can then be set up by underlaying 0.8 ml of 1.7 g/ml CsCl, 2 mg/ml PI beneath 3 ml of 1.4 g/ml CsCl, 2 mg/ml PI. The sample is diluted 1 : 2 with TE and loaded on top of the 1.4 g/ml CsCl layer. The gradients are counterbalanced,

covered with mineral oil, and placed in a SW50.1 rotor and spun at 45,000 rpm for 3.5 hr. The mtDNA will band at the interface of the 1.4 and 1.7 g/ml CsCl layers and can be collected by bottom or side puncture of the tube.

8. Alternatively, secondary gradient purification can be performed immediately after Step 6 or 7 by loading the small amount collected by either procedure onto a 1.56 g/ml CsCl gradient. The gradient is then spun again at 36,000 rpm for 42 hr.

9. The mtDNA band can be collected by side or bottom puncture and carried on to the assay phase.

10. Alternative ultracentrifuge rotors can be used rather than the ones mentioned here. The step gradients suggested above have to be carried out in swinging-bucket rotors, and the times suggested should be followed. The CsCl equilibrium gradients can also be performed less efficiently in fixed-angle or vertical rotors, if smaller volume samples are being processed, in which case run times can be decreased considerably.

IV. Assay Phase

1. The mtDNA band is extracted with butanol to remove the PI by adding an equal volume of butanol to a glass tube containing the sample. The tube is mixed gently, and the phases are allowed to settle on the bench top (~30 sec).

2. The butanol will be in the upper layer, and it is removed with a Pasteur pipette. This extraction step is repeated 5 to 6 times or until most of the reddish color from the PI is removed.

3. The butanol-extracted mtDNA fraction is then dialyzed in 1 × TE for 12–14 hr to remove CsCl and butanol residue.

4. The dialyzed mtDNA is removed from the dialysis bag to a 15-ml Corex tube, and a 1/10 volume of ammonium acetate is added to the dialyzed mtDNA followed by a 2- to 3-fold volume of ethanol. The tube is mixed by inverting and placed at −20° for at least 2 hr. Samples can also be left overnight.

5. The mtDNA is pelleted by spinning in the SS34 rotor at 12,000 rpm for 20 min.

6. The pellet is resuspended in 0.05 to 0.1 ml of TE or water.

7. To assay for the presence of supercoiled, circular mtDNA several approaches can be taken. The simplest method is to run a small part of the entire sample (1/20 of the total), either undigested or digested with a restriction enzyme, on a 0.8% (w/v) agarose gel in 1 × TBE (Tris–borate–EDTA buffer) and to visualize the DNA by staining the gel with ethidium bromide. If the DNA is undigested by a

restriction enzyme, then the mtDNA will band as a large DNA fragment on the gel; if the DNA has been digested, then discrete bands less than 16 kilobases (kb) will be visible. If the preparation has any nuclear contamination, there will be a smear of DNA that will partially or completely obscure the mtDNA bands.

8. The concentration of the sample can be estimated by examining the ethidium bromide gel or by using a spectrophotometer. The products from this particular preparation can then be used for RFLP with end labeling, for cloning, for PCR, or for making radioactive probes for Southern blots or filter lifts. The purity of the sample is essential for all applications except PCR.

Method 2. Crude Total Cellular Preparation: Full Scale

I. Tissue Preparation and Grinding

1. Depending on the starting tissue the grinding procedure is different.[19]

   A. For large amounts of tissue such as vertebrate organs, whole large insects, or "clonal" collections of insects (e.g., isofemale lines), liquid nitrogen can be used to powder the tissue for extraction. Grind between 0.5 and 1.5 ml of tissue to a fine powder in a mortar and pestle. Gentle grinding is accomplished by making sure that none of the liquid nitrogen falls out of the mortar.

   B. For small, soft organisms the tissue can be placed directly into a grinding buffer without liquid nitrogen treatment. The tissue can then be dispersed by grinding in a Teflon Dounce homogenizer.

2. Commonly used extraction buffers include the following:

   TSM buffer: 0.2 M Tris, 0.1 M EDTA, 1% SDS
   Urea grinding buffer (1 liter): 168 g urea, 25 ml of 5 M NaCl, 20 ml of 1 M Tris, 16 ml of 0.5 M EDTA, 20 ml of 20% sarcosine, 190 ml sterile distilled water
   HOM buffer: 80 mM EDTA, 100 mM Tris, 160 mM sucrose

Any of these extraction buffers are adequate for most vertebrate tissues, such as heart, kidney, muscle, and liver, and, in most cases, for insect tissues. HOM buffer is also very good for these tissues. For small, soft organisms with potential nuclease activities, the urea grinding buffer works best. Apparently, the urea decreases or eliminates nuclease activity and

---

[19] T. D. Kocher, W. K. Thomas, A. Meyer, S. V. Edwards, S. Paabo, F. X. Villabanca, and A. C. Wilson, *Proc. Natl. Acad. Sci. U.S.A.* **86,** 6196 (1989).

ensures elimination of any contaminants that might interfere with subsequent enzymatic manipulations such as PCR and restriction digestion.

3. An alternative to proceeding with a crude DNA preparation[19] at this point is to employ a cytoplasmic DNA (CNA) preparation.[18] We present protocols for both.

II. Lysis: Crude Total Cellular DNA

1. After sufficient grinding, buffer should be added so that there is at least a 3:1 buffer to sample ratio.[19] It is not necessary to determine the volume or weight of the sample precisely, but try not to exceed a final total volume of 5 ml. The efficiency of the extraction will be improved by increasing the buffer to sample ratio. This is especially true if the tissue used was not thoroughly minced. Transfer the minced tissue plus buffer to 15-ml Corex tube.

2. Add 1/50 volume of protease K (10 mg/ml) and incubate the sample from 2 hr to overnight at between 55° and 65° with vigorous shaking. The origin of the tissue being processed will dictate the length of the proteinase K digestion. Certain tissues such as kidney, heart, and liver do not require overly long digestions (2–3 hr), whereas skin, muscle, or harder tissues will require longer digestion (8–12 hr). Museum preserved specimens, such as skins and teeth (that have been powdered by a hammer), will require very long digestion periods (1–2 days).

3. Optional, but recommended: Add 1/10 volume of 5 M potassium acetate and mix well by inverting the tubes. Incubate on ice for 30 min.

4. Spin tubes at 10,000 rpm for 10 min in an SS34 rotor, then transfer the supernatant to a new 15-ml Corex tube.

II. Lysis: Enriched Cytoplasmic Nucleic Acid Preparation

1. The minced, ground, powdered, or Dounced tissue is brought up to 5 ml of either HOM or TSM buffer in a 15-ml Corex tube.[18] The homogenized tissue is made 0.2% in Nonidet P-40 or 0.2% in Triton X-100 by the addition of 0.1 ml of a 10% stock solution of either detergent and incubated on ice for 5 min.

2. Nuclei are removed by low-speed centrifugation (2500 rpm) for 5 min in an SS34 rotor. Several low-speed spins may be necessary to remove nuclei to a satisfactory degree.

3. The supernatant contains cytoplasmic nucleic acids and proteins, and extraction of the nucleic acids can be accomplished with phenol (see below). This preparation does not entirely purify cytoplasmic nucleic acids, as some nuclear nucleic acids are also isolated. In our

hands the enrichment factor tends to vary from preparation to preparation.

III. Purification

1. Add an equal volume of TE-buffered phenol–chloroform–isoamyl alcohol (PCI) in a 25:24:1 ratio, then cover the Corex tube with Saran wrap. Mix the solution by inverting the tubes. Be extremely careful not to spill any phenol.

2. Spin tubes at 5000 rpm for 5 min in an SS34 rotor. The resultant solution should have two phases. The lower phase is organic and contains the PCI and extracted proteins and nucleases, whereas the upper phase contains the aqueous solution of DNA. Transfer the top layer to new Corex tube, with either the "wrong" end of a glass pipette or a large-bore Pasteur pipette. The larger the opening of the pipette that is used to transfer the upper phase, the higher the molecular weight of the DNA will be. If a preparation appears to be overly "dirty," another phenol extraction may be necessary. In addition, if a particular sample is valuable or if the supernatant layer is of low volume, "back extraction" of the phenol layer may be needed. This is performed by adding an equal volume of TE to the previously extracted phenol layer, spinning at 5000 rpm for 5 min and removing the resultant aqueous layer. The back-extracted aqueous layer can then be combined with the first aqueous layer isolated above.

3. Add an equal volume of chloroform to the tubes and cover with Saran wrap. Mix solution by inverting the tubes, once again being very careful not to spill the organic phase, this time chloroform.

4. Spin the tubes at 5000 rpm for 5 min in an SS34 rotor. Transfer the top layer to new Corex tubes or to a set of 1.5-ml Eppendorf tubes.

5. If the potassium acetate step (above) was omitted, then add a 1/10 volume of 5 M ammonium acetate and 2 volumes of ethanol, and allow the DNA to precipitate for at least 30 min at $-20°$. If a considerable amount of DNA is present in the solution, addition of ethanol should immediately produce a stringy white precipitate.

6. The DNA can be collected by spinning the Corex tubes at 10,000 rpm for 10 min in the SS34 rotor or, if Eppendorf tubes were used, by spinning in a microcentrifuge at top speed for 5 to 10 min.

7. After the ethanol is gently poured off the pellet, the pellet is resuspended in 0.5 ml of TE and the resuspended DNA transferred to a 1.5-ml Eppendorf tube. Then 0.05 ml of 5 M ammonium acetate and 1 ml of ethanol is added, and the tube mixed by inverting and placed at $-20°$ for at least 20 min. This second ethanol precipitation is essential to ensure that the isolated DNA will be of high enough

purity to manipulate with Taq polymerase and restriction enzymes.

8. The tubes are then spun at highest speed in a microcentrifuge for 5 to 10 min, and the ethanol is gently poured off. The tubes can then be left on their sides on the bench top for 1 to 2 hr to allow the ethanol to evaporate. Alternatively, the tubes can be placed in a Speed-Vac concentrator or under a vacuum for 10 to 20 min to facilitate and speed the evaporation of ethanol.

9. The resultant dried pellet should be whitish in color, but light brown to light red pellets are also sometimes obtained. Discolored pellets are of high enough purity that they are almost always good targets for PCR or RFLP analysis. The pellet should be resuspended in from 0.1 to 0.5 ml of TE or sterile distilled water. The size of the pellet will determine the amount of TE or water the pellet is resuspended in.

IV. Assay Phase

1. The integrity of the isolated DNA can be assayed on a 0.8% agarose minigel run in $1\times$ TBE. The gel should also contain a known quantity of uncut λ DNA to get an estimation of concentration of the samples. From this rough estimation, decisions regarding the dilution of the samples with TE or water can be made. Sometimes visualization of the DNA on minigels can be sufficient to estimate concentration, but we recommend that spectrophotometer readings [optical density (OD) readings] be made on the samples if accurate estimates of concentration are required.

2. The purity of the DNA can be assayed by assessing the susceptibility of the DNA to restriction enzymes. This can be done by running three forms of the isolated DNA on a minigel. The first is untreated DNA, the second is DNA treated with proper amounts of a common restriction enzyme (e.g., *Hin*dIII or *Eco*RI), and the third sample is the DNA plus 0.25 to 0.5 μg of λ DNA treated with a common restriction enzyme. If the DNA is pure enough for RFLP analysis or PCR, the cut DNA will show a smear compared to the uncut DNA. The λ DNA plus isolated DNA sample should show a smear plus the λ pattern overlaid on the smear. If the DNA is not pure, high molecular weight DNA will appear in all three lanes.

Method 3. Crude Total Cellular DNA: Minipreparation

I. Tissue Preparation and Grinding

1. When only small amounts of tissue are available or large numbers of individual samples need to be processed, the following preparation is highly recommended.[20]

[20] E. S. Coen, J. M. Thoday, and G. A. Dover, *Nature (London)* **295,** 564 (1982).

2. Small amounts of vertebrate tissue, small insects, small parts of larger insects, and small amounts of other invertebrate tissue (the equivalent of no more than 0.25 ml) should be placed in a 1.5-ml Eppendorf tube and, after adding 0.2 ml of mini grinding solution (see below), crushed with a minipestle that fits the 1.5-ml Eppendorf tube.

II. Lysis

1. Add 0.2 ml of mini lysis solution (see below) to each sample. The contents are mixed by inverting the tubes.
2. The tubes are placed at 65° for 30 min.
3. Add 0.60 ml of 5 M potassium acetate. The tubes are mixed gently by inverting.
4. The tubes are then placed on ice for 45 min.

III. Purification

1. The tubes are spun at maximum speed in a microcentrifuge for 5 to 10 min, and the supernatant is removed to a new 1.5-ml Eppendorf tube.
2. Add 1 ml of ice-cold ethanol. The tubes are mixed by inverting and allowed to sit at room temperature for 20–30 min.
3. The DNA is pelleted by spinning at maximum speed in a microcentrifuge for 5 to 10 min.
4. The ethanol supernatant is poured off, and the pellet is resuspended in 0.1 ml of 0.5 M ammonium acetate.
5. Add 0.4 ml of ethanol. The tubes are mixed and allowed to sit at room temperature for 10 min.
6. The DNA is pelleted by a maximum-speed run in a microcentrifuge for 5 to 10 min.
7. The ethanol supernatant is poured off, and the pellet is left to dry at room temperature 1 to 3 hr or is dried under a vacuum (10–20 min).
8. The pellet is resuspended in 0.05–0.1 ml of TE or water.

IV. Assay

1. One-tenth of the final resuspended DNA is treated as in the large-scale crude preparation assay section.

Solutions. The following solutions (usually made fresh each time) are used in minipreparations:

Mini grinding solution: 10 mM Tris-HCl, 60 mM NaCl, 10 mM EDTA, 5% (w/v) sucrose

Mini lysis solution: 0.3 M Tris-HCl, 1.1% (w/v) SDS, 0.1 M EDTA, 5% (w/v) sucrose; add 0.04 ml of diethyl pyrocarbonate (DEPC) per 5 ml of solution

Alternatives for Polymerase Chain Reaction

The PCR has revolutionized the way that DNA sequence data are collected. Consequently, certain newer procedures and precautions need to be considered.

Isolation of DNA from Single Hair Follicles

Isolation of small amounts of DNA from single hair follicles that have been stored in 100% ethanol or fresh follicles can be accomplished by incubating the follicles in 0.5 ml of extraction buffer plus proteinase K.[21] Phenol extraction of the DNA is then performed, with careful back extraction. The DNA is concentrated with either ethanol precipitation or as in the next section.

Alternative Rapid DNA Isolation Method for Small Specimens

A second type of rapid preparation that can be used to prepare DNA successfully from small amounts of starting material[22] is to lyse an extremely small amount of cell suspension (usually less than 10 μl) by adding 5 μl of 200 mM potassium hydroxide, 50 mM dithiothreitol and incubating at 65° for 10 min. Five microliters of neutralization solution (900 mM Tris-HCl, pH 8.3, 300 mM potassium chloride, 200 mM hydrogen chloride) is then added. This solution should then be ready for PCR amplification.

Concentration of DNA Samples

Concentration of small amounts of DNA is critical in the PCR. Often a large portion of the DNA is lost in an ethanol precipitation. Microconcentration columns are commercially available that allow the concentration of small amounts of DNA without loss of material. The general procedure for using these microconcentration columns is to load the sample onto the column and spin in a fixed-angle position for 10–15 min to allow the eluent to pass through the column. TE or water is then added to the

[21] L. Vigilant, R. Pennington, H. Harpending, T. D. Kocher, and A. C. Wilson, *Proc. Natl. Acad. Sci. U.S.A.* **86**, 9350 (1989).
[22] L. Zhang, X. Cui, K. Schmitt, R. Hubert, W. Navidi, and N. Arnheim, *Proc. Natl. Acad. Sci. U.S.A.* **89**, 5847 (1992).

column to wash the DNA, and the spin is repeated. Three to four washes ensures that the DNA will be of sufficient purity.

Contamination of DNA Samples

Contamination has been recognized as a potential problem in isolating DNA for PCR amplification. When isolating human DNA, contamination by the researcher can be common unless precautions are taken. These precautions include using positive displacement pipettes, plugged pipette tips, and exercising extreme caution in manipulating all tissue. Below are some suggestions that can be followed to reduce the problems of contamination. In addition, primers designed for mtDNA amplification have been reported to amplify nonfunctional mtDNA sequences that have been incorporated into the nuclear genome,[23] creating the possibility of a second source of contamination. Although this phenomenon is rare, the possible occurrence of nonfunctional mtDNA sequences inserted in the nuclear genome and generated using highly conserved mtDNA primers should be examined.

Negative Controls. Negative controls can be run during isolation of DNA. This requires the use of a dummy tube with no sample in it.

Alternative Homogenization Methods. Rather than processing tissue with a Dounce homogenizer, either fresh tissue or frozen material allowed to partially thaw can be effectively processed by carefully dicing it up using a disposable scalpel blade. The tissue in placed in a small plastic "weigh boat" that is placed in a larger weigh boat filled with crushed ice and chopped up until it is fully homogenized. A small amount of lysis buffer (see above) can be added to facilitate this. The resulting homogenate can be scraped into a tube suitable for carrying out the extraction. Care should be taken to dispose of weigh boats, scalpels, and gloves between each sample that is processed. Mortar and pestles have been traditionally used to powder frozen tissues for extraction. Frozen tissue can be powdered without using a mortar and pestle by placing the tissue in a small plastic bag and crushing it. These two methods are recommended if DNA is being made for PCR amplification as both use disposable materials, reducing the opportunities for contamination.

Miniature Grinding Device. A miniature mortar and pestle can be devised by putting the sample in a 1.5-ml Eppendorf centrifuge tube and using a 0.5-ml tube to crush the sample. The lip and cap of the 0.5-ml tube can be trimmed with a razor blade to facilitate the fit of the two tubes. If the samples are to be processed with PCR amplification, it is desirable to discard each 0.5-ml tube "pestle" after grinding to avoid cross-contamina-

[23] M. F. Smith, W. Kelly Thomas, and J. L. Patton, *Mol. Biol. Evol.* **9,** 204 (1992).

tion with the grinder. This "tube inside of tube" method works well on small amounts of vertebrate tissue or insect tissue. Alternatively, a sterile pipette tip can be used to macerate insect tissue, with each new sample being macerated by a new tip. The tissue is gently crushed between the two tubes, but be careful to avoid overmacerating the tissue.

Alternative Method of Powdering Samples. Powdering samples has been a common approach to dispersing tissues. Conventionally this has been done in a mortar and pestle. The mortar and pestle are cleaned between sample processing. This procedure is tedious, and contamination can often occur from sample to sample. Cross-contamination can be avoided in powdering samples by placing deep-frozen tissues ($-70°$) in individual thick plastic bags. The tissue can then be quickly tapped or even pounded with a hammer, making sure that the hammer does not puncture the bag. Once the tissue is sufficiently powdered, it can actually be poured into a tube and processed. The bag is then discarded, eliminating contamination.

Detection of Nonfunctional Sequences. Detection of nonfunctional mtDNA sequences after amplification with highly conserved PCR primers is accomplished by comparison of sequences to an authentic mtDNA sequence from an organism that is closely related to the study organism. The occurrence of insertions or deletions in the study sequence should be a warning as to the authenticity of the sequence. In addition, if there are deletions or insertions in protein-coding genes that are affected by this phenomenon, then deletions that are not multiples of three (i.e., the length of one codon) should be a further warning signal that the generated sequence is nonfunctional.[23] When a nonfunctional sequence is obtained, there are two strategies for obtaining the authentic sequences. The first involves isolating highly purified mtDNA via the methods outlined above (Table III, method 1). The second method involves the redesigning of primers that take advantage of the changes between the authentic and nonfunctional sequences.

Polymerase Chain Reaction Primers

Primer sequences for many of the regions of the mtDNA have been published. Primer designing, in general, involves the alignment of sequences or examination of sequences from various taxa. Regions of broad similarity in the nucleotide sequences are identified, and oligonucleotides of the proper sequence are designed. The methodology for determining primer sequences is discussed elsewhere.[24,25] The broader the taxonomic

[24] R. K. Saiki, *in* "PCR Protocols" (M. A. Innis, D. H. Gelfand, J. J. Sninsky, and T. J. White, eds.), p. 13. Academic Press, San Diego, 1990.
[25] C. Simon, A. Franke, and A. Martin, *in* "Molecular Techniques in Taxonomy" (G. M. Hewitt, A. W. B. Johnston, and J. P. W. Young, eds.), p. 329. Springer-Verlag, Berlin 1991.

TABLE V

COMMONLY USED POLYMERASE CHAIN REACTION PRIMERS FOR MITOCHONDRIAL DNA ANALYSIS[a]

3' position	Primer sequence (5' → 3')	Use with	Size	Organisms	Refs.[b]
L14841	CTTCCATCCAACATCTCAGCATGATGAAA	H15149	308	Universal	1–4
L14724	CGAAGCTTGATATGAAAAACCATC	H15149	415	Universal	5, 6
L14979	GACGTCAACTACGGCTTGAAT	H15149	170	Mammals	5
L15162	GCAAGCTTCTACCATGAGGACAAAATATC	H15915	753	Mammals	5
L15408	ATAGACAAAATCCCATTCCA	H15915	507	Mammals	5
L15513	CTAGGAGACCCTGACAACTA	H15915	402	Mammals	5
L15775	GACGGCCATACATGAATTGGAGGACAACCAGTC	H15915	140	Mammals	5
H14927	GTGACAGAGGAGAATGCTGT	L14724	203	Mammals	5
H15915	AACTGCAGTCATCTCCGGTTTACAAGAC	—	—	Mammals	5
H15149	AAACTGCAGCCCCTCAGAATGATATTTGTCCTCA	—	—	Universal	2, 3, 5
L14841	CCAACATCTCAGCATGATGAAA	H15163	322	Universal	7, 8
H15163	CTCAGAATGATATTTGTCCTCA	—	—	Universal	7, 8
L15915	AACTGCAGTCATCTCCGGTTTACAAGAC	L14841	1073	Birds	8
L14230	GCTTCCATCCAACATCTCAGCATGATG	H14542	366	Rodents	9
L14115	CGAAGCTTGATATGAAAAACCATCGTTG	H14542	427	Rodents	9, 10
H14542	GCAGCCCCTCAGAATGATATTTGTCCTC	L14230	—	Rodents	9, 10
L15087	TACTTAAACAAAGAAACCTGAAA	H15767	680	Birds	2
L15114	GGAGTCATCCTACTCCTAACCCT	H15767	653	Birds	2
L15299	CGATTCTTCGCCCTGCACTTCCTCC	H15767	468	Birds	2
L15424	ATCCATTCCACCCATACTACTC	H15767	343	Birds	2
L15564	CCACACATTAAACCGAATGATA	H15767	203	Birds	2
L15609	ATTCTACGATCCATCCAAACAAACT	H15767	158	Birds	2
H15547	AATAGGAAGTATCATTCGGGTTTGATG	L14841	706	Birds	2
H15767	ATGAAGGGATGTTCTACTGGTTG	—	—	Birds	2

b

TABLE V continued
COMMONLY USED POLYMERASE CHAIN REACTION PRIMERS FOR MITOCHONDRIAL DNA ANALYSIS[a]

3' position	Primer sequence (5' → 3')	Use with	Size	Organisms	Refs.[b]
12					
L1091	AAAAAGCTTCAAACTGGGATTAGATACCCCACTAT	H1478	387	Universal	1, 11
L1478	TGACTGCAGAGGGTGACGGGCGGTGTGT	—	—	Universal	1, 11
L1067	AAACTAGGATTAGATACCCTATTAT	H1478	411	Universal	12
H1478	AAGAGCGACGGGCGATGTGT	—	—	Universal	12
H1416	AAGGTGGATTTGGTAGTAAA	L1067	349	Insects	12
L1154	ATTCAAAGAATTTGGCGGTA	H1478	324	Insects	12
H651	AAGTTTATTTTGGCTTA	L877	226	Insects	12
L877	GACAAATTCGTGCCAGCAGT	—	—	Insects	12
L1373	CGCTGCAGAGAAATGGCTACATTTTCT	H1478	105	Humans	13
L1435	GGGAAGCTTAAGGAGGATTTAGCAGTAAA	H1478	53	Humans	13
16					
H2491	ATGTTTTTGGTAAACAGGCG	L1067	1200	Insects	12
L2491	CGCCTGTTTATCAAAAACAT	H3058	546	Universal	1, 12
H3058	CCGGTCGAACTCAGATCACGT	—	—	Universal	1, 12
L2492	GAATTCCTATTAAACCAACTTTATT	H3397	900	Insects	14
H3397	CGCCTGTTTAACAAAAACAT	—	—	Insects	14
I					
H7227	AGTATAAGCGTCTGGGTAGTC	L6454	773	Vertebrates	12
L6454	TCGTCTGATCCGTCTTTGTCAC	—	—	Universal	12
H7258	GAACATGATGAAGAAGTGCACCTTCCC	L6454	804	Urchins	12
H7110	CCAGAGATTAGAGGGAATCAGTG	L6569	541	Universal	12
L6569	CCTGCAGGAGGAGGAGACCC	—	—	Universal	12
L5950	ACAATCACAAAGAYATYGG	H7176	1226	Fish	3
L6586	CCTGCAGGAGGAGGAGAYCC	H7176	590	Fish	3
H7086	CCTGAGAATARKGGGAATCAGTG	L5950	1136	Fish	3
H7176	AGAAATGTTGWGGGAARAA	—	—	Fish	3

	Primer	Sequence	Pair	Size	Taxon	
II	H7552	AACCATTTCATAACTTTGTCAA	L8321	769	Primates	*15*
	L8321	CTCTAATCTTTAACTAAAG	—	—	Primates	*15*
	L7953	GGACTAATCTTCAACTCCTACATACT	H8003	50	Humans	*13*
	H8003	ATCGGGAGTACTACTCGATTGT	—	—	Humans	*13*
	L7773	GACGGCTCAGGAAATAGAAAC	H8003	228	Humans	*13*
	L7450	AAAGGAAGGAATCGAACCCC	H8055	605	Fish	*3*
	H8055	GCTCATGAGTGGAGGACGTCTT	—	—	Fish	*3*
	L3018	ATGGCAGATTAGTGCAATGG	H3804	786	Insects	*16*
	H3804	GTTTAAGAGACCAGTACTTG	—	—	Insects	*16*
III	L9490	GGAATAATTTATTTATTTATCAGAAG	H10140	550	Insects	*17*
	H10140	AGTTTACCACTCAAAAAAGAGC	—	—	Insects	*17*
	L9459	TTATTTATTGCATCAGAAGT	H9924	465	Universal	*12*
	H9924	TCAACAAAGTGTCAGTATCA	—	—	Universal	*12*
Con Reg	L15996	CTCCACCATTAGCACCCAAAGC	H16401	405	Humans	*18*
	H16401	TGATTTCACGGAGGATGGTG	—	—	Humans	*18*
	L29	GGTCTATCACCCTATTAACCAC	H408	379	Humans	*18*
	H408	CTGTTAAAAGTGCATACCGCCA	—	—	Humans	*18*
	L625	TCTTCTAGGCATTTCAGTG	L15980	—	Verte-brates	*12*
	L15980	CTACCTCCAACTCCCAAAGC	—	—	Verte-brates	*12*
	L15774	GACGGCCAGTACATGAATTGGAGGACAACCAGT	H16498	724	Verte-brates	*19*
	H16498	CCTGAACTAGGAACCAGATG	—	—	Verte-brates	*19*
	L15910	GAATTCCCGGTCTTGTAAACC	H615	1200	Cetaceans	*19*
	H615	TCTCGAGATTTTCAGTGTCTTGCTTT	—	—	Cetaceans	*20*
	L15926	TCAAAGCTTACACCAGTCTTGTAAACC	H16498	572	Humans	*20*

TABLE V *continued*

COMMONLY USED POLYMERASE CHAIN REACTION PRIMERS FOR MITOCHONDRIAL DNA ANALYSIS[a]

3' position	Primer sequence (5' → 3')	Use with	Size	Organisms	Refs.[b]
4					
L11110	CAGCCACAGAACTAATCATA	H11396	286	Humans	18
L11136	ATCTTCTTCGAAACCACACT	H11396	260	Humans	18
L11230	TCTACACCCTAGTAGGCTCC	H11396	166	Humans	18
H11396	GCTTTAGGGAGTCATAAGTG	—	—	Humans	18
L11574	CTATCCCTATGAGGCATAATTATAAC	H12175	601	Vertebrates	1
L11838	CTTCCACTAATAGGAACCTGATGACT	H12175	337	Vertebrates	1
L11975	TACTCCCTATACATATTT	H12175	200	Universal	12
H12175	AACCAAAACATCAGATTGTGAATCT	—	—	Vertebrates	1
5					
H12650	GAATTCTATGATGATCATGT	L11975	675	Universal	12
6					
H8936	GTGCGCTTGGTGTTCCCGTGG	L9459	523	Urchins	1
8					
L8352	TAAAGATTGGTGACTCCCAACCACC	H8773	421	Fish	6
H8773	GTAGGGAAGTAAGCCAATATGTT	—	—	Fish	6
tRNA					
L4452	TGCTGCAGCCATACCCGAAAATGTTGGTT	H5545	1093	Marsupials	22
H5545	CGAAGCTTTGTACTTGCTTAGGGCTTG	—	—	Marsupials	22

[a] H and L refer to heavy and light strands of the human mitochondrial genome. The position number refers to the 3' base of the primer in the complete human mtDNA sequence (23). Gene letters are explained in Table I. In practice these primer pairs should be compared across several taxa to ensure proper similarity, especially at the 3' end. The following abbreviations are used to refer to degeneracy at a particular position in a primer sequence: R = AG; Y = CT; M = AC; K = GT; S = CG; W = AT; H = ACT; B = CGT; V = ACG; D = AGT; N = ACGT.

[b] Key to references: (1) T. D. Kocher, W. K. Thomas, A. Meyer, S. V. Edwards, S. Paabo, F. X. Villablanca, and A. C. Wilson, Proc. Natl. Acad. Sci. U.S.A. 86, 6196 (1989); (2) S. V. Edwards, P. Arctander, and A. C. Wilson, Proc. R. Soc. London 243, 99 (1991); (3) B. B. Normark, A. R. McCune, and R. G. Harrison, Mol. Biol. Evol. 8, 819 (1991); (4) S. B. Hedges, R. L. Bezy, and L. R. Maxson, Mol. Biol. Evol. 8, 767 (1991); (5) D. M. Irwin, T. D. Kocher, and A. C. Wilson, J. Mol. Evol. 32, 128 (1991); (6) M. A. Grachev, S. Ja. Slobodyanyuk, N. G. Kholodilov, S. P. Fyodorov, S. I. Belikov, D. Yu. Sherbakov, V. G. Sideleva, A. A. Zubin, and V. V. Kharchenko, J. Mol. Evol. 34, 85 (1992); (7) T. D. Kocher and K. Lockwood, personal communication, 1991; (8) S. V. Edwards and A. C. Wilson, Genetics 126, 695 (1990); (9) M. F. Smith, W. Kelly Thomas, and J. L. Patton, Mol. Biol. Evol. 9, 204 (1992); (10) M. F. Smith and J. L. Patton, Mol. Biol. Evol. 8, 85 (1991); (11) A. Meyer and S. I. Dolven, J. Mol. Evol. 35, 102 (1992); (12) S. R. Palumbi, personal communication (1990) from the widely distributed "Simple Fool's Guide to PCR"; C. Simon, A. Franke, and A. Martin, in "Molecular Techniques in Taxonomy" (G. M. Hewitt, A. W. B. Johnston, and J. P. W. Young, eds.), p. 329. Springer-Verlag, Berlin, 1991; (13) S. Paabo, personal communication, 1991; (14) R. DeSalle, Mol. Phyl. Evol. 1, 31 (1991); (15) M. Ruvolo, T. R. Disotell, M. W. Allard, W. M. Brown, and R. L. Honeycutt, Proc. Natl. Acad. Sci. U.S.A. 88, 1570 (1991); (16) H. Liu and A. T. Beckenbach, Mol. Phyl. Evol. 1, 41 (1992); (17) A. Vogler and R. DeSalle, Ann. Entomol. Soc. 86, 142 (1993); (18) L. Vigilant, R. Pennington, H. Harpending, T. D. Kocher, and A. C. Wilson, Proc. Natl. Acad. Sci. U.S.A. 86, 9350 (1989); (19) G. F. Shields and T. D. Kocher, Evolution 45, 218 (1991); (20) A. R. Hoelzel, J. M. Hancock, and G. A. Dover, Mol. Biol. Evol. 8, 475 (1991); (21) A. De Rienzo and A. C. Wilson, Proc. Natl. Acad. Sci. U.S.A. 88, 1597 (1991); (22) S. Paabo, W. Kelly Thomas, K. M. Whitfield, Y. Kumazawa, and A. C. Wilson, J. Mol. Evol. 33, 426 (1991); (23) S. Anderson, A. T. Bankier, B. G. Barrell, M. H. L. de Bruijn, A. R. Coulson, J. Drouin, I. C. Eperon, D. P. Nierlich, B. A. Roe, F. Sanger, P. H. Schreierr, A. J. H. Smith, R. Staden, and I. G. Young, Nature (London) 290, 457 (1981).

range in the alignment, the more "universal" the primers will be. The references to the complete sequences of animal mtDNA listed in Tables I and IV can be of assistance in accomplishing this. In Table V we present a list of primers that have been used in various laboratories and/or published in publications up to 1992.

[15] Comparative Studies of Genome Content

By A. Lane Rayburn

Introduction

Genome content, or genome size, refers to the amount of DNA contained in the genome of an organism. A genome is defined as the basic (monoploid) chromosome set of an organism, consisting of a species-specific number of linkage groups.[1] Therefore, when determining the genome size of an organism, only the DNA contained in the chromosomes (nuclear DNA) is considered. The nuclear DNA content of a cell is usually expressed in terms of the C value of the cell. The letter C is used to distinguish terminology used for genome studies from the terms x and n, reserved for chromosome number.[2] The $1C$ value of an organism is the amount of DNA that resides in an unreplicated haploid nucleus.[3] A diploid nucleus ($2n$) in early interphase contains a $2C$ amount of DNA, while the same $2n$ nucleus at late interphase, after replication, would contain a $4C$ DNA amount.

One must be cautious when comparing DNA amounts in various species. Comparisons may be made at either the C level or genome level. This becomes increasingly important when dealing with polyploid species. For example, hexaploid breadwheat (*Triticum aestivum*: $2n = 6x = 42$) has been reported to have a genome size of approximately 18 picograms (pg) of DNA.[4] Contained within the $1C$ nucleus of *T. aestivum* are haploid complements of three diploid genomes (A, B, and D). Each of these diploid genomes comprises approximately one-third of the $1C$ genome of hexaploid wheat, or 6.0 pg. Hexaploid wheat has been proposed to arise from

---

[1] R. Reiger, A Michaelis, and M. M. Green, "Glossary of Genetics and Cytogenetics," 4th Ed. Springer-Verlag, Berlin, Heidelberg, and New York, 1976.

[2] M. D. Bennett and J. B. Smith, *Philos. Trans. R. Soc. London B* **277**, 201 (1977).

[3] H. Swift, *Proc. Natl. Acad. Sci. U.S.A.* **36**, 200 (1950).

[4] C. E. May and R. Appels, *in* "Wheat and Wheat Improvement" (E. G. Heyne, ed.), p. 166. ASA Monograph, Madison, Wisconsin, 1987.

the hybridization of tetraploid wheat (*T. turgidum*: $2n = 4x = 28$) with diploid *T. monococcum* ($2n = 2x = 14$). The $1C$ value of *T. turgidum* has been reported as 12.0 pg, whereas the $1C$ amount of *T. monococcum* has been reported to be 6.2 pg.[3] *Triticum monococcum* is the putative *A* genome donor species, and the $1C$ size of 6.2 pg fits the predicted 6.0 pg for each of the *A*, *B*, and *D* genomes. In addition, the 18.1 pg obtained for *T. aestivum* is in good agreement with the combined genome size of 12.0 and 6.2 pg. If one compares $1C$ values, the values increase in approximately equal amounts between ploidy levels: diploid ($2x$), 6.2 pg; tetraploid ($4x$), 12.0 pg; and hexaploid ($6x$), 18.0 pg. However, if one compares the DNA content per diploid genome (x), the amount at each ploidy level is very similar: diploid, 6.2 pg; tetraploid, 6.0 pg; and hexaploid, 6.0 pg. It is imperative that the term C is correctly used in genome size studies to avoid confusion while attempting to make comparisons.

The two ways genome content may be reported are in absolute amounts and in arbitrary units. When expressing genome size in absolute terms, various units may be employed. The most common unit is picograms (1 pg is equal to 10^{-12} g). Other units that may be used are daltons and nucleotide pairs. One is able to convert from one unit to another based on the following equations: 1 nucleotide pair $= 660$ daltons, and 1 pg $= 0.965 \times 10^9$ nucleotide pairs. It should be noted that these conversions are only estimates and should be considered accordingly.

The genome size of experimental lines may also be expressed relative to a standard. The standard might be a taxonomically well-defined plant of constant genome size. The nuclear DNA amount of experimental lines would be expressed as percentages of the standard. The standard is defined as 100% and has a genome size of 100 arbitrary units (AU) As AUs are a relative term, they provide no additional information as to absolute genome size. Therefore, although the relative comparisons made within a given experiment are valid, comparisons between studies are difficult to make. There are two methods to overcome this limitation: (1) ensure that all studies use the same identical standard or (2) determine the absolute DNA amount in picograms for the standard. Having the absolute amount for the standard will allow one to convert the AUs of experimental lines to picograms.

Various methods exist for the estimation of genome size. The three most common methods are chemical extraction, densitometry, and flow cytometry. Although each of these techniques has advantages and disadvantages, the method that appears to have the greatest potential is flow cytometry. Therefore, flow cytometric analysis of genome size is presented as the method of choice for genome size comparisons. If, however, one is interested in the other methods, several excellent references provide much of the theoretical and technical aspects of chemical extraction and densitometry.[2,3]

Flow cytometric analysis of genome size requires isolating cells or nuclei in high concentrations. The nuclei are then stained with various DNA-specific fluorochromes. The stained nuclei are subsequently analyzed by a flow cytometer cell sorter. Critical to the technique is the freeing of intact cells or nuclei.

Isolation of Nuclei

Nuclear isolation is a method by which large numbers of free intact nuclei are obtained. The method of isolation resulting in the highest yield of intact nuclei is dependent on the age and type of the particular tissue used. The method presented here[5] has been demonstrated to give high yields of intact nuclei in maize, wheat, barley, oats, sorghum, and soybeans.

For maize, seeds are germinated in a soilless mix of a 1 : 1 mixture of perlite and vermiculite. The plants are exposed to a day length of 14 hr. The irradiance is approximately 100 μM photons m^{-2} s^{-1}. These conditions result in tissue that is most conducive to nuclear isolation. After 2 weeks, the seedlings are harvested, and a 2.5-cm section of the seedling is removed starting at the first node above the mesocotyl. The tissue types that reside in this area are leaf sheaths, rolled up leaf blades, and the coleoptile. The coleoptile is removed. In addition, if the seedling is heavily pigmented with anthocyanins, the first leaf and its sheath are also removed. The removal of these tissues results in a sample that is easier to grind and that is free of pigments which may affect the analysis. The seedling section is then sliced into disks of approximately 3 mm in length and placed in a nuclear isolation buffer previously described,[6] with minor modifications. The buffer consists of 1 M hexylene glycol, 10 mM Tris (pH 8.0), 10 mM MgCl$_2$, and 0.5% Triton X-100. At this and all subsequent steps, the sample is kept at 4°. The sliced tissue is then homogenized with a Tissuetearor (Biospec Products, Bartlesville, OK) for 30 sec at 4500 rpm. The homogenate is sequentially filtered through nylon mesh of sizes 250 and 53 μm. The filtrate is then centrifuged in a 15-ml Corex tube at 500 g for 15 min. The supernatant is aspirated off and the pellet resuspended in 250 μl of fresh isolation buffer. An aliquot of the nuclear suspension is removed and mixed with a staining solution of 0.15% (w/v) pyronin, 0.1%

[5] A. L. Rayburn, J. A. Auger, A. E. Benzinger, and A. G. Hepburn, *J. Exp. Bot.* **40**, 1179 (1989).
[6] J. C. Watson and W. F. Thompson, this series, Vol. 118, p. 57.

(w/v) methyl green. The concentration of nuclei is determined with a hemocytometer. The method described results in large numbers of intact nuclei. The normal number of nuclei obtained from a maize seedling is about 250,000 to 300,000 nuclei per isolation.

Nuclear Staining

After the nuclei are isolated, they may be stained with a variety of DNA fluorochromes. A large number of such fluorochromes exists, each with its own advantages and disadvantages. The three most popular fluorochromes for genome analysis are mithramycin, 4',6'-diamidino-2-phenylindole (DAPI), and propidium iodide (PI). All three fluorochromes have been useful in detecting both inter- as well as intraspecific genome size variation. Depending on the particular species being analyzed, one will find that only one, two, or perhaps all three will give the expected results. The procedures described below have been successfully used in a variety of plant species.

DAPI is the fluorochrome of choice when dealing with small intraspecific differences in maize. DAPI binds to both A-T and G-C base pairs and fluoresces when bound to either type. However, the fluorescence when DAPI is bound to A–T-rich DNA is greater than when bound to G–C-rich DNA. Comparisons of DAPI-stained nuclei to Feulgen-stained nuclei have demonstrated that this base pair specificity does not preclude the use of DAPI in detecting small DNA differences in maize.[5]

After the numbers of nuclei have been determined, the fluorochrome is added to the nuclei. A staining concentration of 2.5 μg of DAPI per 10^6 nuclei (determined by a titration curve) is used. The stock solution of DAPI consists of 1 μg of DAPI per 1 μl of distilled water. Critical to proper staining is the accurate determination of the number of nuclei to be stained (fluorochrome : nuclei ratio). This caveat should be heeded with all fluorochromes. The titration curve of DAPI is such that significant over- as well as underestimates of the number of nuclei result in distortions of the resulting data. Similar results, although not so extreme, have been observed with other fluorochromes. After staining, the samples are kept on ice until examined.

If one chooses PI for DNA staining, the following procedure may be used.[7] After the nuclei are pelleted, they are resuspended in staining solution, which consists of 3% (w/v) polyethylene glycol (PEG) 6000, 50 μg/ml PI, 180 units/ml RNase, 0.1% Triton X-100 in 4 mM citrate buffer. The pH of the staining solution is 7.2. The nuclei are then incubated at 37° for 20 min. After incubation, an equal volume of salt solution is added. The

[7] K. Bauer, Northwestern University, personal communication, 1989.

salt solution consists of 3% (w/v) PEG, 50 μg/ml PI, 0.1% Triton X-100 in 0.4 M NaCl. The final pH of the solution is 7.2. After adding the salt solution, the sample should be placed in the dark at 4° for at least 1 hr. As previously described, one must determine the proper titration curve for PI.

A word of caution should be noted at this juncture. The titration curve for the fluorochrome should be determined on the species of interest. Two species that vary widely with respect to nuclear DNA content may not have the same titration curves. For example, the titration curve of sorghum at 1.63 pg of DNA per $2C$ nucleus is much different than the titration curve of maize at 6.0 pg per $2C$ nucleus. After the proper staining conditions have been determined, the nuclei are ready for analysis.

Flow Analysis

After staining, the sample is passed through a flow cytometer cell sorter. A typical flow cytometer is the Coulter EPICS 751 system (Coulter Electronics, Hialeah, FL). The excitation beam is provided by a 5-W argon ion laser. The laser is tuned to the proper excitation wavelength, which is determined by the fluorochrome used. DAPI excitation is accomplished using the UV lines (351–363 nm) available from the argon laser. For PI, the excitation wavelength is 488 nm (monochromatic blue light). Emitted fluorescence is detected by a PMT (photomultiplier tube). Appropriate optical filters provide the fluorescence detection path. The emission maximum for DAPI is 470 nm. The filter used for DAPI is a 418 nm longpass laser-blocking filter. The emission maximum for PI is 623 nm. For PI, the system uses two optical filters. A 457–502 nm laser-blocking filter is used in conjunction with a 515 nm longpass filter. A minimum of 5000 nuclei are examined per sample. The flow rate for the sample is approximately 300 nuclei/sec.

On analysis a histogram is obtained (Figs. 1–3). The frequency of nuclei is plotted versus the fluorescence channel (or fluorescence intensity) using a linear scale. Owing to stoichiometric binding of dye to DNA, a relationship exists between fluorescence channel and DNA content. The higher the fluorescence channel, the more DNA is present in the nucleus. A histogram of nuclei, obtained from tissue which contains dividing cells, has three components. Each component relates directly to a stage of the cell cycle. The first peak (the peak with the lowest fluorescence channel mean) contains nuclei isolated from cells in the G_0–G_1 portion of the cell cycle and are henceforward referred to as the G_1 peak. The second peak, with the higher fluorescence intensity, contains nuclei from the G_2 portion of the cell cycle. The mean of the G_2 peak should be approximately twice that of the G_1 peak. The ratio of the area under the G_1 peak to the area

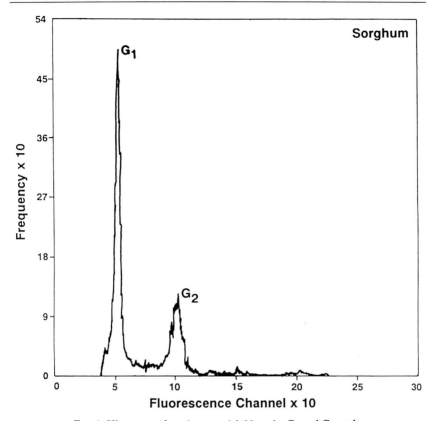

FIG. 1. Histogram of sorghum nuclei. Note the G_1 and G_2 peaks.

under the G_2 peak changes depending on the amount of meristematic activity in the sample. Nuclei isolated from rapidly growing root tips have a much larger G_2 peak than nuclei obtained from tissue that contains almost entirely differentiated cells with only a small amount of meristematic activity (leaf). The area between the two-peaks represent the S phase of the cell cycle.

The coefficient of variation *(CV)* of the G_1 and G_2 peaks may be used to gauge the isolation and staining procedures. When using DAPI, a *CV* of approximately 3.5 to 4.5 appears to reflect proper staining and isolation procedures (A. L. Rayburn, unpublished). If the *CV* is lower than 3.5, the nuclei are generally understained. If the *CV* is higher than 4.5, either the samples are overstained or the isolation procedures resulted in fragmented nuclei. With PI, however, *CV*s of 2.0 to 3.0 are most common, with *CV*s less than 1.0 occurring frequently. Understaining with PI does not result in any appreciable effect on the *CV*. *CV*s above 3.5 indicate suboptimal

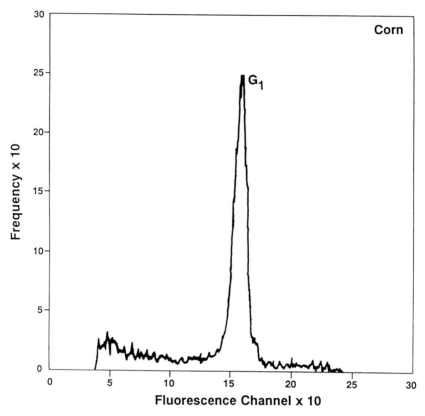

FIG. 2. Histogram of corn nuclei. Note that only the G_1 peak is observed. Owing to the scale used the G_2 peak is not seen.

nuclear isolation conditions. It should be noted that *CV*s can be used in this manner only if either (1) nuclei are isolated from individual plants and are kept separate during analysis or (2) multiple plants are combined from populations which should be homogeneous for genome size such as maize inbred lines. If plants of a population are segregating for genome size, mixing plants will result in increases in the *CV*s of the G_1 peaks independent of staining or isolation conditions.

For comparisons of genome size, the G_1 peaks from the species of interest are compared. Initially, a species should be selected as a standard. The fluorescence channel or fluorescence intensity of nuclei is not a meaningful number unless it is compared relative to a standard. Therefore, each day an analysis is run, at least two samples of the standard should be examined. The flow cytometer parameters are set such that all analysis will

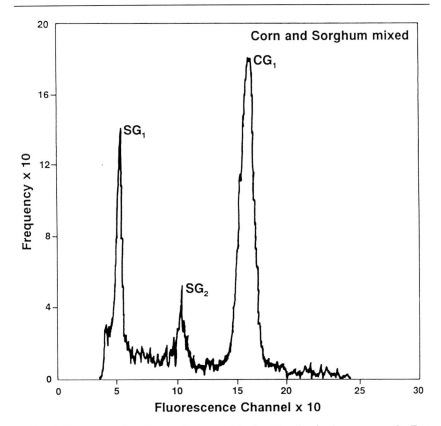

FIG. 3. Histogram of sorghum and corn nuclei mixed together in the same sample. Two sorghum peaks (SG$_1$ and SG$_2$) are observed, but only one corn peak (CG$_1$) is observed.

be comparisons between the experimental and the standard line. The absolute channel number obtained directly from the analysis means little by itself.

Standards may be used in two different manners. First, the standard may be an external standard. It is this method that was described above. The experimental samples and the standard samples are kept separate during the isolation, staining, and flow analysis. Each day the fluorescence intensity of the experimental lines are calculated relative to the standard. The standard is defined as 100% or 100 AU. Second, the standard may be used as an internal standard. In an internal standard, the standard and experimental lines are mixed at the isolation, staining, or analysis stage. This results in two G$_1$ peaks in the resulting histogram (Fig. 3). The relative position of the experimental G$_1$ peak with respect to the G$_1$ peak of the standard is used to determine genome size. It is important that neither the G_1 nor the G_2 peak of the standard overlap with the G_1 peak of the experimental sample. Also the G_2 peak of the experimental sample should not be in close proximity to the G_1 peak of the standard. Overlapping of peaks results in a distortion of the peaks and results in mean estimates that are not true means of the G_1 peaks. This is especially important when trying to detect small amounts of DNA variation. If, however, one picks a standard that is far removed from the experimental G_1 peak, internal standards work quite well. The major disadvantage arises if the staining titration curves for the experimental and the standard nuclei are very different. In this case, external standards should be used. It should be noted that either method can give comparable results. On estimating the genome size of maize using sorghum as a standard, the relative genome size of maize is the same regardless of whether an external or internal standard is used (Figs. 1–3).

The above techniques have been used to detect intra- as well as inter-specific DNA content differences.[8-10] The critical factors in estimating genome content by this method are the isolation of intact stable nuclei and the staining procedure used. If only ploidy differences are of interest, either of the described staining procedures can be used, as well as several others not mentioned. If the differences among the experimental plants are in the 15–25% range, the staining becomes more critical, and the fluorochromes of choice at least in maize are PI and DAPI. If the differences are 3–5% or less, it is extremely important that the fluorochrome and staining parameters are well defined. Not all fluorochromes allow one to detect such small differences. In maize, DAPI is the fluorochrome of choice for this type of analysis. Therefore, if the experiments are critically designed, flow cytometric analysis allows for the rapid and efficient estimation of genome size in a large number of individuals. Accurate genome content comparison studies in plants are now possible with this technology.

---

[8] L. M. McMurphy and A. L. Rayburn, *Crop Sci.* **31**, 63 (1991).

[9] A. L. Rayburn, *Evol. Trends Plants* **4**, 53 (1990).

[10] D. W. Galbraith, K. R. Harkins, J. M. Maddox, N. M. Ayres, D. P. Sharma, and E. Firoozabady, *Science* **220**, 1049 (1983).

[16] Evolutionary Analyses of Repetitive DNA Sequences

By MARK A. BATZER, CARL W. SCHMID, and PRESCOTT L. DEININGER

Introduction

In the mid 1970s, Britten and colleagues developed simple, reliable procedures for renaturing DNA duplexes from dissociated single strands.[1] These investigators immediately recognized that both the rate of cross-hybridization and the thermal stability of resulting interspecies DNA heteroduplexes measure the DNA sequence relatedness of divergent species. Whereas the single-copy sequences are directly suitable for phylogenetic comparisons, repetitive sequences present complications. A single-copy sequence from one species is constrained to cross-hybridize with its ortholog from a divergent species so that the mispairing of the resulting heteroduplex reflects the sequence relatedness of the two species.[2,3] In contrast, a repetitive sequence can hybridize to any one of many potential complements (i.e., paralogous sequences) and will rarely hybridize to its corresponding ortholog. The mispairing of paralogous heteroduplexes more nearly reflects the divergence of these repeated sequences from their most recent common ancestral sequence than the divergence of the species being examined.

For purposes of identification, we refer to the preceding studies as being "genomic hybridizations" in that they involve the cross-hybridization of either total DNA or some large subfraction of total DNA, that is, a mixture of many different sequences. In contrast, specific nucleotide substitutions can be targeted with short oligonucleotide probes to facilitate "sequence-specific" hybridization. Despite the limitations mentioned above, genomic hybridization of repetitive sequences has led to many worthwhile conclusions that have both stood the test of time and have indeed been verified by more direct cloning and sequencing results. These achievements are worth noting as they document the applicability of this approach. We limit this synopsis to studies of human DNA that are relevant to topics we are examining by improved techniques described below.

The rate of hybridization of repetitive human DNA suggested the existence of a single major repetitive DNA sequence family, a prediction

---

[1] E. H. Davidson, G. A. Galau, R. C. Angerer, and R. J. Britten, *Chromosoma* **51,** 253 (1975).
[2] B. H. Hoyer, N. W. van de Velde, M. Goodman, and R. B. Roberts, *J. Hum. Evol.* **1,** 645 (1972).
[3] D. E. Kohne, J. A. Chiscon, and B. H. Hoyer, *J. Hum. Evol.* **1,** 627 (1972).

METHODS IN ENZYMOLOGY, VOL. 224

that was subsequently verified by the identification of human *Alu* repeats.[4,5] The melting temperature of renatured human repeats indicated these sequences to be approximately 20% divergent from each other, a value confirmed and refined by detailed sequence comparisons.[6] Comparisons of the melting temperatures of human–chimpanzee DNA heteroduplexes showed that repetitive and single-copy sequence classes diverge at similar rates, a conclusion that has also been verified by sequence comparisons.[7,8] Although the thermal stability of DNA heteroduplexes accurately indicated the relative divergence of the major family of repeats in divergent primates (i.e., human, chimpanzee, monkey, and galago, a prosimian), these observations did not reveal specific sequence differences that distinguish the major family of repeats in human and galago genomes.[9,10] We think it is likely that today the same questions would be investigated by very different and more incisive procedures and do not review these earlier genomic hybridization methods.

The study of repetitive DNA sequences has been refined by advances in cloning, sequencing, and oligonucleotide synthesis. Whereas the original studies of repeated DNA evolution had to analyze whole families of repeated DNA sequences using genomic hybridization techniques, it is now possible to use cloning and DNA sequence analysis to define subfamilies of repeated DNA sequences. These subfamilies may then be characterized rapidly utilizing other approaches, such as specific oligonucleotide probes and the polymerase chain reaction (PCR).[11] It is these procedures that are covered in more detail here.

Current Approaches

It is difficult to improve on direct DNA sequence comparisons for evolution studies of the repeated DNA sequences. The only drawback is that these studies are relatively labor intensive, limiting the experimental sample to a much smaller one than can be studied using hybridization procedures. As a first step in analyzing any repeated DNA family, however, several independent copies should be sequenced in order to obtain some

---

[4] C. M. Houck, F. P. Rinehart, and C. W. Schmid, *Biochim. Biophys. Acta* **518**, 37 (1978).

[5] C. M. Houck, F. P. Rinehart, and C. W. Schmid, *J. Mol. Biol.* **132**, 289 (1979).

[6] P. L. Deininger, D. J. Jolly, C. M. Rubin, T. Friedmann, and C. W. Schmid, *J. Mol. Biol.* **151**, 17 (1981).

[7] P. L. Deininger and C. W. Schmid, *Science* **194**, 846 (1976).

[8] I. Sawada, C. Willard, C.-K. J. Shen, B. Chapman, A. C. Wilson, and C. W. Schmid, *J. Mol. Evol.* **22**, 316 (1985).

[9] P. L. Deininger and C. W. Schmid, *J. Mol. Biol.* **127**, 437 (1979).

[10] G. R. Daniels, G. M. Fox, D. Loewensteiner, C. W. Schmid, and P. L. Deininger, *Nucleic Acids Res.* **11**, 7579 (1983).

[11] K. B. Mullis and F. A. Faloona, this series, Vol. 155, p. 335.

knowledge of the structure and general variability of the sequences in the family. Sequence analysis of fairly large numbers of sequences may eventually be required to determine the detailed subfamily structure of a repeated DNA family. Traditional subcloning procedures for sequencing are now being supplemented and superseded by a variety of extremely promising PCR approaches[12] and novel cloning vectors such as λ ZAP II.[13] For these reasons, we think it is both likely and desirable that future investigations of phylogenetic relatedness will rely increasingly on DNA sequence comparisons and PCR approaches. However, hybridization techniques continue to be a valuable indirect method of rapidly comparing sequences at a large number of loci or between a large number of individuals or species. Additionally, they provide simple, effective methods of isolating clones for subsequent sequence analysis.

Oligonucleotide synthesis provides precisely defined sequences for use as hybridization probes to specific sequences. Once a family, or subfamily, of repeated sequences is defined in sequence, further DNA sequences are easily determined and sequence data accumulate in readily accessible databanks. As suggested above, genomic hybridization studies involving the cross-hybridization of many different sequences are relatively insensitive to precise differences that may distinguish otherwise closely related sequences. In contrast, the utility of oligonucleotide hybridization probes for this purpose is illustrated by recent findings concerning human *Alu* repeats. The number of human *Alu* repeats that have been sequenced is especially large owing both to the ubiquity of *Alu* repeats in the human genome and to the special emphasis human DNA has received in sequence studies. The nucleotide sequences of individual subfamily members can be aligned and family or subfamily consensus sequences determined (Fig. 1A). The various subfamilies are defined by members which share common nucleotide variants. Independent analysis of *Alu* sequences by six laboratories suggested the existence of distinct *Alu* sequence subfamilies that inserted into the human genome at different times in evolution.[14-19] Examples of the consensus sequences advanced for these subfamilies are shown in Fig. 1B. Whereas six laboratories arrived at similar conclusions

[12] W. Bloch, *Biochemistry* **30**, 2735 (1991).
[13] J. M. Short, J. M. Fernandez, J. A. Sorge, and W. D. Huse, *Nucleic Acids Res.* **16**, 7583 (1988).
[14] V. Slagel, E. Flemington, V. Traina-Dorge, H. Bradshaw, Jr., and P. L. Deininger, *Mol. Biol. Evol.* **4**, 19 (1987).
[15] C. Willard, H. T. Nguyen, and C. W. Schmid, *J. Mol. Evol.* **26**, 180 (1987).
[16] R. J. Britten, W. F. Baron, D. B. Stout, and E. H. Davidson, *Proc. Natl. Acad. Sci. U.S.A.* **85**, 4770 (1988).
[17] J. Jurka and T. Smith, *Proc. Natl. Acad. Sci. U.S.A.* **85**, 4775 (1988).
[18] Y. Quentin, *J. Mol. Evol.* **27**, 194 (1988).
[19] D. Labuda and G. Striker, *Nucleic Acids Res.* **17**, 2477 (1989).

A

```
              110       120       130       140       150       160       170       180       190       200
HS-1 CON  TGAAACCCCGTCTACTAAAAATACAAAAATTAGCGGGCGGCCGTAGGGGCCCCCGCCCTGTAGTCCCAGCTACTTGGGAGGCTGAGGCAGGAGAATGCCGT
PV 92     ..........T.........................................................................................
PV 83     ....................................................................................................
PV 6      ....................................................G..M.............................................
HS C2N4   ..................................................A...T.............M...T............................
HS C3N1   ....................................................A..T.............................................
HS C4N4   ..................................................A...T.............................................
HS C4N6   ...................................................A.............................................T...
HS C4N8   ...................................................A.................................................

PV 71     ...................................................C....N.........T.T................................
TPA 25    .....A.............................................C....N.........C..................................A.
HS C4N5   ...................................................C....N.........N.N................................
```

B

```
              10        20        30        40        50        60        70        80        90        100
PS        GGCCGGGCGCGTGGCTTCACGCCTGTAATCCCAGCACTTTGGGAGGCCGAGGCGGGCGGATCACCTGAGGTCAGGAGTTCGAGACCAGCCTGCCAACAT
AS        ....................................................................................................
CS        ..................................................................................................G.
HS-1      ............................................................................A.......T..T.....C......
HS-2      ............................................................................A.......T.C.T.A.C.......

              110       120       130       140       150       160       170       180       190       199
PS        GGTGAAACCCCGTCTCTACTAAAAATACAAAA--TTAGCCGGGCGTGGTGGCGCGCGCCTGTAGTCCCAGCTACTCGGGAGGCTGAGCGAGGAGAATCGC
AS        ....................................................................................................
CS        ....................................A..........G...............................................G....
HS-1      ....................................A..........G.........................T..........T...............
HS-2      ....................................A-.........G.........................T..........T.............G.

              210       220       230       240       250       260       270       280       290
PS        TTGAACCCGGGAGGCGGAGGTTGCAGTGAGCCGAGATCGCGCCACTGCACTCCAGCCTGGGCGACAGAGCGAGACTCCGTCTCTCAAAAAAA
AS        G.........C..........................................................................
CS        G.........C..............................C..................................
HS-1      ..........C..............................C..................................
```

concerning the existence of *Alu* subfamilies,[14-19] there are naturally some differences in details concerning the number of subfamilies and refinement in the corresponding consensus sequences. There is also no common nomenclature for the *Alu* subfamilies. Because this chapter is not intended to resolve these issues, but rather to describe procedures that can be employed to distinguish between closely related sequences, we arbitrarily adopt the subfamily names identified by the Deininger group[20,21] to provide specific examples for our discussion. Matera *et al.*[22,23] have studied the identical subfamilies, albeit under different names, so that the hybridization procedure and results of our two laboratories can be directly compared. The human-specific (HS-1) subfamily differs by five concerted mutations from the cattarhine-specific (CS) subfamily (Fig. 1B). Two of

[20] M. A. Batzer and P. L. Deininger, *Genomics* **9**, 481 (1991).
[21] M. R. Shen, M. A. Batzer, and P. L. Deininger, *J. Mol. Evol.* **33**, 311 (1991).
[22] A. G. Matera, U. Hellmann, and C. W. Schmid, *Mol. Cell. Biol.* **10**, 5424 (1990).
[23] A. G. Matera, U. Hellmann, M. F. Hintz, and C. W. Schmid, *Nucleic Acids Res.* **18**, 6019 (1990).

FIG. 1. Alignment of several *Alu* subfamily members and comparison of five *Alu* consensus sequences. (A) Partial alignment of the TPA 25 *Alu* family member [S. J. Friezner Degen, B. Rajput, and E. Reich, *J. Biol. Chem.* **261**, 6972 (1986)] and several additional *Alu* HS subfamily members (see the references below). The consensus sequence (CON) is depicted at the top and represents the most common nucleotide found within the subfamily members at each position. Positions in the individual sequences that are the same as the consensus sequence are represented as dots. Substitutions are marked with the appropriate nucleotide, and deletions are indicated with an x or −. The boxed nucleotides represent HS-2 subfamily diagnostic mutations. (B) Representation of the five *Alu* consensus sequences as reported by Shen *et al.* [M. R. Shen, M. A. Batzer, and P. L. Deininger, *J. Mol. Evol.* **33**, 311 (1991)]. Each of the consensus sequences is defined by a number of diagnostic mutations and has been given a biologically relevant name. The PS (primate-specific) *Alu* consensus sequence represents the oldest and largest subfamily of *Alu* sequences found within primate genomes. The AS (anthropoid-specific) *Alu* consensus sequence differs from the PS consensus by a single 2-base pair deletion at position 65. The CS (catarrhine-specific) subfamily consensus sequence shares nine diagnostic mutations that are not found in the AS consensus sequence. The HS-1 (human-specific-1) and HS-2 (human-specific-2) consensus sequences are defined by five and three unique diagnostic mutations from the CS and HS-1 consensus sequences, respectively. The observation that each of the subfamilies has all of the diagnostic changes of the previous subfamily, as well as unique changes, supports the sequential appearance of different subfamilies within the genome. The HS subfamilies represent the most recently amplified *Alu* family members found within the human genome [M. A. Batzer, G. E. Kilroy, P. E. Richard, T. H. Shaikh, T. D. Desselle, C. L. Hoppens, and P. L. Deininger, *Nucleic Acids Res.* **18**, 6793 (1990); A. G. Matera, U. Hellmann, and C. W. Schmid, *Mol. Cell. Biol.* **10**, 5424 (1990); A. G. Matera, U. Hellmann, M. F. Hintz, and C. W. Schmid, *Nucleic Acids Res.* **18**, 6019 (1990); M. A. Batzer and P. L. Deininger, *Genomics* **9**, 481 (1991)].

these five differences, a C at position 91 and an A at position 98, are close enough so that both laboratories experimentally confirmed and extended these sequence predictions using oligonucleotide hybridization probes directed to this region (see Methods).

Special Considerations for Repetitive DNA Evolution

Many of the other chapters in this volume deal with evolutionary analyses of specific genes and unique DNA sequences.[24-28] There are, however, some evolutionary aspects unique to repeated DNA sequences. The most important of these factors is the amplification dynamics. Sequences become repetitive because there are amplification processes that make extra copies of them. These include retroposition and transposition mechanisms that would explain the majority of interspersed repeated DNA sequences, as well as recombination or replication slippage mechanisms that would probably explain most tandem replications. For any given repeated sequence, various factors may combine to increase or decrease the amplification rate of that sequence at various times in the evolutionary process. Thus, the dynamics of the amplification process could greatly affect the observed evolution of the family. This is particularly important in cross-species comparisons, because the amplification dynamics of a specific repeated DNA family may be altered in one species, relative to another.

Once a sequence amplification event occurs, the nature of any selection on the copies is important. In many (or even most) cases, it appears that the majority of repeated DNA sequences represent pseudogenes, which mutate at a neutral rate of evolution.[8] Along with amplification dynamics, the possible removal of repeated sequences must also be considered. Removal does not seem to play a major role with the interspersed repeated DNA elements,[8,29,30] but it is likely to be important in tandemly repeated satellite elements. Other mechanisms might also alter evolution of parts of a repeated DNA sequence. For instance, human *Alu* family copies are initially rich in CpG dinucleotides. These sites appear to be approximately 10-fold more subject to mutation than other sites in the genome,[19,31]

[24] D. Stahl, this volume [27].
[25] E. P. Lessa, this volume [31].
[26] J. M. Chesnick and R. A. Cattolico, this volume [13].
[27] D. B. Stein, this volume [12].
[28] R. DeSalle, A. K. Williams, and M. George, this volume [14].
[29] I. Sawada and C. W. Schmid, *J. Mol. Biol.* **192,** 693 (1986).
[30] B. F. Koop, M. M. Miyamoto, J. E. Embury, M. Goodman, J. Czelusniak, and J. L. Slightom, *J. Mol. Evol.* **24,** 94 (1986).
[31] M. A. Batzer, G. E. Kilroy, P. E. Richard, T. H. Shaikh, T. D. Desselle, C. L. Hoppens, and P. L. Deininger, *Nucleic Acids Res.* **18,** 6793 (1990).

probably because of methylation of these sites in the copies.[32,33] Other sequences, such as regions containing short repeated segments or homopolymeric runs, also seem subject to higher rates of mutation.

Methods

The study of repetitive sequences utilizes many relatively routine techniques in molecular genetics that are described in a number of excellent manuals[34,35] and are not discussed here. Instead, we consider the aspects of experimental design and data analysis that are specific for the study of repeated DNA families.

Cloning Repetitive DNA Sequences

Several special considerations arise when cloning repeated DNA sequences. One consideration is the stability of repeated sequences cloned into *Escherichia coli*. Instability is generally attributed to recombinations between tandemly repeated or inverted repeated sequences. These problems may be minimized by keeping the insert size as small as possible. Several genetic factors also influence instabilities in *Escherichia coli*. These include the general host restriction and modification systems, as well as *RecA* and *RecB* (homologous recombination) and *uvrC* and *umuC* (recombination involving inverted repeats).[36] Methylation has also been found to have a significant effect on the cloning of methylated DNA fragments, with hosts deficient in *mcrA* and *mcrB* host methylation being the best choice.[37]

Second, it is important to consider whether a clone library will be representative of a particular repeated sequence. Besides the genetic factors, above, unusual patterns of restriction sites in some repeated sequences may influence their relative abundance in a library. This would be more likely for a tandemly repeated sequence or a very long repeated sequence than for short, interspersed repeated DNA sequences. Traditional λ or plasmid libraries would be sufficient for most studies, but in certain situations it might be necessary to resort to DNA libraries of randomly frag-

[32] C. Coulondre, J. H. Miller, P. J. Farabaugh, and W. Gilbert, *Nature (London)* **244**, 775 (1978).

[33] A. P. Bird, *Nucleic Acids Res.* **8**, 1499 (1980).

[34] F. M. Ausabel, R. Brent, R. E. Kingston, D. D. Moore, J. G. Seidman, J. A. Smith, and K. Struhl, "Current Protocols in Molecular Biology." Wiley, New York, 1987.

[35] J. Sambrook, E. F. Fritsch, and T. Maniatis, "Molecular Cloning: A Laboratory Manual." Cold Spring Harbor Laboratory, Cold Spring Harbor, New York, 1989.

[36] A. Greener, *Strategies* **3**, 5 (1990).

[37] J. P. Doherty, M. W. Graham, M. E. Linsenmeyer, P. J. Crowther, M. Williamson, and D. M. Woodcock, *Gene* **98**, 77 (1991).

mented DNA derived from sonicated DNA for short fragments[38] or from DNA sheared through a syringe for large fragments.[20,31]

Screening of a library with standard hybridization conditions [42°, 1 M NaCl in 50% formamide, with a final wash at 65° in 0.1 × standard saline citrate (SSC)] will detect sequences having a maximum 20–30% mismatch. For more divergent repetitive sequences, a screening may also be attempted under somewhat lower stringency (e.g., 37° hybridization with a final wash at 50° in 1 × SSC) to determine whether a large number of sequences can be detected. The source of the probe may represent either sequences from a previously isolated member of a repeated DNA family or simply radiolabeled genomic DNA. In the latter screening, only those sequences that are present at a fairly high copy number (i.e., represent greater than 0.1% of that genome) will produce a hybridization signal in this experiment. The sensitivity of this approach could easily be increased by utilizing a $C_o t$ fractionation to isolate various repetitive fractions that could then be utilized to probe the library.[38a,39]

Sequence Determination

Routine sequencing may be carried out using either shotgun or sequential deletion procedures.[40] The latter strategy is particularly useful to help align segments of a long tandemly repeated sequence. However, for experiments involving sequence analysis of multiple members of an interspersed repeated DNA family, sequence determination using sequencing primers from within the repeated DNA sequence can greatly streamline the analysis. The primers are generally 17–20 bases in length. By utilizing primers in both orientations and sequencing both strands of the sequence, many copies of a repeated DNA family, including their immediate flanking regions, can be rapidly and accurately sequenced (Fig. 2). In a repeated DNA family with a great deal of sequence mismatch this may not work well, as it is important that the primer match the sequence reasonably well, particularly at the last several 3′ bases.[41] Difficulties may also arise if more than one copy of the repeated sequence are present in a single recombinant DNA molecule. This could result in determination of a mixed sequence. However, it is also possible to minimize this problem in some cases. For instance, if efforts are being made to sequence members of a specific repeated DNA subfamily (see below), a primer can be made which will only sequence members of that subfamily by placing one of the subfamily

[38] P. L. Deininger, *Anal. Biochem.* **129**, 216 (1983).
[38a] M. S. Springer and R. J. Britten, this volume [17].
[39] P. E. Nisson, P. C. Watkins, J. C. Menninger, and D. C. Ward, *Focus* **13**, 42 (1991).
[40] P. L. Deininger, *Anal. Biochem.* **135**, 247 (1983).
[41] M. A. Batzer and P. L. Deininger, unpublished data (1989).

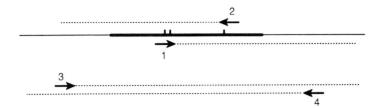

FIG. 2. Schematic representation of a DNA sequencing strategy for the analysis of repetitive DNA sequences. The individual *Alu* (repetitive) family member is depicted by the thick line. DNA sequencing primers are indicated by the arrows (1–4). Diagnostic mutations unique to the particular subfamily are indicated by the tick marks near the 3' end of sequencing primers 1 and 2. Initially primers that anneal within the repetitive element are used to generate DNA sequence information, which begins within the element and proceeds out to unique DNA sequences 3' and 5' of the element. The design of these primers allows exact base pairing only with subfamily members (owing to the 3' unique subfamily diagnostic mutations), permitting the analysis of relatively large clones that may contain more than one repetitive element. Subsequent primers complementary to the unique 5' and 3' flanking sequences (3 and 4, respectively) can be made for generating overlapping nucleotide sequence information or for PCR analysis of the locus.

diagnostic mutations at the 3' end of the sequencing primer. This primer will be very ineffective at sequencing members of the same repeated DNA family that do not have the diagnostic change.

Analysis of Sequences

There are two basic strategies for comparing repeated sequences. The first, and most common, is to align the sequences (as in Fig. 1A) to develop a consensus sequence. Each individual sequence can then be compared to the consensus sequence. Many repetitive DNA sequences will vary from their consensus by 0.5% to more than 30%.[6,14] The consensus sequence provides an improved estimate of what the ancestral or parental repeated DNA sequence looked like prior to accumulation of mutations in the individual copies. In sequence families that have distinct subfamilies, such studies may be somewhat misleading unless the consensus used is that for the appropriate subfamily. However, such an alignment may help detect changes within subgroups of the repeated DNA family members that may represent subfamilies (Fig. 1A). There are multiple alignment programs (e.g., CLUSTAL in the PC/GENE suite from Intelligenetics) which can also help align sequences. However, the alignments almost always will require manual improvement, as these programs tend to include more insertions and deletions than necessary.

The alternate form of analysis would be to carry out pairwise comparisons of individual repeated DNA sequence members.[14,42] Such pairwise comparisons can provide an excellent method for suggesting subfamily structure, as well as dating a subfamily age. However, it is important to check the alignment criteria and manually refine the alignments in these analyses. Other chapters of this[43] and other texts[44] cover phylogenetic tree formation from such alignments in detail.

In making the alignments and determining subfamily structure, one must be aware that some sequence changes may occur in parallel in totally different members of a family. Changes that are held in common do not always indicate a subfamily (see Fig. 1A). Some analyses deal with this by including statistics on the probability of multiple common changes occurring in two family members.[45] One must also consider sequences (such as potential CpG methylation sites and simple sequences) that may be especially prone to specific types of sequence changes and may mimic subfamily relations, when they really represent parallel changes in random family members (Fig. 1A). For example, position 143 in Fig. 1A shows parallel mutations of a CpG dinucleotide to CpA. If subfamily changes are suspected, they may then be confirmed by further sequencing and oligonucleotide hybridization studies as described below.

Specific Sequence Hybridization

Specific sequence hybridizations utilize specific oligonucleotides or longer probes to detect repeated sequence subfamilies (as discussed below). The use of repeated DNA probes to screen recombinant DNA libraries for new sequence members has been discussed in general above. We consider this approach to be one of the best and most direct methods for determination of repeated DNA sequence copy number as well. If the repetitive family is randomly represented in the library, the most direct count of repetitive sequence members can be estimated by screening the library and determining how many hybridizing positive members are obtained relative to the number of plaques screened and the average insert size in the library. Dot blots, Southern blot hybridizations, or traditional $C_o t$ plots are alternatives, but such measurements rely on relative renaturation rates. These rates depend not only on copy number, but also on sequence length and mismatching, potentially necessitating significant corrections to the data. In addition, as a result of the subfamily structure of repeated DNA se-

[42] P. L. Deininger and V. K. Slagel, *Mol. Cell. Biol.* **8**, 4566 (1988).
[43] D. M. Hillis, M. W. Allard, and M. M. Miyamoto, this volume [34].
[44] D. L. Swofford and G. J. Olsen, *in* "Molecular Systematics" (D. M. Hillis and C. Moritz, eds.), p. 411. Sinauer, Sunderland, Massachusetts, 1990.
[45] J. Jurka and A. Milosavljevic, *J. Mol. Evol.* **32**, 105 (1991).

quences, these hybridization techniques probably have a much higher signal-to-noise ratio than library screening owing to the background caused by related but nonidentical sequences.

Hybridization techniques are also the methods of choice to look at the RNA expression of repeated DNA family members. Such studies[46-48] are not discussed here, but they can be an important part of understanding the function and evolutionary mechanisms associated with a repeated DNA family.

Either cloned sequences or chemically synthesized oligonucleotides might be used as specific sequence hybridization probes. However, the most thermally stable region in a long duplex determines the temperature at which denatured single strands separate and the probe elutes from DNA immobilized on the filter, usually the critical parameter for the observations described below. A long duplex consisting of both poorly base-paired and exactly base-paired regions might denature at the same temperature as exact sequence complements. For this reason we generally expect oligonucleotide hybridization probes to be more discriminating than probes using longer cloned sequences and recommend their use whenever possible. This expectation is documented by Southern blot hybridization of a cloned HS subfamily member to a restriction digest of total human DNA (Fig. 3), where the cloned HS subfamily member hybridizes to a prominent *Bam*HI restriction fragment of 1 kilobase (kb). Higher stringency washing eliminates hybridization to both the 1-kb *Bam*HI band and the higher molecular weight smear, so that the stability of the hybrid formed by this band is indistinguishable from that of HS subfamily members. However, sequence analysis of the 1-kb *Bam*HI fragment demonstrates that is not a member of the HS subfamily reported in Fig. 1. Rather, the sequence of this restriction fragment reveals the presence of two complete *Alu* repeats and one partial *Alu* repeat interrupted by the *Bam*HI cloning site. Included within the sequence of one *Alu* repeat is a short (31 nucleotides) GC-rich (66%) sequence that only differs by two mispairs (one of which is a GT mispair) from the cloned *Alu* hybridization probe. Plausibly the high genomic copy number of this fragment, its multiple *Alu* composition, and the excellent sequence match between short regions of the hybridization probe and the fragment might all contribute to their cross-hybridization under stringent conditions. Regardless of the correct explanation, long cloned sequences do not provide the specificity required to identify sequence subfamilies.

Selection of Oligonucleotide Hybridization Probes. The shortest possi-

[46] K. E. Paulson and C. W. Schmid, *Nucleic Acids Res.* **14,** 6145 (1986).
[47] J. B. Watson and J. G. Sutcliffe, *Mol. Cell. Biol.* **7,** 3324 (1987).
[48] J. Skowronski, T. G. Fanning, and M. F. Singer, *Mol. Cell. Biol.* **8,** 1385 (1988).

B BI E H P

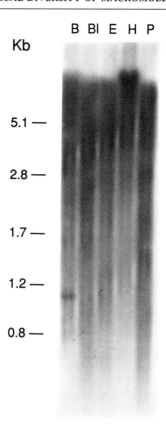

FIG. 3. Low specificity of long hybridization probes. Human DNA was digested with the following restriction enzymes: *Bam*HI (B), *Bgl*II (Bl), *Eco*RI (E), *Hin*dIII (H), and *Pst*I (P). The DNA was then transferred to filters and hybridized to a 600-bp fragment containing the polymorphic *Alu* repeat situated near the human tissue plasminogen activator gene [S. J. Friezner Degen, B. Rajput, and E. Reich, *J. Biol. Chem.* **261**, 6972 (1986)]. The blot was washed at 0.04×SSC and 60° and exposed for 4.5 hr. These washing conditions approximate that of exactly paired sequence complements. The *Bam*HI band persists even after washing at 0.025×SSC and 60°.

ble oligonucleotide that targets the maximum number of diagnostic base changes provides the most selective hybridization probe. As a lower limit on the size of the oligonucleotide, sequences of 16 or fewer nucleotides would occur at random in the human genome, which is about 2.5 billion base pairs in length. In our experience, oligonucleotides that consist of about 20 residues are sufficiently long to target a particular complement but are also sufficiently short to be sensitive to single nucleotide mismatches. To reduce background hybridization, all four nucleotides should

TABLE I
RELATIVE ORDER OF BASE-PAIRING STABILITY[a]

Watson–Crick		Non-Watson–Crick	Noncontributing
			T-T
		G-T	A-A
G-C	A-T	G-A	C-C
		G-G	C-A
			C-T

[a] The noncontributing base pairs are the most disruptive to the hybridization, whereas the Watson–Crick base pairs act as the most positive contributors as originally shown by Ikuta *et al.* [S. Ikuta, K. Takagi, R. B. Wallace, and K. Itakura, *Nucleic Acids Res.* **15**, 797 (1987)].

be represented in the target sequence in an approximately even distribution, and targets that include runs of a particular base should be avoided if possible. However, as shown below, even a run of T residues on the end of an oligonucleotide can be used successfully.

Both the position and type of sequence mismatch determine duplex stability.[49] Pyrimidine–pyrimidine mispairs tend to be maximally destabilizing, whereas mispairs involving G tend to be least destabilizing (e.g., see Table I). By judicious choice of the complementary strand to be targeted for hybridization, the most destabilizing base mispairs can be selected for the oligonucleotide sequence. The mispair provides the maximum effect on thermal stability by being centrally located in the oligonucleotide. The terminal base pairs on the two ends of a DNA duplex are only "half-stacked" so that the duplex ends are already somewhat destabilized compared to the middle; a short duplex effectively melts from its ends. A base mispair, centrally located, essentially destabilizes the region that has the greatest effect on the strand-dissociation temperature.

The thermal stability of a short DNA duplex can be estimated by the simple $4 + 2$ rule; each GC pair contributes $4°$ and each AT pair contributes $2°$ to the duplex melting temperature in $0.9\ M$ NaCl solution.[50] Although more rigorous estimates of duplex stability are possible,[51] this simple method is reasonably accurate, and, in any event, we find it useful

[49] S. Ikuta, K. Takagi, R. B. Wallace, and K. Itakura, *Nucleic Acids Res.* **15**, 797 (1987).
[50] R. B. Wallace, J. Shaffer, R. F. Murphy, J. Bonner, T. Hirose, and K. Itakura, *Nucleic Acids Res.* **6**, 3543 (1979).
[51] C. R. Cantor and P. R. Schimmel, "Biophysical Chemistry, Part I: The Conformation of Biological Macromolecules." Freeman, San Francisco, 1980.

to compare the stabilities of perfectly paired and imperfectly paired duplexes empirically.

Optimization of Washing Conditions. One approach to setting exact hybridization and washing conditions is to determine the temperature at which the oligonucleotide elutes from filter-bound hybrids (Fig. 4). For example, a 22-nucleotide probe melts from its exact HS complement (TPA) at 67° compared to 66° as predicted by the 4 + 2 rule. Incorporating the two mispairs depicted in Fig. 4 lowers the duplex melting temperature by 10° (AFP) or an average of 5° per each base mispair. In a similar calibration experiment involving a different oligonucleotide sequence, we also observe a 10° depression in duplex melting temperature resulting from two base mispairs. As a possible generalization of these observations, there is approximately a 1° depression in DNA melting temperature for each 1%

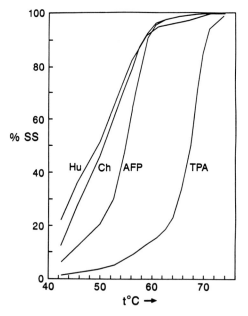

Fig. 4. Melting profiles of a subfamily-specific oligonucleotide. The oligonucleotide (5′ ATCGAGACCATCCCGGCTAAAA 3′) was melted from human (Hu), chimpanzee (Ch), and the TPA [S. J. Friezner Degen, B. Rajput, and E. Reich, *J. Biol. Chem.* **261**, 6972 (1986)] and AFP [P. E. M. Gibbs, R. Zielinski, C. Boyd, and A. Dugaiczyk, *Biochemistry* **26**, 1332 (1987)] *Alu* control DNAs. The underlined bases represent the HS-1 subfamily specific positions. The profile of BLUR 11 was indistinguishable from that of salmon sperm DNA (not shown). Note the high temperature melting component present in the human but not chimpanzee DNA. In this particular set of melting profiles, the filters were charged with 30 μg of each DNA. %SS, Percent single strand. [From A. G. Matera, U. Hellmann, and C. W. Schmid, *Mol. Cell. Biol.* **10**, 5424 (1990).]

TABLE II
EFFECT OF ADDED 3' AT BASE PAIRS[a]

	Added 3' AT pairs		
	0	1	2
Measured $T_d(°)$	40.5	44.5	46.5
Estimated T_d $(4 + 2$ rule,°)	38	40	42

[a] The oligonucleotide 5' AGACTCCGTCTC-TTTT 3' is an exact match to the HS subfamily except for the four T residues situated at the 3' end replacing the A residues normally occupying this position. Measured T_d is the thermal elution temperature of the oligonucleotide (5 × SSPE) from different DNA sequences with exact complements to the first 12 nucleotides and to additional 3' AT pairs as listed. The $4 + 2$ rule [R. B. Wallace, J. Shaffer, R. F. Murphy, J. Bonner, T. Hirose, and K. Itakura, *Nucleic Acids Res.* **6**, 3543 (1979)] predicted values are shown for comparison.

sequence mismatch.[52] Again the exact position and sequence context of a mispair can markedly influence duplex stability, so these generalizations are subject to the peculiarities of any particular oligonucleotide.

In one unfavorable case, we wished to isolate *Alu* members with 3' ends that terminate in four or more T residues rather than the A-rich region which normally occupies this position. Our strategy was to first determine the dissociation temperatures (T_d) of exact complements having no T, one T, and two T residues, which are summarized in Table II. Interestingly, the thermal stability of these structures increase in about 2° increments for each added T residue, as predicted by the $4 + 2$ rule. Using the preliminary calibration shown in Table II and modified hybridization and washing conditions that we use in library screening (see below), we succeeded in isolating *Alu* complements that terminate in four or more 3' T residues. Based on these experiences, we find it useful to preface library screening with simple filter hybrid melts to define the useful temperature range of the selected oligonucleotide and then to make judicious choices for the library screening conditions as described below.

[52] T. I. Bonner, T. D. Brenner, B. R. Neufeld, and R. J. Britten, *J. Mol. Biol.* **81**, 123 (1973).

60°C   65°C

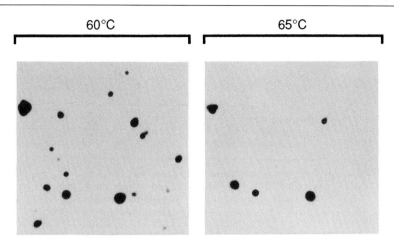

FIG. 5. Effect of varying the stringency of the final washes on the specificity of an oligonucleotide probe. Plaque lifts were performed on a human genomic library. The lifts were hybridized to the human-specific Alu family member oligonucleotide probe (5′ CACCGTTTTAGCCGGGATGG 3′, with the underlined bases representing the HS-1 specific positions) as previously described [M. A. Batzer and P. L. Deininger, *Genomics* **9**, 481 (1991)]. The final washes were then performed using 6× SSC, 0.05% sodium pyrophosphate at 60°, followed by final washes of the same filter at 65°. The autoradiographs were much cleaner after washes at 65° than at 60°, allowing the isolation of exact complements to the oligonucleotide at the higher temperature. Previous studies in our laboratory have shown that the clones which hybridize less intensely at 60° and subsequently disappear result from hybridization of inexact complements to the oligonucleotide.

Alternatively, the exact temperature of stringent washes that facilitate the isolation of perfect complements may be determined using library screening. Using the 4 + 2 rule, the T_d of another oligonucleotide that was used to isolate HS subfamily members (Fig. 5) should be 64°. After hybridization with the human-specific oligonucleotide at 42° overnight to plaque lifts from a human genomic library,[53] a comparison of washes (6× SSC/0.05% sodium pyrophosphate) at two different temperatures was made (Fig. 5). Filters washed at 60° contained both light and dark hybridizing plaques, whereas those washed at 65° contained only dark hybridizing plaques. Experiments in our laboratory have shown that the less intense hybridizations result from imperfect hybrids, whereas clones containing exact complements hybridize more intensely. Using this approach, the ideal temperature of the most stringent wash (65° in this case) to isolate perfect complements for any oligonucleotide probe can be determined.

[53] D. Woods, *Focus* **6**, 1 (1984).

In addition to the washing temperature, the hybridization temperature should also be selected to reduce background.[49] By hybridizing at the highest possible temperature, hybridization to inexact complements is minimized; the subsequent stringent washing then further reduces what is already a diminished background. We typically hybridize at about 5° lower than the elution temperature that was determined in the previously discussed filter hybrid melts (Fig. 4). Whenever possible, positive and negative control lifts of exact and inexact complements should be included in the library screening. After hybridization, usually 4 hr or overnight, we exhaustively wash with several room temperature changes of $5 \times$ SSPE until the wash shows negligible radioactivity compared to the filters, as judged by a hand-held radioactivity monitor. One or more stringent washes are then performed for 5 min with shaking at a temperature just below that of the sharpest rise in the transition for the filter hybrid melting profile of exact sequence complements (e.g., 63° in the example of Fig. 4). Stringent washes are always followed by room temperature washes to dilute any residual radioactivity on the damp filters. Both the positive and negative controls are directly followed by a hand-held monitor during these procedures to ensure the selectivity of the stringent washing. If the background is too high, more stringent washing can be subsequently employed. Theory suggests that multiple stringent washings reduce background more than the signal, although the authentic hybridization signal is also diminished. Again, the internal positive and negative controls provide confidence that the procedures are being appropriately executed. We routinely perform our most stringent final washes at, or even below, the thermal elution temperature of exact sequence complements (e.g., 67° in the example of Fig. 4) The radioautograph for the exact complements following this most stringent wash should be noticeably less intense than that resulting from the previous less stringent washings, and it is hoped that there is no radioautographic exposure resulting for the negative control.

Analysis of Orthologous Loci

An alternative to the analysis of random copies of a repetitive DNA family is to study the evolution of a single repetitive DNA family member at a given locus. Such studies have proved very important in eliminating factors such as gene conversion and excision of repetitive sequences as being important considerations in *Alu* evolution.[8,29,30] They are also the most direct measure of the divergence rate seen for repetitive sequence family members. Traditionally these experiments have involved the cloning of a given genetic locus from a number of species and sequence analysis of that region to allow comparison. The PCR approach described below now makes such studies much more rapid and capable of being carried out easily through a wide range of species.

Choice of Primers. The development of the PCR has facilitated the exponential amplification of specific DNA sequences. This technique may be applied to the analysis of orthologous repetitive loci as described below. Initially oligonucleotide primers complementary to unique DNA sequences flanking any repetitive DNA element of interest are chosen manually[11] or with the aid of a computer program such as OLIGO.[54] These primers generally are 25 bases long, contain equal numbers of A, G, C, and T nucleotides, have about the same T_d as calculated by the 4 + 2 rule, and are manually compared to each other to preclude primer–dimer amplification. The primers are then searched against the EMBL/GenBank database (using a program such as QGSEARCH in the PC/GENE suite from Intelligenetics) to determine whether they reside in a previously described region of the genome. For efficient evolutionary PCR, the match of the primers with target DNA at the most 3′ nucleotides is critical for successful amplification.[55] We have previously found that the inclusion of an inosine residue at the 3′ terminal nucleotide mitigates mismatches at the 3′ terminal nucleotide, thereby enhancing the range and reproducibility of evolutionary PCR.[56]

Reaction Conditions and Optimization of Annealing Temperature. Amplification of repetitive loci is typically carried out in a 100-μl reaction consisting of 100 ng of target DNA, 750 ng of each primer, 2.5 units of *Taq* DNA polymerase, a 10× reaction buffer (generally supplied by the manufacturer of the *Taq* polymerase) and 200 μM deoxynucleoside triphosphates (dNTPs). Reactions are carried out for 30 cycles, with each cycle consisting of 1 min at 94° (denaturation), 2 min at an experimentally determined annealing temperature, and 2 min at 72° (extension). One-fifth (20 μl) of the reaction products are subsequently fractionated on a 2% agarose gel containing 0.5 μg/ml ethidium bromide and visualized directly by UV fluorescence. The optimal annealing temperature for any set of primers is determined by amplifying target DNA using different annealing temperatures beginning at 5°–10° below the T_d of either member of the primer pair. The specificity of the reaction increases with increasing temperature, with the reaction products proceeding from a smear of nonspecific amplification products to the amplification of one or a few specific bands.

Amplification of Orthologous Loci. Once the optimal annealing temperature is determined (generally the highest temperature that provides

[54] W. Rychlik and R. E. Rhoads, *Nucleic Acids Res.* **17**, 8543 (1989).
[55] G. Sarker, J. Cassady, C. D. K. Bottema, and S. S. Sommer, *Anal. Biochem.* **186**, 64 (1990).
[56] M. A. Batzer, J. E. Carlton, and P. L. Deininger, *Nucleic Acids Res.* **19**, 5081 (1991).

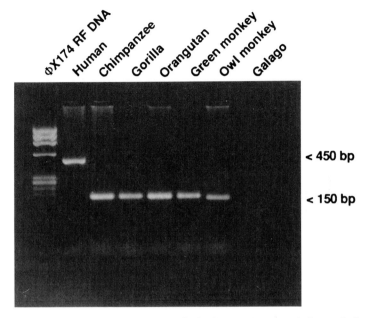

FIG. 6. PCR analysis of an individual *Alu* family member at orthologous loci within primate genomes. PCR amplification was carried out with unique primers (schematically demonstrated in Fig. 2, primers 3 and 4) flanking the HS C4N4 *Alu* family member [M. A. Batzer and P. L. Deininger, *Genomics* **9**, 481 (1991)]. Products resulting from the amplification of *Alu* subfamily member HS C4N4 were run on an agarose gel containing ethidium bromide and visualized by UV fluorescence. A 450-bp band is present if the *Alu* family member is located within the chromosome, whereas heterozygotes produce both bands; genomes that do not contain an *Alu* family member insertion produce only the 150-bp band. The marker was φX174 RF DNA digested with *Hae*III. The analysis shows that this *Alu* family member is located only within the human genome and is absent from the genomes of chimpanzee, gorilla, orangutan, green monkey, and owl monkey. No fragment was amplified from the galago genome, indicating that the galago was too divergent for the effective use of primers predicted from a gene located within the human genome.

sufficient specific product), orthologous loci can generally be amplified. The amplification of an *Alu* HS subfamily member (HS C4N4) locus is shown in Fig. 6. We can see that a 450-base pair (bp) fragment (indicating that a 300-bp *Alu* family member inserted between the two oligonucleotide primers) is present only in human DNA. Amplification of other ape (chimpanzee, gorilla, and orangutan), Old World monkey (green monkey), and New World monkey (owl monkey) DNAs resulted in the amplification of only a 150-bp fragment (no *Alu* family member present). The amplification of DNA from the prosimian galago resulted in no specific amplification products at this temperature. These data suggest that this *Alu* family member amplified sometime after the human/great ape divergence.[20] We have routinely found that the amplification of orthologues within humans,[57] and as far back as New World monkeys (diverged from humans approximately 30–40 million years ago), is possible.[21] However, we have not been able to amplify any orthologous loci within the galago genome.[21] The effectiveness of this procedure will be dependent on the molecular clock in the species of interest, as well as the random location of mutations. Again, mutations occurring at, or near, the 3' end of one of the primers will be most detrimental to amplification.

Conclusion

Before cloning and routine sequence analysis, DNA renaturation provided the only method for comparing nucleic acid sequences. Naturally, for a time, the importance of this approach decreased as the newer methods became routine. However, the availability of oligonucleotides for use as hybridization probes and PCR primers has reinvigorated the usefulness of nucleic acid hybridization as a complement to DNA sequence analysis.

Acknowledgments

We thank Drs. Morris Goodman and Jerry Slightom for providing orangutan DNA, Ms. Utha Hellmann-Blumberg for the use of her unpublished data, and the department of photographic services at the Louisiana State University Medical Center for assistance. This research was supported by U.S. Public Health Service Grants GM-21346 (C.W.S.) and RO1 HG 00340 (P.L.D.) as well as a grant from the Cancer Crusaders (M.A.B.). Work by M.A.B. was also performed by Lawrence Livermore National Laboratory under the auspices of the U.S. Department of Energy Contract No. W-7405-ENG-48.

[57] M. A. Batzer, V. A. Gudi, J. C. Mena, D. W. Foltz, R. J. Herrera, and P. L. Deininger, *Nucleic Acids Res* **19**, 3619 (1991).

[17] DNA–DNA Hybridization of Single-Copy DNA Sequences

By MARK S. SPRINGER and ROY J. BRITTEN

Introduction

Single-copy DNA hybridization techniques have seen widespread application to problems in systematics. Most notably, Sibley and Ahlquist[1] have produced a phylogeny for many of the birds of the world. Other taxa

[1] C. G. Sibley and J. E. Ahlquist, "Phylogeny and Classification of Birds." Yale Univ. Press, New Haven, Connecticut, 1990.

that have been studied include carnivores, rodents, marsupials, sea urchins, sea stars, sand dollars, cave crickets, and dipterans. In addition, single-copy DNA hybridization has been used to investigate aspects of genome organization and evolution. Examples of rate disparity in the single-copy genome are appearing in a wide variety of taxa.[2–5]

The purpose of this chapter is 3-fold: (1) to review basic principles of hybridization and the kinetics of reassociation, (2) to provide an overview that summarizes and compares different techniques used in single-copy hybridization, and (3) to examine different estimates of distance derived from melting curves. Most of the topics in this chapter are discussed in greater detail elsewhere, and readers are referred to Britten *et al.*,[6] Sheldon and Bledsoe,[7] and Werman *et al.*[8] for consideration of specific topics. Also, we do not discuss the application of DNA hybridization data to problems in phylogenetic reconstruction, but rather refer readers to Springer and Krajewski[9] and Werman *et al.*[8]

General Principles of DNA Hybridization

Native DNA is isolated and purified to remove RNA and protein. Long-stranded DNA is then sheared to short fragments to permit the separation of repetitive and single-copy DNA and to reduce viscosity and gel formation.[6] Reassociation kinetic techniques developed by Britten *et al.*[6] are used to fractionate single-copy DNA from repetitive sequences. $C_o t$ plots may then be constructed that present the percentage of single-stranded DNA versus the log of $C_o t$ ($C_o t$ is the initial concentration of DNA in moles of nucleotides per liter multiplied by time of incubation in seconds). An appropriate $C_o t$ value can then be chosen that allows separation of single-stranded, single-copy DNA by hydroxyapatite column chromatography.[10] Fractionated single-copy DNA from one species is then radioactively labeled (tracer) and hybridized with unlabeled DNA (driver) from the same species (homoduplex reaction) and from different species

[2] R. J. Britten, *Science* **231**, 1393 (1986).
[3] F. H. Sheldon, *Mol. Biol. Evol.* **4**, 56 (1987).
[4] M. S. Springer and J. A. W. Kirsch, *Mol. Biol. Evol.* **6**, 331 (1989).
[5] S. D. Werman, E. H. Davidson, and R. J. Britten, *J. Mol. Evol.* **30**, 281 (1990).
[6] R. J. Britten, D. E. Graham, and B. R. Neufeld, this series, Vol. 29E, p. 363.
[7] F. H. Sheldon and A. H. Bledsoe, *J. Mol. Evol.* **29**, 328 (1989).
[8] S. D. Werman, M. S. Springer, and R. J. Britten, *in* "Molecular Systematics" (D. M. Hillis and C. Moritz, eds.), p. 204. Sinauer, Sunderland, Massachusetts, 1990.
[9] M. S. Springer and C. Krajewski, *Q. Rev. Biol.* **64**, 291 (1989).
[10] D. E. Kohne and R. J. Britten, *in* "Procedures in Nucleic Acid Research" (G. L. Cantoni and D. R. Davies, eds.), p. 500. Harper & Row, New York, 1971.

(heteroduplex reactions). When the hybridizations are complete, melting profiles and the extent of reassociation are determined for each hybridization reaction. Median and/or modal melting temperatures may then be determined from melting profiles, and differences in these parameters between homoduplex and heteroduplex curves provide estimates of genetic distance (ΔT_m and ΔT_{mode}). The normalized percentage of hybridization *(NPH)* may also be obtained for interspecies heteroduplex measurements by dividing the extent of hybridization for an interspecies heteroduplex by the homoduplex control and multiplying by 100. Finally, ΔT_m and *NPH* are sometimes combined into another distance measure, $\Delta T_{50}H$. Different measures of distance may then be used in phylogenetic reconstruction and in comparing rates of single-copy DNA evolution.

Kinetics of Reassociation

Rates of reassociation of DNA are influenced by several factors including genome complexity, DNA concentration, fragment size, reassociation temperature, and cation concentration. The complexity of the sheared genome is defined as the length in base pairs (bp) of the longest nonrepeating sequence that could be produced by splicing together fragments in the population.[11] The rate of reassociation is inversely proportional to the complexity of the genome. Because genome complexity ranges from as low as 10^4 bp for some viruses up to 10^{10} bp for some eukaryotes, it is clearly significant in affecting the rate of reassociation and needs to be taken into account. Additionally, DNA reassociation is approximately a bimolecular second-order reaction that depends on the concentration of single-stranded DNA.

Fragment size is important because second-order rate constants for reassociation vary in proportion to the square root of fragment length. For example, rate constants for 700-bp fragments are higher by a factor of 1.32 than for 400-bp fragments. Cation concentration is important because cations decrease the intermolecular repulsion of negatively charged DNA strands. Finally, the optimal temperature for reassociation is about 25° below the melting temperature (T_m) of the native duplex DNA.

Given these constraints that impinge on the reassociation of DNA, experimental designs must be constructed that facilitate (1) the separation of single-copy DNA to be used as tracer DNA from repeated sequences and (2) the hybridization of single-copy tracer DNA with driver DNA from the same and different species.

---

[11] L. E. Hood, J. H. Wilson, and W. B. Wood, "Molecular Biology of Eukaryotic Cells." Benjamin/Cummings, Menlo Park, California, 1974.

Isolation of Single-Copy DNA

Repeated sequences are a universal feature of the eukaryotic genome. Moreover, they typically show large amounts of divergence within the genome of an individual and often have biased AT to GC ratios. Partly for these reasons, hybridization of repetitive DNA has not been used effectively for resolution of systematic issues. Therefore, these sequences should be removed from labeled single-copy DNAs. Because repeats are distributed throughout the genome, their removal requires fragmentation of long-stranded DNA to fragments several hundred base pairs long.

Removal of almost all of the repeats is possible because they are present at a higher concentration than single-copy sequences and therefore reassociate faster. Single-stranded DNA will not adsorb to hydroxyapatite at 50° in 0.12 M phosphate buffer (PB), whereas double-stranded DNA will adsorb, permitting fractionation of the single-stranded, single-copy fraction from the double-stranded, repetitive fraction (see Werman et al.[8] for details). However, separation of single-copy sequences from low-frequency repeats is never absolute, and small numbers of copies of each repeat family will remain in the single-copy fraction.

The fraction of DNA that remains single-stranded in hydroxyapatite fractionation can be calculated as follows:

$$S = 1/(1 + kC_0t)$$

where k is a second-order rate constant that depends on cation concentration, temperature, fragment size, and sequence complexity. At standard experimental conditions (60°, 0.18 N cation, DNA fragments 500 bp long), k is 10^6 divided by the genome size in base pairs. It is generally both practical and satisfactory to reassociate fragmented DNA to a C_0t value at which the single-copy DNA is 10% reassociated before fractionation over hydroxyapatite. To achieve this C_0t value, rates of reassociation can be accelerated by increasing both DNA concentration (up to about 10 $\mu g/\mu l$) and cation concentration. Acceleration factors (relative to 0.12 M PB at 60°) for different cation concentrations are given in Britten et al.[6]

Hybridization of Tracer DNA with Driver DNA

The range of precision of duplexes permitted between tracer and driver DNA is controlled by the temperature and ionic strength of the incubation buffer. Together, they establish the criterion, or stringency, or reassociation. This is usually described as the difference between the T_m of perfect duplexes in the incubation buffer and the temperature of incubation. If criterion is too stringent, only a subset of sequences can hybridize, and the resolving power of the method is reduced. Conversely, if stringency conditions are too low, unrelated sequences may hybridize and complicate the interpretation of hybridization data as a record of evolutionary history. Hybridizations are frequently carried out at 60° in 0.48 M PB and subsequently diluted to 0.12 M PB for thermal chromatography. This criterion provides for reasonable acceleration of the reaction and leads to the formation of duplexes that are thought to involve orthologous sequences.

DNA Hybridization Methods

Most DNA hybridization studies have been performed in a phosphate buffer/hydroxyapatite system. One of the disadvantages of this system is that the base composition of duplexes can affect their individual melting temperatures: DNA duplexes that are (G + C)-rich will denature at higher temperatures than (A + T)-rich duplexes because G-C base pairs share three hydrogen bonds and A-T pairs share only two hydrogen bonds. As an alternative to phosphate buffer, some workers have employed tetraethyl-ammonium chloride (TEACL) to eliminate the effect of base composition. However, TEACL interferes with hydroxyapatite chromatography, and single-stranded DNA is generally removed by S1 nuclease digestion rather than hydroxyapatite.

In this section, we review basic procedures that are used in single-copy DNA hybridization studies. In addition to the PB/hydroxyapatite and S1/TEACL systems, we also discuss other methods such as the PERT system and the PT system that combines TEACL with hydroxyapatite.

DNA Isolation and Purification

Protocols for the isolation and purification of DNA have been presented by a number of different workers. Most are modifications of the method developed by Marmur.[12] Differences between protocols reflect the diversity of organisms and tissues from which DNA is isolated as well as the amount of DNA that is being purified. For hybridization studies it is important that DNA be free from contaminants such as protein, glycogen, and metal ions. Briefly, purification involves repeated extractions with phenol and chloroform and treatment with proteinase K/pronase and RNase. A general method for the isolation and purification of DNA from tissues is given by Werman et al.[8] Krajewski[13] presents a method for extracting DNA from bird blood, and Kirsch et al.[14] present a protocol for

[12] J. Marmur, J. Mol. Biol. **3**, 208 (1961).
[13] C. Krajewski, Auk **106**, 603 (1989).
[14] J. A. W. Kirsch, M. S. Springer, C. Krajewski, M. Archer, K. Aplin, and A. W. Dickerman, J. Mol. Evol. **30**, 434 (1990).

mammal tissues. Protocols for DNA extraction from small organisms such as fruit flies are given by Hillis *et al.*[15]

Preparation of Sheared DNA

DNA must be fractionated into 500-bp fragments for both tracer and driver DNA because repeats are dispersed throughout the genome. Methods of shearing include homogenization in a tissue blender[6,16] and the commonly used procedure of sonication. Sonication protocols vary depending on the model of sonifier used. Kirsch *et al.*[14] describe a protocol using a Branson sonifier/cell disrupter that is useful for large sample volumes (15 ml). Werman *et al.*[8] describe a protocol for a Kontes (Vineland, NJ) Micro ultrasonic cell disrupter that is useful for smaller samples (500 μl). DNA samples should be passed over a chelating resin column and through cellulose acetate filters after sonication to filter out any metal ions or particles introduced during sonication.[8] The size of sonicated samples is monitored by gel electrophoresis. If sonicated DNAs are too long, they can be additionally sonicated; if they are too short they should not be used in hybridization reactions. Thus, DNA should not be oversheared at first. It is well to examine the size distribution with alkaline gel electrophoresis.

Tracer Preparation and Repeat Removal

Most of the DNA hybridization studies carried out on birds and mammals have made use of [125]I. Protocols for iodination are given by Sibley and Ahlquist,[17] Kirsch *et al.,*[14] and Werman *et al.*[8] These protocols, in turn, are derived from more general protocols for iodination.[18–21] Iodination is carried out on single-stranded single-copy DNA that has already been fractionated on hydroxyapatite (see below). In contrast, enzymatic labeling of DNA such as nick translation and random priming is carried out on native DNA with subsequent isolation of single-copy DNA on hydroxyapatite. Werman *et al.*[8] present a nick-translation procedure adapted to DNA hybridization studies. In this protocol, unincorporated nucleotides are

[15] D. M. Hillis, A. Larson, S. K. Davis, and E. Z. Zimmer, *in* "Molecular Systematics" (D. M. Hillis and C. Moritz, eds.), p. 318. Sinauer, Sunderland, Massachusetts, 1990.

[16] J. A. Hunt, T. J. Hall, and R. J. Britten, *J. Mol. Evol.* **17**, 361 (1981).

[17] C. G. Sibley and J. E. Ahlquist, *in* "Evolution Today" (G. G. E. Scudder and J. L. Reveal, eds.), p. 301. Carnegie-Mellon Univ., Pittsburgh, Pennsylvania, 1981.

[18] S. L. Commorford, *Biochemistry* **10**, 1993 (1971).

[19] D. M. Anderson and W. R. Folk, *Biochemistry* **15**, 1022 (1976).

[20] H.-C. Chan, W. T. Ruyechan, and J. G. Wetmur, *Biochemistry* **15**, 5487 (1976).

[21] W. Prensky, *in* "Methods in Cell Biology" (D. M. Prescott, ed.), p. 121. Academic Press, New York, 1976.

eliminated using the glass powder elution procedure of Davis *et al.*[22] rather than the spin-column technique.

An advantage of ^{125}I is its long half-life (60 days), but the iodination procedure is somewhat difficult to establish. Nick translation and random priming with ^{32}P and ^{3}H are easier to control, but ^{32}P does have a shorter half-life (14 days) than [125]I. We prefer nick translation to random priming because it is easier to produce longer tracers.

Fractionation and Sizing of Single-Copy DNA

Single-copy tracer DNA must be fractionated from repeated sequences. For iodinated tracers, fractionation is performed prior to labeling. Nick-translated and random-primed tracers, in turn, are fractionated after labeling. In all cases, an appropriate $C_o t$ must be chosen to reassociate the repeat DNA while leaving the single-copy component single-stranded so as to separate these components using hydroxyapatite. $C_o t$ can be calculated by the following formula:

$$C_o t = [\mu g \text{ DNA/sample volume } (\mu l)] \times 10 \times AF \times time \text{ (hr)}$$

where AF is the acceleration factor due to an increase in PB concentration over 0.12 M PB. The acceleration factor for 0.48 M PB is 5.65. Estimates of $C_o t$ values sufficient for repeat removal have been made for *Drosophila* ($C_o t$ 10–50), sea urchins ($C_o t$ 100), and primates ($C_o t$ 300).[8]

It is important to estimate the size of denatured single-copy tracer DNA. Relative to unsheared DNA, fragmented DNAs have a T_m that is lower by 500/duplex length in degrees Celsius. The size of single-stranded tracer DNA can be estimated using alkaline gel electrophoresis.[8,23] A practical consideration is that the capacity of hydroxyapatite is about 100 μg of DNA per 400 mg of hydroxyapatite.

Hybridization with Hydroxyapatite and Phosphate Buffer

Both manual procedures[6,24] and automated methods[17,25] have been used to perform thermal chromatography with hydroxyapatite and phosphate buffer. The efficiency and accuracy of automated methods make them desirable for large-scale taxonomic projects.

[22] L. G. Davis, M. D. Dibner, and J. F. Battey, "Basic Methods in Molecular Biology." Elsevier, New York, 1986.
[23] M. J. Sambrook, E. F. Fritsch, and T. Maniatis, "Molecular Cloning: A Laboratory Manual." Cold Spring Harbor Laboratory, Cold Spring Harbor, New York, 1989.
[24] D. E. Kohne, J. A. Chiscon, and B. H. Hoyer, *J. Hum. Evol.* **1**, 627 (1972).
[25] J. A. W. Kirsch, R. J. Ganje, K. G. Olesen, D. W. Hoffman, and A. H. Bledsoe, *BioTechniques* **8**, 505 (1990).

A simple, manual procedure for the preparation of thermal elution of hybrids is described in detail by Werman *et al.*[8] The DNAlyzer of Sibley and Ahlquist[17] was the first custom-built, automated thermal elution device. The DNAlyzer allowed simultaneous thermal elution of 25 different hybrids. Kirsch *et al.*[25] described a thermal elution device (TED) patterned after the DNAlyzer that also allows simultaneous elution of 25 hybrids. The temperature is maintained within 0.15° with the TED, and the equilibration and elution are controlled by a Neslab MTR-5 programmer working with a Neslab DCR-4 digital temperature controller.

Regardless of the method used, several considerations are relevant. First, driver DNA should be from 1000- to 10,000-fold in excess over tracer DNA. Second, concentrations of PB stock solutions, the hybridization reaction mix, and the elution buffer are critical and must be known with accuracy. Third, reactions should be allowed to go at least 90% completion of the single-copy DNA. Estimates of reassociation rates for interspecies hybrids should also take into account sequence divergence and its effect on rate constants; that is, the rate constant for interspecies hybrid formation is retarded by a factor of $2^{\Delta T_m/10}$ owing to sequence mismatch. It is convenient to reassociate DNAs in 0.48 M PB to accelerate their reassociation. Fourth, when hybrid DNAs are diluted from 0.48 to 0.12 M PB and loaded onto hydroxyapatite columns, they should not be diluted with pure water or premature denaturation may take place.[8] Finally, those contemplating work with hydroxyapatite should consult papers by Martinson,[26] Martinson and Wagenaar,[27] and Fox *et al.*[28] regarding the binding properties of hydroxyapatite. These authors have constructed phase diagrams for perfectly and imperfectly base-paired DNA which show that there is a very narrow window of buffer concentrations at which the elution temperature of the duplex DNA corresponds to the melting temperature. Moreover, the size of this window decreases for imperfectly matched DNA. A good choice for most studies is 0.12 M PB, but for comparisons involving distantly related taxa 0.10 M PB may in fact be better for thermal elution.[28]

S1 Nuclease and Tetraethylammonium Chloride

An alternative to the hydroxyapatite procedure for discrimination between single-stranded and duplex DNA is the S1 nuclease procedure.[29,30]

[26] H. G. Martinson, *Biochemistry* **12**, 145 (1973).
[27] H. G. Martinson and E. B. Wagenaar, *Anal. Biochem.* **61**, 144 (1974).
[28] G. M. Fox, J. Umeda, R. K.-Y. Lee, and C. W. Schmid, *Biochim. Biophys. Acta* **609**, 364 (1980).
[29] R. E. Benveniste and G. J. Todaro, *Nature (London)* **261**, 101 (1976).
[30] R. E. Benveniste, *in* "Molecular Evolutionary Genetics" (R. J. MacIntyre, ed.), p. 359. Plenum, New York, 1985.

According to this method, DNA is treated with S1 nuclease after hybridization and then precipitated. S1 nuclease digests single-stranded DNA but not double-stranded DNA. The rigor of enzyme treatment can have a dramatic effect on the effective criterion, and it is important to establish strict control in S1 nuclease assays.

The extent of hybridization is never as great with this procedure as with hydroxyapatite because all unduplexed regions are digested. In contrast, single-stranded tails of fragments containing duplexes bind to hydroxyapatite. The kinetics of S1-digested reassociated DNA are described by the equation $S = (1 + kC_0 t)^{-0.44}$, where S is the fraction of single-stranded DNA.

S1 nuclease can be used in conjunction with NA$^+$ or TEACL, but not phosphate. TEACL has the desirable property of eliminating the effect of base composition on hybrid melting temperature at a concentration of 2.4 M.[31,32] The width of a melting curve for precise duplexes is about 1.5° in TEACL compared to about 14° in 0.12 M PB. Unlike PB, however, TEACL interferes with hydroxyapatite chromatography. Therefore, it has always been employed in conjunction with S1 nuclease although recent advances partially mitigate the problems of TEACL with hydroxyapatite. Details of the S1 nuclease/TEACL method are given by Powell and Caccone[33] and Werman et al.[8]

Other Methods

Studies indicate that TEACL and hydroxyapatite can be combined over a limited temperature range. However, in the presence of high-molarity TEACL, the PB concentration that permits double-stranded DNA to bind but single-stranded DNA to pass through hydroxyapatite is much reduced. Specifically, a buffer (PT) containing 2.0 M TEACL and 13 mM PB can be used with hydroxyapatite. This solvent is stable from 4° to 75°, and precise duplexes bind hydroxyapatite from room temperature upward. Native DNA melts at 68° with a width of about 3°, and 500-bp fragments melt at 65°. Unfortunately, single strands bind below 50°, so the method is only useful for closely related species.

One other technique that may be useful in hybridization work is the phenol emulsion reassociation technique (PERT). This is a high-stringency system that achieves high $C_0 t$ values at low DNA concentrations and room temperature by using a phenol emulsion phase to accelerate the rate of reassociation as much as 10,000-fold.[34] A major advantage of the PERT

[31] W. B. Melchior and P. H. Von Hippel, *Proc. Natl. Acad. Sci. U.S.A.* **70**, 298 (1973).
[32] J. R. Hutton and J. G. Wetmur, *Biochemistry* **12**, 558 (1973).
[33] J. R. Powell and A. Caccone, *J. Mol. Evol.* **30**, 267 (1990).
[34] D. E. Kohne, S. A. Levison, and M. J. Byers, *Biochemistry* **16**, 5329 (1977).

system is that only small amounts of driver DNA (5 μg) are needed for each reaction. However, its value for systematic studies remains to be explored. A basic protocol for setting up hybrids for the PERT system is given by Werman *et al.*[8]

Distance Estimates

Several estimates of divergence may be derived from melting curves and the extent of reassociation. Those that are most commonly employed are ΔT_m, ΔT_{mode}, *NPH,* and $\Delta T_{50}H$. The latter combines ΔT_m and *NPH* into a single measure. T_m is the interpolated temperature at which 50% of the hybrids that formed remain in duplexes. T_{mode} is the interpolated temperature at which the maximum number of hybrids melt. *NPH* is the normalized percentage of hybridization, namely, the fraction of hybridization in a heteroduplex reaction divided by the fraction of hybridization in a homoduplex reaction multiplied by 100. Finally, $T_{50}H$ is an estimate of the temperature at which 50% of the potentially hybridizable DNA remains in duplexes.

Sibley and Ahlquist,[17] Sheldon and Bledsoe,[7] and Werman *et al.*[8] discuss different methods for the computation of different indices as well as characteristics of these distances. Some of their most salient properties are as follows.

T_m. The Melting Temperature (T_m) is a useful measure of thermal stability for small and intermediate divergences. The decrease in T_m with increasing divergence is steady until it reaches about the halfway point between criterion and the T_m of precise duplexes. At greater divergences, the fraction of DNA that hybridizes continues to drop steadily but the T_m decreases very little. This results in compression of ΔT_m values at greater distances and is discussed elsewhere.[7,9,35]

Precision of T_m values is generally excellent. Also, different techniques for calculating T_m give highly congruent estimates.

T_{mode}. Like T_m, T_{mode} is moderately precise for small and intermediate distances. However, different methods for calculating T_{mode} sometimes give incongruent estimates. In contrast to T_m, T_{mode} does not appear to be bounded from below by criterion and does not suffer as much from compression. Even so, the melting curve flattens out at increased distances, and the mode becomes extremely difficult to locate; false modes may result from the imprecision of replicate measurements if the curve is broad and flat.

NPH. In general, *NPH* is not as accurately determined as T_{mode} and T_m.[14,17] However, *NPH* continues to drop steadily after T_m becomes com-

---

[35] M. S. Springer and J. A. W. Kirsch, *Syst. Zool.* **40**, 131 (1991).

pressed and T_{mode} becomes difficult to locate. Therefore, at larger evolutionary distances, *NPH* may be the only measure of distance with any resolving power. On the other hand, the interpretation of *NPH* is additionally complicated because it is not yet known how much of the reduction in *NPH* is due to kinetic effects and thus is not a direct measure of the fraction of the DNA capable of forming hybrids under the conditions. In other words, rates of reassociation differ in homoduplex versus heteroduplex reactions, and the DNA available for hybridization may be used up before the tracer–driver sequences can completely hybridize.

Kinetic effects are probably small for closely related species,[36] but one study suggests a large kinetic effect for larger distances.[37] Additional studies will be required to resolve this issue. In any case, the kinetic effect can be reduced by incubation to high C_0t because the concentration of the single-stranded part of the driver falls at a rate of $(1 + kC_0t)^{-0.44}$ and there should be some single-stranded driver available very late in the incubation.[38]

$T_{50}H$. Kohne et al.[24] devised $T_{50}H$ to remedy the compression of T_m values that is caused by criterion conditions and to correct for reduced *NPH* values that occur even for closely related species. Hall et al.[39] interpret $\Delta T_{50}H$ as a measure of median sequence divergence between species. Smith et al.[40] showed that $T_{50}H$ values are relatively constant even at different incubation temperatures. It is clearly useful to have such a criterion-independent measure of distance for comparisons made at different incubation temperatures. However, $T_{50}H$ is not immune to all the flaws of its constituent parts. Most importantly, the imprecision that is typical of *NPH* measurements gets incorporated into $T_{50}H$. Also, $T_{50}H$ will be affected by kinetic effects just as is *NPH*.

Conclusions

Single-copy DNA hybridization has been and continues to be an important technique for addressing problems in systematics and genome evolution. One advantage of this technique for systematic studies is that it encompasses the entire single-copy genome. Rates of evolution for the single-copy genome provide an important yardstick for gauging rates of

[36] T. Bonner, R. Heinemann, and G. J. Todaro, *Nature (London)* **286,** 420 (1980).

[37] G. A. Galau, M. E. Chamberlin, B. R. Hough, R. J. Britten, and E. H. Davidson, *in* "Molecular Evolution" (F. J. Ayala, ed.), p. 200. Sinauer, Sunderland, Massachusetts, 1976.

[38] R. J. Britten and E. H. Davidson, *in* "Nucleic Acid Hybridisation: A Practical Approach" (B. D. James and S. J. Higgins, eds.), p. 3. IRL Press, Oxford, 1985.

[39] T. J. Hall, J. W. Grula, E. H. Davidson, and R. J. Britten, *J. Mol. Evol.* **16,** 95 (1980).

[40] M. J. Smith, R. Nicholson, M. Stuerzl, and A. Lui, *J. Mol. Evol.* **18,** 92 (1982).

evolution among individual genes. Technical advances are improving the precision of the method, and theoretical studies are providing a strong underpinning for the use of DNA hybridization data in phylogenetic reconstruction. Finally, the feasibility for large-scale taxonomic comparisons has been greatly enhanced by the construction of automated thermal elution devices in several different laboratories.

[18] Comparative Studies of Gene Copy Number

By SCOTT O. ROGERS, GREGORY C. BEAULIEU, and ARNOLD J. BENDICH

Introduction

For determination of gene copy number, slot blotting is the method of choice. When compared with Southern blotting, the slot blotting procedure is simpler and more accurate, so that large numbers of samples can be analyzed. Four steps are involved in the process. First, the DNA is quantified by spectrophotometry or densitometry. Second, samples are loaded onto a membrane. Third, the DNA is hybridized to labeled probe DNA. Fourth, gene copy number is calculated. Copy numbers of some gene families can change rapidly. Individuals within a species, and even tissues within an individual, can possess different copy numbers.[1] The data, therefore, are not always of phylogenetic significance and should be interpreted cautiously.

Step 1: Quantitation of DNA

Two methods for DNA quantitation are described: spectrophotometric and densitometric. We have used both methods for analyzing hundreds of samples. The first method is simpler and more precise; however, more DNA is required. The second method requires much less DNA (several nanograms) and is the only option when little DNA is available. A method for rapidly extracting DNA from many samples of milligram (and larger) amounts of plant,[2,3] algal,[3a] and fungal[4] tissues has been described. We do

[1] S. O. Rogers and A. J. Bendich, *Genetics* **117**, 285 (1987).
[2] S. O. Rogers and A. J. Bendich, *Plant Mol. Biol.* **5**, 69 (1985).
[3] S. O. Rogers and A. J. Bendich, "Plant Molecular Biology Manual" (S., B. Gelvin and R. Schilperoot, eds.), p. A6:1. Kluwer Academic Publishers, Boston, 1989.
[3a] M. S. Shivji, S. O. Rogers, and M. J. Stanhope, *Mar. Ecol. Prog. Ser.* **84**, 197 (1992).
[4] S. O. Rogers, S. A. Rehner, C. Bledsoe, G. J. Mueller, and J. F. Ammirati, *Can. J. Bot.* **67**, 1235 (1989).

METHODS IN ENZYMOLOGY, VOL. 224

not recommend quantitation on agarose gels, because it is not sufficiently accurate.

Spectrophotometric Method

Procedure

1. Treat the sample with 0.3 N NaOH at 37° for about 18 hr. This hydrolyzes RNA to mononucleotides, which are then removed in the next step. Treat enough DNA not only for the quantitation but also for the subsequent blotting.

2. Precipitate the single-stranded DNA with 2 volumes of 95% (v/v) ethanol at room temperature.

3. Centrifuge at 10,000–15,000 g for 5 min at room temperature.

4. Wash the pellet in 80% ethanol, dry, and rehydrate in TE (10 mM Tris, pH 8.0, and 1 mM EDTA, pH 8.0).

5. Determine the absorbance at 260 nm. A 40 μg/ml solution of single-stranded DNA has an A_{260} of 1.0. To denature DNA that may have reassociated during storage (which will reduce light absorbance), the sample can be heated briefly to 95° and quickly cooled on ice before the measurement is made.

We obtain a coefficient of variation of 1–2% for replicate determinations using this spectrophotometric method.

Densitometric Method

In the densitometric method, DNA is stained with ethidium bromide, drawn into microcapillaries, and photographed; the photographic negative is then scanned with a densitometer.[5] There are several points of caution to observe. (a) One should use a standard DNA with a similar G + C content as the unknown DNA because base composition affects ethidium fluorescence.[5] The standard DNA is purified DNA that has been quantified by spectrophotometry. It can be from the organism under study or, alternatively, from a linearized plasmid or a phage such as λ. (b) The unknown DNA should be digested with one or more restriction enzymes. This prevents clumping of the DNA and ensures uniform fluorescence in the microcapillaries. (c) RNA in the sample makes a small contribution to ethidium fluorescence and can be removed with RNase (DNase-free) for 15 min at 37°. The resulting oligoribonucleotides still fluoresce, but to a

[5] S. P. Moore and B. M. Sutherland, *Anal. Biochem.* **144**, 15 (1985).

much reduced degree. As long as their concentration is below 500 ng/μl and the DNA concentration is above 5 ng/μl, they will not pose a problem. (d) The range of DNA concentration to use is between 5 and 50 ng/μl. Above 60–70 ng/μl the DNA becomes saturated with ethidium bromide, and the relationship between DNA concentration and fluorescence becomes nonlinear.

Procedure

1. Make a series of dilutions of the standard DNA that covers the range 5–150 ng/μl.

2. Make a dilution of the unknown DNA that falls in the range 10–100 ng/μl. An estimation of the initial concentration of each unknown can be made by either ethidium bromide staining on an agarose gel, a preliminary run-through of the densitometric procedure, or the knowledge of the yield of DNA usually obtained from the source material.

3. For each standard and unknown DNA sample, mix the following: 1.5 μl distilled water, 1.0 μl ethidium bromide (50 μg/ml), and 2.5 μl DNA. Three replicates per sample are recommended. The mixing can be done in 0.5-ml microcentrifuge tubes, or on a piece of Parafilm taped to the bench top. If the latter, mix only 5–8 samples at a time so that evaporation does not alter the DNA concentration. Note that, after mixing, DNA concentrations are now one-half the initial values.

4. Break several 5- or 10-μl glass microcapillaries in half. Draw each sample into a microcapillary so that approximately equal volumes of air are present at either end. The microcapillaries can be cleanly broken by scoring them first with a diamond-tipped pencil or similar instrument. We use glass capillaries marked at the 5- or 10-μl point by the manufacturer (VWR, San Francisco, CA).

5. Flame-seal the ends of the microcapillary, taking care not to boil the solution inside.

6. Attach to the laboratory bench two parallel pieces of tape, sticky side up. They should be about 4 cm apart. Line up the sealed microcapillaries on the tape in a parallel array, with a spacing of at least 4–5 mm between each pair of capillaries.

7. Place another piece of tape, sticky side down, on top of the two initial pieces. Label the array for future reference. Wipe fingerprints off the tubes with a Kimwipe tissue.

8. Photograph the microcapillaries much as you would an ethidium bromide-stained gel, but use f 8 or higher rather than a lower f-stop. This reduces lens distortion and gives sharper images. The frame should allow a generous border around the microcapillary array for the same reason; do not fill the frame with arrays. Record all camera settings exactly (f-stop,

height of the lens, exposure time) and maintain those values, so that comparisons between negatives will be meaningful. Some preliminary photographs should be taken of standard sets to determine the exposure time that best corresponds to the linear range of the film. The correct exposure time depends on the equipment used and the age of the UV light source. We use a Fotodyne 300 nm UV light source (four bulbs), a Polaroid MP-4 camera, and Polaroid Type 55 film, which has a fine-grain negative that is excellent for densitometry. With this equipment the best exposure times at f8 at a distance of 60 cm are 60–90 sec. Finally, use a fresh set of standards in each photograph when you do this preliminary determination, because UV irradiation bleaches the ethidium bromide fluorescence during exposure and the signals will be reduced in subsequent photographs of the same standard set.

9. Scan the dry negatives on a densitometer. Scan the standard set first and adjust the "zero" and "gain"; then scan the unknowns without changing these settings. If the images of the microcapillaries are all of equal sharpness, density is directly proportional to peak height. A meaningful standard curve can be constructed from a standard set that either was present in the same photograph as the unknown set or was photographed by the same equipment at exactly the same settings as the unknown set.

We obtain a coefficient of variation of 10–15% for replicate determinations using the densitometric method.

Step 2: Slot Blotting

In the slot-blotting procedure, DNA is applied to a membrane through slots in a manifold. The membrane is then probed by one of the many variants of the Southern hybridization procedure.[6] Slot blotting is similar, but superior to dot blotting because the signals on the slot-blot autoradiograph have parallel edges that provide uniformity and facilitate densitometric scanning. The manifold we use is the Manifold II apparatus made by Schleicher and Schuell (Keene, NH).

Choice of membrane type is important. Nylon-based membranes are stronger, have higher binding capacities, and thus allow detection of sequences present in low copy number. However, complete removal of probe is difficult.[7] Nitrocellulose membranes are brittle and have lower binding capacities, but they can be stripped and reprobed more easily. In general,

---

[6] E. M. Southern, *J. Mol. Biol.* **98**, 503 (1975).
[7] G. Cannon, S. Heinhorst, and A. Weissbach, *Anal. Biochem.* **149**, 229 (1985).

quantitation of the signal on a reprobed blot is not as accurate as on the initial probing. We recommend making several duplicate blots on nylon-based membranes, using one blot for each probe.

The correct amount of DNA must be loaded into each slot. The following points will affect the amount chosen. (a) The smallest amount of hybridizing DNA that can be detected in a slot using contemporary methods is between 0.1 and 10 pg. Usually a higher signal is desired, to increase the signal-to-background ratio. (b) The larger the genome size of the organism, the more DNA must be loaded. Genome sizes are given for bacteria,[8] fungi,[9-11] algae,[9] plants,[2,12-14] protozoa,[9] and animals.[15] (c) The lower the expected copy number, the more DNA must be loaded. (d) There is no point in trying to load an amount of DNA that exceeds the binding capacity of the membrane. Nytran, for example, has a binding capacity of 400 $\mu g/cm^2$, or 24 μg per slot (6 mm^2); the capacity of nitrocellulose is 80 $\mu g/cm^2$, or 4.8 μg per slot.

Procedure

1. Make a map relating slot numbers (which are identified on the manifold) with sample numbers. We recommend three replicates per sample. The positions of the replicates on the blot should be randomized for the results to be statistically meaningful.

2. Soak the membrane in 10 × SSC (1.5 M NaCl, 0.15 M sodium citrate) at least 30 min before blotting.

3. For each replicate of each DNA sample to be loaded, place the desired amount of DNA into a 0.5-ml microcentrifuge tube.

4. Add an equal volume of 1 N NaOH and mix. Incubate at room temperature for 5–10 min.

5. Add an equal volume (i.e., twice the original DNA volume) of 1 M Tris (pH 7.5), 3 M NaCl and mix. This solution should not sit for hours before loading, since highly repetitive DNA will have a chance to reanneal, thus reducing DNA binding to the membrane.

[8] S. Krawiec and M. Riley, *Microbiol. Rev.* **54**, 502 (1990).
[9] A. H. Sparrow, H. J. Price, and A. G. Underbrink, "Evolution of Genetic Systems" (H. H. Smith, ed.), p. 451. Gordon and Breach, New York, 1972.
[10] H. Brody and J. Carbon, *Proc. Natl. Acad. Sci. U.S.A.* **86**, 6260 (1989).
[11] D. H. Griffin, *in* "Handbook of Applied Mycology: Fungal Biotechnology" (D. K. Arora, R. P. Elander, and K. G. Mukerji, eds.), Vol. 4, p. 445. Dekker, New York, 1991.
[12] M. D. Bennett and J. B. Smith, *Proc. R. Soc. London B* **334**, 309 (1992).
[13] D. Ohri and T. N. Khoshoo, *Plant Syst. Evol.* **153**, 119 (1986).
[14] S. S. Dhillon, "Cell and Tissue Culture in Forestry," (J. M. Bonga and D. J. Durzan, ed.), p. 298. Martinus Nijhoff, Dordrecht, The Netherlands, 1987.
[15] B. John and G. Miklos, "The Eukaryote Genome in Development and Evolution," p. 150. Allen and Unwin, London, 1988.

6. Place two wetted pieces of Whatman (Clifton, NJ) 3MM paper and the wetted membrane into the manifold. Load each well with $10 \times$ SSC and adjust the vacuum so that the wells take at least 30 sec to empty. If the vacuum is strong enough to draw the solution through in $1-10$ sec, the result can be asymmetrical signals on the autoradiograph, which cannot be scanned accurately.

7. Load the DNA into the wells according to the map.

8. When all the wells have emptied, remove the membrane, then soak it for 30 sec in 0.4 N NaOH and for 30 sec in 0.2 M Tris (pH 7.5), $2 \times$ SSC.

9. Dry the membrane completely according to the manufacturer's directions.

10. Place the membrane in a plastic hybridization bag. The membrane can be stored like this for months.

For absolute copy numbers to be determined, a blot with varying amounts of probe DNA must be prepared to construct the standard curve. The range covered by the standards must include the copy number values present in the unknowns. Four points should be made here. (a) The blotting and labeling of the probe must be done to avoid artifactual signals. For example, if both the standards on the blot and the labeled probe contain cloning vector plus insert, the presence of the vector in the standards will result in more hybridizing sequences (per insert copy) in the standards than in the unknowns, which might lead to an underestimation of unknown copy number. This can be avoided by blotting insert plus vector and then probing with the insert alone. (b) The standards can be on the same blot as the unknowns or on a different blot, but they must be present in the same hybridization reaction as the unknowns. (c) Add heterologous carrier DNA to the standards in roughly the same proportion as heterologous DNA is present in the unknowns. (d) Frequently an autoradiographic signal from a slot containing a large amount of standard will obliterate weaker signals in neighboring slots rendering densitometry impossible. Therefore, standard DNA samples can be loaded in alternate slots, with the slots between left empty.

Step 3: Hybridization

Two alternative hybridization procedures are described: one is a modified sodium phosphate method performed at high temperature,[16] and the

---

[16] G. M. Church and W. Gilbert, *Proc. Natl. Acad. Sci. U.S.A.* **81,** 1991 (1984).

other is performed at lower temperature in formamide solution. Other alternative procedures that give low background may also be used.

Procedure A

1. Label 10–200 ng of probe DNA using a random primer labeling system and 10–100 μCi of an α-^{32}P-labeled nucleotide (e.g., [^{32}P]-dCTP at 3000 Ci/mmol). Incubate at 37° for 30 min. Alternative radiolabeling systems may be used. Percent incorporation of label can be tested with a Geiger–Mueller counter. If the amount of radioactivity remaining in the column is much more than that in the probe, then a new labeling reaction should be done. (Typical sources of trouble are the Klenow enzyme, contaminants in the DNA to be labeled, and radioactive or unlabeled nucleotides that have not been kept frozen.) Generally, 30–80% incorporation is desirable. If less DNA is used in the labeling reaction, percent incorporation will be lower but the specific activity of the probe will be higher, and vice versa.

2. During the above incubation, prehybridize the blot at 65° for at least 5 min in 30 ml hybridization buffer [1 mM EDTA, 7% sodium dodecyl sulfate (SDS), 0.5 M NaHPO$_4$, pH 7.2]. A 1 M stock of NaHPO$_4$ contains 134 g Na$_2$HPO$_4 \cdot 7$H$_2$O and 4 ml of 85% H$_3$PO$_4$ per liter.[16]

3. Separate unincorporated nucleotides from the labeled probe with a Sephadex G-50 column (using a homemade apparatus[17]) or a commercial device designed for this purpose.

4. Denature the probe DNA at 95° for 5 min. Chill quickly on ice.

5. Remove the hybridization solution from the bag. Add 10 ml fresh hybridization buffer and the probe.

6. Incubate with agitation at 65° for 8–16 hr. If a heterologous hybridization probe is used, the temperature of incubation may have to be lowered (by 1° for each 1% change in sequence) because of base mispairing in the hybrids formed.

7. Wash the blot twice, 5 min each with agitation, at 65° in wash buffer 1 (1 mM EDTA, 5% SDS, 40 mM NaHPO$_4$, pH 7.2). It is most convenient to heat the buffer to 65°C in a microwave oven and do the wash in a plastic container, on a shaker, at room temperature. Use 125 ml wash buffer for each 1200 cm^2 of membrane (one slot-blot sheet).

8. Wash the blot 8 times, 5 min each with agitation, at 65° in wash buffer 2 (1 mM EDTA, 1% SDS, 40 mM NaHPO$_4$, pH 7.2).

9. Do a final wash in wash buffer 2 for 20 min at 65° in a shaker–incubator.

10. Do several quick rinses, several seconds each, in 1 × SSC.

[17] J. Sambrook, E. F. Fritsch, and T. Maniatis, "Molecular Cloning: A Laboratory Manual," 2nd Ed., p. E.34. Cold Spring Harbor Laboratory, Cold Spring Harbor, New York, 1989.

11. Wrap the blot in plastic wrap and perform autoradiography using X-ray film [e.g., Kodak (Rochester, NY) XAR-2].

Procedure B

1. Prehybridize the blot for 4–12 hr at 65° in 30 ml of 50% formamide, 1% SDS, 1 *M* NaCl. We have obtained low background and high sensitivity using BDH (Poole, UK) AnalaR grade and Fisher (Pittsburgh, PA) molecular biology grade formamide (without further purification). Higher backgrounds were found when lower grades of formamide were used. If low amounts of hybridization are expected (less than 1–2 pg), nonhomologous unlabeled DNA should be added (e.g., 100 μg/ml sheared salmon sperm DNA). This will decrease the amount of nonspecific hybridization, which becomes significant at these low hybridization levels. For some probes, higher percentages (2–5%) of SDS are suggested (for steps 1 and 4) to lower background. Concentrations of SDS lower than 1% may cause increased background with nylon-based membranes.

2. Add the probe (prepared as in steps 1–4 of Procedure A) to the bag and incubate with agitation at 42° for 6–16 hr. The hybridization and washing temperatures in steps 2 and 4, respectively, should be lowered if a heterologous probe is used. When wash temperatures are below 50–55°, however, higher backgrounds may result.

3. Wash the blot twice, 2–5 min each with agitation, in 2× SSC at room temperature.

4. Wash the blot twice, 30 min each with agitation, in 2× SSC, 1% SDS at 65°.

5. Wash the blot twice, 30 min each with agitation, in 0.1× SSC at room temperature.

6. Wrap the blot and perform autoradiography (as in step 11 of Procedure A).

Step 4: Calculation of Copy Number

1. Calculate the proportion of total DNA that hybridizes with the probe (i.e., picograms of hybridized probe DNA divided by picograms of DNA loaded in that slot). If the genome size of the organism is unknown, no further calculations are possible, but useful comparisons among samples can still be made.

2. Multiply the calculated value by the genome size of the organism.

3. Multiply this value by 9.11×10^5 kilobases (kb)/pg [which equals $(6.03 \times 10^{11} \text{ daltons/pg})/(6.62 \times 10^5 \text{ daltons/kb})$]. This gives the number of kilobases of DNA that hybridized.

4. Divide by the length of the probe sequence to obtain the copy number.

For replicate copy number determinations of ribosomal RNA genes in maize and broad bean, we obtain a coefficient of variation of 8–15% using the spectrophotometric method of DNA quantitation with hybridization Procedure A. The coefficient of variation is about 50% using the densitometric method of quantitation with hybridization Procedure B.

[19] Optimal Preparative Methods for Producing Comparative Gene Libraries

By JERRY L. SLIGHTOM

Introduction

Recombinant DNA cloning techniques, for the isolation of a specific region(s) of the genome of an organism, and nucleotide sequencing methods have been extremely productive in providing information regarding the organization and structure of many genes. Analyses of coding and noncoding orthologous DNA regions from many different mammalian species have also provided a new dimension for the detailed analyses of phylogenetic relationships.[1–5] The results from molecular type analyses have for the most part paralleled the phylogenetic relationships derived from the fossil record.[6–10] In some cases, however, such investigation has been useful in resolving long-standing phylogenetic questions [e.g., the

[1] B. F. Koop, M. Goodman, P. Xu, K. S. Chen, and J. L. Slightom, *Nature (London)* **319**, 234, (1986).

[2] M. M. Miyamoto, J. L. Slightom, and M. Goodman, *Science* **238**, 369 (1987).

[3] M. M. Miyamoto, B. F. Koop, J. L. Slightom, M. Goodman, and M. R. Tennant, *Proc. Natl. Acad. Sci. U.S.A.* **85**, 7627 (1988).

[4] B. F. Koop, D. Tagle, M. Goodman, and J. L. Slightom, *Mol. Biol. Evol.* **6**, 580 (1989).

[5] W. J. Bailey, D. H. A. Fitch, D. A. Tagle, J. Czelusniak, J. L. Slightom, and M. Goodman, *J. Mol. Biol. Evol.* **8**, 155 (1991).

[6] A. C. Wilson, S. S. Carlson, and T. J. White, *Annu. Rev. Biochem.* **46**, 573 (1977).

[7] P. Andrews, *in* "Molecules and Morphology in Evolution: Conflict or Compromise" (C. Paterson, ed.), p. 23. Cambridge Univ. Press, Cambridge, 1987.

[8] D. Pilbeam, *Am. Anthropol.* **88**, 295 (1986).

[9] M. Goodman, J. Czelusniak, B. F. Koop, D. A. Tagle, and J. L. Slightom, *Cold Spring Harbor Symp. Quant. Biol.* **51**, 875 (1987).

[10] M. Goodman, B. F. Koop, J. Czelusniak, D. H. A. Fitch, D. A. Tagle, and J. L. Slightom, *Genome* **3**, 316 (1989).

placement of orangutan *(Pongo pygmaeus)* into the genealogical family Hominidae].[11] These analyses have also been extremely valuable in determining how gene families have evolved from an ancestral gene (e.g., the evolutionary history of the β-globin gene cluster),[12-19] and in investigating intergenic exchanges (e.g., gene conversions that have occurred between β-type globin genes).[20-27]

The ability to clone and purify specific DNA regions became feasible on a large scale with the development of vectors derived from *Escherichia coli* phage λ, which is capable of growth after cloning of a portion of foreign DNA into its genome. Lambda phage was first used for cloning following partial purification of the target gene region.[28,29] However, the development of improved λ vectors, capable of cloning larger regions of the target DNA[30] [15–20 kilobases (kb)], together with rapid screening proce-

[11] B. F. Koop, M. M. Miyamoto, J. E. Embury, M. Goodman, J. Czelusniak, and J. L. Slightom, *J. Mol. Evol.* **24,** 94 (1986).

[12] A. Efstratiadis, J. W. Posakony, T. Maniatis, R. M. Lawn, C. O'Connell, R. A Spritz, J. K. DeRiel, B. Forget, S. M. Weissman, J. L. Slightom, A. E. Blechl, O. Smithies, F. E. Barralle, C. C. Shoulders, and N. J. Proudfoot, *Cell (Cambridge, Mass.)* **21,** 653 (1980).

[13] F. S. Collins and S. M. Weissmann, *Prog. Nucleic Acid Res. Mol. Biol.* **31,** 315 (1984).

[14] R. C. Hardison, *Mol. Biol. Evol.* **1,** 390 (1984).

[15] J. B. Margot, G. W. Demers, and R. C. Hardison, *J. Mol. Biol.* **205,** 15 (1989).

[16] W. R. Shehee, D. D. Loeb, N. J. Adey, F. H. Burton, N. C. Casavant, P. Cole, C. J. Davies, R. A. McGraw, S. A. Schichman, D. M. Severynse, C. F. Voliva, F. W. Weyter, G. B. Wisely, M. H. Edgell, and C. A. Hutchison III, *J. Mol. Biol.* **205,** 41 (1989).

[17] B. F. Koop and M. Goodman, *Proc. Natl. Acad. Sci. U.S.A.* **85,** 3893 (1988).

[18] L. B. Giebel, V. Van Santen, J. L. Slightom, and R. A. Spritz, *Proc. Natl. Acad. Sci. U.S.A.* **82,** 6984 (1985).

[19] D. A. Tagle, B. F. Koop, M. Goodman, J. L. Slightom, R. T. Jones, and D. L. Haas, *J. Mol. Biol.* **203,** 439 (1988).

[20] J. L. Slightom, A. E. Blechl, and O. Smithies, *Cell (Cambridge, Mass.)* **21,** 627 (1980).

[21] J. L. Slightom, L.-Y. E. Chang, B. Koop, and M. Goodman, *Mol. Biol. Evol.* **2,** 370 (1985).

[22] J. L. Slightom, T. Theisen, B. F. Koop, and M. Goodman, *J. Biol. Chem.* **262,** 7472 (1987).

[23] J. L. Slightom, B. F. Koop, P.-L. Xu, and M. Goodman, *J. Biol. Chem.* **263,** 12427 (1988).

[24] B. F. Koop, D. R. Siemieniak, J. L. Slightom, M. Goodman, J. Dunbar, P. C. Wright, and E. L. Simons, *J. Biol. Chem.* **264,** 68 (1989).

[25] D. H. A. Fitch, C. Mainone, M. Goodman, and J. L. Slightom, *J. Biol. Chem.* **265,** 781 (1990).

[26] D. A. Tagle, J. L. Slightom, R. T. Jones, and J. L. Slightom, *J. Biol. Chem.* **266,** 7469 (1991).

[27] D. H. A. Fitch, W. J. Bailey, D. A. Tagle, M. Goodman, L. Sieu, and J. L. Slightom, *Proc. Natl. Acad. Sci. U.S.A.* **88,** 7396 (1991).

[28] P. Leder, S. M. Tilghman, D. C. Tiemeier, F. I. Polisby, J. G. Seidman, M. L. Edgell, L. W. Enquist, A. Leder, and B. Norman, *Cold Spring Harbor Symp. Quant. Biol.* **42,** 915 (1977).

[29] S. Tonegawa, C. Brock, N. Hozumi, and R. Schuller, *Proc. Natl. Acad. Sci. U.S.A.* **74,** 3518 (1977).

[30] F. R. Blattner, B. G. Williams, A. E. Blechl, K. Denniston-Thompson, H. E. Faber, L.-A. Furlong, D. J. Grunwald, D. O. Kiefer, D. D. Moore, J. W. Schumm, E. L. Sheldon, and O. Smithies, *Science,* **196,** 161 (1977).

dures[31,32] allowed for the construction of complete genomic recombinant λ phage libraries[32,33] (random shotgun collections) from which single-copy gene regions could be efficiently isolated from an initial background of several million recombinant phage clones.

With the development of direct genome sequencing using the polymerase chain reaction (PCR) procedure,[34] many have questioned whether there is still a need for the construction of recombinant DNA libraries. Direct genomic sequencing with the PCR has already added greatly to our ability to obtain homologous or orthologous nucleotide sequence information rapidly from a large number of individuals or species (see Kocher[35]). With the existence of two uniquely different procedures to obtain an orthologous series of nucleotide sequences for comparative purposes, researchers must decide which method is best suited to obtain their goals. The answer to this question is neither easy nor straightforward because each situation contains many different variables. A general rule may be to ask two questions before starting: (1) What is the amount of nucleotide data that must be collected from each individual and/or species? (2) How many individuals and/or species will be examined? These questions are somewhat diametrically opposed because in most cases the nucleotide sequencing part remains the most labor-intensive step. If the nucleotide sequence region of comparison is small, say, 1 to 2 kb, and a large number of individuals or species are involved, the PCR approach may be the most suitable. If the nucleotide sequence region is large, say, above 10 kb, and a limited number of individuals and species are involved, say, fewer than 10, then a gene isolation approach may be more suitable. The answer to the questions above also depend on the method with which one has the most experience, because time will not be consumed to learn new techniques. However, for the purpose of this chapter it will be assumed that the decision has been made to clone extensive regions from one or several species.

The major goal in constructing recombinant λ phage libraries is to obtain a sufficient number of recombinant λ phage clones to ensure that the entire genome of a species has been cloned (see below). The library must also be in a form that can be easily stored over long periods of time, is not subject to rearrangements (mostly deletions) within the cloning vector,

---

[31] W. D. Benton and R. W. Davis, *Science* **196**, 180 (1977).

[32] F. R. Blattner, A. E. Blechl, K. Denniston-Thompson, H. E. Faber, J. E. Richards, J. L. Slightom, P. W. Tucker, and O. Smithies, *Science* **202**, 1279 (1978).

[33] R. M. Lawn, E. F. Fritsch, R. C. Parker, G. Blake, and T. Maniatis, *Cell (Cambridge, Mass.)* **15**, 1157 (1978).

[34] R. K. Saiki, D. H. Gelfand, S. Stoffel, S. Scharf, R. Higuchi, G. T. Horn, K. B. Mullis, and H. A. Erlich, *Science* **239**, 487 (1988).

[35] W. K. Thomas and T. D. Kocher, this volume [28].

and can be used to extend genetic analyses outward from a specific cloned genetic region (chromosome walking). In addition, the construction and screening of the library should be relatively easy and straightforward to allow first-time users to succeed without spending large amounts of time troubleshooting technical steps. Two major types of cloning vector systems have been developed that have retained their viability because they satisfy most of these conditions, namely, the above-mentioned bacteriophage λ and the related cosmid vector systems.[36,37] The major advantage of the cosmid system over the λ system is a larger insert cloning capacity (35 to 45 kb versus a maximum of about 20 kb for λ). Thus, fewer cosmid clones would be needed to walk across a larger region of genomic DNA. However, the cosmid system requires a higher degree of skill for screening (the difference between screening phage plaques versus bacterial colonies), and the larger insert size increases the potential to clone more genomic DNA elements that are unstable in *E. coli,* making them more prone to deletions during growth in *E. coli.* Because of these disadvantages it is recommended that experience be first gained using the λ vector system. For this reason and because the comparative primate libraries described here were made in λ phage vectors, this chapter describes procedures for the use of λ vectors for the construction of genomic libraries. Those interested in using cosmid vector systems should consult other references.[36-38]

The construction of a recombinant λ phage genomic DNA library requires the isolation of genomic DNA in the size range of about 50 kb (nearly 100 kb for cosmid vectors), followed by a series of partial restriction enzyme digestions to obtain a complete or pseudorandom (depending on the restriction enzyme used) fragmentation of the target genome. Randomization of the partial digest can be increased by using different time points or restriction enzyme concentrations in a series of digestions ranging from a minimum of about 10 to 20% to a maximum of about 60 to 70% of completion (see below). A series of digests is useful because not all restriction enzyme recognition sites (for a specific restriction enzyme) are equally susceptible to cleavage. After obtaining the digestion series, the individual digests are pooled and size fractionated on a velocity sedimentation gradient. Target DNA fragments between 15 and 20 kb in length are selected for λ cloning, and those between 35 and 55 kb can be used for cosmid cloning. The selection of target DNA fragments between 15 and 20 kb for λ cloning is useful because it reduces the chance of cloning multiple frag-

[36] J. Collins, this series, Vol. 68, p. 309.
[37] P. F. R. Little and S. H. Cross, *Proc. Natl. Acad. Sci. U.S.A.* **82**, 3159 (1985).
[38] J. Sambrook, E. F. Fritsch, and T. Maniatis, "Molecular Cloning: A Laboratory Manual," 2nd Ed. Cold Spring Harbor Laboratory, Cold Spring Harbor, New York, 1989.

ments, which is not desirable because they will generally be noncontiguous in the target species genome and therefore interrupt chromosome walking continuity. The cloning into λ of two fragments between 15 and 20 kb in size will produce a recombinant λ molecule that is too large (> 52) to be encapsidated.[39]

The selection of which restriction enzyme to use for preparing the target DNA for cloning is an important decision because of its impact on whether the library will contain overlapping fragments that theoretically represent the complete target species genome. The construction of some of the first λ libraries, using λ replacement vectors (see below), involved use of a partial digest by multiple frequently cutting enzymes [with 4-base pair (bp) recognition sites] followed by the addition of EcoRI linkers.[33] Libraries constructed using only an EcoRI digest of the target DNAs are incomplete because the 6-bp recognition site of EcoRI (or most any other enzyme that recognizes a 6-bp site) is not represented at a high enough frequency that ensures complete randomness of the digested target genomic DNA. The EcoRI site is represented at random every 4^6 or 4096 bp; however, in generating a partial EcoRI digest not all sites are cut, and skipping just a few sites (five or more) would result in fragments that exceed the cloning capacity of a λ vector. In addition, the distribution of EcoRI sites is not truly random; thus, there are some regions of a target species genome in which even a complete EcoRI digest will yield fragments that exceed the capacity of a λ vector.

The construction of completely random λ libraries can be achieved using the procedure described by Lawn et al.[33] in which the target DNA is digested with a restriction enzyme(s) whose recognition site is only 4 bp, which is present on average every 4^4 or 256 bp. However, the library construction method described by Lawn et. al.[33] required the use of additional enzymatic steps (EcoRI site methylation and the ligation of linkers). However, these additional enzymatic steps were eliminated by the identification of restriction enzymes with 4- and 6-bp recognition sites that share identical single-stranded core sequences. The most frequently used restriction enzyme recognition site combinations utilize λ vectors that contain a BamHI cloning site, which has a 6-bp recognition site,

[39] N. E. Murray, in "Phage Lambda and Molecular Cloning" (R. W. Hendrix, J. W. Roberts, F. W. Stahl, and R. A. Weisberg, eds.) p. 395. Cold Spring Harbor Laboratory, Cold Spring Harbor, New York, 1983.

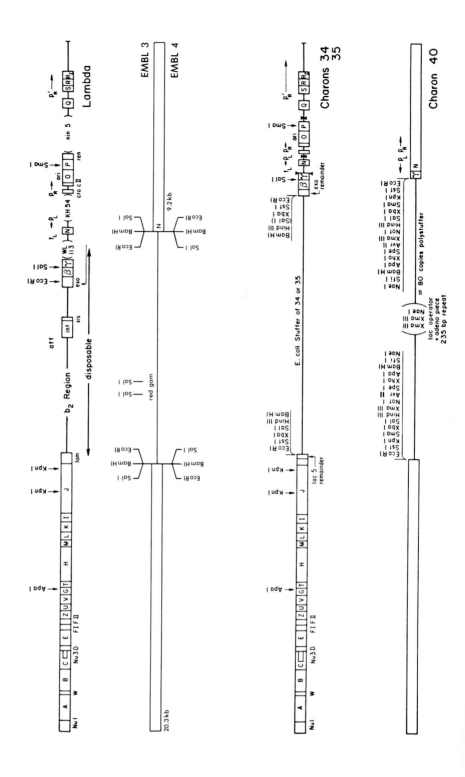

that can be used to clone target genomic DNA digested with either *Sau*3A or *Mbo*I, which has the following 4-bp recognition site:

$$
\begin{array}{cccc}
| & & & \\
G & A & T & C \\
C & T & A & G \\
& & & |
\end{array}
$$

Both recognition sites share the identical single-stranded sequence GATC. Many λ vectors have been engineered to contain *Bam*HI cloning sites, with the most widely used λ vectors being EMBL3 and EMBL4[40] and Charon 35 and 40 vectors.[41,42] Although there is only a one in four chance that a cloned *Sau*3A- or *Mbo*I-cut fragment will regenerate a *Bam*HI site, analysis of the cloned insert is not difficult because these vectors contain additional restriction enzyme recognition sites flanking the *Bam*HI sites (see the maps of EMBL and Charon vectors in Fig. 1).

Assuming that the complete target species genome is clonable, the number of λ clones needed to contain the complete genome can be calculated using Eq. (1), described by Clark and Carbon[43]:

$$N = \ln(1 - P)/\ln(1 - x/y) \tag{1}$$

where the number of clones needed *(N)* can be calculated for a set probability *(P)*, knowing the size of the cloned fragments *(x)* and the total size of the target species genome *(y)*. In using this calculation two assumptions must be made: first, that fragmentation of the target genomic DNA is completely random; second, that the exact size of the cloned inserts is

[40] A.-M. Frischauf, H. Lehbach, A. Poustra, and N. Murray, *J. Mol. Biol.* **170**, 827 (1983).
[41] W. A. M. Loenen and F. R. Blattner, *Gene* **26**, 171 (1983).
[42] I. S. Dunn and F. R. Blattner, *Nucleic Acids Res.* **15**, 2677 (1987).
[43] L. Clark and J. Carbon, *Cell (Cambridge Mass.)* **9**, 91 (1976).

FIG. 1. Genetic and restriction enzyme site maps of wild-type λ and the λ vectors EMBL3, EMBL4, and Charon 34, 35, and 40. Maps of wild-type λ and Charon vectors are from Dunn and Blattner,[41] and maps of the EMBL vectors are from Frischauf *et al.*[40] The DNA region between the indicated polylinkers in the λ cloning vectors is the disposable stuffer that can be replaced by the target DNA for library construction. The stuffer region of the EMBL vectors includes the loci for *gam* and *red* gene function, which are used for Spi⁺ phenotype selection by growth on an *E. coli* (P2) host. The stuffer region of Charon 40 consists of 80 copies of a specific 235-bp fragment; it is referred to as a polystuffer. Removal of the polystuffer and preparation of Charon 40 arms require first digestion with *Nae*I followed by a second digest with any of the 15 other restriction enzymes that have recognition sites within the polylinker. Charon 40 clones are *gam*⁺; thus, they can be grown on *E. coli* hosts that are *recA*⁻ such as ED8767. [From J. L. Slightom and R. F. Drong, *in* "Plant Molecular Biology Manual" (S. B. Gelvin and R. A. Schilperoort, eds.), p. A8, Kluwer Academic Publishers, Dordrecht, The Netherlands, 1988. Reprinted by permission of Kluwer Academic Publishers.]

known. Although the pooled fragments selected for cloning are between 15 and 25 kb, analysis of cloned inserts from many libraries (see Table I) suggests that the average insert size is skewed toward the lower end of this range; the value of x appears to be closer to about 17 kb. The genome size of most mammals is relatively constant, being about 3×10^9 bp.[44] Many plant genomes are of similar complexity.[44a] Assuming that the target genome library will contain inserts of 17 kb, the number of clones needed to obtain a complete library at a probability of $P = 0.99$ is nearly 1×10^6. The techniques for library construction described below use specific primate examples but should be applicable to most eukaryotic organisms.

Bacteriophage λ Vector Systems

Bacteriophage λ is a useful cloning vector system because the center third of the viral genome, the region between the J and N genes (Fig. 1), is not essential for lytic growth; λ vectors designed for removal of this region are referred to as λ replacement vectors. This region can be replaced by foreign genetic information without greatly affecting propagation of the phage, provided that the size of the recombinant λ genome satisfies the specific size restrictions (being larger than about 40 kb but not larger than about 52 kb).[39] As a result of many genetic and recombinant DNA manipulations, a large number of λ replacement type vectors have been developed that contain single or multiple restriction enzyme sites surrounding different dispensable center fragments (also called stuffer fragments); many of these λ vectors are still in use. Many different stuffer fragments have been used to facilitate their later removal (by either physical or genetic means, see below) from the essential λ genes that remain within the flanking DNA region. These flanking regions are linked through the λ *cohesive end (cos)* site and are referred to as the λ right and left arms (Fig. 1).

Construction of a recombinant λ library first requires removal of the stuffer fragment by physical or genetic means so that it, or part of it, will not end up in the library. The physical removal method utilizes restriction enzyme sites located within the stuffer, to fragment it. This is followed by separation of the λ arms from the stuffer fragments on a salt or sucrose velocity sedimentation gradient.[20,38,44] This method is used for the preparation of Charon 4A (*Eco*RI digest)[30] and Charon 35 (either *Eco*RI or *Bam*HI digest)[41] λ arms. In the case of Charon 40, the stuffer fragment contains multiple copies (80 repeats) of a 235-bp fragment flanked by the *Nae*I restriction enzyme site (Fig. 1), and on *Nae*I digestion the stuffer

[44] T. Maniatis, E. F. Fritch, and J. Sambrook, "Molecular Cloning: A Laboratory Manual." Cold Spring Harbor Laboratory, Cold Spring Harbor, New York, 1986.
[44a] A. L. Rayburn, this volume [15].

fragment is essentially destroyed.[42] In the genetic method the stuffer fragments are not removed prior to cloning the target DNA fragments, because the stuffer fragment contains a genetic locus that prevents the phage from growing and thus from being represented within the library. This is done by using stuffer fragments that include functional *gam* and *red* loci such as the λ EMBL vector series (Fig. 1). Lambda clones that contain the stuffer fragment will produce *gam*+ phages, the growth of which can be suppressed by propagating in an *E. coli* (P2) host such as Q359 or P2392.[39] Sensitivity to P2 interference (the Spi+ phenotype) appears to be due in part to inactivation of the *E. coli recBC* nuclease by the *gam* gene product. Infection of a P2 lysogen by *gam*+ λ produces a *recBC*⁻ phenotype that leads to inhibition of both protein synthesis and DNA replication by the P2 prophage. However, infection by *gam*⁻ phage permits replication provided that they are Chi+, which is the case for the EMBL λ vectors.

Instead of using the Spi⁻ phenotype to select EMBL3 and EMBL4 recombinant phage, the stuffer region of these λ vectors can be removed by two other methods that rely on physical techniques. The first method involves digesting either EMBL3 or EMBL4 with *Sal*I (two *Sal*I sites are located within the stuffer) and *Bam*HI, followed by separation of the annealed arms (31 kb) from the fragmented stuffer by a velocity sedimentation gradient, similar to that used for the isolation of Charon 4 arms.[20] The second method, which is applicable only to EMBL3, involves the digestion of the vector with both *Bam*HI and *Eco*RI, which releases a short 10-bp fragment (Fig. 1). The intact stuffer fragment is flanked by exposed *Eco*RI sites that will not ligate to the *Bam*HI sites on the EMBL3 arms; the efficiency of this approach can be increased by using a 5% polyethylene glycol (PEG) precipitation step to separate the large DNAs from the 10 bp fragment.

The EMBL vector systems have been subjected to many modifications to enhance their usefulness. The most useful modifications involve the addition of more restriction enzyme recognition sites flanking the stuffer fragment and the addition of T3 and T7 RNA polymerase promoters. The T3 and T7 RNA polymerase promoters are used to synthesize ^{32}P-labeled RNAs that can be used as probes for the identification of overlapping λ clones. Many of the λ vectors described above are available from or have been developed by commercial vendors. As in the case with many basic molecular biology techniques, the components needed to construct a specific recombinant phage library (λ vector arms, *in vitro* packaging extracts, etc.) and even the library can be obtained from commercial sources.

Although the Charon 32 and 40 λ vector systems have not been improved on since they were developed,[41,42] they still offer several advantages, such as a 16 restriction enzyme site polylinker and the *gam* locus (Fig. 1).

Recombinant *gam*$^+$ Charon 35 and 40 phage are capable of growth in *recA*$^-$ *E. coli* hosts, such as ED8767 *(supE, supF, hsdS*$^-$*, met*$^-$*, recA56).*[45] Charon 35 or 40 λ vectors for the construction of recombinant libraries have been found to be extremely useful because there are many regions of mammalian genomes that are unstable in λ vectors and must be grown in a *recA*$^+$ *E. coli* host. For example, the DNA region that encodes the duplicated fetal globin genes (a 5-kb direct repeat) of higher primates has been very difficult to clone in λ vectors grown in a *recA*$^+$ host[46] but has been readily cloned into λ Charon 32 and 40 vectors.[21,22]

The availability of many different λ vector systems from commercial sources has essentially removed the need to prepare λ vector arms. However, procedures for the preparation of λ arms from Charon 4A and 40 are described here to aid those who need the experience or who want to use noncommercially available vectors, such as Charon 40. The Charon series λ vectors have been used to construct many recombinant λ phage libraries that contain primate DNAs, and these libraries have been used for the isolation of much of the β-globin gene region. The use of these libraries has resulted in the addition of a vast amount of information concerning the organization, function, and evolution of the primate β-globin gene cluster (Table I). These libraries have been widely distributed and have been used for the isolation of many other DNA regions and specific genes.

Purification of λ Vector Arms

Charon 4A. The λ vector Charon 4A was one of the first vectors constructed capable of cloning large DNA fragments in the range of 8 to 22 kb, and it has been widely used for the construction of genomic libraries containing insert DNAs from many different organisms. The procedure for growing and isolating Charon 4A DNA in the *E. coli* host DP50*sub*F has been described.[20] The stuffer fragment of Charon 4A includes the *bio* (7.8 kb) and *lac* (6.9 kb) *Eco*RI fragments that can be separated from the 31-kb annealed arms by passage through a 5 to 20% NaCl gradient. NaCl is used to form the gradient because of its higher capacity to fractionate DNA without smearing. The following procedure and conditions are used for the preparation of Charon 4A arms.

1. Completely digest about 300 μg of Charon 4A phage DNA with *Eco*RI; ensure that the digest is complete by analysis on an agarose gel.

2. If the *Eco*RI digest is complete, inactivate the enzyme by adding diethyl pyrocarbonate (DEPC) to a final concentration of 0.1% and EDTA to 10 mM. Heat the sample to 65° for 10 min.

[45] N. E. Murray, W. J. Brammar, and K. Murray, *Mol. Gen. Genet.* **150**, 53 (1977).
[46] L.-Y. E. Chang and J. L. Slightom, *J. Mol. Biol.* **180**, 767 (1984).

TABLE I
PRIMATE LIBRARIES CONSTRUCTED IN VARIOUS VECTORS AND
GENES ISOLATED FROM β-GLOBIN GENE CLUSTER

Species	Vector	Gene region isolated	Refs.
Homo sapiens			
Hsa 563	Charon 4A (*Eco*RI)	ϵ, $^G\gamma$, $^A\gamma$, $\psi\eta$	Slightom *et al.*,[20] Miyamoto *et al.*[2]
Hsa 267 (maternal parent of 563)	Charon 32 (*Eco*RI)	$^G\gamma$, $^A\gamma$	J. L. Slightom (unpublished)
Hsa $^T\gamma$	Charon 32 (*Eco*RI)	$^G\gamma$, $^A\gamma$	J. L. Slightom (unpublished)
Pan troglodytes			
Chimpanzee (#1)	Charon 30 (*MBo*I) Charon 4A (*Eco*RI)	γ^1, γ^2, $\psi\eta$	Chang and Slightom[46] Slightom *et al.*,[21] Miyamoto *et al.*[2]
Pan paniscus			
Pygmy chimpanzee	Charon 40 (*Bam*HI)	γ^1, γ^2, $\psi\eta$	W. Bailey and J. L. Slightom (unpublished)
Gorilla gorilla gorilla			
Gorilla	Charon 34 (*Mbo*I)	γ^1, γ^2, $\psi\eta$	Chang and Slightom,[46] Miyamoto *et al.*[2]
Pango pygmaeus			
Orangutan YO-1	Charon 32 (*Eco*RI)	γ^1, γ^2, $\psi\eta$, δ	Slightom *et al.*,[22] Koop *et al.*,[1] Miyamoto *et al.*[2]
Orangutan NZ-1	Charon 35 (*Sau*3A)	ϵ, γ^1, γ^2, $\psi\eta$	Slightom *et al.*,[22] Koop *et al.*[11]
Hylobates lar			
Gibbon	Charon 40 (*Sau*3A)	γ^1, γ^2, $\psi\eta$, δ	Fitch *et al.*,[25] Bailey *et al.*[5]
Macaca mulatta			
Rhesus	Charon 30 (*Mbo*I) Charon 32 (*Eco*RI) Charon 40 (*Eco*RI)	γ^1, γ^2, $\psi\eta$	Slightom *et al.*[23] Miyamoto *et al.*[3]
Papio cynocephalus			
Baboon	Charon 30 (*Mbo*I) Charon 4A (*Eco*RI)	Not screened	J. L. Slightom (unpublished)
Ateles geoffroyi			
Spider monkey	Charon 35 (*Sau*3A)	γ^1, γ^2, $\psi\eta$, δ, β	Giebel *et al.*,[18] Fitch *et al.*[27]
Cebus albifrons			
Capuchin	Charon 4A (*Eco*RI)	ϵ, γ^1, γ^2, $\psi\eta$	J. L. Slightom and K. Hayasaka (unpublished)
Saimiri sciureus			
Squirrel monkey	Charon 34 (*Sau*3A)	Not screened	J. L. Slightom (unpublished)

TABLE I *(continued)*

Species	Vector	Gene region isolated	Refs.
Tarsius syrichta			
Tarsius	Charon 35 (*Sau*3A)	γ^1, $\psi\eta$, δ, β	Koop et al.,[4,24]
Galago crassicaudatus			
Galago	Charon 35 (*Mbo*I)	ϵ, γ, $\psi\eta$, α, β	Tagle et al.[19,26]

3. From a linear gradient maker, pump 5 ml of 5% and 5 ml of 20% NaCl solution for each gradient. Note that these volumes are appropriate when using a Beckman (Palo Alto, CA) SW-41 rotor as each tube holds a total volume of 11 ml. Form the gradient by pumping in the less dense NaCl solution into the bottom of the tube followed by the more dense NaCl solution. The less dense solution will rise above as the more dense solution is added.

4. Gently pipette the *Eco*RI-digested Charon 4A DNA onto each gradient tube; the volume placed on top of each tube should not exceed 300 μl, and the amount of Charon 4A DNA should not exceed 300 μg per gradient tube. Spin at 35,000 rpm, at 20°, for at least 6 hr but not longer than 7 hr.

5. After centrifugation, collect about 20 fractions (0.5 ml) from each gradient tube, load 5 to 10 μl from each fraction directly onto a 0.7% agarose gel, and electrophorese.

6. The gradient profile should appear similar to that shown in Fig. 2, with the Charon 4A arms located nearer the bottom and the *bio* and *lac* fragments located nearer the top of the gradient.

7. Pool the fractions that contain the major portion of annealed Charon 4A arms and measure the volume. Then add an equal volume of sterile double-distilled water, mix, and add 2.5 volumes of 100% ethanol. Precipitate the Charon 4A arms by first chilling the sample at −70° for at least 2 hr, then centrifuge in a SS34 rotor at 10,000 rpm for 2 hr. Decant the supernatant from the DNA pellet.

8. Resuspend the DNA pellet in 0.5 ml of 0.3 M sodium acetate and reprecipitate in a 1.5-ml microcentrifuge tube. Decant the supernatant from the DNA pellet and dry under vacuum.

9. Resuspend the Charon 4A arms in TEN buffer (see reagents list at the end of this chapter), measure the DNA concentration (OD_{260}), and adjust the volume to a final DNA concentration of about 1 mg/ml. Analyze the Charon 4A arms by electrophoresis through a 0.7% agarose gel; if

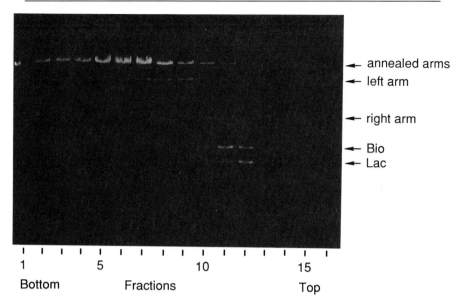

←— annealed arms
←— left arm

←— right arm

←— Bio
←— Lac

```
I   I   I   I   I   I   I   I   I   I   I   I   I   I   I   I
1           5                   10                  15
Bottom              Fractions                Top
```

FIG. 2. Separation of annealed Charon 4A arms from internal *bio* and *lac* fragments using a 5 to 20% NaCl gradient. Charon 4A DNA (300 μg) was first digested with *Eco*RI followed by layering onto a 5 to 20% linear NaCl gradient and centrifuging as described in the text. The gradient was fractionated into 0.5-ml aliquots, from which 5 μl was directly loaded onto a 0.7% agarose gel. After electrophoresis and ethidium bromide staining, fractions containing the majority of the annealed Charon 4A arms (near the bottom of the gradient, 31 kb) could be distinguished from those containing the internal *bio* and *lac* fragments (near the center of the gradient). Fractions containing annealed Charon 4A arms were pooled and treated as treated as described in the text. [From M. T. Sung and J. L. Slightom, *in* "Genetic Engineering in the Plant Sciences" (N. Panapoulos, ed.), p. 31. Praeger Publishers, Santa Monica, California, 1981. Reprinted by permission of Praeger Publishers.]

the sample still contains a large amount of the stuffer fragments, repeat the gradient steps.

10. Before using the Charon 4A arms for constructing a recombinant library, determine the number of background phage (phage resulting from cloning the stuffer fragments) by ligating a small fraction (about 1 μg) of the arms and packaging *in vitro* (see below) followed by determining the titer of the phage resulting from packaging the purified Charon 4A arms. If the titer of the purified and ligated Charon 4A arms is greater than 5×10^4 plaque-forming units (pfu) per microgram of arms, these Charon 4A arms should be repurified by a second NaCl gradient.

Charon 40. The replacement λ phage Charon 40 avoids the use of a velocity sedimentation gradient. The stuffer region of Charon 40 can be cut into 80 fragments, each 235 bp in length, by digestion with the restriction enzyme NaeI (Fig. 1). The small NaeI fragments can be separated from the Charon 40 arms by differential precipitation. The method described is from Dunn and Blattner.[42]

1. Digest Charon 40 DNA at a concentration of 100 μg/ml with NaeI. Use NaeI at about 1 unit/μg of Charon 40 DNA in the appropriate enzyme salts. Incubate at 37° for 2–3 hr, followed by analysis of 1 μg on a 0.7% agarose gel (after heating the sample to 65° for 10 min to separate the annealed cos sites). The NaeI digest should show λ left (19 kb) and right (10 kb) arms and a single intense band at about 235 bp (Fig. 3). If multiple bands with a repeat periodicity of 235 bp are present the NaeI digestion is not complete. Add more NaeI (0.1 unit/μg) and digest for an additional hour, followed by agarose gel analysis.

2. After completion of the NaeI digestion, heat-inactivate NaeI by incubating at 65° for 10 min. Cool the sample on ice and adjust the enzyme buffer for digestion with BamHI. Add BamHI to a concentration of about 1 unit/μg of Charon 40 DNA and incubate at 37° for 1 hr. Heat-inactivate the enzyme and then place on ice.

3. Raise the NaCl concentration in the DNA solution by adding $\frac{1}{10}$ volume of salt adjustment buffer [5 M NaCl, 100 mM Tris-HCl (pH 7.4), 10 mM EDTA]. Mix this solution, spin down briefly, and then add $\frac{1}{8}$ volume of sterile 40% PEG and repeat the mixing. Incubate the mixture at 25° (room temperature) for about 2 hr.

4. Precipitate the annealed Charon 40 arms by spinning in a microcentrifuge at about 25°, 10,000 rpm, for 5 min. Carefully remove the supernatant solution and save for later analysis. Resuspend Charon 40 arms in 200 μl of 0.2 M NaCl, then add 500 μl of 100% ethanol and precipitate to remove traces of PEG. Wash the pellet with 70% ethanol, dry briefly under vacuum, then resuspend in TEN buffer to obtain a final concentration of Charon 40 arms near 500 μg/ml.

5. Analyze the PEG supernatant solution and purified Charon 40 arms by electrophoresis through a 0.7% agarose gel to determine the purity of the Charon 40 arms and the amount of arms remaining in the PEG supernatant. Theoretically, the PEG supernatant solution should contain only the stuffer fragments while the pellet should contain only Charon 40 arms (Fig. 3). If the arms contain a large amount of stuffer fragments the PEG precipitation steps should be repeated.

6. Determine the efficiency of NaeI and BamHI digestions and PEG precipitation by ligating 1 μg of purified Charon 40 arms and packaging in vitro (see below) into particles. Titer the in vitro packaged phage to determine the number of viable (background) phage that can be expected. If the

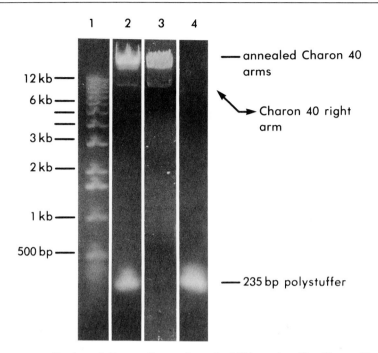

FIG. 3. Purification of Charon 40 arms from the 235-bp polystuffer. Charon 40 phage DNA (100 μg) was digested with *Nae*I; after this digest was determined to be complete, the DNA was subjected to digestion with *Bam*HI for preparation of the vector cloning site used for cloning *Sau*3 A-cut and size-selected target DNA (see text). The annealed Charon 40 arms (31 kb) were differentially precipitated away from the polystuffer using 5% PEG. The efficiency of the PEG precipitation step was analyzed by electrophoresis through a 0.7% agarose gel. Lane 1, DNA size standard (1-kb ladder from Bethesda Research Laboratories, Gaithersburg, MD); lane 2, Charon 40 DNA after digestion with *Nae*I and *Bam*HI; lane 3, Charon 40 annealed arms purified by 5% PEG precipitation; lane 4, supernatant solution from the 5% PEG precipitation (contains only polystuffer fragments). This analysis shows that annealed Charon 40 arms can be effectively purified from the polystuffer fragments by the 5% PEG differential precipitation step. [From J. L. Slightom and R. F. Drong, *in* "Plant Molecular Biology Manual" (S. B. Gelvin and R. A. Schilperoort, eds.), p. A8. Kluwer Academic Publishers, Dordrecht, The Netherlands, 1988. Reprinted by permission of Kluwer Academic Publishers.]

background titer is less than 5×10^4 pfu/μg of purified Charon 40 arms, they can be used for the construction of recombinant λ clone libraries. If the titer exceeds this value, repeat the PEG precipitation step; if this does not help it indicates that arms contain a small amount of *Nae*I and/or *Bam*HI uncut Charon 40, which would be effectively packaged. At this point one can redigest or initiate a new preparation.

Preparation of Target Genomic DNA for Cloning into λ Vectors

The construction of completely random recombinant λ genomic libraries can only be accomplished if the target DNA is fragmented by a sequence-independent method such as mechanical shearing. However, the cloning of sheared DNAs requires the additional steps of linker addition and methylation prior to cloning. As suggested above, it is much more convenient to rely on the use of restriction enzymes that possess 4-bp recognition sites, specifically enzymes such as Sau3A and MboI, which can be used with many λ vectors (Fig. 1). To achieve the goal of constructing a random genomic library, the strategy involves using a series of partial Sau3A or MboI digests of the target DNA, followed by size fractionation to select DNA fragments of optimal size for cloning into a λ vector digested with BamHI. This strategy does not guarantee the construction of a λ library that contains all regions of the target genome because some regions may not be stable in E. coli.[46] The Charon 35 to 40 series of λ vectors has reduced this stability problem because of their ability to grow in recA⁻ E. coli hosts. The example presented here involves the cloning of Hylobates lar (common gibbon) genomic DNA into Charon 40.

Sau3A Partial Digestion and Size Selection of Gibbon Genomic DNA

The purity of the target genomic DNA is very important to any cloning experiment because impurities can greatly affect restriction enzyme digestion and, more importantly, the λ *in vitro* packaging system. General methods for the isolation of mammalian DNAs, such as that described by Blin and Stafford,[47] yield DNAs of sufficient purity for restriction enzyme digestion and cloning. However, if contaminating proteins and/or carbohydrates (glycogen, if the tissue source is liver) are present they can inhibit many DNA-modifying enzymes and *in vitro* packaging enzymes. These contaminants can be removed by subjecting the DNA to either a neutral or ethidium bromide–CsCl gradient centrifugation. The amount of target DNA needed to initiate library construction is generally between 200 and 300 μg. Although a library can be constructed with considerably less DNA, this also demands more skill.

1. Check the Sau3A digestibility of the target DNA sample by titering it with the enzyme. Prepare a DNA mix sample by first adding 10 μg of target DNA, $\frac{1}{10}$ volume of enzyme buffer, and water to obtain a final volume of 150 μl in a 1.5-ml microcentrifuge tube.

2. Dispense 30 μl of the DNA mix into a sterile 1.5-ml microcentri-

---

[47] N. Blinn and D. W. Stafford, *Nucleic Acids Res.* **3**, 2303 (1976).

fuge tube (tube 1) and set on ice, then add 15 μl of the DNA mix into each of six other tubes (tubes 2 to 7). Chill all tubes on ice and then add 4 units of Sau3A to tube 1, mix, and briefly spin down the solution. Remove 15 μl and add to tube 2, mix well, and continue the 2-fold serial dilution process through tube 6. Tube 7 is the nondigested control. Incubate tubes 1 to 6 at 37° for 30 min, stop the reaction quickly by adding DEPC and EDTA to a final concentration of 0.1% and 10 mM, respectively, and continue by incubating the samples at 65° for 5 min. Analyze the digestion results by electrophoresis through a 0.7% agarose gel.

3. Evaluate the digestion results and estimate the enzyme concentration and incubation times needed to generate a series of partial digest samples in the range of 10, 20, 30, 40, and up to 60% of completion. The dilution series will yield the following ratio of Sau3A enzyme (1 hr units) per microgram of DNA: tube 1, 1 unit/μg; tube 2, 0.5 unit/μg; tube 3, 0.25 unit/μg; tube 4, 0.125 unit/μg; tube 5, 0.06 unit/μg; and tube 6, 0.03 unit/μg. The amount of enzyme required for digestion will also be an indication of the purity of the target DNA. If no digestion is observed except at the higher enzyme concentrations, then the target DNA is not pure and could be difficult to clone.

4. Using the digestion data, prepare a series of large-scale digests to obtain digested samples in the range of 10, 20, 30, 40, and 60% of completion. Depending on the availability of the target DNA, digest between 20 and 50 μg per reaction sample. Incubate the reaction at 37° for the prescribed times, then rapidly inactivate the enzyme by adding 0.1% final volume of DEPC and 100 mM EDTA and incubating at 65° for 5 min.

5. Analyze the digestion results by removing between 0.5 and 1.0 μg of the DNA and electrophresing through a 0.7% agarose gel. After staining with ethidium bromide and photographing, estimate the extent of each digestion, as was done with the primate target DNAs shown in Fig. 4. If the expected digestion profiles were obtained, proceed to the next step. However, if the samples are underdigested, add fresh 10 × enzyme buffer and Sau3A enzyme and repeat the digest. (The DEPC will not inhibit the freshly added enzyme because it rapidly decomposes at 65° to CO_2 and ethanol.) If the samples are overdigested, keep the digest nearest 40 to 60% of completion and set up three new digestions using lower concentrations of Sau3A enzyme or reduced time.

6. If the samples are within the expected digestion ranges, extract the samples with phenol and precipitate the DNA with ethanol. Resuspend each sample in 50 μl of TEN buffer and then pool; the volume of the pooled samples should not exceed 300 μl, and they should not contain more than 300 μg of DNA.

7. Using a linear gradient maker, form a 5 to 20% NaCl gradient in

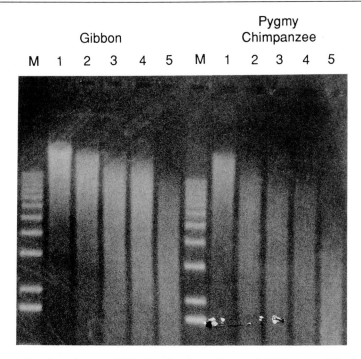

Fig. 4. Pseudorandom partial *Sau*3A digestions of genomic DNA isolated from gibbon *(Hylobates lar)* and pygmy chimpanzee *(Pan paniscus)*. For each species DNA, five aliquots containing about 50 μg of DNA were subjected to a set limit digestion with *Sau*3A (see text). After the allocated digest was completed, the restriction enzyme was rapidly deactivated (by the addition of DEPC and EDTA, see text), and about 1 μg of DNA from each aliquot was electrophoresed through a 0.7% agarose gel to determine the extent of digestion. The lanes containing the designated DNAs (gibbon or pygmy chimpanzee) were digested for specified times and amounts of *Sau*3A as follows: lanes 1, 3 units for 15 min; lanes 2, 3 units for 30 min; lanes 3, 6 units for 30 min; lanes 4, 12 units for 15 min; and lanes 5, 12 units for 30 min. Lanes labeled M contain the 1-kb DNA size standards. The digestion series for DNAs isolated from each species shows the expected degree of *Sau*3A digestion, ranging from about 30% of completion in lanes 1 to about 60 to 70% of completion in lanes 5. Digestion samples were pooled and size fractionated as shown in Fig. 5 for the digested gibbon DNA.

Beckman polyallomer tubes (1.43 × 8.9 cm; used in a SW41 rotor). Before loading the gradient, heat the pooled samples to 37° for 10 min to ensure that the enzyme sites are free and not annealed. Centrifuge at 35,000 rpm for 6 to 7 hr at 20°.

8. After centrifugation, fractionate the gradient from the bottom using tubing connected to a peristaltic pump. Carefully place the capillary tube in from the top until it reaches the bottom of the tube, then turn on the pump and collect 0.3- to 0.4-ml samples in sterile 1.5-ml microcentri-

fuge tubes. Remove between 5 and 10 μl from each fraction and analyze by electrophoresis through a 0.7% agarose gel. The result of fractionating *Sau*3A-digested gibbon DNA is shown in Fig. 5.

9. Pool the fractions that contain *Sau*3A-cut target DNA between 15 and 25 kb, measure the volume, and then add an equal volume of sterile water and mix. Then add 2.5 volumes of 100% ethanol, transfer the sample to siliconized Corex tubes, and incubate at −70° for at least 2 hr or overnight.

10. Precipitate the DNA by centrifuging in an SS34 roto, at 10,000 rpm and −10°, for 2 hr. After spinning, carefully remove the supernatant solution and save. Resuspend the DNA pellet in 0.4 ml of 0.3 *M* sodium acetate, transfer the sample to a sterile 1.5-ml microcentrifuge tube, and add 1 ml of 100% ethanol. Chill the sample at −70° for at least 10 min, then microcentrifuge at 4° for 10 min. Carefully remove the supernatant solution and dry the DNA sample under vacuum. Resuspend the dried DNA pellet in TEN buffer to obtain a final concentration between 0.5 and 1.0 mg/ml.

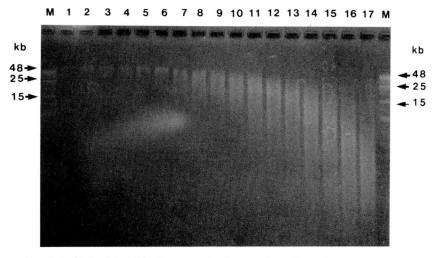

Fig. 5. Analysis of the DNA size separation from a salt gradient using the pseudorandom partial digested (*Sau*3A) gibbon DNA shown in Fig. 4. The gibbon digestion aliquots were pooled and loaded onto a 5 to 20% linear NaCl gradient. After centrifugation, fractions were collected (0.5 ml), and 5 μl from each fraction was loaded directly onto a 0.7% agarose gel. The lanes are labeled according to the tube fractions, and the flanking lanes labeled M contain the high molecular weight DNA markers obtained from Bethesda Research Laboratories. Fractions containing *Sau*3A-cut gibbon DNA fragments in the range of 15 to 25 kb were located; the highest concentration appears to be located in lanes (fractions) 7 to 10. These fractions were pooled, ethanol precipitated, and then cloned into Charon 40 to obtain the library listed in Table I.

11. Remove about $\frac{1}{20}$ of the size-fractionated target DNA and electrophroses through a 0.7% agarose gel alongside a DNA size standard; estimate the average size of the fragments and the DNA concentration of the sample. The *Sau*3A-cut target DNA is ready to be cloned into *Bam*Hi-cut Charon 40 arms (see below).

Cloning and Screening a Recombinant λ Genomic DNA Library

The high efficiency of the λ vector cloning systems is in part due to the λ infection mechanism that is used to transfer the recombinant λ DNA molecules into the host *E. coli* cells. However, to take advantage of this efficient transformation mechanism, the recombinant λ molecules must first be packaged into λ phage particles, which is done reassembling *in vitro* the necessary phage components. Several procedures for the isolation of λ packaging components have been described,[44,48] and the λ *in vitro* packaging components isolated are capable of yielding packing efficiencies in the range of 10^7 to 10^8 pfu/µg using wild-type λ DNA. The preparation of these λ packaging components is a time-consuming operation, and each component must be subjected to quality control testing prior to use. However, this can now be avoided because many different commercial vendors supply the λ *in vitro* packaging components, and they have been able to increase packaging efficiencies up to 10^9 pfu/µg using wild-type λ DNA. Such high efficiencies are desirable since it allows one to construct a complete recombinant λ library for a mammalian species from as little as 1 µg of size-selected DNA. Because the λ *in vitro* packaging components, with high efficiencies, can be obtained from commercial sources, their isolation is not covered in this chapter; those interested in isolating these components should refer to the appropriate references.[44,48]

Ligation of the λ arms and size-selected *Sau*3A-cut target genomic DNA is done by mixing them together at a concentration that favors the formation of concatenated λ DNA. Cocatenated λ DNA are the most efficient substrate for the *in vitro* packaging reaction; monomeric λ DNA molecules are not efficiently packaged. Two important parameters to consider when setting up λ ligation reactions are (1) the ratio of λ arm cloning sites to insert DNA cloning sites and (2) the concentration of each DNA species. Optimal values for these parameters can be theoretically estimated, assuming that all the DNA molecules will participate in the ligation reaction.

To obtain the best substrate for the *in vitro* packaging reaction (λ left arm – insert – λ right arm)$_N$ the type of available cloning sites *(cs)* must be

---

[48] B. Hohn, this series, Vol. 68, 299.

considered. Each λ arm (right and left) contains two different types of cohesive termini, a *cos* site that is compatible only with its complement on the opposite arm and one *cs* that is compatible with both termini of the size-fractionated target DNA (also referred to as *cs*) and the *cs* of the opposite λ arm. The ligation reaction should contain equimolar concentrations of the *cs* ends for each of the three different types of DNA molecules present, and, because each λ arm contains only one *cs* end, the ratio of molecules in a ligation reaction should be 2 : 1 : 2 (λ left arm – insert – λ right arm) or a molar ratio of 2 : 1 for annealed λ arms to size-selected target DNA. As a first approximation, ligation reactions should contain about 4 μg of annealed λ arms (31 kb) for every 1 μg of size-selected target DNA, assuming an average size-selected value of 17 kb. However, if *in vitro* packaging efficiencies are not as expected (because of damaged *cos* or *cs* sites, errors in estimating DNA concentrations, etc.), a series of pilot ligation and *in vitro* packaging reactions should be done to determine the best ratio of λ arms to size-selected target DNA for each particular batch of λ arms and target DNAs.

The second parameter, the total DNA concentration in the ligation mixture, is important because the DNA concentration must be high enough to ensure that intermolecular ligations, which lead to concatamers, are favored over self-ligation, which reduces the yield of viable recombinant λ DNA molecules. A theoretical discussion concerning the effects of DNA concentration and size on the ligation reaction has been presented previously.[44] The results of the calculations are that the concentration of annealed arms should be about 135 μg/ml, and the size-selected target DNA concentration should be about 43 μg/ml.

Ligation and in Vitro Packaging of Recombinant λ Molecules

Prior to ligating λ arms with size-selected target DNA, the λ arms should be tested by self-ligation followed by *in vitro* packaging to determine the amount of background phage that can be expected per microgram of arms. Purified λ arms should not yield more than 5 × 10⁴ background pfu/μg. For the procedure outlined below, Charon 40 arms were purified as shown in Fig. 3 and ligated with the size-selected gibbon DNAs shown in Fig. 5.

1. Charon 40 arms and size-selected target DNA should be resuspended in TEN buffer at a concentration of about 500 μg/ml. Add 10 μl (8 μg) of Charon 40 arms and 4 μl (2 μg) of size-selected gibbon DNA to a sterile 1.5-ml microcentrifuge tube. Incubate at 37° for 5 min, then cool on ice. Add 2.5 μl of 10× ligase salts (see reagents list below) and adjust the total volume to about 25 μl. Remove a 2-μl aliquot prior to adding ligase

and store at 4° for later agarose gel analysis. Add ligase, generally about 1 μl for a ligase concentration of about 10 units/μl.

2. Incubate at 37° for 1 hr, then at 4° overnight. After ligating, heat-inactivate the ligase by incubating at 65° for 5 min, then remove another 2-μl aliquot and analyze, along with the aliquot removed in Step 1, by agarose gel electrophoresis. The agarose gel analysis should show few DNA fragments in the size range of the original size-selected target DNA for the ligated sample; most of the DNAs in this sample should be larger than wild-type λ phage (50 kb). Store the ligation mix at 4° while waiting for gel analysis results.

3. Obtain λ *in vitro* packaging extracts from a commercial source, and *in vitro* package one-half of the ligation mix as instructed by the vendor. After packaging, add 1 ml of λ dilution buffer and 20 μl of chloroform. Mix gently and titer the supernatant solution at dilutions of 10^{-3} and 10^{-4} using *E. coli* host ED8767 (use this host only for *gam*$^+$ λ vectors, see above). If the ligation and *in vitro* packaging reactions are successful, the titer should be between 10^5 and 10^6 pfu/ml. Titers as high as 10^7 can be expected.

4. The next step depends on the titer obtained and the goals of the experiment. If the titer is barely sufficient to contain the complete genome at a probability of 0.99, it should be amplified directly before screening (see below). However, if the library contains 5 to 10 genome equivalents, one genome equivalent can be used for direct screening and amplification (see below), with the remaining part being stored for later amplification or distribution.

Screening of a Recombinant λ Library with ^{32}*P-Labeled Probes*

As mentioned above, the λ vector system is a highly efficient system for obtaining a large number of recombinant clones; however, its immense power lies in the efficiency with which it can be screened for clones that contain DNA regions (genes) represented at single-copy levels within the target species genome. Benton and Davis[31] were the first to describe a highly effective and efficient *in situ* phage plaque screening protocol, which involves the simple contact between the phage plaques and nitrocellulose (more recently, nylon membranes have been preferred). Treatment of the filters with 0.5 M NaOH causes the phage to lyse, and the exposed recombinant λ DNAs are denatured and fixed *in situ* on the filters. Contact between the filter and the λ plaques does not destroy nor greatly distort the location of each λ plaque; thus, the λ plaques remaining on the surface of the agar are viable and can be recovered. After fixing the recombinant λ DNAs onto the filter, the filter is hybridized against ^{32}P-labeled DNA or

RNA probes, and the resulting film exposures can be used to locate the recombinant λ plaque responsible for a specific hybridization signal. After a series of plaque purification steps (repeated plating and plaque isolation), an individual λ recombinant clone can be purified to homogeneity.

The number of recombinant λ phage plaques that must be screened to obtain a specific DNA region depends on the size of the target genome, the nature of the DNA region (i.e., single copy or repetitive), and the background of nonrecombinant λ phage contained within the library. For example, to isolate a single-copy DNA region from a recombinant λ phage gibbon library requires screening at least 10^6 phage plaques; however, to compensate for background nonrecombinant λ phage and other discrepancies during library construction, 2 to 3 times the minimum number of phage plaques should be screened. At this point you may be asking, How would one screen 2×10^6 recombinant λ phage plaques? This must require many plates. The numbers of phage plaques that can be effectively plated and screened on different size culture plates are listed in Table II. The smaller plates should not be used for the first round of screening for a single-copy DNA region; they are more suited for the isolation of highly repetitive DNA or for later purification steps (see below). The largest plates that are convenient for screening are the 23×24 cm plates supplies by Nunc (Vanguard International, Inc, Neptune, NJ) plates, which are referred to as kiloplates (KPs) in the procedure presented below. These plates are extremely convenient because they are square (easy to fit with the filter

TABLE II
AGAR PLATE SIZES AND NUMBER OF PHAGE PLAQUES PLATED[a]

| Size of plate | Total area (cm²) | Volume (ml) | | | Maximum number plaques/plate |
		Bottom agar	Indicator bacteria (ml)	Top agar	
9 cm round plate	64	25–30	0.1	3.0	15,000
9 × 9 cm square plate	81	35–40	0.1	4.0	20,000
15 cm round plate	177	70–80	0.3	8.0	50,000
23 × 24 cm square plate	552	300	4.0	40	250,000
44 × 32 cm cafeteria-type tray	1408	900	12.0	100	1,000,000

[a] From J. L. Slightom and R. F. Drong, in "Plant Molecular Biology Manual" (S. B. Gelvin and R. A. Schilperoort, eds.), p. A8. Kluwer Academic Publishers, Dordrecht, The Netherlands, 1988. Reprinted by permission of Kluwer Academic Publishers.

membrane) and because, after exposure of the hybridized filter membrane to film, hybridizing plaques can be identified by placing the agar plate over the exposed film while sitting on a light box.

The following procedure is for the screening of a recombinant λ library that has been constructed in Charon 40 and is being screeened after plating on the recA⁻ E. coli host ED8767. For the composition of solutions and other reagents, see the list following the protocol.

1. Prior to starting, prepare the following: (a) an overnight culture (50 ml) of the appropriate E. coli host (in the case of plating Charon 40 use ED8767 cells); and (b) agar, both top and bottom, for pouring the necessary number of KPs [2×10^6 phage can be screened on 8 KPs, which requires about 2 liters of bottom agar (1.5%), and phage spreading requires about 400 ml of top agar (1%).]. Also, the titer of the recombinant λ library should already be known.

2. Pour the required number of KPs prior to starting the E. coli overnight growth. Allow the plates to sit overnight at room temperature to ensure that the KPs are free of excess moisture. If necessary place in a biological hood, remove the cover, and wipe away excess moisture.

3. Set up preinfections of the E. coli host and λ library; for each KP add the following to a sterile 50-ml polystyrene tube: 4 ml of overnight growth of E. coli host ED8767, 4 ml of λ 10 mM NaCl and CaCl solution, and the appropriate volume of the recombinant λ library expected to produce 250,000 pfu. Incubate the preinfection mixture at 37° for 15 min.

4. Add 40 ml of top agar (stored in a 50° water bath) to the preinfection tubes (one at a time), gently invert 2 to 3 times, and then pour evenly over the surface of a KP. Repeat this step until all of preinfection tubes have been plated. Allow the top agar to harden before moving (15–20 min should be adequate).

5. Place the KPs in a 37° incubator (do not invert); to avoid condensation on the lids to not stack the KPs. Use as large an incubator as possible or a warm room if available. Allow the E. coli and phage to grow overnight to obtain full plaque development.

6. Carefully remove the KPs from the incubator; if condensation is present on the lids, remove the lid and either shake off the condensation or wipe it off with a Kimwipe. Allow the KPs to cool to room temperature, followed by storage in a cold room. Prior to placing filters over the plaques place the KPs on ice (use a large smooth-bottom photo tray filled with ice) and cool the plates for about 30 min.

7. While cooling the KPs, set up two smooth-bottom photo trays with denaturation and neutralization solutions by lining with two sheets of Whatman (Clifton, NJ) 3MM paper and saturating with the appropriate solution.

8. Cut filter membranes to 22×23 cm and treat according to the vendor's instruction. Generally, filters are labeled in permanent ink with the KP number, λ vector used, and the name of target species. First, soak filters in distilled water followed by soaking in a solution of 1 M NaCl. Blot the filters dry on Whatman 3MM paper to remove excess moisture just prior to use.

9. Carefully place the filter membrane on top of the appropriate KP and mark the position of the filter by using a blunt needle dipped in India ink (water resistant) and stabbing it through the membrane into the agar below. Traces of ink should remain in the agar and on the filter. The marks should be placed about every 4 to 5 cm to provide easy alignments for later picking of the phage plaque responsible for a particular hybridization signal.

10. Allow the filter to remain in contact with the λ phage plaques for about 4 min, then carefully remove (avoid lifting the top agar) and place the filter, plaque side up, onto Whatman 3MM paper previously saturated with denaturation solution. Remove trapped air bubbles and leave for 4 min; then transfer onto Whatman 3MM paper previously saturated with neutralization solution; and leave for 4 min. After neutralization, place the filter on dry Whatman 3MM paper and allow it to air-dry. Repeat these steps for each KP used and for each replica filter obtained from the same KP.

11. Treatment of the filter membrane to ensure fixing of the recombinant λ DNAs depends on the type of membrane used. Nitrocellulose requires only heating (68°, overnight), whereas nylon requires UV light treatment (see vendor's instructions).

12. After fixing DNA onto the filters, prehybridize them in a smooth-bottom photo tray (25×25 cm). For eight filters, prehybridize in about 600 ml of prehybridization solution and incubate at 65°–68°, with shaking (50 rpm), for at least 2 hr. Cover the photo tray with plastic wrap and a glass plate to prevent evaporation.

13. Hybridize the eight filters in about 200 ml of hybridization solution that contains the denatured ^{32}P-labeled probe [specific activity, 1×10^8 counts/min (cpm)/μg] of at least 10^5 cpm/ml. Denature the probe by adding $\frac{1}{20}$ volume of 5 M NaOH, then incubate at 68° for 15 min followed by the addition of $\frac{1}{20}$ volume of 5 M HCl and $\frac{1}{10}$ volume of 1 M Tris-HCl (pH 7.4). Mix, then quickly add filters to hybridization solution that contains 50 μg/ml poly(A) (Sigma, St. Louis, MO, P-9403), which is used to reduce nonspecific binding of the probe to the filter. Pour off the prehybridization solution and add the hybridization solution; incubate at 68° with shaking for at least 12 hr or overnight.

14. Decant the hybridization solution and rinse briefly with 200 ml of $3 \times$ SSC plus 0.5% sodium dodecyl sulfate (SDS). Replace with 300 ml of

fresh $3 \times$ SSC solution and incubate at $68°$ with shaking for 30 min to 1 hr.

15. Remove the filters from the wash solution and place on Whatman 3MM paper. Allow the filter to dry and then scan the filter with a Geiger counter. If the counts are below 200 cpm, obtain film exposures of the filters. Expose filters to film using enhancer screens (DuPont Quanta III) at low temperatures $(-70°)$. If the initial scan or later film exposures indicate high background signals, rewash the filters in a solution of lower ionic strength (e.g., $1 \times$ SSC and, if necessary, down to 0.1X SSC).

16. After obtaining film exposures, align the corresponding films and filters and transfer the reference ink marks onto the film. Placing the film on a light box and the corresponding KP on top of the film, align the reference marks on the film with those in the agar. Align the reference marks closest to a particular hybridization signal (λ clone candidate) and pick the plaque region using the large end of a sterile Pasteur pipette. The agar plug can be excised from the plate by placing a finger over the small end of the pipette, loosing the plug with rotary movement, and lifting the plug free using the "built-up" vacuum. Phage plugs are placed in 1 ml of phage storage buffer and numbered with the KP number followed by the pick number from a particular KP. Thus, the first plaque pick from KP 1 is referred to as KP 1.1.

17. The recombinant λ phage responsible for the hybridization signal must be purified from the contaminating phage also contained within the agar plug. Purification of signal-producing phage is done by using a series of successive plating and hybridization reactions similar to that just described for the KPs, but on smaller agar plates (see Table II). An agar plug can contain well over 100 independent phage clones. The next purification plating should involve the plating of about 500 pfu. This can usually be obtained by plating a 10^{-4} dilution from the original pick (KP 1.1, etc.). Plating is done using the same technique used to obtain a phage titer: 100 μl of E. coli host cells, 100 μl of λ infection salts, and the appropriate dilution volume of the phage sample. The sample is preincubated at $37°$ for 10 min followed by the addition of top agar and plating (use 3 ml of top agar for round 9-cm plates and 4 ml for square 9-cm plates).

18. Plaque purification plates are screened using procedures similar to those described above for KPs, but, instead of using the large end of the Pasteur pipette for picking plaques, the small end is used to reduce the number of contaminating phage carried along with the hybridizing λ phage plaque. The small phage plugs are placed into 1 ml of phage storage buffer and numbered. The secondary purified phage pick for KP 1.1 is referred to as KP 1.1.1, and so on for the other KPs and multiple picks from each KP. The procedure of successive plating, screening, and picking is continued

until a plating is obtained where every phage plaque yields a hybridization signal with the ^{32}P-labeled probe. For each successive plating step, add an additional number (.1) to distinguish it from any of the previous purification phage plaque picks. Once the recombinant λ clone(s) that hybridizes to the ^{32}P-labeled probe is plaque-pure, a primary phage growth should be started from a single hybridizing plaque; this is referred to as the primary growth (1°). From this primary growth, a large-scale λ growth can be initiated and the recombinant λ DNA purified.[20,42,49]

Reagents

NZC medium: 9 g NZ amine (Gibco BRL, Grand Island, NY), 4.5 g NaCl, 4.5 g casamino acids, water to 0.9 liter; mix, then autoclave

NZC bottom agar: 10 g NZ amine, 5 g NaCl, 5 g casamino acids, 15 g agar, water to 1 liter

NZ top agar: 10 g NZ amine, 5 g NaCl, 10 g agar, water to 1 liter

ø80 buffer: 0.1 M NaCl, 10 mM Tirs-HCl (pH 7.4)

ø storage buffer: ø80 buffer plus 0.05% gelatin and 10 mM MgCl$_2$; saturate with chloroform after autoclaving

λ infection salts: 10 mM MgCl$_2$, 10 mM CaCl$_2$

Salt adjustment buffer: 5 M NaCl, 100 mM Tris-HCl (pH 7.4), 10 mM EDTA

TEN buffer: 20 mM Tris-HCl (pH 8.0), 10 mM NaCl, 1 mM EDTA

10X ligase salts: 800 mM Tris-HCl (pH 8.0), 200 mM MgCl$_2$, 150 mM dithiothreitol (DTT), 10 mM ATP; store at $-20°$ and heat to room temperature prior to use

λ dilution buffer: 10 mM Tris-HCl (pH 7.4), 10 mM MgSO$_4$, 0.01% gelatin

Denaturation solution: 0.5 M NaOh, 1.5 M NaCl

Neutralization solution: 3.0 M NaCl, 0.5 M Tris-HCl (pH 7.2), 1 mM EDTA

Prehybridization solution: 6 × SSC (1 × SSC is 0.15 M NaCl, 15 mM sodium citrate, pH 7.2), 0.02% (w/v) bovine serum albumin (BSA), 0.02% (w/v) Ficoll (M_r 400,000), 0.02% (w/v) polyvinylpyrrolidone (PVP, M_r 36,000), 1% (w/v) SDS

Hybridization solution: Prehybridization solution plus 50 μg/ml poly (A) (Sigma, P-9403), note that 20 μg/ml denatured sonicated calf thymus or salmon sperm DNA can be used in place of poly(A).

[49] J. L. Slightom and R. F. Drong, *in* "Plant Molecular Biology Manual" (S. B. Gelvin and R. A. Schilperoort, eds.), p. A8. Kluwer Academic Publishers, Dordrecht, The Netherlands, 1988.

Acknowledgments

I thank Drs. Allan Wilson, Elizabeth Zimmer, Sandy Martin, and Barb Chapman for gifts of the gorilla SF-4 (San Francisco Zoo), chimpanzee #1, orangutan YO-1 (Yerkes Regional Primate Center), rhesus, capuchin, and squirrel monkey DNAs and for encouraging me to construct primate recombinant phage libraries. I also thank Drs. Frederick Blattner and Oliver Smithies for providing me with the opportunity to learn the λ biology needed for the construction of phage libraries. Special thanks go to the many individuals who have worked with me in the Department of Anatomy and Cell Biology, Wayne State University, Detroit, Michigan, in the comparative analyses of the primate β-globin gene cluster: Drs. Morris Goodman, John Czelusniak, Ben Koop, Michael Miyamoto, Dan Tagle, David Fitch, Wendy Bailey, and Kenji Hayasaka. The work was supported by research grants from The National Institutes of Health (HD 16595 to J.L.S. and HL 33940 to M.G. and J.L.S.) and the National Science Foundation Award BSR-8607202 to M.G. and J.L.S.

[20] Surveying Plant Genomes for Variable Number of Tandem Repeat Loci

By Steven H. Rogstad

Introduction

A delineated class of DNA, minisatellite DNA, has been found to be moderately to extremely variable in certain eukaryote genomes.[1-3] Minisatellite DNA is composed of tandem repeats of a "core" or "consensus" sequence reiterated a low to moderate number of times (relative to satellite DNA, where consensus motifs, or variants thereon, may be repeated tens of thousands of times). For convenience, minisatellite DNA as defined here includes simple sequence[4] and microsatellite[5] DNA. Thus, minisatellite consensus sequences range from 2 to approximately 70 base pairs (bp). Several different minisatellite families (members of a family have consensus sequence similarities) have been described.[6]

The different alleles at a minisatellite locus are thought to vary in the number of repeats of the consensus sequence, hence the name variable number of tandem repeats (VNTR) locus.[7] For some human VNTR loci

---

[1] A. J. Jeffreys, V. Wilson, and S. L. Thein, *Nature (London)* **314**, 67 (1985).
[2] T. Burke and M. W. Bruford, *Nature (London)* **327**, 149 (1987).
[3] G. Vassart, M. Georges, R. Monsieur, H. Brocas, A. S. Lequarre, and D. Christophe, *Science* **235**, 683 (1987).
[4] D. Tautz, *Nucleic Acids Res.* **17**, 6463 (1989).
[5] M. Litt and J. A. Luty, *Am. J. Hum. Genet.* **44**, 397 (1989).
[6] J. Pemberton, B. Amos, and S. Patton, *Fingerprint News* **3**(2), 22 (1991).
[7] Y. Nakamura, M. Leppert, P. O'Connell, R. Wolff, T. Holm, M. Culver, C. Martin, E. Fujimoto, M. Hoff, E. Kumlin, and R. White, *Science* **235**, 1616 (1987).

surveyed across random populations, the number of alleles available is so high that heterozygosities in excess of 90% (hypervariable loci) have been found.[1] In humans, probes are now known that reveal alleles at single hypervariable loci[7] or at several variable–hypervariable loci simultaneously.[1,3]

These highly variable genetic markers have proved useful in a number of biological applications including forensics,[8] pedigree analysis,[9] linkage studies and mapping,[10] population studies,[11,12] clonal analysis,[13] cell line and tumor typing,[14] and *in situ* hybridization for the unique marking of chromosomes.[15] For a comprehensive list of minisatellite DNA analyses, see Pemberton *et al.*[6] Inspection of this bibliography reveals that studies utilizing VNTR genetic markers with plants are relatively few. Two contributing factors to this deficiency are as follows: (1) restriction enzyme/ Southern blot minisatellite analyses require relatively large amounts of DNA per individual, and (2) more minisatellite probes for plants are needed. Here, methods are described that, it is hoped, will contribute to alleviating these problems.

Plant DNA Extraction and Purification

Previously described methods for plant DNA extraction (e.g., see Hillis *et al.*[16] and references therein) have usually either required CsCl purification or resulted in a final DNA product that often contained other contaminating substances. For restriction site analysis of chloroplast or mitochondrial genomes, where usually less than 0.5 μg of DNA per individual is required, any procedure was appropriate and practical. However, for minisatellite restriction enzyme analyses, typically 5–10 μg of DNA is required per individual per trial. For such analyses, obtaining enough DNA from an individual for several trials using CsCl is costly and time-consuming;

[8] P. Gill, A. J. Jeffreys, and D. J. Werrett, *Nature (London)* **318**, 577 (1985).

[9] A. J. Jeffreys, V. Wilson, S. L. Thein, D. J. Weatherall, and B. A. Ponder, *Am. J. Hum. Genet.* **39**, 11 (1986).

[10] M. Georges, M. Lathrop, P. Hilbert, A. Marcotte, A. Schwers, S. Swillens, G. Vassart, and R. Hanset, *Genomics* **6**, 461 (1990).

[11] J. H. Wetton, R. E. Carter, D. T. Parkin, and D. Walters, *Nature (London)* **327**, 147 (1987).

[12] D. A. Gilbert, N. Lehman, S. J. O'Brien, and R. K. Wayne, *Nature (London)* **344**, 764 (1990).

[13] S. H. Rogstad, H. Nybom, and B. Schaal, *Plant Syst. Evol.* **175**(3–4), 115 (1991).

[14] J. A. L. Armour, I. Patel, S. L. Thein, M. F. Fey, and A. J. Jeffreys, *Genomics* **4**, 328 (1989).

[15] P. R. Simpson, M. A. Newman, D. R. Davies, T. H. Noel Ellis, P. M. Matthews, and D. Lee, *Genome* **33**, 745 (1990).

[16] D. M. Hillis, A. Larson, S. K. Davis, and E. A. Zimmer, *in* "Molecular Systematics" (D. M. Hillis and C. Moritz, eds.), p. 318. Sinauer, Sunderland, Massachusetts, 1990.

moreover, plant DNA preparations lacking a CsCl step are often so contaminated that complete restriction enzyme digestion analyses are inhibited and/or interference of electrophoresis occurs. For a variety of plants, many of the contaminating substances coprecipitating with DNA in isolation protocols lacking the CsCl step appear to be separated from the DNA on agarose gel electrophoresis.

DNA Purification by Agarose Gel Electrophoresis: Protocol

1. Grind fresh tissue in liquid nitrogen or according to the method described by Tai and Tanksley.[17] Although it is possible to use a number of different initial extraction buffers followed by 2-propanol precipitation, typically 0.5–5 g of ground tissue is added to 8 ml of 2× hexadecyltrimethylammonium bromide (CTAB) extraction buffer[16] with 1% 2-mercaptoethanol added just prior to tissue addition. Thirty-milliliter centrifuge tubes with caps are used for this step. Alternatively, for plants that yield relatively high amounts of DNA with little contamination, 0.2–0.6 g of powdered tissue can be added to 800 μl of 2× CTAB extraction buffer with 1% 2-mercaptoethanol in microcentrifuge tubes, and the following protocol scaled accordingly.

2. Incubate closed tubes at 65° for 1–15 hr. Note that, for most plants extracted, DNA and contaminant yields increase with the time of incubation. For some plants that have especially high production of contaminants, impurities can be attenuated by incubation at lower temperatures for shorter periods (e.g., incubation with chilled CTAB on ice for 1 min to 1 hr), although this usually decreases the DNA yield (by up to 90%). After incubation, and only if it proves necessary, some of the contaminants can be removed by centrifugation (retain the supernatant) at 2500 g for 1–5 min, although some of the DNA can be lost to the pellet (up to 60% loss).

3. After incubation (and centrifugation if needed) add 0.6 volumes of chloroform and mix gently but thoroughly. Centrifuge at 8000 g for 10 min and transfer the upper aqueous layer to a new 30-ml centrifuge tube. Precipitate the DNA (and contaminants) with 0.6 volumes of 2-propanol for 20 min to overnight at −20°. Centrifuge at 8000 g for 10 min and gently dispose of the supernatant. Air-dry the pellet; then add 0.5–1 ml of 1× Tris–EDTA (TE) buffer[18] (the amount of 1 × TE will depend on prior experience with the steps given below). When the pellet is more or less dissolved, add ribonuclease A to a final concentration of 10 μg/ml. Incubate at 37° for 30 min.

4. Aliquot 0.5–1 ml of the above solution into the pockets of a 0.45–0.55%, 1× Tris–borate–EDTA (TBE) buffer[18] (sterile), 2.7 μg/ml ethi-

[17] T. H. Tai and S. D. Tanksley, *Plant Mol. Biol. Rep.* **8**, 297 (1991).

dium bromide (sterile), low-gelling-temperature agarose gel prepared as follows. Typically, a "minigel" size gel bed is used (e.g., gel bed 8 × 10 cm). A comb is constructed (using plastic blocks glued to a support) to make wells measuring 1.5 cm (on the side where DNA will enter the gel) by 5–8 mm. Teeth should be separated by at least 5 mm to ensure separation of specimens and to ensure that side walls are strong when the comb is removed. The gel is poured with at least 2 mm of gel below the wells and with the top gel surface at least 5 mm above the bottom of the wells. With this low concentration of agarose, the gel should be cooled at 4° for at least 1 hr. Remove the comb with gentle rocking, using a spatula to tease the gel free from the teeth. DNA preparations that are especially contaminated can be centrifuged for 2 sec on a microcentrifuge before adding them to the wells (this may cause some loss of DNA).

5. Add 1 × TBE buffer to the electrode chambers on both sides of the gel until it is approximately 1–2 mm below the top of the gel. Do not cover the gel with buffer at any time. Run the gel at 70 mA for 2.5 min, then at 60 mA, usually for 15–20 min (higher amperages may shear the DNA or adversely affect its detectability). The electrophoresis of the DNA should be observed carefully. Ideally, the DNA enters the gel in the first 5 min and moves as a clear, unified, thin wave that should be visible without UV light. Once the wave has moved 1–2 mm into the gel, turn off the power, remove the leads, and remove most of the 1 × TBE from the electrode chambers. Dispose of the now DNA-free solutions in each pocket and rinse the pockets with sterile distilled water.

Note that the rate at which the DNA moves through the gel and the form of the wave can be adversely affected by a DNA concentration that is too high or by contaminants. In the case of the former, add less sample to the wells and/or dilute the samples. In the latter case, species, individuals within a species, and even leaves from one individual can differ as to contaminants produced. One way to reduce impurities is to remove the well solution after the DNA has fully entered the gel (usually after 5 min; turn off the power), replace it with distilled sterile water, and then run the gel normally. This will affect the DNA wave form and movement, however. Also, in some cases, contaminants moving through the gel ahead of the DNA may absorb ethidium bromide to the degree that the DNA wave is difficult or impossible to see (even under UV light). DNA is probably still present behind such materials and can often be detected by a slight wrinkling of the gel surface that always forms as the DNA moves through the gel. If in doubt about where the DNA is, treat the gel region with a wrinkled surface behind (cathodal to) the contaminants as indicated below (Step 6). In fact, it is possible to run purification gels without ethidium bromide, excising only the region where the wrinkle forms. However, ethidium bromide makes DNA move more slowly through the gel, and it seems to give better separation between contaminants and the DNA wave. Reducing the ethidium bromide concentration below 2.7 μg/ml may reduce yields.

It is sometimes obvious that contaminants are present in the region of the DNA wave after the period of electrophoresis indicated above. In such a case, the time of electrophoresis at 60 mA can be extended. Also, the well contents can be replaced with distilled sterile water as indicated above. However, the further the DNA moves into the gel, the bigger the gel slice needed and/or the lower the yield, so run the DNA into the gel as little as possible.

6. Being mindful of the possibility of cross-contamination of specimens, use clean razor blades to excise carefully the DNA wave for each specimen, transferring each to its own marked 1.5-ml microcentrifuge tube. If possible, remove the DNA–gel slices using only visible light (a sheet of white paper underneath the gel will help) as UV light may cause damage. If necessary, use long-wave length UV light at low intensity (i.e., a hand-held unit). Proper care should be taken to avoid unwanted distribution of ethidium bromide.

7. Close the tubes and incubate in a 68° water bath for 20 min, then transfer them to a 38° water bath for 8 min. The melted gel volume should be approximately 400–600 μl. Keeping the tubes in the 38° water bath as much as possible, add 15 units of agarase (Sigma, St. Louis, MO) to each tube. Close the tubes and mix thoroughly. Incubate at 38° for 12–18 hr, with occasional mixing. Note that protocols exist for removal of low gelling temperature agarose using phenol followed by chloroform[18] (chloroform alone is not effective). However, phenol treatment usually reduces yields by up to 50% compared to those obtained with the present protocol, and phenol treatment may increase the shearing of the DNA.

8. Precipitate the enzyme and undigested materials by bringing each tube to 0.5 M NaCl, cooling on ice 30 min, and then microcentrifuge at 12,000 rpm for 15 min at 4°. The precipitate is often loose at the bottom of the tube, so carefully withdraw the supernatant to a new tube. At this stage, the solution can be either precipitated immediately with 0.6 volumes of 2-propanol or, if the restriction enzymes to be used are salt sensitive, diluted with distilled water to 900 μl followed by addition of 0.6 volumes of 2-propanol. Incubate for at least 30 min at −20°

9. Microcentrifuge the tubes at 12,000 rpm for 8 min. Carefully discard the supernatant and rinse the pellet with 1 ml of 70% (v/v) ethanol. Spin the tubes again for 6 min at 12,000 rpm and discard the supernatant (repeat this rinse to further dilute salts if necessary). Air-dry the pellets at room temperature. Add the appropriate amount of 1 × TE for the desired

concentration (50–100 µl) and quantitate. The DNA is now ready for restriction enzyme digestion using 4-bp cutters.

Note that a small white residue often remains after agarase digestion. For many species, this residue is usually many times smaller than the precipitate that develops from other protocols that do not use CsCl. If a large precipitate does develop, it may indicate that the agarase digestion was not complete (try running the gel for a longer or shorter time to make sure that the DNA is free of contaminating materials).

Plant Minisatellite DNA Probes

It was previously demonstrated that particular minisatellites might be taxonomically widespread[19-22] and, further, that numerous different minisatellite families with utilizable genetic variation might be distributed variously throughout a wide variety of living organisms.[20,21] Although numerous minisatellite families seem to exist, obtaining particular minisatellite probes has not always been a straightforward process. Initially, probes were discovered more or less by accident when DNA sequences cloned for other purposes were seen to reveal complex endonuclease fragment profiles for humans and other organisms.[1] Such probes were subsequently derived by screening cloned DNA specifically with the intention of seeking variation.[7,23] This approach involves cloning genomic DNA and screening the clones, a time-consuming process for each cloned probe obtained.

Some of these multilocus probes are now available commercially (e.g., the so-called Jeffreys probes[1] can be obtained from CellMark Diagnostics (Germantown, MD); Molecular Biosystems, Inc. also offers minisatellite probes). The American Type Culture Collection (Rockville, MD) is a repository for cloned sequences and primers. Many of these sequences have been shown to be polymorphic (some with heterozygosities in excess of 80%) in humans and other organisms. It seems likely that some of the more highly variable sequences include minisatellites that might be used as probes in other organisms. Although some minisatellite probes may be

[18] J. Sambrook, E. F. Fritsch, and T. Maniatis, "Molecular Cloning." Cold Spring Harbor Laboratory, Cold Spring Harbor, New York, 1989.
[19] J. F. Dallas, *Proc. Natl. Acad. Sci. U.S.A.* **85**, 6831 (1988).
[20] S. H. Rogstad, J. C. Patton, and B. A. Schaal, *Proc. Natl. Acad. Sci. U.S.A.* **85**, 9176 (1988).
[21] S. H. Rogstad, B. L. Herwaldt, P. H. Schlesinger, and D. J. Krogstad, *Nucleic Acids Res.* **17**, 3610 (1989).
[22] A. P. Ryskov, A. G. Jincharadze, M. I. Prosnyak, P. L. Ivanov, and S. A. Limborska, *FEBS Lett.* **233**, 388 (1988).
[23] Z. Wong, V. Wilson, A. J. Jeffreys, and S. L. Thein, *Nucleic Acids Res.* **14**, 4605 (1986).

utilized across wide taxonomic expanses, there is no guarantee that any one will be informative with any particular untested species.

M13 Repeat Probe: Polymerase Chain Reaction Protocol

The M13 repeat probe[3] has been shown to reveal polymorphisms in a wide variety of organisms including plants.[20,22] Previous methods for amplification of this sequence utilized standard clonal growth of the M13 bacteriophage, with random priming used to label the sequence obtained by restriction digestion and isolation of the proper electrophoresed fragment. An improved method brings about both amplification and labeling using the polymerase chain reaction (PCR).

1. Obtain DNA from any strain of the M13 bacteriophage (commercially available) that has the minisatellite sequence indicated by Vassart et al.[3]

2. Using the M13 sequence (GenBank), synthesize primers for the complementary strands that can be used to amplify the minisatellite sequence (e.g., 20-mers 5' TGTAGTTTGTACTGGTGACG and 5' CCTTATTAGCGTTTGCCATC).

3. For amplification, follow the protocol of Saiki,[24] except use approximately 3 ng template DNA, maintain the $MgCl_2$ concentration at 0.2 mM, and amplify only through 10–15 cycles. With increased amplification, rather than obtaining a sharp band product when checked on an agarose gel, DNA sequences of a very broad range of sizes are produced (nonspecific amplification).

4. For PCR labeling of M13, follow the directions in Step 3 except modify the deoxynucleotide triphosphate mix to include three unlabeled nucleotides and one labeled nucleotide (yielding a final concentration for each at 10 μM). Ten cycles of amplification should suffice.

Creation of Synthetic Tandem Repeats via Polymerase Chain Reaction

Subsequent to the description of the first minisatellite probes, it was determined that simple repeats, used in the form of labeled oligomers, could be used to reveal genetic variation in a number of organisms.[25] A possible drawback to this method is that such oligomer probes, owing to

[24] R. K. Saiki, D. H. Gelfand, S. Stoffel, S. J. Scharf, R. Higuchi, G. T. Horn, K. B. Mullis, and H. A. Erlich, *Science* **239**, 487 (1988).

[25] K. Weising, F. Weigand, A. J. Driesel, G. Kahl, H. Zischler, and J. T. Epplen, *Nucleic Acids Res.* **23**, 10128 (1989).

their small size, cannot be used effectively in hybridizations with Southern blot membranes[18] and must instead be used directly with wet or dried agarose gels.[25] Vergnaud[26] showed that "random," short sequences (14 bp) could be ligated, the ligation products then being cloned into a vector. The cloned synthetic tandem repeats (STRs) were then clonally amplified and labeled by standard techniques to be used as probes. Although these random STRs revealed variation in humans, they still required cloning steps.

A method is described here that generates labeled STRs using the polymerase chain reaction, bypassing some of the disadvantages of the previously discussed methods of probe creation. A few applications of PCR–STR probes designed after probes previously shown to detect minisatellite variation in humans and other organisms are demonstrated. The existence of variable minisatellite DNA genetic markers has been more amply demonstrated for vertebrates than for other organisms, and, to supplement previous findings that such systems are widespread throughout the eukaryotes, taxonomically distant plant species were chosen for this study.

Two angiosperms are among the species investigated: (1) two sibling individuals (seedlings derived from the same "maternal" tree) of *Polyalthia glauca* (Hassk.) Mueller (Annonaceae) were compared for differences in fragment profiles with different PCR–STR probes; and (2) several randomly chosen, commercially available individuals of *Raphanus sativus* L. (Cruciferae; radish) were selected based on preliminary results showing the presence of minisatellites in this species. Two individuals of *Equisetum arvense* L. (Equisetaceae; field horsetail), one each from two stands separated by 100 m near Georgetown, Colorado, and three individuals of the walking-fern (*Asplenium rhizophyllum* L.; Aspleniaceae) collected within 20 m of one another near Festus, Missouri, were also analyzed for minisatellite variation as described below.

Only healthy green tissues free of epiphytes or debris were sampled. Tissues were stored on ice until they could be ground (within 48 hr) in liquid nitrogen for storage at −70°. Details of DNA extraction, agarose gel electrophoresis, Southern blotting, and autoradiography are the same as described previously[20] with the exceptions noted here. Hybridization conditions of Westneat *et al.*[27] were used, and filters were stripped for subsequent rehybridization following Sambrook *et al.*[18] With the *Polyalthia glauca* seedlings, replicate samples of both seedlings were run in multiple

[26] G. Vergnaud, *Nucleic Acids Res.* **19,** 7613 (1989).
[27] D. F. Westneat, W. A. Noon, H. K. Reeve, and C. F. Aquadro, *Nucleic Acids Res.* **16,** 4161 (1988).

lanes on one gel, the resulting Southern filter then being cut into strips that were subjected to different probes.

PCR–STR probes are generally prepared as follows, although slight variations will be described for specific applications. For each probe, two oligomers are synthesized, arbitrarily designating one the template strand and the other the complementary strand. Template oligomers are designed to be 16–20 bases long and include at least two tandem copies of the repeat sequence (i.e., the sequence to ultimately be tandemly repeated numerous times in the product DNA). All template oligomer repeat sequences, with one exception, are chosen based on sequences shown previously in some other organism to reveal minisatellite variation or are related variant sequences. The exception is a randomly chosen oligomer repeat sequence.

Templates used include ACTGGG (the *Drosophila* Per gene probe[28]); CAC, GACA, GATA (repeat sequences used with vertebrates[29]); GATGTGGG (human derived); and GAGGT (sequence selected at random). None of these repeat sequences has *Hae*III or *Taq*I recognition sequences when in tandem.

The creation of a PCR–STR probe proceeds by using an equimolar mixture of the template and complementary oligomer strands in otherwise standard PCR amplifications. When the two oligomers are mixed under annealing conditions, some of them will anneal unevenly in the proper configuration for 3′ terminus polymerase activity, adding an additional copy (or copies) of the repeat sequence to each oligomer. When such products are heated to denaturation and then allowed to reanneal, some of them will again anneal unevenly, permitting 3′ end extension and adding more copies of the repeat sequence to each strand. After numerous cycles, very long STRs composed of many repeats are created. The initial steps of this process are shown in Fig. 1.

Starting with template and complementary oligomers, each having two copies of either the repeat sequence or its complement, respectively, and given that at each annealing step at least one template–complement pair will undergo 3′ extension, only the last 3′ end copy of the repeat sequence for each strand will anneal (as shown in Fig. 1). The longest product strands will have $2^n + 1$ repeats of the repeat sequence, where n is the number of PCR cycles. Thus after 10 cycles, the longest strands will theoretically have 1025 repeats, and after 20 cycles they will have 1,048,577 repeats.

Because the hybridization–stripping characteristics of such long re-

[28] M. Georges, P. Cochaux, A. S. Lequarre, M. W. Young, and G. Vassart, *Nucleic Acids Res.* **17**, 7193 (1987).
[29] J. T. Epplen, *J. Hered.* **79**, 409 (1988).

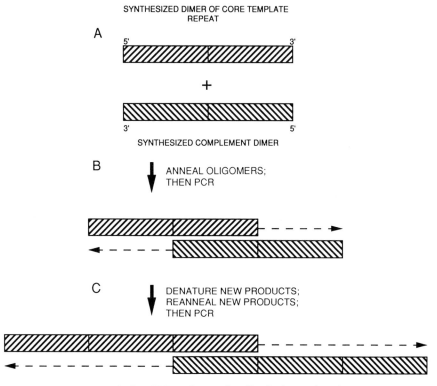

SYNTHESIZED DIMER OF CORE TEMPLATE
REPEAT

A

SYNTHESIZED COMPLEMENT DIMER

B   ANNEAL OLIGOMERS;
      THEN PCR

C   DENATURE NEW PRODUCTS;
      REANNEAL NEW PRODUCTS;
      THEN PCR

FIG. 1. Initial steps of PCR–STR probe creation. For further explanation, see text.

peats are unknown, after 15 PCR cycles the total PCR–STR product is electrophoresed on 1.5% agarose gels, and, by comparison with a size standard (λ DNA cut with HindIII or DraI), strands of approximately 200–500 bp are isolated by electroelution[18] (the agarase digestion method described above can also be used). Approximately 20–40 ng of these size-selected strands are then used in asymmetric PCR amplifications (5–12 cycles with cycling parameters after Saiki et al.[24]) using a ^{32}P-labeled nucleotide for one of the four nucleotides, resulting in PCR products that are labeled.

Asymmetric PCR, as opposed to symmetric PCR, is used in the labeling step because, although the former creates strands at a slower arithmetic rate, it will produce an abundance of labeled single strands lacking complementary strands with which to reanneal. The PCR–STR template and complementary strands are repetitive and have a high potential to reanneal. With symmetric PCR to make labeled strands, the more balanced

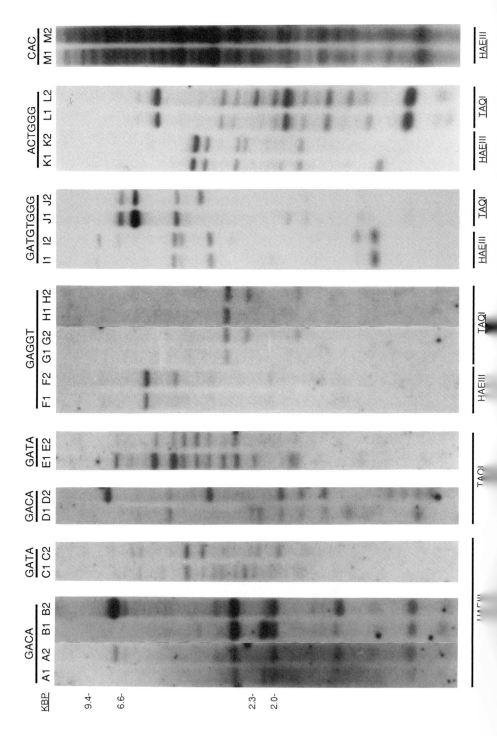

numbers of template and complementary product strands might lead to reannealing of these strands prior to, and hence rendering them useless for, hybridization with the genomic target DNA fragments bound to the Southern blots. Furthermore, asymmetric PCR with relatively low concentrations of PCR–STR template strands and high concentrations of the one primer will promote product strands in the general size range of the size selected strands. In contrast, symmetric PCR will produce, by the mechanisms shown in Fig. 1, some strands that are much longer than the size selected strands, a trend that will increase with increasing cycles. The primer chosen for the asymmetric strand is the strand with the repeat sequence having the highest frequency of the labeled nucleotide used.

The PCR mixture conditions in both the PCR–STR production and the PCR–STR labeling steps follow Saiki et al.[24] except as noted. The volume of the production reaction is 100 μl; that of the labeling step is 30 μl. Ten picomoles of each oligomer primer is added to the PCR–STR production reaction, and 5 pmol of a single primer is added to the labeling reaction.

Thermocycling parameters are generally as follows: 95° for 1 min 10 sec (it is imperative to separate long, repetitive strands); 47°–50° for 1 min 10 sec (at least for the early cycles a slightly lower annealing temperature seems to improve productivity, perhaps owing to the difficulty of annealing small oligomers in the desired configurations); and 63°–67° for 2 min without a serial time increase for the extension step. From 15 to 20 cycles are used for PCR–STR production. To test whether changes in these protocols would affect the endonuclease fragment detection characteristics of the PCR–STR products, some of the above parameters are changed in comparative experiments. These changes are described below.

Examples of Polymerase Chain Reaction–Synthetic Tandem Repeat Probes

Figure 2 shows the *Hae*III and/or *Taq*I endonuclease fragment profiles resulting from probing digested genomic DNA from two siblings (seeds derived from one maternal parent) of *Polyalthia glauca* with six different PCR–STR probes. As siblings (each is designated uniquely as 1 or 2 in Fig.

FIG. 2. Autoradiographs from PCR–STR probings (repeat sequences given above the top lines) of the genomic DNA from two siblings (designated as 1 and 2) of *Polyalthia glauca* that was digested with endonucleases (see lines below lanes), electrophoresed, and blotted onto nylon. Replicate lanes for each individual were run on one gel, although in some cases probed lanes were stripped and probed with a new probe (e.g., the I and J lanes are reprobed F and G lanes). Letters under the top lines indicate different treatments as discussed in the text; KBP, kilobase pairs.

2), fragment profiles are expected to be similar for each probe–endonuclease combination.

Lanes A1 and A2 in Fig. 2 show the *Hae*III fragment patterns revealed for the two siblings when the PCR–STR GACA probe is prepared as described above. Differences in their fragments are apparent. Lanes B1 and B2 (Fig. 2; run on the same gel and blotted onto the same membrane as lanes A1 and A2) are produced with the exact same protocol except that three initial rounds of polymerization with the Klenow fragment are undertaken in the same reaction solution subsequently used for PCR production of the labeled probe. The solution is heated to 96° for 4 min before each Klenow round, for which fresh enzyme is always added when the solution reaches room temperature and according to the manufacturer's instructions (Promega, Madison, WI). This Klenow treatment is used for two reasons. First, there is some question as to whether the short oligomers employed in the PCR–STR would anneal efficiently at elevated reaction temperatures in the proper configuration for elongated STR production to proceed (Fig. 1). The Klenow fragment is known to polymerize in random hexamer primed reactions,[18] so its efficacy in polymerizations from short fragments is known. Second, thermophilic polymerases may have a terminal transferase activity[30] and thus may potentially alter the desired repeat sequence pattern irregularly with PCR–STR creation. Treatment with the Klenow fragment does not appear to produce this transferase activity,[30] and thus initial rounds of amplification with Klenow would establish longer templates without irregular sequences, thereby yielding more uniform repeats in subsequent PCR–STR production.

Comparisons of lanes A1 with B1 and A2 with B2 of Fig. 2 show that similar fragment patterns are produced from these two different PCR–STR preparation protocols, the major difference being in intensity. Thus, the hybridization of these probes appears to be little affected by these differences in protocol.

The DNA of lanes C1 and C2 (Fig. 2) was digested with the same enzyme (*Hae*III), run on the same gel, and blotted on the same filter (which was subsequently cut into smaller strips) as lanes A1, A2, B1, and B2 (Fig. 2), but the former two lanes have been probed with PCR–STR GATA rather than the GACA probes used for the latter four lanes. Note that the two siblings of *P. glauca* have differing fragment patterns with the GATA probe; furthermore, the one nucleotide difference between the repeat sequence of the GATA and GACA probes produces very different fragment profiles.

---

[30] D. Denney, Jr., and I. Weissman, *Amplifications* **4**, 25 (1990).

Differences between the two siblings of *P. glauca* are also revealed with endonuclease *Taq*I and PCR–STR probes GACA (lanes D1 and D2 in Fig. 2) and GATA (lanes E1 and E2). The one nucleotide difference between the GACA and GATA probes makes a marked difference in the fragments revealed, as can be seen by comparing the D and E lane sets (Fig. 2).

*Hae*III (lanes F1 and F2, Fig. 2) and *Taq*I (lanes G1 and G2) digested DNA from the two siblings of *P. glauca* probed with the PCR–STR GAGGT probe revealed relatively simple, although different, patterns between the two siblings. Lanes H1 and H2 (Fig. 2) are replicate filters of the G1 and G2 lanes. The former were probed with a PCR–STR GAGGT probe labeled with 12 cycles of asymmetric PCR, the GAGGT probe for the latter being labeled with only 7 cycles. This difference in cycle number has no detectable effect on the fragments revealed.

The I through M lanes of Fig. 2 are comparisons of fragment profiles for the *P. glauca* siblings produced using *Hae*III or *Taq*I as enzymes and PCR–STR probes GATGTGGG (I and J lanes), ACTGGG (K and L lanes), and CAC (M lanes). For the first two probes, differences between the siblings are apparent for both enzymes. The profiles produced with the CAC probe were very complex, with no differences between siblings being readily detectable, although probes based on this repeat sequence have detected intraspecific variation in other organisms.[29]

Figure 3A demonstrates how different probes have different capacities for revealing fragments and variation in different organisms (*Hae*III is the only enzyme used in Fig. 3). Lanes N1 and N2 (Fig. 3) are two different individuals of *Raphanus sativus,* whereas O1 and O2 are different *Equisetum arvense* individuals, all probed with PCR–STR GACA. Only very faint patterns are discernible for *R. sativus,* in contrast to *E. arvense* for which hybridization is intense. These patterns are quite different from those produced for *P. glauca* individuals (Fig. 2, B and D lanes), even though these filters were all probed in the same hybridization solution.

PCR–STR CAC was used to probe the same *R. sativus* and *E. arvense* filters (Fig. 3A, P and Q lanes, respectively). In contrast to results with the GACA probe, the CAC probe reveals fragment differences between individuals for both species. Four randomly selected individuals of *R. sativus* were probed with the PCR–STR ACTGGG probe (Fig. 3A, lanes R1–R4), revealing several differences among individuals.

In probings of three individuals of *Asplenium rhizophyllum* collected within 20 m of one another, the PCR–STR CAC probe did not hybridize to any significant degree (Fig. 3B, lanes S1–S3). The GATA probe, however, detected several fragment differences (Fig. 3B, lanes T1–T3).

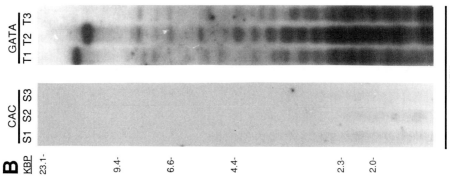

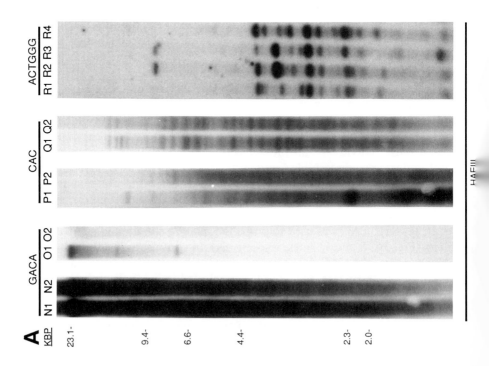

Conclusions

It has been proposed that different minisatellite DNA systems are probably taxonomically widely distributed, and that variation in such systems, where it exists, can be used to investigate various biological questions.[20,21] Numerous examples of repetitive sequences revealing intraspecific genetic variation have now been described,[6] and, in fact, it appears that randomly chosen repeats will often reveal variation in a variety of organisms (Vergnaud[26]; the PCR–STR GAGGT probe used here). Furthermore, intraspecific genetic variation of repetitive sequences can occur whether the repeat sequences are 2–3 bp (simple sequences or microsatellites[4,5,25,29]) or relatively large sequences.[31,32] Thus, it seems likely that the variation of repetitive sequences is a general characteristic of such sequences regardless of size or composition, rather than a special characteristic of particular repetitive sequences. The results here further confirm that repetitive sequences in general can be used to seek variable genetic markers for a wide array of organisms. In the case of the *P. glauca* siblings, variable markers were revealed with five of the six probes used. However, it is clear that no one repeat sequence will be universally optimal.

Given that for any particular organism it will probably be necessary to survey with a variety of minisatellite probes to select those most appropriate for the specific biological issues to be investigated, methods that facilitate such surveys are needed. Advantages of the PCR–STR approach include the following: (1) cloning manipulations are not required such as clone creation and screening, vector maintenance, and repeated cloned DNA preparation; (2) amplification with PCR to produce PCR–STRs can proceed from minute quantities of initial, short oligomers, in contrast to using ligation of oligomers where much larger quantities must be expended; and (3) using asymmetric PCR to make labeled strands produces an abundance of labeled single-strand probes (of high specific activity)

---

[31] A. G. Matera, U. Hellmann, M. F. Hintz, and C. W. Schmid, *Nucleic Acids Res.* **18**, 6019 (1990).
[32] E. W. Jabs, C. A. Goble, and G. R. Cutting, *Proc. Natl. Acad. Sci. U.S.A.* **86**, 202 (1989).

FIG. 3. Autoradiographs from PCR–STR probings (repeat sequences given above the top lines) of genomic DNA digested with endonucleases (see lines below) and electrophoresed for (A) two individuals of *Equisetum arvense* (designated 1 and 2 for the N and P lanes), two individuals of *Raphanus sativus* (designated 1 and 2 for the O and Q lanes), four individuals of *Raphanus sativus* (R1 and R4), and (B) three individuals of *Asplenium rhizophyllum* (designated 1, 2, and 3 for the S and T lanes). The N and O lanes are from the same autoradiograph of one filter. The P, Q, and T lanes are reprobings of the N, O, and S lanes, respectively. KBP, Kilobase pairs.

lacking complementary strands and thus free to hybridize with target genomic DNA.

In conclusion, these results further open the door to the creation of numerous minisatellite probes to survey for VNTR loci. Such probes can be used not only for restriction fragment length polymorphism (RFLP) analyses of the many biological applications outlined in the introduction, but they can also be employed to screen cloned libraries with the ultimate goal of determining the sequences flanking VNTR loci. The latter information provides perhaps the most accurate and rapid means of utilizing the variation at such loci via the PCR analysis of alternate alleles.

Acknowledgments

I thank Rebecca German, William Hamilton, Roger Ruff, and Earl Whitson. This research was funded by National Science Foundation Grant BSR 90-96317.

[21] Use of Randomly Amplified Polymorphic DNA Markers in Comparative Genome Studies

By Brunella Martire Bowditch, Darrilyn G. Albright, John G. K. Williams, and Michael J. Braun

Introduction

Detection of genetic variation is essential to a wide range of comparative genetic research endeavors. These include studies as diverse as gene mapping, individual identification, parentage determination, population genetics, and molecular phylogenetics. The speed and accuracy of comparative genetic research often depend on the methodology for detecting variation.

The advent of allozyme electrophoresis and restriction fragment length polymorphism (RFLP) analysis of DNA made it possible to detect many polymorphisms in most organisms at the protein or DNA level.[1] These genetic polymorphisms are crucial in examining the many different fields noted above. Specifically, polymorphisms can aid in the determination of relatedness of groups of taxa, in the analysis of parentage in domestic and wild animal species, in the identification of individuals for captive breeding programs in endangered species, in the comparison of wild and cultivated plants, and in the estimation of levels of inbreeding or outbreeding in

[1] C. Chang and E. M. Meyerowitz, *Curr. Opin. Genet. Dev.* **1**, 112 (1991).

populations. They have also been essential, for example, both in linkage mapping quantitative trait loci in cultivated plants and in mapping the human genome.[2,3]

Although modern techniques for studying genetic polymorphisms have provided may new vistas of research, they do have some limitations. For instance, RFLP analysis requires fairly large amounts of genomic DNA (often unavailable from rare, ancient, and/or field-collected materials) and cloned probes that may be specific to an organism or group. Targeted PCR (polymerase chain reaction) assays, although requiring much less genomic DNA, depend on DNA sequence knowledge of the organism or gene under study. Williams *et al.*[4] have described a technique for detecting DNA polymorphisms which they call RAPD (random amplified polymorphic DNA). The technique requires only small amounts of DNA, and no prior knowledge of the genome in question is necessary. It is based on random amplification of DNA fragments, via PCR, using short primers of arbitrary sequence. The method detects abundant polymorphism in most organisms. We have used this method in our own systematic studies of genetic variation in natural populations of plants and animals. It clearly has advantages in certain situations, and promises to become a basic tool for genetic studies of closely related organisms.[5]

Outline of Method

As stated above, the RAPD technique makes use of PCR technology, allowing geometric amplification of DNA templates. With standard PCR, it is first necessary to determine the sequence of the DNA to be analyzed. Then two specific primers (commonly 18–25 bases long), complementary to sequences flanking the target segment, are synthesized to prime the DNA amplification reaction. In contrast, the RAPD method uses a single short primer (e.g., 9 or 10 bases) of arbitrary sequence and a lower annealing temperature than the average PCR reaction (Fig. 1). These two modifications lower the specificity of the reaction so that a number of anonymous but reproducible fragments can be amplified from most complex genomes. Reaction products are analyzed by electrophoresis on agarose gels and ethidium bromide staining, so radioactive isotopes are unnecessary. Mutations that inhibit primer binding or otherwise interfere with amplification can be detected as the absence of the pertinent band in those individuals.

[2] D. Nelson, *Curr. Opin. Genet. Dev.* **1,** 62 (1991).

[3] S. J. O'Brien, *Curr. Opin. Genet Dev.* **1,** 105 (1991).

[4] J. G. K. Williams, A. R. Kubelik, K. J. Livak, J. A. Rafalski, and S. V. Tingey, *Nucleic Acids Res.* **18,** 6531 (1990).

[5] H. Hadrys, M. Balick, and B. Schierwater, *Mol. Ecol.* **1,** 55 (1992).

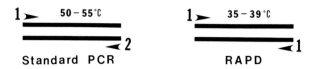

FIG. 1. Schematic showing the use of two primers of common length 18–25 bases in standard PCR versus the use of a single primer of common length 9–11 bases in RAPD reactions.

Because the short oligonucleotides are able to prime reproducible amplification from a variety of genomes under the specified conditions, it is possible to develop a panel of "universal" primers that can be used to detect polymorphism directly in practically any organism using only nanogram quantities of genomic DNA.

Detailed Protocol for Random Amplified Polymorphic DNA

The final concentrations of all components contained in the RAPD reaction are listed below along with the composition of stock solutions. This is followed by a detailed procedure for mixing a RAPD reaction of 25 μl, which is our usual reaction volume.

Reaction Components

11 mM Tris-HCl (pH 8.3)
50 mM KCl
1.9 mM MgCl$_2$
0.001% (w/v) Gelatin or 0.1 mg/ml bovine serum albumin (BSA)
0.1 mM each of dATP, dCTP, dGTP, and TTP
0.2 μM primer (equals 5 pmol in a 25-μl reaction)
0.02–0.04 units/μl *Taq* DNA polymerase (Perkin-Elmer Cetus, Norwalk, CT)
0.1–4.0 ng/μl of genomic DNA

Stock Solutions

10× RAPD buffer: For 1 ml of stock combine:
100 μl 1 M Tris-HCl, pH 8.3 (100 mM final concentration)
500 μl 1 M KCl (500 mM)
19 μl 1 M MgCl$_2$ (19 mM)
100 μl 10 mg/ml BSA (1 mg/ml)
281 μl Distilled water
10× dNTPs: For 1 ml of stock combine:
10 μl 1 M Tris, pH 8.3 (10 mM final concentration)

10 μl 0.1 M dATP (1 mM)
10 μl 0.1 M dGTP (1 mM)
10 μl 0.1 M dCTP (1 mM)
10 μl 0.1 M TTP (1 mM)
950 μl Distilled water

Procedure

1. Prepare stock solutions of reaction buffer (10× RAPD buffer) and deoxynucleotides (10× dNTPs). Larger quantities of both solutions can be prepared, subdivided into aliquots, and stored frozen at $-20°$.

2. Dilute primer and template DNAs in TLE (10 mM Tris-HCl, 0.1 mM EDTA, pH 8.0) such that 1 μl of the primer dilution contains 5–6 pmol of primer and 1 μl of the template dilution contains the desired amount of DNA for the reaction (between 5 and 15 ng of template DNA per 25-μl reaction has worked best in our hands).

3. Prepare a mix of those reagents common to all tubes in the experiment. Prepare a quantity sufficient for the total number of samples and controls to be run, plus one extra aliquot to ensure adequate volume for pipetting. In a single tube combine the following amounts **per sample:** 2.5 μl of 10× dNTPs, 2.5 μl of 10× RAPD buffer, and 0.5–1 unit *Taq* polymerase (volume will depend on supplier's concentration). Add distilled water to final volume of 23 μl per sample and *mix well.* Making a reaction mix greatly reduces the chances of pipetting errors for individual samples.

4. Label the reaction tubes (600-μl Eppendorf tubes) and add 23 μl of reaction mix (Step 3), 1 μl diluted template DNA (5–100 ng), and 1 μl diluted primer DNA (5–6 pmol). The advantage of adding the reaction mix first is that the surface tension will help pull down the 1 μl volumes, which would otherwise be hard to pipette into the tubes. If using only one template DNA with a number of primers or one primer DNA with a number of templates, the single primer or template DNA can be combined with the mixture in Step 3 instead of adding it separately in Step 4.

5. Mix the 25-μl reactions by flicking the capped tubes with a finger, then spin the tubes briefly in a microcentrifuge to remove any air bubbles.

6. Overlay the reactions with mineral oil to avoid evaporation in the thermocycler (20 μl of oil is usually sufficient) and place the tubes in the thermocycler. Turn on the machine 15 min before starting the reactions to warm it up, as suggested by the instruction manual.

7. Record the location of each of the reaction tubes in the thermocycler block. This is helpful in troubleshooting poor reactions to rule out effects of any temperature inconsistencies within the block itself.

8. Program the thermocycler for 40–45 cycles with the following parameters: denature for 1 min at 94°, anneal for 1 min at 35°–36° (55° for 16-mers), and extend for 2 min at 72°. The total reaction time is approximately 5 hr in a Perkin-Elmer Cetus thermocycler using minimum ramping times between incubation temperatures.

9. Place completed reactions at 4° until gel analysis is performed.

The ranges listed for the amount of template DNA, the amount of primer DNA, the amount of *Taq* polymerase, the number of PCR cycles, and the annealing temperature used do not seem to significantly affect the results of the RAPD reaction. Likewise, substituting BSA for gelatin in the RAPD buffer does not affect the results.

Detailed Protocol for Gel Analysis of Randomly Amplified Polymorphic DNA Reactions

1. Remove 23 μl of the RAPD reaction to a new Eppendorf tube by placing the pipette tip to the bottom of the reaction tube and drawing the reaction up from the bottom. This avoids most of the mineral oil.

2. Add 2.5 μl of 10X gel loading buffer [30% (w/v) Ficoll, 200 mM EDTA (pH 8.0), 0.5% (w/v) bromphenol blue, dissolved in water and stored at $-20°$] to each sample.

3. Load the samples on 1.4–1.5% (w/v) agarose gels. We use 50-ml minigels made with TBE electrophoresis buffer (89 mM Tris base, 89 mM boric acid, 2 mM EDTA, pH 8.3) and add 5 μl of 5 mg/ml ethidium bromide directly to the molten agarose before pouring the gel. Normally, we use 12-well combs (1.5 mm thick) in the minigels, which hold about 20 μl of sample volume total, but 16-well combs can also be used. Because the ethidium bromide-stained RAPD bands appear sharper the smaller the well size used, larger well sizes are not recommended. DNA length standards in the range of 200–5000 base pairs (bp) should also be included on each gel to facilitate determination of band sizes.

4. Electrophorese the gels at 50 V in TBE buffer for approximately 3 hr. We have found that longer runs at lower voltages yield clearer images of band patterns than shorter runs at a higher voltage. It is crucial to obtain the best separation of bands possible since the gel results are documented only by a photograph.

5. Photograph the gel on an ultraviolet transilluminator using an orange UV filter for the camera lens. For good gels, it is useful to use film which gives both a positive and a negative of the gel so that additional prints of that gel can be made at a later time.

Considerations for Optimizing Randomly Amplified Polymorphic
DNA Reactions

Primers

Although the primers used in this technique are random sequences, there are several criteria that must be met to produce results. Williams *et al.*[4] found that the minimum primer length to detect amplification in ethidium bromide-stained agarose gels was 9 bases, and a minimum of 40% GC content was also required to produce amplification (50–80% GC content is generally used). These criteria were determined at an annealing temperature of 36° but also held true at annealing temperatures as low as 15°.

Primer dimer formation can be a potential problem in PCR amplifications. Although we generally avoided the use of self-complementary sequences when designing primers, we found that dimer formation was not a problem at an annealing temperature of 35° even when 6 bases of a 10-bp primer could potentially hybridize.

Primers can be purchased or synthesized using standard phosphoramidite chemistry. Operon Technologies Inc. (Alameda, CA) has available 25 kits of 20 different 10-base primers.

Stringency

The stringency limits for RAPD reactions are fairly straightforward. Annealing temperatures above 40° prevented amplification by many of the 10-base primers.[4] Annealing temperatures of 35° or 36° are routinely used for 9-, 10-, and 11-base primers, while an annealing temperature of 55° is used for 16-base primers.[6,7] D. Fong (personal communication, 1991) has observed that an annealing temperature of 39° for 10-base primers tends to decrease the total number of amplification products for any given primer, but the bands that do appear are highly reproducible.

Number of Cycles, DNA Quality, and Concentration

DNA samples purified by either stringent methods (equilibrium centrifugation in cesium chloride density gradients) or simple methods (organic extraction) have yielded reproducible RAPD patterns. However, in plants (e.g., members of the evening primrose family, Onagraceae) and

[6] J. G. K. Williams, M. K. Hanafey, J. A. Rafalski, and S. V. Tingey, this series, Vol. 218 p. 704 (1993).
[7] M. Arnold, *Proc. Natl. Acad. Sci. U.S.A.* **88**, 1398 (1991).

animals (e.g., molluscs) that have high concentrations of polysaccharides, alternative methods of DNA purification were necessary for successful amplification (Bult et al.,[8] L. Adamkewicz, personal communication, 1992). General conditions of the PCR protocol are also important to consider. For RAPD reactions, the number of cycles appears to be more critical than the starting concentration of the template DNA. We have used between 0.0002 and 4.0 ng/μl of template DNA from a single individual titmouse (Parus bicolor) in a 45-cycle RAPD reaction. Equivalent band patterns, as viewed on an ethidium bromide-stained agarose gel, were observed for template DNA concentrations between 0.004 and 0.8 ng/μl. Template DNA concentrations both higher and lower than this range, however, produced only a subset of the original band pattern (Fig. 2A). These data are supported by the findings of Williams et al.,[6] that RAPD band patterns were generally preserved until the template DNA concentration fell below 0.1 ng in a 25-μl reaction (0.004 ng/μl). However, when testing 0.6 ng/μl of the same template DNA at 45, 40, 35, 30, 25, and 20 cycles, some bands in the pattern were poorly reproducible at 30 and 25 cycles, and all bands were absent in 20-cycle runs (Fig. 2B).

Taq Polymerase Concentration

We have found that 0.02 units/μl of Taq polymerase is often sufficient for RAPD reactions using 0.2–0.6 ng/μl of template DNA with 0.24 pmol/μl of primer DNA, though the optimal quantity of Taq polymerase may depend on the supplier used. Excessive amounts of Taq polymerase, or in some cases template DNA, can cause the reaction to appear as a smear of amplified DNA rather than as discrete bands.[4]

Ultraviolet Light Range for Photography

Higher contrast photographs of ethidium bromide-stained agarose gels are obtained if the gel is illuminated with short-wavelength UV radiation (254 nm), whereas longer wavelengths (312 nm) yield less satisfactory results. This enhanced contrast may allow some polymorphisms to be scored under short-wavelength UV light that could not reliably be scored under longer wavelengths. Note that wavelengths shorter than 254 nm should not be used to view a gel band that is to be excised for cloning or direct sequencing since such wavelengths cause breaks in the DNA backbone much more frequently than longer wavelengths do.

---

[8] C. Bult, M. Kallersjo, and Y. Suh, Plant Mol. Biol. Rep. **10**(3), 269 (1992).

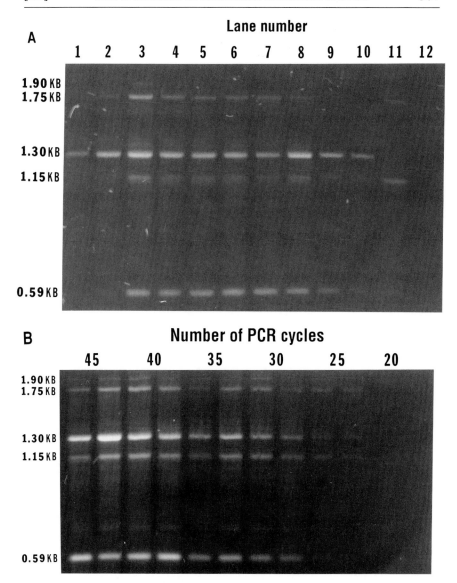

Fig. 2. Effect of DNA concentration and number of PCR cycles on RAPD analysis, shown on ethidium bromide-stained agarose gels, using template DNA from a single individual titmouse *(Parus bicolor)* and the 10-base primer AP5a[4] (5′ CTGTTGCTAC 3′). **(A)** Various concentrations of template DNA amplified through 45 cycles of PCR. Lanes 1–11 contain 100, 50, 20, 10, 5, 2, 1, 0.1, 0.05, 0.01, and 0.005 ng of template DNA, respectively, in 25-μl reactions. Lane 12 contains no template DNA. **(B)** Constant amount of template DNA (0.6 ng/μl) amplified with primer AP5a between 20 and 45 cycles. All reagents for the experiment were combined in a single tube, then aliquoted into twelve 25-μl reactions. Duplicate reactions were performed for each cycle length variation.

Use of Polyacrylamide Gels

As compared to ethidium bromide-stained agarose gels, many more amplified bands can be visualized by separating the products of the reaction on a polyacrylamide gel and staining with silver.[9] This gel system also permits detection of bands amplified weakly by primers as short as 5 nucleotides.

Characteristics of Randomly Amplified DNA Polymorphisms

A typical RAPD amplification will yield from 2 to 10 visible bands, of which most are reproducible, when performed by the methods described herein. For any particular genome, a certain percentage of RAPD primers will not produce satisfactory amplification products. These primer/template combinations either yield no amplification whatsoever or yield very light or hazy bands that are hard to reproduce in replicate reactions. However, the same primers may work perfectly well on the next genome tested, so the failure of a particular primer/template combination is more likely due to lack of readily amplifiable segments in the template DNA rather than a systematic problem with the primer.

A large majority of the strong reaction products fall in the range of 400–2000 bp. This is presumably due to limitations in the resolving power of the agarose gels at lower molecular weights as well as inefficiency of the extension reaction under the described PCR conditions at higher molecular weights. Although optimization of the PCR protocol for higher molecular weight markers is possible, it is probably neither time nor work efficient considering the number of scorable bands obtained in the lower molecular weight range.

Most RAPD markers show dominant/recessive inheritance in diploid organisms, that is, a DNA segment of a certain molecular weight is amplified from some individuals but not from others. In general, it is not possible to distinguish heterozygous individuals from those homozygous for the dominant allele at such loci; both have the "band present" phenotype. Presumably, insertion/deletion events of moderate size (perhaps 50–1500 bp) will result in codominant alleles detectable as amplification products of different size, assuming the mutation has no major effect on the kinetics of amplification. Experience to date, however, indicates that such codominant size polymorphisms are relatively rarely detected in RAPD assays. Williams *et al.*[4] found only four codominant markers among 88 RAPD polymorphisms mapped in the *Neurospora crassa* ge-

---

[9] G. Caetano-Annoles, B. J. Bassam, and P. M. Gresshoff, *Bio/Technology* **9**, 553 (1991).

nome. The frequency may be greater in genomes where short interspersed repetitive elements are common.

When a small number of individuals is analyzed, the statistical inference of allelism may be weak for a particular pair of bands showing a pattern of inheritance consistent with codominance. In this case, homology between the bands can be established by excising them from the gel for sequencing or for use as probes on Southern blots.

Nature of Randomly Amplified DNA Polymorphisms

Although the RAPD technique can potentially detect all classes of mutations (substitutions, insertions, deletions, inversions), the basis of most polymorphism is presumably substitution. However, substitutions apparently need not occur in the primer binding sites themselves to be detected by the RAPD assay. Braun and Albright[10] describe evidence indicating that the sequence of the primer binding sites is frequently unchanged in RAPD polymorphisms. Primers containing restriction sites were used to detect RAPD polymorphisms on the assumption that such polymorphisms would also be frequently detected as RFLPs when the RAPD bands were used as probes in Southern transfers. Although the predicted restriction fragments were observed, these fragments were rarely polymorphic. The most likely explanation appears to be that the RAPD assay can detect substitutions outside the primer binding sites themselves.

Priming of RAPD assays is carried out at relatively low temperatures that may allow the formation of secondary structure within the single stranded template at or around the primer binding sites. Such secondary structure could inhibit the binding of primer to template sufficiently to prevent amplification. Alternatively, secondary structure might inhibit elongation by *Taq* polymerase. Mutations which stabilize secondary structure may thus be detectable as RAPD polymorphisms even though they do not reside within the primer binding site. Under this hypothesis, the RAPD assay potentially can detect mutations in a much larger region than that covered by the primer binding sites. This hypothesis may help explain the high degree of polymorphism observed in RAPD band patterns for most organisms tested.

Reproducibility and Experimental Strategies

Because the RAPD assay is often used to identify individuals or populations, reproducibility of individual RAPD reactions must be high. Once

[10] M. J. Braun and D. G. Albright, in preparation (1993).

ideal conditions for a certain reaction are found, it is imperative to keep them constant in order to be able to compare results from different reactions. The same is true for the thermocycler; once a study is begun on a certain instrument it is best to run all samples on the same instrument.

With careful attention to laboratory technique, the reactions are quite reproducible from run to run within the same thermocycler and within different wells of a thermocycler on the same run. The strongest PCR products are the most highly reproducible, with fainter products tending to come and go on multiple runs. For this reason, we suggest that strong products be chosen over faint products when scoring individuals or populations.

An example of band scoring using a particular RAPD primer in marsh wrens *(Cistothorus palustris)* is shown in Fig. 3. Although all bands in the schematic are visible on the original gel photograph, the accuracy with which they might be scored as present or absent is quite variable. Bands d, f, h, and k (Fig. 3) are dominant PCR products that can be easily scored as to their presence or absence. Band e, present in individuals 1 and 4 in Fig. 3, is very close in size to band f but is most likely distinct since individual 5 appears to have both band e and band f. Bands a and c in Fig. 3 vary so much in intensity between individuals that they would be hard to score reliably across a population. The remaining bands (b, g, i, and j, Fig. 3) are quite faint so their reproducibility would be suspect. Thus, only bands d, e, f, h, and k (Fig. 3) are good candidates for scoring within this population.

Because the number of polymorphisms detected is primarily limited by the number of primers tested, it is often possible to identify more polymorphisms than are needed for a particular study. Therefore, the best

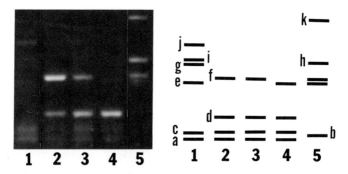

FIG. 3. RAPD analysis on an ethidium bromide-stained agarose gel using 0.6 ng/μl template DNA from several individual marsh wrens *(Cistothorus palustris)* and the 10-base primer AP4c[4] (5' TCTCGATGCA 3'). Both the gel photograph and a schematic of visible bands are shown.

strategy to be used with RAPD technology is to screen many primers and select only those that give highly reproducible bands for scoring, rather than trying to optimize every primer/template combination.

By the same rationale, the presence of a band in the pattern of an individual can be more reliably scored than absence of a band. For this reason, we have on several occasions repeated the RAPD assay on all individuals with the band-absent (negative) phenotype for particular polymorphisms, and we recommend this precaution, especially for those investigators becoming familiar with the technique. In our experience, repeating the assay has rarely resulted in a changed score when the positive phenotype consists of a strong PCR product, but replication of experiments serves as a gauge of the accuracy of results and is a useful aid to choosing polymorphisms that are easily reproducible.

A significant source of false-negative scores stems from DNA samples containing impurities that inhibit PCR amplification. Such samples are readily detected by noting the relative intensities of PCR products in an RAPD pattern other than the particular polymorphism being scored (Fig. 4, lane 9). If bands that are monomorphic throughout a population are weak or absent in an individual sample, it is likely that this sample contains some inhibitory contaminant. Further purification of the sample by any of a number of standard protocols normally corrects this problem.

Finally, band patterns seem fairly reproducible between individual Perkin-Elmer Cetus thermocycler units (especially the strong products), but we have had trouble using the precise protocol described here with other brands of thermocyclers. This is most likely due to the fact that different brands of thermocyclers have quite different temperature cycling profiles even when programmed identically. The largest variations occur in ramping times between the various cycle temperatures and in the way the machine determines those temperatures (i.e., temperature probe in the thermocycler block versus temperature probe in a sample itself). The RAPD technique can likely be adapted to any brand of thermocycler by optimizing the programmed reaction times for the machine used.

Use of Randomly Amplified Polymorphic DNA Technology in Comparative Genome Studies

With RAPD technology, polymorphisms can be detected in closely related organisms such as those that compose a species complex, different populations of a single species, or individuals within a population. For this reason, the RAPD technique is likely to find wide application in gene mapping, in individual and strain identification, and in those issues in ecology and population biology requiring genetic analysis of relatedness or

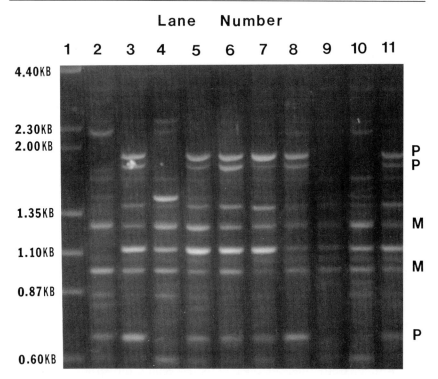

Fig. 4. Typical RAPD analysis on an ethidium bromide-stained agarose gel. **Lane 1** contains DNA length standards. **Lanes 2–11** contain 0.6 ng/μl each of genomic DNA from different individual marsh wrens *(Cistothorus palustris)* amplified with the 10-base primer AP5a[4] (5' CTGTTGCTAC 3'). Note that most amplified products lie between 0.6 and 2.3 kb, and there are about 8 to 10 well-defined PCR products per lane. Some products are monomorphic in this population (M), whereas others are polymorphic (P). Amplification of all bands in lane 9 is weak, and testing of this individual should be repeated before polymorphic fragments are scored as present or absent.

identity. We believe that RAPD technology will be very useful, for example, in studying breeding/mating systems for both plant and animal species, such as determining the number of individuals that have originated from outcrossing versus self-crossing in plants. We have successfully applied the RAPD method to several questions involving geographical "species" or populations in both avian and plant systems.

Avian Systems

Two different species of birds were analyzed using the RAPD technique. Five individuals each from an eastern and western population of marsh wrens *(Cistothorus palustris)* were compared, and five individuals each of two subspecies of titmice *(Parus bicolor)* were also compared. In both species, 25 of 31 primers tested were able to amplify genomic DNA. In several cases, primers that failed to produce scorable results in one avian species were successful in the other species, supporting the idea that lack of amplification from a particular primer/genome combination is more often due to chance than nonsuitability of the primer. For the wrens, 17 of the 25 usable primers showed some individual polymorphisms. The 25 primers produced 128 scorable bands, 50 of which were polymorphic. Thus, 55% of all primers tested detected polymorphisms, for an average of 1.6 polymorphic bands per primer tested. Similarly, in titmice, 18 of the 25 usable primers showed some individual polymorphisms. A total of 116 bands was produced, 53 of which were polymorphic. Thus, 58% of primers tested detected an average of 1.7 polymorphic bands per primer tested.

Plant Systems

Ninety primers were screened on a single individual from the weedy *Euphorbia esula* L. complex. Of those tested, 60 primers amplified the DNA. Subsequently, the 60 primers were tested on an individual from each of six morphologically indistinguishable populations within the complex. Forty of the 60 primers used produced fragment patterns which differentiated at least four of the individuals, and 23 of the 40 usable primers gave a unique pattern for every geographical population screened. Of the 231 scorable bands present in the fragment patterns produced by the primers giving population-specific patterns, 82 bands were unique to one of the six populations, and 8 bands were shared by all six populations.

RAPD analysis was also performed on DNA extracted from the monocots *Heliconia latispatha* Bentham and *Phenakospermum guianense* L. C. Rich. Endl. et Miq. Of a total of 35 primers tried, 28 amplified DNA from *Heliconia latispatha,* and 27 amplified DNA from *Phenakospermum guianense.* RAPD technology has been used to examine interspecific gene flow between two species of Louisiana irises, *Iris fulva* Ker-Gawl. and *Iris hexagona* Walt.[7] Six of the seven primers tested detected differences between the two species. Three of the primers were then used to confirm the hybridity of *Iris nelsonii* and its origin from *Iris fulva* and *Iris hexagona.* Finally, Williams *et al.*[4] used RAPD analysis to construct genetic maps in soybeans (*Glycine max* L. Merr. and *Glycine soja* Sieb. et Zucc.), corn (*Zea mays* L. Off.), and the mold *Neurospora crassa.*

Other Systems

Currently, RAPD analysis is being used to study a wide variety of organisms. Successful amplifications have been obtained from mosquito

DNA[11], isopods (D. Fong, personal communication, 1991), molluscs (L. Adamkewicz, personal communication, 1992), humans, and bacteria.[4] It seems likely that the technique can be adapted to virtually any organism.

Welsh and McClelland[12] have used an approach similar to the RAPD technique that they call arbitrarily primed PCR (AP–PCR). In this technique, longer primers (which may have been made for some other purpose) are used as random primers by beginning with two cycles of low stringency PCR followed by further amplification at higher stringencies. The results are repeatable, and have permitted them to distinguish genetically 24 strains within 5 species of *Staphylococcus,* 11 strains of *Streptococcus pyogenes,* and 3 varieties of rice *(Oryza sativa).*

Although RAPDs promise to be useful in parentage analysis and other DNA typing studies, the occasional appearance of nonparental bands in offspring of known parentage has raised concern in some cases.[13] Such anomalous bands may be amplification artifacts caused by heteroduplex formation between alternative parental alleles[5] or between repetitive elements from distinct loci in the genome. Repetition of the analysis may resolve some anomalies. In a study of domestic horses, 5 cases of nonparental bands were observed among 59 sire–dam–offspring triplets analyzed. In all 5 cases, on repetition, either the anomalous band disappeared in the offspring or appeared in one of the parents (E. Bailey, personal communication, 1992). It may be possible to account for most or all nonparental bands by the analysis of "synthetic offspring," namely, an equal mixture of DNAs from putative parents.[5]

Applications of Randomly Amplified Polymorphic DNA Technology in Systematics

The use of RAPD technology in phylogenetics is limited to extremely closely related organisms. In this instance, the systematic characters are defined as the RAPD fragments of a certain molecular weight, while the character states are the presence or absence of that band. It must be noted that the absence of a band can potentially be the phenotype of many different alleles at a RAPD locus, while the presence of a band demonstrates an amplifiable sequence of a specific length. In other words, the character state "band absent" may in reality encompass many different character states such as inversion, different secondary structure, or any number of point mutations. This implies that the likelihood of band loss through mutation will generally be greater than the likelihood of regaining

[11] R. C. Wilkerson, T. J. Parsons, D. G. Albright, T. A. Klein, and M. J. Braun, *Insect Mol. Biol.* **1** (4), in press (1993).

[12] J. Welsh and M. McClelland, *Nucleic Acids Res.* **18**, 7213 (1990).

[13] M. F. Riedy, W. J. Hamilton, and C. F. Aquadro, *Nucleic Acids Res.* **20**, 918 (1992).

the same band: the transition probabilities between character states are asymmetrical. This situation is analogous to that which arises when restriction site data are used for phylogenetic inference[14-16] and may require similar precautions in analysis.

It is very important to realize that in coding characters we are making a statement of homology. The great advantage of RAPD technology is the ability to obtain DNA polymorphisms without having to sequence or otherwise characterize the genomic DNA of interest. On the other hand, because the amplifications are random, we cannot be sure that the comigrating gel bands we see are homologous in every sample analyzed. The inference of homology is strong when total sequence divergence between taxa is low and many RAPD bands are shared. At higher taxonomic levels, it is likely that only a few shared bands would be generated and their homology would be highly questionable. To avoid gross errors, it is important to limit this type of study to closely related organisms and not infer systematic relationships at higher taxonomic ranks.

[14] V. A. Albert, B. D. Mishler, and M. W. Chase, *in* "Molecular Systematics of Plants" (P. S. Soltis, D. E. Soltis, and J. J. Doyle, eds.), p. 369. Chapman and Hall, NY, 1992.
[15] D. M. Hillis, M. W. Allard, and M. M. Miyamoto, this volume [34].
[16] W. Wheeler, *Syst. Biol.,* submitted for publication.

[22] Molecular Approaches to Mammalian Retrotransposon Isolation

By SANDRA L. MARTIN and HOLLY A. WICHMAN

Introduction

Interspersed repetitive DNA, which accounts for a significant proportion of the DNA found in mammalian genomes, appears to arise largely from the replicative dispersal of transposable elements. Because of the high copy number of these elements and their mobility within the genome, it is reasonable to propose that their sequences play an important role in the evolution of mammalian genomes and thus of mammals themselves, yet the extent of that role remains to be elucidated. From an evolutionary point of view, it is interesting to ask why mammalian genomes tolerate so much (apparently) excess DNA. One view is that transposable elements persist despite neutral or even deleterious effects on their host because they have evolved selfish mechanisms to maintain themselves as genomic parasites.[1-3] An alternative view is that, although they may evolve as selfish

[1] W. F. Doolittle and C. Sapienza, *Nature (London)* **284,** 601 (1980).
[2] L. E. Orgel and F. H. C. Crick, *Nature (London)* **284,** 604 (1980).
[3] D. Hickey, *Genetics* **101,** 519 (1982).

parasites, they confer some advantage for the genome and hence the organism in which they reside. Among the many possible roles for interspersed repeated sequences are their ability to change patterns of gene expression by landing within or near genes, and their interspersed nature providing target sites for recombination. Recombination between homologous sequences in mispaired interspersed repeats may lead to gene duplication, loss, and rearrangements including exon shuffling. Finally, the potential to aid or block chromosome pairing could impact the occurrence and rate of speciation.[4-6]

All of the mammalian transposable elements that have been characterized to date seem to be the result of transpositions that proceeded through an RNA intermediate. This process is known as retrotransposition or retroposition. Three classes of these retrotransposable elements are known in mammals: (1) SINEs, or short interspersed repeated sequences such as the human *Alu* family and rodent B1; (2) LINEs, or long interspersed repeated sequences such as L1 in a variety of mammalian species; and (3) retrovirus-like elements, such as THE 1 in humans and *mys* and IAP in rodents. Retrovirus-like elements have long terminal repeats (LTRs) that often surround two open reading frames (ORFs) like those of retroviruses, but they lack the ability to leave one cell and enter another. LINEs also have two ORFs, but have no LTRs. SINEs have no LTRs and no ORFs. Transposition of all of these elements must involve reverse transcription of the RNA intermediate; in some cases the required reverse transcriptase is apparently encoded by the element itself.

Although a number of mammalian retrotransposons have been discovered by chance, a variety of approaches have been developed specifically to survey the genome for the presence of retrotransposons as well as to facilitate their isolation and characterization. Of the three methods outlined here in detail, two (conserved restriction sites and phylogenetic screening) could lead to isolation of all three types of retrotransposons, whereas the third method [polymerase chain reaction (PCR) amplification of reverse transcriptase] will exclude the SINEs. An unrelated method that has been particularly useful for isolation of SINEs[7] will not be detailed.

[4] C. A. Hutchison III, S. C. Hardies, D. D. Loeb, W. R. Shehee, and M. H. Edgell, *in* "Mobile DNA" (D. E. Berg and M. M. Howe, eds.), p. 593. American Society for Microbiology, Washington, D.C., 1989.

[5] P. L. Deininger, *in* "Mobile DNA" (D. E. Berg and M. M. Howe, eds.), p. 619. American Society for Microbiology, Washington, D.C., 1989.

[6] J. Brosius, *Science* **251**, 753 (1991).

[7] K.-I. Matsumoto, K. Murakami, and N. Okada, *Biochem. Biophys. Res. Commun.* **124**, 514 (1984).

Isolation of Interspersed Repetitive DNA Sequences Using Conserved Restriction Sites

When mammalian genomic DNA is cleaved with restriction endonucleases and sized-fractionated by electrophoresis through agarose gels, a smear of DNA extending from the top to the bottom of the gel is always seen. With some restriction enzymes, however, there are discrete bands superimposed on the smear. These bands are due to a large number of fragments of exactly the same size, the hallmark of repetitive DNA. Although both tandem and interspersed repeated DNA may be visualized by this method, a single band or a small number of bands with any given enzyme usually will be due to an interspersed repeat family. In the case of interspersed repeats, at least two restriction sites must lie within the repeat unit in order to generate a discretely sized restriction fragment. That fragment can be studied as a subset of the entire element, or it can be used to isolate the complete element for further study. The methods below detail the specific application of standard procedures in molecular biology for studies of interspersed repeated DNA. The detailed molecular biology methods themselves can be found elsewhere.[8,9]

Method

1. Prepare high molecular weight genomic DNA from the species of choice.

2. Cleave several 1-μg aliquots of the DNA with different restriction endonucleases, following the manufacturer's recommendations. Fractionate the digested DNA by electrophoresis through 0.8% (w/v) agarose gels, then visualize the DNA by ethidium bromide staining and transillumination.

3. Choose the restriction endonuclease that gives a strong, discrete band(s) above the smear of DNA, as is characteristic of repeated DNA.

4. Cut 20 μg of DNA with the appropriate enzyme, fractionate the digest by electrophoresis through 0.8% agarose, and stain with ethidium bromide. Locate the band of interest by placing the gel on a transilluminator, then either electroelute the DNA from the band or cut it from the gel using a clean razor blade and extract the DNA from the slice of agarose.

5. Prepare a plasmid cloning vector, preferably choosing one that contains the same restriction site as the one used for the genomic DNA digest so that the insert may be excised cleanly from the plasmid for subsequent

---

[8] S. L. Berger and A. R. Kimmel (eds.), this series, Vol. 152.

[9] J. Sambrook, E. F. Fritsch, and T. Maniatis, *in* "Molecular Cloning: A Laboratory Manual," 2nd Ed. Cold Spring Harbor Laboratory, Cold Spring Harbor, New York, 1989.

use. Cut 10 μg of plasmid with the appropriate restriction enzyme, then treat the digested DNA with alkaline phosphatase to prevent the vector from religating. Extract with phenol to remove all of the enzymes and with chloroform to remove the phenol. Add the purified repetitive fragment to the prepared vector and ligate. Transform the DNA into a suitable *Escherichia coli* host strain, and select for the presence of the plasmid using plates containing the appropriate antibiotic for that plasmid.

6. Pick colonies onto two fresh antibiotic-containing plates into a grid pattern for screening. Transfer the colonies from one of the plates onto a nitrocellulose filter. Screen the clones for repeated DNA inserts by colony hybridization using a ^{32}P-random-prime labeled probe prepared from total genomic DNA. Because each single-copy sequence occurs at low frequency in the probe, only repetitive sequences exhibit detectable hybridization under these conditions. The intensity of hybridization is a direct reflection of the copy number of the particular element in the DNA used as a probe. It is likely that more than half of the clones will have the repeated DNA fragment as an insert and will hybridize strongly to the probe.

7. Prepare DNA minipreparations from the positive clone(s). Prepare a hybridization probe from the cloned DNA using ^{32}P-labeled deoxynucleotide and random-prime labeling. Hybridize this probe to Southern blots of genomic DNA digested with the cloning enzyme and several others to confirm that the clone contains interspersed repetitive DNA. Interspersed repeats can be distinguished from tandem repeats (e.g., ribosomal genes and satellite sequences) at this stage. As mentioned above, a discrete fragment arising from an interspersed repeat must come from cleavage at both of two internal restriction sites for that enzyme. If one or both of those sites are lost and there are no other sites for the enzyme within the element, the probe will hybridize to a number of fragments with different sizes, leading to a smear (Fig. 1). If the isolated repeated DNA fragment was instead from a tandemly repeated unit, the loss of one or both sites would lead to a ladder of differently sized fragments rather than a smear.

This method is useful for the isolation of interspersed repetitive DNA elements whose copy number is on the order of tens of thousands, but it will not be practical for the isolation of less abundant elements. As long as the copy number is high enough, SINEs, LINEs, and the retroviral-like elements can all be isolated using this approach.

Characterization of Elements

Once a clone for an interspersed repetitive sequence has been isolated, it is useful to use it as a hybridization probe to isolate additional, full-length copies of the element from a genomic DNA library. In this way, the entire

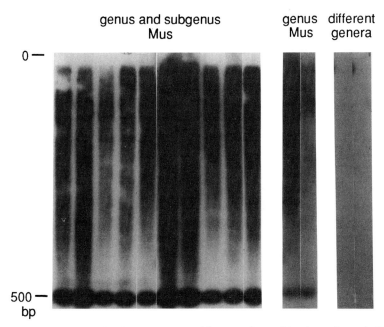

FIG. 1. Southern blot of an interspersed repetitive DNA in murid rodents. Genomic DNA was digested with *Bam*HI, fractionated by electrophoresis through 0.8% agarose, and then blotted to nitrocellulose. A 0.5-kb *Bam*HI fragment that lies within the L1 structure of *Mus domesticus* [S. L. Martin, C. F. Voliva, S. C. Hardies, M. H. Edgell, and C. A. Hutchison III, *Mol. Biol. Evol.* **2,** 127 (1985)] was used as a hybridization probe after labeling with ^{32}P. The resulting autoradiogram shows that a homologous *Bam*HI fragment is present as part of an interspersed repeated structure in all of the species belonging to the genus *Mus* (note the smear extending upward from the 0.5-kb fragment). Under the stringent conditions used in the hybridization (50% formamide, 42°, washes at 52° in 0.1 × SSC), no homolog is detected in the two species from different genera, *Rattus* and *Acomys*. The diminished intensity of the signal in the two more distantly related *Mus* species [*M. (Coelomys) pahari* and *M. (Pyromys) saxicola*] could be due either to reduced overall homology to the probe or loss of *Bam*HI sites (both because of accumulated sequence divergence) or to a lower copy number of the element in these species.

element will be available for further study. After several clones from a particular repetitive sequence family have been mapped and the boundaries of the repetitive portions of each clone have been identified, sequence analysis of a typical clone can be used to further determine the characteristics of the family. In addition to being used in a GenBank or EMBL database search for similar elements, the sequence should be examined for the presence of direct or inverted repeats, open reading frames, and regions of amino acid similiarity to coding regions of known retroelements, especially reverse transcriptase.[10] The phylogenetic distribution of the element can be studied using genomic DNA from a variety of species and Southern blots (Fig. 1). Alternatively, if a number of individual clones are isolated from a given subset of the element, their DNA sequences can be determined and used to study the evolution of the element in lieu of the entire element.[11,12]

Isolation of Rapidly Evolving Repetitive Sequences Using Phylogenetic Screening

The phylogenetic screening procedure, or phi-screen, was developed as a method for isolating rapidly evolving repetitive sequences, and specifically for isolating transposable elements that have been active in the recent evolutionary history of an organism.[13] The application of this procedure for the isolation of active retroelements is based on the premise that elements which are actively moving in the genome by retrotransposition change more rapidly both in copy number and sequence than elements which are static in the genome.

Method

1. Construct a genomic library from the species of interest (the ingroup). It is not necessary to construct a complete library for purposes of phi-screening, although it will be useful to have a complete genomic library for further characterization of retroelement families. The library to be used for phi-screening should be constructed using standard procedures from partially digested, size-selected genomic DNA from the species of interest, with the following considerations. A plasmid vector such as pUC or pBluescript is generally used because of the ease of small-scale plasmid DNA isolation compared to λ DNA isolation. Some thought should be given to the insert size in a library constructed specifically for phi-screening. For organisms with a short-term interspersion of repetitive sequences throughout the genome (including mammals), two approaches have been used. When small inserts [1–3 kilobases (kb)] are used to construct the library, it is expected that multiple repetitive sequence families will seldom be present in a single clone. Alternatively, larger inserts (10–20 kb) can be

[10] Y. Xiong and T. H. Eickbush, *EMBO J.* **9**, 3353 (1990).
[11] S. L. Martin, C. F. Voliva, S. C. Hardies, M. H. Edgell, and C. A. Hutchison III, *Mol. Biol. Evol.* **2**, 127 (1985).
[12] S. C. Hardies, S. L. Martin, C. F. Voliva, C. A. Hutchison III, and M. H. Edgell, *Mol. Biol. Evol.* **2**, 109 (1986).
[13] H. A. Wichman, S. S. Potter, and D. S. Pine, *Nature (London)* **317**, 77 (1985).

used, in which case SINEs will be present in most clones. The choice of insert size will determine the method used in the phi-screening procedure. For organisms (e.g., birds) with a long-term interspersion pattern,[14] a library with an insert size of 20 kb or more can be used, because most clones will lack repetitive sequences, and the incidence of multiple repetitive elements in a single clone should be rare.

2. Make duplicate blots from a number of arbitrarily selected clones isolated from the ingroup library. In phi-screens of mammalian genomes, we have screened about 0.1% of the genome, which is about 1000 3-kb clones or 200 15-kb clones. It is important to have equivalent amounts of DNA on each filter; hence, screening of library lifts results in a large number of false positives. Therefore, DNA is isolated from each clone, and duplicate blots are made for all clones to be screened. For clones with small insert size, or for organisms with long-term interspersion, dot and slot blots can be used, since each insert is likely to have only a single repetitive element. Clones with larger insert size from organisms that have short-term interspersion may contain multiple repetitive elements; therefore, the signal from one element, especially a highly repetitive SINE, may mask the signal from a second element in the same clone. For these clones, Southern blots are made instead of dot and slot blots. Each clone is digested with three restriction enzymes so that the insert is generally cut into a number of small fragments. We use *Eco*RI, *Hind*III, and *Bam*HI because these enzymes are inexpensive. Electrophoresis is carried out on duplicate 0.8% agarose gels, which are photographed and subjected to Southern blotting.

3. Hybridize duplicate blots to total genomic DNA probes from the ingroup and an outgroup. Hybridization probes are labeled with ³²P then used at 5×10^6 to 1×10^7 counts per minute (cpm) per hybridization under conditions of reduced stringency. As explained above, only repetitive sequences will be detected. One of the duplicate blots is hybridized to total genomic DNA probe prepared from the same sample used to construct the library (the ingroup). The duplicate blot is hybridized to total genomic DNA probe prepared from a related taxon (the outgroup). Selection of the appropriate outgroup is empirical. A taxon is selected which is related closely enough to the ingroup that most repetitive sequences will cross-hybridize, but distantly enough that some differences are detected. A divergence time of about 30 million years for the two species provides a good starting point for selection of an outgroup. If too many or too few differences are detected, the filters may be stripped and additional outgroup probes prepared for phi-screening with little additional effort.

---

[14] R. A. Bouchard, *Int. Rev. Cytol.* **76**, 113 (1982).

4. Compare autoradiographs to identify rapidly evolving repetitive sequences. In a successful phi-screen, most repetitive sequences will have a similar hybridization intensity between the ingroup and the outgroup. Clones of interest are those which hybridize differentially to the two probes. These repetitive sequences are evolving more rapidly than the majority of repetitive sequences in the genome. This may be due either to a sequence or a copy number change in a repetitive sequence family that is shared between the two species or to the acquisition of a new repetitive sequence family since their divergence. Once a group of clones are identified which appear to contain repetitive sequences that differ markedly in the intensity of hybridization, these differences should be verified by repeating the screening procedure.

5. Characterize families of rapidly evolving repetitive sequences. The first step in characterizing the rapidly evolving repetitive sequences identified by phi-screening will be to sort them into cross-hybridizing families, keeping in mind that clones may contain multiple families of repetitive sequences. If the goal is to isolate retroelements, the most promising clones are those which contain both repetitive and nonrepetitive sequences, indicative of interspersion. Clones which contain only repetitive sequences are usually satellite DNAs. One good experiment for initial characterization is to use clones from each cross-hybridizing family (or bands from clones containing multiple elements) to probe genomic digestions from a range of taxa selected to span the phylogenetic distance between the ingroup and the outgroup. The best restriction enzymes for these studies can be identified in a preliminary analysis in which 15 to 20 enzymes are used to digest genomic DNA from the ingroup, and Southern blots are probed with a clone from each family to identify enzymes which yield discrete bands. This experiment will give a good indication of the phylogenetic distribution of the family, as well as changes in restriction sites or copy number over evolutionary time. Clones are further characterized as described above, in order to differentiate between interspersed and tandem repeats and to determine the boundaries of the repetitive DNA within the clone, its structure, and whether it is related to other, known retroelements.

Phylogenetic screening has been used successfully to isolate retroelements that have been active in recent evolutionary time. In a phi-screen of the white-footed mouse, *Peromyscus leucopus,* the retrovirus-like element *mys* was identified by virtue of its hybridization to *Peromyscus* DNA but its lack of hybridization to *Mus* (the outgroup) DNA.[13] In a phi-screen of human DNA using *Galago* as an outgroup, both the 5′ end of the L1 element and the retrovirus-like element THE 1 hybridized differentially, but no significant differences were found in the initial screening of the human library when African green monkey was used as an outgroup.[15] This method is not, however, specific for retroelements. In all phylogenetic screening procedures conducted on mammals to date, satellite DNAs have been detected as hypervariable repetitive sequences.[15-17] The other disadvantages of this method are that it is labor intensive, requires a genomic library for each species to be screened, and will only detect elements which are repetitive and are either recently introduced into the genome or are rapidly evolving. However, it has the advantage of requiring only standard techniques, so that it is a feasible approach even for the novice molecular biologist. Evolutionary information is also built into the isolation process, so that repetitive elements isolated by this technique are likely to be of interest to an evolutionary biologist.

Isolation of Reverse Transcriptase-Encoding Elements Using Polymerase Chain Reaction

The PCR is an expedient method for isolating defined sequences from total genomic DNA, but it requires that specific primers be designed to flank the sequence of interest. The concept of universal primers,[18] which takes advantage of evolutionarily conserved regions of a gene to design primers that will amplify specific sequences from a range of taxa, can be applied to the search for mammalian retrotransposons. Because reverse transcriptase is the most conserved region in the retroviral genome and has been intensively studied in an evolutionary context,[10,19] it is the PCR target of choice for screening mammalian genomes for retrotransposons. Two different strategies have been used to screen for retrotransposons via PCR,[20,21] but other approaches are possible. Here we discuss some general considerations for primer design and some specific details to consider when using PCR with highly degenerate primers.

[15] J. A. Lloyd, A. N. Lamb, and S. S. Potter, *Mol. Biol. Evol.* **4,** 85 (1987).
[16] H. A. Wichman, C. T. Payne, and T. W. Reeder, *in* "Molecular Evolution" (M. T. Clegg and S. J. O'Brien, eds.), Wiley–Liss, New York, 1990.
[17] H. A. Wichman, C. T. Payne, O. A. Ryder, M. J. Hamilton, M. Maltbie, and R. J. Baker, *J. Hered.* **82,** 369 (1991).
[18] T. D. Kocher and T. J. White, *in* "PCR Technology" (H. A. Erlich, eds.), p. 137. Stockton, New York, 1989.
[19] R. F. Doolittle, D.-F. Feng, M. S. Johnson, and M. A. McClure, *Q. Rev. Biol.* **64,** 1 (1989).
[20] A. Shih, R. Misra, and M. G. Rush, *J. Virol.* **63,** 64 (1989).
[21] H. A. Wichman and R. A. Van Den Bussche, *BioTechniques* **13,** 258 (1992).

```
MULV   ...   LDQGILVPCQSPWNTPLLPVKKPGTNDYRPVQDLREVNKRVDEIHPTVPNPYNLLSGLPPSHQWYTVLDLKDAFFCLRLHPTSQPLFAFEWRDPEM-
RSV    ...   LQLGHIEPSLSCWNTPVFVIRKASGS-YRLLHDLRAVNAKLVPFGAVQQGAPVLSA--LPRGWPLMVLDLKDCFFSIPLAEQDREAFAFTLPSVNNQ
MMTV   ...   LQLGHLEESNSPWNTPVFVIKKKSGK-WRLLQDLRAVNATMHDMGALQPGLPSPVA--VPKGWEIIIDLQDCFFNIKLHPEDCKRFAFSVPSPNFK
BLV    ...   LEAGYISPWDGPGNNPVFPVRKPNGA-WRFVHDLRATNALTKPIPALSPGPPDLTA-IPTHPPHIICLDLKDAFFQIPVEDRFRFYLSFTLPSPGGL
VISNA  ...   EGKVGRAPPHWTCNTPIFCIKKKSGK-WRMLIDFRELNKQTEDLAEAQLGLPHPGG--LQRKKHVTILDIGDAYFTIPLYEPYRQYTCFTMLSPNNL
HSRV   ...   LKQGVLTPQNSTMNTPVYPVPKPDGR-WRMVLDYREVNKTIPLTAAQNQHSAGILA-TIVRQKYKTTLDLANGFWAHPITPESYWLTAFTWQ------

             N+P++ + K       R + D R N              +D+ ++  +            +F+


GISGQLTWTRLPQGFKNSPTLFDEALHRDLADFRIQHPDLILLQYVDDLLLAATSELD-CQQGTRALLQTLGNLGYRASAKKAQICQKQVKILGYLL ... MULV
APARRFQWKVLPQGMTCSPTICQLVVGQVLEPLRLKHPSLCMLHYMDDLLLAASSHDG-LEAAGEEVISTLERAGFTISPDKVQR-EPGVQYLGYKL ... RSV
RPYQRFQWKVLPQGMKNSPTLCQKFVDKAILTVRDKYQDSYIVHYMDDILLAHPSRSI-VDEILTSMIQALNKHGLVVSTEKIQK-YDNLKYLGTHI ... MMTV
QPHRRFAWRVLPQGFINSPALFERALQEPLRQVSAAFSQSLLVSYMDDILYASPTEEQ-RSQCYQALAARLRDLGFQVASEKTSQTPSPVPFLGQMV ... BLV
GPCVRYYWKVLPQGWKLSPAVYQFTMQKILRGWIEEHPMIQFGIYMDDIYIGSDLGLEEHRGIVNELASYIAQYGFMLPEDKRQE-GYPAKWLGFEL ... VISNA
GKQYCWTRLPQGFLNSPALFTADVVDLLKEIPN-----VQVYVDDIYLSHDDPKE-HVQQLEKVFQILLQAGIVVSLKKSEIGQKTVEFLGFNI ... HSRV

++ +P G  +P+   +          Y+DD+++ +      +      G     K         +LG +
```

Method

1. Identify regions that will serve as targets for PCR primers and design degenerate primers for these regions. Examples of regions chosen for PCR amplification of retroviral reverse transcriptase are shown in Fig. 2. The design of each primer takes advantage of an evolutionarily conserved domain. The YXDD domain is the most conserved region of reverse transcriptase, and it is included in both primer pairs (Fig. 3). Although the examples here are based on the sequence of retroviral reverse transcriptase, degenerate primers also could be designed to conserved domains of other retroviral genes or to LINE ORFs. In choosing the specific region to be used in designing PCR primers, leucine (L) and arginine (R) are generally avoided because each has six possible codons, whereas the use of phenyl-alanine (F), methionine (M), tyrosine (Y), histidine (H), glutamine (Q), asparagine (N), lysine (K), aspartic acid (D), glutamic acid (E), cysteine (C), tryptophan (W), and serine (S) is favored because each has only one or two possible codons. The amino acid sequences of the chosen regions are reverse-translated into DNA sequences to design the primers, and degenerate sites are included to conform to the degenerate nature of the genetic code. Remember to use the reverse complement in constructing the downstream primer. Different cloning sites are added to the 5′ end of each primer just inside a short GC clamp. In choosing restriction sites, consider the vector to be used for cloning and make sure that the digestion buffers are compatible. Although primers as short as 12 base pairs (bp) (exclusive

---

FIG. 2. Amino acid sequence alignments of the reverse transcriptase domains of six distinct retroviruses. MULV, Moloney murine leukemia virus [T. M. Shinnick, T. A. Lerner, and J. G. Sutcliffe, *Nature (London)* **293**, 543 (1981)]; RSV, Rous sarcoma virus [D. E. Schwartz, R. Tizard, and W. Gilbert, *Cell (Cambridge, Mass.)* **32**, 853 (1983)]; MMTV, mouse mammary tumor virus [I.-M. Chiu, A. Yaniv, J. E. Dahlberg, A. Gazit, S. F. Skuntz, and S. R. Tornick, *Nature (London)* **317**, 366 (1985)]; BLV, bovine leukemia virus [N. Sagata, T. Yasunage, J. Tsuzuku-Kawamura, K. Ohishi, Y. Ogawa, and Y. Ikawa, *Proc. Natl. Acad. Sci. U.S.A.* **82**, 677 (1985)]; VISNA, ovine visna virus [P. Sonigo, M. Alizon, K. Staskus, D. Klatzmann, S. Cole, O. Danos, E. Retzel, P. Tiollais, A. Haase, and S. Wain-Hobson, *Cell (Cambridge, Mass.)* **42**, 369 (1985)]; HSRV, human spumaretrovirus [B. Maurer, H. Bannert, G. Darai, and R. M. Flugel, *J. Virol.* **62**, 1590 (1988)]. Dashes within the amino acid sequence indicate gaps introduced to facilitate alignment. Symbols below the sequence of HSRV indicate amino acid sites that are highly conserved (letters) or undergo chemically conservative substitutions (+) among viruses and LTR-containing retrotranspo-sons, excluding copia and Ty [Y. Xiong and T. H. Eickbush, *Mol. Biol. Evol.* **5**, 675 (1988)]. These regions are suitable sites to consider when designing primers for PCR. The dashed line above the amino acid sequences indicates the regions used to design primers for PCR amplification of endogenous reverse transcriptase related to that encoded by human T-cell leukemia virus I and II (HTLV I and II).[20] The solid line above the amino acid sequences indicates the location of the primers shown in Fig. 3.

Degenerate primers:

Left: G C T C T A G A T T Y S Y N R T N A A R A A

Right: G C G A A T T C R T C R T C C A Y R T A

Degenerate primers with inosine:

Left: G C T C T A G A T T Y S Y I R T I A A I A A

Right: G C G A A T T C I T C R T C C A Y I T A

Y = C+T; S = G+C; R = A+G; N = A+T+G+C

FIG. 3. Universal primers for PCR amplification of endogenous retroviral reverse transcriptase. The primers anneal within the retroviral reverse transcriptase gene at the position indicated by the thick lines in Fig. 2. The degenerate primer pair has been used successfully,[21] but it gives a number of nonspecific products. Substitution of inosine at some of the degenerate positions increases the effective concentration of primer. *Xba*I and *Eco*RI restriction endonuclease cleavage sites are underlined.

of the cloning site and clamp) and as degenerate as 512-fold have been used successfully to amplify endogenous reverse transcriptase from mammalian genomes,[21] it is important that the 3' end of the primer be as conservative as possible.

2. Carry out the PCR under conditions appropriate for degenerate primers. PCR amplification should be carried out under standard conditions except as noted. The concentration of the primer may need to be increased up to 5-fold over standard conditions if the primers are very degenerate. The appropriate template concentration must be determined empirically, and various concentrations of template from 1 ng to 1 μg should be tested for each new primer–template pair. For very short or degenerate primers, the increase in temperature from the primer binding step to the synthesis step should be gradual in the initial PCR cycles so that the primer does not fall off the template before strand elongation occurs. Ten cycles in which this temperature increase was ramped over 3 min, followed by 25 cycles in which this step was ramped over 1 min, with a primer binding temperature of 50°, has been a successful profile with short, degenerate primers.[21] The ideal thermal profile will depend on the specific primers and template, and it must be determined empirically.

3. Clone and sequence the PCR product. Because of the degenerate primers used in these amplifications, a number of nonspecific products are expected. This, coupled with the possibility of numerous target sequences for amplification in the genome being screened, means that direct se-

quencing of the PCR product is not feasible. Therefore, the product must be cloned, using the restriction sites designed into the primers, for further characterization. The total PCR product should be chloroform extracted, digested with the two enzymes of choice, and fractionated by size using agarose gel electrophoresis. Although multiple bands are likely, the most reliable predictor as to whether a band would prove to be a reverse transcriptase (and thus warrant further study) in our hands was its size.[21] Product of the predicted size class should be cloned into pBluescript for ease of sequencing. The number of clones that are chosen for sequencing can be narrowed to repetitive DNA by hybridization with a probe prepared from total genomic DNA. Once the sequence of individual clones is obtained, positive clones can sometimes be identified by searching the DNA database. In addition, novel reverse transcriptase-like sequences can be identified by the presence of conserved amino acid domains between the primers.[21] Retrotransposons may be quite degenerate, and not all copies will necessarily have a complete ORF for reverse transcriptase.

4. Characterize retrotransposons. Once a positive clone has been identified, it can be used as a probe to isolate full-length copies of the element from a genomic library. These clones can also be characterized by phylogenetic screening.

One advantage of this method is that it does not require library construction unless one wishes to use the PCR product to isolate the entire element. Additionally, this method can be used to detect both repetitive and low-copy elements. Using these primers, however, it will not be possible to detect retroposons such as SINE elements or retrovirus-like elements such as THE 1 that do not have a region of amino acid similarity to reverse transciptase.

Conclusion

The presence of retrotransposon-derived repetitive DNA in such abundance in mammalian genomes is a mystery in terms of its origin as well as its role in the evolution of the genomes it inhabits. These mysteries will best be solved by detailed comparative studies of a number of elements in an evolutionary context. Already, phylogenetic analysis has led to an improved understanding of the structure and dynamics of retrotransposition of LINE and SINE elements of vertebrate genomes.[4,11,12,22–24] We have detailed three useful methods for the isolation of interspersed repetitive

[22] E. Pascale, E. Valle, and A. Furano, *Proc. Natl. Acad. Sci. U.S.A.* **87,** 9481 (1990).

[23] Y. Kido, M. Aono, T. Yamaki, K.-I. Matsumoto, S. Murata, M. Saneyoshi, and N. Okada, *Proc. Natl. Acad. Sci. U.S.A.* **88,** 2326 (1991).

[24] B. A. Rikke, L. D. Garvin, and S. C. Hardies, *J. Mol. Biol.* **219,** 635 (1991).

DNA/retrotransposons from mammalian genomes. In addition, methods to characterize and study the phylogenetic distribution of these elements are presented. These methods are applicable to screening any mammalian genome and should facilitate the much needed examination of genome evolution in a wide variety of species.

Acknowledgment

This work was supported by National Institutes of Health Grants GM 40367 (S.L.M.) and GM 38727 (H.A.W.).

[23] Detection and Characterization of Transposable Elements

By INGRID FELGER and JOHN A. HUNT

Introduction

The first characterization of transposable genetic elements was the genetic analysis of the unstable mutations in *Zea mays* (maize) by McClintock.[1] Plant geneticists have since made extensive use of transposons, such as activator *(Ac)* and *Tam3,* for cloning genes of interest.[2] However, the first recombinant DNA clones of eukaryotic transposable elements were discovered serendipitously. This occurred during the screening of the first *Drosophila* genomic DNA library for gene sequences, using a probe of cDNA to poly(A) RNA, when *copia* and other similar elements were isolated because they make an abundant poly(A) RNA.[3] Other serendipitous isolations occurred during the cloning of specific genes when the elements were found within the same clonal isolate. Later, new elements were identified by their mutagenic effect on previously cloned genes, especially in *Drosophila*[4] and maize.[5]

In general the definition of a transposable element is not that any one particular element has been proved to transpose, rather that a particular repetitive DNA sequence can be demonstrated to be interspersed in genomic DNA and have sequence similarity with established classes of trans-

---

[1] B. McClintock, *Cold Spring Harbor Symp. Quant. Biol.* **21,** 197 (1955).
[2] L. Balcells, J. Swinburne, and G. Coupland, *Trends Biotechnol.* **9,** 31 (1991).
[3] D. J. Finnegan, G. M. Rubin, M. W. Young, and D. S. Hogness, *Cold Spring Harbor Symp. Quant. Biol.* **42,** 1053 (1977).
[4] P. M. Bingham, R. Levis, and G. M. Rubin, *Cell (Cambridge, Mass.)* **25,** 693 (1981).
[5] N. Federoff, S. Wessler, and M. Shure, *Cell (Cambridge, Mass.)* **35,** 235 (1983).

posable elements. Transposons, such as *Ac* and *Spm,* which appear to be inactive in maize are active when introduced into other plants such as potato, tomato, tobacco, and *Arabidopsis.*[2] This may be widely true as a principle for many eukaryotic transposable elements.

Genetic Methods for Isolation of Transposable Elements

The *white* locus of *D. melanogaster,* which is mostly used as the locus for isolation of transposable elements by their mutagenic affect, was itself first isolated by a mutation caused by the insertion of a *copia* element. Several clones from different chromosomal origins containing *copia* elements were isolated from the genomic DNA library of this strain by using a *copia* DNA probe. The *white* locus was identified by *in situ* hybridization to polytene chromosomes using the nonrepetitive DNA of different copia containing clones.[4]

In maize and other plants, the properties of the transposable elements which form unstable mutations have helped in isolating the gene affected by the mutations caused by the elements. In turn, new elements are characterized by clonal isolation of the DNA from genomic libraries of strains containing the mutations (see Döring and Starlinger[6]). Because mutations can be maintained in heterozygous plants and recovered in the homozygous state by self-crossing F_1 progeny, large numbers of individual isolates can be quickly recovered. This makes population genetic analysis easy in plants, compared to other eukaryotes.

Nongenetic Methods for Isolation of Transposable Elements

Random Isolation

Young[7] screened 80 randomly isolated plasmids from a genomic library of *D. melanogaster* DNA for repetitive elements. Of these 23 were found to be repetitive by hybridization to Southern blots of *D. melanogaster* DNA, and 17 of these were interspersed repetitive elements as shown by *in situ* hybridization with polytene chromosomes.

Isolation by Hybridization to Repetitive DNA

Genomic DNA of most eukaryotes has about 10–20% of moderately repetitive DNA.[8] Sequences of this fraction are represented in the genome

[6] H.-P. Döring and P. Starlinger, *Annu. Rev. Genet.* **20,** 175 (1986).
[7] M. W. Young, *Proc. Natl. Acad. Sci. U.S.A.* **76,** 6274 (1979).
[8] R. J. Britten, D. E. Graham, and B. R. Neufeld, this series, Vol. 29E, p. 363.

in 10–1000 copies. This middle repetitive portion of the genome can be used to screen a genomic library for middle repetitive clones by hybridization to plaque lifts of bacteriophage plates.[9] Labeled total genomic DNA can be used for screening, provided that the level of labeled single-copy DNA is insufficient to allow the detection of plaques containing single-copy DNA sequences. By contrast, the amount of label deriving from a single repetitive element is, according to its copy number, 10–1000 times more abundant. Therefore, it is possible to determine concentrations of labeled probe that will readily discriminate between plaques on bacteriophage plate lifts that contain single-copy and repetitive DNA. The presence of highly repetitive tandem repeated "satellite" sequences in the genomic libraries may cause some difficulty when using unfractionated DNA as a probe. In most genomic libraries, which are made using restriction enzyme digestion, highly repetitive DNA is usually absent or underrepresented because of the lack of restriction sites in these simple sequence tandem repeats. In libraries made using randomly sheared DNA, highly repetitive DNA is expected to be present, and therefore it is necessary to remove the highly repetitive DNA from the probe DNA by denaturation and reannealing of the DNA to low C_0t values followed by removal of the double-stranded DNA by chromatography on hydroxylapatite.[8]

Using a partial *Eco*RI bacteriophage library from *D. silvestris,* no evidence was found for a highly repetitive clone. In a similar library from *D. picticornis,* 4 plaques out of 200 were found with very high levels of hybridization, indicating the presence of a highly repetitive DNA sequence.[10] For screening plaque lifts from two λ bacteriophage Charon 4 libraries, the following conditions were used. The probe DNA was made from genomic DNA of both *D. silvestris* and *D. picticornis* using nick translation of 0.25 μg of DNA with 50 μCi [^{32}P]dCTP to yield DNA with a specific activity of $1.8-2.5 \times 10^8$ counts per minute (cpm)/μg. Hybridization of replica filters of lifts from plates containing 2–400 plaques from each of the two species with each of the probes used 90 ng of probe for each hybridization. The hybridization was performed in 50% formamide, $5\times$ standard saline citrate (SSC) in the presence of $5\times$ Denhardt's solution and 10% dextran sulfate at 37° for 16 hr. The filters were washed 3 times in 0.1% sodium dodecyl sulfate (SDS) and $0.2\times$ SSC at 42° for 10 min each. Under these conditions, using *Drosophila* DNA, all of the plaques were visible after autoradiography, but with those which contained repetitive elements showed much stronger hybridization.

[9] J. Sambrook, E. F. Fritsch, and T. Maniatis, "Molecular Cloning: A Laboratory Manual," 2nd Ed., Cold Spring Harbor Laboratory, Cold Spring Harbor, New York, 1989.
[10] I. Felger and J. A. Hunt, *Genetica* **85,** 119 (1992).

Initial Characterization of Elements

Elements Discovered Either as Causing Mutation within Known Gene or Found Close to Gene in Clonal Isolate

The first indication that a part of the clone contains repetitive DNA is obtained by hybridization of the isolated recombinant DNA to genomic DNA. This can be done by labeling the whole bacteriophage, cosmid, or plasmid clone. The presence of repetitive DNA is shown by a greater than expected level of hybridization with Southern blots of restriction enzyme-digested genomic DNA (as compared to that found for a single-copy gene such as the *Adh* gene in *Drosophila*[11]). If the repetitive DNA is a tandem repeat, then a simple restriction pattern containing few bands would be expected for the majority of the 6-base cutting enzymes used. For an interspersed repetitive sequence, many bands or a smear will be seen. Other evidence for interspersion of repetitive elements is simple to obtain if the organism being studied is *Drosophila,* or another dipteran with large polytene chromosomes, since the cloned DNA containing the repetitive element can be used for *in situ* hybridization.[12] The finding of well-distributed bands of hybridization on the chromosomes indicates the presence of an interspersed repetitive element. Low-copy elements may be detectable in metaphase chromosome spreads.

Elements Found by Random Isolation of Clones or by Hybridization of Genomic DNA to Genomic Library

The clones that are chosen at random have to be screened to determine if they contain a repetitive element. DNA is isolated from the clone and hybridized to Southern blots of genomic DNA.

The alternate method is screening a genomic library by hybridization with total genomic DNA as described above. With this procedure 70 positive clones were identified from 800 recombinant bacteriophage plaques from a *D. silvestris* genomic DNA library.[10] Sixteen clones were plaque-purified by replating and rescreening. Of these 9 were found to be repetitive, both by hybridization of their DNA to an *Eco*RI digest of genomic DNA and by *in situ* hybridization to polytene chromosomes from *D. silvestris.* Five clones were identified as interspersed repetitive elements hybridizing to 10 to 60 sites. One clone hybridized at more than a hundred sites and showed strong centromeric hybridization. Two repetitive clones appear to contain tandem repeated DNA sequences, and one cross-hybridized with the histone gene cluster of *D. melanogaster.*

[11] J. A. Hunt, J. G. Bishop III, and H. L. Carson, *Proc. Natl. Acad. Sci. U.S.A.* **81**, 7146 (1984).
[12] M. L. Pardue and J. Gall, *Methods Cell Biol.* **10**, 1 (1975).

DNA Sequence Determination of Repetitive Element

For all elements, except those that are inserted into a gene whose DNA sequence is already known, it is first necessary to determine the limits of the element in the clonal isolate. This can be done by hybridizing Southern blots of the clone with a genomic DNA probe. Restriction fragments containing repetitive DNA produce a stronger signal, whereas signals from single-copy DNA are weak. Because only a small fraction of the labeled genomic DNA contains the repetitive sequence, care should be taken that this DNA is not preferentially hybridized to the largest of the restriction fragments containing the element by using an excess of the labeled probe. Restriction fragments that contain only a short sequence of the repetitive element will also give a weak signal, so that it is important to have a complete restriction map of the clonal isolate in order to recognize the restriction fragments flanking those containing most of the sequence.

Once a plasmid containing the repetitive sequence is recognized or subcloned from the restriction fragments of the larger clone, it can be used to isolate other elements of the same family from the genomic library. Additional isolates may be necessary to determine the ends of the element.

The plasmids or M13 subclones of the repetitive regions can then be used for DNA sequencing by the Sanger dideoxy method.[13] In previous studies we have experienced extreme difficulty in maintaining subclones of a single repetitive element in fragments of 2–3 kilobases (kb) in M13 owing to instability, presumably caused by portions of the transposable element, and have chosen to use direct sequencing on double-stranded DNA or generation of single-stranded DNA from the chimeric plasmids pZf18U and pZf19U.[14] The ease of obtaining reliable sequences using Sequenase (U.S. Biochemical Corp., Cleveland, OH) on double-stranded plasmids isolated by the rapid boiling method makes this the method of choice in our laboratory.[15]

Recognition of Transposable Element Sequence

The elements so far known fall into two major classes: those containing terminal repeats which may be direct or inverted and those that lack terminal repeats. There are two methods to look for terminal repeats.

---

[13] F. Sanger, S. Nicklen, and A. R. Coulson, *Proc. Natl. Acad. Sci. U.S.A.* **74**, 5463 (1977).
[14] D. A. Mead and B. Kemper, *Bio Technology* **10**, 85 (1988).
[15] F. Toneguzzo, S. Glynn, E. Levi, S. Mjolsness, and A. Hayday, *BioTechniques* **6**, 460 (1988).

Dot Matrix Analysis

Dot matrix analysis is most useful for determining tandem repeats of any sequence. The program for these analyses can be found in the University of Wisconsin GCG package and the Staden package, which run on VAX VMS computers, or in the MacVector program from IBI, which is available for both IBM and Macintosh microcomputers.[16] A public domain program for the Macintosh called DottyPlot is available through the EMBL file server on Bitnet.[17] Basically any dot matrix program compares the sequence against itself using a defined window size and places a dot in the matrix whenever the sequence within the window finds a match with a preset degree of fit. For DNA the matches will be typically 67–73% with a window of 11 to 21 nucleotides. This produces a matrix that is not too dense and where the diagonal matches of the repeats, offset from the diagonal of perfect match, can be easily seen, and it gives the length and separation of the matches for direct repeats. For inverted repeats it is necessary to compare the sequence against its complement. Some programs will do this automatically since it is also a way of finding secondary structure in a single-stranded molecule. Here the presence of the inverted repeat is seen as a row of dots along a diagonal, with the spacing from the opposite diagonal being the distance between the repeats. The dot matrix analyses for the *copia*[18] element from *D. melanogaster* showing a direct repeat of 144 nucleotides and the *Uhu*[19] element from *D. heteroneura* which has an inverted repeat of 49 nucleotides are shown in Fig. 1. Some dot matrix programs, such as that in MacVector, are capable of zooming into regions of interest and will show the actual alignment of the repeats. If this is not available then it is necessary to examine the sequences themselves for the homology. Retroviral elements similar to *copia* are also found in maize and yeast as well as in mammals.[20]

---

[16] MacVector is available from IBI, P.O. Box 9558, New Haven, CT 06535; University of Wisconsin GCG programs are available from the Genetics Computer Group, University Research Park, Madison, WI 53711; and the Staden programs are available from the Medical Research Council Laboratory for Molecular Biology, Hills Road, Cambridge, UK.

[17] Public domain programs for microcomputers are available by electronic mail from NETSERV@EMBL-HEIDELBERG.DE on Bitnet by sending a one-line message HELP. The GenBank data-base is also available via Bitnet by sending the HELP message to BLAST@NCBI.NLM.NIH.GOV.

[18] S. M. Mount and G. M. Rubin, *Mol. Cell. Biol.* **5**, 1630 (1985).

[19] L. Brezinsky, G. V. L. Wang, T. Humphreys, and J. Hunt, *Nucleic Acids Res.* **18**, 2053 (1990).

[20] A. M. Weiner, P. L. Deininger, and A. Efstratiadis, *Annu. Rev. Biochem.* **55**, 631 (1986).

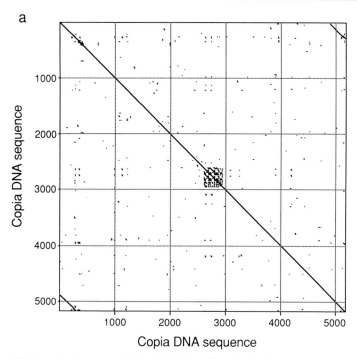

Fig. 1. Dot matrix plots created using MacVector[16] and edited on MacDraw II to obtain square diagrams. (a) Analysis of the *copia* sequence against itself using a window of 21 nucleotides and 67% matching. The diagonal from the top left to the bottom right shows the perfect match, and the two diagonal lines at the top right and bottom left show the match of the terminal direct repeats. (b) Analysis of the *Uhu* sequence against its complement using the same 21-nucleotide window and 67% matching as for the *copia* sequence. Here there is no perfect match on the diagonal, but the presence of the inverted terminal repeats is shown by the diagonal lines at the top right and bottom left corners of the plot.

Use of Local Homology Programs

Local homology programs set up a matrix of the sequences to be compared much as in the dot matrix analysis. The window size and weighting for deletions and insertions can be set by the operator. The number of fits is selected by choosing a probability level that the match found is better than random. For a reasonable match larger than 20 base pairs (bp) the probability can be set to less than 1×10^{-9}. The program SEQH from Kanehisa[21] is available for the VAX VMS. It is written in Fortran and can be ported to other computers with a Fortran compiler. It does require a large memory since a 5000×5000 matrix will use 25

---

[21] M. Kanehisa, *Nucleic Acids Res.* **12**, 203 (1984).

b

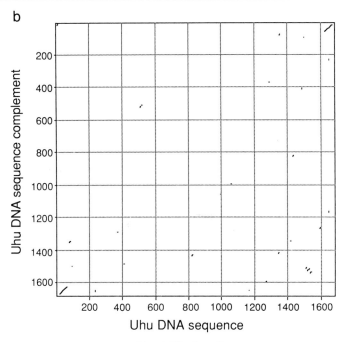

FIG. 1 (*Continued*)

megabytes of memory, and usually a comparison must be done by breaking down one of the sequences into smaller units. For such a comparison the program is quite slow. The comparisons are done as for the dot matrix analysis using the sequence compared with itself or its complement. The GCG suite of programs has a program that will find repeat sequences within a sequence, but it does not appear to work for inverted repeats.

If either of the methods gives evidence for a terminal repeat it is then relatively easy to discover the flanking sequences at the point of insertion. In addition, there is usually a short direct repeat of the genomic DNA sequence at the insertion site. To confirm the terminal repeats, and whether the total element has been determined, a second element has to be isolated from the genomic library using the repetitive sequence as a probe. DNA sequence analysis of the second element will confirm the completeness of the element first isolated.

Elements without Terminal Repeats

The major forms of transposable elements in eukaryotes that lack terminal repeats belong to the class of retrotransposons. The LINEs and SINEs in mammals (long interspersed elements and short interspersed elements, respectively) occur in very high copy number. Evidence for their existence is found in restriction digests of genomic DNA, where prominent bands are seen using ethidium bromide staining of the digested DNA in the digests whereas the rest of the DNA is seen as a smear. The first element isolated from such a band was the *Alu* element from humans.[22] This SINE is made up of dimers of the 7SL small nuclear RNA. Other SINEs are derived from tRNA. The method of transposition is presumably by reverse transcription of an RNA transcript of the element.

The LINEs are much larger and seem to be derived from retroviral sequences. However, they do not have terminal repeats, and many of the elements found in the genome are truncated at the 5' end. Most SINEs and LINEs have a duplication of the genomic DNA at the insertion site. The LINEs generally have long open reading frames which share sequence similarity with retroviral genes. In *Drosophila* several middle repetitive LINE-like sequences have been characterized such as the *I* element, *F* element, and the ribosomal insertion sequences.[20]

Characterization of a LINE-like element can be quite difficult. First, the repetitive region of the isolated clone is sequenced. Then a computer analysis is performed to determine the characteristics of the sequence, such as the presence of inverted or direct terminal repeats. If terminal repeats are not found it is still possible that the element is one of the terminal repeat class which has a deletion at one end. Further clones are then obtained from the library by using the repetitive portion of the clone as a probe. By way of cross-hybridization, regions of homology between the two isolates can be delineated. The sequence of portions of other isolates should then be determined to search for the ends of the element.

The next search is within the DNA sequence. In general LINE-like elements have A-rich sequences at the 3' terminus.[20] If these are found in other isolates of the repetitive sequence, this is a strong presumption for the classification of the element as a LINE. By comparing different isolates, deletions within the element can be determined as well as truncation of the 5' end. The sequence is also analyzed for open reading frames. In our analysis of the *Loa* element, a retrotransposon from *D. silvestris* lacking a long terminal repeat (LTR), we needed to sequence three isolates of the element and then construct a consensus sequence to eliminate the deletions found in all of the isolates. Two long open reading frames spanning 7 kb were identified in the consensus sequence.

---

[22] C. W. Schmid and C.-K. J. Shen, *in* "Molecular Evolutionary Genetics" (R. J. MacIntyre, ed.), p. 323. Plenum, New York, 1986.

Identification of Protein Sequences Encoded by Open Reading Frames

Many retrotransposons encode proteins that are homologous with the *gag* and *pol* genes of retroviruses.[23] Because the function of the reverse transcriptase enzyme may be necessary for the transposition of these elements, it is important to search for sequence similarities in the repetitive elements that are newly isolated.

The procedure here is both easy and difficult. It is easy to translate the DNA sequence into polypeptides using each of the six reading frames or, if the element shows strong evidence for a long open reading frame, just this frame. Usually all of the reading frames are used in the first analysis since it is relatively easy to make deletion and insertion mistakes in the sequencing process. The next step is to search a protein database for similarities with other sequences. This can be done using the GCG package or any other sequence analysis package which allows access to the most up-to-date nucleic acid and protein sequence databases. These are the EMBL and GenBank nucleic acid libraries and the NBRF protein library. The most common program for searching these databases is FASTA (a public domain program written by Pearson and Lipman[24] which is included in these packages). This will search both protein and nucleic acid databases, but for transposable elements the NBRF protein database should be searched because the nucleic acid similarity is too low (< 30%) to obtain meaningful matches. The output of the search is sometimes confusing, and two types of matches are found, both of which give scores that are over 80. The first is from a high degree of similarity with over 40% match over a small number of amino acids (typically 10–20). This usually means that you have identified a motif which is found in a particular protein or proteins associated with some function, but which might not be the same type of protein as is being used for the search. The other similarity will vary from less than 20 to 100% (if the element has been sequenced by someone else!) covering 100 amino acids and more. Such a match is presumptive evidence that the protein sequence from the open reading frame has a homolog in some other species, or even a distant or close relative in the same species. The output from FASTA gives the alignment of the sequences and the identity of the sequence for the best 40 fits, which is usually more than necessary.

In the case of the *Uhu* element,[19] which was shown to contain inverted direct repeats and was about 1500 bp long, a high degree of similarity was found with the amino acid sequence of the putative transposase from the

[23] Y. Xiong and T. H. Eickbusch, *Mol. Cell. Biol.* **8,** 114 (1988).
[24] W. R. Pearson and D. J. Lipman, *Proc. Natl. Acad. Sci. U.S.A.* **85,** 2444 (1988).

```
(Peptide) FASTA of: Uhu1.Aa  from: 1 to: 311  March 27, 1991  11:51

REFORMAT of: Uhu.1aa  check: 556  from: 1  to: 311  March 27, 1991  11:25
(No documentation)

TO: NBRF:*  Sequences: 25,814  Symbols: 7,348,950  Word Size: 2

The best scores are:                                    init1 initn opt..

New:S06303  Hypothetical protein 174 (transposon Tc1) - C... 296   302   337
Protein:Qqkwta  Tc1 hypothetical protein A-273 - Caenorha... 296   296   345
New:A35062  *Kex2 protein homolog precursor - Human          64    94    76

N;Alternate names: probable transposase
C;Species: Caenorhabditis elegans
C;Accession: S06303
R;Plasterk, R.H.A.
Nucleic Acids Res. 15, 10050, 1987 . . .

SCORES     Init1: 296 Initn: 302 Opt: 337
           40.2% identity in 174 aa overlap

             100        110        120        130      ' 140        150
Uhu1.A VFNIIRRFVDENRIEDKGRKAPNKIFTEQEERRIIRKIRENPKLSAPKLTQQVQDEMGKK
                                  :|:|:|:  ||:|: :|::: : : :|
S06303                            MNRNILRSAREDPHRTATDIQMIISSPKGPL
                                          10        20        30

             160        170        180        190        200        210
Uhu1.A CSVQTVRRVLHNHDFNARVPRKKPFISTKNKGTRMTFAKTHLDKDLEFWNTIIFEDESKF
        | :||||  |::  ::::| |  ||||||:||: :|:::||:|| : : |:: |::|||||
S06303 PSKRTVRRRLQQAGLHGRKPVKKPFISKKNRMARVAWAKAHLRWGRQEWAKHIWSDESKF
            40         50         60         70         80         90

             220        230        240        250        260        270
Uhu1.A IIFGSDGRNYVRRQSNTELNPKNLKATVKHGGGSVMVWACISAASVGNLVCIETTTDRNV
        :|||||:::|||: ::  :||   :|||||||||||:|:::::|:| |  |::: ||
S06303 NLFGSDGNSWVRRPVGSRYSPKYQCPTVKHGGGSVMVWGCFTSTSMGPLRRIQSIMDRFQ
            100        110        120        130        140        150

             280        290        300        310
Uhu1.A DLSILKENLLQSAEKLGIRRTFRFYQDNDQDNNQAZ
        :|:::::: : | : :: |:| |
S06303 YENIFETTMRPWALQ-NVGRGFVF
            160        170
```

FIG. 2. Edited portion of the FASTA search of the NBRF database (release date September 1990) using the amino acid translation of the large open reading frame of the *Uhu* element from *D. heteroneura*. This shows that the first two matches are nearly identical and that they are clearly superior to any other matches found. The alignment to the *Tc1* element transposase is shown.

Tc1 element of *Caenorhabditis elegans* (Fig. 2). No similarity with other transposable elements from *D. melanogaster* was found in the search. This is because the NBRF database does not contain all translations of open reading frames in DNA sequences. In fact it is possible to find similarity between the translated open reading frames of two *Hb* elements of *D.*

melanogaster, Uhu and *Tc1*. There is no doubt from all the analyses that the *Uhu* element is a transposon of the *P* and *Tc1* type.

The computer search of the NBRF protein database with the *Loa* element identified the ribosomal insertion element *R1* from *D. melanogaster* as the best fit. Other significant fits were obtained with the *Jockey*

```
              L       PG D                        F       G   P
     440  DTCVARLKSRRSPGLDGINGTICKAVWRAIPEHLASLFSRCIRLGYFPAE 489
              ::  ::: :||||:|||: :: ::    : ::: : : ::: ||::|::
    1180  NWAIGTFQPYKSPGMDGISPAFLQTGQDILLSRIRKALVSSLALGHIPSA 1229

                       K            YRP  L      K           R
     490  WKCPRVVSLLKGPDKDKCEPSSYRGICLLPVFGKVLEAIMVNRVREVLPE 539
              : :|||  |: :||  :|:|:| |:| : : |:|| ::   ::|   | :
    1230  CRRARVVFIPKAGKKDITDPKSFRPISLTSFLLKTLEKMVDYKIRSTLLK 1279

                   Q  GF     T                         LD   AFD
     532  GCRW---QFGFRQGRCVEDAWRHVKSSVGASAA--QYVLGTFVDFKGAFD 584
              :     |  ::|  ||::::|   :::::: :|: :   : :|  :|:|::||||
    1280  QRPLHPAQHAYRVGRSTDTALYQLQRTZSAAIDYKEVALCAFLDIEGAFD 1329
                                                        D      F

                                                            G
     585  NVEWSAALSRLADLG-CREMGLW-QSFFSGRRAVIRSSSGTVLVPVTRGC 632
              |:: :|  : |:  |  :: : |   :::::|:: :  ::|| | :|:||
    1330  NTSHDAIKDTLSRRGLDPTTSRWILALLRSRQVTASVHDSTVTVLTTKGC 1379

              QG       L         L            ADD
     633  PQGSISGPFIWDILMDVLLQRL-QPYCQLSAYADDLLLLVEGNSRAVLEE 681
              |||:: :|::|::|:| || ||  :  :|:::::  :  :
    1380  PQGGVLSPLLWSLLVDELLNRLTNSGIQCQGYADDIVIIARGKFEESLCD 1429
              P  G   P                       Y  DD

     682  KGAQLMSIVETWGAEVGDCLSTSKTVIMLL--KGALRRAPTVRFAGANLP 729
              : :  ::|: :|  ||| |:::||||: :  :  |:| : :::::|: |:
    1430  MVQSGLRITYDWCKEVGLNLNPTKTVIVPFTRRHKLQRMRQIWLSGTPLE 1479
                                  G     K

              LGV
     730  YVRSCRYLGITVSEGMKFLTHIASLRQRMTGVVGALARVLRADWGFSPRA 779
              |: :|||::  :: ::| ||::: : : :: :  :: :|| ||:
    1480  RSREVKYLGVIFDSKLNFGTHVQNAMLKCSRALYTCRSIAGKSWGTSPKI 1529
              G

     780  RRTIYDGLMAPCVLFGAPVWYDTAEQVAAQRRLASCQRLILLGCLS--VC 827
              | :|  :: | : :|: :| | |  :::::::|:: ||:  :  :|  |
    1530  VRWLYLMVVRPMLTYGVIAWGDRARLITVKKQLQKLQRMACV-CMTGVMC 1578

     828  RTVSTVALQVLGGAPPLDLAAKLLAIKYKLKRG 860
              | :|:||::| : :||:  :| :    |::::
    1579  -TCPTMALEALMELTPLHHIIRLKQKATLLRMS 1610
```

FIG. 3. Comparison of the deduced amino acid sequence of the consensus element *Loa* from the Hawaiian *Drosophila* (*D. silvestris,* lower line) and the ribosomal insertion sequence, *R1,* element from *D. melanogaster* (upper line). The amino acids above the sequences at that site are found in the putative reverse transcriptase of at least five of six non-LTR retrotransposons.[23] The amino acids below the sequences are invariant in the reverse transcriptase of retroviruses.[25]

and *I* elements of *D. melanogaster* and the *R1* element of *Bombyx mori*. The central part of the second open reading frame has a similarity spanning more than 800 amino acids with 22–25% identity. Although the similarity is low, it is strongly supported by the presence of shared invariant amino acid residues at specific locations (Fig. 3).[25] This pattern of conserved amino acids has been established by comparisons of reverse transcriptase sequences from retroviruses and retrotransposons.[25] In addition, similarities between *Loa* amino acid sequences in the first open reading frame are found with the gag protein as indicated by characteristically spaced histidine and cysteine residues ("Cys" motifs). An exonuclease-specific "Cys" motif is also found in the second open reading frame.

Evolutionary Consequences

In searching for new transposable elements, it is generally not possible to use DNA cross-hybridization to detect related elements from the same class, because the similarity is detectable only in the amino acid sequences. Instead, the features that are diagnostic of some transposable elements are the presence of terminal repeats, whether inverted or direct, and the amino acid sequence of the proteins they encode. Before a new transposable element can be classified, it is necessary to determine the DNA sequence of the element.

Once a new element is isolated, it is possible to find homologous DNA sequences in closely related species by DNA hybridization. It is unlikely, however, that this method would detect homologous DNA in *Drosophila* species which diverged more than 10–20 million years ago, because transposable elements have a more rapid rate of evolutionary change. An exception to this rapid divergence rate is seen in the *copia* and *P* elements, which are transmitted horizontally, and therefore cross-hybridization is the method of choice for isolating these elements in new species.

[25] H. Toh, R. Kikuno, H. Hayashida, T. Miyata, W. Kugimura, S. Inouye, S. Yuki, and K. Saigo, *EMBO J.* **4**, 1267 (1985).

[24] DNA–DNA Hybridization Approaches to Species Identification in Small Genome Organisms

By CLETUS P. KURTZMAN

Introduction

Until the late 1960s, the primary means for defining species of microorganisms lay with the subjective appraisal and weighting of phenotypic properties, primarily cellular morphology and growth responses on various sugars and other compounds. Although it was known that the assimilation and fermentation of many compounds were controlled by one or only a few genes[1] and, therefore, could result in an erroneous definition of species, practical genetic means for defining species often did not exist.

With the development of methods for measuring DNA duplex formation came the opportunity to define all species on the basis of genetic similarity. The theoretical and operational aspects of DNA hybridization were elegantly presented by Britten and Kohne,[2] and their paper has served as a primer for those now using DNA reassociation techniques.

The definition of a species, whether of higher plants, animals, or microorganisms, has as its basis the principle of genetic isolation. Simply put, members of a species are considered interfertile, whereas genetically separate species are not.[3] When measurements of DNA reassociation are used to define species, the extent of DNA relatedness expected between members of a species must first be determined. The discussion at the conclusion of this chapter addresses this issue and offers guidelines for predicting genetic relatedness from the extent of DNA complementarity.

Isolation of DNA

Most methods for DNA hybridization require substantial amounts of DNA, and it is not uncommon to process 5–50 g (wet weight) of cells. Although numerous methods have been developed for isolation of DNA from microorganisms, many are modifications on basic themes, and the following account is intended to be an overview of the procedures. Cells are generally grown to log phase to ensure that each has undergone complete

---

[1] C. C. Lindegren and G. Lindegren, *Proc. Natl. Acad. Sci. U.S.A.* **35**, 23 (1949).

[2] R. J. Britten and D. E. Kohne, *Science* **161**, 529 (1968).

[3] T. Dobzhansky, *in* "Molecular Evolution" (F. J. Ayala, ed.), p. 95. Sinauer, Sunderland, Massachusetts, 1976.

chromosome replication.[4] However, some fungi represent a practical exception because much greater yields of DNA may be obtained from young mycelia.[5]

For microorganisms with cell walls, wall breakage may be effectively accomplished either mechanically or enzymatically. Mechanical breakage of the cell walls of yeasts and filamentous fungi is easily done in the Braun cell homogenizer (B. Braun Biotech, Allentown, PA) using 0.5-mm glass beads.[5,6] Breakage should be kept to 25–50% to prevent large-scale shearing of the DNA. The French pressure cell [SLM Instruments (AMINCO), Urbana, IL] is another mechanical means for cell breakage that is sometimes effective. Generally, sonication has proved inferior to the preceding two methods. Because some fungal nucleases seem quite active even in the presence of EDTA, Specht et al.[7] developed a protocol for obtaining high molecular weight DNA from freeze-dried cells. Bacterial cell walls are usually degraded enzymatically,[8,9] and this is also successful for many fungi.[10]

Once cell wall breakage or enzymatic dissolution has occurred, a variety of procedures can be used for isolation and purification of the DNA. These generally rely on an initial deproteinization using phenol or a mixture of perchlorate, sarcosine, and chloroform–isoamyl alcohol.[6,9] The crude DNA is further purified by enzyme treatments, alcohol precipitation and spooling, hydroxylapatite chromatography, cesium chloride gradient centrifugation, or any combination of these methods.

DNA Shearing

Once the DNA has been purified, it is used unsheared when attached to membranes; for free solution reactions and for membrane probes, however, it must be sheared to satisfy reaction kinetics and to maintain specificity. Strand lengths of 400–500 base pairs (bp) are usually employed.[2,8] The French pressure cell is commonly used for shearing, and a double

---

[4] R. J. Seidler and M. Mandel, *J. Bacteriol.* **106,** 608 (1971).

[5] C. P. Kurtzman, M. J. Smiley, C. J. Robnett, and D. T. Wicklow, *Mycologia* **78,** 955 (1986).

[6] C. W. Price, G. B. Fuson, and H. J. Phaff, *Microbiol. Rev.* **42,** 161 (1978).

[7] C. A. Specht, C. C. DiRusso, C. P. Novotny, and R. C. Ullrich, *Anal. Biochem.* **119,** 158 (1982).

[8] J. L. Johnson and E. J. Ordal, *J. Bacteriol.* **95,** 893 (1968).

[9] J. L. Johnson, *in* "Manual of Methods for General Bacteriology" (P. Gerhardt, R. G. E. Murray, R. N. Costilow, E. W. Nester, W. A. Wood, N. R. Krieg, and G. B. Phillips, eds.), p. 450. American Society for Microbiology, Washington, D.C., 1981.

[10] C. P. Kurtzman and H. J. Phaff, *in* "The Yeasts, Volume 1: Biology of Yeasts" (A. H. Rose and J. S. Harrison, eds.), p. 63. Academic Press, London, 1987.

passage at 10,000–30,000 psi followed by filtration through a 0.45-μm membrane filter gives fragments that are predominantly in the specified range. Shearing may also be done with a sonifier or even by multiple passages through a 27-gauge hypodermic needle.[4]

Following shearing, dialyze the DNA at approximately 5° against 0.001× SSC (standard saline citrate: 0.15 M NaCl, 15 mM sodium citrate, pH 7.0) and 1 mM EDTA.[11] The preparation can then be taken to dryness in a freeze-dryer, redissolved at 1–5 μg/μl in distilled water, and frozen at −20° until needed.

Isotope Labeling

Most hybridization methods require radioisotope-labeled DNA. With the development of *in vitro* techniques, *in vivo* labeling is now seldom used, but the methodologies have been described by Johnson.[9] *In vitro* methods rely on the binding of ^{125}I to DNA or the incorporation of ^{3}H-, ^{32}P-, or ^{35}S-labeled nucleotides into the DNA strands by nick translation. Commerford[12] has described ^{125}I labeling. Nick-translation kits are available from isotope suppliers and represent the most expedient means for using this effective labeling technique, which is based on generating single-strand nicks with pancreatic DNase I followed by the incorporation of radiolabeled nucleotides with *Escherichia coli* DNA polymerase I.[9,13,14]

Determination of Guanine Plus Cytosine Contents

The percent guanine plus cytosine (G + C) in DNA represents an exclusionary descriptor for separating species. If the G + C contents of strains differ by more than 1.5% when determined by buoyant density or more than 2.5% when estimated from thermal melts, the strains can be expected to represent different species.[10] The determination of G + C contents is relatively rapid and, when the values are different, allows exclusion of strains from the more time-consuming DNA hybridization experiments.

In addition to the exclusionary value, it is necessary to know the percent G + C content to maximize DNA reassociation kinetics. Most faithful pairing of complementary nucleotide sequences occurs at T_m − 25°, that is, a temperature 25° below the midpoint of the thermal melt

[11] C. P. Kurtzman, M. J. Smiley, C. J. Johnson, L. J. Wickerham, and G. B. Fuson, *Int. J. Syst. Bacteriol.* **30**, 208 (1980).
[12] S. L. Commerford, *Biochemistry* **10**, 1993 (1971).
[13] R. B. Kelly, N. R. Cozzrelli, M. P. Deutscher, I. R. Lehman, and A. Kornberg, *J. Biol. Chem.* **245**, 39 (1970).
[14] T. Maniatis, A. Jeffrey, and D. G. Kleid, *Proc. Natl. Acad. Sci. U.S.A.* **72**, 1184 (1975).

curve.[2,9] The G + C contents of DNAs are generally determined by thermal melt or from buoyant density in cesium chloride gradients. Unsheared DNAs are used for these determinations. High-performance liquid chromatography (HPLC) of enzymatically hydrolyzed DNA is another method that may be effective.[15]

Thermal Melt

Application of heat to solutions of double-stranded DNA causes a disruption (melt) of hydrogen bonding, and the resulting strand separation can be detected as increased absorbance at 260 nm. A linear relationship exists between progressively higher melt temperatures and increasing G + C contents. Pure DNA shows approximately a 40% increase in A_{260} following complete strand separation. Another aspect is that as the molarity of the reaction buffer increases, melting temperature increases. As a consequence of equipment limitations, DNAs having a G + C content under 55% are generally melted in 1× SSC; those over 55% are placed in 0.1× SSC.

Necessary equipment includes a spectrophotometer with a 4- or 6-place thermally controlled cuvette holder, quartz cuvettes with Teflon stoppers, an automatic cuvette changer, and a recorder with offset to exclude the absorbance of the initial DNA concentration. Other arrangements of equipment can be made to work.

1. Based on a 4-place cuvette holder, make up the following four cuvettes: (1) blank to contain 1× SSC; (2) 25 μg/ml (A_{260} 0.5) of reference DNA, such as *E. coli* (G + C 51 mol%), in 1× SSC; (3) and (4) 25 μg/ml of the undetermined DNAs in 1× SSC.
2. After making the required equipment adjustments, raise the temperature to 50°, then begin a programmed temperature increase of approximately 0.25°/min until maximum hyperchromicity is reached.
3. Determine the T_m as follows. With a pencil and ruler, mark the beginning and concluding curves of each melt profile and determine the point equidistant between them. The temperature of the melt at this point is the T_m. Calculate G + C contents from Eqs. (1) and (2).[16] For 1× SSC,

$$\text{Mol\% G} + \text{C} = 2.44T_m - 169.00 \tag{1}$$

[15] J. Tamaoka and K. Komagata, *FEMS Microbiol. Lett.* **25**, 125 (1984).
[16] J. Marmur and P. Doty, *J. Mol. Biol.* **5**, 109 (1962).

For 0.1× SSC,

$$\text{Mol\% G} + \text{C} = 2.08T_{\text{m}} - 106.40 \tag{2}$$

Buoyant Density

When a cesium chloride solution is spun in an ultracentrifuge, a density gradient is formed. The position of DNA in the gradient is determined by its G + C content, which can be calculated from the relative position of a second DNA of known density. Determinations are generally made in an analytical ultracentrifuge, but a preparative ultracentrifuge will also serve for this purpose.

1. Make a stock cesium chloride solution in the following manner: Add 130 g CsCl to 75 ml of 10 mM Tris buffer, pH 8.5. Dissolve and treat for 20 min with 2 g activated charcoal to remove any material absorbing at 260 nm. Filter. Determine the exact concentration of CsCl from the refractive index. Solutions are usually approximately 1.87 g/ml, and about 405 μl is used in a total volume of 500 μl.
2. Combine the required amount of CsCl stock solution, 1 μg of undetermined DNA, 1 μg of *Micrococcus luteus* reference DNA (buoyant density 1.7311 g/ml),[6] and bring to a final volume of 500 μl with distilled water. Load centrifuge cells. Centrifugation is generally for 20 hr at 44,000 rpm.[17]
3. Determine distances of peak center points for unknown and reference DNAs and determine the G + C content using Eqs. (3) and (4).[17]

$$\rho = \rho_0 + 4.2\omega^2(r^2 - r_0^2) \times 10^{-10} \text{ g/cm}^3 \tag{3}$$

where ρ is the density of unknown DNA, ρ_0 the density of known DNA, ω the radians/sec (2π radians/revolution), r the distance of unknown from the center of rotation, and r_0 the distance of standard from the center of rotation.

$$\text{Mol\% G} + \text{C} = (\rho - 1.66)/0.098 \times 100 \tag{4}$$

Procedures for Measuring DNA–DNA Hybridization

Methods for measuring DNA relatedness fall into two general categories: (1) the free-solution technique, in which all of the reactants are solubilized, and (2) filter membrane technique, in which the DNA of one strain is immobilized on nitrocellulose or other filter materials and DNA

[17] C. L. Schildkraut, J. Marmur, and P. Doty, *J. Mol. Biol.* **4**, 430 (1962).

from the other strain is solubilized in the buffer surrounding the membrane. Each method has its strengths, and, when properly done, each provides the same measure of relatedness.[4,11] The standard error for the determinations is usually around 5%.

Free-Solution Hybridizations

Spectrophotometric Method. The spectrophotometric (optical) method is a convenient nonisotopic procedure that has the advantage of simultaneously providing data for estimates of genome size. The rationale for this method is based on the observation that DNA reassociation is a concentration-dependent, second-order reaction.[2] As a result, if 50 μg/ml of DNA reassociates at a certain rate, 25 μg/ml will take twice as long. Consequently, if a mixture of two DNAs reassociates at the same rate as an equivalent concentration of unmixed DNA, the organisms providing this DNA belong to the same species. If the reaction time of the mixture is the sum of that of the two unmixed DNAs, the organisms are different species. Because the genome size of microorganisms is relatively small, the midpoint of the reaction, which is often used as a reference, is reached within 30 min to about 12 hr, depending on reaction conditions and actual genome sizes.

For a typical determination, cuvettes contain 75 μg/ml of DNA (A_{260} 1.5) which, following melting, still maintains the absorbance in a linear range for most spectrophotometers. Reactions are carried out in SSC, and greater concentrations of SSC are used for larger genome sizes in order to accelerate the reassociation to a reasonable recording time.[4,5,18,19] For bacteria use 3× SSC, and for yeasts, molds, and actinomycetes use 5× SSC. Because the higher concentrations of SSC raise the melting temperature to over 100°, 20% (v/v) dimethyl sulfoxide (DMSO) is added to the reaction mixture to depress the melting temperature. For 5× SSC, 20% DMSO depresses the temperature by about 12°.[11] The appropriate incubation temperature can be determined experimentally from a melt with 25 μg/ml of DNA under the reaction conditions selected.

Formamide depresses melting temperatures, but it has high absorbance at 260 nm. Phosphate buffer, which is often used for DNA reassociations, should not be used with DMSO because of the possibility of a violent reaction at higher temperatures, at least in the presence of metals.[11]

Equipment for determinations includes a spectrophotometer with a 4- or 6-place thermally controlled cuvette holder, quartz cuvettes with Teflon

---

[18] R. J. Seidler, M. D. Knittel, and C. Brown, *Appl. Microbiol.* **29**, 819 (1975).
[19] J. De Ley, H. Cattoir, and A. Reynaerts, *Eur. J. Biochem.* **12**, 133 (1970).

stoppers, an automatic cuvette changer, and a recorder with offset to exclude the absorbance of the initial DNA concentration.

1. Based on a 4-place cuvette holder, use the following protocol. To two 1-ml cuvettes add 500 μl of 10× SSC and 200 μl DMSO. Each cuvette then receives 75 μg of sheared double-stranded DNA in 0.001× SSC/1 mM EDTA from one member of the pair to be reassociated. Distilled water is added to each cuvette to bring the volume to 1000 μl. Verify the absorbance of each to be 1.5 at 260 nm and adjust if necessary. Use an aqueous solution of 5× SSC and 20% DMSO as the blank. Remove 350 μl from each cuvette containing DNA and mix in another cuvette. The four cuvettes will now contain the following: (1) blank, (2) 75 μg/ml DNA from the first strain, (3) 75 μg/ml DNA from the second strain, and (4) 37.5 μg/ml DNA from each of the two strains (total 75 μg/ml). Transfer to thermal microcuvettes, cover the filling ports with one to several layers of Teflon tape (~0.09 ml thickness), and tightly insert stoppers. The remainder of the preparations from the first set of cuvettes can be frozen and used for a second determination.

2. Read A_{260} for each cuvette at the incubation temperature (usually 50°–60°), then raise the temperature to 90°–95° and hold for 10 min to denature the DNA. Cool to the reassociation temperature at 3°/min. Record the absorbance until each cuvette reaches $C_0 t_{0.5}$, that is, one-half the absorbance resulting from hyperchromicity as seen at the start of the $T_m - 25$° incubation. The relatively slow cooling rate allows reassociation of multicopy DNA and any contaminating mitochondrial DNA.[11]

3. At the conclusion of $C_0 t_{0.5}$, remelt the preparations at 90°–95°. If the hyperchromicity is greater than the initial melt, the cuvettes have leaked and the $C_0 t_{0.5}$ values will be overestimated.

4. Calculate percent relatedness using Eq. (5).[4] $C_0 t_0$ is the A_{260} at which $T_m - 25$° is first reached, not the maximum hyperchromicity at 90°–95°.

% Relatedness

$$= \{1 - [\text{obs } C_0 t_{0.5}^{\text{mix}} + (C_0 t_{0.5}^{100} - C_0 t_{0.5}^{0})]/C_0 t_{0.5}^{100}\} \times 100 \quad (5)$$

The term obs $C_0 t_{0.5}^{\text{mix}}$ is the observed $C_0 t_{0.5}$ of a renatured mixture, $C_0 t_{0.5}^{100}$ is the $C_0 t_{0.5}$ of the mixture expected if the two DNA molecules are identical in sequence, and $C_0 t_{0.5}^{0}$ is the $C_0 t_{0.5}$ of the mixture expected for no sequence similarity (complete additivity of the independently measured $C_0 t_{0.5}$ values). The term $C_0 t$ is used to denote the product of the DNA concentration and the time of incubation for the reaction and is expressed in moles of nucleotides times seconds per liter.[2] The midpoint of a reassociation is expressed as $C_o t_{0.5}$, and the time required to reach this point is a reflection of the genome size of an organism: the larger the genome, the fewer DNA copies at a given DNA concentration. Genome sizes are calculated as a proportionality of $C_o t_{0.5}$ values using a species of known genome size for reference. The $C_o t_{0.5}$ values must be corrected for the effect of G + C content.[4,20]

5. De Ley et al.[19] proposed Eq. (6) for calculating spectrophotometric reassociation data. Procedural aspects of the reassociations are the same as above.

$$D = \frac{4v'_m - (v'_A + v'_B)}{2(v'_A v'_B)^{1/2}} 100 \tag{6}$$

The initial straight line portion of each cuvette plot on the recorder is used in the calculation, and the reaction velocities (v'_A and v'_B for individual DNAs and v'_M for the mixture of the two DNAs) are treated as decreases in vertical distances (decreasing absorbance) over a fixed, arbitrary horizontal distance (time). This method is much quicker than $C_o t_{0.5}$ determinations and works well for bacteria, but results for fungi have been less reliable, perhaps because of the presence of a greater number of multicopy genes.

Hydroxylapatite Hybridization. Hydroxylapatite (HA) is a form of calcium phosphate that preferentially binds double-stranded DNA when in an appropriate concentration of phosphate buffer.[21,22] This specificity provides a means for separating renatured DNA duplexes from free-solution reassociation reactions. Brenner et al.[21] developed a batch procedure that allows handling of up to 10 samples at the same time, thus greatly increasing the utility of HA for DNA hybridizations. The procedure given here was derived from the work of Brenner et al.[21] and Price et al.[6]

1. Place in sealable 0.5-dram culture tubes a mixture of 200 μg of sheared, unlabeled target DNA and 0.2 μg of sheared, labeled probe DNA in a total volume of 500 μl of 280 mM phosphate buffer (PB).

2. Heat vials in boiling water for 10 min to denature the DNA, transfer to a water bath at $T_m - 25°$, and incubate for 20 hr. The T_m can be calculated from the data of Gruenwedel et al.[23] Higher G + C DNAs

[20] M. Gillis, J. De Ley, and M. De Cleene, *Eur. J. Biochem.* **12**, 143 (1970).
[21] D. J. Brenner, G. R. Fanning, A. V. Rake, and K. E. Johnson, *Anal. Biochem.* **28**, 447 (1969).
[22] G. Bernardi, M. Faures, G. Piperno, and P. P. Slonimski, *J. Mol. Biol.* **48**, 23 (1970).
[23] D. W. Gruenwedel, C. Hsu, and D. S. Lu, *Biopolymers* **10**, 47 (1971).

can first be heat-denatured in low concentrations of PB and then brought to 280 mM.

3. The samples may be frozen following incubation or processed immediately. Process by diluting each to yield 12 ml of 140 mM PB containing 0.2% sodium lauryl sulfate.

4. Add 2.4 g of freshly washed (in 140 mM PB) HA (BioGel HTP, Bio-Rad Laboratories, Richmond, CA) to each 12-ml sample and shake for 20 min at 60° at a rate to maintain a slurry. Incubation at 60° prevents adsorption of single-stranded DNA. Centrifuge to pellet the HA and remove the supernatant. Repeat with two additional washes as before and determine the counts in the supernatants by liquid scintillation. A clinical-style centrifuge used in a 60° incubator prevents temperature drops during processing. Elute double-stranded DNA from the HA with two 12-ml washes of 400 mM PB at 95°–100°. Determine the radioactivity in the supernatants. Calculate relatedness by dividing the amount of heterologous reassociation by the amount of homologous reassociation and multiplying by 100.

It is sometimes necessary to remove rapidly renaturing sequences from DNA samples before measuring the extent of relatedness, and this can be done on HA columns as described by Price et al.[6] Self-annealing of the radiolabeled probe DNA can be minimized if the ratio of target to probe is kept to at least 1000:1. It may also be useful to determine the amount of self-annealing to be certain that this does not bias the calculated relatedness. Finally, a simplified version of the HA procedure was described by Lachance[24] in which HA-containing minicolumns maintained under controlled temperatures are used to separate the duplexed DNAs.

S1 Nuclease Method. Sutton[25] and Crosa et al.[26] demonstrated that DNA hybridization mixtures could be freed of single-stranded DNA by hydrolysis with S1 endonuclease. Double-stranded hybrids are then removed from the reaction mix by precipitation with trichloroacetic acid (TCA) or by collection on Whatman (Clifton, NJ) DE-81 filters.[27]

1. Place in sealable 0.5-dram culture tubes a mixture of 200 μg of sheared, unlabeled target DNA and 0.2 μg of sheared, labeled probe DNA in a total volume of 500 μl of 2× SSC.

2. Heat-denature the DNAs by immersing the reaction tubes in boiling water for 10 min and then incubate at $T_m - 25°$ for 20 hr.

[24] M. A. Lachance, *Int. J. Syst. Bacteriol.* **30**, 433 (1980).
[25] W. D. Sutton, *Biochim. Biophys. Acta* **240**, 522 (1971).
[26] J. H. Crosa, D. J. Brenner, and S. Falkow, *J. Bacteriol.* **115**, 904 (1973).
[27] M. Popoff and C. Coynault, *Ann. Microbiol.* **131A**, 151 (1980).

3. Following reassociation, place a 100-μl sample from each hybridization vial into a polypropylene tube containing 1 ml of 50 mM sodium acetate–0.3 M NaCl–0.5 mM ZnCl$_2$ buffer (pH 4.6) and 50 μl of 0.5 mg/ml denatured sheared calf thymus DNA. Mix the contents and add 50 μl (1 unit/μl) of S1 nuclease. Mix and incubate for 1 h at 50°.

4. Following incubation, add an equal volume of 10% TCA to the reaction mixture. Mix and cool to about 5° for 1 hr. Collect the precipitates, which are the S1-resistant duplexed DNA fragments, on nitrocellulose filters, dry, and count for radioactivity.

5. As an alternative to TCA precipitation, transfer 100 μl of the S1-digested preparation to a dry 2.4-cm Whatman DE-81 filter disk in a vial. Wash the disk twice with 5-ml volumes of 5% Na$_2$HPO$_4$ and twice with 5-ml amounts of distilled water. Dry the filter and measure the radioactivity.

Membrane Filter Hybridization

Filter Preparation. The following protocol for immobilizing single-stranded DNA on nitrocellulose membrane filters is essentially that of Gillespie and Spiegelman.[28] Different filter sizes may be used, but it is more efficient to immobilize DNA on large filters which can be subdivided rather than to use many smaller filters. Generally, the amount of DNA immobilized is 25 μg/cm^2.

1. Dilute unsheared DNA stock in 0.1$\times$ SSC to 50 μg/ml, starting with an amount sufficient to give 25 μg/cm^2 for the filter size to be used.

2. Denature the DNA by heating in a boiling water bath for 10 min and then quickly cool by pouring into 9 volumes of ice-cold 6.7$\times$ SSC to give 5 μg of DNA/ml in 6$\times$ SSC. Maintain this solution on ice. The combination of low temperature and low DNA concentration keeps the DNA single-stranded.

3. Float the nitrocellulose filter on distilled water until evenly wet, and then place in a filter holder. Wash with cold 6$\times$ SSC using about 3 ml/cm^2 of filter and a flow rate of 0.2 ml/cm^2/min.

4. Pass the denatured DNA solution through the filter and then wash with 6$\times$ SSC. Flow rates and the second wash should be as before.

5. Dry the filter overnight at room temperature and then for 2 hr at 80° in a vacuum oven. Label the edge of the filter with a pencil, place between the membrane filter shipping papers, and store in a desiccator at room temperature until needed.

[28] D. Gillespie and S. Spiegelman, *J. Mol. Biol.* **12,** 829 (1965).

Hybridization Procedures. Filters are placed in small leakproof vials, preincubated with a medium to block all adsorptive sites not occupied by DNA, then placed within reaction vials containing buffer and the labeled probe. On completion of the incubation, the filters are washed and dried, and the radioactivity is measured.

1. Cut individual filters from the master membrane filter using a paper punch or other device. The filter sizes commonly used carry from 5 to 25 μg of DNA.
2. Preincubate the filters in Denhardt's[29] medium (0.02% bovine serum albumin, 2% polyvinylpyrrolidone, 0.02% Ficoll 400, in 2× SSC) for 30 min to 2 hr at the incubation temperature to be used for hybridization. Agitate the filters occasionally during preincubation.
3. Blot the membranes on absorbent paper and place in reaction vials.
4. To each reaction vial add 1 μg of labeled [2000–10,000 counts/min (cpm)], single-stranded DNA probe and just enough 2× SSC (e.g., 200 μl) to constantly wet the filter. Incubate at $T_m - 25°$ for 12–20 hr.
5. After incubation, wash each filter twice with 2–3 ml of 2× SSC at the incubation temperature; blot and allow to dry.
6. Place dried filters in scintillation vials and determine the radioactivity. The counts from heterologous DNA filters are divided by counts from homologous DNA filters and the ratio multiplied by 100 to get percent relatedness. Because of variability among filters, it is generally necessary to have 4–6 replicates.

One of the major problems with DNA hybridizations on nitrocellulose filters is that DNAs tend to elute from the filters. The higher incubation temperatures required for high G + C DNAs accentuate the problem. Formamide is often used to lower incubation temperatures in filter hybridization experiments. The T_m of DNA is lowered by 0.6° for each percent of formamide added to buffers that contain 0.035 to 0.88 M NaCl.[30]

Competition Reactions

Competition reactions are carried out as outlined above except that the reaction mixture contains an excess of unlabeled, sheared, single-stranded DNA that competes with the labeled probe for hybridization sites on the filter-bound DNA. If the supplemental heterologous DNA shows the same interference to probe binding as is found in vials containing excess homologous DNA, the two DNAs are considered homologous. If more probe

[29] D. T. Denhardt, *Biochem. Biophys. Res. Commun.* **23,** 641 (1966).
[30] J. R. Hutton, *Nucleic Acids Res.* **4,** 3537 (1977).

binds in the presence of heterologous DNA, the extent of divergence between the two DNAs can be calculated from a ratio of the scintillation counts. Johnson[9] has provided additional details of this procedure. The strength of competition reactions for DNA hybridization studies is that many different DNAs can be compared with a reference species, yet only the probe and filters need to be made with DNA from the reference species.

Thermal Stability of DNA Hybrids

An estimate of the extent of base-pair mismatch between hybridized DNA strands can be made by measuring the thermal stability of the reassociated duplexes. Place hybridized filters in fresh incubation buffer and heat at 5° increments to 90°. At each 5° increase, remove a small sample of the buffer and count. Controls are reassociated homologous DNAs. The amount of nucleotide mispairing has been estimated at 1% for each degree the T_m of the heterologous duplex is lowered (ΔT_m).[2,31]

Nylon Membrane Filters

Various formulations of nylon membrane filters have appeared on the market in recent years.[32] The filter material is less brittle than nitrocellulose, and the stable linkages formed with DNA allow stripping off probes by heat while still retaining the target DNA on the filter, thus allowing reuse. The membranes have proved particularly effective for Southern blot hybridizations and are quite amenable to use in slot-blot and dot-blot devices, which allow application of up to 96 DNAs on a single filter. The membranes can be incubated in thermostable plastic pouches that require only a small volume of buffer. However, quantitation of DNA hybridizations on these filters has been no better than for nitrocellulose (C. P. Kurtzman, unpublished, 1987).

Interpretation of DNA Hybridization Data

DNA reassociation data require interpretation in the context of the species concept. Dobzhansky[3] has championed the idea that species can be described in terms of genetics and that, among sexually reproducing and outbreeding organisms, species can be defined as Mendelian populations or arrays of populations that are reproductively isolated from other population arrays. This concept seems apparent for mammals, but its application to microorganisms is less straightforward because not all are known to have sexual cycles and some that do show little or no outbreeding. Further,

[31] D. J. Brenner, *Int. J. Syst. Bacteriol.* **23**, 298 (1973).
[32] J. L. McInnes, S. Dalton, P. D. Vize, and A. J. Robins, *Bio/Technology* **5**, 269 (1987).

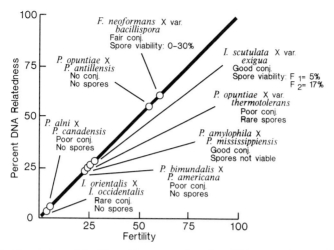

FIG. 1. Correlation of fertility with extent of nuclear DNA complementarity among heterothallic yeasts showing low to intermediate DNA relatedness. Mating responses parallel measurable DNA reassociation. These and other data suggest that strains showing 70% or greater DNA relatedness will, in most cases, represent members of the same biological species.[10,37] P., Pichia; I., Issatchenkia; F., Filobasidiella.

species formation does not usually leave a clear-cut separation between groups comprising the new species, and genetically intermediate populations may initially survive. Consequently, in order to use DNA reassociation data to define species, there must be some concept of the extent of DNA divergence to be found among members of a microbial species.

For various groups of eubacteria, correlation of DNA relatedness with phenotypic clustering has suggested that strains showing 70% or greater nucleotide similarity represent the same species.[9,31,33,34] Similar conclusions have been drawn for defining species of actinomycetes.[35,36]

Yeasts represent the first microbial group in which DNA relatedness has been correlated with sexual fertility. Some of these comparisons are presented in Fig. 1 as summarized from Kurtzman[37] and Kurtzman and Phaff.[10] What is apparent from these data (Fig. 1) is that mating responses can occur among strains showing as little as 10% DNA relatedness, even

[33] L. K. Nakamura, Int. J. Syst. Bacteriol. 32, 43 (1982).

[34] L. G. Wayne, D. J. Brenner, R. R. Colwell, P. A. D. Grimont, O. Kandler, M. I. Krichevsky, L. H. Moore, W. E. C. Moore, R. G. E. Murray, E. Stackebrandt, M. P. Starr, and H. G. Truper, Int. J. Syst. Bacteriol. 37, 463 (1987).

[35] M. Modarski, M. Goodfellow, S. T. Williams, and P. H. A. Sneath, in "Biological, Biochemical, and Biomedical Aspects of Actinomycetes" (G. Szabo, S. Biro, and M. Goodfellow, eds.), p. 517. Akademiai Kaido, Budapest, 1986.

[36] D. P. Labeda and A. J. Lyons, Syst. Appl. Microbiol. 14, 158 (1991).

[37] C. P. Kurtzman, Stud. Mycol. 30, 459 (1987).

though no fertile progeny result. In general, strains showing 70% or greater DNA relatedness are considered to be members of the same species because, if crossed, they would be expected to form a reasonable number of fertile progeny. Isolates showing approximately 50–70% DNA relatedness are considered taxonomic varieties unless there is genetic evidence that crosses are completely infertile. One exception concerns species of *Saccharomyces*. *Saccharomyces cerevisiae* and *S. bayanus* exhibit only about 10% DNA relatedness (background), but *S. pastorianus* shows 60–70% complementarity with each.[38] It appears that *S. pastorianus* is a partial amphidiploid, which was formed by a chance hybridization between *S. cerevisiae* and *S. bayanus,* and that it represents a biologically separate species despite the high relatedness shown between it and its putative progenitors.

Few data are available correlating fertility and DNA complementarity among the filamentous fungi. Preliminary work with species of *Fusarium* suggests that strains showing 70% or greater DNA relatedness are conspecific.[39,40] Comparisons among microbial groups, other than those discussed, remain to be done.

From the preceding studies, it appears that, for a majority of bacteria and yeasts, strains showing greater than 70% DNA relatedness represent members of the same biological species. This may also apply to filamentous fungi and other microbial groups, but there are few comparisons. Despite a generally good correlation between extent of DNA duplex formation and fertility, the exceptions noted make it prudent to view measurements of DNA complementarity as only predictive of a certain amount of genetic relatedness. Finally, all studies indicate that, among microorganisms, DNA relatedness resolves only to the genetic distance of closely related species and not beyond. For resolution of greater genetic distances, comparisons of conserved genes, such as those coding for ribosomal RNA, need to be made.[41,42]

Acknowledgments

I thank L. K. Nakamura and S. W. Peterson for helpful discussions. The mention of firm names or trade products does not imply that they are endorsed or recommended by the U.S. Department of Agriculture over other firms or similar products not mentioned.

[38] A. Vaughan Martini and C. P. Kurtzman, *Int. J. Syst. Bacteriol.* **35,** 508 (1985).
[39] J. J. Ellis, *Mycologia* **81,** 307 (1989).
[40] S. W. Peterson and A. Logrieco, *Mycologia* **83,** 397 (1991).
[41] C. R. Woese, *Microbiol. Rev.* **51,** 221 (1987).
[42] S. W. Peterson and C. P. Kurtzman, *Syst. Appl. Microbiol.* **14,** 124 (1991).

[25] Direct Ribosomal RNA Sequencing for Phylogenetic Studies

By JEAN-PIERRE BACHELLERIE and LIANG-HU QU

Principle of Method

Primer extension with reverse transcriptase[1] in the presence of chain-terminator dideoxynucleotides[2] is a straightforward approach for direct RNA sequencing. A variation[3] of this method largely extends its usefulness and range of application, particularly for systematic studies in the field of molecular evolution. Selection of the primer sequence within a highly conserved region of the RNA allows for direct RNA sequencing on numerous organisms with the same primer. In principle, this sequencing method can be applied to any molecular species of RNA, provided a strongly conserved region can be identified in its sequence. Obviously in the case of least abundant RNAs, a prior enrichment has to be performed (e.g., through hybrid selection).[4] Ribosomal RNAs (rRNAs) constitute choice targets for this sequencing approach. Their high relative abundance in cellular RNA permits direct determinations on total unfractionated RNA,[3] and their complex mosaic pattern of variation[5,6] provides the basis for refined evaluations of phylogenetic distances. Universal primers for rRNA sequencing can be selected within a set of almost perfectly conserved sequence motifs scattered along both the small and large subunit molecules. Each provides a starting point for the derivation of the sequence of about 300–350 nucleotides for any species.[3] Although most of the length of the rRNA molecule can be sequenced in this way, partial determinations focused on well-selected regions may suffice for reliable distance evaluations.[7]

This approach has been extensively applied to small subunit rRNA,[8] and sequences of synthetic oligodeoxynucleotides that can serve as universal primers for this molecule have been reported previously.[9] Owing to its

[1] D. Zimmern and P. Kaesberg, *Proc. Natl. Acad. Sci. U.S.A.* **75,** 4257 (1978).

[2] F. Sanger, S. Nicklen, and A. R. Coulson, *Proc. Natl. Acad. Sci. U.S.A.* **74,** 5463 (1977).

[3] L. H. Qu, B. Michot, and J. P. Bachellerie, *Nucleic Acids Res.* **11,** 5903 (1983).

[4] R. Jagus, this series, Vol. 152, p. 567.

[5] M. Salim and B. E. H. Maden, *Nature (London)* **291,** 205 (1981).

[6] N. Hassouna, B. Michot, and J. P. Bachellerie, *Nucleic Acids Res.* **12,** 3563 (1984).

[7] L. H. Qu, M. Nicoloso, and J. P. Bachellerie, *J. Mol. Evol.* **28,** 113 (1988).

[8] D. J. Lane, B. Pace, G. J. Olsen, D. A. Stahl, M. L. Sogin, and N. R. Pace, *Proc. Natl. Acad. Sci. U.S.A.* **82,** 6955 (1985).

[9] D. J. Lane, K. G. Field, G. J. Olsen, and N. R. Pace, this series, Vol. 167, p. 138.

larger size and the presence of some extended highly variable domains,[6] the large subunit rRNA molecule potentially represents a more sensitive indicator with a wider range of applications, even if its utilization may be rendered somewhat more complex by the substantial degree of diversification of secondary structure during evolution.[10] In addition to rapidity (sequence information for several organisms can be generated within 3 days), direct RNA sequencing has important advantages. The method ensures derivation of the actual consensus sequence for functional genes without the limitations associated with cloning of a single gene copy. It may also allow the detection of occasional sequence polymorphisms, provided they concern large subsets of the gene family.[11] Because sequencing band patterns generated from the same primer can be examined in parallel for different organisms, the procedure is particularly well-suited for systematic comparative analyses. Errors arising from band compression effects in measurements of nucleotide differences between species are reduced to a minimum.

The susceptibility of RNA to hydrolysis does not usually pose a serious problem, and sequences can frequently be derived from RNA samples in which only a minor fraction of rRNA molecules have retained their integrity. Nevertheless, a few nucleotide positions may occasionally remain ambiguous, either because of the frequent occurrence of nicks at some particular sites in the RNA molecule or because of reverse transcriptase pausing at very stable secondary structures in the template. As discussed below, such problems can often be solved by adjusting the conditions of the reverse transcriptase reaction or by introducing a further purification of the RNA template. Difficulties arising from prematurely terminated cDNAs can be eliminated when using the dye-labeled dideoxynucleotides introduced for automated nucleic acid sequencing (Dye-Deoxy terminators, Applied Biosystems, Foster City, CA). The parasitic bands in the sequencing pattern are no longer seen using this procedure.

It must be stressed that in this approach sequence determinations can still be carried out on preparations heavily contaminated by RNAs from different origins, provided the primer is selected accordingly. In such cases universal primers have to be replaced by more specific priming sequences. The presence of several phylogenetic group-specific motifs[10,12] along the sequences of both small and large subunit rRNAs allows for the selection of a variety of discriminating primers, chosen according to the particular problem and nature of the contaminating species.

[10] B. Michot and J. P. Bachellerie, *Eur. J. Biochem.* **188,** 219 (1990), and references therein.
[11] L. H. Qu, M. Nicoloso, and J. P. Bachellerie, *Nucleic Acids Res.* **19,** 1015 (1991).
[12] E. F. DeLong, G. S. Wickham, and N. R. Pace, *Science* **243,** 1360 (1989).

Finally this sequencing method is not restricted to species for which abundant biological material is available: 2–3 μg of unfractionated cellular RNA is usually sufficient to perform a reliable sequence determination from a given primer.[13] In the protocol described below, which essentially applies to rRNA, radioactive cDNAs are generated by utilization of a 5′ end-labeled primer rather than by incorporation of an α-labeled deoxynucleoside triphosphate (dNTP), in order to eliminate potential problems arising from self-primed reverse transcriptions.

Materials

Oligonucleotide primers are synthesized with an Applied Biosystems 380A synthesizer by the phosphoramidite method

T4 polynucleotide kinase (10 U/μl) is obtained from Amersham (Amersham, UK)

Avian myeloblastosis virus (AMV) reverse transcriptase (AMV RT-XL; 17 U/μl) is purchased from Life Sciences Inc. (St. Petersburg, FL); the enzyme is aliquoted and stored at −20°

Nucleotides: 2′-deoxynucleoside triphosphates (dNTPs) are purchased from Sigma (St. Louis, MO), and the 2′,3′-dideoxynucleoside triphosphates (ddNTPs) from Boehringer-Mannheim Biochemicals (Mannheim, Germany)

5′-[γ-^{32}P]ATP (3000 Ci/mmol; 10 μCi/μl) and 5′-[γ-^{35}S]ATP (1000 Ci/mmol; 10 μCi/μl) are obtained from Amersham

5 M Ammonium acetate, pH 5.5

3 M Sodium acetate, pH 5.0

1 M Acetic acid

5 M KOH

1 mg/ml Yeast tRNA (Sigma, NR 9001)

RE buffer: 3 M LiCl, 6 M urea, 10 mM sodium acetate (pH 5.0), 0.1% sodium dodecyl sulfate (SDS)

4 M LiCl, 8 M urea

TE buffer: 10 mM Tris-HCl, pH 7.6, 1 mM NaEDTA, pH 7.6

10× PL buffer: 500 mM Tris-HCl, pH 7.6, 100 mM MgCl$_2$, 50 mM dithiothreitol (DTT), 1 mM spermidine, 1 mM NaEDTA, pH 7.6

5× RT buffer: 250 mM Tris-HCl, pH 8.3, 200 mM KCl, 30 mM MgCl$_2$

TEN 1 buffer: 50 mM Tris-HCl, pH 7.5, 150 mM NaCl, 5 mM NaEDTA

TEN 2 buffer: 50 mM Tris-HCl, pH 7.5, 1.5 M NaCl, 5 mM NaEDTA

---

[13] L. H. Qu, N. Hardman, L. Gill, L. Chappell, M. Nicoloso, and J. P. Bachellerie, *Mol. Biochem. Parasitol.* **20**, 93 (1986).

Phenol–chloroform (1 : 1, v/v); Phenol (redistilled) is equilibrated with TE buffer

$2\times$ dNTPs, ddNTP mixes are prepared from 10 mM stock solutions of each dNTP (in 10 mM Tris-HCl, pH 7.5) and from 1 mM stock solutions of each ddNTP (in 10 mM Tris-HCl, pH 7.5)

$2\times$ dNTPs, ddA mix: 10 μM ddATP, 50 μM dATP, 200 μM dCTP, 200 μM dGTP, 200 μM dTTP

$2\times$ dNTPs, ddG mix: 10 μM ddGTP, 50 μM dGTP, 200 μM dATP, 200 μM dCTP, 200 μM dTTP

$2\times$ dNTPs, ddC mix: 5 μM ddCTP, 50 μM dCTP, 200 μM dATP, 200 μM dGTP, 200 μM dTTP

$2\times$ dNTPs, ddT mix: 10 μM ddTTP, 50 μM dTTP, 200 μM dATP, 200 μM dGTP, 200 μM dCTP

Another set of $2\times$ dNTPs, ddNTP mixes in which ddNTPs concentrations are doubled compared to the standard values indicated above is also used for improved sequence determinations in the vicinity of the primer 3′ end.

$2\times$ dNTPs mix: 200 μM dATP, 200 μM dGTP, 200 μM dCTP, 200 μM dTTP

Gel loading mix: 80% (v/v) deionized formamide, 10 mM NaOH, 1 mM NaEDTA, 0.1% (w/v) xylene cyanol, 0.1% (w/v) bromphenol blue

$10\times$ Electrophoresis buffer: 0.5 M Tris base, 0.5 M H$_3$BO$_3$, 10 mM NaEDTA, pH 8.3

Polyacrylamide gels: 6 or 8% acrylamide (N,N'-methylenebisacrylamide: acrylamide, 1 : 20) gels containing 7 M urea are run in 1$\times$ electrophoresis buffer (gel dimensions are 38 $\times$ 23 $\times$ 0.04 cm)

Gel fix: 20% ethanol, 10% acetic acid in water

Protocols

Purification of RNA Template

In most cases any of the widespread techniques[14–16] for isolating RNA from intact cells, tissues, or subcellular fractions can be used successfully.

---

[14] D. M. Wallace, this series, Vol. 152, p. 33.
[15] S. L. Berger, this series, Vol. 152, p. 215.
[16] R. J. MacDonald, G. H. Swift, A. E. Przybyla, and J. M. Chirgwin, this series, Vol. 152, p. 219.

Mechanical homogenization steps may have to be introduced for improving cell disruption for bacteria[9] or fungi.[7,17] The final choice has to be made on an empirical basis, according to the yield of the preparation and the degree of preservation of high molecular weight rRNA from nucleolytic activity. However, the procedure[18,19] relying on the utilization of 3 M LiCl/6 M urea has proved efficient and convenient for a variety of eukaryotic cell types.

In a typical case for the analysis of an eukaryotic sample, 0.01 – 1 g of tissue or packed cells is lysed at 4° in 10 ml RE buffer in the presence of 200 μg/ml heparin. The mixture is homogenized with an Ultra-Turrax homogenizer (Janke and Kunkel, Staufen, Germany) until the viscosity is drastically reduced (usually less than 1 min) and kept overnight at 4°. RNA is then recovered by centrifugation (15,000 g, 20 min) and the pellet washed with 10 ml of cold 4 M LiCl, 8 M urea before recentrifugation (15,000 g, 20 min). After redissolution in water, the RNA solution is adjusted to 0.1 M sodium acetate, pH 5.0, and 0.1% SDS before two phenol – chloroform extractions, several ether extractions, and ethanol precipitation. After ethanol washings, the RNA is dissolved in TE buffer and its concentration determined spectrophotometrically. The RNA may then be directly utilized for sequence determination (it can also be stored for months at − 20° as an ethanol precipitate) or submitted to additional extraction steps, depending on the indications of the UV absorption spectra. Analysis of a small aliquot of the RNA preparation (1 – 5 μg) by agarose gel electrophoresis[20] followed by ethidium bromide staining is recommended. In some cases, the introduction of a sucrose gradient purification step[3] may be necessary, to ensure the utilization of a high molecular weight rRNA template. This purification step may also be required in the case of plant RNAs in order to get rid of the bulk of shorter chloroplast rRNA molecules, which may generate a parasitic sequencing band pattern with some universal primers.

Selection and Preparation of Primer

Synthetic deoxyribonucleotides in the size range 20–25 nucleotides have proved to be effective primers and provide a high degree of specificity.

[17] J. Guadet, J. Julien, J. F. Lafay, and Y. Brygoo, *Mol. Biol. Evol.* **6**, 227 (1989).
[18] C. Auffray and F. Rougeon, *Eur. J. Biochem.* **107**, 303 (1980).
[19] M. Le Meur, N. Glanville, J. L. Mandel, P. Gerlinger, R. Palmiter, and P. Chambon, *Cell (Cambridge, Mass.)* **23**, 561 (1981).
[20] R. C. Ogden and D. A. Adams, this series, Vol. 152, p. 61.

TABLE I
PRIMERS FOR DIRECT SEQUENCING OF LARGE SUBUNIT RIBOSOMAL RNAs[a]

Primer	Specificity	Sequence	Hybridization site
P1	EK	TTTMACTCTCTCTTCAAAGTNCTTTTC	*Mm*, 382; *Sc*, 370
	EB, AB	TTTCACTCCCYYTTTGGGGTNCTTTTC	*Ec*, 476; *Mv*, 481
P2	EK	TCCTTGGTCCGTGTTTCAAGACGGG	*Mm*, 1126; *Sc*, 635
	EB	TCGCNGGCTCATTCTTCAANAGG	*Ec*, 562
	AB	TCCCTGGCTCGTGTTTCAAGACG	*Mv*, 615
P3	U	CTTCGGRGRGAACCAGCTG	*Mm*, 1464; *Sc*, 934; *Ec*, 803; *Mv*, 886
P4	EK	GCTTACCAAAAGTGGCCCAC	*Mm*, 1652; *Sc*, 1112
P5	U	GTGAGYTRTTACGCAYTYYTT	*Mm*, 1799; *Sc*, 1258; *Ec*, 1084; *Mv*, 1171
P6	EK	TTGCTACTRCYRCCRAGATCT	*Mm*, 2101; *Sc*, 1432
	EB	TCGTTACTYANRTCRGCATTC	*Ec*, 1252
	AB	GCTGYTACTRYNRCCRGGAT	*Mv*, 1335
P7	EK	ACCTGMTGCGGTTATNGGTACG	*Mm*, 2532; *Sc*, 1830
	EB, AB	ACCTGTGTCGGWTYNSGGTACG	*Ec*, 1600; *Mv*, 1693
P8	EK	ACCGCCCCAGNCAAACTCCCC	*Mm*, 3833; *Sc*, 2604
	EB, AB	ACCGCCCCAGYCAAACTGCCC	*Ec*, 2237; *Mv*, 2293

[a] Eight sets of priming sequences (P1–P8) are among the most generally useful. Some are of universal utilization (U), whereas others are more specific for one of the primary kingdoms (EK, eukaryotes; EB, eubacteria; AB, archaebacteria). Locations of the hybridization site along the sequence are given for some reference species: eukaryotes, *Mus musculus (Mm)* and *Saccharomyces cerevisiae (Sc)*; eubacteria, *Escherichia coli (Ec)*; archaebacteria, *Methanococcus vanniellii (Mv)*. The coordinates correspond to the large rRNA position complementary to the 3′-terminal nucleotide of the primer. Sequences of primers are given in the 5′ → 3′ direction. Nucleotide M = C or A; N = A, G, C, or T; R = A or G; S = C or G; W = A or T; Y = C or T.

A list of primer sequences that may be of universal applicability for the large subunit rRNA molecule is given in Table I. Primers having a more restricted range of utilization may also be selected, particularly when analyzing portions of the rRNA molecule that have already been subjected to an intensive comparative analysis. It is noteworthy that the very large collection of partial sequences[7,21,22] (more than 100) for the 5′ terminal

[21] R. Perasso, A. Baroin, L. H. Qu, J. P. Bachellerie, and A. Adoutte, *Nature (London)* **339**, 142 (1989).
[22] R. Christen, A. Ratto, A. Baroin, R. Perasso, K. G. Grell, and A. Adoutte, *EMBO J.* **10**, 499 (1991).

region of eukaryotic 28 S rRNAs makes this region a particularly well-calibrated index for evaluations of phylogenetic relationships.

The deprotected oligonucleotide synthetic product is precipitated twice in ethanol, and a 0.5 $\mu g/\mu l$ solution in water is prepared (concentration is measured from a UV absorption spectrum). One microliter of the oligodeoxynucleotide solution is mixed with 2 μl of 10× PL, 5 μl of [γ-^{32}P]ATP (or [γ-^{35}S]ATP), 1 μl of T4 polynucleotide kinase, and 11 μl water. After incubation at 37° (for 45 min with [γ-^{32}P]ATP or for 2 hr with [γ-^{35}S]ATP), the reaction is stopped by the addition of 150 μl of 5 M ammonium acetate, pH 5.5, and 130 μl water and 10 μl of the yeast tRNA solution are added to the mixture before precipitation with 1 ml ethanol. After chilling at −70° for at least 15 min, the precipitate is collected by centrifugation (12,000 g, 15 min), redissolved, and submitted to two additional cycles of precipitation–redissolution. Finally, the precipitate is redissolved in 20 μl of gel loading mix and the mixture analyzed on a 8% acrylamide–7 M urea slab gel in 1× electrophoresis buffer, until the bromphenol blue has reached the middle of the gel.

After the gel is exposed to X-ray film for 10 min, the band containing the labeled primer is excised and the oligodeoxynucleotide eluted by 1–2 ml TE buffer (for 1 hr at room temperature with frequent vortexing). After removal of the supernatant (5 min in a microcentrifuge), gel pieces are resuspended in 0.5 ml TE buffer. After this reextraction, the second supernatant is pooled with the first, and the labeled oligodeoxynucleotide is recovered by ethanol precipitation and further purified by DEAE-cellulose (DE-52) chromatography. DEAE-cellulose (DE-52) is saturated with *Escherichia coli* DNA in TEN 1 buffer and submitted to several washings in TEN 2 buffer before utilization. The labeled oligonucleotide in TEN 1 buffer is fixed to the column, and, after several washings in TEN 1 buffer, elution is achieved in TEN 2 buffer. After ethanol precipitation, the labeled oligonucleotide is redissolved in TE buffer and reprecipitated with ethanol. Finally the nucleic acid pellet is dissolved in 50 μl TE buffer and can be stored in small aliquots [typically 5 × 10⁵ to 10⁶ counts/min (cpm) per tube] at −20° for up to 3 weeks.

Sequencing Reactions

For sequencing, in parallel to the four base-specific reactions (each containing one of the dideoxy chain terminators), a fifth reaction is always performed in the presence of only the four dNTPs. It detects sites of premature termination of cDNAs along the RNA template, resulting from either nicks in rRNA or from secondary structures blocking the progression of reverse transcriptase. Accordingly, this reaction serves as an index when attempting to solve problems created by such nonspecific bands in the sequencing patterns by modifying some parameters of the standard procedure.

For a set of five reactions, 5 μl of the labeled primer solution in TE buffer (typically $10^5 - 10^6$ cpm) is mixed with 5 μl of RNA template solution (0.5 – 15 μg in water) and 10 μl water. The solution is heated at 65° for 5 min before addition of 5 μl of 5× RT buffer. The primer–RNA mix is then immediately allowed to cool to room temperature for at least 10 min.

Reverse transcription is performed at 37° for 30 min after adding 4–8 units of enzyme to 10 μl of a solution made up by the addition of 5 μl of 2× dNTPs, ddNTP to 5 μl of primer–RNA mix. The reaction is stopped by the addition of 90 μl of 0.33 M KOH, and hydrolysis of the RNA template is achieved by an overnight incubation at 37°. Complementary DNAs are recovered by ethanol precipitation after the addition of 200 μl of 0.3 M acetic acid in the presence of 0.5 μg yeast tRNA carrier. After redissolution in 6 μl gel loading mix, 1.5-μl aliquots of the cDNA solutions are analyzed on 6% acrylamide–7 M urea gels run at high temperature (50°). For each set of reactions, a minimum of three migration times is used. Typically, the shortest run is stopped when the bromphenol blue reaches the bottom of the gel, and the two other runs are usually 2 and 3 times longer, respectively. After electrophoresis, gels are dried under vacuum after fixation for 30 min. in gel fix solution and are autoradiographed with XM X-ray films (3M, Cergy-Pontoise, France) using an intensifying screen for 1–15 days. To reduce problems posed by secondary structures in an RNA template, the temperature of the reverse transcription may be increased up to 50°. Adjustments in the ddNTP concentrations may also be attempted in order to solve ambiguities in sequencing band patterns due to prematurely terminated cDNAs. This is done by increasing the relative proportion of dideoxynucleotide-terminated molecules at the ambiguous positions.

Analysis of Sequence Data

Parallel analysis on the same sequencing gel of several sets of reactions for different species allows for a reliable assessment of actual sequence differences. Inclusion in the panel of sequenced RNAs of a control sample, corresponding to a species for which a sequence has already been determined on a rDNA clone, may also help to assess potential problems arising from band compression effects.

It is noteworthy that the accuracy of a major fraction of sequence differences can be tested by an independent criterion, namely, by reference to the highly conserved secondary structure of the molecule.[23] A nucleotide change in a base-paired segment must be generally confirmed by the detection of a compensatory base change at the opposite position of the helix.

Procedures involved in the analysis of rRNA sequence data for inferring phylogenetic relationships have been extensively discussed.[24,25] A critical prerequisite is that sequence alignments be carefully examined by reference to rRNA secondary structure.[23,26]

[23] B. Michot, N. Hassouna, and J. P. Bachellerie, *Nucleic Acids Res.* **12**, 4259 (1984).
[24] G. J. Olsen, this series, Vol. 164, p. 793.
[25] C. R. Woese, *Microbiol. Rev.* **51**, 221 (1987).
[26] R. R. Gutell and G. E. Fox, *Nucleic Acids Res.* **16**, r175 (1988).

[26] *In Vivo* Analysis of Plant 18 S Ribosomal RNA Structure

By Julie F. Senecoff and Richard B. Meagher

Introduction

The ribosomal RNAs are involved in multiple molecular interactions with both protein factors and other RNAs in addition to having an extensive intramolecular secondary structure of their own. These interactions involve not only particular sequence domains but also recognition of secondary and tertiary structural features of the RNA molecule.[1–12] Un-

[1] D. A. Brow and H. F. Noller, *J. Mol. Biol.* **163**, 27 (1983).
[2] N. M. Chapman and H. F. Noller, *J. Mol. Biol.* **109**, 131 (1977).
[3] A. E. Dahlberg, *Cell (Cambridge, Mass.)* **57**, 525 (1989).
[4] D. Moazed and H. F. Noller, *Cell (Cambridge, Mass.)* **47**, 985 (1986).
[5] D. Moazed, B. J. Van Stolk, S. Douthwaite, and H. F. Noller, *J. Mol. Biol.* **191**, 483 (1986).
[6] M. Mougel, F. Eyermann, E. Westhof, P. Romby, A. Expert-Bezancon, J.-P. Ebel, B. Ehresmann, and C. Ehresmann, *J. Mol. Biol.* **198**, 91 (1987).
[7] H. F. Noller, *Biochemistry* **13**, 4694 (1974).
[8] H. F. Noller, *Annu. Rev. Biochem.* **53**, 119 (1984).
[9] D. A. Peattie, S. Douthwaite, R. A. Garrett, and H. F. Noller, *Proc. Natl. Acad. Sci. U.S.A.* **78**, 7331 (1981).
[10] D. A. Peattie and W. Herr, *Proc. Natl. Acad. Sci. U.S.A.* **78**, 2273 (1981).
[11] J. B. Prince, B. H. Taylor, D. L. Thurlow, J. Ofengand, and R. A. Zimmerman, *Proc. Natl. Acad. Sci. U.S.A.* **79**, 5450 (1982).
[12] S. Stern, T. Powers, L.-M. Changchien, and H. F. Noller, *Science* **244**, 783 (1989).

derstanding the functional evolution of rRNA sequence as a consequence of the conservation of distinctive structures with or without sequence conservation has presented a unique problem to molecular biology.

DNA, RNA, and protein sequence comparisons have been used to infer the phylogenetic relationships and time of divergence of a variety of species. The small subunit ribosomal RNA has been most useful in these studies, allowing wide-ranging comparisons at the species level even among eukaryotic and prokaryotic kingdoms.[13] However, for ribosomal RNA, evolutionary differences may be more directly reflected in alteration or conservation of specific structures than in general RNA sequence changes. To determine the structural features of rRNA in its native state, a number of *in vitro* chemical modification methods have been developed.[14-18] We have extended these procedures to the *in vivo* chemical modification of rRNA within the leaves of higher plants. This rapid procedure requires only bulk purification of RNA from tissues following modification and avoids more time-consuming procedures for isolation and purification of ribosomes and ribosomal subunits. Modified and control rRNA is then examined by primer extension and sequencing. Using this method, a partial comparison of 18 S rRNA sequence and structure from four plant species has been performed. Experiments utilizing this protocol will potentially yield information regarding structural changes that occur as a consequence of sequence diversity across species. Thus, a more complete determination of relatedness can be gained using both sequence and structural information. This method should be useful in the analysis of native structures of other RNA molecules.

Principle of Method: Technology of Dimethyl Sulfate Modification

Dimethyl sulfate (DMS) has been used extensively as a reagent for the chemical sequencing of DNA[19] and in the analysis of protein–DNA

[13] J. A. Lake, *Cell (Cambridge, Mass.)* **33**, 318 (1983).
[14] T. R. Cech, N. K. Tanner, I. Tinoco, B. R. Weir, M. Zucker, and P. S. Perlman, *Proc. Natl. Acad. Sci. U.S.A.* **80**, 3903 (1983).
[15] C. Ehresmann, F. Baudin, M. Mougel, P. Romby, J.-P. Ebel, and B. Ehresmann, *Nucleic Acids Res.* **15**, 9109 (1987).
[16] L. Lempereur, M. Nicoloso, N. Riehl, C. Ehresmann, B. Ehresmann, and J.-P. Bachellerie, *Nucleic Acids Res.* **13**, 8339 (1985).
[17] D. Moazed, S. Stern, and H. F. Noller, *J. Mol. Biol.* **187**, 399 (1986).
[18] A. Rairkar, H. M. Rubino, and R. E. Lockard, *Biochemistry* **27**, 582 (1988).
[19] A. M. Maxam and W. Gilbert, this series, Vol. 65, p. 499.

interactions.[20-22] These procedures detect reactions between the N-7 atom in the purine ring of guanine residues and DMS. Adenosine residues are also a target for modification by DMS, but they react less efficiently than do guanosine residues. Cleavage at a modified base is effected by heating DMS-treated DNA in piperidine.

Protein binding to DNA may result in the protection of specific nucleotide residues from chemical and enzymatic cleavages. Conversely, treatment of DNA with chemical modifiers, such as DMS, prior to protein binding can identify the residues required for specific protein–DNA interactions. These methylation interference assays have been used to map the specific binding region of proteins on DNA.[22] Experiments analyzing protein–nucleic acid interactions using chemical modification reagents in general give more precise definitions of the specific molecular interactions involved in binding than do analogous experiments performed with nonspecific nuclease probes.

In addition to modification at the N-7 position of guanine, DMS also reacts with specific elements available only within single-stranded regions of nucleic acids. Thus, this reagent has been used to identify single-stranded regions in a variety of RNA molecules[16-18,23-25] as well as the protein–RNA interactions involved in ribosome assembly in vitro.[2,4,6,12,26] Similar procedures have been used to demonstrate the sites within the 16 S rRNA that are involved in tRNA recognition.[4] Residues within nucleotide pairs involved in hydrogen bonding are inaccessible to modification by dimethyl sulfate. These positions are also resistant to modification if protein binds directly to these sites within the bases. A summary of the modified bases produced from the reactivity of DMS on single-stranded RNA is shown in Fig. 1. Previous work has established the extent of reactivity at each position.[16] The primary reaction observed on RNA bases is at the N-1 position of adenine followed by the N-3 of cytosine. Weaker reactions occur at N-1 of guanine and N-3 of uracil.

All the modifications described above for RNA interfere with Watson–Crick base pairing and thus impede the progress of reverse transcriptase

[20] W. Gilbert, A. Maxam, and A. Mirzakekov, in "Control of Ribosome Synthesis" (N. Kjeldgaard and O. Maaloe, eds.), p. 139 Munksgaard, Copenhagen, 1976.
[21] W. Ross and A. Landy, Cell (Cambridge, Mass.) 74, 560 (1983).
[22] U. Siebenlist and W. Gilbert, Proc. Natl. Acad. Sci. U.S.A. 77, 122 (1980).
[23] J. Atmadja and R. Brimacombe, Nucleic Acids Res. 12, 2649 (1984).
[24] S. Stern, D. Moazed, and H. F. Noller, this series, Vol. 164, p. 481.
[25] B. J. Van Stolk and H. F. Noller, J. Mol. Biol. 180, 151 (1984).
[26] S. Stern, R. C. Wilson, and H. F. Noller, J. Mol. Biol. 192, 101 (1986).

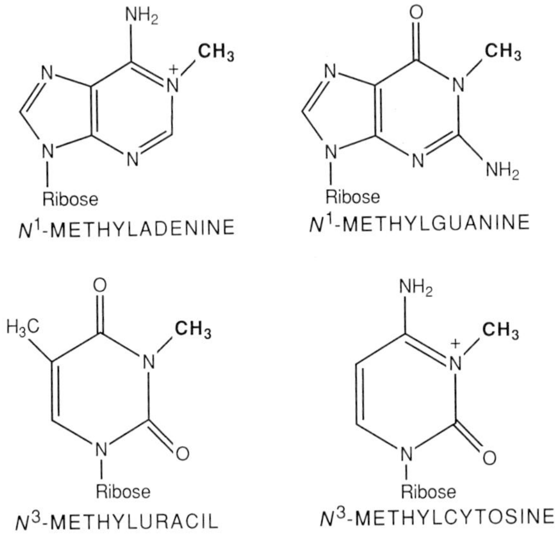

FIG. 1. Structure of nucleotides and site of modification by dimethyl sulfate in single-stranded RNA. DMS-modified structures of the four normally occurring nucleotides in RNA are shown. Indicated in bold type are the methyl groups added by DMS which inhibit the progress of reverse transcriptase during primer extension reactions.

during primer extension reactions on these templates. This results in the appearance of a new band on sequencing gel electrophoresis of a primer extension of treated RNA as compared to untreated RNA. The extension reaction on a modified RNA template terminates one base prior to the methylated position. Therefore, when interpreting sequencing gels, one residue must be added to the terminal position in order to identify the reacted base. A schematic diagram of this analysis is presented in Fig. 2.

A principle of the *in vivo* treatment which has been developed is that the reagents used must be able to penetrate the plant cell wall. Inclusion of low concentrations of detergent are necessary for *in vivo* modification of RNA by DMS. Vacuum infiltration also has been used successfully to introduce DMS into leaf cells; however, this procedure results in less consistent modification patterns among different leaf samples and developmental stages than does the detergent procedure described below. Modification of RNA *in vivo* with kethoxal (Research Organics, Cleveland, OH) and 1-cyclohexyl-3-(2-morpholinoethyl)carbodiimide (CMCT, Sigma, St. Louis, MO) has also been attempted. Although both of these reagents react with purified plant RNAs *in vitro,* no evidence of reaction was observed on RNA isolated from leaves treated with these chemicals under a number of reaction conditions (J. F. Senecoff and R. B. Meagher, unpublished obser-

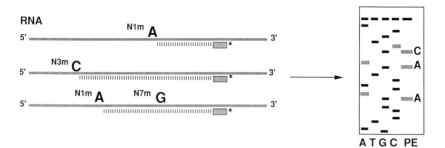

FIG. 2. Primer extension assay to detect DMS-modified nucleotides in single-stranded RNA. RNAs modified with DMS are examined by primer extension. Bases at which modifications occur are unable to form complementary hydrogen bonds with the cognate nucleotide. Thus, chain termination results. Primer extension products (PE) terminate one nucleotide before the modified residue compared to a sequencing ladder.

vations, 1989). However, as dimethyl sulfate modifies all four bases to some extent,[16] this analysis has resulted in an extensive amount of information concerning the secondary structure of RNA *in vivo*. A portion of these data is the subject of this chapter.

Method of Analysis

Dimethyl Sulfate Treatment

Leaf tissue (1–5 g) is harvested from 8-day-old soybean (*Glycine max* var. Pella 86) or 5- to 8-day-old maize (*Zea maize* var. Early Sunglow) seedlings. Experiments with *Arabidopsis thaliana* var. Colombia are performed using leaves from mature plants. In experiments using *Lemna minor,* the entire plant is treated. The tissues are placed into a beaker containing 100 ml of DMS modification buffer [80 mM potassium cacodylate buffer, pH 7.4; 300 mM potassium chloride; 20 mM magnesium acetate, and 0.1% (v/v) nonidet P-40 (NP-40)]. An empirically determined amount of DMS (concentrations vary from 0.1 to 1.0%; treatments are commonly performed at 0.5%) is then added to the leaf suspension; control tissue is treated with buffer only. The leaves are swirled periodically for 5 min, filtered through cheesecloth, and washed with 2–3 volumes of distilled water. The tissue is immediately frozen in liquid nitrogen and stored at −70° until use. For some leaf samples such as soybean, the DMS treatment is allowed to proceed for as long as 20 min. To prevent overmodification of *Lemna* RNA it is necessary to use lower DMS concentrations (0.2%).

RNA Isolation

RNA is prepared from tissue following a standard protocol for the isolation of RNA from leaves.[27] RNA is stored in 70% ethanol without sodium acetate at $-70°$ and dried immediately prior to use.

Primer Extension Analysis

Primer extension analysis of RNA is carried out following modifications of published procedures for sequencing and primer extension reactions.[28] Sequencing primers are end labeled as follows. Ten picomoles primer is incubated in 10 μl total reaction volume containing 50 μCi [γ-^{32}P]ATP, 20 mM Tris-Cl (pH 7.5), 10 mM MgCl$_2$, and 1 mM dithiothreitol (DTT). Reactions are initiated with 10 units of T4 polynucleotide kinase and incubated at 37° for 60 min. The kinase is inactivated by heating at 65° for 5 min. There is no need to purify the primer following the kinasing reaction.

In general, 2 μg of total RNA isolated from plant tissue is dried and resuspended in 10 μl of an 8:2 mix of hybridization buffer [60 mM NaCl, 50 mM Tris-Cl (pH 8.3), 10 mM DTT] and ^{32}P-end-labeled rRNA primer. The annealing reactions are incubated at 50° or $T_m - 5°$ (i.e., 5° below the melting temperature), whichever is lower, for 30–60 min. Longer annealing does not adversely affect the primer extension reactions. Following hybridization, MgCl$_2$ and reverse transcriptase are added to the primer–template mixture [2 μl of 36 mM MgCl$_2$, 60 mM NaCl, 50 mM Tris-Cl (pH 8.3), 10 mM DTT containing 8–10 units avian myeloblastosis virus (AMV) reverse transcriptase]. Extension mixes contain 375 μM of each deoxynucleoside triphosphate (dNTP) in hybridization buffer. Reactions are initiated as follows: 2 μl of the hybridization reaction containing reverse transcriptase is mixed with 3 μl of the dNTP mix for both the control and modified templates. Reactions are incubated at the same temperature as the annealing reactions for 30 min. Sequencing reactions using unmodified templates are performed in an analogous manner with the inclusion of 100 μM dideoxynucleoside triphosphate (ddNTP).[28] Termination solution (5 μl of 98% v/v formamide, 1 mM EDTA, 10 mM NaOH, 0.03% bromphenol blue, and 0.03% xylene cyanol) is added, the reactions are heated to 90° for 3 min, and 3 μl of each reaction is loaded onto a 6 or 8% polyacrylamide–8.3 M urea gel. Gels are run at constant power (35–50 W) for various lengths of time, fixed in 10% methanol–10% acetic acid,

[27] C. M. Condit and R. B. Meagher, *Mol. Cell Biol.* **7,** 4273 (1987).
[28] T. Inoue and T. R. Cech, *Proc. Natl. Acad. Sci. U.S.A.* **82,** 648 (1985).

dried, and exposed for various times to X-ray film without an intensifying screen.

Sequence and Structure Analysis

The structures of the rRNAs were generated and sequences compared using the University of Wisconsin Genetics Computer Group Nucleic Acid sequence analysis package[29] and the University of Georgia Biological Structure and Sequence Computation Facility.

Results and Discussion

The sequence and structure of a short region (111 nucleotides) of the molecule within the 3' minor domain of the 18 S rRNA have been examined in detail in four distant plant species. This region is hypothesized to have extensive and stable secondary structure, based on data available for *Escherichia coli* 16 S rRNA[8,17,24] and for rabbit 18 S rRNA.[18] Furthermore, this region is known to contain the peptidyl or P site, which mediates tRNA binding.[30] This region may also mediate mRNA binding during scanning by the 40 S subunit.[31] These features are essential to translational mechanisms and thus must be conserved throughout evolution.

Sequence

Comparison of the sequences obtained from dideoxy primer extension of total RNA (Fig. 3) reveals a high degree of similarity in this segment among the four plant species examined. Included in this alignment are the previously determined sequences from rabbit,[18] *Xenopus laevis,*[32] and *Saccharomyces cerevisiae*[33] 18 S rRNAs and a consensus sequence from all seven RNAs. Among the plant species the greatest diversity in nucleotide sequences is observed between *Lemna* and soybean 18 S rRNAs where only 99 of 111 nucleotides are shared between the two RNAs in this region. *Arabidopsis* and soybean 18 S rRNA share 103 of 111 residues, as do maize and soybean 18 S rRNAs. However, two of the differences between the maize and soybean sequences are transversion changes, which have the potential to alter secondary structure more severely than transitional changes.

[29] J. Devereaux, P. Haeberli, and O. Smithies, *Nucleic Acids Res.* **12**, 387 (1984).
[30] J. Ofengand, P. Gornicki, K. Chakraburtty, and K. Nurse, *Proc. Natl. Acad. Sci. U.S.A.* **79**, 2817 (1982).
[31] M. Kozak, *Cell (Cambridge, Mass.)* **15**, 1109 (1978).
[32] M. Salim and B. E. H. Maden, *Nature (London)* **291**, 205 (1981).
[33] P. M. Rubstov, M. M. Musakhanov, V. M. Zakharyev, A. S. Krayev, K. G. Skryabin, and A. A. Bayev, *Nucleic Acids Res.* **8**, 5779 (1980).

```
              1          11         21         31         41         51
SOYBEAN       CUCCUACCGA UUGAAUGGUC CGGUGAAGUG UUCGGAUuGC GGCGACGuGa GCGGUuCGCu

ARABIDOPSIS   CUCCUACCGA UUGAAUGaUC CGGUGAAGUG UUCGGAUCGC GGCGACGuGG GuGGUuCGCC

MAIZE         CUCCUACCGA UUGAAUGGUC CGGUGAAGUG UUCGGAUCGC GGCGACGGGG GCGGUuCGCC

LEMNA         CUCCUACCGA UUGAAUGGUC CGGUGAAGcG cUCGGAUCGC GGCGACGaGG GCGGUcCcCC

RABBIT        CUaCUACCGA UUGgAUGGUu uaGUG.gGcc cUCGGAUCGg ccCGcCGGGG uCGGccCaCg

YEAST         CUagUACCGA UUGAAUGGcu uaGUGAgGcc UcaGGAUCug c....uaGaG aaGGgggcaa

XENOPUS       .UaCUACCGA UUGgAUGGUu uaGUGAgGUc cUCGGAUCGg ccCcgCcGGG GucGgcCaCg

CONSENSUS     CUCCUACCGA UUGAAUGGUC CGGUGAAGUG UUCGGAUCGC GGCGACGGGG GCGGU.CGCC


              61         69         79         89         99         109
SOYBEAN       GCCC..GCGA CGUuGuGAGA AGUCCAcUGA ACCUUAUCAU UUAGAGGAAG GAG

ARABIDOPSIS   GCCC..GCGA CGUCGCGAGA AGUCCAcUaA ACCUUAUCAU UUAGAGGAAG GAG

MAIZE         GCCC..cCGA CGUCGCGAGA AGUCCAUUGA ACCUUAUCAU UUAGAGGAAG GAG

LEMNA         GCCC..GCGA CGUCGCGAGA AGUCCgUUGA ACCUUAUCAU UUAGAGGAAG GAG

RABBIT        GCCCugGCGg aGcgcuGAGA AGaCggUcGA ACuUgAcuAU cUAGAGGAAG uAa

YEAST         cuCCaucucA gagCG.GAGA AuuUggAcaA ACuuggUCAU UUAGAGGAAc uAa

XENOPUS       GCCCugGCGg aGcgcCGAGA AGaCgAUcaA ACuUgAcuAU cUAGAGGAAG uAa

CONSENSUS     GCCC..GCGA CGUCGCGAGA AGUCCAUUGA ACCUUAUCAU UUAGAGGAAG GAG
```

FIG. 3. Comparison of 18 S rRNA sequences from four plant species, yeast, rabbit, and *Xenopus*. Sequences of a portion of the 3' minor domain of 18 S rRNA are compared. Also shown is a consensus derived from the comparison. Lowercase letters indicate where the sequence differs from the consensus. Numbers correspond to the coordinates for the four plant rRNAs determined in this study. Gaps introduced by sequence alignment are ignored in the numbering.

In Vivo *Dimethyl Sulfate Modification*

Modification experiments and primer extension analyses were performed on RNA from all four plant species. Representative gels are shown in Fig. 4. A large number of stops occur in the native RNA without DMS treatment. This phenomenon has been observed reproducibly in multiple RNA preparations using a variety of polymerization temperatures, and it may be due to a strong secondary structure within the rRNA that cannot be sequenced by reverse transcriptase. Also, there are many naturally occurring modified base residues within the rRNA sequences that can impede reverse transcriptase.[23] Another possibility is that a portion of the RNA in each preparation is degraded at specific sensitive sites, and thus chain terminations result. Similar patterns of termination have been ob-

SOYBEAN   ARABIDOPSIS   MAIZE

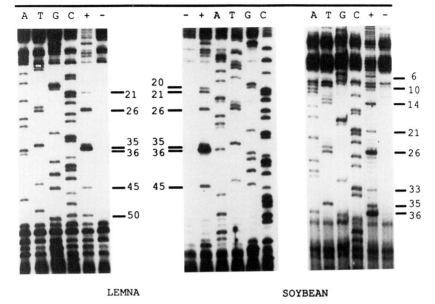

LEMNA   SOYBEAN

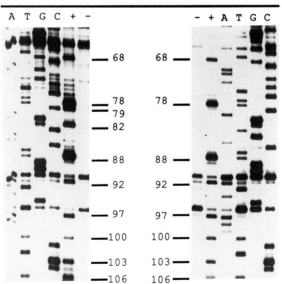

FIG. 4. Primer extension analysis of rRNAs isolated from four plant species treated with dimethyl sulfate. Sequencing lanes are indicated as A, T, G, and C. A plus sign (+) over the lane denotes RNA prepared from DMS-treated tissues; a minus (−) denotes RNA isolated from mock-treated tissues. The nucleotide numbers are the coordinates at which primer extension reactions terminate (see Figs. 3 and 5A). The position of the modified base in the sequence of the 18 S rRNA is determined by comparing novel bands in the DMS-reacted RNA with the sequencing ladder and adding one nucleotide (see Fig. 2 for details). Various regions of the RNA are shown for the different plant species. Several sites of chain termination arising from nascent structure are also evident.

A

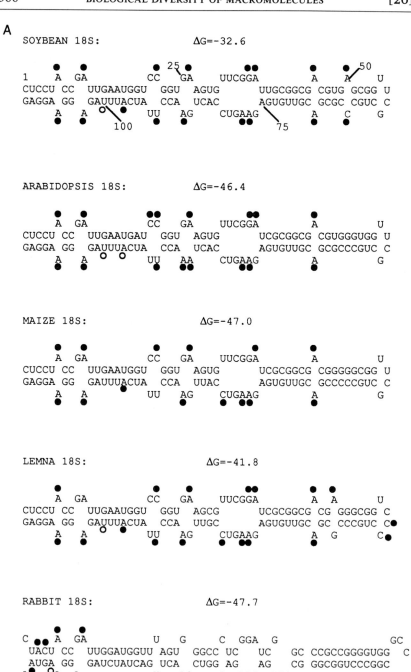

B

MAIZE 18S: ΔG=-49.8

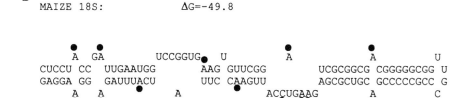

FIG. 5. (A) Structure of four plant rRNA segments determined by DMS modification and primer extension. The structures of the 3′ minor domain for the four plant species and rabbit 18 S rRNA are shown. Filled circles indicate sites of strong modification and open circles sites of weak or variable modification. The free energy of each structure is indicated in Kcal/mol. The structure and modification data presented for rabbit were published previously.[18] Structures presented for the plant RNAs are based on DMS data. Numbering indicated for soybean RNA is as in Fig. 3. (B) Optimal energy structure for maize 18 S rRNA 3′ minor domain. A computer-generated structure of maize 18 S rRNA within the region studied used the parameters of Zuker and co-workers.[35,36] DMS-modified residues within double-stranded regions in the structure are indicated.

served in other rRNA sequences.[34] Thus, the sequence and structure of small regions of the rRNA molecule cannot be determined. Treatment with DMS results in a new pattern of chain termination during primer extension. Treatment with DMS can be controlled such that only a few modifications occur on each RNA molecule. Comparison of primer extension reactions on control and treated RNAs reveals significant increases in band intensity or the appearance of a new band in the modified RNA. This observation indicates that a particular base is single-stranded within the structure of the molecule. A summary of the modifications observed for each RNA in the region examined is shown in Fig. 5A.

Secondary Structures

Proposed secondary structures for all four plant RNAs based on the information gained experimentally are presented in Fig. 5A. It is clearly evident that the secondary structures within this region of RNA from the four species are highly homologous. The predicted RNA structures from the two dicot species, soybean and *Arabidopsis,* differ only in that an adenine at position 50 in soybean is a guanine in *Arabidopsis,* thus promoting formation of an additional base pair with the cytosine at position 63 in *Arabidopsis.* Other sequence differences between the two dicot species have no effect on the secondary structure of this region. The predicted

[34] J. F. Connaughton, A. Rairkar, R. E. Lockard, and A. Kumar, *Nucleic Acids Res.* **12,** 4731 (1984).

structure of this segment in maize 18 S rRNA is indistinguishable from that of *Arabidopsis*, containing the GC pair between positions 50 and 63. The formation of this base pair is substantiated by the experimental data. In the soybean sequence, both the adenine at position 50 and the cytosine at 63 are subject to modification by DMS. No modification is observed at cytosine 63 in *Arabidopsis*, maize, and *Lemna* RNAs, indicating that it is now base paired. Several nucleotide differences between the maize and soybean sequences result in an increased stability of this segment in maize (see below). Finally, for *Lemna*, which has the highest degree of sequence difference from the other three plant sequences, only minor changes in secondary structure are observed. In particular only the change to an adenine at position 48 alters the structure from that in *Arabidopsis* and maize. That this position now becomes single-stranded in *Lemna* is confirmed by DMS modification of the adenine residue.

One indication that supports assigning helicity to a region is the presence of compensatory base changes, that is, where both nucleotides participating in the formation of a base pair are altered. This observation may also help determine the degree of divergence of two species. Within the sequences examined here, the two dicot species have no compensatory base changes. Furthermore, only base transition differences are evident between the two species. One instance of compensatory nucleotide differences was observed between the maize and soybean RNA sequences, where position 48 is a uracil in soybean but a guanine in maize and position 65, a guanine in soybean, is a cytosine in maize. Thus, a U-G base pair is converted to a G-C base pair. These changes preserve the integrity of the putative double helical region and also increase the stability of this structure. An analogous situation is observed between *Lemna* and soybean RNAs involving the base pair between nucleotides 29 and 84. Thus, although based on sequence comparison alone maize and *Arabidopsis* are equally homologous to soybean, the information gained by examining where these base changes reside within the secondary structure of the molecule may provide additional information on how the elements of structure are reserved between species.

Overall Stability of Structures

One might assume that the overall stability of this region had a selective advantage or disadvantage to the organism. Several nucleotide differences among the four species result in variable stabilities of the secondary structures. For example, the structure within the soybean RNA contains nine G-U base pairs, which are only singly hydrogen bonded. These base pairs contribute negligibly to the stability of the secondary structure. The structures of *Arabidopsis,* maize, and *Lemna* RNAs all have fewer G-U base pairs, owing to nucleotide differences that alter these to G-C or A-U base pairs. These changes result in an increased stability, evidenced by the increased negative free energy change (ΔG). In comparing the sequence and structure of soybean and *Arabidopsis* RNAs, which are the least and most stable structures, respectively, based on calculated free energies, several differences result in an increased stability predicted for the *Arabidopsis* RNA structure. In particular, the U/C differences at positions 38, 60, 72, and 74, the G/A at position 18 (changing the G^{18}-U^{95} base pair to A^{18}-U^{95}), and the A/G difference at position 50 (creating an additional base pair) all result in increased stability. Only one difference, C/U at position 52, would decrease the stability. Similar comparisons can be made with maize and *Lemna* RNAs relative to soybean. Briefly, for soybean and *Lemna* RNAs, six nucleotide differences which decrease the stability of the soybean RNA structure are observed, whereas only two differences would result in increased stability. For maize, five differences increase the stability and only one decreases. Whether these increased stabilities are significant in terms of environmental adaptation or reflective of the G + C composition of the specific plant genome is unclear.

It is likely that protein binding and other macromolecular interactions also play a role in stabilizing the rRNA structures. A recent analysis[12] reveals that the binding of specific ribosomal protein to rRNA *in vitro* results in changes within the rRNA modification patterns. Because the experiments detailed herein examine a population of rRNA within the cell, and thus may be analyzing a variety of RNA conformations, it is remarkable that very consistent results are observed. However, the consequence of polysome assembly and translation on the structure of the rRNA and how these structures might be distinct are not addressed in this analysis. *In vitro* DMS modification of soybean RNA has revealed some differences in base reactivity relative to that observed on RNA modified *in vivo* as described herein.[36a]

Computer-generated structures have been determined for these RNA segments. The program utilized gives the lowest energy structure if no constraints are imposed during the operations.[35] An optimal energy structure for the minor domain of maize 18S rRNA is presented in Fig. 5B.[36] The structure presented in Fig. 5A is more consistent with DMS modification data since the structure in Fig. 5B places three modified A residues within double-stranded regions. However, the computer-predicted struc-

---

[35] M. Zuker and P. Stiegler, *Nucleic Acids Res.* **9**, 133 (1981).

[36] M. Zuker, *Science* **244**, 48 (1989).

[36a] J. F. Senecoff and R. B. Meagher, *Plant Molec. Biol.* **18**, 219 (1992).

ture within this region is very similar to the one presented in Fig. 5A. Within this small region only a few alternate foldings are available. The problem in structure predicton arises when large sequences capable of conforming to many high-stability structures are analyzed.[36] In these cases, the usefulness of direct conformational analysis of specific nucleotides within RNA molecules can be invaluable. The DMS-reactive nucleotides within the entire sequence of the 18 S rRNA from soybean have been evaluated.[36a] This technique has also been applied to the analysis of soybean ribulose-1,5-bisphosphate carboxylase small subunit mRNA secondary structure.[36a]

Evolutionary Significance

Extensive structural analyses have been performed on E. coli 16 S rRNA in vitro.[8] This RNA has been examined in free form, in fully assembled 30 S, subunits and at intermediate stages of ribosome assembly. Comparative sequence analysis of a variety of eubacterial rRNAs has resulted in a generalized model for the structure of 16 S like rRNAs.[37,38] Zwieb and co-workers have expanded this study in the analysis of 18 S rRNA sequences from six eukaryotic species.[39] Based on the information available for the E. coli 16 S rRNA secondary structure, structures for the 18S molecules were proposed and analyzed for compensatory mutations within putative helical regions. Although this method provides strong evidence for conservation of helical regions, no direct evidence supporting the predicted secondary structures of these molecules was presented. The technique described herein will help to confirm the predictions made using comparative sequence analysis in addition to providing information in regions where little sequence homology exists between various species.

The 3' minor domain (domain IV) of the small subunit ribosomal RNA contains many elements which mediate interactions with other molecules during translation. However, despite the conservation of function within this domain, the homology between the plant 18 S rRNAs and E. coli small subunit RNA within this 3' region is limited.[34,40,41] This segment within the E. coli molecule contains only 83 nucleotides as compared to 111 nucleotides for the four plant species examined. The additional nucleotides within the plant RNA have been constrained to conform to a

[37] R. R. Gutell, B. Weiser, C. R. Woese, and H. F. Noller, Prog. Nucleic Acids Res. Mol. Biol. 32, 155 (1985).
[38] C. R. Woese, R. Gutell, R. Gupta, and H. F. Noller, Microbiol. Rev. 47, 621 (1983).
[39] C. Zwieb, C. Glotz, and R. Brimacombe, Nucleic Acids Res. 9, 3621 (1981).
[40] Y.-L. Chan, R. Gutell, H. F. Noller, and I. G. Wool, J. Biol. Chem. 259, 224 (1984).
[41] V. K. Eckenrode, J. Arnold, and R. B. Meagher, J. Mol. Evol. 21, 259 (1985).

structure which is very similar to that contained within eubacterial rRNA sequences. Presumably, these structural features mediate important interactions with other macromolecules.

A structural analysis of the 18 S rRNA from rabbit has also been published[18]; the sequence within this region is only 60% similar to that of the plant RNAs (Figs. 3 and 5A). However, one of the lowest energy structures proposed for this segment is very similar to that proposed herein for the four plant rRNAs (Fig. 5A). The *in vivo* DMS modification data reported for the plant RNAs support the proposed secondary structures of the four plant RNA segments, and, in fact, they are more consistent with the proposed structure than the *in vitro* DMS modification data on rabbit rRNA for this region. However, for the analysis of RNA secondary structure to have its full impact on studies in molecular evolution, it must be possible to quantify observed differences.

Phylogenetic reconstructions based on sequence comparisons of 18 S rRNAs have in some cases weighted transversions more strongly than transitions.[42] This is done for two reasons. First, the ratio of transversions to transitions observed between two nuclear DNA sequences is usually about 4 times less than predicted by random mutation.[42,43] Second, transversions inevitably disrupt secondary structures that are essential to the integrity of the rRNA molecule. Conservation of 18 S structure is recognized to supersede the conservation of sequence, and thus the weighting of transversions has been the subject of some debate.[42] However, only a fraction of transversions occur in base-paired regions and hence have any impact on secondary structure.

When sequences involved in base pairing are clearly identified, the weighting of a transversion in these regions in rRNA can be more fully evaluated. Furthermore, when the Watson and Crick strands of each base-paired region are both known, then the impact of transversion on structure and stability can be taken into account. For example, based only on numbers of sequence differences in the rather limited 3′ region analyzed for the four plant rRNAs, the dicot sequence from soybean appears as closely related to the sequence from the monocot (maize) as it does to that from a distant dicot, *Arabidopsis.* The aquatic monocot *Lemna* is the most distant in sequence. An overview of secondary structure suggests that *Arabidopsis* and maize are more closely related (compare the terminal stems in Fig. 5A, nucleotides 46–67). However, as discussed earlier, a

[42] W. M. Brown, E. M. Prager, A. Wang, and A. C. Wilson, *J. Mol. Evol.* **18,** 225 (1982).
[43] E. A. Zimmer, R. K. Hamby, M. L. Arnold, D. A. LeBlanc, and E. C. Theriot, *in* "Hierarchy of Life: Proceedings of the 70th Nobel Symposium" (B. Fernholm, K. Bremmer, and H. Jornvall, eds.), p. 205, 1989. Excerpta Medica, New York.

significant pair of transversions has occurred in this region since the divergence of *Arabidopsis* and maize from a common ancestor. In the *Arabidopsis* sequence nucleotide 48 is a U and is paired with a G in position 65. This U-G pair is substituted by a G-C pair in the maize sequence. The soybean sequence has a U-G pair in this position as does *Arabidopsis,* the other dicot in this study. Based on any of several taxonomic schemes for these plants, *Arabidopsis* and soybean shared ancestors more recently with each other than they did with maize. Only by heavily weighting this transversion would the expected phylogeny be obtained in a comparison of these short sequences. In light of the structures presented for these molecules and their partial confirmation by DMS modification, weighting this transversion for its impact on secondary structure may be warranted.

It is interesting to note that the *Lemna* sequence differs in this same base-paired position by only one transversion each from the two dicot sequences and from the maize sequence. *Lemna* is an aquatic monocot. This subclass split from other monocots very early in their evolution.[44] Extensive comparisons of other sequences have placed its divergence very close to the divergence of monocots and dicots.[45] Further analysis and comparison of *in vivo* 18 S rRNA structures should help to establish guidelines for quantifying both sequence and structural changes in RNA and measuring their evolutionary significance.

Acknowledgments

We thank Dr. Elizabeth Zimmer for proposing an evolutionary comparison of RNA secondary structures in plants and Dr. Harry Noller for encouragement and communications during the course of this study. We also thank Dr. Michael Arnold for constructive criticism and comments on the manuscript. This work was supported by a grant from the U.S. Department of Energy to R.B.M. and an National Institutes of Health postdoctoral fellowship to J.F.S.

[44] A. Cronquist, "An Integrated System of Classification of Flowering Plants." Columbia University Press, New York, 1981.
[45] R. B. Meagher, S. L. Berry-Lowe, and K. Rice, *Genetics* **123**, 845 (1989).

[27] Comparison of Nucleic Acids from Microorganisms: Sequencing Approaches

By DAVID STAHL

Introduction

This chapter has several purposes: to provide an overview of existing nucleic acid sequencing approaches, to indicate areas of rapid technical development, and to serve as an introduction to nucleic acid sequence determination. Also, it may provide a slightly different perspective for those already familiar with sequencing routines. Thus, this chapter is not intended to serve as a source of all available techniques.

Current practitioners of nucleic acid sequencing are divided into two camps: the old and the new (or low versus high technology). Newer technology is primarily concerned with automation and increased throughput. The demands and resources available to large sequencing projects, most notably bacterial and eukaryotic genome projects, has fostered automation. However, much of this is evolving technology and is presently not suited for the smaller scale sequencing projects discussed in this chapter. Although automated technology is not central to this chapter, a brief commentary is offered. The primary audience for this chapter is the scientist approaching sequencing projects of less than 100 kilobases (kb) or so of sequence. Also beyond the scope of this chapter is detailed discussion of cloning and isolation strategies. The reader is directed to a variety of techniques volumes for more detailed discussion of such issues as vector and host selection, screening, and methods for RNA and DNA isolation/fractionation.[1-3]

The exception to limited technical coverage concerns the trend toward direct sequencing of plasmids and polymerase chain reaction (PCR) products. This is a reflection of the increasing need for ease and rapidity in sequence determination. However, many published protocols lack consistency. As discussed at greater length below, there are few or no (as yet) tried-and-true favorites. Thus, the techniques presented in this chapter comprise detailed protocols for only one fairly robust rapid isolation and

---

[1] C. J. Howe and E. S. Ward, *in* "Nucleic Acid Sequencing: A Practical Approach" (C. J. Howe and E. S. Ward, eds.), p. 99. IRL Press, Oxford, 1989.

[2] J. Sambrook, E. F. Fritsch, and T. Maniatis, "Molecular Cloning: A Laboratory Manual," 2nd Ed. Cold Spring Harbor Laboratory, Cold Spring Harbor, New York, 1989.

[3] F. M. Ausubel, R. Brent, R. E. Kingston, D. D. Moore, J. A. Smith, J. G. Seidman, and D. Struhl (eds.), "Current Protocols in Molecular Biology." Wiley, New York, 1989.

METHODS IN ENZYMOLOGY, VOL. 224

sequencing technique and variations that have yet to be fully evaluated. These should not be viewed as ultimate protocols, but rather as workable options.

General Considerations

Choice of Molecule: DNA versus Message (versus Protein)

Because most sequences are determined in the study of the biology of well-circumscribed biochemical or genetic systems (eukaryotic or prokaryotic), the question of what or which molecule(s) to analyze is generally not an issue. However, in addition to the selection of an appropriate biopolymer, the choice of sequencing method may be less straightforward in more strictly comparative studies. Here, the need for rapid sequence determination, issues of fidelity of technique, or modification of DNA sequence associated with posttranscriptional processing events may be primary considerations in deciding sequencing strategy.

Until recently the rules for making these decisions were thought to be relatively straightforward and essentially defined by the central dogma of molecular biology: DNA makes RNA makes protein. The recognized complexities of posttranscriptional modification of RNA (e.g., capping, splicing, and polyadenylation) were variations on this basic theme. However, deviations from the standard genetic code were initially observed in mitochondria and later in the nuclear genes of certain prokaryotes and eukaryotes. This presented the first uncertainty about the rules for translation of nucleic acid sequences into amino acid sequences.[4,5] For example, the observation of variable termination codon assignment among lower eukaryotes suggests that the rules defining termination were still in flux at the time of ciliate diversification.[5] That additional coding information can be added to the gene following transcription (via insertion, deletion, or substitution of nucleotides) is a process now generally referred to as RNA editing (reviewed in Refs. 6–8). In fact, some of the apparent deviations from the standard genetic code are now recognized to be corrected by RNA editing.[8]

RNA Editing

The most dramatic examples of RNA editing, so far observed, occur in certain mitochondrial genes of kinetoplastid protozoa. Here editing creates

[4] T. D. Fox, *Annu. Rev. Genet.* **21,** 67 (1987).
[5] F. Meyer, H. J. Schmidt, E. Plumper, A. Hasilik, G. Mersmann, H. E. Meyer, A. Engstrom, and K. Heckmann, *Proc. Natl. Acad. Sci. U.S.A.* **88,** 3758 (1991).
[6] R. Benne, *Trends Genet.* **6,** 177 (1990).
[7] L. Simpson and J. Shaw, *Cell (Cambridge, Mass.)* **57,** 355 (1989).
[8] J. Scott, *Curr. Opin. Cell Biol.* **1,** 1141 (1989).

open reading frames via the frequent insertion (insertional editing) and occasional deletion of uridine residues. The most spectacular editing results in creation of message from a precursor mRNA (for cytochrome-*c* oxidase III) completely lacking an open reading frame. Although the first observations of editing concerned addition or deletion of uridine residues, studies of the α subunit of mitochondrial ATP synthase of *Physarum polycephalum* have revealed that insertion of nonencoded cytidine residues may also occur.[9] Substitutional editing, represented by C to U conversions (or T to C conversions), observed in certain plant mitochondrial mRNAs and during developmentally regulated editing of mammalian apolipoprotein B mRNA, are apparently mechanistically distinct (e.g., deamination of specific cytidine residues[10]). Other possibly mechanistically distinct types of RNA editing involve insertion of guanosine or adenosine residues in viral RNAs (e.g., paramyxovirus P protein[11]).

Elucidating mechanisms of RNA editing is an area of intensive study, and the reader is referred to reviews of this subject.[6-8] However, editing may complicate comparative sequence analyses. Although most examples of RNA editing are of mitochondrial (or kinetoplastid) origin, they are not exclusively so. The generality of RNA editing remains to be established. Thus, with regard to comparative sequence analysis, a basic issue is the translation of sequence established at the DNA level. For example, because editing appears to be a general phenomenon in plant mitochondria, it is now necessary to reevaluate all protein sequences inferred from DNA sequences. More generally, are there rules for predicting RNA editing? Is what appears to be a pseudogene necessarily a pseudogene? Given that comparative sequencing studies often span large phylogenetic distances, the potential for organism-specific (idiosyncratic) editing of homologous transcripts must also be considered.

The biology of RNA editing is an open arena for comparative studies. A variety of immediate questions concern the evolution of the editing process: questions of function, origins, mechanisms, and distribution. For example, are these primitive or degenerate processes? Such issues must be evaluated comparatively, emphasizing sequence determination at the various informational levels of DNA, RNA, and/or protein. However, a more practical concern for comparative sequencing studies is the influence of the editing process on gene sequence. For example, editing might impose greater constraints on the primary coding sequence of a gene destined for editing. Thus, although comparative studies are based on sequence divergence of the functional biopolymer (protein or RNA), the editing process might influence both the rate and character of accepted mutation.

[9] R. Mahendran, M. R. Spottswood, and D. L. Miller, *Nature (London)* **349**, 434 (1991).
[10] H. Wintz and M. R. Hanson, *Curr. Genet.* **19**, 61 (1991).
[11] S. M. Thomas, R. A. Lamb, and R. G. Paterson, *Cell (Cambridge, Mass.)* **54**, 891 (1988).

Choice of Method: Templated versus Nontemplated Sequencing Protocols

There are two general strategies for nucleic acid sequence determination, here referred to as templated and nontemplated sequencing methods. Nontemplated methods use nucleotide-specific cleavage of end-labeled (generally radioactively) molecules (either RNA and DNA) followed by size fractionation of cleavage products on high-resolution polyacrylamide gels (sequencing gels) to generate the now-familiar sequencing ladder. Base-specific cleavage is accomplished either by selective chemical modification (DNA or RNA) or partial nuclease digestion (RNA). Nontemplated sequencing of DNA is now the less popular technique. Nevertheless, although somewhat cumbersome relative to templated sequencing methods, chemical methods offer certain advantages and should not be immediately discounted (see below).

Templated sequencing protocols rely on the enzymatic synthesis of a DNA strand complementary to a defined template (RNA or DNA). The basic approach has been well described in both general texts and specific methods manuals.[1,2] Briefly, synthesis of the complementary strand is initiated from the 3' terminus of a priming oligonucleotide (the primer) annealed to a specific sequence (generally 15–20 nucleotides) flanking the region of DNA (or RNA) to be sequenced. The primer–template hybrid serves as a specific initiation site for the polymerase (e.g., Klenow, reverse transcriptase, Sequenase). Thus, all polynucleotide chains have the same 5' terminus. Termination of the growing chains at specific 3' nucleotides is controlled by the inclusion of one of four dideoxynucleotides in each of four reaction mixes (C, A, T, and G). The DNA fragments from each reaction mix (differing chain extension from a common 5' terminus) are resolved by size on acrylamide gels and visualized using either radioactive or fluorescent labels. Automated sequencers (e.g., Applied Biosystems, Foster City, CA) use either fluorescent primers (having a different fluorescent dye-labeled primer for each nucleotide termination mix) or, alternatively, fluorescent-dye labeled dideoxynucleotides (e.g., Dye Deoxy terminators, Applied Biosystems). Radioactive methods generally introduce the radiolabel during polymerization via α-labeled triphosphates (^{32}P or ^{35}S) or, alternatively, by labeling the primer prior to addition to the reaction mix. Primers are most commonly labeled by using polynucleotide kinase to introduce a ^{32}P at the 5' terminus of the oligonucleotide.

A greater variety of classic and contemporary methods are available for the sequencing of RNA. A thorough description of existing methodologies of more current utility are found in Stahl *et al.*[12] However, as for DNA sequencing, the two most commonly applied methods for RNA sequenc-

---

[12] D. A. Stahl, G. Krupp, and E. Stackebrandt, *in* "Nucleic Acids Sequencing: A Practical Approach" (C. J. Howe and E. S. Ward, eds.), p. 137. IRL Press, Oxford, 1989.

ing are (1) sequencing of end-labeled RNA (nontemplated sequencing) and (2) primer extension sequencing (templated sequencing). Primer extension sequencing uses the same basic experimental approach applied to templated DNA sequence determination. The main difference (other than the use of an RNA template) is the use of reverse transcriptase.

Nontemplated Sequencing

Nontemplated DNA Sequencing. Advantages of direct DNA sequencing include the minimization of cloning and subcloning procedures as well as possible instabilities of certain DNA inserts in M13 vectors.[13] Modifications in chemical sequencing protocols made since initial publication of the method by Maxam and Gilbert[14] have shortened sample processing time and substituted less hazardous chemical. This technique has remained the method of choice for many investigators.[13] The reader is referred to publications that address modifications in chemical methodology for DNA sequence determination.[13,15]

Nontemplated RNA Sequencing. Direct RNA sequencing remains an important tool for biological and comparative studies. The RNAs are end-labeled at a common 3'- or 5'-terminal position by a variety of techniques and subjected to base-specific enzymatic or chemical cleavage. Generally, RNA sequence analysis makes use of standard acrylamide sequencing gels, although chromatographic techniques also have been used extensively for short sequence determinations (e.g., of oligonucleotides). Using standard acrylamide gels, approximately 120 nucleotides of a sequence can be deduced from a single end-labeled molecule. In general, fewer positions can be read from the terminus of an RNA molecule, than for DNA, on sequencing gels. Presumably, this in part is a consequence of incomplete denaturation of RNA fragments on sequencing gels, resulting from the greater stability of RNA structure relative to DNA. Therefore, fragments must be generated, isolated, and sequenced for the complete sequence determination of larger RNAs. Such fragments have been generated by mild nuclease treatment in high-salt buffers[16] or preferentially by site-specific cleavage with RNase H and DNA oligonucleotides.[17]

[13] R. F. Barker, *in* "Nucleic Acid Sequencing: A Practical Approach" (C. J. Howe and E. S. Ward, eds.), p. 117. IRL Press, Oxford, 1989.

[14] A. M. Maxam and W. Gilbert, *Proc. Natl. Acad. Sci. U.S.A.* **74**, 560 (1977).

[15] G. M. Church and S. Kieffer-Higgins, *Science* **240**, 185 (1988); sequence autoradiographs readers developed by Bio-Rad, Hitachi, and Seiko are mentioned in *Nature (London)* **325**, 771 (1987), and *Nature (London)* **328**, 460 (1987).

[16] H. J. Gross, G. Krupp, H. Domdey, M. Raba, P. Jank, C. Lossow, H. Alberty, K. Ramm, and H. L. Sanger, *Eur. J. Biochem.* **121**, 249 (1982).

[17] S. Shibahara, S. Mukai, T. Nishikara, H. Inoue, E. Ohtsuka, and H. Morisawa, *Nucleic Acids Res.* **15**, 4403 (1987).

A notable example of direct RNA sequencing is the sequence determinations of small RNAs (guide RNAs) implicated in kinetoplastid RNA editing.[18] In the study, guanylyltransferase in combination with [α-^{32}P]GTP was used to label specifically the 5′ terminus of primary RNA transcripts (containing 5′-di- or 5′-triphosphates). Sequences of 5′-terminal fragments (derived from complete RNase T1 digestion of the end-labeled RNAs) were determined using enzymatic sequencing protocols. The reader is referred to Stahl *et al.*[12] for detailed methods and protocols for end-labeling and sequencing of RNA.

Templated Sequencing

As already noted, templated sequencing protocols are now the most generally used methods for nucleic acid sequence determination. The use of the highly processive modified bacteriophage T7 polymerase (Sequenase; U.S. Biochemical Corp., Cleveland, OH) has eliminated many earlier problems associated with variable intensity of individual bands on sequencing gels.

DNA Templates and Use of M13

All templated sequencing protocols require a single-strand template. Although double-strand sequencing is now a generic term for the sequencing of recombinant plasmids and PCR products, all such techniques require the availability of single-strand template. Single strands are generated either via denaturation of duplex structures or by generating an excess of one strand (the template). By far the most standard (and robust) techniques for generating the necessary single-strand template make use of cloning/sequencing vectors derived from the filamentous bacteriophage M13. Here the single-strand template is generated as part of the normal reproductive cycle of the phage. Thus, large amounts of very clean single-strand template are readily generated. The reader is referred to Messing and Bankier[19] for detailed descriptions of cloning in M13 vectors and plasmid–phage chimeric vectors.

In general, primary cloning in M13 is not recommended owing to biased representation of genomic fragments and instability of large inserts. Insert instability may also interfere with sequencing strategies that require subcloning in M13. Although the use of chimeric vectors (phagemids) offers additional subcloning strategies and, in principle, obviates some of

[18] V. W. Pollard, S. P. Rohrer, E. F. Michelotti, K. Hancock, and S. L. Hajduk, *Cell (Cambridge, Mass.)* **63,** 783 (1990).

[19] J. Messing and A. T. Bankier, *in* "Nucleic Acids Sequencing: A Practical Approach" (C. J. Howe and E. S. Ward, eds.), p. 1. IRL Press, Oxford, 1989.

the problems associated with insert instability, production of single-stranded DNA (ssDNA) in chimeric vectors is very dependent on the cloned DNA fragment. Also, fragment orientation and the strain of helper phage influence single-strand production.[20,21]

Templated Sequencing of Double-Stranded DNA

Plasmids. The direct sequencing of recombinant plasmid (or PCR product) DNA is an increasingly popular approach to sequence analysis. Direct sequencing eliminates the need for subcloning in an M13 vector or *in vitro* generation of single-strand template (see Sequencing of Plasmid DNA below). However, as is evident by the great variety of "improved methods" for rapid plasmid isolation and direct sequencing, these techniques are not (as yet) robust; that is, template is of variable sequencing quality and/or techniques appear not to be transferable from laboratory to laboratory (or from worker to worker). However, these qualifiers are not intended to discourage the use of this technique, but rather to prepare the reader for the variation inherent in current methods. When procedures are working well, sequencing gels generated from direct plasmid sequencing cannot be distinguished from those derived from single-strand templates.

As for all sequencing protocols, template purity is generally regarded as the first essential. The requirement for high-purity template cannot be overemphasized. Techniques for rapid plasmid isolation are therefore usually a compromise between rapidity of the isolation technique and template quality. CsCl-purified plasmid is generally regarded as the gold standard in comparisons of double-strand sequencing protocols. The reader is referred to a collection of standard methods for plasmid isolation using CsCl.[2] Yet in some instances the quality of plasmid-templated sequence derived from rapid isolation protocols is superior to that obtained from CsCl-isolated material. Such differences have in part been attributed to greater nicking of the supercoiled plasmid resulting from the greater handling required in preparation. However, because the drive toward increased use of double-strand sequencing protocols can be mostly attributed to the relative rapidity of sequence determination, each investigator should evaluate the scope of the intended sequencing project; this evaluation should balance greater throughput with sequence fidelity (i.e., unidentified or misidentified nucleotide positions).

Progressive Deletion Method of Sequencing Plasmid Inserts. Direct plasmid (and M13-based) sequencing may also be coupled to a number of directed deletion protocols, for example, using ExoIII/mung bean nu-

[20] R. Zagursky and K. Baumeister, this series, Vol. 155, p. 139.
[21] N. A. Straus and R. J. Zagursky, *BioTechniques* **10**, 376 (1991).

clease[22] or ExoIII–ExoVI for creating progressive nuclease deletions of the cloned fragment.[19,23] When methods are working well, the generation of a group of plasmids containing an appropriate set of nested deletions of the cloned insert eliminates the need for time-consuming subcloning or the synthesis of additional sequencing primers. In our experience the success of this technique is very much dependent on the character and orientation of the cloned insert. Given this variability, and the time-consuming steps required, the progressive deletion technique is better suited to smaller sequencing projects.[19]

Double-Stranded Polymerase Chain Reaction Products. A second direct sequencing strategy makes use of DNA specifically amplified by the PCR. This DNA can be cloned (and sequenced as described in other sections of this chapter) or used to generate single-strand template using *in vitro* techniques (e.g., asymmetric amplification,[24] magnetic bead separation of strands,[25,26] or exonuclease digestion of one strand[21,27]). Alternatively, amplified product is directly used as a template in chain-termination sequencing reactions. As for plasmid sequencing, the use of double-stranded DNA requires a preliminary denaturation step to initiate primer-directed polymerization. A relatively straightforward technique that has offered reasonable (but not general) success is snap-freezing. This technique imposes a rapid freezing step immediately following denaturation to retard reannealing of the denatured DNA. Optimization of snap-freezing in relationship to other reaction conditions for double-stranded DNA sequencing using Sequenase has been investigated by Casanova *et al.*[28] However, in our experience, neither direct sequencing (without thermal cycling, below) nor asymmetric amplification are robust sequencing strategies. Discussion with other investigators currently examining alternative direct sequencing strategies suggests that either magnetic bead separation or exonuclease digestion may provide more consistent sequencing results. The reader is referred to Thomas and Kocher[29] for more detailed discussion and protocols for sequencing PCR-amplified DNAs.

[22] Stratagene pBluescript Exo/Mung DNA Sequencing System, Stratagene Systems, La Jolla, CA (1988).
[23] L.-H. Guo and R. Wu, *Nucleic Acids Res.* **10,** 2065 (1982).
[24] U. B. Gyllensten and H. A. Erlich, *Proc. Natl. Acad. Sci. U.S.A.* **85,** 7652 (1988).
[25] T. Hultman, S. Stahl, E. Hornes, and M. Uhlen, *Nucleic Acids Res.* **17,** 5864 (1989).
[26] M. Espelund, R. A. P. Stacy, and K. S. Jakobsen, *Nucleic Acids Res.* **18,** 6157 (1990).
[27] R. G. Higuchi and H. Ochman, *Nucleic Acids Res.* **17,** 5865 (1989).
[28] J.-L. Casanova, C. Pannetier, C. Jaulin, and P. Kourilsky, *Nucleic Acids Res.* **18,** 4028 (1990).
[29] W. K. Thomas and T. D. Kocher, this volume [28].

Temperature Cycling Sequencing

An approach having increased application is the use of multiple cycles of the sequencing reaction to increase the signal. This technique is particularly useful for the sequencing of double-stranded templates; repeated denaturation steps reduce problems associated with template reannealing and premature termination/displacement of the growing strand. The basic format is much the same as that used for PCR amplification. Temperature cycling has been used for sequencing small amounts of double-stranded DNA in combination with dye-labeled terminators on automated sequencers[29] and in combination with end-labeled primers and standard techniques.[30]

Templated Sequencing of RNA

The use of RNA templates in primer extension dideoxynucleotide sequencing reactions was introduced soon after the inception of chain-termination (dideoxy) sequencing protocols for DNA.[31-33] The technique differs little from DNA-templated reactions except for the requirement for reverse transcriptase. The greatest (single) advantage of RNA-templated sequencing is elimination of cloning. Disadvantages include the need for prior knowledge of the priming site sequence and the limitation to single-strand sequence determination. However, the combination of cDNA synthesis followed by PCR amplification now provides more immediate access to the sequence of the complementary strand if necessary for validation of sequence or resolution of sequencing gel ambiguities (e.g., band compression or nonuniform dideoxynucleotide incorporation). Also, given the greater susceptibility of RNA to degradation by nuclease, considerable care must be taken to isolate intact template. Modified nucleotides (e.g., within the ribosomal RNAs) may cause premature truncation of the transcript. Nevertheless, the technique is rapid, given the above provisos, and of generally high fidelity.

Although applicable to virtually any RNA, the most concerted application of templated sequencing of RNA has been to sequencing the transcripts of homologous genes. For such studies the identification of a common priming site or sites among transcripts provides access to rapid sequence comparisons of adjacent variable regions. This has been used to compare immunoglobulin mRNA sequences, taking advantage of constant

[30] A. M. Carothers, G. Urlaub, J. Mucha, D. Grunberger, and L. A. Chasin, *BioTechniques* **7**, 494 (1989).
[31] G. G. Brownlee and E. M. Cartwright, *J. Mol. Biol.* **114**, 93 (1977).
[32] G. R. Baralle, *Cell (Cambridge, Mass.)* **12**, 1085 (1977).
[33] N. J. Proudfoot, *Cell (Cambridge, Mass.)* **10**, 559 (1977).

region priming sites adjacent to variable regions,[34,35] and viral RNAs.[36] Probably the most widespread application of direct RNA sequencing has been in the comparative sequencing of ribosomal RNAs.[37,38] As for the immunoglobulin mRNAs, the rRNAs offer variable regions (organism to organism sequence variation) adjacent to conserved regions of sequence. The conserved regions within the rRNAs have been used as PCR primer sites for the selective amplification of these genes, using either DNA from pure culture or directly recovered from environmental or medical samples.[39-41] Amplified DNA is either cloned for subsequent sequence analysis or alternatively sequenced directly (taking advantage of conserved priming sites as noted above). These and associated techniques are well described by Stackebrandt and Goodfellow.[42]

Selected Protocols

The intent of this limited protocols section is not to provide the reader with an optimal protocol for sequence determination. The protocols listed are presented because they work reasonably well; they give the reader a starting point. However, they are not the final word. This is an area of rapid development, as reflected by frequent publication of "improved" methods. However, there have been few side-by-side comparisons of protocols. Until such comparisons are made, there can be no specific recommendation.

Rapid Plasmid Isolation and Direct Sequencing

There are many published variations of the two most popular approaches to rapid plasmid isolation using either alkaline lysis[43] or boiled

---

[34] P. H. Hamlyn, M. J. Gait, and C. Milstein, *Nucleic Acids Res.* **9**, 4485 (1981).

[35] P. F. Robbins, E. M. Rosen, S. Haba, and A. Nisonoff, *Proc. Natl. Acad. Sci. U.S.A.* **83**, 1050 (1986).

[36] D. Marc, G. Drugeon, A. L. Haenni, M. Girard, and S. Van der Werf, *EMBO J.* **8**, 2661 (1989).

[37] D. L. Lane, B. Pace, G. J. Olsen, D. A. Stahl, M. L. Sogin, and N. R. Pace, *Proc. Natl. Acad. Sci. U.S.A.* **82**, 6955 (1985).

[38] C. R. Woese, *Microbiol Rev.* **51**, 221 (1987).

[39] L. Medlin, H. J. Elwood, S. Stickel, and M. L. Sogin, *Gene* **71**, 491 (1988).

[40] K. H. Wilson, R. B. Blitchington, and R. C. Greene, *J. Clin. Microbiol.* **28**, 1942 (1990).

[41] S. Giovannoni, R. B. Britschgi, C. L. Moyer, and K. G. Field, *Nature (London)* **345**, 60 (1990).

[42] E. Stackebrandt and M. Goodfellow (eds.), "Nucleic Acid Techniques in Bacterial Systematics." Wiley, Chichester, 1991.

[43] C. Birnboim and J. Doly, *Nucleic Acids Res.* **7**, 1513 (1979).

lysis[44] for cell breakage and release of plasmid DNA. Both protocols yield plasmid DNA of acceptable template quality. Although the boiled lysis method is generally considered to yield plasmid of somewhat better template quality (i.e., generating fewer sequencing artifacts such as banding in all four lanes of the sequencing gel), this has not been a consistent observation. Variations in published protocols are primarily associated with those steps used to further purify the plasmid following initial fractionation from total cellular material. Secondary purification steps include RNase A digestion, ethanol or 2-propanol precipitation at high ionic strength, organic extraction, and selective precipitation with the cationic detergent cetyltrimethylammonium bromide[45] or polyethylene glycol.[35] Although removal of RNA is generally considered important, some methods publications (e.g., using high-temperature alkaline denaturation[46]) (see below) report that the presence of contaminating RNA has no effect on either the reaction backgrounds or the amount of sequence obtained.

General Comments. The first reference to plasmid sequencing was by Chen and Seeburg.[47] Wang[48] and Toneguzzo *et al.*[46] report that RNase digestion and organic extraction are not necessary for quality sequencing gels (using modified rapid boiling technique). These authors claim that direct sequencing of plasmids isolated by rapid techniques is indistinguishable from single-strand DNA or CsCl-purified plasmid. Both investigators note that denaturation at 85° improves sequencing results. Wang also reports that DNA isolated from *Escherichia coli* strain HB101 is unstable when stored at 4° (in contrast to a stability of several months reported for other strains). This difference in host strains was earlier reported by Chen and Seeburg.[47] A report by Taylor *et al.* (1993) discusses the influence of different host strains on the quality of small scale plasmid DNA preparations for sequencing.[48a] *E. coli* strains DH5, JM109, and SURE were reported to consistently produce the highest yields of template quality DNA. These authors also noted that plasmid DNA isolated from endonuclease A (endA+) *E. coli* strains was highly susceptible to degradation if care was not taken to remove the endonuclease by phenol : chloroform extraction. Thus, these reports suggest that host strain can significantly influence the quality of plasmid sequencing results.

[44] D. S. Holmes and M. Quigley, *Anal. Biochem.* **114,** 193 (1981).

[45] G. Del Sal, G. Manfioletti, and C. Schneider, *BioTechniques* **7,** 514 (1989).

[46] F. Toneguzzo, S. Glynn, E. Levi, S. Mjolsness, and A. Hayday, *BioTechniques* **5,** 460 (1988).

[47] E. Y. Chen and P. H. Seeburg, *DNA* **4,** 165 (1985).

[48] Y. Wang, *BioTechniques* **6,** 843 (1988).

[48a] R. G. Taylor, D. C. Walker and R. R. McInnes. *Nucleic Acids Res.* **21,** 1677 (1993).

Rapid Alkaline Lysis Isolation Protocol

Our laboratory has found that the inclusion of an additional purification step involving ethanol precipitation from an aqueous solution made 2.5 M in ammonium acetate greatly improves template quality. Such a step should be examined as an amendment to any protocol that yields suboptimal template. The following protocol is a modified alkaline lysis protocol[2] that in our experience routinely yields plasmid of reasonable template quality. All centrifugation steps are performed in an Eppendorf microcentrifuge (e.g., Model 5414) unless otherwise specified. The reader is referred to Sambrook *et al.*[2] for media formulations.

1. Inoculate a single colony into 1.5 ml of L broth (include 50 μg/ml ampicillin if necessary) and grow overnight.

2. Transfer to 1.5-ml Eppendorf tube and centrifuge at 14K rpm for 2 min at 4°. Remove all media by aspiration. Note that it is very important to remove all media since it can interfere with subsequent sequencing reactions.

3. Spin briefly (5 sec) and remove the remaining supernatant with a Pipetman (screw up by thumb to remove last traces of medium).

4. Add 100 μl glucose buffer solution I [50 mM glucose, 25 mM Tris-Cl, (pH 8.0), 10 mM EDTA (pH 8.0)] and resuspend cells by vortexing.

5. Add 1 μl of lysozyme (50 mg/ml; freshly prepared or aliquoted in small quantities of 10–20 μl and stored at −20° until needed); mix and incubate for 10 min at room temperature.

6. Transfer the tubes to wet ice and add 200 μl of freshly made solution II [5 ml 0.4 M NaOH, 1.0 ml 10% (w/v) sodium dodecyl sulfate (SDS), 4.0 ml water]. Mix by inverting 3 times (do not vortex).

7. Add 150 μl of solution III (3 M potassium acetate, 5 M acetic acid). Solution III is made by combining 60 ml of 5 M potassium acetate, 11.5 ml glacial acetic acid, and 28.5 ml water. Mix thoroughly and incubate on ice–water for 10 min.

8. Centrifuge at 14K rpm for 10 min at 4°.

9. Transfer the supernatant to a clean 1.7-ml Eppendorf tube and add 500 μl of phenol–chloroform–isoamyl alcohol (25:24:1, v/v), pH 8.0. Vortex and centrifuge at 14K rpm for 5 min at 4°.

10. Transfer the aqueous phase to a clean 1.7-ml Eppendorf tube and add 800 μl room temperature absolute ethanol. Incubate for 15 min at room temperature, then centrifuge at 14K rpm for 10 min at room temperature.

11. Remove the supernatant and air-dry the pellet for approximately 5 min.

12. Dissolve the pellet in 100 μl water. Add 2.0 μl RNase solution (10 mg/ml stock boiled to inactivate DNase) and incubate for 30 min at 37°.

13. Add 260 μl water and 400 μl phenol–chloroform–isoamyl alcohol (25:24:1, v/v), pH 8.0. Vortex and centrifuge at 14K rpm for 5 min at 4°.

14. Remove the aqueous phase to a clean Eppendorf tube. Add 0.5 volume of 7.5 M ammonium acetate. To amended volume add 1 volume absolute ethanol. Vortex and precipitate overnight at −20°.

15. Centrifuge at 14K rpm for 30 min in microcentrifuge at 4°.

16. Decant the supernatant and wash 3 times with 70% ethanol (centrifuge at 14K rpm for 10 min between each wash addition).

17. After the final wash invert the tube on a Kimwipe and remove residual fluid on the tube walls using a cotton-type swab. Vacuum dry and resuspend the pellet in TE [10 mM Tris-HCl (pH 7.2), 1 mM EDTA]. Transfer to a clean Eppendorf tube.

Optional Steps

18. Reprecipitate with 0.1 volume of 3 M sodium acetate and 2.0 volumes absolute ethanol. Precipitate on dry ice for 5 min or overnight at −20°.

19. Centrifuge at 14K rpm for 30 min at 4°.

20. Remove the supernatant and wash 3 times with 70% (v/v) ethanol as above. Dry the pellet for 5 min at 37° and then resuspend in 50 μl TE.

Boiled Lysis Isolation Protocols

The following are bare bones protocols of the general format specified by Toneguzzo et al.[46] and Liszewski et al.[49] They are primarily offered for comparative reference, and the reader is referred to additional indicated references that specify more elaborate isolation techniques.

Boiled Lysis Protocol Adapted from Toneguzzo et al.[46]

1. Chill 1.5 ml of an overnight culture (with antibiotic) on ice. Harvest by 5 min of centrifugation in microcentrifuge.

2. Resuspend the pellet rapidly in 300 μl STET buffer [8% (w/v) sucrose, 5% (v/v) Triton X-100, 50 mM Tris-HCl (pH 8.0), 50 mM EDTA (pH 8.0)] and add 25 μl freshly prepared lysozyme at 10 mg/ml. Vortex briefly (2 sec) and immediately boil for 45 sec. Centrifuge at 14K rpm for 15 min at room temperature.

3. Remove approximately 210 μl to clean tube and add 230 μl of 2-propanol. Incubate for 5 min at −20°, then centrifuge at 14K rpm for 10 min at 4°.

4. Wash the pellet 2 times with 70% (v/v) ethanol and resuspend in 25 μl TE [10 mM Tris-HCl (pH 7.4), 1 mM EDTA]. For RNA hydrolysis add 1 μl of 10 mg/ml RNase A (DNase-free) and incubate for 15 min at 37°.

[49] M. K. Liszewski, V. Kumar, and J. P. Atkinson, *BioTechniques* **7**, 1079 (1989).

Boiled Lysis Protocol Adapted from Liszewski et al.[49]

1. Grow 50 ml of an overnight culture ($\sim$18 hr) in Luria broth, NZYM, or Circle Grow medium (BIO 101 Labs, La Jolla, CA) with shaking (225 rpm). Harvest by centrifuging at 7000 g for 5 min.

2. Resuspend the pellet in 5 ml STET [8% sucrose, 5% Triton X-100, 50 mM EDTA, 50 mM Tris-HCl (pH 7.5)]. Add RNase A (boiled) to 10 μg/ml final. Add lysozyme to 0.5 mg/ml final (e.g., 250 μl of freshly made 10 mg/ml solution). Incubate for 10 min at room temperature.

3. Boil for 1 min (excess boiling time increases chromosomal contamination), then hold on ice for 10 min. Centrifuge at 12,000 g for 10 min at 4°.

4. Carefully remove the supernatant (discard pellet). Add 0.6–1.0 volume ice-cold 2-propanol ($\sim$4 ml) and mix by inversion. Immediately centrifuge (10,000 g for 10 min at 4°).

5. Resuspend the pellet in 600 μl TE [10 mM Tris-HCl (pH 7.4), 1 mM EDTA] and transfer to clean microcentrifuge tube.

6. Extract 2 times with an equal volume (1:1) of phenol–chloroform (centrifuge at 10,000 g for 7 min at room temperature) and 1 time equal volume of chloroform (centrifuge at 10,000 g for 2 min at room temperature).

7. Transfer to 15 ml centrifuge tube. Add an equal volume of 7.5 M ammonium acetate ($\sim$500 μl) and 2.5 volumes of $-20°$ ethanol ($\sim$2.5 ml). Hold on ice for 15 min, then centrifuge at 10,000 g for 10 min at 4°.

8. Wash the pellet 2 times with 70% ethanol and resuspend in 300 μl TE. The average yield is 150 μg (range 80–300 μg).

Plasmid Kits

A number of kits for the reported isolation of template quality plasmid DNA are currently on the market. These generally use modifications of the above protocols in combination with additional purification, for example anion exchange or glass adsorption and elution. A partial list of suppliers includes: 5 Prime → 3 Prime, Inc. (Boulder, CO), Qiagen (Chatsworth, CA), Pharmacia (Piscataway, NJ), Clontech Laboratories, Inc. (Palo Alto, CA). However, this author knows of no side-by-side comparison of these products.

Sequencing of Plasmid DNA

The modified phage T7 DNA polymerase marketed as Sequenase (U.S. Biochemical Corp.) has become the standard enzyme for both single- and double-strand DNA sequence analysis. Attributes of this enzyme that favor its use include high processivity, no 3′ → 5′-exonuclease activity, and less discrimination against nucleotide analogs. Thus, bands on sequencing gels are more uniform in intensity and demonstrate less background than with the large fragment of *E. coli* polymerase I (large Klenow fragment) or avian myeloblastosis virus (AMV) reverse transcriptase. A convenient kit is provided by U.S. Biochemical Corp., and the reader is referred to the product manual for detailed protocols.

Toneguzzo et al.[46] compare Sequenase, standard Klenow, modified Klenow, and reverse transcriptase sequencing of single-strand and double-strand templates. This is one of the few papers that provides side-by-side comparisons of parameters examined, and general observations include the following. Reverse transcriptase is not suitable for double-strand sequencing and is unable to provide sequences from AT-rich templates. This was previously reported by Pfeffer and Mierendorf.[50] Denaturing supercoiled DNA in 0.2 N sodium hydroxide at 85° rather than at room temperature greatly improved the sequence information obtained from a GC-rich template and eliminated several cross-banding regions. Factors identified as contributing to poor labeling of short transcription products using the Sequenase protocol include (1) the molar ratio of template to primer, (2) the amount of template used in the labeling reaction, (3) the deoxynucleoside triphosphate (dNTP) concentration in the labeling reaction, and (4) the incubation time of the termination reaction. Termination reaction time varied with template (attributed to pause sites that required extension of the incubation time from 15 min to 20 or 30 min). The presence of contaminating RNA had no influence on either the background reactions or the amount of sequence information obtained.

Sequenase Parameter Optimizations. The tabulation below provides a summary of Sequenase parameter optimizations.[46]

Primer–template ratio	5 : 1 ratio required for uniform labeling of small fragments
Amount of template	At least 0.6 μg of single-strand DNA (1 μg optimal) and 2–4 μg supercoiled DNA
dNTP concentration in the labeling reaction	2.5 μM (corresponding to 1 : 3 dilution of the Sequenase kit) gave better results than the standard reaction mixture; for greater amounts of template, higher dNTP concentrations were required (up to the standard 7.5 μM concentration)
Length of termination reaction	15 min at 37°

High-Temperature Alkaline Denaturation and Primer Annealing[48]

1. Combine in 1.7-ml Eppendorf tube 2.0–4.0 μl plasmid (~2–4 μg), 20 μl water, and 20 μl of 0.4 M NaOH, 4 mM EDTA (pH 8.0). Vortex and incubate 10 min at room temperature or 85° for 5 min. *Note:* Denaturation at elevated temperature has been reported to improve sequencing reactions (see below).[45,46,51]

[50] D. Pfeffer and R. Mierendorf, *Promega Notes* **5,** 1 (1986).
[51] G. Gabriel and P. Burbelo, *Editorial Comments, U.S. Biochemical Corp.* **17,** 21 (1991).

2. Add 4.0 μl neutralization solution (2 M ammonium acetate, pH 4.5).

3. Immediately add 100 μl cold absolute ethanol and precipitate on dry ice (5 min) or overnight at $-20°$.

4. Centrifuge for 30 min in microcentrifuge at 4°. Decant the supernatant, wash the pellet 3 times with 70% ethanol (spin after each wash), and then vacuum dry.

5. Resuspend in 10.0 μl water, transfer to clean 0.65-ml Eppendorf tube, and add 2.0 μl primer (1–2.5 pmol) and 3.0 μl of 5× Sequenase reaction buffer. Heat for 2 min at 65°. Cool for 10 min (on bench top).

6. Continue as specified by Sequenase standard protocols.

Variations on Denaturation/Annealing Steps Adapted from Del Sal[45]

1. Combine 1 μg plasmid DNA in 7 μl TE, 2 μl primer (4.4 ng/μl), and 1 μl of 1 M NaOH. Incubate for 10 min at 68°.

2. Add 4 μl TDMN solution [0.28 M TES (free acid, Sigma, St. Louis, MO, T-1375), 0.12 M HCl, 50 mM dithiothreitol (DTT), 80 mM MgCl$_2$, and 0.2 M NaCl]. Incubate for 10 min at room temperature.

3. Add 6 μl elongation premix consisting of 3 μl labeling mix (Sequenase kit), 1 μl [α-^{35}S]dATP, and 2 μl of T7 polymerase (3 units). Incubate for 5 min at room temperature.

4. Add 4 μl of the above reaction to each of four termination mixes (Sequenase kit; tubes C, A, T, and G). Incubate for 10 min at 37°.

5. Add 5 μl stop solution (Sequenase kit). Heat for 3 min at 90° before loading sequencing gels.

Variations on Denaturation/Annealing Steps Adapted from Gabriel and Burbelo[51]

1. Add 1 μl primer (50 ng) to 5 μl of plasmid preparation (see above).

2. Add 1 μl of 0.5 N NaOH. Heat to 80° and slow cool (in small/minimum volume of water) to 37°.

3. Neutralize by the addition of 1 μl of 0.5 N HCl.

4. Follow the standard Sequenase protocol.

Troubleshooting and Technical Suggestions

Primer Design and Storage. The primers are synthetic DNA oligonucleotides containing free 3′-hydroxyl groups. Although a minimum primer length based on probability considerations is 11 to 12 nucleotides, in practice most range from 15 to 20 nucleotides. Even within this size range, however, priming at multiple or alternative sites sometimes occurs. If possible, primers should be of greater than 50% G + C content so they will compete well with secondary structure within the template, and they should have one or more C or G residues near the 3′ (priming) end so that they "lay down" well on the template. Extensive primer extension sequencing studies of ribosomal RNAs have shown considerable variation in priming efficiency between primers of similar size and composition, presumably reflecting competing template structure. In these instances, "sliding" the target site several nucleotides to the 5' or 3' direction has sometimes improved priming efficiency. Thus, primer design remains largely empirical.

Resolution of Sequencing Gel Ambiguities and Troubleshooting. Sequence information is commonly lost as a result of two different anomalies, namely, band compression and premature termination of the transcript. Band compression is a consequence of undenatured secondary structure (usually GC-rich hairpin helices). The inclusion of helix-destablizing analogs (inosine or deazaguanine) frequently eliminates or reduces this effect.[1] Premature termination of the transcript results in a band across all lanes of the sequencing gel, obscuring the identity of the dideoxynucleotide truncation. For templated RNA sequencing, coincident termination may be a consequence of nicked template (generally caused by RNase contamination), secondary structure, or modified nucleotides. However, because the obscuring transcripts still retain 3'-hydroxyl groups, they are substrates for chain elongation and can be chased out (of a unique gel position) by chain extension using terminal deoxynucleotidylexotransferase (TdT).[52] This technique was originally developed for direct sequencing of RNA, but a protocol for direct DNA sequencing has also been described.[53] Since some spurious banding can result from the inclusion of the TdT extension step, this should be used as a secondary cleanup reaction.

Automation and Nonstandard Formats

An issue addressed by any laboratory involved in routine sequence determination is automation. At what point, measured in nucleotides per year, is it appropriate to consider automation? The definition of automation used here does not include equipment used to facilitate the reading of sequencing films by hand. Rather, automation eliminates or significantly reduces the amount of direct inspection of raw data (films or scans). There are two formats of automation, the first being automated sequencers used for on-line readout during the fractionation of reaction products on acrylamide gels [Applied Biosystems Inc. (ABI), Pharmacia (Piscataway, NJ), and LI-COR Inc. (Lincoln, NE)]. The second avenue is to couple conventional DNA sequencing (or multiplexing techniques; see below and Ref. 15) with automatic or semiautomatic DNA sequence film reading [Millipore (Bedford, MA), PDI (Huntington, Station, NY), Bio-Rad (Rich-

[52] D. C. DeBorde, C. W. Naeve, M. L. Herlocher, and H. F. Maassab, *Anal. Biochem.* **157,** 275 (1986).

[53] D. W. Fawcett and S. F. Bartlett, *Editorial Comments, U.S. Biochemical Corp.* **17,** 19 (1991).

mond, CA), Hitachi (Yokohama, Japan), IntelliGenetics (Mountain View, CA), and Scanalytics (Billerica, MA)]. The direct sequencers are high end equipment, whereas the film readers are a factor of five or so less in price. The existing film readers are not completely automated and require significant operator interaction. Their advantage and market appeal are to provide a significant reduction in the fatigue (and error) associated with standard "by hand" read and verify approaches. Given relative affordability, these units definitely merit consideration. Although user satisfaction with the ABI system is generally quite high, none of the systems should be considered to have satisfied all user expectations (although possibly meeting or exceeding product specifications). An issue that remains somewhat open is that of acceptable error frequency. This number that may range from 1 : 2000 to 1 : 10,000 depending on project and personality.

Multiplexing and Nonradioactive Detection

The number of repetitive steps required for nucleic acid sequencing can be reduced by multiplexing techniques. Multiplex DNA sequencing, first described by Church and Kieffer-Higgins,[15] involves the mixing of cloned fragments prior to chemical cleavage (nontemplated sequencing). Each cloned fragment in the mix is identified by a unique tag. As originally described, the tag was a unique flanking sequence within the plasmid vector used for cloning. Following standard acrylamide gel fractionation, the separated DNA fragments are transferred to a support membrane for indirect detection of clone-specific tags (originally detection was via hybridization of radioactive probes complementary to the clone-specific sequences). DNA fragments specific to each recombinant plasmid are visualized by serial probing of the membrane for each plasmid-specific sequence (repeated washings and probings of the membrane). The number of probings is essentially defined by the number of unique plasmid vectors.

A variation on the multiplexing approach, possibly more suited to small sequencing projects, uses analog-specific tagging of sequencing primers. Using this variation, the reaction products from each analog mix (A, G, C, and T) are combined and electrophoresed in a single lane of a sequencing gel. The separated DNA fragments from each base-specific reaction mix are visualized by repeated probing of the membrane. The individual images derived from a single lane on the sequencing gel are then superimposed for reading the sequence. In principle, this approach is more suited to automated film readers because no correction for lane-to-lane variation is needed. The general multiplexing approach has been modified for use with templated sequencing protocols and nonradioactive detection, for example, using sequencing primers labeled with biotin or digoxigenin. A comparative evaluation of these techniques for use with multiplex sequencing has been presented by Richterich and Church.[54]

To conclude, automation probably should be considered if the sequencing effort exceeds about 100,000 nucleotides per year. No equipment recommendations are made since this remains both an area of rapid development and one of large capital investment.

Acknowledgments

I thank Bruce Roe, Pat Gillavet, George Church, and Mitch Sogin for helpful discussions and acknowledge the United States Environmental Protection Agency, the Office of Naval Research, and the National Science Foundation for supporting our research related to the methods described in this chapter.

---

[54] G. Richterich and F. M. Church, in press (1992).

[28] Sequencing of Polymerase Chain Reaction-Amplified DNAs

By W. KELLEY THOMAS and THOMAS D. KOCHER

Introduction

The enzymatic amplification of a specific segment of DNA (target sequence) by the polymerase chain reaction (PCR) has several advantages over traditional, *in vivo* methods of cloning DNA (Table I). The PCR is much more rapid than traditional library construction and screening because it involves far fewer manipulations. In addition, smaller regions of sequence similarity can be used to retrieve a particular sequence, and it is possible to amplify from relatively crude preparations of DNA as well as from degraded DNA samples.[1] The sensitivity of PCR allows amplification of as little as a single molecule of DNA, which has permitted the recovery

TABLE I
COMPARISON OF DIFFERENT APPROACHES TO SEQUENCING

Feature	PCR and direct sequencing	PCR and cloning/sequencing	Traditional cloning/sequencing
Speed	Rapid	Semirapid	Slow
Sensitivity	1 molecule	1 molecule	1 ng
Accuracy	High	Low	Moderate
Cloning artifacts	None	Yes	Yes
Maximum length	<2 kb	<5 kb	<1000 kb
Repair	None	None	Yes

[1] W. K. Thomas, S. Pääbo, F. X. Villablanca, and A. C. Wilson, *J. Mol. Evol.* **31,** 101 (1990).

of sequences from unculturable bacteria[2] and even single sperm.[3] Finally, in many cases, direct sequencing of PCR products offers the opportunity to determine DNA sequences with greater accuracy than traditional cloning methods.

This chapter describes various methods for generating templates suitable for sequencing from PCR products. Three general approaches are discussed: (1) the cloning of PCR products, (2) preparation of single-stranded templates from PCR products, and (3) direct use of double-stranded PCR products as sequencing templates.

Principles

Template Preparation Using Polymerase Chain Reaction

Synthesis of a specific DNA segment via the PCR is directed by two oligonucleotide primers. Multiple rounds of extension from the primers amplify the target sequence exponentially, with termini defined by the 5′ ends of the two primers. This process allows the determination of sequences from a very small number of target molecules. Unique sequences can be amplified from a complex mixture of nucleic acids because of the high specificity of the primer–template hybridization.

Error Rates in Polymerase Chain Reaction Cycling

During the polymerization of DNA, misincorporation of nucleotides leads to errors in the resulting daughter molecules. The error rate for the most commonly employed polymerase *(Taq)* has been estimated to be as low as 1 error in 9000 nucleotides incorporated, which is somewhat higher than that of the Klenow fragment of polymerase (1 of 36,000 nucleotides). The error rate of *Taq* polymerase is dependent on several factors including the quality of the target DNA[4] and the reaction conditions, particularly the concentrations of the deoxynucleoside triphosphates (dNTPs).[5] The relevance of PCR errors and cloning are presented below (Fig. 1).

[2] S. J. Giovannoni, T. B. Britschgi, C. L. Moyer, and K. G. Field, *Nature (London)* **345**, 60 (1990).

[3] H. Li, U. B. Gyllensten, X. Cui, R. K. Saiki, H. A. Erlich, and N. Arnheim, *Nature (London)* **335**, 414 (1988).

[4] S. Pääbo, D. M. Irwin, and A. C. Wilson, *J. Biol. Chem.* **265**, 4718 (1990).

[5] M. A. Innis, K. B. Myambo, D. H. Gelfand, and M. A. D. Brow, *Proc. Natl. Acad. Sci. U.S.A.* **85**, 9436 (1988).

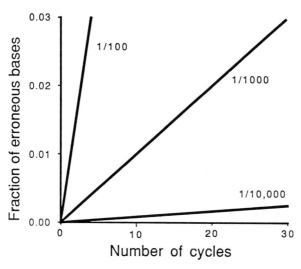

FIG. 1. Fraction of bases expected to be in error in a cloned PCR product as a function of the number of cycles. Three potential misincorporation rates are shown (solid lines).

Rationale of Direct Sequencing

For most applications the error rate of the polymerase has no effect on the quality of the sequence information obtained. This is because the amplified product (sequencing template) is a population of molecules with only a relatively small fraction of the copies containing misincorporated nucleotides at any one position (Table II). The fraction of template molecules that will contain a misincorporated base is dependent on the starting number of target molecules and the cycle in which the error occurred. The fraction is very small, undetectable in most applications, if the number of starting molecules is at least 10.

TABLE II

FRACTION OF MUTANT MOLECULES IN FINAL AMPLIFICATION PRODUCT

Number of starting molecules[a]	Cycle of PCR in which mutation first appears			
	1	2	5	10
1	0.5	0.25	0.003	0.001
10	0.05	0.025	0.0003	0.0001
100	0.005	0.0025	0.00003	0.00001
1000	0.0005	0.00025	0.000003	0.000001

[a] Number of starting molecules refers to the number of intact single-stranded template molecules.

The error rate of the polymerase has no effect on the quality of the sequence information obtained by direct sequencing. Errors during amplification become problematic only when the product of the PCR is cloned prior to sequencing. An apparent exception is homopolymer/homodimer sequences (i.e., long runs of a single base or of a dimer).[6] Direct sequences from PCR products of such regions are clear from either sequencing primer to the homopolymer, yet unreadable beyond the homopolymer. The problem appears to be the result of length variation in the sequencing template at the homopolymer stretch (Fig. 2). As few as 10 T residues appear to promote this problem (Fig. 2A), with longer runs being even more problematic (Fig. 2B). This phenomenon might arise from misincorporation during PCR, or it may reflect *in vivo* heterogeneity in the target sequence which is missed by traditional cloning. An important point is that the sequences are unreadable but do not give a wrong sequence.

Cloning Amplified Products

Direct cloning of PCR products is made more difficult by the propensity of *Taq* polymerase to incorporate an A beyond the template strand. This leaves a single-base 3′ overhang.[7] Cloning of PCR products can be facilitated by several means such as the inclusion of restriction sites in the primer sequences,[8] blunt-end cloning after removal of a 3′ A overhang,[9] or the use of cloning vectors compatible with the ends of PCR products.[10]

The major drawbacks to cloning PCR products prior to sequencing are the extra preparation time and the need to sequence multiple independent clones. This is necessary because, unlike direct sequencing of PCR products, each clone represents a single molecule form the PCR reaction and can contain a sequence that is different from the target molecule. The percentage of bases in a clone that will be erroneous owing to misincorporation during the PCR is dependent on the error rate and the number of cycles employed prior to cloning (Fig. 1). For example, under typical PCR conditions we might expect less than 1 in 1000 nucleotides to be misincorporated. If we amplified a 1-kilobase (kb) target sequence for 20 cycles, the percentage of bases in each clone that we would expect to be erroneous would be less than 2%. Therefore, to ensure that we determine the se-

[6] P. Goloubinoff, S. Pääbo, W. K. Thomas, and A. C. Wilson, manuscript in preparation.
[7] J. M. Clark, *Nucleic Acids Res.* **16**, 9677 (1988).
[8] S. J. Scharf, G. T. Horn, and H. A. Erlich, *Science* **233**, 1076 (1986).
[9] S. J. Scharf, *in* "PCR Protocols" (M. A. Innis, D. H. Gelfand, J. J. Sninsky, and T. J. White, eds.), p. 84. Academic Press, San Diego, 1990.
[10] D. Marchuk, M. Drumm, A. Saulino, and F. S. Collins, *Nucleic Acids Res.* **19**, 1154 (1991).

A   B

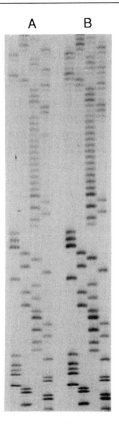

FIG. 2. DNA sequences across a homopolymer stretch in the control region of the mitochondrial genomes from two salmon. (A) Ten T residues in pink salmon *(Oncorhynchus gorbuscha)* and (B) 14 T residues in rainbow trout *(Oncorhynchus mykiss).* Lanes for each species are from left to right G, A, T, and C. Sequencing templates were prepared by asymmetric amplification directly from purified mitochondrial DNA. Sequencing reactions were performed with Sequenase and the products separated in 6% polyacrylamide, 7 *M* urea gels.

quence in the original target segment, we would need to sequence at least three clones.

Single-Stranded Templates

The method we have used most for generating DNA sequencing templates is unbalanced priming, which utilizes asymmetric ratios of primer concentration to synthesize an excess of one strand during the PCR.[11] This

[11] U. B. Gyllensten and H. A. Erlich, *Proc. Natl. Acad. Sci. U.S.A.* **85,** 7652 (1988).

method is especially suitable for short products [<500 base pairs (bp) in length] and has the advantage of not requiring special modifications of the oligonucleotide primers.

Other methods for obtaining single-stranded DNA for sequencing include binding of a biotinylated primer to streptavidin–agarose beads.[12] Another method involves exonuclease digestion of one strand.[13] Alternatively, the complementary strands can be isolated from agarose strand-separation gels, especially if the two strands have different base compositions.[14]

Double-Stranded Templates

In some cases the efficient generation of single-stranded DNA molecules proves difficult even though double-stranded products amplify readily. In light of this, a variety of methods have been developed to sequence double-stranded PCR products. These linear DNAs are more difficult to sequence than supercoiled plasmid DNAs. Difficulties that must be overcome are a low priming efficiency (because the complementary strands tend to reanneal) and spurious priming on both strands. One of the most successful approaches has been the addition of dimethyl sulfoxide (DMSO), at a final concentration of 10% (v/v), to the sequencing reaction prior to the primer annealing step.[15] The DMSO tends to prevent renaturation of the double-stranded product.

An alternative approach is to perform multiple cycles of sequencing to increase the sequencing signal.[16] Especially when done with *Taq* polymerase, cycle sequencing allows much more sequencing signal to be obtained from small quantities of double-stranded DNAs. This technique becomes extremely useful when dye-labeled terminators are available.

Methods

Protocol 1: Single-Stranded Amplifications

Synthesis of a single DNA strand for sequencing can be generated by asymmetric amplification. Asymmetric amplifications can be performed either on PCR products or directly from unamplified DNA. Typically

---

[12] L. G. Mitchell and C. R. Merril, *Anal. Biochem.* **178**, 239 (1989), see also B. H. Bowman and S. R. Palumbi, this volume [29].

[13] R. G. Higuchi and H. Ochman, *Nucleic Acids Res.* **17**, 5865 (1989).

[14] T. Maniatis, E. F. Fritsch, and J. Sambrook, *in* "Molecular Cloning" p. 179. Cold Spring Harbor Laboratory, Cold Spring Harbor, New York, 1982.

[15] P. R. Winship, *Nucleic Acids Res.* **17**, 1265 (1989).

[16] V. Murray, *Nucleic Acids Res.* **17**, 8889 (1989).

double-stranded PCR products are gel-purified before reamplification. Asymmetric amplifications are very similar to symmetric ones, except that one of the primers in the reaction is at a low, limiting concentration. This results in copies of one strand being amplified (linearly) relative to the other after the limiting primer has been effectively used up. A typical asymmetric amplification (50-μl reaction volume) involving the production of template for both strands is carried out as follows.

1. Add primers to each tube, including a control without template. Tube 1 contains 5 μl primer A (10 μM) and 5 μl primer B (10 μM) (no-template control). Tube 2 contains 5 μl primer A (10 μM) and 5 μl of a 1 : 100 dilution of primer B (0.1 μM). Tube 3 contains 5 μl primer B (10 μM) and 5 μl of a 1 : 100 dilution of primer A (0.1 μM).

2. Prepare a master mixture, including enzyme: 5 μl of 10× Taq buffer (670 mM Tris-HCl, 20 mM MgCl$_2$, 100 mM 2-mercaptoethanol, 0.1% Tween 20, v/v), 1 μl of 40 mM dNTP stock, 33.5 μl of water, and 0.25 μl (1 – 2 units) of Taq polymerase. The total volume is 40 μl per reaction tube. Mix well by pumping the pipette.

3. Aliquot 40 μl to each of the three tubes (tubes 1 – 3 in Step 1).

4. Add template to the two noncontrol tubes (tubes 2 and 3). This is usually 1 μl of the double-stranded amplification product purified from a gel. Mix well.

5. Add two drops of mineral oil to each tube to prevent evaporation and spin briefly.

6. Cycle the temperature of the reaction tubes 30 to 35 times, using annealing conditions at least as stringent as those used in the double-stranded amplification.

7. Analyze 5 μl on an agarose gel. Migration of the single-stranded bands relative to the double-stranded products varies with gel type and with the product being amplified. It is not necessary to remove the mineral oil from the tube.

8. Salts and nucleotides are most easily removed by centrifugal dialysis in a microconcentrator such as the Millipore (Bedford, MA) Ultrafree 30,000 NMWL. Do not use if the mineral oil has been removed with chloroform. Pipette the reaction from under the oil, wiping the pipette tip.

8a. Add 0.25 ml distilled water and centrifuge for 10 min at 2000 g in a fixed-angle rotor at room temperature.

8b. Repeat this washing procedure 2 additional times.

8c. Collect the DNA from the filter in about 15 – 40 μl of distilled water.

9. Use 7 μl for sequencing according to standard protocols (e.g., Sequenase, U.S. Biochemical, Cleveland, OH).

Protocol 2: Sequencing Double-Stranded Amplification Products on Applied Biosystems 373A Automated Sequencer

Isolation of Template

1. Run 50 μl of a double-stranded PCR amplification in a low melting point agarose gel. After electrophoresis, cut the desired double-stranded product from the gel with as little excess agarose as possible. The gel band should contain approximately 200 ng of DNA (a good strong band by ethidium bromide fluorescence).

2. Purify the DNA. We prefer extracting with hot phenol. Dilute the gel band with 400 μl of 0.2 M NaCl–TE (10 mM Tris, 1 mM EDTA), pH 8.0. Add 500 μl of hot buffered phenol (pH 8.0).[14] Place into a 65° water bath for 5 min. Vortex briefly to break up the agarose plug, then return the sample to the water bath for an additional 5 min.

3. Remove the tube from the water bath. Vortex until an emulsion is formed throughout the tube. Spin in a microcentrifuge at maximum speed for 5 to 10 min at room temperature. Transfer to an ice bath for 5 min and microcentrifuge at 4° at maximum speed for 10 min.

4. Recover the aqueous supernatant from the top of the tube. Avoid the thick white interface, which contains most of the agarose.

5. Extract with a solution containing 0.25 ml of phenol and 0.25 ml of chloroform.

6. Extract with 0.5 ml of chloroform.

7. Place the aqueous layer into a clean tube. The solution volume should be approximately 0.5 ml. Add 5 μl of 1 M MgCl$_2$ and 1 ml of 2-propanol. Place on ice for 20 min, then pellet the DNA by centrifugation in a microcentrifuge for 30 min at 4°. Because relatively small quantities of DNA are being precipitated, the precipitation must be efficient. Wash the pellet with 1 ml of 70% (v/v) ethanol and centrifuge in a microcentrifuge at 4° for 5 min. Decant the ethanol and dry the pellet.

Sequencing Reactions

8. Resuspend the DNA in 19 μl of water. Avoid the use of EDTA, as this can interfere with the PCR cycling to follow. Set up a sequencing reaction with each primer, using the reagents supplied by Applied Biosystems, Inc. (Foster City, CA), as follows:

Reaction premix:

5× TACS buffer	4 μl
dNTP mix	1 μl
Dye terminator A	1 μl
Dye terminator T	1 μl
Dye terminator G	1 μl

Dye terminator C   1 μl
Taq polymerase   4 units

9. Prepare the individual reactions to contain 12 μl DNA template, 7.25 μl reaction premix, and 1 μl primer (10 μM).

10. Cover with one drop of mineral oil and thermal cycle as follows: denature at 96° for 30 sec, anneal at 50° for 15 sec, and extend at 60° for 4 min.

11. Repeat for a total of 25 cycles. *Note:* The thermal cycle of this protocol is optimized for traditional M13 forward and reverse sequencing primers. Other primer sequences may work most effectively at a different annealing temperature. Also, the extension temperature is lower than the temperature optimum of *Taq* polymerase, to allow the incorporation of the highly modified dye-labeled dideoxy terminators.

12. After the cycling, remove the dNTPs and decomposed dye via a spin column (Centrisep, Princeton Separations, Adelphia, NJ) according to the instructions of the manufacturer. Add 4 μl of a 5:1 solution of formamide and 50 mM EDTA. Vortex and heat to 90° for 2 min. Apply to sequencing gels.

Acknowledgments

We thank J. Conroy, E. M. Prager, T. W. Quinn, and A. Sidow for helpful discussions and the National Science Foundation and the National Institutes of Health for financial support.

[29] Rapid Production of Single-Stranded Sequencing Template from Amplified DNA Using Magnetic Beads

By Barbara H. Bowman and Stephen R. Palumbi

Introduction

We present a powerful, general method for producing a pure preparation of single-stranded DNA (ssDNA) sequencing template from double-stranded DNA (dsDNA) amplified using the polymerase chain reaction (PCR). The amplified DNA is synthesized using one biotinylated and one nonbiotinylated primer and is bound to streptavidin beads via the biotin molecule. The two DNA strands are then dissociated, and the nonbiotinylated strand is released and recovered from the supernatant. The procedure presented here is a modification and simplification of the method originally proposed by Mitchell and Merril.[1] It produces ssDNA from templates of any amplifiable length, without strand bias and with sufficient yield for several sequencing reactions. This method utilizes the sensitivity and yield of double-strand PCR reactions, and it provides pure templates that give optimal sequencing results.

The double-stranded amplified DNA to be sequenced is generated in two separate, reciprocal reactions. In the first, the 5' primer is biotinylated[2] and the 3' primer is not; in the second, the reverse is true. This enables either of the two strands to be recovered in single-stranded form for sequencing. Each double-stranded DNA contains a biotin molecule at one end (Fig. 1a) and is captured on streptavidin-coated magnetic beads (Fig. 1b). Other PCR reaction components are removed by washing. The dsDNA is denatured by the addition of base (Fig. 1c), whereupon the strand lacking biotin is separated from the bound strand. The released strand can be recovered from the supernatant, while the biotinylated strand remains attached to the streptavidin-coated magnetic beads. Typical yields of ssDNA are sufficient for five sequencing reactions, which may be performed using either the PCR primer (Fig. 1d) or internal sequencing primers.

The streptavidin capture method is applicable to any PCR product, in that the efficiency of recovery of ssDNA does not appear to be affected by the length, base composition, or secondary structure of the amplified region. The method allows direct sequencing of amplified DNA, which for many applications is preferable to sequencing individual, cloned PCR products because direct sequencing obscures any random errors induced by *Taq* polymerase. The technique requires only a single round of amplification to produce each ssDNA sequencing template, thus limiting both the opportunity for contamination and the costs of a second amplification reaction. To minimize the number of primers that must be biotinylated, a single set of primers that anneal to highly conserved regions of the target molecule can be used to amplify DNAs from a wide variety of taxa.

This method is of particular use for sequencing long templates such as small subunit ribosomal DNA (rDNA). Such long molecules [over 1700 base pairs (bp) in the case of eukaryotic small subunit rDNA] exceed the length at which asymmetric amplification[3] and double-stranded sequencing[4] are reliable alternatives. For several reasons, sequencing ribosomal RNA (rRNA) genes from an amplified product is strongly preferred over

[1] L. G. Mitchell and C. R. Merril, *Anal. Biochem.* **178**, 239 (1989).
[2] C. Levenson and C. Chang, *in* "PCR Protocols" (M. A. Innis, D. H. Gelfand, J. J. Sninsky, and T. J. White, eds.), p. 99. Academic Press, San Diego, 1990.
[3] U. B. Gyllensten and H. A. Ehrlich, *Proc. Natl. Acad. Sci. U.S.A.* **85**, 7652 (1988).
[4] B. Bachmann, W. Lüke, and G. Hunsmann, *Nucleic Acids Res.* **18**, 1309 (1991).

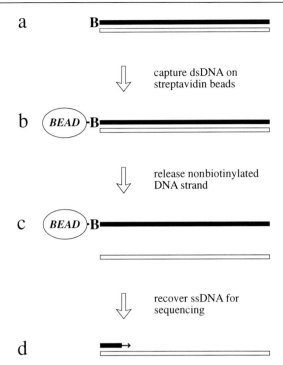

FIG. 1. Preparation of ssDNA from biotinylated PCR-amplified DNA. (a) DNA is amplified using one primer that has a biotin molecule (B) at the 5′ end and one primer that lacks biotin. (b) Double-stranded amplified DNA is captured on a streptavidin-coated magnetic bead via the biotin molecule. (c) The dsDNA is denatured, with the biotinylated strand remaining attached to the streptavidin-coated magnetic bead and the nonbiotinylated strand being released into the supernatant. (d) The nonbiotinylated strand is used as an ssDNA sequencing template.

direct, dideoxy sequencing of an rRNA template using reverse transcriptase. Both DNA strands are available for sequencing, the rDNA will not contain new information incorporated by RNA editing, and considerably less starting material is required. If PCR is performed starting from a preparation of whole RNA[5] instead of DNA, the present method shares with direct sequencing the advantage that only expressed genes are amplified and sequenced.

The streptavidin capture method is also of advantage to population biologists, because it provides for rapid, repeatable, and reliable sequences

[5] T. W. Myers and D. H. Gelfand, *Biochemistry* **30**, 7661 (1991); E. S. Kawasaki, *in* "PCR Protocols" (M. A. Innis, D. H. Gelfand, J. J. Sninsky, and T. J. White, eds.), p. 21. Academic Press, San Diego, 1990.

from large numbers of individuals. When the same DNA region must be sequenced from hundreds or even thousands of individuals, this method provides the best compromise among accuracy, speed, and cost.

Procedure

Handling Magnetic Beads

For each step below, add liquid to the streptavidin paramagnetic beads in a microcentrifuge tube, agitate the mixture gently by hand or on a rotator, and then remove the supernatant. Before removing each supernatant, pulse-spin the tube in a microcentrifuge very gently (< 1000 rpm) and briefly, so as to remove bead slurry from the top of the tube without pelleting the magnetic beads. If necessary, use a micropipettor to dislodge beads from the bottom of the tube and resuspend them.

To concentrate the beads before removing the supernatant, lay the microcentrifuge tube on the laboratory bench and place a small magnet by its side. (Alternatively, tube and magnet can be placed into a holder that will keep them side by side.) When the beads have congregated on the side of the tube, lift the tube, holding the magnet in place, and pipette off the liquid. Summoning beads to the side rather than to the bottom of the tube makes it possible to remove all the liquid without concern that some magnetic beads will accidentally be pipetted into the supernatant.

Notes. The different brands of streptavidin magnetic beads have different molarities of attached streptavidin, different resuspension protocols, and different capture properties. Bead volumes given in this protocol are for Promega (Madison, WI) paramagnetic particles (No. Z5241) prepared as indicated below. If other beads are used, appropriate volumes may differ by a factor of 3 or more from those presented.

Bead Preparation

Remove the storage buffer from 0.6 ml magnetic bead slurry (one tube) and wash 3 times with bead buffer (0.2 M NaCl, 20 mM Tris-HCl, 1 mM EDTA, pH 7.6). Resuspend beads in 625 μl of this buffer and store at 4°.

Amplification

Prepare a 100-μl PCR mix of the target to be sequenced, using one primer that has been biotinylated[2] and one that has not. The strand initiated with the nonbiotinylated primer will become the single-stranded sequencing template. Because the yield of dsDNA directly determines the amount of ssDNA recovered, procedures that maximize the efficiency of the PCR reaction, such as the addition of empirically determined concentrations of glycerol or dimethyl sulfoxide (DMSO),[6] are recommended. Determine the yield of amplified DNA by evaluating 3–5 μl of the dsPCR product on an agarose gel stained with ethidium bromide. Another 3–5 μl may be saved for a later gel to evaluate the recovery of ssDNA.

All of the following procedures are done at room temperature.

Optional Reduction of Primer Concentration

1. Reduce the concentration of excess primers relative to product by prespinning the remaining product plus 1.9 ml sterile water in a Centricon-100 tube (Amicon, Beverly, MA, 4211 or 4212). Centrifuge for 25 min in a fixed-angle rotor (e.g., Sorvall, Norwalk, CT, SA600 or SS34) at the highest speed recommended by the manufacturer for the Centricon tubes.

2. Cap the tube with the Centricon retentate cup, then backspin the retentate (about 40 μl) into the cup in a swinging-bucket rotor by bringing the speed to 1000 rpm and then braking immediately. The retentate contains the biotinylated dsDNA for use in the capture step below.

3. If desired, 100–250 μl of the liquid in the Centricon filtrate cup may be concentrated by evaporation, and selective elimination of primers can be evaluated on an agarose gel.

4. Centricon tubes may be rinsed and reused for the same preparations later in the procedure. Fill the top of the tube (Centricon concentrator) with sterile water, cap with the retentate cup, and shake gently. Repeat a total of 3 times. The tube need not be dried, but the retentate cup must be pipetted dry before reusing it to capture the final ssDNA, so that it will not be diluted. Rinsing can conveniently be done during incubation in the capture step, below.

Notes. A Centricon-30 tube may be substituted for use with very small DNAs that may not be retained by the Centricon-100 tube. Millipore (Bedford, MA) MC 100,000 MWL filter units (UFC3 THK 00) may be substituted if desired. Because of their smaller volume, additional dilution steps may be required to desalt the ssDNA before sequencing. For brands of streptavidin magnetic beads whose binding efficiency for biotinylated DNA may decrease as the length of the DNA increases, it may be essential to reduce the primer concentration via this prespin step.

---

[6] K. T. Smith, C. M. Long, B. H. Bowman, and M. M. Manos, *Amplifications* **5**, 16 (1990).

Capture of Double-Stranded Amplification Product

In a 2.0- or 1.5-ml microcentrifuge tube, combine the amplified DNA, or prespun product from the optional step, with 100 μl of well-suspended magnetic bead slurry. Place on a rotator or agitate gently by hand for 15–30 min. Remove the liquid. Save at least a 10-μl sample for later analysis of capture efficiency on an agarose gel. (Typically, some dsDNA is visible in this supernatant fraction. Only a very bright band, combined with low yield of ssDNA, should be of concern.)

Washing

Wash with 500 μl TE (10 mM Tris-HCl, 1 mM EDTA, pH 8.0) or bead buffer by shaking gently or placing on a rotor for about 1 min. Remove and discard liquid. Repeat.

Denaturation of Double-Stranded DNA, Recovery of Single-Stranded DNA, and Neutralization

1. Dilute 5 M NaOH to 0.2 M just before use. Add 150 μl of 0.2 M NaOH to streptavidin magnetic beads to which dsDNA is bound. Agitate or rotate for 6 min.

2. During the incubation, add 100 μl of 5 M ammonium acetate, pH 6.8, to a Centricon-100 tube.

3. After the 6-min incubation, remove the NaOH solution (which contains the released, nonbiotinylated ssDNA strand) from the magnetic beads and add it to the ammonium acetate in the Centricon-100 tube. It is very important not to pipette any magnetic beads into the NaOH fraction. Most ssDNA will be recovered in this single NaOH wash. If desired, Steps 2 and 3 can be repeated.

Desalting and Concentration of Single-Stranded DNA

1. Add sterile water to the neutralized ssDNA solution in the Centricon-100 tube to a total volume of 2.0 ml. Spin for 25 min in a fixed-angle rotor at the highest speed recommended for the Centricon tubes. Add 2.0 ml of sterile water and repeat the spin.

2. Discard the water in the filtrate cup of the Centricon-100, add 2.0 ml more of sterile water, and spin a third time. This spin may be extended to 45 min, if this is necessary to bring the volume of the retentate down to 40–60 μl.

3. Remove the filtrate cup and backspin the retentate into the retentate cup in a swinging-bucket rotor by raising the speed to 1000 rpm and braking immediately.

Evaluation of Yield

Test the ssDNA yield (as well as the effectiveness of intermediate steps in the procedure, if desired) by agarose gel electrophoresis of the following samples: (1) retained PCR product, (2) concentrated filtrate from the prespin, (3) supernatant containing the primers and dsDNA that were not captured on the magnetic beads, and (4) 3–5 μl of ssDNA. (The ssDNA band will not run at the same speed as dsDNA.) A larger volume of ssDNA may be necessary to visualize very short products on an agarose gel.

Sequencing

Standard sequencing reactions can be performed on 7.5 μl of the resulting ssDNA. Possible sequencing primers include the nonbiotinylated version of the biotinylated primer or any primer that anneals to the same strand. If the ssDNA yield is low, the ssDNA can be concentrated by evaporation under vacuum before sequencing.

Sequencing reactions have been performed using as a template a 350-base biotinylated ssDNA attached to Dynabeads M-280 (Dynal, Great Neck, NY).[7] The beads must be washed with TE or bead buffer to remove residual NaOH and then resuspended in low-EDTA TE (10 mM Tris-HCl, 0.1 mM EDTA, pH 8.0). The bead slurry then can be used in a sequencing reaction just as other ssDNAs. Primers may include the original, nonbiotinylated primer or other primers that anneal to the attached strand. Beads must be resuspended during sequencing reactions, but after final denaturation they must be excluded from the fraction loaded on the sequencing gel. Sequences using the template attached to the beads have not been as clean as those using the released strand as a template; optimization of this step will greatly increase the desirability of the procedure.

Alternative Low-Cost Procedure

The following procedure reduces the cost and time requirements for ssDNA preparation and yields enough template for one or two sequencing reactions. It is designed to speed analysis of a short DNA segment from many different individuals in a population. The PCR reaction is performed in a smaller volume, the Centricon prespin is omitted, and concentration steps are replaced by 2-propanol precipitation.

1. Amplify the DNA to be sequenced in a 50-μl volume. After evaluating yield on an agarose gel, add 45 μl of the PCR reaction to 50 μl of paramagnetic bead slurry and agitate for 15–30 min.

[7] T. Hultman, S. Bergh, T. Moks, and M. Uhlen, *BioTechniques* **10**, 84 (1991).

2. Remove the supernatant and wash with 200 μl bead buffer or TE.

3. Add 60 μl of 0.2 M NaOH and agitate 6 min. Remove the NaOH, which contains the released ssDNA, to a fresh microcentrifuge tube.

4. To the NaOH fraction, add: 70 μl of 5 M ammonium acetate (pH 6.8), 140 μl of 2-propanol, and 1 μl of 10 μg/μl tRNA *of* 1 μl of 1.25% linear polyacrylamide. (The tRNA or linear polyacrylamide is necessary to achieve sufficient yield for sequencing.)

5. Incubate at $-20°$ for at least 1 hr.

6. Spin at high speed in a microcentrifuge for 10 min at 4°.

7. Pipette or decant off the supernatant, taking care not to dislodge the pellet.

8. Add 100 μl of 70% (v/v) ethanol. Tap the tube lightly to wash the pellet. Pipette or decant off the ethanol.

9. Air- or vacuum-dry the pellet completely. Resuspend in 7–14 μl TE.

Notes. Initial experiments show that Steps 3–9 of the above procedure can be eliminated in some cases. Washed beads from Step 2 can be resuspended in an annealing reaction designed for sequencing double-stranded templates.[4] Following boiling, instead of the usual rapid-freezing step, the beads are concentrated immediately and the supernatant is removed. The remainder of the sequencing procedure is conducted on the supernatant as described.[4] Our experiments show good sequencing yields from templates less than 500 bp in length. Longer templates have not yet been tested.

Acknowledgments

We thank P. Clyne, S.-Y. Chang, K. S. Greisen, and T. J. White for helpful comments on the manuscript and W. E. Rainey, K. S. Greisen, and C. Orrego for feedback and suggestions on the method described. B.H.B. is supported by a grant from the National Institutes of Health (RO1-A128545). S.R.P. holds National Science Foundation Grant BSR-9000006.

[30] DNA Sequences from Old Tissue Remains

By W. Kelley Thomas and Svante Pääbo

Introduction

One aspect of molecular evolution concerns the structural change of macromolecules over time. Traditionally, inferences about such changes are made from information about the differences within and between extant species. The development of the polymerase chain reaction (PCR) has made it possible to retrieve DNA sequences from tissues found in museum collections and at archaeological excavations. This has opened up the possibility of including ancestral taxa in molecular evolutionary studies and examining genetic changes in populations over time.

Such an approach is possible in contexts where collections exist that include extinct forms or where large numbers of individuals are preserved, making it possible to evaluate levels of variation within ancestral populations. This is often the case in museum collections, particularly of smaller organisms, as well as in some archaeological contexts. In this chapter we discuss the major sources of material for diachronic molecular studies, review some commonly used methods for extracting and enzymatically (PCR) amplifying ancient DNA, discuss some technical problems, and offer a few theoretical considerations regarding the design of projects involving museum specimens and archaeological remains.

Sources of Material

Museum collections represent the single largest source of tissues readily available to the investigator. Such collections are of particular value because of provenance data and habitat information on the specimens. In addition, many of the collections include large series representing single localities.

Zoological collections are composed predominantly of skeletal remains, dried skin, and specimens fixed in formalin or alcohol. Of these, alcohol-fixed material and dried untanned skins that have been prepared rapidly after the death of the animals can be expected to yield DNA sequences on a routine basis.[1,2] Skeletal material makes up a large proportion of vertebrate collections but is more likely to be devoid of amplifiable DNA. Nevertheless, Hagelberg and colleagues have amplified DNA from archeological bone samples,[3] and methods developed for protein extractions may be particularly useful in this regard (see Tuross and Stathoplos, [9] in this volume). In clinical pathology, paraffin-embedded tissues prepared for histology represent a large and valuable resource for retrospective studies.[4] Fixation conditions have been shown to affect the recovery of

[1] C. E. Greer, J. K. Lund, and M. M. Manos, *PCR Methods Appl.* **1,** 46 (1991).
[2] W. K. Thomas, S. Pääbo, F. X. Villablanca, and A. C. Wilson, *J. Mol. Evol.* **31,** 101 (1990).
[3] E. Hagelberg, B. Sykes, and R. Hedges, *Nature (London)* **342,** 485 (1989).
[4] D. K. Shibata, N. Arnheim, and W. J. Martin, *J. Exp. Med.* **167,** 225 (1988).

amplifiable DNA. These studies indicate that the best fixative is 95% (v/v) ethanol.[1]

Archaeological remains are limited to skeletons in most areas of the world. However, where climatic or local conditions permit, dried tissues may be preserved in the form of mummies. Furthermore, wet sites such as peat bogs often yield macroscopically well-preserved material. However, the likelihood of retrieval of DNA is dependent on factors such as the pH of the water. Thus, acid peat bogs of Europe have yet to yield any DNA from human remains, whereas two samples from the neutral peat bogs of Florida[5,6] have shown that DNA may be preserved in the presence of persistent standing water. The above materials yield DNA that goes back in time approximately 40,000 years. Theoretical considerations indicate that should be about the upper limit for the preservation of DNA when water is present.[7] However, under some circumstances DNA may survive for several millions of years in plant compression fossils (the interested reader is referred to Refs. 8 and 9 for information on DNA from plant fossils).

State of Preservation of Old DNA

The major problem with DNA extracted from old tissues is that the average length of intact template molecules is much reduced owing to various forms of DNA damage. In general, the average size of the DNA from ancient soft tissue, as visualized by ethidium bromide staining in agarose gels, is reduced to less than 200 base pairs (bp).[10] The higher molecular weight DNA seen occasionally has been shown to be due to bacterial or fungal growth by Southern blotting or by amplification with bacteria-specific primers.[9] The reduced molecular size of ancient DNA demonstrates that one form of damage which affects it is strand breakage. A large fraction of the damage may be caused by autolytic processes that occur rapidly after death.[11] However, an equally important source of damage may be oxidative and hydrolytic effects that either directly break

[5] G. H. Doran, D. N. Dickel, W. E. Ballinger, Jr., O. F. Agee, P. J. Laipis, and W. W. Hauswirth, *Nature (London)* **323**, 803 (1986).

[6] S. Pääbo, J. A. Gifford, and A. C. Wilson, *Nucleic Acids Res.* **16**, 9775 (1988).

[7] S. Pääbo and A. C. Wilson, *Curr. Biol.* **1**, 45 (1991).

[8] E. M. Golenberg, D. E. Giannasi, M. T. Clegg, C. J. Smiley, M. Durbin, D. Henderson, and G. Zurawski, *Nature (London)* **344**, 656 (1990).

[9] A. Sidow, A. C. Wilson, and S. Pääbo, *Philos. Trans. R. Soc. London B* **333**, 429 (1991).

[10] S. Pääbo, *Proc. Natl. Acad. Sci. U.S.A.* **86**, 1939 (1989).

[11] L. B. Rebrov, V. L. Kozeltdev, S. S. Shishkin, and S. S. Debov, *Vestn. Akad. Med. Nauk SSSR* **10**, 82 (1983).

phosphodiester bonds or induce chemical modifications that labilize the bonds. Damage in the form of oxidized pyrimidine residues has been demonstrated by enzymatic assays in DNA extracted from dry soft tissue remains of animals.[10] Also, cross-links and other, unknown, damage are undoubtedly present. If high molecular weight noncontaminating DNA is found, the possibility that it may represent cross-linked molecules should be kept in mind.

An interesting finding is that plant remains may be less susceptible to damage than animal remains. This is illustrated by the fact that ancient corn cobs yield DNA of a size that, in some cases, allows for determination of restriction fragment length polymorphisms[12] as well as the amplification of nuclear single-copy genes.[13]

Enzymatic Amplification

The polymerase chain reaction (PCR) is an ideal tool for the study of ancient DNA because it has the ability to amplify a small number of intact DNA molecules that exist in a complex mixture of large amounts of partially degraded and modified templates. Of crucial importance for the use of ancient DNA extracts is the extent to which the damage limits or inhibits the enzymatic reaction. An observation often made is that the maximum sizes of amplifiable products are reduced in old, damaged DNA compared to modern DNA extracts.[14] This is also true for DNA from ancient bones, which seem in many cases to allow for longer amplifications than soft tissues. For example, we have determined the maximum size of amplifiable DNA from 3500-year-old moas found at a dry cave site in New Zealand and found that, whereas soft tissues allowed the amplification of pieces only up to 120 bp, bone extracts from the same individual yielded products of up to 380 bp. However, DNA extracted from a modern ratite bird easily allowed the amplification of pieces of over 1000 bp.[15]

Damage of ancient DNA reduces the number of undamaged molecules that can serve as templates for a DNA polymerase. In addition, damage to the templates may cause the polymerase to stall and thus slow down the

---

[12] T. Helentjaris, *Maize Genet. Cooperation News Lett.* **62**, 104 (1989).

[13] P. Goloubinoff, S. Pääbo, and A. C. Wilson, *Proc. Natl. Acad. Sci. U.S.A.* **90**, 1997 (1993).

[14] S. Pääbo, in "PCR Protocols: A Guide to Methods and Applications" (M. A. Innis, D. H. Gelfand, J. J. Sninsky, and T. J. White, eds.), p. 159. Academic Press, San Diego, 1990.

[15] A. Cooper, C. Mourer-Chauviré, G. K. Chambers, A. von Haeseler, A. C. Wilson, and S. Pääbo, *Proc. Natl. Acad. Sci. U.S.A.* **89**, 8741 (1992).

initial rounds of amplification of the old DNA. Consequently, a few, modern, undamaged DNA molecules that contaminate a specimen may be preferentially amplified. Because the contaminating DNA will permit longer amplifications than the old DNA, one way to discriminate against nonremovable contaminating DNA that stems from the specimen itself is to use primers that are very closely spaced. These will amplify short fragments in which the ancient DNA may quantitatively predominate, so that the contaminant is not detected. For example, in a study of marsupial wolf specimens, amplifications of 120 bp yielded human sequences, whereas amplifications 60 bp long from the same extract yielded marsupial sequences.[16] An alternative approach, when animal remains are studied and the contamination is of human origin, is to design primers specific for the animal in question.

Direct Sequencing

Once ancient mitochondrial (mt) or chloroplast DNA sequences are amplified, no further problems need be anticipated from damage in the template DNA because only one type of sequence is expected in the individual. This allows for direct sequencing of the amplification product. When that is performed, the average sequence of all amplified molecules is determined, reflecting the consensus sequence of all molecules from which the PCR was initiated. In most cases, the number of initial template molecules is substantial, and thus any misincorporation or recombination event that affected a single molecule even during the first cycle of the PCR will not be seen in the directly sequenced product. However, if the amplification is initiated from less than about 10 template molecules, then an ambiguous sequencing reaction may be observed. The problem may be overcome by the use of more template DNA, using high fidelity amplification conditions, or by performing multiple amplifications.

When nuclear sequences are determined, the presence of two or more alleles will make separation of the alleles necessary in order to assign the observed sequence differences to each allele. Ideally, this is performed by a technique that still allows direct sequencing to be performed from the alleles, for example, denaturing gradient gel electrophoresis (see [31] by Lessa, in this volume). However, if cloning is performed, a high error rate at the primary sequence level is observed. For example, a sequence 45 bp long determined from four clones of a mitochondrial amplification product from heavily damaged DNA from a 4000-year-old Egyptian mummy

---

[16] R. H. Thomas, W. Schaffner, A. C. Wilson, and S. Pääbo, *Nature (London)* **340,** 465 (1989).

contained erroneous bases at almost 4% of the positions.[10] In such cases, a large number (at least three per mtDNA or nuclear allele) of cloned sequences need to be determined so that the allelic sequence positions can be distinguished from chimeric clones and errors caused during polymerization by template damage or spontaneous misincorporation.[13,17]

Jumping Polymerase Chain Reaction

The reconstruction of ancient alleles needs to take into account one other effect of DNA damage on amplification via the PCR. Premature termination of the extension from primers may result in *in vitro* recombination events. This effect is due to the fact that partially extended primers may, with their 3' ends, anneal to other template molecules with sequence similarity to the initial template.[18] Resulting molecules may represent chimeras of both alleles at the locus being amplified or may arise from alleles at other loci when multigene families are investigated. For example, when major histocompatibility class I heavy chain genes were amplified from a 7000-year-old human, five chimeric clones were found among 89 clones that represented 14 sequences.[17] Similarly, when alcohol dehydrogenase 2 genes were sequenced from 4500-year-old maize from Peru, 2 of 12 clones proved to be recombination products of the two alleles present in the plant.[13] Based on these observations, it is imperative that multiple clones representing each allele be sequenced.

Basic Protocol for Polymerase Chain Reaction of Ancient DNA

A sample protocol is given following the general scheme outlined in Fig. 1. For each step detailed precautions are discussed.

Removal of Samples

Before extraction of DNA, a sample needs to be taken from a tissue specimen. In general, because of the potential for contamination from previous handling, it is advisable to remove a sample that is large enough to allow for the superficial part of it to be cut away, trimmed off, or avoided. Typically 2–3 mm² is sufficient for dried soft tissue remains.[2] Gloves should be worn as well as face masks; alternatively, speaking during handling of the samples should be strictly avoided to prevent saliva aero-

[17] D. A. Lawlor, C. D. Dickel, W. W. Hauswirth, and P. Parham, *Nature (London)* **349**, 785 (1991).
[18] S. Pääbo, D. M. Irwin, and A. C. Wilson, *J. Biol. Chem.* **265**, 4718 (1989).

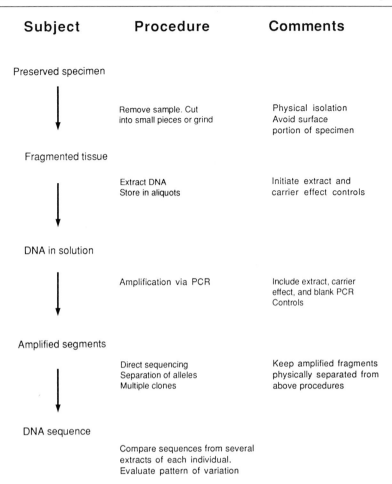

Subject   **Procedure**   **Comments**

Preserved specimen

Remove sample. Cut
into small pieces or grind

Physical isolation
Avoid surface
portion of specimen

Fragmented tissue

Extract DNA
Store in aliquots

Initiate extract and
carrier effect controls

DNA in solution

Amplification via PCR

Include extract, carrier
effect, and blank PCR
Controls

Amplified segments

Direct sequencing
Separation of alleles
Multiple clones

Keep amplified fragments
physically separated from
above procedures

DNA sequence

Compare sequences from several
extracts of each individual.
Evaluate pattern of variation

FIG. 1. Schematic of the overall procedure of determining DNA sequences from old tissue samples. The state of each sample is given, and the procedure and special comments are also shown. For detailed procedures and precautions, see the text.

sols. For cutting the specimens, sterile surgical blades are generally useful. For bone specimens, exchangeable saw blades washed in 1 N HCl and discarded after one use are appropriate. At this time an extract control should be initiated. Ideally, such a control would follow the same procedure as other samples without a sample being taken. Although not always employed, an additional control extraction of DNA from an unrelated organism may be useful to control for carrier effects.

Fragmentation of Tissue

The tissue sample needs to be fragmented into small pieces to allow the extraction buffer access to the cell remnants. This has to be done in a way that minimizes the risk of contamination. For soft tissues it is generally sufficient to cut the sample into fragments as small as possible with sterile surgical blades. For bone samples a Spex mill (Spex Industries, Inc., Edison, NJ) allows grinding the bone into a fine powder. Between samples, the steel cylinders of the mill are cleaned in 1 N HCl for about 10 min in order to hydrolyze carryover DNA without overly corroding the instrument. The acid treatment is followed by several washes in double-distilled water. All washings are performed in glass vessels that have been acid-washed as above. For the disruption of particularly tough formalin-fixed tissues, sonication in the presence of micro-sized glass beads has been useful.[19]

DNA Extraction

A DNA extraction protocol that has proved useful for most ancient tissues is a modification of the protocol initially published by Blin and Stafford.[20] Approximately 0.1 g of small pieces of soft tissue is added to 5 ml of extraction buffer containing 10 mM Tris-HCl (pH 8.0), 2 mM ethylenediaminetetraacetic acid (EDTA), 10 mM NaCl, 1% (w/v) sodium dodecyl sulfate (SDS), 10 mg/ml dithiothreitol (DTT), and 0.5 mg/ml proteinase K. Incubation at 37° with gentle agitation overnight will allow most or all of the tissue to go into solution. An equal volume of phenol, equilibrated with 1 M Tris-HCl (pH 8.0), is added. When the phenol is being equilibrated, care should be taken to use uncontaminated Tris buffer and to measure the pH only on aliquots that are removed from the water phase and then discarded. Two phenol extractions and one chloroform extraction are performed, and the water phase is concentrated and purified on a Centricon 30 microconcentrator (Amicon, Danvers, MA). The retentate can be stored frozen, preferably in a few aliquots. In all cases solutions should be manipulated with DNA-free positive displacement pipettes.

Bone samples can also be extracted by the above protocol, with yields similar to more specialized protocols. However, the most successful protocols involve the chelation of the hydroxyapatite by 0.5 M EDTA for prolonged periods of time.[3] The remaining decalcified tissue is made up of collagen and contains the trapped DNA. The near future will undoubtedly

[19] M. J. Heller, L. J. Burgart, C. J. TenEyck, M. E. Anderson, T. C. Greiner, and R. A. Robinson, *BioTechniques* **11,** 372 (1991).
[20] N. Blin and D. W. Stafford, *Nucleic Acids Res.* **3,** 2303 (1976).

see the development and evaluation of other methods for extraction of DNA from ancient bone. The use of Chelex[21] as well as the elution of DNA bound to hydroxyapatite by phosphate buffer deserve special attention. Alternative methods of DNA extraction may prove useful for many old tissue sources. Given the irreplaceable nature of archeological and museum samples, whatever procedure for DNA isolations is followed, a premium must be placed on the efficiency of extraction of amplifiable DNA, with little if any emphasis placed on ease of isolation.

Analysis of Extracted DNA

An aliquot of the extracted DNA can be run on an agarose or polyacrylamide gel and visualized by ethidium bromide staining. It should be noted that most ancient extracts contain compounds that fluoresce spontaneously with a blue-green fluorescence under UV light. This most likely represents Maillard products of reducing sugars that are formed over prolonged periods of time.[22] The presence of Maillard products in DNA extracts seems to correlate with inhibition of amplification, but it is not known if the inhibition is due to the products themselves. It should be further noted that ethidium bromide staining is not sensitive enough to detect traces of DNA that may be sufficient for the PCR. In many applications we therefore go ahead and analyze the extract by the PCR. Alternatively, the DNA can be labeled with radioactive nucleotides via DNA polymerase and visualized in a gel by autoradiography.[23]

Amplification of Sequences from Ancient DNA

To minimize the risk of contamination, we have eliminated all unnecessary components from the PCR buffer, such as ammonium sulfate and 2-mercaptoethanol. A typical 25 μl double-stranded amplification mixture contains 67 mM Tris (pH 8.8), 2 mM MgCl$_2$, bovine serum albumin (20 μg/ml), 1 mM of each deoxynucleoside triphosphate (dNTP), 1 μM of each primer, template DNA (10–1000 ng), and *Taq* polymerase (2 units, Perkin-Elmer Cetus, Norwalk, CT). Amplifications always include two important controls: an extract control, which is a blank extraction to control for contamination of extraction components as explained above, and several negative PCR controls, which should control for contamination of PCR components during preparation of reagents or setup of the amplification reactions. The controls should be made identical to the

[21] P. S. Walsh, D. A. Metzger, and R. Higuchi, *BioTechniques* **10,** 506 (1991).
[22] T. M. Reynolds, *Adv. Food Res.* **14,** 167 (1965).
[23] S. Pääbo, *Cold Spring Harbor Symp. Quant. Biol.* **51,** 441 (1986).

experimental amplifications by adding an equal volume of blank extract to the extract control and of water to the negative PCR controls.

Sequence Determination

Following successful double-stranded amplification, sequences of the product are preferably determined by direct sequencing or, alternatively, by cloning and sequencing. For a more thorough discussion of sequencing PCR products, see Thomas and Kocher.[24]

Contamination Precautions

As explained above, the amplification of nontarget templates (contamination) is the most pervasive problem faced in retrospective studies employing PCR. Precautions against contamination during sampling include avoiding biological contact with the specimen through the use of gloves, masks, and DNA-free instruments. Because specimens are often contaminated prior to sampling for DNA extraction, it is useful either to limit the sample to uncontaminated sections or to clean external surfaces.[16,25]

Carryover of Amplification Products

In our experience, most cases of contamination stem from carryover of previous PCR amplifications performed in the laboratory. This type of contamination can be curbed by physical separation of the area where DNA extractions are carried out and PCR experiments are set up from the area where PCR products are handled and analyzed after amplification. To be effective this separation has to be absolutely rigorous and include isolation of pipettes, plasticware, reagents, water, etc. It should be noted that micropipettors used to aliquot amplification products or concentrated DNA solutions may during aliquotting contaminate specimen extracts or new PCRs by aerosols for substantial times after the initial contamination occurred.

An elegant way to curb carryover of amplified DNA is to replace thymidine in the PCR with deoxyuridine, which can be eliminated from subsequent PCR amplifications by treating the reaction mixture exclusive of template DNA with uracil–DNA glycosylase, which is inactivated by heating before the PCR is done.[26] However, because the digestion with uracil–DNA glycosylase only reduces amplification of contaminating

[24] W. K. Thomas and T. D. Kocher, this volume [28].
[25] E. Hagelberg and J. B. Clegg, *Proc. R. Soc. London B* **244,** 45 (1990).
[26] M. C. Longo, M. S. Berninger, and J. L. Hartley, *Gene* **93,** 125 (1990).

templates from previous amplifications, this method cannot be regarded as more than a useful complement when working with ancient DNA; it does not eliminate the need to separate physically the pre- and post-PCR laboratory procedures, and it does not eliminate contamination caused by contemporary specimens or dTMP-containing amplified DNA.

Reagent Contamination

The second most common source of contamination is contemporary DNA present in chemicals used for DNA extraction or for the PCR. All reagents should be stored in small aliquots to facilitate replacing them when contamination occurs. Solutions and chemicals should be screened by performing control extracts and PCR amplifications as discussed above. Sometimes extensive testing is required before, for example, a source of uncontaminated water can be found. Once a clean source is identified, it should be aliquotted to avoid future contamination of the entire batch and used exclusively for PCR of ancient DNA.

Note that autoclaving will just decrease the size of DNA and is therefore not a suitable means of removing contaminants. Irradiation with UV light has been suggested and used as a method of destroying contaminating DNA.[8] This has the advantage that all components of the PCR except *Taq* polymerase and the DNA extract can be treated because the single-stranded primers are much less sensitive to ultraviolet irradiation than is double-stranded DNA. However, as with the uracil–DNA glycosylase method, because low levels of contamination may survive to be partially extended by jumping PCR or may appear in the PCR performed on extracts due to carrier effects (see below), such irradiation will only complement the physical separation methods.

Carrier Effects

The insidious nature of the contamination problem is illustrated by the fact that some contamination fails to be detected by the canonical controls. We (A. Cooper, M. Höss, and S. Pääbo, unpublished observations, 1990) have observed PCR experiments where controls did not show any traces of contamination and the ancient extract gave a clearly visible band. Yet these products proved (on sequencing) to be the result of carryover from a previous amplification. Even multiple controls failed to reveal traces of contamination, and this phenomenon has been observed on several occasions in different laboratories. These observations are consistent with either very low levels of contamination or ancient extracts serving as carriers for low-level contaminants that are not detected in the controls. One control that may be of use is to include an extract of unrelated ancient tissue for which the primers should not work. For example, one could use ancient plant extracts to control for mitochondrial DNA sequences in ancient animal remains. Further work is needed to elucidate the nature of the extract components responsible for the carrier effect.

Criteria of Authenticity

In experiments where museum skins and frozen liver samples were compared from the same individuals, in all cases identical sequences were obtained from both tissues.[27] Nevertheless, from the above discussion it is evident that even with the use of all controls and precautions discussed, some contaminations may occur. Therefore, it is important to evaluate the authenticity of the sequences that are determined from ancient sources. The following possibilities are offered to the investigator who has obtained a putative ancient DNA sequence.

Multiple Extracts

One of the obvious procedures that should be used is to remove more than one sample from different portions of a specimen and to extract the DNA from those samples on different occasions, preferably with different batches of reagents. The sequences determined should be identical in samples that stem from the same specimen.

Pattern of Variation

The positions at which sequences differences occur when compared to other ancient and modern individuals can be evaluated with regard to what is known about the functional constraints acting on a sequence. For example, substitutions in protein-coding genes that result in stop codons or unlikely amino acid replacements may invoke suspicion until verified by additional extracts.[28]

The ultimate reason for amplifying ancient DNA is to learn more about the history of species or populations. This is mostly done by the use of comparative analysis of DNA sequences. Such studies do, in themselves, provide indications as to the authenticity of the sequences obtained. For example, when ancient or old populations are studied, extensive knowledge of the present-day descendant populations is necessary not only to make inferences about the history of the populations, but also to allow the investigator to evaluate the authenticity of the sequences by comparing the

[27] M. F. Smith and J. L. Patton, *Trends Genet.* **7,** 4 (1991).
[28] S. Pääbo and A. C. Wilson, *Nature (London)* **334,** 387 (1988).

pattern of variation in the old and new samples.[2] Similarly, when extinct species are studied, phylogenetic analyses are expected to show that the species fall within the relevant families or higher orders. For example, the sequence that was determined for the marsupial wolf[16] fell within the radiation of extant marsupials but was not identical to any other marsupial that had been studied. Such observations can be taken to support the claim that a particular sequence is of ancient origin.

Error Frequency in Amplification Products

It is often observed that when PCR products from ancient remains are cloned, they contain a large number of random substitutions which are believed to be due to damage present in extracted DNA.[10] When this error frequency is compared to that of a control template of modern, presumably undamaged DNA, the increased number of errors itself may indicate that the PCR products stem from a damaged template and thus that it likely represents an old sequence.[13] However, it should be remembered that if the presumed ancient sequence in reality stems from a contaminating modern sequence present in only a few copies, the PCR product will have gone through many more cycles of polymerization before the plateau phase of the amplification is reached. For example, if an amplification product contains twice as many errors as a control DNA amplified for 30 cycles, that result could be due to errors in the initial template or to the presence of template in a 10- to 100-fold lower concentration (A. von Haeseler and S. Pääbo, unpublished, 1991). Thus, because the concentration of ancient DNA is hard or impossible to determine accurately, error rates in amplification products may be difficult to use as a criterion of authenticity.

Sample Size

A problem always faced in designing an evolutionary investigation is judging how many individuals need to be analyzed from a particular species or population. This problem manifests itself across the range of questions, from inferring species trees from gene trees to comparing variation within and between populations.

The necessary number of samples, of course, varies depending on the question being addressed. For populations, a useful measure of diversity at the nucleotide level is the number of nucleotide differences per site between randomly chosen genotypes. This estimate is related to many population parameters, such as the mean time to a common ancestor for genotypes in a population, and is a basis for evaluating differences between populations. As expected, when the number of individuals increases, the accuracy of the estimate increases. Relevant to the discussion, the variance of the estimate does not improve significantly beyond 10, or even 5, individuals.[29] This is not to say that an effective sample size is reached at 10 samples, only that the accuracy is not significantly improved. Therefore, when investigating variation at the population level, a sample size of 10 may be a good rule of thumb but may ultimately prove inadequate for the particular comparison. By contrast, for interspecific and especially higher-level taxonomic questions, levels of divergence often far exceed the intraspecific differences. In such cases, single specimens can suffice to test particular hypotheses.

Acknowledgments

We are especially grateful to A. Cooper, O. Handt, M. Höss, E. Prager, N. Tuross, T. White, and E. Zimmer for comments and advice on earlier versions of this chapter. This work received support from National Science Foundation grants to Allan C. Wilson.

[29] F. Tajima, *Genetics* **105**, 437 (1983).

[31] Analysis of DNA Sequence Variation at Population Level by Polymerase Chain Reaction and Denaturing Gradient Gel Electrophoresis

By ENRIQUE P. LESSA

Introduction

The study of DNA sequence variation at the population level is of much interest to population geneticists, molecular evolutionists, and systematists. By now, researchers in all these fields have realized that the polymerase chain reaction (PCR) and direct sequencing have greatly simplified the process of gathering data at the DNA level. The additional opportunities provided by methods for the physical separation of PCR products, however, have remained largely unappreciated. Here, we show how one such technique, denaturing gradient gel electrophoresis (DGGE),[1-3] can be combined with PCR amplification and direct sequencing to assess efficiently variation at the population level.

[1] R. M. Myers, T. Maniatis, and L. S. Lerman, this series, Vol. 155, p. 501.
[2] R. M. Myers, V. C. Sheffield, and D. R. Cox, *in* "PCR Technology: Principles and Applications for DNA Amplification" (H. A. Erlich, ed.), p. 71. Stockton, New York, 1989.
[3] V. C. Sheffield, D. R. Cox, and R. M. Myers, *in* "PCR Protocols: A Guide to Methods and Applications" (M. A. Innis, D. H. Gelfand, J. J. Sninsky, and T. J. White, eds.), p. 206. Academic Press, San Diego, 1990.

There are many procedures for retrieving or generating useful indicators of DNA sequence variation in natural populations. However, the full analytical power of population genetics can only be applied to a subset of such indicators. In general, population genetic analyses can be best utilized if the information gathered has the following characteristics. (1) Samples are sufficiently large and encompass several local populations across the area of interest. (2) The data consist of clearly defined orthologous sets, for example, haplotypes of specified regions of the mitochondrial genome and alleles of nuclear loci. (3) Information includes frequencies and distributions of alleles (and haplotypes) and of genotypes, as well as the underlying sequence differences. (4) Several loci are usually required to make inferences about the structure and history of populations. In animal systems, for example, comparisons of several nuclear loci, and comparing these with a mitochondrial locus, are desirable. Only through such comparisons can one make inferences about the population structure and history of the organisms from the idiosyncratic patterns of individual genes. (5) To be informative, the target loci should be variable at the population level.

In the discussion that follows, we focus on the application of PCR and DGGE to the study of nuclear loci in mammalian populations. Extensions to other organisms and genes, both nuclear and cytoplasmic, should be straightforward.

Denaturing Gradient Gel Electrophoresis: Basic Principles

Using a variety of approaches, DGGE is a versatile technique for the physical separation of DNA sequences that have only minor differences.[1] Only the application of this technique to populational analyses of PCR products is discussed here.

Two types of gels are used in DGGE. Perpendicular gels have linear gradients of denaturants from side to side, that is, the gradient is perpendicular to the direction of migration during electrophoresis. Parallel gradient gels have linear gradients of denaturants from top to bottom, that is, parallel to the direction of migration during electrophoresis.[1] Perpendicular DGGE is used at the beginning of a project to determine the "melting profile" of the amplified DNA segment. A single PCR product is loaded across the entire gel, which is cast with a full range of denaturants (from 0 to 80%). The results of perpendicular DGGE tell whether the PCR product behaves as one or more melting domains. Sharp inflections in the melting profile correspond to the melting points of the domain or domains. Because the gradient is linear, the concentration of denaturants of the observed inflection points can be easily measured on the gel. This information is used for narrowing the range of parallel gradient gels, which are used to compare multiple samples run side by side (Fig. 1).

Several conditions have to be met to achieve the full resolving power of DGGE. First, denaturing gradient gels are run immersed in an aquarium filled with buffer and kept at a constant temperature (usually 60°). The temperature is therefore the same across the entire gel, as well as between runs. Second, it has been shown that the resolving ability of DGGE is maximal when the amplified DNA segment behaves as a single melting domain, and consequently shows a single inflection point in a perpendicular gradient gel. This condition can be met by targeting segments of small to moderate size [up to about 500 base pairs (bp)] for amplification and by

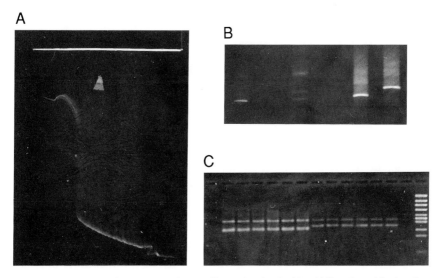

Fig. 1. (A) Perpendicular denaturing gradient gel stained with ethidium bromide showing a single inflection point. The gel is shown lying between a glass plate and a layer of plastic wrap. The plate may be lifted for accurately measuring the position of the inflection point. (B) Parallel denaturing gradient gel. Samples were heated at 95° for 2 min and allowed to cool before loading. Homozygotes show a single band, with differences in mobility denoting different alleles. A single specimen turned out to be a heterozygote and consequently shows several bands: the bottom two correspond to the homoduplex, allelic strands, and the top, fuzzy band groups are the heteroduplex molecules. Trailing smears in two of the lanes result from overloading. (C) Agarose minigel showing asymmetric amplifications of selected samples. Each set of six represents amplification of six samples with a different limiting primer. In each set, the first four samples were reamplified from agarose plugs, whereas the last two were reamplified from alleles cut from denaturing gradient gels. Direct sequencing of all products was equally successful.

attaching a GC clamp (a 40-base sequence of G and C residues) to the 5′ end of one of the primers.[4]

Finally every substitution affects the melting point of its domain, so that different alleles will have slightly different melting points. This allows the separation of PCR products differing by one or more substitutions on the basis of slight differences in melting point. To this end, the range of urea used in parallel DGGE should be centered about the melting point determined by perpendicular DGGE and extended 10 to 15 points in each direction. For example, for a melting point of 50% denaturants, parallel gels might be cast with gradients from 40 to 60% denaturants.

Application of Denaturing Gradient Gel Electrophoresis to Population-Level Studies

PCR-based studies of sequence variation generally begin by amplifying double-stranded copies of a target sequence. Then each strand is sequenced directly after either asymmetric amplifications or physical separation of the two strands. Under the right conditions, DGGE is sufficiently sensitive to allow the separation of double-stranded products differing by as little as a single base substitution.

The screening of PCR products by DGGE before sequencing has distinct advantages. (1) As an analytical technique, DGGE suffices to determine the genotypes of all sampled individuals and to estimate genotypic and allelic frequencies within and between populations. These data can be subjected to population genetic analyses and can be of great value in their own right. Also, heterozygotes are hard to determine unequivocally on sequencing gels, but easily identified by DGGE. (2)The information gathered by DGGE should help decide which samples should be used for sequencing. In principle, one homozygous representative of each allele should suffice, but replicas are in order. In any case, the number of alleles usually is much lower than the number of individuals, so that the efforts saved in sequencing may be significant. (3) DGGE can also be used as a preparative technique. Alleles are in effect gel-purified during DGGE, and samples taken from the gels can be reamplified and sequenced (Fig. 1).[5] Some alleles present in low frequencies may not be found in homozygous conditions but may be of much biological significance (e.g., rare alleles in hybrid zones or as indicators of gene flow). Their identification and sequencing are greatly simplified by DGGE.

[4] V. C. Sheffield, D. R. Cox, L. S. Lerman, and R. M. Myers, *Proc. Natl. Acad. Sci. U.S.A.* **86**, 232 (1989).
[5] E. P. Lessa, *Mol. Biol. Evol.* **9**, 323 (1992).

Primer Design and Experimental Approach

No sequence data are available for most living species. Therefore, one must rely on data for other species to choose the genes to be studied and to design oligonucleotide primers for PCR amplifications and sequencing. The endeavor is more likely to succeed and the data more likely to be interpretable if the target genes are chosen on the following bases. (1) Single-copy genes or members of well-known gene families with limited copies should be chosen to ensure that alleles are not confused with paralogous sequences. (2) Ideally, the genes should include conserved regions for designing primers interspersed with variable regions to be screened for variation. (3) To allow maximal resolution in DGGE, the total size of the amplified products should be no greater than about 500 bp. The known intron/exon structure of many eukaryotic genes provides an easy, but not unique, way of meeting these requirements. Introns are less constrained than exons and, consequently, are good candidates to be variable within populations. At the same time, flanking exons are likely conserved, so that primers may be designed on the basis of available sequences.

Figure 2 illustrates the use of these ideas in the study of a single locus, namely, intron 1 of the adult β-globin of pocket gophers.[5] The fact that this is a member of a multigene family might create some complications. The PCR might amplify more than one adult β-globin (many mammals have two of these globin genes), the embryonic β-globins, or even some of the pseudogenes in the family. Furthermore, the conserved regions in exons are very similar in all functional genes of the family. The procedures outlined in Fig. 2 are designed to overcome these potential problems and effectively target the gene of interest. Note that the GC clamp is eliminated from asymmetric amplifications and sequencing reactions by this protocol.

Under the conditions outlined in the sections above, DGGE should resolve the vast majority of alleles. However, it still is possible for two alleles to have converged to very similar, indistinguishable, melting domains. Two precautions can be taken to ensure even greater resolution. First, "heteroduplex analysis"[6] should be used routinely as follows. The double-stranded products should be briefly denatured at 95° and allowed to renature before being loaded on the gels. In heterozygous samples, this will regenerate the two original alleles and also create hybrid, heteroduplex molecules. Because these molecules have at least one mismatch, they are guaranteed to have lower melting points and will separate from the homoduplex, allelic molecules during DGGE (Fig. 1). Such additional bands will indicate the presence of hidden alleles in those individuals and should prompt their sequencing in search for variable sites.

[6] E. S. Abrams, S. E. Murdaugh, and L. S. Lerman, *Genomics* 7, 463 (1990).

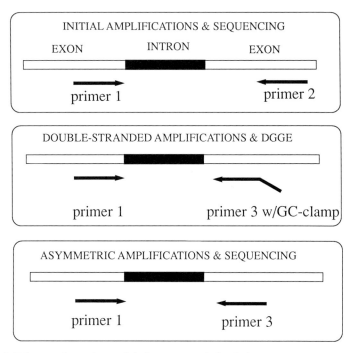

Fig. 2. Primer and experimental design to use variation in intron sequences for the study of population structure. Primers 1 and 2 are designed on the basis of alignments of available sequences of taxa related to the one under study. They are used in a pilot project to generate initial sequences of the target gene. The identity of the amplified product(s) should be verified by comparisons with published sequences. Primer 3 is designed on the basis of the initial sequences and used instead of primer 2 to ensure specificity of subsequent PCR amplifications. One version of primer 3 has a 40-base GC clamp on the 5′ end and is used with primer 1 for double-stranded PCR amplifications. The double-stranded products are screened for allelic variation by DGGE. Reamplifications of selected samples (agarose plugs of homozygous individuals or alleles purified during parallel DGGE) and direct sequencing are carried out with primer 1 and a version of primer 3 without the GC clamp.

Second, common sense should be applied in choosing which specimens to use to generate replicate sequences of alleles identified as identical by DGGE. For example, specimens representing the extremes of the observed geographic distribution of the alleles might be chosen.

Even though DGGE has been fine-tuned to be a very powerful method, organismal and molecular biologists alike often express concerns about the lack of absolute guarantees of complete resolution. This is a mistake that can be corrected by thinking in terms of sampling efficiency. One could spend all available resources in an exhaustive analysis of a few samples. The resulting sense of security is a false one. Such data are definitive only about the samples in question but are probably a crude representation of the population of interest. On the other hand, a significantly larger sample will provide a more refined picture of the population even if a small margin of uncertainty is introduced into the method. Obviously, if some samples are absolutely crucial for a certain project, nothing short of a full analysis should be accepted.

Extensions: Surveying Gene Families and Mitochondrial DNA Haplotypes

In principle, it is possible to relax the precautions directed at obtaining amplifications from a single locus in order to gather data on several paralogous genes simultaneously. This may be done without losing the ability to discriminate loci if the number of target genes remains limited. Larger gene families can be screened for variation by a similar approach, but the locus-by-locus assessment will be difficult. Choosing among these alternatives will depend on the relative merits, for the project in question, of (1) having clearly identified allelic variation amenable to population genetic studies, (2) simply finding "markers" of genetic differentiation, and (3) focusing on diversity of the gene family itself. Regardless of the option taken, it is always advisable to check the sequences against both orthologous and paralogous sequences of other organisms when working with nuclear genes.

Like nuclear alleles, mitochondrial haplotypes can be sorted out on denaturing gradient gels. DGGE can be very useful in detecting heteroplasmy. Otherwise, the main applications to mitochondrial PCR products are screening for variability and minimizing sequencing efforts. The savings can range from trivial, if hypervariable regions are being studied, to substantial for relatively conserved genes.

Alternative Methods

In determining how best to screen for allelic variation in PCR products, the following alternatives to DGGE should also be considered. Temperature gradient gel electrophoresis[7] (TGGE) is conceptually similar to DGGE, except that it works by establishing gradients based on temperature, rather than urea. In principle, most of the issues addressed here apply to TGGE.

Single-strand conformation polymorphism[8] (SSCP) analysis identifies

---

[7] R. M. Wartell, S. H. Hosseini, and C. P. Moran, Jr., *Nucleic Acids Res.* **18**, 2699 (1990).
[8] M. Orita, Y. Suzuki, T. Sekiya, and K. Hayashi, *Genomics* **5**, 874 (1989).

allelic variation on several types of acrylamide gels at various temperatures. It is based on the fact that even minor differences in sequence may change the conformation of single-stranded products and allow the separation of variants by gel electrophoresis. However, the conditions leading to the separation of alleles are difficult to understand and control. Also, the sizes of the segments must be under 200 bp to allow for efficient separation and to minimize the number of alternative conformations each single-stranded product may take. One advantage of SSCP is that it works with regular sequencing gel boxes. If a project can benefit from the identification of variants and can afford missing alleles, SSCP is a good alternative. The original description of the method incorporates radioactive labeling into the PCR protocol, but this can probably be avoided by using ethidium bromide staining.

Heteroduplex analysis can also be used in combination with long (overnight) runs on Hydrolink sequencing gels to detect allelic variation.[9] The main appeal of this approach, as for SSCP, is its simplicity, but the factors controlling the level of resolution of this technique are unknown.

Finally, restriction fragment analysis of PCR products can sometimes compensate for its inability to uncover all substitutions by allowing one to survey larger DNA segments (up to several kilobases in length). It must be noted that DGGE can be coupled with restriction fragment analysis to allow the detection of a fraction of the mutations that do not affect restriction sites.[3]

In spite of the unquestionable appeal of working at the DNA level, population biologists should avoid discarding "old" techniques without serious consideration of the advantages and disadvantages of the various options. With all its limitations, allozyme electrophoresis still has no rival in allowing for the rapid and inexpensive surveying of large numbers of loci and individuals.

Procedures

Detailed descriptions of equipment, stock solutions, and protocols are available elsewhere.[1-3,10] Here, we provide an outline of the procedures used to survey allelic variation in a single-copy nuclear gene within and between populations of wild mammals. The procedures can easily be modified for other purposes.

[9] J. Keen, D. Lester, C. Inglehearn, A. Curtis, and S. Bhattacharya, *Trends Genet.* 7, 5 (1991).
[10] R. M. Myers, V. C. Sheffield, and D. R. Cox, *in* "Genome Analysis: A Practical Approach" (K. Davies, ed.), p. 95. IRL Press, Oxford, 1988.

Perpendicular Denaturing Gradient Gel Electrophoresis

1. Prepare a 100-μl reaction for double-stranded amplification of a single specimen for perpendicular DGGE. Use the same primer combination planned for subsequent comparisons of multiple samples. One of the primers should include a GC clamp.

2. Cast a perpendicular gradient gel with 0–80% denaturants from side to side. The gel is poured sideways and should not be moved until the acrylamide polymerizes.

3. Mix 40–80 μl of the amplified product with 10 μl of the DGGE dye. Load samples evenly across the entire upper surface of the gel. Run at 60° and 150 V for about 5 hr.

4. Separate the glass plates. Place the one with the gel on a tray. Gently flush the gel with water for about 5 min.

5. Pour 1 ml of ethidium bromide solution (1 μg/ml) on top of the gel. Cover with plastic wrap or with a plastic plate transparent to ultraviolet light.

6. After about 15 min, invert on a transluminator. Remove the glass plate after verifying that the DNA has been stained. If necessary, increase the contrast by flushing the excess ethidium bromide with water. Determine the inflection point by measuring its position along the gradient directly on the gel or on an instant picture.

Parallel Gradient Gels

1. Obtain double-stranded amplifications of all samples with the same pair of primers used for perpendicular DGGE (one should include a GC clamp). Use total reaction volumes of 25 or 12.5 μl. Run on agarose minigels and take plugs by puncturing with clean, positive displacement pipettes. Save the plugs in 250 μl of LTE buffer.

2. Cast a parallel gradient gel, with the gradient centered about the estimated inflection point and spanning a total range of 20–30 percentage points. Gently insert an inverted sharkstooth comb (as for any acrylamide gel), making sure the gradient is not disturbed. Do not move the gel before it polymerizes.

3. Invert the comb, place the gel box in the aquarium at 60°, and fill the top chamber with buffer.

4. In Eppendorf tubes, mix 2–6 μl of each double-stranded amplified product with 1 μl of DGGE dye. Heat at 95° for 1–2 min, then allow samples to cool. Flush the wells with buffer and load the samples.

5. Run gels at 60° and 150 V. Running times vary and must be optimized by running an experimental gel, loading samples at 1.5-hr intervals. Time is deemed to be sufficient if the samples reach the center of the gel

(i.e., the melting point) and/or allele separation is observed. Exceeding minimal running times usually is not a problem, since allele migration slows dramatically once the melting point is reached.

6. Stain and visualize gels as above. If needed, take samples of individual alleles by cutting small pieces of the gel with a clean scalpel or razor blade. Avoid cross-contamination. Place cut pieces in 250 μl of LTE buffer.

Sequencing Alleles Identified by Denaturing Gradient Gel Electrophoresis

1. Determine which samples need to be sequenced. Place tubes with alleles sampled from agarose or denaturing gradient gels in LTE in a heating block at 65° for 15 min. Even though the acrylamide pieces never dissolve, enough DNA diffuses into the buffer to allow reamplification.

2. Use 25 μl of the melted products to prepare 50-μl reactions for asymmetric amplifications by PCR. Use primer versions without GC clamps.

3. Evaluate amplifications on an agarose minigel, then proceed to sequence the alleles directly.[11]

Acknowledgments

I am grateful to Cristián Orrego, James Patton, Kelley Thomas, and Allan Wilson for discussions of DGGE. Financial support was provided by National Science Foundation Grant BSR-88-22345 (to J. L. Patton) and by an Alfred P. Sloan Postdoctoral Fellowship in Molecular Studies of Evolution.

[11] W. K. Thomas and T. D. Kocher, this volume [28].

[32] Detection of Variability in Natural Populations of Viruses by Polymerase Chain Reaction

By SHENG-YUNG CHANG, ANDY SHIH, and SHIRLEY KWOK

Introduction

The retroviruses associated with AIDS (autoimmune deficiency syndrome), namely, human immunodeficiency virus types 1 and 2 (HIV-1, -2), display considerable genome sequence variability not only between isolates from different individuals but also between isolates from the same individual over time.[1-5] This diversity is due largely to the extensive misincorporation of nucleotides by the viral reverse transcriptase, which lacks the proofreading $3' \rightarrow 5'$-exonuclease activity.[6] Understanding the nature of sequence changes in the HIV-1 genome may be central to understanding the pathogenesis of the virus and the immune response to infection. For example, is there any correlation between the viral strain and the clinical course of infection? Are there variants with envelope glycoprotein sequences that will serve as more effective antigenic targets for destruction by the immune system?

Conventional methods for molecular characterization of viral isolates involve cultivation of the agent, extraction of nucleic acid, and analysis by restriction enzyme mapping. Higher analytical resolution generally requires more time-consuming procedures such as molecular cloning and sequencing of entire genomes. The polymerase chain reaction (PCR)[7,8] provides a rapid means for analyzing genomic diversity.[4,5] There are two major advantages in using the PCR to study viral variants. First, it obviates the need to propagate the virus in culture, a procedure which is expensive, time-consuming, and biohazardous. Second, the PCR allows detection of naturally occurring isolates and circumvents the viral selection which often occurs during cultivation.[5]

To increase the validity of PCR results, reaction conditions must be carefully designed to minimize biased selection and amplification artifacts. For example, the selected primers should amplify all variants present. The cloning and selection of clones for sequencing should be unbiased. Moreover, the intrinsic misincorporation rate of *Taq* DNA polymerase during the PCR should be maintained at a level significantly lower than the assumed/expected variation within the sequence to be analyzed. Finally, the frequency of PCR-mediated recombination should be kept to a minimum. In the following sections, we address amplification strategies and strategies to maximize *Taq* polymerase fidelity and minimize PCR-me-

[1] B. H. Hahn, M. A. Gonda, G. M. Shaw, M. Popovic, J. A. Hoxie, R. C. Gallo, and F. Wong-Staal, *Proc. Natl. Acad. Sci. U.S.A.* **82**, 4813 (1985).

[2] M. S. Saag, B. H. Hahn, J. Gibbons, Y. Li, E. S. Parks, W. P. Parks, and G. M. Shaw, *Nature (London)* **334**, 440 (1988).

[3] S. Peng, Y. Shlesinger, E. S. Daar, T. Moudgil, D. D. Ho, and I. S. Y. Chen, *AIDS* **6**, 453 (1992).

[4] M. Goodenow, T. Huet, W. Saurin, S. Kwok, J. Sninsky, and S. Wain-Hobson, *JAIDS* **2**, 344 (1989).

[5] A. Meyerhans, R. Cheynier, J. Albert, M. Seth, S. Kwok, J. Sninsky, L. Morfeldt-Manson, B. Asjo, and S. Wain-Hobson, *Cell (Cambridge Mass.)* **35**, 901 (1989).

[6] D. A. Steinhauer and J. J. Holland, *Annu. Rev. Microbiol.* **41**, 409 (1986).

[7] K. B. Mullis and F. Faloona, this series, Vol. 155, p. 335.

[8] R. K. Saiki, S. Scharf, F. Faloona, G. T. Horn, H. A. Erlich, and N. Arnheim, *Science* **230**, 1350 (1985).

diated recombination. Protocols for amplification, cloning, and sequencing of naturally occurring variants are presented. Although initially designed for HIV-1 detection, these strategies should be applicable to all variable gene targets.

Amplification Strategies

Primer Selection

Given the inherent genetic variability of the HIVs, detection by PCR requires the identification of primers that will amplify all variants. Therefore, only highly conserved nucleic acid regions should be considered as primer sites. Restriction endonuclease sites can be incorporated into the 5' ends of the primers to facilitate cloning. Primers are typically positioned 300–350 base pairs (bp) apart as products of this size can be analyzed in their entirety on a single sequencing gel. The suitability of a selected oligonucleotide to serve as a primer for PCR amplification is dependent on several factors, including (1) the effects of mismatched bases on primer–template duplex stability, (2) the location of the mismatched bases within the primer, (3) the kinetics of association and dissociation of primer-template duplexes at the annealing and extension temperatures, and (4) the efficiency with which the polymerase can recognize and extend a mismatched duplex.

Although the number and location of mismatched bases will affect PCR amplification, mismatches at the 3' terminus are expected to have the greatest effect on the PCR. Kwok et al.[9] reported on the effects of 3'-terminal mismatches on amplification of HIV sequence. In their system, A–G, G–A, C–C and A–A mismatches were detrimental while all other mismatches had minimal effect. Most notably, oligonucleotides with a 3'-terminal T served efficiently as primers even when mismatched with T, C, or G. The concentration of deoxynucleoside triphosphates (dNTPs) in the reaction mix also affects *Taq* polymerase extension. For example, whereas most 3'-terminal mismatches with the exception of A–G, G–A, C–C, and A–A amplified efficiently in the presence of 800 μM dNTPs, only T–G and G–T mismatches and perfectly matched sequences amplified in the presence of 6 μM dNTPs. Thus, 3'-terminal mismatches can be more efficiently extended if the primers terminate in a T and if amplifications are carried out at higher dNTP concentrations.

Oligonucleotide primers that have 25–30 bases of complementarity

[9] S. Kwok, D. E. Kellogg, N. McKinney, D. Spasic, L. Goda, C. Levenson, and J. J. Sninsky, *Nucleic Acids Res.* **18,** 999 (1990).

with the target sequence can better accommodate mismatches than shorter oligonucleotides. An HIV-1 primer with as many as 5 mismatches within a 28-mer was efficiently extended when a relatively relaxed annealing temperature of 55° was employed.[9] At 60°, however, amplification with the primer was dramatically reduced. To minimize amplification bias, the annealing temperature should be sufficiently relaxed so that minimally mismatched templates can be amplified, yet sufficiently stringent to prevent nonspecific amplifications from predominating the reactions. The optimal amplification condition needs to be empirically determined for each primer–pair system.

Strategies to Improve Specificity of Amplifications

To facilitate cloning and to increase the probability that all variants will be represented, amplifications should be as specific and efficient as possible. Ideally, the amplified product should be visible by ethidium bromide staining of agarose gels.

Several strategies can be used to enhance amplification specificity. The most commonly used method is the "nested" PCR.[7] The potential drawbacks with nested PCRs are 2-fold. First, successful amplification requires conservation of four primer binding regions and may compromise the range of variants detected. Second, the nested PCR, as commonly performed, lends itself to contamination as products from the first amplification are openly transferred to a new tube for the second amplification.

"Hot start" PCRs can also greatly enhance PCR specificity.[10,11] In this procedure, one of the critical components for amplification (commonly *Taq* polymerase) is added to the tube only after the reaction temperature reaches above the T_m (melting point) of the primers. This allows for primers that annealed nonspecifically at the lower temperatures to be melted off prior to addition of *Taq* polymerase. As a consequence, unique amplification products can often be obtained from low copy number targets that are present in a vast background of genomic DNA.

The incorporation of dUTP/ung (uracil *N*-glycolase)[12,13] into amplification reactions has two advantages. In addition to preventing dU-carryover products from serving as templates in the PCR, the incorpora-

[10] F. Faloona, S. Weiss, F. Ferre, and K. Mullis, *Abstract: Sixth International Conference on AIDS* **2**, 318 (1990).
[11] R. T. D'Aquila, L. J. Bechtel, J. A. Videler, P. Gorcyca, and J. C. Kaplan, *Nucleic Acids Res.* **19**, 3749 (1991).
[12] M. C. Longo, M. S. Berninger, and J. L. Hartley, *Gene* **93**, 125 (1990).
[13] D. E. Kellogg, C. Gates, B. Dragon, K. Kung, J. Wang, S. Kwok, and J. J. Sninsky, *Abstract: Seventh International Conference on AIDS* **2**, 59 (1991).

tion of dUTP/ung also yields more specific amplifications (S. Kwok, unpublished results, 1991). Specifically, incubating the preamplification reactions with ung for 2 min at 50° prior to temperature cycling dramatically reduces nonspecific amplifications. Presumably, both specific and nonspecific dU extensions that occur during PCR setup are cleaved by ung at 50°. Because the temperature in subsequent cycling is maintained at or above the annealing temperature, a pretreatment with ung mimics "hot start" in that all products are cleaved prior to the initial cycling at 95°. Initial studies suggest that ung pretreatment may be a productive avenue to explore if an increase in PCR specificity is required.

Cloning and Sequencing of Amplified Product

Depending on the extent of variability, results from direct sequencing of amplified products may be difficult to interpret, and sequencing of cloned products may be required. Cloning can be expedited by adding different restriction enzyme recognition sequences to the 5' end of each primer. It should be noted, however, that variants which harbor either of these restriction endonuclease sites within the amplified region may not be successfully cloned. Such biases can be avoided by cloning of blunt-ended PCR products.

Probes are often used to identify clones of interest. However, just as primers can bias the amplification of variants, so too can the sequence of the oligonucleotide probe influence selection of candidate variants. A long probe that spans the region flanked by the primers coupled with low to moderate hybridization stringencies will minimize exclusion of divergent isolates. In the event of suspected extreme divergence, clones that contain inserts of the correct size should be sequenced even if hybridization to the probe is not observed.

A preferred strategy is to purify the amplified products from the region of expected size(s) from the agarose gel and clone them into the polylinker sites of a high copy number plasmid vector. The size(s) of candidate clones is confirmed by PCR with primers that flank the polylinker region in the vector. We have observed that at least 95% of clones contain inserts of interest. This method circumvents potential selection by the probe.

Fidelity of *Taq* DNA Polymerase

When using the PCR to study variability in a natural population of viruses, an important consideration is the fidelity of *in vitro* DNA synthesis catalyzed by *Taq* polymerase. Unlike enzymes such as T4 polymerase and the Klenow fragment of *Escherichia coli* DNA polymerase I, *Taq* polymerase lacks $3' \rightarrow 5'$ proofreading exonuclease activity.[14] Misincorporations with this enzyme may lead to either chain termination or an elevated error rate (misincorporation per nucleotide per cycle) or frequency (accumulative misincorporation per nucleotide). Chain termination is usually benign to sequence interpretation since partial extension products will not contain both primer sites and will not be propagated in subsequent reaction cycles. However, if the misincorporated bases are extended, especially in the early rounds of amplification, the observed variability within a viral population will be overestimated.

The error rate of *Taq* polymerase during the PCR was originally estimated to be about 1×10^{-4}/cycle for substitution events[14,15] and 3×10^{-5}/cycle for frame-shift events,[15] with most point mutations being transitions rather than transversions. More recently, it has been shown that the observed error rate of *Taq* polymerase can range between 1×10^{-3} and less than 1.2×10^{-5} for a given PCR amplification, depending on the reaction conditions and the nature of the template sequence.[4,16,17] Reaction parameters such as $MgCl_2$ concentration, pH, and dNTP concentration all have significant effects on enzyme error rate. For instance, a 60-fold reduction in polymerase fidelity was observed simply by changing the reaction pH from 5 to 8.2. Similar compromises in fidelity were seen when the $MgCl_2$ concentration was in molar excess relative to the total dNTP concentration of the reaction. In general one can expect, at a reaction pH of 7 to 8, an accumulative error frequency of approximately 1 in 5000 as the molar ratio of $MgCl_2$ to dNTPs approaches 10. In fact, according to Eckert and Kunkel,[17] the polymerase error rate can be kept to a minimum ($< 10^{-6}$/cycle) only when the $MgCl_2$ concentration is equimolar to the total reaction dNTP concentration and the reaction pH is maintained at 5 or 6 (70°).

Minimizing the polymerase error rate by maintaining certain reaction parameters may not always be experimentally feasible, but some optimization of reaction conditions to maximize fidelity should always be considered when determining variability within a viral population by the PCR. In the absence of such optimization, knowing the expected error frequency of *Taq* polymerase under a given reaction condition will allow a more realistic assessment of viral population variance. To this end, the amplification and sequencing of an appropriate plasmid control template will serve as an

[14] K. R. Tindall and T. A. Kunkel, *Biochemistry* **27**, 6008 (1988).
[15] R. K. Saiki, D. H. Gelfand, S. Stoffel, S. J. Scharf, R. Higuchi, G. T. Horn, K. B. Mullis, and H. A. Erlich, *Science* **239**, 487 (1988).
[16] P. Keohavong and W. G. Thilly, *Proc. Natl. Acad. Sci. U.S.A.* **86**, 9253 (1989).
[17] K. S. Eckert and T. A. Kunkel, *Nucleic Acids Res.* **18**, 3739 (1990).

indicator of *Taq* polymerase fidelity for a given PCR experiment and aid in defining the level of confidence in sequence interpretation.

The fidelity of *Taq* polymerase under the conditions described in the section on experimental procedures is summarized in Table I. Fifty molecules of linearized plasmid DNA pSYC1857[18] (rearranged, inactivated HIVZ6 which was cloned between the *Nru*I and *Bam*HI sites of pBR322) were amplified with primers to the conserved regions flanking the hypervariable VI/V2 region of the *env* gene. Fifty picomoles of each primer in PCR buffer containing 1.5 mM MgCl$_2$, either 0.4 or 0.8 mM total dNTPs, and 2.5 units *Taq* polymerase were used for each reaction. After two rounds of 35 cycles of nested PCR, the products were cloned and sequenced. A total of 4.6 and 3.6 kilobases (kb) of PCR products from amplifications performed in the presence of 0.4 and 0.8 mM dNTPs, respectively, were sequenced. The relatively low error rate is consistent with results of other studies that used similar conditions.[15-17]

Jumping Polymerase Chain Reaction

The PCR is known to generate certain undesirable products, such as "hybrid," or "recombinant," or "shuffled," or "heteroduplex" clones if the target DNA has more than one type of sequence.[15,19-23] These PCR artifacts come from (1) partial PCR products of the first few PCR cycles

TABLE I
FIDELITY OF *Taq* DNA POLYMERASE

Total dNTP concentration (mM)	Mutations		Error frequency	Error rate/cycle
	Number	Type		
0.4	5	Transition	1.3×10^{-3}	3.7×10^{-5}
	1	Single-base deletion	(1/770)	
0.8	5	Transition	1.4×10^{-3}	4.0×10^{-5}
			(1/720)	

[18] C. Hart, S.-Y. Chang, S. Kwok, J. Sninsky, C.-Y. Ou, and G. Schochetman, *Nucleic Acids Res.* **18,** 4029 (1990).
[19] S. J. Scharf, A. Friedmann, C. Brautbar, F. Szafer, L. Steinman, G. Horn, U. Gyllensten, and H. A. Erlich, *Proc. Natl. Acad. Sci. U.S.A.* **85,** 3504 (1988).
[20] A. R. Shuldiner, A. Nirula, and J. Roth, *Nucleic Acids Res.* **11,** 4409 (1989).
[21] S. Paabo, D. M. Irwin, and A. C. Wilson, *J. Biol. Chem.* **265,** 4718 (1990).
[22] A. Meyerhans, J. Vartanian, and S. Wain-Hobson, *Nucleic Acids Res.* **18,** 1687 (1990).
[23] R. Jansen and F. D. Ledley, *Nucleic Acids Res.* **18,** 5153 (1990).

hybridizing to different target sequences in the subsequent PCR cycles and (2) *in vivo* repair of heteroduplex DNA after cloning. We use the term "jumping PCR" here to describe these two phenomena.

DNA templates damaged by freeze–thaw cycling during storage, by nuclease contamination of the sample, or by a short extension time of the *Taq* polymerase at 72° can all contribute to the formation of partially extended PCR products. To minimize the jumping of the partial PCR products between different templates, long extension times should be used, and a two-step PCR (i.e., only cycling between the denaturing and annealing temperatures) should be avoided. Freshly prepared DNA templates should be used if possible. The exponential accumulation of PCR products plateaus when the concentration of the products reaches 10^{-8} to 10^{-7} M.[24] At that point, 5–10 pmol of short products (several hundred base pairs) are present in a typical 100-μl reaction. If the amplification is not stopped after the plateau stage, heteroduplex DNA would start to form from the separation and reannealing of different sequences in the subsequent cycles. On uptake into the host cells by transformation, the heteroduplex DNA molecules may be repaired before replication takes place, resulting in the generation of recombinant molecules. Therefore, the amount of input DNA target and the number of cycles should be carefully monitored. It is difficult to eliminate jumping PCR totally, but there are many parameters (i.e., number of cycles, amount of input DNA, duration of *Taq* polymerase extension) that can be optimized to lower the possibility of creating more artificial variabilities during PCR.

Experimental Procedures

Preparation of Samples

Fast and simple methods for extraction of DNA or RNA from cells, whole blood, and other fluids have been reviewed by Bloch[24] and Kawasaki.[25] In general, target DNA is released from the cells or virus by proteinase K digestion in the presence of detergent followed by heat inactivation of the proteinase K. A small aliquot of the cell lysate is then used directly as the template in the amplification reaction. For cell culture supernatant fractions, the proviral DNA released from the lysed cells can be directly

[24] W. Bloch, *Biochemistry* **30**, 2735 (1991).
[25] E. S. Kawasaki, *in* "PCR Protocols: A Guide to Methods and Applications" (M. A. Innis, D. H. Gelfand, J. J. Sninsky, and T. J. White, eds.), p. 146. Academic Press, San Diego, 1990.

amplified after heating the supernatant to 65° for 30 min to inactivate the nucleases.[26]

To extract DNA and RNA from serum samples or from culture supernatant fractions which contain a low level of proviral DNA, the IsoQuick extraction kit (a modified guanidinium salt–organic solvent extraction method) by MicroProbe Corporation (Bothell, WA) has been used successfully by our group. Following the manufacturer's recommended procedures for total nucleic acid extraction, the final DNA–RNA suspension can be amplified directly to detect proviral DNA. Using cDNA templates synthesized from total nucleic acids with random hexamer priming generally increases the PCR product yield.

A protocol for cDNA synthesis[27] from RNA extracted from a 20- to 50-μl serum sample using the "total nucleic acids extraction" procedures of the IsoQuick kit is as follows. Resuspend the DNA–RNA pellet in 20 μl of RNase-free water. Use one-half for cDNA synthesis by combining 10.00 μl of DNA/RNA or RNA suspension, 5.00 μl of 25 pmol/μl random hexamer primer, and 18.75 μl of RNase-free water. Heat to 63° for 3 min, then put on ice. Add the following: 5.00 μl of 10 × PCR buffer [100 mM Tris (pH 8.3), 500 mM KCl, 15 mM MgCl$_2$, 0.1% (w/v) gelatin], 5.00 μl of 10 mM dithiothreitol (DTT), 2.50 μl of 10 mM dNTPs (total), 1.25 μl of 40 units/μl RNase inhibitor (RNasin, Promega, Madison, WI), and 2.50 μl of 200 units/μl Moloney murine leukemia virus (MMLV) reverse transcriptase (Bethesda Research Laboratories, Gaithersburg, MD). Incubate at 37° for 1 hr, boil for 5 min, and put the tube on ice.

Amplification

Nested Polymerase Chain Reaction. For the first round of amplification, 20 to 50 μl of cell lysate, 2 μl of DNA–RNA or 10 μl DNA–cDNA is amplified for 30–35 cycles in an 100 μl-reaction volume. As products from the first amplification are usually not visible on the gel with ethidium bromide staining, 5 μl of the first amplification products is then used to seed a second 100 μl reaction and amplified for another 35 cycles.

Hot Start. To minimize nonspecific PCR, hot start can be incorporated into the amplification protocol.[11] Reaction mixes containing DNA template, primer pair and dNTP in PCR buffer are heated to 80°, and an aliquot of diluted *Taq* polymerase is then added to the mixture to initiate the amplification reaction.

---

[26] S. Wain-Hobson, J. P. Vartanian, M. Henry, N. Chenciner, R. Cheynier, S. Delassus, L. Pedroza Martins, M. Sala, M. T. Nugeyre, D. Guetard, D. Klatzmann, J. C. Gluckman, W. Rozenbaum, F. Barre-Sinoussi, and L. Montagnier, *Science* **252**, 961 (1991).

[27] G. T. Gerard, *Focus* **9**, 5 (1987).

Cloning and Sequencing of Amplification Products

Preparation of Sticky Ended Amplification Fragments. Combine 100 μl of PCR products, 100 μl of 4 M ammonium acetate, and 200 μl of 2-propanol. Mix the contents of the tube well and incubate at room temperature for at least 10 min. Centrifuge to pellet the DNA for 10 min, remove the supernatant, wash the pellet with 70% (v/v) ethanol, and dry. Digest the resuspended PCR products with restriction enzymes in a total volume of 30 μl. Load one-fifth of the reaction into one lane of a minigel that consists of 2 or 3% NuSieve GTG agarose (FMC, Rockland, ME) plus 0.5% agarose and electrophorese. Excise the PCR fragments from the gel at the region of the expected size(s) and place into a 0.5-ml microcentrifuge tube that is punctured with a 23-gauge needle at the bottom and plugged with siliconized glass wool. Collect the liquid from the gel piece into a 1.5-ml microcentrifuge tube by centrifuging at 10,000 rpm for 10 min at room temperature. Add 25 to 50 μl of water to the gel fragment and spin again to collect the liquid to increase the recovery. Ethanol-precipitate the DNA fragments, wash the DNA pellet with 70% ethanol, and dry.

Preparation of Blunt-Ended Amplification Fragments. Kinased primers are used in the PCR, and the ends of PCR fragments are "flushed" with *E. coli* DNA polymerase I and dNTPs to increase the efficiency of ligation. The fragments are then purified as described for the sticky ended fragments.

Vectors. High copy number plasmid vectors like the pUC series can be used for the cloning of PCR products. Standard ligation and transformation procedures are used for cloning.[28]

Confirmation of Candidate Clones with Polymerase Chain Reaction. M13 or pUC sequencing and reverse sequencing primers that flank the polylinker region are used to confirm, by PCR, the size(s) of a large number of the transformants. First, 50-μl PCR mixtures containing 10 pmol of each primer, 0.1 mM of each dNTP, and 1.25 units of *Taq* polymerase in PCR buffer are aliquoted into tubes. Then, cells from a single colony are transferred with a toothpick into the PCR mixture after first patching the cells on a fresh master plate, which is saved for future use. The amplification is carried out for 20 cycles; the size of the insert of each individual clone can be confirmed by loading one-tenth of the PCR product on the agarose gel.

Sequencing. Plasmid DNA from 1.5 ml of an overnight culture is prepared using a modified Holmes–Quigley method.[28] The pellets are resuspended in 10 μl of TE buffer (10 mM Tris, pH 8.0, and 1 mM EDTA) and

---

[28] J. Sambrook, E. F. Fritsch, and T. Maniatis, *in* "Molecular Cloning: A Laboratory Manual." Cold Spring Harbor Laboratory, Cold Spring Harbor, New York, 1989.

treated with 50 μg/ml of RNase A for 10–15 min at 37°. One-half of the DNA is used for one set of sequencing reactions. The nucleotide sequence of double-stranded DNA is determined using the dideoxy–chain-termination procedure with Sequenase (U.S. Biochemical Corp., Cleveland, OH).

Discussion

The ability of the PCR to amplify rapidly sequences directly from small amounts of clinical specimens provides a means to study naturally occurring viral variants without introducing the selective pressures of virus cultivation. Some of the concerns with applying the PCR to study viral heterogeneity are the introduction of errors during the amplification process and potential biased selection of variants during amplification, cloning, and sequencing. In addition, there is a tradeoff between the sensitivity/specificity of a reaction and the fidelity of the *Taq* DNA polymerase. For example, conditions that allow for amplification of all viral variants, such as lower annealing temperatures and higher concentrations of dNTPs (200–800 μM total), may reduce the fidelity of *Taq* polymerase. Conversely, conditions that favor high *Taq* fidelity may compromise variant detection. Ideally, amplification conditions which minimize biased selection and maximize *Taq* fidelity are desirable. Different reaction conditions can and different template sequences may influence the fidelity of the *Taq* polymerase. Thus, the fidelity of the *Taq* polymerase must be determined for each experiment by amplifying control templates under the conditions used. In situations where extremely low error rates are required, DNA polymerases that have proofreading functions such as T4 DNA polymerase can be used in the PCR.

The PCR may also underestimate the genetic variability of a viral population. Highly divergent isolates may not be amplified or detected. Because the number and location of primer/template mismatches determine the efficiency of extension, the relative proportion of each isolate may be skewed. Finally, unless large numbers of isolates are sequenced, rare variants in the biological sample may be missed.

Despite the potential limitations, the PCR has provided an invaluable means to study not only viral variants but also gene polymorphisms and evolution. The construction of phylogenetic trees based on comparing the sequences of homologous genes in many different species has been significantly facilitated by PCR techniques.

[33] Phylogenetic Analysis of Restriction Site Data

By Kent E. Holsinger and Robert K. Jansen

Introduction

Restriction site variation has provided a wealth of data to address questions in systematic and evolutionary biology. In animals, studies have used variation in mitochondrial DNA or nuclear ribosomal DNA to assess patterns of population differentiation within species, to study the dynamics of hybrid zones, and to reconstruct the biogeographic history of closely related species.[1,2] In plants, studies have used variation in chloroplast DNA or nuclear ribosomal DNA to determine systematic relationships among taxa at levels ranging from species within a genus to genera and tribes within a family.[3] Although some studies, particularly in animals, have focused primarily on the pattern of genetic diversity within and among populations, most have included an attempt to reconstruct some aspect of phylogenetic history. Indeed, analysis of restriction site variation in chloroplast DNA may now be the most widely used tool for determining phylogenetic relationships within families of flowering plants.

Restriction site variation has several characteristics that make it particularly appropriate for phylogenetic reconstruction. Three of these characteristics it shares with nucleotide sequence data: (1) character states can be scored unambiguously, (2) a large number of characters can be scored in each individual, and (3) it provides information on both the extent and the nature of divergence between two sequences. Unlike DNA–DNA hybridization techniques, which provide information only on the extent of divergence between two sequences, restriction site and sequence analyses allow us to determine whether changes arise from nucleotide substitution, insertion/deletion events, or genome rearrangement.[4] Although restriction site comparisons offer less direct information on the evolution of nucleotide sequences than comparison of the underlying sequences, they may sometimes be the preferred method of analysis. It is (at present) simpler and less expensive to do a restriction site survey than to perform a sequence survey.

[1] J. C. Avise, J. Arnold, R. M. Ball, E. Bermingham, T. Lamb, J. E. Neigel, C. A. Reeb, and N. C. Saunders, *Annu. Rev. Ecol. Syst.* **18**, 489 (1987).
[2] C. Moritz, T. E. Dowling, and W. M. Brown, *Annu. Rev. Ecol. Syst.* **18**, 269 (1987).
[3] J. D. Palmer, R. K. Jansen, H. J. Michaels, M. W. Chase, and J. R. Manhart, *Ann. Mo. Bot. Gard.* **75**, 1180 (1988).
[4] T. E. Dowling, C. Moritz, and J. D. Palmer, *in* "Molecular Systematics" (D. M. Hillis and C. Moritz, eds.), p. 250. Sinauer, Sunderland, Massachusetts. 1990.

More importantly, it is sometimes more informative. A very long nucleotide sequence may be required to provide more information than a restriction site survey. For example, a phylogenetic analysis of 22 taxa from the sunflower family used both restriction site variation in chloroplast DNA and nucleotide sequence variation in the chloroplast-encoded gene rbcL.[5] The restriction site survey identified 583 variable sites of which 169 were shared between two or more taxa, whereas the sequence survey found only 109 shared variable positions. Even though the restriction site survey included only 11 enzymes, nearly 3500 nucleotides are included in the variable sequences surveyed, whereas rbcL is only 1428 base pairs (bp) in length.

In spite of its usefulness, restriction site variation does pose some problems for phylogenetic analysis. Consider a 6-bp stretch of DNA. If it is the recognition sequence for a restriction enzyme, any of 18 different nucleotide substitutions will remove that restriction site. If it differs from the recognition sequence in one position, there is only one substitution that will create a restriction site. Thus, a genome having a restriction site at a particular position is far more likely to lose it than a genome lacking that site is to gain it. As a consequence, convergent losses of a restriction site are more likely than convergent gains, and the ratio of convergent losses to convergent gains increases as the taxa become more divergent.[6] This chapter reviews different methods that have been proposed for phylogenetic analysis of restriction site variation and illustrates how these methods can be used by applying them to restriction site variation among genera within a subtribe of the sunflower family.[7]

Methods

Restriction Site Mapping

Restriction fragment patterns can be characterized either as a simple list of fragment lengths or as mapped restriction sites. When comparing closely related taxa with a low level of sequence divergence, it may be possible to infer restriction site mutations by inspecting fragment profiles. This approach has been used successfully in restriction enzyme studies of chloroplast DNA at the inter- and intraspecific levels, where sequence

[5] K.-J. Kim, R. K. Jansen, R. W. Wallace, H. J. Michaels, and J. D. Palmer, *Ann. Mo. Bot. Gard.* **79,** 428 (1992).
[6] A. R. Templeton, *in* "Statistical Analysis of DNA Sequence Data" (B. S. Weir, ed.), p. 151. Dekker, New York, 1983.
[7] R. K. Jansen, R. S. Wallace, K.-J. Kim, and K. L. Chambers, *Am. J. Bot.* **78,** 1015 (1991).

divergence is in the range of 0.5–1.0%.[8,9] At higher taxonomic levels, however, mapping the location of both restriction site and length mutations enables one to make a much more accurate determination of character homology. Bremer[10] suggested that mapping and aligning restriction sites may not be worth the extra time and effort it entails, but in our view mapping restriction sites is the only way we can be sure that we have correctly identified the characters. Without mapping studies we might regard two fragment patterns as identical when the underlying pattern of restriction sites is very different. Thus, every restriction fragment difference used in a phylogenetic analysis should be fully characterized. Furthermore, accurate estimates of the degree of sequence divergence depend on a knowledge of the number of variant and invariant sites, information that can be obtained only with a detailed knowledge of the location of restriction sites.

Restriction fragment differences can be the result either of nucleotide substitutions in the recognition sequences or of length mutations. Mapping will allow these changes to be distinguished from one another, but should both types of differences be included in a phylogenetic analysis? Most of the phylogenetic methods we discuss below assume that changes in restriction sites are the result of nucleotide substitution. Obviously, including length variation in such analyses is inappropriate. Only Wagner parsimony does not assume that all restriction site changes result from nucleotide substitution. Thus, length mutations should be included in Wagner parsimony analyses, but only when the homology of those mutations can be confidently established, for example, through the presence of discrete, readily distinguishable size classes.

Choice of Restriction Enzymes

The choice of restriction enzymes for a study is usually based on practical matters like cost and the number of fragments generated,[4] but the recognition sequence of the enzymes should also be considered. Many restriction enzymes recognize multiple sites. For example, *Ban*II (GUG-CYC) can have any purine (U) or any pyrimidine (Y) at the second and fifth position, respectively, in the recognition sequence. This enzyme will cleave DNA with any one of four different nucleotide sequences (GGGCCC = *Apa*I, GAGCCC, GGGCTC, and GAGCTC = *Sac*I, *Sst*I). Because *Ban*II cuts DNA at the same places as *Apa*I, *Sac*I, and *Sst*I, the latter enzymes should not be included in any surveys that use *Ban*II.

[8] J. D. Palmer, and D. Zamir, *Proc. Natl. Acad. Sci. U.S.A.* **79**, 5006 (1982).
[9] M. T. Clegg, A. D. H. Brown, and P. R. Whitfield, *Genet. Res.* **43**, 339 (1984).
[10] B. Bremer, *Plant Syst. Evol.* **175**, 39 (1991).

Clearly, SacI and SstI should not be included in the same survey, since their recognition sequences are identical. In addition, the available maximum-likelihood methods rely on calculated probabilities of restriction site change for enzymes with a single recognition sequence. Thus, it is better to avoid enzymes with multiple recognition sequences when likelihood analyses are planned.

Finally, the evolution of any restriction sites recognized by enzymes that differ at only one base in their recognition sequence will not be completely independent, violating one of the critical assumptions for use of maximum-likelihood and bootstrapping techniques. In short, we recommend that the following types of enzymes be avoided in restriction site surveys: (1) those with multiple recognition sequences, (2) those with a recognition sequence identical to that of another enzyme included in the survey, (3) those with a recognition sequence that differs from that of another enzyme being used by a single nucleotide, and (4) those with a recognition sequence included in that of another enzyme being used.

Distance Methods

The character state data of a restriction site survey can be summarized in a distance matrix that relates all pairs of sequences. To be useful for phylogenetic analysis, however, a distance measure must do more than measure the similarity in restriction maps. It should reflect the expected amount of change between the sequences.[11] Nei and Li[12] developed just such a measure for mapped restriction sites. The distance measure they propose is expressed in terms of the average number of substitutions per nucleotide since two genomes diverged from a common ancestor. Although they assumed that all recognition sites have the same number of nucleotides, their method can be generalized for surveys including multiple enzymes that differ in the length of their recognition sequence.[13] This distance measure assumes that all differences in the restriction maps are a result of nucleotide substitution. Thus, any length mutations that are detected must be excluded from the calculation.

In describing the relationship between two genomes as a single number, some information is necessarily lost,[14] but such a description also has certain advantages. First, the distance measure of Nei and Li gives a direct

[11] J. Felsenstein, *Evolution* **38,** 16 (1984); but see also the exchange between Farris and Felsenstein: J. S. Farris, *Cladistics* **1,** 67 (1985); J. Felsenstein, *Cladistics* **2,** 130 (1986); J. S. Farris, *Cladistics* **2,** 144 (1986).

[12] M. Nei, and W.-H. Li, *Proc. Natl. Acad. Sci. U.S.A.* **76,** 5269 (1979).

[13] M. Nei, and F. Tajima, *Genetics* **105,** 207 (1983).

[14] D. Penny, *J. Theoret. Biol.* **96,** 129 (1982).

estimate of the amount of evolutionary change between two genomes since they diverged from the most recent common ancestor. Second, it includes a correction for multiple substitutions at a single nucleotide site, whereas parsimony methods consistently underestimate the number of nucleotide substitutions along any given branch.[15] Finally, it may be possible to assess the reliability of phylogenetic hypotheses directly, because variances on the branch points in the phylogeny can be computed from the variances of the distance estimates.[16,17]

A variety of methods have been proposed for phylogenetic analysis of distance data. The most widely used are UPGMA cluster analysis (unweighted pair group method using arithmetic averages),[18] Fitch–Margoliash (a form of weighted least squares),[19] and neighbor joining.[20] UPGMA appears to work well only if the rate of substitutions is constant and the distances among the taxa are large,[21] assumptions that are not reasonable for many data sets. Fitch–Margoliash and neighbor joining analyses do not require the assumption of a constant rate of substitutions, but simulations have suggested that neighbor joining is better at obtaining the correct phylogeny.[22] It is also more computationally efficient than the Fitch–Margoliash method. Thus, neighbor joining appears to be the method of choice for phylogenetic analysis of distance data.[23]

Parsimony Methods

Although distance methods have certain advantages for the analysis of restriction site data, notably their computational efficiency, they also have an important disadvantage. By reducing the data to pairwise distances, important information on the evolutionary history of individual restriction sites is lost. In chloroplast DNA, for example, individual sites often evolve at very different rates. A parsimony analysis of 328 restriction site mutations shared among two or more taxa in a survey of 57 taxa from the sunflower family showed that 186 of the sites had only a single mutation, whereas 6 sites showed nine or more mutations.[24] A conservative statistical

[15] N. Saitou, *Syst. Zool.* **38,** 1 (1989).
[16] W.-H. Li, *Mol. Biol. Evol.* **6,** 424 (1989).
[17] N. Takahata, and F. Tajima, *Mol. Biol. Evol.* **8,** 494 (1991).
[18] R. R. Sokal and C. D. Michener, *Univ. Kans. Sci. Bull.* **28,** 1409 (1958).
[19] W. M. Fitch and E. Margoliash, *Science* **155,** 279 (1967).
[20] N. Saitou and M. Nei, *Mol. Biol. Evol.* **4,** 406 (1987).
[21] Y. Tateno, M. Nei, and F. Tajima, *J. Mol. Evol.* **18,** 387 (1982).
[22] N. Saitou and T. Imanishi, *Mol. Biol. Evol.* **6,** 514 (1989).
[23] See also the discussion by D. M. Hillis, M. W. Allard, and M. M. Miyamoto, this volume [34].
[24] R. K. Jansen, K. E. Holsinger, H. J. Michaels, and J. D. Palmer, *Evolution* **44,** 2089 (1990).

test for rate constancy among the sites showed that these six sites were undergoing substitutions more rapidly than other sites detected in the survey. Clearly, those sites that change only once are more likely to be a reliable guide to phylogenetic relationships than those that change many times, but existing distance measures do not allow us to determine which sites are which.

Three different parsimony methods have been used for the analysis of restriction site data: Wagner parsimony,[25] Dollo parsimony,[26] and weighted parsimony or generalized parsimony.[27] As the names suggest, they all invoke a parsimony principle to select the best tree (or trees) from the set of all possible trees. Specifically, the trees selected are those that require the minimum amount of evolutionary change. The methods differ from one another in how the amount of evolutionary change is calculated.

Wagner parsimony, as applied to restriction site data, counts each change of restriction site, from presence to absence or absence to presence, as a single step, and minimizing the length of the tree under the Wagner criterion corresponds to minimizing the total number of steps on the tree. Thus, it presumes that gain of a restriction site is as likely as loss of a restriction site. Wagner parsimony produces an unrooted tree, which is usually rooted by use of an outgroup. Dollo parsimony, on the other hand, presumes that gain of a restriction site is so unlikely relative to its loss that any taxa sharing a particular site must have inherited it from a common ancestor. Thus, a restriction site can be gained only once, but it may be lost many times; minimizing the length of a tree under the Dollo criterion corresponds to minimizing the total number of restriction site losses while allowing each site to be gained only once. Dollo parsimony, as originally proposed, produces a rooted tree, but an equivalent unrooted method exists.[28]

Weighted parsimony is intermediate between these two extremes. Each restriction site loss is counted as a single step, but each restriction site gain is counted as M steps, where M can be any positive number greater than 1. M can be estimated from an initial Wagner parsimony analysis that relates it to the observed probabilities of site gains and losses,[29] or it may be chosen

[25] J. S. Farris, *Syst. Zool.* **19,** 83 (1970).

[26] J. S. Farris, *Syst. Zool.* **26,** 77 (1977).

[27] D. Sankoff, and R. J. Cedergren, *in* "Time Warps, String Edits and Macromolecules: The Theory and Practice of Sequence Comparison" (D. Sankoff and J. B. Kruskal, eds.), p. 253. Addison-Wesley, Reading, Massachusetts, 1983.

[28] D. L. Swofford and G. J. Olsen, *in* "Molecular Systematics" (D. M. Hillis and C. Moritz, eds.), p. 411. Sinauer, Sunderland, Massachusetts, 1990.

[29] R. K. Jansen, K. E. Holsinger, and R. Olmstead, *Am. J. Bot.* **78,** (Suppl.) 193 (1991).

independently of the observed variation.[30] The higher the probability of a site loss relative to a site gain, the higher M is. Weighted parsimony produces a rooted tree. If the ancestral condition is specified as all site absences, then weighted parsimony becomes equivalent to a rooted Dollo method as M is increased.[28] If the ancestral condition is not specified, then weighted parsimony produces total loss topologies as M is increased.[30] In these topologies taxa are grouped only by shared restriction site losses.

Choosing the appropriate parsimony method is a difficult task. Wagner parsimony obviously errs in treating restriction site gains and losses as equally likely. Dollo parsimony errs in the opposite direction, by completely disallowing parallel gains. Thus, weighted parsimony seems to be the obvious choice. Unfortunately, weighted parsimony is much more computationally intensive than other parsimony methods, which may prevent its use with large data sets. In practice it is best to apply all three methods to a set of data. If weighted parsimony is impractical, because of the size of the data set, both Wagner and Dollo analyses should still be performed. Groups that are found in each of the analyses may be regarded as the most robust phylogenetic hypotheses, since many different assumptions about the pattern of character state change are consistent with their monophyly. Notice that if length mutations are included among the characters for a parsimony analysis, the presumption of asymmetry does not necessarily apply. Thus, length mutations should be treated as unordered characters.

With weighted parsimony we recommend that a range of weights be used and that these weights be chosen based on calculations from the observed rate of character state change as estimated from an initial Wagner parsimony analysis. In addition, it seems best to root the tree by specifying the ancestral states as site absences. Using weights derived from observed character state changes allows the observed asymmetry between site gains and site losses to determine the appropriate weight, which is necessary because the degree of asymmetry differs from data set to data set. Rooting by specifying the ancestral states as all absent provides a parsimony method that grades continuously between Wagner parsimony and Dollo parsimony. It is reasonable to require that a parsimony method for restriction site data converge to Dollo parsimony as the weight is increased, because the increasing weight is intended to reflect the decreasing likelihood of parallel site gains. Furthermore, Dollo parsimony is the maxi-

[30] V. A. Albert, B. D. Mishler, and M. W. Chase, *in* "Molecular Systematics in Plants" (D. E. Soltis, P. S. Soltis, and J. Doyle, eds.), p. 369. Chapman & Hall, New York, 1992.

mum-likelihood estimate of phylogeny in the case when rates of change are low and the probability of site loss is much greater than that of a site gain.[31,32]

Maximum-Likelihood Model

Parsimony methods, although they are preferable to distance methods in some ways, are by no means the ideal method for analysis of restriction site variation in a phylogenetic context. Even the most appropriate of the parsimony methods, weighted parsimony, does not correctly account for multiple changes at a restriction site along a branch. Furthermore, parsimony methods are known to fail when branch lengths are very unequal.[33] In addition, parsimony methods do not use all of the information that is available for phylogenetic reconstruction. Specifically, only sites that are present (or absent) in two or more of the taxa included in the survey are phylogenetically informative in a parsimony context. Even though a taxon that has had a long history of independent evolution is expected to have many unique changes, these changes have no effect on the phylogenetic estimates obtained from a parsimony analysis.

Perhaps the most promising approach for the phylogenetic analysis of restriction site data is to use a statistical model of restriction site change to construct a maximum-likelihood estimate. Smouse and Li[34] showed how to use such an approach when only four taxa are involved, and Felsenstein[35] has extended this approach to an arbitrary number of species. As currently implemented the statistical model underlying the maximum-likelihood model assumes that (1) substitutions among all nucleotides are equally likely, (2) each restriction site evolves independently, (3) the recognition sequences of all enzymes used in the survey are of equal length, (4) the rate of nucleotide substitution is the same at all sites, and (5) all differences in the restriction map of species are a result of nucleotide substitutions, not length mutations. The maximum-likelihood model does not assume that nucleotide substitutions accumulate in a clocklike manner. A more general model of nucleotide substitution could be used, one that allows different rates of transitions and transversions, for example, but such a method appears to be computationally infeasible. In any case, studies of nucleotide sequence data have repeatedly shown that sequence divergence values depend little on the model of nucleotide substitution

[31] J. Felsenstein, *Syst. Zool.* **28**, 49 (1979).
[32] R. W. DeBry and N. A. Slade, *Syst. Zool.* **34**, 21 (1985).
[33] J. Felsenstein, *Syst. Zool.* **27**, 401 (1978).
[34] P. E. Smouse and W.-H. Li, *Evolution* **41**, 1162 (1987).
[35] J. Felsenstein, *Evolution* **46**, 159 (1992).

when the probability of substitution per site is 20% or less.[36] Thus, the completely symmetric substitution model currently assumed in the maximum-likelihood method is unlikely to lead to gross errors.

One of the great advantages of likelihood methods is that their statistical properties are well understood, since they are the basis for many classic statistical methods. This means that relatively straightforward statistical tests on phylogenetic hypotheses are possible. For example, an approximate test of the null hypothesis of zero branch length is performed by setting that branch length equal to 0 and recomputing the log-likelihood. If the difference between the log-likelihood with and without the zero branch length is L, then $-2L$ is approximately distributed as χ^2 with one degree of freedom. Determining whether the difference between two topologies is statistically significant is more difficult, but Kishino and Hasegawa[37] have developed a paired-sites method that seems broadly applicable. It would be difficult to use their method to enumerate all topologies that are statistically indistinguishable from the maximum-likelihood estimate, but it can be used to test for significant differences between topologies arrived at through other means (e.g., a priori hypotheses of relationship based on existing taxonomy or preliminary analyses of restriction site data using distance or parsimony methods).

There is one difficulty in applying the maximum-likelihood method to large data sets, especially those with a large number of taxa. It is the least computationally efficient of any of the methods we discuss, making it impractical to compute the likelihood of a large number of alternative topologies. With a small number of taxa, say, fewer than 10 or 20, this may not be a problem. With larger numbers of taxa, however, the computational burden quickly becomes impractical. Thus, this method may be most useful with large data sets as a means of evaluating tree topologies found using one of the other methods.

Bootstrapping

Systematists have tended to pay much more attention to the methods used to reconstruct phylogenies than to methods that can be used to assess the reliability of phylogenetic hypotheses. Attempts to determine the reliability of phylogenetic hypotheses have often involved either assessing the congruence of the hypothesis with groups that have already been recognized or testing its robustness to perturbations of the character state data matrix. Neither of these approaches is entirely satisfactory. The first tends only to reconfirm previous hypotheses, not to evaluate them. The second

[36] M. Nei, "Molecular Evolution." Columbia Univ. Press, New York, 1987.
[37] H. Kishino and M. Hasegawa, *J. Mol. Evol.* **29**, 170 (1989).

provides no measure of "robustness" beyond the simple intuitive feel that much of the data are (or are not) consistent with the hypothesis. Estimates of reliability derived from bootstrap resampling are preferable to these qualitative approaches because bootstrapping is based on well-understood statistical principles and because its results are directly comparable to confidence intervals in classic statistics. Furthermore, bootstrapping can be used with any of the methods discussed above to give a statistical indication of the amount of confidence that should be placed in any particular branch point.

To employ the bootstrap we need only assume that the restriction sites in our sample evolve independently of one another. We do not need to assume that they all evolve at the same rate, only that the rate of evolution at each site is determined independently from the same underlying probability distribution. To put it another way, all we need assume is that the character state distribution we find among extant taxa is the realization of an underlying substitution process that can be described by a stochastic process.[38] Bootstrap resampling draws sites at random (with replacement) to construct a series of new character state data matrices whose distribution of character states will reflect the underlying statistical substitution process. If sites vary in their rate of change, the compound statistical distribution is faithfully represented by the independent sampling with replacement to form each new bootstrap sample data set, assuming we have a large enough sample of characters. If the sites do not vary in their rate of change, a smaller sample of characters will suffice to represent the underlying statistical distribution of character states.

For each of the character state data matrices produced by bootstrap resampling a phylogenetic estimate is made, using one of the above methods. How are the results of this analysis to be presented and interpreted? There is some disagreement about the best way to summarize the results of a bootstrap analysis of phylogeny,[39] but let us assume for the moment that the results have been summarized as a majority-rule consensus tree. If a particular group appears in 70% of the bootstrap sample trees, then we can conclude that this group will appear in 70% of the trees constructed using this tree construction method on new data collected in the same way. If the tree construction method used were known to be unbiased, we could make the even stronger statement that 70% of the groups identified in this way will reflect the true phylogeny.

[38] J. Felsenstein, *Evolution* **39**, 783 (1985).
[39] M. Sanderson, *Cladistics* **5**, 113 (1989).

An Example From the Sunflower Family

Data

To illustrate how the methods described can be used in the analysis of real data, we have selected as an example a survey of restriction site variation within a subtribe of the sunflower family, the Microseridinae.[7] As currently circumscribed the Microseridinae includes seven genera: *Agoseris, Krigia, Microseris, Nothocalais, Phalacroseris, Picrosia,* and *Pyrrhopappus.* Representatives of all genera except *Picrosia* were included in the survey, as were representatives of other genera in the remaining subtribes of the tribe Lactuceae, *Hypochaeris, Lactuca, Leontodon, Stephanomeria,* and *Tolpis.* In all, 20 species were represented in the survey. A total of 914 restriction sites were mapped using 17 different restriction enzymes. Of the 17 enzymes used, 11 have recognition sequences of 6 bases, 5 have recognition sequences of 6 bases with two of the positions requiring only the presence of a purine or a pyrimidine, and 1 has a recognition sequence of 5 bases in which either an adenine or a thymine is recognized at one position. Thus, sequence variation at approximately 5400 nucleotides was surveyed in the study. Of the 914 restriction sites mapped, 541 were unvaried (i.e., the site was present in every species). Of the remaining 373 variable sites, 193 showed differences that were unique to a single species, and 180 showed differences that were shared among two or more species.

Wagner, Dollo, and weighted parsimony analyses of the restriction site data were done on a Macintosh IISE microcomputer using the PAUP program (Phylogenetic Analysis Using Parsimony, Version 3.0q, developed by D. Swofford). We used the branch-and-bound algorithm to identify all equally parsimonious trees for Dollo and Wagner analyses. We used tree bisection–reconnection branch swapping to identify trees with weighted parsimony, and we rooted the weighted parsimony trees by assuming that all restriction sites were absent from the ancestor. Bootstrap analyses using both Wagner and Dollo parsimony were done on an IBM 3090 using PHYLIP (Phylogenetic Inference Package, Version 3.2, developed by J. Felsenstein). The neighbor joining analysis and evaluation of likelihoods for each of the tree topologies were done on a Zenith Z-386 microcomputer using PHYLIP (Version 3.41). Although some enzymes with recognition sequences of different length were used, distance estimates for the neighbor joining analysis and likelihoods of alternative tree topologies were obtained assuming that all enzymes used had 6-base recognition sequences. This will lead to underestimates of the amount of nucleotide sequence divergence but should not seriously bias the phylogenetic estimates obtained.

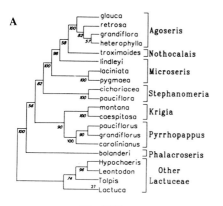

A

ln L = -5060.4

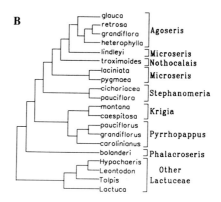

B

ln L = -5062.67

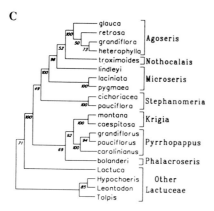

C

ln L = -5057.96

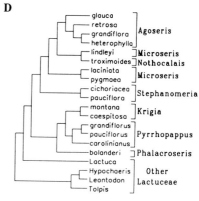

D

ln L = -5060.08

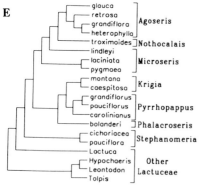

E

ln L = -5064.21

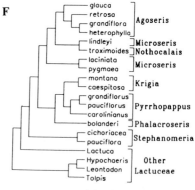

F

ln L = -5066.34

Results

The results of our phylogenetic analyses are summarized in Fig. 1. The topologies include all of the alternatives found among the equally parsimonious Wagner and Dollo parsimony trees, the weighted parsimony trees, and the neighbor joining tree. The two equally parsimonious Wagner parsimony trees (Fig. 1A,B) differ only in the position of *Microseris lindleyi* and *Nothocalais troximoides*. The topology of the Wagner parsimony bootstrap majority-rule consensus tree is that of Fig. 1A. The topology of the neighbor joining tree is that of Fig. 1B. The four equally parsimonious Dollo parsimony trees (Fig. 1C–F) show substantial congruence with the Wagner parsimony trees, but they differ in two significant ways. First, in the Wagner trees *Phalacroseris* is the sister group of the remaining members of the Microseridinae, whereas in the Dollo trees it is found in a clade including *Krigia* and *Pyrrhopappus*. Second, *Stephanomeria* is included within the Microseridinae in only two of the four Dollo trees (although the bootstrap includes it in all replicate trees), whereas it is included in the Microseridinae in both Wagner trees. The topology of the Dollo parsimony bootstrap majority-rule consensus tree is that of Fig. 1C.

As we have argued above, neither Wagner nor Dollo parsimony is entirely appropriate for the phylogenetic analysis of restriction site variation. Weighted parsimony methods are preferable. To estimate the weights we count the number of times each character undergoes a gain or a loss, as estimated from the Wagner parsimony analysis, and make a maximum-likelihood estimate of the probability of character state change along a branch.[29] In our experience, which of the several topologies is selected for this count is not particularly important, because the number of changes required does not differ between topologies. More importantly, the number of gains and losses differs little between topologies, and a large difference in the pattern of gains and losses is required to produce markedly different weight estimates. The data and the estimated weights derived from them are presented in Table I. For weights ranging from 1.3 to 1.45, the topology shown in Fig. 1A is the only one produced. For a weight of

FIG. 1. Neighbor joining, Wagner parsimony, Dollo parsimony, and weighted parsimony trees of the Microseridinae based on chloroplast DNA restriction site mutations. The numbers below the nodes on (A) and (C) indicate the number of times that the group to the right was monophyletic in 100 bootstrap samples. The number below each of the trees is its log-likelihood. (A) One of two Wagner parsimony trees, the Wagner parsimony bootstrap majority-rule consensus tree, and the weighted parsimony tree for $1.3 \leq M \leq 1.5$. (B) One of two Wagner parsimony trees and the neighbor joining tree. (C) One of four Dollo parsimony trees, the Dollo parsimony bootstrap majority-rule consensus tree, and the weighted parsimony tree for $1.5 \leq M \leq 1.8$ (D–F) Dollo parsimony trees.

TABLE I
PATTERNS OF RESTRICTION SITE CHANGE IN
MICROSERIDINAE[a]

Number of changes	Gains	Losses
1	90	70
2	9	24
3	3	9
4	—	2

[a] Weight assuming all sites evolve at same rate:
1.28. Weight assuming gamma distribution of
rates: 1.45.

1.5, both the topology in Fig. 1A and the one in Fig. 1C are produced. For weights from 1.55 to 1.8, the topology in Fig. 1C is produced. The model that assumes a gamma distribution of substitution rates among sites fits the data better than one assuming a constant rate, and the weight estimate calculated under this model is 1.45. Thus, the weighted parsimony analysis would suggest that the topology in Fig. 1A is the best estimate of phylogenetic relationships. It should not be too strongly preferred to that in Fig. 1C, however, since a small change (<5%) in the weight estimate could reverse this conclusion.

The best way to choose among the competing topologies is to compare their likelihoods. When we do this we see that the topology in Fig. 1C has the highest (least negative) likelihood. According to the paired-sites test proposed by Kishino and Hasegawa,[37] none of the tree topologies shown is significantly worse than the one in Fig. 1C. There is, however, another way to look at it. The next most likely topology, according to these calculations, is the one in Fig. 1D. Its log-likelihood is −5060.08 versus −5057.96 for the topology in Fig. 1C. The likelihood of one of these topologies is simply the probability of getting the observed data given that topology and the underlying assumptions about the substitution process. Thus, a difference of 2.12 units in log-likelihood means that the data we observed are 8 times more likely if the true relationships are given by the topology in Fig. 1C than if they are given by the topology in Fig. 1D. Similarly, the data are 11 times more likely given the topology in Fig. 1C than the one in Fig. 1A. Although such comparisons are not equivalent to the significance tests of classic statistics, they do provide a measure of support for alternative hypotheses, and these differences have been used to develop an alternative approach to statistical hypothesis testing.[40] Thus, there seems to be reason-

[40] A. W. F. Edwards, "Likelihood." Cambridge Univ. Press, Cambridge, 1972.

ably strong support for the topology in Fig. 1C compared to the alternatives. In short, the topology produced as the majority-rule consensus tree from the bootstrapped Dollo analysis, which is also one of the equally parsimonious Dollo topologies, has a higher likelihood than either of the equally parsimonious Wagner parsimony topologies.

Conclusions

Wilson and colleagues have argued that direct sequencing via the polymerase chain reaction (PCR) has made restriction mapping obsolete in molecular analyses of variation in animals,[41] and they may well be right. A survey of restriction site variation in hominids that used 18 6-base enzymes and 1 4-base enzyme, for example, was able to examine only about 50 sites per mitochondrial genome, or about 300 bp.[42] High-resolution mapping using 12 enzymes, 10 of which recognize 4-base sequences, allowed detection of about 370 restriction sites per mitochondrial genome, or about 1500 bp,[43] but interpretation of the complex fragment patterns was possible only because a complete reference sequence was available. Surveys of nuclear genes will reveal even fewer restriction sites, simply because the regions analyzed are typically of the order of 1 to 2 kilobases (kb) while the mitochondrial genome is about 15 kb in length. Because direct sequencing with the PCR allows regions known to be highly variable to be targeted, direct sequencing is rapidly becoming the method of choice for molecular analyses of animal species.

Direct sequencing is not yet the method of choice for many analyses of plant species. Restriction site analyses in plants have primarily focused on variation in the chloroplast genome. Because the chloroplast genome is roughly 10 times the size of the animal mitochondrial genome,[3] it should not be surprising that these restriction site surveys have been able to examine far more variable sites than have comparable surveys of animal species. Table II shows, for six representative surveys in flowering plants, the number of restriction enzymes used, the average number of sites recognized per genome, the number of variable sites mapped, and the approximate number of nucleotides included in the variable regions that were mapped. Even the survey in *Clarkia*, which identified the fewest variable sites, studied over 700 nucleotides at variable sites. If the proportion of unvaried to variable sites in the *Clarkia* data set is comparable to

[41] A. C. Wilson, E. A. Zimmer, E. M. Prager, and T. D. Kocher, *in* "The Hierarchy of Life" (B. Fernholm, K. Bremer, and H. Jornvall, eds.), p. 407. Elsevier, Amsterdam, 1989.

[42] S. D. Ferris, A. C. Wilson, and W. M. Brown, *Proc. Natl. Acad. Sci. U.S.A.* **78,** 6319 (1981).

[43] R. L. Cann, M. Stoneking, and A. C. Wilson, *Nature (London)* **325,** 31 (1987).

TABLE II
NUMBER OF NUCLEOTIDES EXAMINED IN REPRESENTATIVE CHLOROPLAST DNA SURVEYS

Group surveyed	Number of enzymes used	Number of sites per genome	Number of variable sites	Nucleotides in variable sites	Ref.[a]
Glycine subgenus *Glycine*	29	543–561	157	940	*1*
Clarkia	18	246	122	730	*2*
Microseridinae	17	570	373	2200	*3*
Asteraceae	11	400	927	5500	*4*
Rubiaceae	8	213	268	1600	*5*
Solanaceae	10	560	880	5300	*6*

[a] *Key to references:* *(1)* J. J. Doyle, J. L. Doyle, and A. H. D. Brown, *Evolution* **44**, 371 (1990); *(2)* K. J. Sytsma, J. F. Smith, and L. D. Gottlieb, *Syst. Bot.* **15**, 280 (1990); *(3)* R. K. Jansen, R. S. Wallace, K.-J. Kim, and K. L. Chambers, *Am. J. Bot.* **78**, 1015 (1991); *(4)* R. K. Jansen, K. E. Holsinger, H. J. Michaels, and J. D. Palmer, *Evolution* **44**, 2089 (1990); *(5)* B. Bremer and R. K. Jansen, *Am. J. Bot.* **78**, 198 (1991); *(6)* R. Olmstead and J. D. Palmer, *Ann. Mo. Bot. Gard.* **79**, 346 (1992).

that in the Microseridinae data set (541 unvaried sites of 914 total sites mapped), then approximately 1800 nucleotides were surveyed in total.

Of course, the utility of restriction site variation for phylogenetic reconstruction is limited by our ability to score shared presence or shared absence of a restriction site as a homologous character. At high taxonomic levels or when the genomes being analyzed are highly divergent, making that determination of homology becomes increasingly difficult. Thus, for phylogenetic studies above the family level in angiosperms, we expect nucleotide sequence variation to be a far more useful approach than restriction mapping. Nevertheless, for the large number of studies within families, tribes, and genera, mapped restriction site variation is likely to be a powerful tool for phylogenetic analysis in plants for many years to come.

We suggest an eclectic approach to the phylogenetic analysis of the data that these surveys will produce. Neighbor joining, because of its computational efficiency, may allow the analysis of certain very large data sets that could otherwise not be analyzed, but our experience has shown that Wagner and Dollo parsimony analyses of very large data sets (57 species, 328 sites) can be done on a microcomputer in 1 hr or less. Furthermore, simulations suggest that the tree produced by neighbor joining is often included among the most parsimonious Wagner or Dollo trees.[22] Thus, we would never recommend relying solely on the neighbor joining tree, unless the data set were enormous. More problematical is the choice among competing parsimony methods. We would always recommend that both Wagner and Dollo parsimony analyses be performed, with bootstrapping whenever computationally feasible. Weighted parsimony analyses should be done with a series of weights, though for large data sets this may be impractical.

For choosing among the alternative topologies produced by these analyses, we recommend that the likelihoods of each of the topologies be computed and the one with the highest likelihood be taken as the best estimate of phylogeny based on the data. For small data sets, say, less than 10 or 20 taxa, it is even possible to use the search strategies currently implemented in PHYLIP to identify a unique maximum-likelihood estimate for the phylogeny. If likelihood analysis of the data is impossible, we recommend caution in choosing among the alternative topologies, if indeed a choice is to be made. In our experience, well-supported groups in a Wagner parsimony analysis are also well supported in Dollo or weighted parsimony analyses. The differences among them are found only in parts of the the the tree where none of the alternatives are strongly supported.

In short, by using a variety of complementary approaches to examine the phylogenetic information found in restriction site variation, considerable insight into the relationships of species and higher taxa can be obtained. To restrict ourselves to any one of the methods discussed here would be to limit our vision unnecessarily. There is clearly an asymmetry in the way restriction sites evolve, but the importance of that asymmetry varies with the taxonomic level of the comparison. At low taxonomic levels, it is almost nonexistent. At high taxonomic levels, asymmetry may be extremely important. None of the methods currently available for phylogenetic analysis is ideal. Even the best of them, the maximum-likelihood method, requires that we assume all restriction sites evolve at the same rate. Nevertheless, it is reasonable to expect that well-supported groups identified by techniques whose assumptions about the evolutionary process differ dramatically are likely to reflect the actual evolutionary history of the taxa involved. Although direct sequencing of PCR products may soon replace restriction site mapping in studies of animal phylogeny, we expect analyses of restriction site variation to be an important source of phylogenetic hypotheses in plants for many years to come.

Acknowledgments

This work was supported, in part, by grants from the National Science Foundation (BSR-9107330 to K.E.H. and BSR87-08246 and BSR90-20171 to R.K.J.). We are indebted to the University of Connecticut Computer Center and the University of Connecticut Biotechnology Center for access to computing facilities and to M. J. Spring for assistance in preparing the figure. We are particularly grateful to Joe Felsenstein, Dick Olmstead, Linda Raubeson, and Chris Simon for comments on early versions of this chapter.

[34] Analysis of DNA Sequence Data: Phylogenetic Inference

By DAVID M. HILLIS, MARC W. ALLARD and MICHAEL M. MIYAMOTO

Introduction

Comparison of biological attributes among organisms or genes in a meaningful manner requires an understanding of the evolutionary connections among the respective taxa or alleles. Thus the emphasis in systematic biology is on phylogeny, namely, the evolutionary history of lineages. Methods for inferring phylogeny from DNA sequences have proliferated greatly in the last few years. Unfortunately, decisions concerning which of many described methods will be used in a given study are rarely made by weighing the advantages and disadvantages of each approach; instead, issues of availability or historical inertia often dictate such choices. In part, this is because each method is advocated in a separate paper, so comparisons among methods are often difficult. Our goal in this chapter is to present a practical guide to selecting a set of methods for phylogenetic analysis of nucleic acid sequences. We focus on the assumptions, advantages, disadvantages, and limitations of the various approaches. Space does not permit a description of each of the algorithms, but many of these are described in an excellent review paper by Swofford and Olsen.[1]

There are five basic steps in the phylogenetic analysis of DNA sequences, although some of the steps are excluded or deemphasized by some investigators. A flowchart that includes these steps is presented in Fig. 1. The sequences under study must first be aligned so that positional homologs (the units of comparison) may be analyzed. Alignment may be straightforward if pairwise differences are small and most differences result from substitutions, but it becomes increasingly difficult as the sequences become more divergent and insertion/deletion events become more common. All phylogenetic analyses assume correct alignment of positional homologs.

Once sequences have been aligned, some assessment of the presence of phylogenetic signal is necessary. If all the sequences are identical, there is obviously no point in additional analysis. At the other extreme, the sequences may be so divergent that they have been randomized with respect

---

[1] D. L. Swofford and G. Olsen, *in* "Molecular Systematics" (D. M. Hillis and C. Moritz, eds.), p. 411. Sinauer, Sunderland, Massachusetts, 1990.

METHODS IN ENZYMOLOGY, VOL. 224

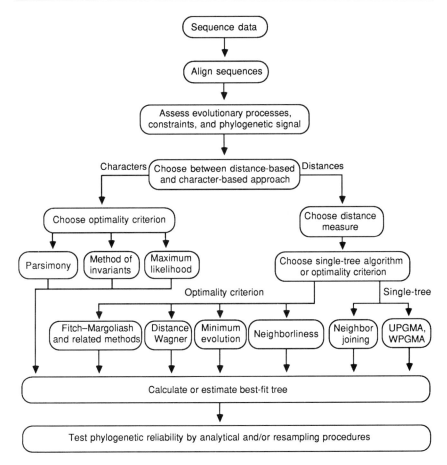

FIG. 1. Flowchart of steps from obtaining the sequence data to assessing the reliability of the final phylogenetic result.

to phylogenetic history. Analysis of the latter sequences will result in an inferred phylogeny, but the phylogenetic hypothesis might as well be selected at random. Many sequence analyses fall between these two extremes: some positions are highly conserved (perhaps invariant among the species), whereas other positions are randomized with respect to phylogenetic history (perhaps the third positions of codons). Thus, assessment of phylogenetic signal requires more than casual inspection of the sequences.

If phylogenetic signal is present in a matrix of sequences, then the third step is selecting a method of phylogenetic inference. Some of the following questions must be answered to make an informed choice among the methods: (1) Are a few broad assumptions preferable to many detailed assumptions about evolution? (2) What parameters of sequence evolution have been examined for the sequences of interest? (3) How variable are rates of change among the study taxa? (4) Are the primary goals of the study to reconstruct accurate branch lengths, reconstruct branching relationships, examine details of character evolution, or some combination of the above? (5) Is combination or comparison of data sets (now or in the future) a goal of the study? (6) Is a particular analysis feasible given the size of the data set and the limitations of computer time?

Once a method has been selected and the appropriate software has been obtained, a strategy must be developed for finding the best tree under the selected optimality criterion. The number of distinct tree topologies (ignoring for the moment the infinite number of possible branch lengths) for even a modest number of taxa is very great.[2] For instance, with just 50 taxa, there are over 2.8×10^{74} distinct, labeled, bifurcating trees, or roughly 10,000 times as many trees as there are atoms in the universe! If one could develop a computer program capable of analyzing 1 trillion trees a second (well beyond the capability of any existing computer), it would still require 8.9×10^{54} years to evaluate all the possible trees for 50 taxa, or about 2×10^{45} times the age of the Earth. Therefore, methods must be selected that estimate the best-fit tree under these circumstances.

Finally, once a tree (or trees) has been obtained, some statement of confidence in the results is desirable. How much better is the tree obtained than the next-best alternative? How does the tree compare to a previous hypothesis of relationships? Which nodes of the tree are well-supported by the data, and which are not?

Alignment

Although alignment of DNA or RNA sequences is often quite simple among closely related taxa, it becomes very difficult as the sequences become more divergent. Alignment is one of the most troublesome aspects of phylogenetic analysis, and it is an area of intensive research and refinement. Details of the commonly used methods have been treated elsewhere in this series,[3] so our comments are limited to some of the practical aspects of producing aligned sequences for phylogenetic analysis.

Most methods of sequence alignment are designed for pairwise comparisons, although alignments among all taxa under study are necessary before phylogenetic analysis can begin. Many of the pairwise approaches

[2] J. Felsenstein, *Syst. Zool.* **27**, 27 (1978).
[3] R. F. Doolittle (ed.), this series, Vol. 183.

are variations on the algorithm described by Needleman and Wunsch,[4] in which matches are scored as positive (e.g., Ref. 1), mismatches as 0, and gaps (corresponding to insertion/deletion events) as negative. The negative scores for gaps can be weighted to account for the size of the gaps. Usually, gaps of more than one position are not weighted in direct proportion to the size of the gap, because it is likely that the adjacent nucleotides were inserted or deleted simultaneously in a single event. The penalty for gaps is typically greater than the positive score assigned to a match, but there are no clear guidelines for assigning the relative weights. One common approach is to assign a weight of -2 to gaps, so that gaps are introduced only if doing so results in the reduction of at least two substitutions.

Modifications of the Needleman–Wunsch criteria can also be used to align multiple sequences,[3] although no efficient algorithms exist to ensure optimal alignments beyond a relatively few sequences if insertion/deletion events are common or substitution rates are high. Therefore, most investigators restrict comparisons to regions in which alignments are relatively obvious. This has the effect of restricting analyses to regions that are likely to have the highest signal-to-noise ratio, because regions of difficult alignment are likely to be evolving at rates too high for effective phylogenetic analysis. Even in regions of high signal-to-noise ratio, however, alternative alignments are likely to be dependent on the weights assigned to gaps. As an example, consider the sequences of ribosomal DNA shown in Fig. 2. The upper alignment requires twenty-two substitutions (at twenty positions), and initial inspection might not indicate the necessity of introducing gaps. However, if matches are assigned a score of 1 and gaps are assigned a penalty between -1 and -3, the lower alignment is favored, which includes four gaps and seven substitutions (at six positions).

A practical method of aligning multiple sequences is to align all pairs of taxa using the Needleman–Wunsch algorithm, then enter the sequences into a text processor for multiple alignment "by hand." A less desirable alternative is to align all taxa to a single reference taxon, but this may be necessary if the number of taxa is great. Another alternative is to use sequence similarity scores to determine the order of alignments (i.e., align the most similar pairs of taxa first). All pairwise alignments can be consulted for possible arrangements, and global alternatives can be evaluated using the Needleman–Wunsch criteria. It is important to establish *a priori* rules for weighting gaps, weighting sizes of gaps, and breaking ties so as not to bias the alignments. Currently, it is not feasible to ensure that the optimum alignment has been achieved unless the number of taxa are few or gaps are uncommon. For this reason, areas of questionable alignments

[4] S. B. Needleman and C. D. Wunsch, *J. Mol. Biol.* **48,** 443 (1970).

```
             22|00              22|20              22|40
Mus          GTCAGCCAGGACTCTCTACCCGCTCACGGCAAGGCTTCCCTGCCCGCTACCGGAGGCAAC
Rattus       GTCAGCCAGGACTCTCTACCCGCTCACGGCAAGGCTTCCCTGCCCGCTACCGGAGGCAAC
Homo         GTCAGCCAGGACTCTCTACCCGCTCGCGGCAAGGCTTCCCTGCCCGCTACCGGAGGCAAC
Rhineura     GTCAGCCAGGATTCTCTATCCGCTCGCGGCAAGGCTTCCCTGCCCGCTACCGGAGGCAAC
Cacatua      GTCAGCCAGGATTCGCTATCCGCTCGCGGCAAGCCTTCCCTGCCCGCTACCGGAGGCAAC
Xenopus      GTCAGCCAGGATTCTCTACCCGCTCGCGGCAAGCCTTCCCTGCCCGCTACCGGAGGCAGC
Rhyacotriton GTCAGCCAGGATTCTCTATCCGCTCGCGGCAAGCCTTCCCTGCCCGCTACCGGAGGCAAC
Typhlonectes GTCAGCCAGGATTCTCTATCCGCTCGCGGCAAGCCTTCCCTGCCCGCTACCGGAGGCAAC
Latimeria    GTCAGCCAGGATTCTCTACCCGCTTGCGGCAAGGCTTCCCTGCCCGCTACCGGAGGCAGC
Cyprinella   GTCAGTCCAGGATTCCTACCCGCTGGCGGTCAAGCCTTCCCTCCGGCTACCGGAGGCAGC
             *  ** ** **   *    **   ** ** * *   **  *             *
```

```
             22|00              22|20              22|40
Mus          GTCAG-CCAGGACTCTCTACCCGCTCACGG-CAAGGCTTCCCTGCCCGCTACCGGAGGCAAC
Rattus       GTCAG-CCAGGACTCTCTACCCGCTCACGG-CAAGGCTTCCCTGCCCGCTACCGGAGGCAAC
Homo         GTCAG-CCAGGACTCTCTACCCGCTCGCGG-CAAGGCTTCCCTGCCCGCTACCGGAGGCAAC
Rhineura     GTCAG-CCAGGATTCTCTATCCGCTCGCGG-CAAGGCTTCCCTGCCCGCTACCGGAGGCAAC
Cacatua      GTCAG-CCAGGATTCGCTATCCGCTCGCGG-CAAGCCTTCCCTGCCCGCTACCGGAGGCAAC
Xenopus      GTCAG-CCAGGATTCTCTACCCGCTCGCGG-CAAGCCTTCCCTGCCCGCTACCGGAGGCAGC
Rhyacotriton GTCAG-CCAGGATTCTCTATCCGCTCGCGG-CAAGCCTTCCCTGCCCGCTACCGGAGGCAAC
Typhlonectes GTCAG-CCAGGATTCTCTATCCGCTCGCGG-CAAGCCTTCCCTGCCCGCTACCGGAGGCAAC
Latimeria    GTCAG-CCAGGATTCTCTACCCGCTTGCGG-CAAGGCTTCCCTGCCCGCTACCGGAGGCAGC
Cyprinella   GTCAGTCCAGGATTC-CTACCCGCTGGCGGTCAAGCCTTCCCT-CCGGCTACCGGAGGCAGC
             *     *  *   *    **   *    *        *               *
```

FIG. 2. Alignment of a section of the 28 S ribosomal RNA gene of 10 species of vertebrates. The upper alignment requires no gaps but does require twenty-two substitutions (at twenty positions, marked with asterisks). The lower alignment is favored if gaps are weighted between −1 and −3, which encompasses the usual range of weighting for gaps. The lower alignment requires four gaps and seven substitutions. [Adapted from "Molecular Systematics" (D. M. Hillis and C. Moritz, eds.), p. 368. Sinauer, Sunderland, Massachusetts, 1990.]

are often removed from consideration prior to phylogenetic analysis, because of the likelihood that positional homology has not been correctly established.[1]

A worthwhile but little-used approach to alignment is to combine alignment and phylogenetic analysis in an iterative process.[5] This results in an alignment and a phylogeny that are both "best fits" according to an optimality criterion such as parsimony. Unfortunately, it is computationally intensive to carry out such an analysis for even a few sequences simultaneously. An initial phylogenetic estimate may be used to break ties in multiple-alignment methods: if two alignments are equally good according to the defined weighting criteria, the alignment that requires fewer changes on the initial phylogenetic estimate is preferred. However, if this tie-breaking procedure is used, it is critically important to use an unbiased method of breaking ties in initial alignments. If this is not done, then preconceived ideas of relationships may be favored at the expense of globally optimal solutions. The danger in such iterative procedures is that

[5] D. Sankoff, C. Morel, and R. J. Cedergren, Nature (London) 245, 232 (1973).

they may be highly sensitive to the initial alignments, and any bias that favors a particular alignment/phylogeny is likely to be magnified through the iterations.

Assessing Phylogenetic Signal

Many molecular systematists assume that obvious alignment of sequences is sufficient to justify the use of these sequences in phylogenetic analysis. However, sequences can be aligned with ease if a majority of the positions are invariant and the remaining positions are randomized by high mutation rates. Analysis of such sequences may produce an optimal tree with a given method, but if the variable sites are truly randomized, there is no reason to expect that the optimal tree is a good estimate of phylogeny. In addition, it is often desirable to partition regions of DNA sequences that are evolving at different rates (e.g., introns versus exons) and to identify the regions that are most likely to be informative for a given problem. Thus, some means of assessing phylogenetic signal in a given set of sequences is necessary.

Many investigators have used pairwise comparisons of the sequences to evaluate the potential phylogenetic importance of their data. For example, the transition/transversion ratios for sequence pairs can be compared to those expected for the sequences at equilibrium, given the observed base compositions (i.e., relative frequencies of A, C, G, and T).[6] DNA sequences that are largely free of homoplasy (parallel fixations and reversals) will have transition/transversion ratios greater than those for sequences that are saturated by change, but similar to those observed for closely related taxa (which remain highly structured). In a similar way, pairwise divergences have been used to assess the potential phylogenetic value of DNA sequences, by plotting percent divergence against time.[7] In such plots, regions of sequences or categories of character state transformations (e.g., transitions) that are saturated by change do not show a significant positive relationship with time. Both the transition/transversion ratio and sequence divergence are influenced by homoplasy, and, as such, both can provide insights into the potential phylogenetic value of sequence data.

Another way of detecting the presence of phylogenetic signal in a given data set is to examine the shape of the tree-length distribution that results from a parsimony analysis of all possible trees or a random subset of all

---

[6] A. Larson, in "Phylogenetic Analysis of DNA Sequences" (M. M. Miyamoto and J. Cracraft, eds.), p. 221. Oxford Univ. Press, New York, 1991.

[7] M. M. Miyamoto and S. M. Boyle, in "The Hierarchy of Life: Molecules and Morphology in Phylogenetic Analysis" (B. Fernholm, K. Bremer, and H. Jörnvall, eds.), p. 437. Elsevier Science Publ., Amsterdam, 1989.

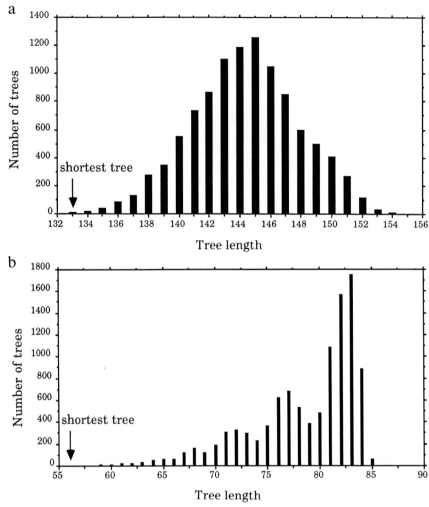

FIG. 3. (a) Nearly symmetrical tree-length distribution, based on an analysis of α-hemo-globin sequences from eight orders of mammals. Such distributions indicate that little or no phylogenetic signal is present in the data set. (b) Strongly skewed tree-length distribution, based on an analysis of a α-crystallin sequences from eight orders of mammals. This distribution indicates that the data are significantly nonrandom and, therefore, potentially informative about phylogeny. (Based on data from Refs. 8 and 9.)

trees.[8-10] Distributions that are close to symmetrical (Fig. 3a) indicate little or no structure in a data set; random sequences produce nearly symmetrical tree-length distributions. A strongly left-skewed tree-length distribution

[8] W. M. Fitch, *Syst. Zool.* **28,** 375 (1979).

(Fig. 3b) is an indication of the presence of correlated characters, which are expected if phylogenetic signal is present (the correlation is the result of the shared history of the taxa). If there is no indication that the data are more structured than random sequences, there is little point in pursuing further phylogenetic analysis of the data. Skewness is measured by the g_1 statistic, and tables of critical values of this statistic[11] for various numbers of taxa and characters should be consulted to test for the presence of nonrandom sequence variation. Skewness is calculated automatically in exhaustive and/or random-tree searches of two phylogenetic analysis software packages (PAUP and MacClade; see Implementation, below).

Figure 4 shows why tests for structured data are important, and that

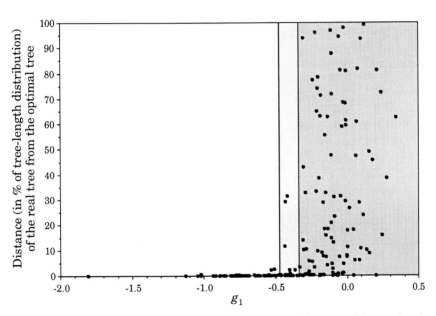

FIG. 4. Relationship between length of the correct tree and skewness of the tree-length distribution in simulated phylogenies. The optimal (most parsimonious) tree is likely to be the correct tree only in analyses of data sets that produce tree-length distributions which are significantly more skewed than expected from random data. The shaded regions correspond to the 95% (dark) and 99% (light) confidence limits for g_1 (the skewness statistic) for random sequence data. (Adapted from Ref. 12).

[9] M. Goodman, J. Czelusniak, and G. W. Moore, *Syst. Zool.* **28**, 379 (1979).
[10] D. M. Hillis, *in* "Phylogenetic Analysis of DNA Sequences" (M. M. Miyamoto and J. Cracraft, eds.), p. 278. Oxford Univ. Press, New York, 1991.
[11] D. M. Hillis and J. P. Huelsenbeck, *J. Hered.* **83**, 189 (1992).

skewness of tree-length distributions is a useful indicator of data sets that are likely to be phylogenetically informative. In simulated phylogenies (in which the true tree is known),[12] data sets that produce significantly skewed tree-length distributions are also likely to produce the correct tree topology in phylogenetic analysis (in this case, using parsimony). However, data sets that produce distributions not significantly different from those obtained from random data (because of high mutation rates) are unlikely to yield trees that resemble the true phylogeny.

Other tests for assessing phylogenetic signal using trees involve repeatedly randomizing characters within data matrices, rather than comparing a given data set to results obtained from random sequences.[13,14] These methods are thus less sensitive to base-compositional or other biases, because the original data are randomized among taxa. However, they also require much greater computational time and are thus less suited for initial assessments of phylogenetic signal than for assessment of confidence in results (see below).

Choosing a Method of Phylogenetic Inference

Assumptions

The first aspect of choosing a method of phylogenetic inference is deciding which assumptions and models one is willing to accept. The choice is important because whenever assumptions of a model are not met by the real patterns of nucleotide substitutions, errors may be introduced into the tree construction. Models of evolutionary processes must reflect biological reality, and the extent to which they fulfill this goal will influence the phylogenetic inferences they provide. It is not always predictable which assumptions, when violated, will affect the phylogenetic estimate. Many different kinds of macromolecules exist, and models that do not take into account this huge amount of variability are bound to fail for some molecular systems. As discussed in the previous section, there is no reason to believe that all molecules or all regions of a single molecule will reflect phylogenetic history. To carefully practice phylogenetic inference one must know more about the specific molecule being examined before it is used to reconstruct phylogeny. It is best if the evolutionary models are chosen based on the molecules that are being studied.

Assumptions are directly related to the evolutionary process of nucleo-

[12] J. P. Huelsenbeck, *Syst. Zool.* **40,** 257 (1991).
[13] J. W. Archie, *Syst. Zool.* **38,** 239 (1989).
[14] D. P. Faith and P. S. Cranston, *Cladistics* **7,** 1 (1991).

tide substitution. One must decide what general assumptions are acceptable for the particular molecule being examined and then choose among the models of phylogenetic inference by the assumptions incorporated in the different tree construction procedures. Most methods share certain assumptions: the characters are evolving independently; the comparisons involve orthologous genes; positional homology has been inferred correctly; and, in many cases, the nucleotide changes examined are neutral.[1,15]

Sequence data are naturally a character-based information source comprising four bases (e.g., A, C, G, and T for DNA) and gaps (insertions/deletions). Multiple mutation events (Fig. 5) can effectively randomize a particular nucleotide position with respect to phylogenetic history. There are 12 possible ways that bases can be substituted, and phylogenetic approaches differ in their treatment of these (Fig. 6). The various methods of phylogenetic inference differ in their assumptions of the pattern of evolutionary change. The observed mutations are analyzed using an explicit or implicit model of nucleotide substitution. Thus, whether one uses a character-based or a distance-based approach to phylogenetic reconstruction, assumptions about the evolutionary process must be made. For a general approach, the most realistic models are limited to a few assumptions, well supported by available evidence. This is important because the more assumptions made, the more likely some of them will be incorrect for the specific macromolecule being examined.

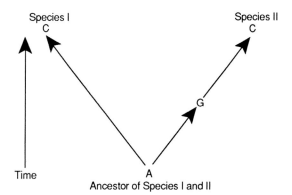

FIG. 5. Example of unobserved multiple substitution events between two distantly related species (I and II). Although three mutations have occurred since the common ancestor, species I and species II show no differences.

[15] J. Felsenstein, *Annu. Rev. Genet.* **22,** 521 (1988).

$$N$$

		A	C	G	T
	A	w	a	b	c
M	C	g	x	d	e
	G	h	i	y	f
	T	j	k	l	z

FIG. 6. The 12 possible pathways of substitution (base M to base N) for the four nucleotides of DNA (A, C, G, and T). Lowercase letters *(a – l)* represent the individual frequencies of each substitution. The symbols $w – z$ therefore correspond to the probability of a base remaining unchanged [e.g., $w = 1 - (a + b + c)$].

Weighting

By examining the way in which nucleotide substitutions are weighted by the various phylogenetic methods, one can understand the evolutionary assumptions that must be accepted if one is to utilize any particular method. Nucleotide substitutions can be subdivided either across sites or across mutations. A natural division of nucleotide site change in protein-coding genes is based on codon structure. One can treat sites preferentially by whether the mutation causes the amino acid to change (nonsynonymous change) or remain the same (synonymous change). Further subdivision exists in the base position of the codon (first, second, or third position) and the number of codons that code for the same amino acid (redundancy, multiplicity class, or degenerate sites). The amino acid code is not universal; thus, different rules may be needed in these cases with regard to redundancy. Codon structure is only relevant to gene sequences that are translated.

The 12 possible types of nucleotide substitution can be treated differently (assuming nonsymmetry of change, e.g., the frequency of A to C does not equal that for C to A) or treated equally, or any combination of these substitutions can be grouped. One obvious division of base substitutions is to treat transitions (changes of purine to purine or pyrimidine to pyrimidine) separately from transversions (change of purine to pyrimidine or vice versa). Insertion/deletion events can also be treated as a separate type of mutation. Additionally, nucleotide substitutions can be preferentially treated by a combination of position and mutation (e.g., transversions occurring in the first and second codon positions).

There are several large classes of DNA sequences which are not translated, including those for structural RNAs [ribosomal RNAs (rRNAs) and transfer RNAs (tRNAs)], pseudogenes, and repetitive DNAs [e.g., short and long interspersed repeated sequences (SINES and LINES)]. Ribosomal

RNA and tRNA genes are constrained by the secondary structures of their products. Their double-stranded (stem) and single-stranded (loop) regions could be treated as a major division of nucleotide site change.[16] Multigene families and repetitive DNAs are influenced by several different mechanisms including unequal crossing-over, gene conversion, and sequence transposition which could affect the types and rates of nucleotide substitution observed. At present few general guidelines are available to account for the particular structure and molecular evolution of these macromolecules.

A Priori and A Posteriori Weighting

All methods weight nucleotide change either equally or selectively, before the analysis *(a priori)* and/or after an initial phylogeny is built *(a posteriori)*. When phylogeneticists pick and choose among the available characters by weighting them, they must justify their rationale; otherwise, one is left with subjective interpretations and considerable confusion for phylogenetic inference. Although there is a great diversity in the way that one can subdivide molecular mutation (see above), it is often unclear which types of characters will consistently provide the greatest signal for phylogenetic inference. Any *a priori* selection may bias the results in favor of a preconceived notion of the evolutionary process. However, weighting of characters can also be approached by *a posteriori* methods, in which one judges the relative importance of characters by the levels of homoplasy observed (rather than expected as done *a priori*). In *a posteriori* weighting, an approximation of a phylogeny is first made, homoplasy is then measured for each character on the tree, and weighting is applied to characters based on the amounts of homoplasy observed. Possible weighting schemes (both *a priori* and *a posteriori*) are limited only by the finite ways in which macromolecules are organized, the possible mutations by which nucleotides can change, and the infinite number of relative weights that could be applied to these substitutions.

When one chooses weights for particular changes, it is usually on the basis of the assumed prevalence of the mutation, with more rapidly evolving changes given less weight and more conservative ones greater weight. One must be willing to accept the hypothesis that where more mutations are taking place there will also be greater chances for homoplasy. One assumes that within a "conservative" gene region, all mutations have occurred slowly, incorporating little homoplasy, thereby making these characters more reliable.

[16] S. A. Gerbi, *in* "Molecular Evolutionary Genetics" (R. J. MacIntyre, ed.), p. 482. Plenum, New York, 1985.

Evolutionary processes that should be closely monitored include whether one mutation is favored over another and whether one gene region evolves differently from another. This can take the form of differential mutation rates for different gene positions or mutations. To add to the complexity, observed differential rates are known for various organisms over the same positions and mutations.[17] Despite this complexity, Nei[18] has stated that "the pattern of evolutionary change is well understood," but, in our opinion, only in broad terms. If selection is invoked, this further complicates understanding the evolutionary process to the extent that Lewontin[19] has remarked that "sequences with significant but intermediate constraints on amino acid replacements are in principle unanalyzable." For phylogeneticists to have a hope of accurately reconstructing history, one must be able to make some predictions about the processes by which macromolecules evolve. Choosing a macromolecule to address a specific phylogenetic question and choosing an appropriate inference method with which to analyze the comparative sequence data are two manifestations of this problem; the latter is addressed in our next sections.

Phylogenetic Reconstruction

Deciding which evolutionary processes are affecting the molecule under study is only the first step toward the resolution of a phylogeny. Numerous alternative methods are currently available (Fig. 1), and each makes different assumptions about the molecular evolutionary process. Some methods are more general and may apply to a wider range of macromolecules and phylogenetic questions, whereas others are restricted to specific types of phylogenetic problems. To assume that any one method can solve all problems is naive, given the complexity of genomes and their evolution. The various phylogenetic methods are interconnected, and we have provided one interpretation of their linkages (Fig. 1). The assumptions, weaknesses, and strengths of each are discussed in more detail below.

Distances and Sequence Divergence

Phylogenetic analyses of sequences can be conducted by analyzing discrete characters (i.e., the nucleotides themselves) or by making pairwise comparisons of whole sequences (the distance approach). Deciding whether to use a distance-based or a character-based method depends on

[17] R. J. Britten, *Science* **321**, 1393 (1986).
[18] M. Nei, *in* "Phylogenetic Analysis of DNA Sequences" (M. M. Miyamoto and J. Cracraft, eds.), p. 90. Oxford Univ. Press, New York, 1991.
[19] R. C. Lewontin, *Mol. Biol. Evol.* **6**, 15 (1989).

the assumptions one is willing to accept and the goals of the study. If one chooses a distance method for phylogenetic inference, then an assumption that a single coefficient of sequence similarity or dissimilarity provides an accurate measure of evolutionary divergence has been accepted. Distance-based approaches may incorporate the various types of change in estimating a single divergence value, leading to a matrix of all pairwise comparisons of the taxa studied. In one sense, the transformation of nucleotide sequence variation to a distance value reduces the available information. However, others have noted that distance approaches may use more of the available information than some character methods such as parsimony procedures which rely only on "phylogenetically informative positions" (see Table III) and ignore variation unique to single taxa.[20] An estimate of nucleotide divergence often uses a model of substitution to "correct" for (unobserved) multiple substitution events occurring between the more divergent pairs of taxa (Fig. 5). Weighting of substitutions is usually done *a priori*. One can sort the various estimates of divergence (Table I)[21] by the number of parameters incorporated in the algorithms to calculate these divergence values. The more complicated models attempt to use numerous parameters of substitution in their calculations.

Both the type and the position of a mutation have been incorporated into the parameters of divergence values. For example, a one-parameter model treats all nucleotide substitutions as equal, whereas a two-parameter model subdivides nucleotide change into transitions and transversions. Three-, four-, six-, and twelve-parameter models have been proposed[22,23] although one could envision seven- to eleven-parameter models depending on which classes of nucleotide change are grouped. The simplest estimates of nucleotide difference count up the total number of substitutions (sometimes including gaps) and divide by the number of base pairs examined, making no attempt to "correct" the distance value.[24,25] For closely related taxa (when substitution events are relatively low, sequence differences < 10%) different estimates of uncorrected and corrected divergence have been shown to give similar values. As distance increases, so does the underestimation of divergence by many methods. When divergence values are very large between taxa, all estimates become suspect.[26] Proponents of

[20] D. Penny, M. D. Hendy, and M. A. Steel, *in* "Phylogenetic Analysis of DNA Sequences" (M. M. Miyamoto and J. Cracraft, eds.), p. 155. Oxford Univ. Press, New York, 1991.
[21] T. Gojobori, E. N. Moriyama, and M. Kimura, this series, Vol. 183, p. 531.
[22] C. Lanave, G. Preparata, C. Saccone, and G. Serio, *J. Mol. Evol.* **20**, 86 (1984).
[23] C. Saccone, C. Lanave, G. Pesole, and G. Preparata, this series, Vol. 183, p. 570.
[24] M. M. Miyamoto, J. L. Slightom, and M. Goodman, *Science* **238**, 369 (1987).
[25] M. Nei, "Molecular Evolutionary Genetics." Columbia Univ. Press, New York, 1987.
[26] T. Gojobori, K. Ishii, and M. Nei, *J. Mol. Evol.* **18**, 414 (1982).

TABLE I
Estimates of Divergence[a]

Method	Comments	Refs.[b]
Nucleotide-based methods		
Uncorrected approaches	Only account for observed differences	
p (difference)	Accounts for observed substitutions and/or gaps	1, 2
Corrected approaches	Attempt to account for unobserved parallel substitutions and reversals in addition to observed differences. For d (divergence), corrections are added for multiple substitution events by adopting some distribution of nucleotide change (e.g., Poisson distribution) and by weighting substitutions	2
One-parameter	All nucleotide substitutions treated as equal	3
Two-parameter	Transitions treated differently than transversions	4
Three-parameter	Two classes of transversion with all transitions treated equally	5
Four-parameter	Two classes of transversion and two classes of transition	6, 7
Six-parameter	Four classes of transversion and two classes of transition	8, 9
Codon-based methods		
Unweighted pathway	Synonymous and nonsynonymous changes, with each further subdivided into three categories of nucleotide substitution	10
Miyata and Yasunaga's weighted pathway	Pathways from one amino acid to another are weighted by biochemical similarity of amino acid replacement	11
Nei and Gojobori's unweighted pathway	Takes into account codon position (first, second, or third) and whether change is synonymous or nonsynonymous (unweighted version of Miyata and Yasunaga's method)	2, 12
Li, Wu, and Luo's weighted pathway	Changes weighted as nondegenerate, 2-fold degenerate, or 4-fold degenerate sites, and by expected to observed frequencies of base pair mutations	13
Four-parameter	Multiplicity classes for 2-, 3-, 4-, and 6-fold degenerate codon groups are used to estimate number of synonymous substitutions per codon. Several options are available for representing constraints on amino acid replacements	14

[a] Formulas for each estimate can be obtained from Gojobori et al.[21]

[b] Key to references: (1) D. L. Swofford and G. Olsen, in "Molecular Systematics" (D. M. Hillis and C Mortiz, eds.), p. 411. Sinauer, Sunderland, Massachusetts, 1990 (see p. 428 for generalized formulas); (2) M. Nei, in "Molecular Evolutionary Genetics" (M. Nei, ed.), p. 64. Columbia Univ. Press New York, 1987; (3) T. H. Jukes and C. R. Cantor, in "Mammalian Protein Metabolism III" (H. N Munroe, ed.), p. 21. Academic Press, New York, 1969; (4) M. Kimura, J. Mol. Evol. 16, 111 (1980); (5) M. Kimura, Proc. Natl. Acad. Sci. U.S.A. 78, 454 (1981); (6) F. Tajima and M. Nei, Mol. Bio Evol. 1, 269 (1984); (7) N. Takahata and M. Kimura, Genetics 98, 641 (1981); (8) T. Gojobori, H Ishii, and M. Nei, J. Mol. Evol. 18, 414 (1982); (9) M. Hasegawa, H. Kishino, and T. Yano, J. Me Evol. 22, 160 (1985); (10) F. Perler, A. Efstratiadis, P. Lomedico, W. Gilbert, R. Kolodner, and Dodgson, Cell (Cambridge, Mass.) 20, 555 (1980); (11) T. Miyata and T. Yasunaga, J. Mol. Evol. 1 23 (1980); (12) M. Nei and T. Gojobori, Mol. Biol. Evol. 3, 418 (1986); (13) W.-H. Li, C.-I. Wu, ar C.-C. Luo, Mol. Biol. Evol. 2, 150 (1985); (14) R. C. Lewontin, Mol. Biol. Evol. 6, 15 (1989).

distance encourage the use of "appropriate distance measures." Divergence measures that are often regarded as "inappropriate" include approaches which add gaps to the calculation (owing to the difficulties in alignment and modeling of these mutations), equally weighted procedures (except when used for closely related taxa) which generally do not "correct" for multiple substitution, and methods that attempt to take selection factors into account. The simplest of the "appropriate" divergence measures is the two-parameter method of Kimura.[27] This model appears to estimate divergence as well as the more complicated algorithms, over a broad range of divergences, and without the need for additional specifics about the evolutionary process.

Clustering Algorithms for Distance Data

Methods for clustering distance data can be broken down into those that rely on clocklike mutation rates and those that are less sensitive to this assumption. Two commonly used algorithms that rely on clocklike behavior are the unweighted and weighted pair group methods using arithmetic means (UPGMA and WPGMA, respectively). These algorithms assume that the data are ultrametric (i.e., clocklike), a property that is often not satisfied by sequence data. Owing to this assumption, UPGMA and WPGMA have largely been replaced by alternative methods that do not rely on this assumption (Table II). The neighbor joining method does not depend on ultrametric data, although it may rely on the assumption of additivity[1,18] (i.e., the evolutionary distance between any two taxa is equal to the sum of the branches that join them). Nonreliance on ultrametric data is a desirable quality for any algorithm as many molecules are not clocklike, even among closely related taxa. The neighbor joining method, because of its ability to handle unequal rates, its connection to minimum-length trees (see below), and its ease of calculation with regard to both topology and branch lengths, has become a popular approach for analyzing sequence distances.

Multiple-tree methods (Table II) rely on a defined criterion of optimality, unlike the single-tree algorithms above which give an answer (usually a single tree topology), but do not select it according to some objective measure of fit or provide a method for ranking alternatives. Optimality is defined as an objective quantity, measuring the conformity of the original data (distance or character) to a tree. Each algorithm is designed to find the best tree given its optimality criterion and different searching methods are employed for this purpose.[1] If one does not agree with the criterion to be optimized, then neither the approach nor the searching method for identi-

---

[27] M. Kimura, *J. Mol. Evol.* **16,** 111 (1980).

TABLE II
CLUSTERING ALGORITHMS USING DISTANCE DATA

Method	Comments	Refs.[a]
Single-tree algorithms	Methods that provide single topology by following specific series of steps. No specific optimality criterion is used to select tree	
UPGMA and WPGMA	Groups taxa in order of decreasing similarity (or increasing dissimilarity); assumes constant rate of evolution	1
Neighbor joining method	Heuristic approach for estimating minimum evolution phylogeny (see below)	2
Multiple-tree algorithms	Methods that use optimality criterion to compare alternative topologies and to select final tree	
Fitch–Margoliash method	Best tree chosen as that which maximizes fit of observed (original) versus tree-derived (patristic) distances, as measured by % standard deviation statistic. Branch lengths are determined by linear algebraic calculations of observed distances among three different taxa interconnected by common node	3
Distance Wagner procedure	Assumes that patristic distances must be greater than or equal to observed distances (e.g., negative branch lengths are not permitted). Best tree chosen is that with shortest overall tree length	4
Neighborliness	Maximizes the four point condition for an additive tree (see text); different quartets of taxa are examined one at a time when five or more taxa are represented	5, 6
Minimum evolution	Computes sum of all branch lengths for each tree, considering all possible topologies, and chooses phylogeny which minimizes total overall length. A Fitch–Margoliash approach is used to calculate branch lengths	7, 8

[a] *Key to references: (1)* P. H. A. Sneath and R. R. Sokal, "Numerical Taxonomy." Freeman, Sa Francisco, 1973; *(2)* N. Saitou and M. Nei, *Mol. Biol. Evol.* **4**, 406 (1987); *(3)* W. M. Fitch and I Margoliash, *Science* **155**, 279 (1967); *(4)* J. S. Farris, *Am. Nat.* **106**, 645 (1972); *(5)* S. Sattath and A Tversky, *Psychometrika* **42**, 319 (1977); *(6)* W. M. Fitch, *J. Mol. Evol.* **18**, 30 (1981); *(7)* L. I Cavalli-Sforza and A. W. F. Edwards, *Am. J. Hum. Genet.* **19**, 233 (1967); *(8)* N. Saitou and T Imanishi, *Mol. Biol. Evol.* **6**, 514 (1989).

fying the optimal tree will satisfy its opponents. In turn, the algorithms used to calculate optimality and to search for optimal trees limit the ability of the investigator to satisfy the original criterion (see below). Quantitative phylogeneticists are continually upgrading their software to improve the speed and accuracy of finding the optimal solution.

Several criteria of optimality are used for building distance trees (Table II). Each approach permits unequal rates and assumes additivity. The Fitch–Margoliash method minimizes the deviation between the observed pairwise distances and the path length distances for all pairs of taxa on a tree. This fit is measured by percent standard deviation (one calculation of the least-squares method). For the distance Wagner method, the observed distance values impose a minimum bound on the branch lengths, thereby ensuring that the tree-derived distances are always greater than or equal to the original ones. The selected topology becomes the one of minimum length, where length is determined as the sum of lengths over all branches of the tree. The neighborliness method compares the distances (d) of the three possible groupings of four taxa (A, B, C, D). Under the assumptions of the four-point condition, two relationships must be satisfied for A–B and C–D to be clustered: (i) $d(A, B) + d(C, D) < d(A, C) + d(B, D)$; and (ii) $d(A, B) + d(C, D) < d(A, D) + d(B, C)$. For larger phylogenetic problems, four-taxon comparisons are conducted for all possible subclusters, and paired taxa are clustered by their arithmetic means. Minimum evolution is the exhaustive implementation of the neighbor joining method (a single-tree heuristic approach). Branch lengths are optimized using the Fitch–Margoliash method and are added to determine the overall length of the tree. The tree with the minimal overall length is then selected.

Character-Based Approaches

Rather than reducing all of the individual variation to a single divergence value, character-based methods treat each substitution separately. By counting each mutation event, one determines the relationships among organisms by the distribution of mutations observed (Table III). These methods are preferred for studying character evolution, for combining

TABLE III
CHARACTER-BASED METHODS OF TREE CONSTRUCTION

Method	Comments	Refs.[b]
Parsimony	Selects phylogeny that minimizes number of evolutionary changes for data set. Approach relies on phylogenetically informative characters (i.e., those with two or more states shared by two or more taxa)	1
Maximum likelihood	Calculates probability of data set, given particular model of evolutionary change and specific topology	2
Method of invariants[a]	Counts number of transversion events supporting phylogeny after adjusting for homoplastic change. Designed for four-taxon problems where homoplasy is expected to be abundant (e.g., for distantly related taxa with unequal rates of evolution)	3, 4

[a] Commonly known as evolutionary parsimony.

[b] *Key to references: (1)* See review by D. L. Swofford and G. Olsen, *in* "Molecular Systematics" (D. M. Hillis and C. Mortiz, eds.), p. 411. Sinauer, Sunderland, Massachusetts, 1990; *(2)* J. Felsenstein, *J. Mol. Evol.* **17,** 368 (1981); *(3)* J. A. Lake, *Mol. Biol. Evol.* **4,** 167 (1987); *(4)* R. Holmquist, M. M. Miyamoto, and M. Goodman, *Mol. Biol. Evol.* **5,** 217 (1988).

multiple data sets or sequentially adding data, and for inferring ancestral genotypes. All sequence information is retained through the analyses; no information is lost in the conversion to distances. Therefore, character-based methods are often preferred when they are feasible. The chief disadvantages are the greater computational time they require and the greater difficulty of correcting for multiple substitutions.

Parsimony

Parsimony is the principle of logic that simple explanations should be preferred over more complex explanations. In the context of phylogenetic inference, the most parsimonious tree is the tree that requires the fewest evolutionary changes to explain the data. Parsimony remains the most popular character-based approach for sequence data. This popularity is due to its logical simplicity, its ease of interpretation, its prediction of both ancestral character states and amount of change along branches, the availability of efficient and powerful programs for its implementation, and its flexibility in terms of maximizing weighting strategies and conducting character analyses. Parsimony procedures search for the phylogeny that minimizes the number of evolutionary events required to explain the original data. Parsimony, which permits unequal rates, assumes that homoplasy occurs at levels that do not interfere with phylogenetic inference. When more than one of the taxa in a study is connected to the tree by an excessively long branch and rates of mutation are relatively high, parsimony procedures can be expected to converge onto the wrong tree even as more data are added.[28] However, inconsistency under these conditions is a property of many tree-building procedures. The excessive homoplasy may be avoided by assigning greater weight to the more conservative sites or gene regions (e.g., functional domains) and/or by giving more weight to the slower types of nucleotide change (e.g., transversions). With parsimony procedures, a wide range of weighting schemes is possible (Table IV).

Weighting strategies must be considered for parsimony analyses as they are for all phylogenetic methods. Weighting is practiced even when weights are not specified, in that all changes are uniformly counted. Thus, no attempt to weight nonetheless carries an assumption about the evolutionary process. Current parsimony programs are well designed to implement sophisticated weighting schemes, and, as such, weights based on at least general patterns of molecular evolution are encouraged (e.g., first and second codon positions versus third), as long as they are explicitly presented and defended.

---

[28] J. Felsenstein, *Syst. Zool.* **27,** 401 (1978).

TABLE IV
METHODS OF CHARACTER WEIGHTING

Weighting method	Comments[a]	Refs.[b]
Uniform weighting	All characters and changes are given equal weight	
Nonuniform weighting	Selective weighting of particular characters and/or changes	
Across positions	Emphasizes structural/functional differences between gene regions or base positions	
Codon positions	Selective weighting of first, second, and third codon positions in translated genes, because of redundancy of genetic code. A general rule is that third-codon positions are under less selective constraint than first and second and, as such, are more likely to change than the latter	
Stems and loops	Selective weighting of double-stranded (stem) versus single-stranded (loop) regions of structural RNAs (tRNA and rRNA), reflecting constraints on stem regions to maintain secondary structure through base pairing	*1*
Within positions	Emphasizes mutational bias	
Transitions versus transversions	Weighting of transition bias which is most evident in vertebrate mitochondrial DNA but apparent in other systems as well. The general rule is that transitions occur more frequently than transversions and, as such, deserve less weight than the latter	
Relative substitution frequencies	The 12 possible substitutions (Fig. 6) are weighted differently according to relative frequencies. Different combinations of the 12 substitutional types can be recognized, with transition/transversion categorizations representing one extreme (see above)	
Weighting by base composition	Weighting schemes based on either observed or expected base compositions of sequences being examined. This approach assumes that base frequencies reflect substitutional frequencies	
Synonymous versus nonsynonymous change	Unlike synonymous mutations, nonsynonymous changes alter primary sequence of a polypeptide and, as such, are under greater selective constraint and occur less frequently. Nonsynonymous mutations therefore warrant greater weight	
Within and across positions	Refers to weighting for both positional effects and mutational bias. A large number of combinations are possible	
Successive approximations	*A posteriori* weighting of characters according to levels of homoplasy, as judged with an initial estimate of topology. Subsequent reiterations of weighting and tree construction are performed until topology stabilizes. Dynamic weighting uses successive approximations approach to weight sequence data both across and within positions	*2–4*

[a] See Swofford and Olsen[1] for review.
[b] *Key to references: (1)* M. J. Dixon and D. M. Hillis, *Mol. Biol. Evol.* **10,** 256 (1993); *(2)* J. S. Farris, *Syst. Zool.* **18,** 374 (1969); *(3)* D. Sankoff and R. J. Cedergren, *in* "Time Warps, String Edits and Macromolecules: The Theory and Practice of Sequence Comparisons" (D. Sankoff and J. B. Kruskal, eds.), p. 253. Addison-Wesley, London, 1983; *(4)* P. L. Williams and W. M. Fitch, *in* "The Hierarchy of Life: Molecules and Morphology in Phylogenetic Analysis" (B. Fernholm, K. Bremer, and H. Jörnvall, eds.), p. 453. Elsevier Science Publ., Amsterdam, 1989.

Maximum-Likelihood Methods

Statistical models have been developed for character-based nucleotide change. By considering each site separately, one determines the likelihood of these changes in the data, given a particular topology and model of molecular evolution. The maximum likelihood method therefore depends heavily on the model chosen and on how well it reflects the evolutionary properties of the macromolecule being studied. Because of questions about the accuracy of the models, coupled with the computational complexities of the approach, maximum-likelihood methods have not received the attention that they probably deserve. The more recent versions of maximum likelihood rely on models of evolution that are quite sophisticated, taking into account the possibility of unequal rates of change among lineages, site-specific rate variability, and the random distribution and/or clustering of variable sites. However, it remains unclear whether these more complex models will be better at phylogenetic inference than the more general ones since the former demand specific insights about the evolutionary process (information which is not typically available).

Methods of Invariants

Phylogenetic inference methods have been developed for selecting the correct topology when large amounts of homoplasy exist, as when rate heterogeneities occur among distantly related branches. By relying on a few specific patterns of nucleotide variation that represent the most conservative changes, one can avoid the abundant homoplasy while recognizing signal. In evolutionary parsimony, quantities called "operator invariants" are calculated for the three possible topologies of four taxa. Each invariant reflects specific patterns of shared transversions corrected for homoplastic similarity. These calculations are based on the variable positions with two purines and two pyrimidines. Zero-value invariants represent cases in which random multiple mutation events have canceled each other out, and, as such, a chi-squared (χ^2) or binomial test is used to identify the correct topology as the one with an invariant significantly greater than zero. Evolutionary parsimony assumes that the transversion rates between the two types of transversions for each given base are equal (e.g., the frequency of A to C is the same as A to T). A recent modification of the procedure has been proposed which corrects such inequalities by taking into account base compositional differences.[29]

The method of invariants recognizes that there are 36 patterns of transitions/transversions (called spectral components) for four taxa. Dif-

[29] A. Sidow and A. C. Wilson, *J. Mol. Evol.* **31**, 51 (1990).

ferent methods of phylogenetic inference rely on various combinations of these components, with 12 used to calculate the three operator invariants of evolutionary parsimony. Unlike parsimony, the method of invariants does not construct intermediate ancestors, and it is limited to direct comparisons of only four taxa at any one time. When more than four taxa are considered, all possible quartets are typically analyzed and a composite tree constructed from the individual results. Relatively long sequences are needed by the procedure to obtain enough transversions for its statistical tests.

Searching for Optimal Trees

Once an optimality criterion has been selected, it is necessary to calculate or estimate the best tree for the given criterion. For relatively few taxa (up to as many as 20 or 30, depending on the level of homoplasy present in the data), it is possible to use exact algorithms that will be certain to find the optimal tree. For greater numbers of taxa, one must rely on heuristic algorithms (i.e., useful and efficient algorithms that approximate the exact solutions but may not give the optimal solution under all conditions). When heuristic algorithms are used, it is always a possibility that a better solution exists. Therefore, the use of heuristic algorithms should be described exactly so that alternative procedures can be explored by other workers who wish to search for better solutions.

Variations of two exact algorithms (algorithms that will always find the optimal solution) are commonly used. The first is to search exhaustively through all possible tree topologies for the best solution(s). This method is computationally simple for 9 or fewer taxa (for which there are $\leq 135{,}135$ labeled, unrooted, bifurcating trees) and is only moderately time-consuming for 10 or 11 taxa (2,027,025 and 34,459,425 trees, respectively).[2] For 12 taxa, the evaluation becomes laborious (654,729,075 trees), and for 13 or more taxa ($\geq 13{,}749{,}310{,}575$ trees) the calculations are usually impractical. The chief advantages of exhaustive searches are (1) the optimal tree(s) is always found and (2) all other possibilities can be ranked with respect to the optimal solution(s).

If an exhaustive search is impractical for a given data set, another exact algorithm can be used that is generally much faster, namely, the branch-and-bound algorithm.[30] Most implementations of this algorithm calculate an initial upper bound for a tree (using one of the heuristic methods described below) and then search exhaustively along paths that lead to all possible trees by sequentially adding taxa. If the upper bound is reached

[30] M. D. Hendy and D. Penny, *Math. Biosci.* **59,** 277 (1982).

before all the taxa have been added to the tree, then no more trees along that search path need be examined, because any further addition of taxa could only increase the length of the tree. If all taxa are added and the upper bound has still not been reached, then a shorter tree has been found. The upper bound is reset to this new score, and the search is continued. In this way, the shortest tree can always be found, even though many trees (all of which must be longer than the upper bound) are never examined by the algorithm.

Although the branch-and-bound algorithm will always find the shortest tree, it cannot rank suboptimal solutions if implemented as above (its usual form). However, the best implementations of this algorithm (e.g., as in the Phylogenetic Analysis Using Parsimony or PAUP program[31]) allow an investigator to save all trees that are shorter than or equal to a specified bound. In this way, it is possible to look at the lower end of a tree-length distribution, even for relatively large numbers of taxa, and thereby rank all alternatives near the optimal solution(s).

If the exact algorithms described above are not feasible for a given data set (the limitation is usually number of taxa), then various heuristic approaches can be tried. The heuristics used should be described in sufficient detail that they can be replicated, and so that alternative searches can be attempted. It is also worthwhile to discuss the number of alternative solutions examined, to give a sense of the thoroughness of the search.

Most heuristic techniques start by finding a reasonably good estimate of the optimal tree(s) and then attempting to find a better solution by examining structurally related trees. The initial tree is usually found by a stepwise addition algorithm.[1,32] These algorithms add taxa sequentially to a tree, in each step adding the new taxon at the optimal place in the growing tree. Once a taxon is added to the tree, however, the tree is constrained for the next round of addition. Therefore, it is likely that the solution that is optimal when only a few taxa are joined together will not be globally optimal for these taxa when the tree is complete (hence, the "inexact" nature of stepwise addition algorithms). The various stepwise addition algorithms differ primarily in the order in which taxa are added to the tree. The simplest (and usually least efficient) algorithms simply add taxa in the order in which they appear in the matrix. Other implementations base the addition of sequences on their distance to a reference taxon or on the number of steps they add to the growing tree.[1]

After an initial tree has been obtained (either by stepwise addition or

---

[31] D. L. Swofford, "PAUP: Phylogenetic Analysis Using Parsimony, Version 3.0." Illinois Natural History Survey, Champaign, Illinois, 1990.

[32] J. S. Farris, *Syst. Zool.* **34,** 21 (1970).

user input), it can often be improved by examining related topologies by a family of procedures known as branch swapping. Several alternatives are commonly implemented and are described by Swofford and Olsen.[1] All involve rearranging branches of the initial tree to search for a shorter alternative (or one equal to or shorter than a specified limit). Because any of these methods may be the most efficient under certain circumstances, it is often necessary to try as many options as are available to be reasonably sure of finding the optimal solution.

Another strategy that may be used for large data sets is to reduce the number of possible topologies by constraining the analyses to look at a subset of trees with exact algorithms. This approach is useful if a study is designed to address specific questions. For instance, assume the relationships among 10 families of angiosperms are in debate, but no one questions the monophyly of each of the 10 families. Also assume that orthologous sequences are available for three species of each of the 10 families. Under such circumstances, it may be desirable to conduct at least one analysis in which the 10 families are each constrained to be monophyletic, because an exact solution is thereby possible (there are only $\sim 1.2 \times 10^{11}$ trees if the 10 families are constrained to be monophyletic but $\sim 8.7 \times 10^{36}$ trees if they are not). Of course, the prior hypotheses of monophyly are being assumed rather than tested, but this is appropriate under the conditions described.

Assessing Confidence in Results

Testing How Well Sequence Data Support Trees

Once the data are collected and a topology constructed, it is necessary to evaluate the reliability of those data and the supported tree. It is important to keep in mind that even randomly generated data can lead to a single, best result. Therefore, several methods exist for testing the robustness of the final topology using analytical and resampling procedures (Table V).

A problem in assessing confidence when there are more than four taxa is that the number of trees available for testing increases dramatically. Thus, few methods can reliably compare the more complex phylogenies. As the size of individual trees increases, the stringency of the tests increases as well. It is generally recommended that subsets of taxa be examined instead, from within the more complex topologies, focusing on specific major questions targeted before the analysis (see below). Alternatively one could limit the comparisons to just those topologies deemed plausible for biological reasons (e.g., a previous hypothesis of relationship based on independent data).

TABLE V
METHODS FOR ASSESSING CONFIDENCE IN RESULTS

Method	Comments	Refs.[a]
Analytical techniques		
For parsimony procedures		
Wilcoxon rank-sum test, sign test, winning sites method	Determines whether significant character support exists for one tree relative to a second. Wilcoxon rank-sum test allows one to assign mutations different weights (i.e., transversions favoring one tree are given greater importance than transitions). For six or fewer taxa and no ordering as above, Wilcoxon rank-sum test reduces to simpler sign test. In winning sites method, binomial test is used to determine whether a greater number of phylogenetically informative positions (*sensu* parsimony) supports one tree versus a second	1–3
Confidence limits without clock	Assumes worst-case scenario for four taxa (two unrelated taxa with fast rates of evolution, with other two and common stem experiencing virtually no change). Under these conditions, two unrelated taxa are expected to share 3/16 of their positions by chance alone. Thus, to be statistically significant, a tree must be supported by more than 3/16 of its characters	4
Confidence limits with clock	Here, polytomy (star phylogeny) for four taxa is taken as worst-case situation. Thus, probability that a phylogenetically informative site supports a tree is same for all three resolutions of polytomy, 1/3	5
Williams/Goodman confidence limits	Similar to approach just described, except that a clock is not assumed. Method is based on a worst-case situation whereby support for correct tree is $\geq 1/3$ and $\leq 2/3$ for the two incorrect topologies combined	6
For evolutionary parsimony	A chi-square or binomial test is used to determine which phylogenetic invariants deviate significantly from zero and which do not	7
For maximum likelihood		
Likelihood ratio test	Ratio of likelihood scores for selected tree and star phylogeny is treated as a chi-square statistic with one degree of freedom. Alternatively, standard normal test of the mean and variance of the difference of their likelihood scores can be used to compare one tree to another	2, 8, 9
For distance approaches		
Branch length variances	An internal branch length is considered significant only if its length plus or minus two standard errors exceeds zero	10–1
Resampling techniques	Characters of original data set are randomly sampled and a tree is produced from new matrix. Many resampled matrices are analyzed (usually ≥ 100). Frequency of replication of a group is taken as measure of its statistical reliability or, at least, its stability	
Booststrapping	Characters are randomly sampled with replacement, leading to new data set of same size as original	13, 1

TABLE V *continued*

Method	Comments	Refs.[a]
Jackknifing	Characters are randomly sampled without replacement, leading to new data set smaller than original one. Jackknifing of taxa is sometimes done instead of characters	15, 16

[a] *Key to references: (1)* A. R. Templeton, *Evolution* **37**, 221 (1983); *(2)* J. Felsenstein, *Annu. Rev. Genet.* **22**, 521 (1988); *(3)* E. M. Prager and A. C. Wilson, *J. Mol. Evol.* **27**, 326 (1988); *(4)* J. Felsenstein, *in* "Statistical Analysis of DNA Sequence Data" (B. Weir, ed.), p. 113. Dekker, New York, 1983; *(5)* J. Felsenstein, *Syst. Zool.* **34**, 152 (1985); *(6)* S. A. Williams and M. Goodman, *Mol. Biol. Evol.* **6**, 325 (1989); *(7)* J. A. Lake, *Mol. Biol. Evol.* **4**, 167 (1987); *(8)* H. Kishino and M. Hasegawa, *J. Mol. Evol.* **29**, 170 (1989); *(9)* J. Felsenstein, *J. Mol. Evol.* **26**, 123 (1987); *(10)* M. Nei, J. C. Stephens, and N. Saitou, *Mol. Biol. Evol.* **2**, 66 (1985); *(11)* M. Hasegawa, H. Kishino, and T. Yano, *J. Mol. Evol.* **22**, 160 (1985); *(12)* W.-H. Li, *Mol. Biol. Evol.* **6**, 424 (1989); *(13)* J. Felsenstein, *Evolution* **39**, 783 (1985); *(14)* D. M. Hillis and J. J. Bull, *Syst. Biol.* **42**, 182 (1993); *(15)* S. Lanyon, *Syst. Zool.* **34**, 397 (1985); *(16)* D. Penny and M. Hendy, *Mol. Biol. Evol.* **3**, 403 (1986).

Groups to be evaluated need to be specified *a priori,* otherwise problems of multiple testing can lead to an unreasonably high probability of accepting some group as significantly supported. In addition, the following tests assume that each nucleotide substitution is independent and derives from a large sample, assumptions which often are not met. Despite these limitations, many systematists have argued for the importance of placing phylogenetic inference in a statistical framework and for improving the "primitive state" of testing its reliability.[33]

Analytical Methods

Analytical procedures for testing phylogenetic reliability operate by comparing the support for one tree to that for another, under the assumption of randomly distributed data. These methods have been extensively developed for parsimony procedures, with one of the earliest approaches using the Wilcoxon rank-sum test to compare the number of unique changes favoring one topology over a second. When fewer than six taxa are considered, this test reduces to the simpler sign test and binomial test (the latter being the winning sites method of Prager and Wilson[34]).

Another approach for testing parsimony results has been to compare the support for the best tree against that expected for a worst-case situation. If no molecular clock is assumed, then the worst-case scenario for four taxa occurs when two unrelated lineages evolve randomly and rapidly, coupled

[33] W.-H. Li and M. Gouy, *in* "Phylogenetic Analysis of DNA Sequences" (M. M. Miyamoto and J. Cracraft, eds.), p. 249. Oxford Univ. Press, New York, 1991.
[34] E. M. Prager and A. C. Wilson, *J. Mol. Evol.* **27**, 326 (1988).

with virtually no change in the other two lineages or the central branch. Under these conditions, one expects two unrelated taxa to share 3 of 16 nucleotide sites by random chance alone. If a molecular clock is assumed, the worst-case situation becomes a trichotomy, with the probability of a phylogenetically informative site (*sensu* parsimony) supporting any one resolution being 1 in 3. Tables have been calculated for each of these cases, summarizing the number of unique changes and extra steps needed to favor statistically one tree over another, relative to the availability of sequence data and phylogenetically informative positions. A recent development of the above tests is the Williams–Goodman approach which does not assume a molecular clock. Instead, this approach assumes that the correct tree will be supported by one-third or more of the informative positions, whereas the two incorrect topologies together will be supported by a total of two-thirds or fewer of the informative sites. Each of the above procedures is largely restricted to four taxa, although at least one heuristic test has been developed to extend this type of approach to five taxa or more.[33] The null model used by these tests assumes equal support for the trees being compared. A significant departure from this expectation implies that more support, greater than expected by chance alone, exists for the best tree relative to the alternatives.

The method of invariants[35] uses a χ^2 test or binomial test to determine which phylogenetic invariants deviate significantly from zero and which do not. Significant departure of an invariant from zero indicates that the associated topology is well supported. In likelihood techniques, one tests the significance of the internal branch length of a tree for four taxa against the null model of an unresolved trichotomy. Here, the logarithm of the ratio of the maximum likelihood scores for the best tree and unresolved phylogeny is treated as a χ^2 statistic, with one degree of freedom. This approach can be extended to more than four taxa. Alternatively, one can compare two maximum likelihood trees in a heuristic way by the mean and variance of the difference of their likelihood scores.[36] Analytical tests of tree reliability for distance approaches have also been developed, with the most popular relying on tests of the variances of internal branch lengths. If an internal branch length plus or minus two times its standard error is greater than zero, then it is considered well-supported at the $\alpha = 0.05$ level.

Resampling Techniques

Resampling procedures estimate the reliability of a phylogenetic result by bootstrapping or jackknifing the characters of the original data set. In bootstrapping, a new data set of the original size is created by sampling the available characters with replacement (Fig. 7). Thus, some characters become represented more than once, others only once, and others not at all in each bootstrapped data set. In contrast, jackknife methods randomly drop one or more data points (or taxa[37]) at a time, thereby creating smaller data sets by sampling without replacement. In either case, a phylogeny is then reconstructed from the resampled data set, and the replication of individual nodes is tallied. The frequency at which a node reappears among different permutations is taken as a measure of its reliability or, at least, of its relative stability (but see below).

Resampling procedures for testing phylogenetic reliability (particularly, the bootstrapping approach) are currently popular, primarily because of the wide availability of powerful and efficient algorithms for their implementation. These approaches have largely been used in conjunction with parsimony procedures, but they can be used in combination with other methods as well (e.g., bootstrapping of sequence positions prior to a distance analysis using the neighbor joining method). The interpretation of bootstrap proportions varies among authors; the values provide unbiased but highly imprecise estimates of repeatability (the probability that the result would be found again given a new sample of characters from the same distribution) and biased, but usually conservative, estimates of phylogenetic accuracy (the probability that the result represents the true phylogeny).[38] However, the degree of bias in the accuracy of estimates varies from node to node in a given tree, as well as from study to study, so bootstrap proportions are not directly comparable with each other. Nonetheless, they are sometimes used as relative measures of confidence among nodes within a single phylogenetic estimate.[39]

Faith and Cranston[14] have developed a procedure (the cladistic permutation tail probability) that randomizes the assignment of character states to taxa at individual sites, while retaining the original configuration of variation at each position. A tree is then constructed and the process repeated to yield a distribution of tree lengths given separate randomizations of the data. The length of a tree is then compared to this distribution to test its departure from lengths expected from randomness. Trees with original overall lengths less than or equal to the shortest 5% of the randomized trees are taken to have significant cladistic structure.

[35] J. A. Lake, *Mol. Biol. Evol.* **4**, 167 (1987).
[36] H. Kishino and M. Hasegawa, *J. Mol. Evol.* **29**, 170 (1989).
[37] S. Lanyon, *Syst. Zool.* **34**, 397 (1985).
[38] D. M. Hillis and J. J. Bull, *Syst. Biol.* **42**, 182 (1993).
[39] M. J. Sanderson, *Cladistics* **5**, 113 (1989).

Original data matrix

Taxa	Characters							
	A	B	C	D	E	F	G	H
One	0	0	1	1	1	1	1	1
Two	0	0	1	1	1	1	1	1
Three	1	1	1	1	1	0	0	0
Four	1	1	0	0	0	0	0	0
Outgroup	0	0	0	0	0	0	0	0

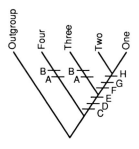

Randomly sample data matrix with replacement; recalculate best tree from new matrix

Resampled data matrix

Taxa	Characters							
	A	B	B	E	E	F	F	H
One	0	0	0	1	1	1	1	1
Two	0	0	0	1	1	1	1	1
Three	1	1	1	1	1	0	0	0
Four	1	1	1	0	0	0	0	0
Outgroup	0	0	0	0	0	0	0	0

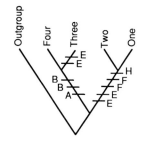

Repeat resampling 1000 times; compute majority-rule consensus tree

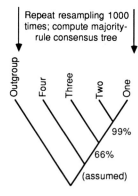

FIG. 7. Bootstrap analysis among characters in a parsimony analysis. The tree to the right of each matrix is the most parsimonious tree for that matrix. The final results of the bootstrap analysis are shown in the tree at the bottom. The number of times each branch was supported in the bootstrap replication is shown as a percentage. Outgroup rooting carries the assumption of ingroup monophyly, so no confidence interval can be assigned to the branch that unites the ingroup.

Statistical versus Phylogenetic Significance

A statistically significant result does not guarantee that an accurate reflection of the phylogeny has been achieved. Rather, it only suggests that a specific topology is strongly supported by a particular method of analysis of the sequence data. This discrepancy can occur because of two different reasons: (1) the statistical significance reflects systematic error that caused the chosen procedure to converge on the wrong topology; and (2) the current topology is correct, but it constitutes a gene tree, which may differ from that for the species owing to allelic polymorphism and lineage sorting, gene duplications or conversion, horizontal transfer, or other molecular evolutionary phenomena. Because of these two possibilities, one cannot accept a tree as correct, given only one set of sequences, even if the current results are considered statistically significant.

The ultimate criterion for determining phylogenetic reliability rests instead on tests of congruence among independent data sets representing both molecular and nonmolecular information.[40,41] Different character types and data sets are unlikely to suffer from the same evolutionary biases, and, as such, congruent results supported by each are more likely to reflect convergence onto the single, correct tree. In the absence of *a priori* knowledge of the truth, congruence remains the final arbiter. Systematists have always relied on concordance in this way to test their hypotheses, and it is therefore not surprising that a similar role for congruence in molecular phylogenetics is starting to emerge as well.

Implementation

No single computer program or package will allow an investigator to conduct all of the analyses described herein. We do not address alignment programs, which were the subject of a recent volume in this series.[3] Many programs for phylogenetic analysis have been described in the literature and are available by writing the authors of the original papers. The following programs are widely used and easily available, either for free or for a small fee. This list is not exhaustive but provides a starting point for conducting most of the analyses described in this chapter.

The PAUP (Phylogenetic Analysis Using Parsimony) program was written by David L. Swofford (Smithsonian Institution, Washington, D.C.; program available from Illinois Natural History Survey, 607 E. Peabody Drive, Champaign, IL 61820). It is a highly versatile, interactive program

---

[40] D. M. Hillis, *Annu. Rev. Ecol. Syst.* **18**, 23 (1987).
[41] M. M. Miyamoto and J. Cracraft, *in* "Phylogenetic Analysis of DNA Sequences" (M. M. Miyamoto and J. Cracraft, eds.), p. 3. Oxford Univ. Press, New York, 1991.

for character-based analyses that allows a wide variety of weighting schemes and modifications of parsimony and methods of invariants. It also conducts bootstrapping, random sampling of trees, analyses of tree-length distributions, consensus analyses, and character analyses, and it has routines for producing camera-ready output of trees. The full-featured program currently is available only for Macintosh computers; an earlier version that lacks many of the advanced features is available for MS-DOS machines (an update is planned). Some versions are also available as C source code for use on workstations and mainframes.

Hennig86 was written by James S. Farris (American Museum of Natural History, New York, NY 10024). It is a fast and effective parsimony program. It is often faster than PAUP but has many fewer features and options. However, Hennig86 does contain a routine for successive approximation *a posteriori* character weighting.

Phylip (Phylogenetic Inference Package) was written by Joseph Felsenstein (Department of Genetics, SK-50, University of Washington, Seattle, WA 98195). The package includes a diverse collection of programs, including routines for calculating estimates of divergence and programs for both distance-based and character-based phylogenetic analyses. The parsimony programs are much slower and less efficient than in PAUP or Hennig86, but Phylip implements many methods that are not widely available elsewhere (e.g., maximum likelihood, many of the distance-based approaches). The package is distributed in Pascal source code or is available in precompiled versions for most computers.

The MacClade program, written by Wayne P. Maddison and David R. Maddison [Department of Ecology and Evolution (WPM) and Department of Entomology (DRM), University of Arizona, Tucson], is another program for parsimony analyses. However, it is primarily designed for interactive tree manipulation and studies of character evolution, rather than for finding most-parsimonious trees. It contains many features especially designed for analysis of DNA sequences and numerous features for the production of camera-ready tree output. It is completely compatible with PAUP, so the two programs are effectively used in combination. MacClade is available only for Macintosh computers; it is available from Sinauer Associates (Sunderland, MA 01375).

NJTREE, UPGMA, and TDRAW were written by Li Jin, J. W. H. Ferguson, N. Saitou, and J. C. Stephens (contact Li Jin, Center for Demographic and Population Genetics, University of Texas Health Science Center at Houston, P.O. Box 20334, Houston, TX 77225). These programs build neighbor joining and UPGMA trees. They are written in FORTRAN-77; precompiled versions are available for MS-DOS computers.

NJDRAW, NJBOOT, and related programs and available from M. Nei and T. S. Whittam (Institute of Molecular Evolutionary Genetics, Penn State University, 328 Mueller Laboratory, University Park, PA 16802-5303). These programs are available precompiled for MS-DOS computers. They are used for computing various DNA distances and for constructing and testing neighbor joining trees.

ANCESTOR, WTSUBS, AUTSUBS, and ALLTOPS were written by P. L. Williams and W. M. Fitch (contact W. M. Fitch, Department of Ecology and Evolutionary Biology, University of California, Irvine, CA 92717). These FORTRAN programs are available in uncompiled form or precompiled for MS-DOS computers. The various programs are designed for generating ancestral states, choosing among tree topologies, and performing various aspects of dynamically weighted parsimony procedures.

Acknowledgments

Our work has been supported by the National Science Foundation (DEB 91-22823 and DEB 92-21052 to D.M.H.; BSR 88-57264 and BSR 89-18606 to M.M.M.). We thank Cliff Cunningham, Winston Hide, David Swofford, Elizabeth Zimmer, and the Smithsonian Molecular Phylogenetics Discussion Group for comments on the manuscript.

Section III

Comparing Macromolecules: Exploring Evolutionary Pattern and Process

Section III. Comparing Macromolecules: Exploring Evolutionary Pattern and Process

Section III describes experimental strategies for examining evolutionary pattern and process in natural and model systems. Pattern determination strategies are presented for probing nucleic acid [35] and protein [36] higher order structure and for detecting more complex features of genomes, such as gene duplication, differentiation, and divergence [37]–[40] and gene recruitment [41]. Chapters [42] and [43] provide examples of how determination of evolutionary patterns from comparative studies can be used to design experimental tests of protein structure and function. These chapters form a bridge to the systems presented for studying some of the processes such as selection [44] and [45], recombination [46], and gene shuffling [47] that have been proposed as fundamental by evolutionary biologists. Recent advances in transgenic technology and in regeneration technology for plants and some animals may soon result in the availability of similar experimental systems for higher eukaryotes.

[35] Experimental Approaches for Detecting Self-Splicing Group I Introns

By BARBARA REINHOLD-HUREK and DAVID A. SHUB

Introduction

Genes can be interrupted by noncoding intervening sequences (introns) that must be removed from initial transcripts to give rise to functional tRNA, rRNA, or mRNA molecules (RNA splicing). Introns have been classified into four different groups, according to the mechanism of their removal. Splicing (of tRNA precursors in eukaryotes and *Archaea*) is mediated by enzymes or (in the case of mRNA precursors in eukaryotes) by protein–RNA complexes (spliceosomes). Alternatively, introns self-splice, with the nucleophile in the first transesterification reaction being the 2'-OH of an internal adenosyl residue (group II) or the 3'-OH of a guanosine cofactor (group I). This guanosine cofactor is covalently bound to the 5' end of excised group I introns (reviewed by Cech[1]). Group I introns that are capable of self-splicing can thus be labeled *in vitro* with [α-^{32}P]GTP.

Introns were initially thought to be widespread in eukaryotic genomes and absent in bacteria. Later they were found in *Archaea*[2] and bacteriophage T4.[3] More recently, the GTP end-labeling reaction has been used to identify additional group I introns in phage T4,[4] in phages of gram-positive eubacteria,[5] and in a tRNA gene in cyanobacteria.[6,7]

Although the radioassay is restricted to group I introns that self-splice efficiently *in vitro,* one advantage of the assay is its lack of bias. Because the reaction can be performed with RNA extracted from any organism, group I introns can be detected without prior information on gene or protein sequences. Although it is not clear how the different intron types are related, further knowledge about patterns of distribution of introns in different organisms and genes may give insights into their evolutionary origin and function.

[1] T. A. Cech, *Annu. Rev. Biochem.* **59**, 543 (1990).
[2] B. P. Kaine, R. Gupta, and C. R. Woese, *Proc. Natl. Acad. Sci. U.S.A.* **80**, 3309 (1983).
[3] J. K. Chu, G. F. Maley, F. Maley, and M. Belfort, *Proc. Natl. Acad. Sci. U.S.A.* **81**, 3149 (1984).
[4] J. M. Gott, D. A. Shub, and M. Belfort, *Cell (Cambridge, Mass.)* **47**, 81 (1986).
[5] H. Goodrich-Blair, V. Scarlato, J. M. Gott, M.-Q. Xu, and D. A. Shub, *Cell (Cambridge, Mass.)* **63**, 417 (1990).
[6] M.-Q. Xu, S. D. Kathe, H. Goodrich-Blair, S. A. Nierzwicki-Bauer, and D. A. Shub, *Science* **250**, 1566 (1990).
[7] M. G. Kuhsel, R. Strickland, and J. D. Palmer, *Science* **250**, 1570 (1990).

Detection of Self-Splicing Group I Introns by [^{32}P]GTP Labeling Assay

Deproteinized RNA extracts of the organism under study are incubated *in vitro* with [α-^{32}P]GTP, then analyzed for occurrence of radioactively labeled RNAs by denaturing polyacrylamide gel electrophoresis (adapted from Garriga and Lambowitz[8]).

Reagents

GTP-labeling buffer (5×): 250 mM NH$_4$Cl, 150 mM MgCl$_2$, 150 mM Tris-HCl (pH 7.5), 5 mM spermidine

Label: [α-^{32}P]GTP, 3000 Ci/mmol, 10 mCi/ml (e.g., Amersham, Arlington Heights, IL, No. PB10201)

Stop solution: 2.5 M ammonium acetate, 25 mM NaEDTA (pH 8.0), 300 μg/ml of yeast RNA

Loading solution: 0.1% bromphenol blue, 0.1% xylene cyanol, 10 mM EDTA, 95% (v/v) formamide

Tris–borate (TBE) buffer: 5.4 g/liter Tris base, 2.75 g/liter boric acid, 1 mM EDTA, pH 8.0

Assay Procedure. Set up the reaction mix in a final volume of 10 μl with 2 μl of RNA (10 μg), 2 μl of GTP labeling buffer (5×), 0.25 μl of 200 mM dithiothreitol (DTT), 0.33 μl of RNase inhibitor (30 U/ml, Promega, Madison, WI), and 2 μl (20 μCi) of label. Incubate at 37° for 1 hr and stop the reaction by adding 150 μl of stop solution chilled on ice and 320 μl of ethanol (−20°). Precipitate the RNA for 30 min at −70°, spin it down for 15 min at 4°, and wash the pellet twice with 1 ml of 70% (v/v) ethanol. Vacuum dry the pellet and dissolve it in 8 μl of water just before loading the gel. Add 8 μl of loading solution and denature the RNA for 15 min at 65°. For RNA samples of 100–600 nucleotides, load the samples on a denaturing 5% polyacrylamide gel (acrylamide–bisacrylamide ratio 30:0.8) made up in TBE buffer with 50 g of urea per 100 ml final volume. On a 40-cm gel, run samples at 200 V overnight. Details on polyacrylamide gels are given by Ogden and Adams.[9] Fix the gel in methanol–glacial acetic acid (5%, v/v, each) for 1 hr, rinse it briefly in water, dry it on a Whatman (Clifton, NJ) 3MM paper for 1 hr under vacuum, and expose on X-ray film. Exposure of up to 1–2 weeks might be necessary to detect weak signals.

Comments. As an example of anticipated results, Fig. 1 shows signals of

[8] G. Garriga and A. M. Lambowitz, *Cell (Cambridge, Mass.)* **39,** 631 (1984).
[9] R. C. Ogden and D. A. Adams, this series, Vol. 152, p. 61.

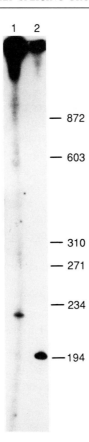

FIG. 1. End labeling of eubacterial RNA with [α-^{32}P]GTP *in vitro.* Deproteinized RNA preparations were incubated for 1 hr at 43° (lane 1) and 50° (lane 2) with [α-^{32}P]GTP as described, then fractionated on a 5% polyacrylamide–8 M urea gel. Labeled fragments were visualized by autoradiography. Lane 1, *Agrobacterium tumefaciens* A136; lane 2, *Azoarcus* sp. BH72. Scale at right indicates position of *Hae*III restriction fragments of phage ΦX174 DNA in nucleotides.

two newly discovered eubacterial introns.[10] An intermediate splicing product, which consists of intron plus 3′-exon after the first step in splicing, can sometimes be seen as a second labeled band of lower electrophoretic mobility (e.g., see Xu *et al.*[6]) The assay can fail to detect group I introns, however, (1) if they require cellular components to facilitate splicing *in vitro* or (2) if the abundance of unspliced transcripts in extracts is low because introns are located in rare transcripts or because they splice very

[10] B. Reinhold-Hurek and D. A. Shub, *Nature (London)* **357,** 173 (1992).

readily *in vivo.* Failure to detect end-labeled RNA in this assay *cannot* be considered proof of the absence of group I introns.

Because RNA is readily degraded, precautions have to be taken against contaminating RNase during all steps (wear gloves; Sambrook *et al.*[11] give hints for general precautions when working with RNA). If the splicing reaction has to be optimized (see following section), several parameters can be taken into consideration. (1) Regarding the temperature of the splicing assay, we incubate routinely at 37° to start with, but several introns show better activity at higher temperatures (e.g., 50° for *Azoarcus* sp. intron). (2) The Mg^{2+} concentration in the labeling buffer is important; for assays of *Tetrahymena* intron it has been described to be as low as 2 mM.[12] (3) The concentration of GTP, under the assay conditions described above, is 0.66 μM; saturating concentrations for GTP can be determined by measuring the intensity of the labeled intron band at varying concentrations of labeled GTP in the reaction mix. An example is given in Fig. 2 for the *Azoarcus* sp. intron, indicating a half-saturating concentration of GTP of 2.7 μM.

Detection of Genomic Locations of Introns and Exons for Cloning

To characterize introns and intron-harboring genes, the respective genomic fragments can be cloned and sequenced. Various approaches have been used to identify these fragments.

Hybridization with End-Labeled Introns from [³²P]GTP-Labeling Assay

After self-splicing of group I introns *in vitro* with [α-³²P]GTP, the 5' end of the excised intron will be end labeled (see previous section). The intron can therefore be used as a radioactive RNA probe to detect homologous DNA sequences. The DNA under question is subjected to restriction digestion, fractionated on agarose gels, and bound to nitrocellulose or nylon filters by Southern transfer according to standard protocols (e.g., Ausubel *et al.*[13]). Hybridization of the filter-bound DNA to the intron probe is carried out as follows.

[11] J. Sambrook, E. F. Fritsch, and T. Maniatis, "Molecular Cloning: A Laboratory Manual." Cold Spring Harbor Laboratory, Cold Spring Harbor, New York, 1989.
[12] C. A. Grosshans and T. R. Cech, *Biochemistry* **28,** 6888 (1989).
[13] F. M. Ausubel, R. Brent, R. E. Kingston, D. D. Moore, J. G. Seidman, J. A. Smith, and K. Struhl, "Current Protocols in Molecular Biology," and supplements. Wiley, New York, 1987.

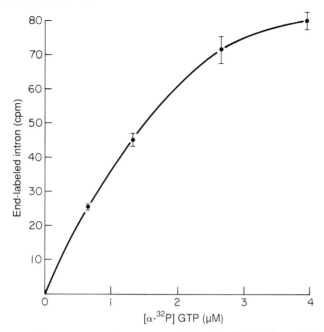

FIG. 2. Effect of GTP concentration on accumulation of end-labeled group I intron during self-splicing *in vitro*. Various amounts of [α-^{32}P]GTP were dried down under vacuum and redissolved in GTP-labeling reaction mixtures containing deproteinized RNA from *Azoarcus* sp. BH72. After incubation at 50°, RNA was separated on a 5% polyacrylamide–8 *M* urea gel and dried, then the activity of the labeled fragments was quantified by direct counting in a Betascope 603 Blot Analyzer (Betagen). Standard deviations are from different measurements of the same gel.

Reagents

Prehybridization/hybridization buffer: 50% deionized formamide, 5× SSC (standard saline citrate), 8× Denhardt's solution, 50 m*M* sodium phosphate buffer (pH 6.5), 0.5% sodium dodecyl sulfate (SDS), 250 μg/ml denatured herring testes DNA, and 500 μg/ml yeast RNA (for the composition of SSC buffer and Denhardt's solution, see, e.g., Ausubel *et al.*[13])

Hybridization Procedure. Formamide is deionized by adding 1 g of mixed bed ion-exchange resin (e.g., AG 501-X8, Bio Rad, Hercules, CA) to 10 ml of formamide, which is stirred on a magnetic stirrer for at least 1 hr and then filtered twice through filter paper. Store at −20°. Prehybridize filters in the above buffer for 4–8 hr at 45°, replace the solution with fresh buffer, and add 0.25 volume of 50% dextran sulfate and labeled probe (denatured at 65° before addition). Hybridize at 45° overnight with shaking.

After removal of the probe, wash filters at 37° four times for 15 min in 100 ml wash solution I (2× SSC, 0.1% SDS) and once for 30 min in 100 ml of wash solution II (0.2× SSC, 0.1% SDS). Expose on film. To obtain a sufficiently hot probe, the splicing reaction might have to be optimized and scaled up. This approach was used successfully to detect intron-containing genes in bacteriophages T4 (see Gott et al.[4]) and SPO1,[5] but it failed to detect the intron in *Azoarcus* sp.[10] Probes can be used in a similar way to screen genomic libraries by colony or plaque hybridizations.

Heterologous Hybridization with Intron Probes

Related introns with considerable sequence similarity may be detected by heterologous hybridization with labeled probes. Restriction fragments of suitable length to cover almost the entire sequence of a cloned intron are labeled radioactively or by alternative, nonradioactive methods. We use the following protocol for "random primed" DNA labeling, derived from Feinberg and Vogelstein.[14]

Reagents

10× Primer buffer: 0.5 M Tris-HCl (pH 7.2), 0.1 M MgCl$_2$, 1 mM dithioerythritol (DTE), 2 mg/ml bovine serum albumin (nuclease free), 60 A$_{260}$ units/ml random primers for 100 μl of buffer. Random primers are synthetic hexadeoxyribonucleotides of random sequence, available, for example, from Pharmacia (Piscataway, NJ, No. 27-2166-01)

Deoxynucleoside triphosphate (dNTP) mix: Mix equal volumes of 0.5 mM dATP, 0.5 mM dGTP, and 0.5 mM dTTP

Heterologous Hybridization Procedure. The DNA fragment to be labeled is denatured in a small volume of distilled water (up to 12 μl) at 100° for 10 min, quick-cooled on ice, and spun down to collect the condensate. Set up a labeling reaction with 2 μl of 10× primer buffer, 3 μl of dNTP mix, 2 μl (20 μCi) of [α-^{32}P]dCTP (3000 Ci/mmol, aqueous solution), and 12 μl of denatured DNA, on ice. Start the reaction by adding 1 μl of 2 U/μl Klenow enzyme (DNA polymerase I, large fragment) and incubate at 37° for 1 hr. Stop the reaction by adding 2 μl of 0.2 M EDTA (pH 8.0). It is optional to separate the labeled probe from unincorporated nucleotides, for example, on a 1-ml Sephadex G-50 column.

[14] A. P. Feinberg and B. Vogelstein, *Anal. Biochem.* **132**, 6 (1983).

After Southern transfer of restriction-digested genomic DNA, filters are prehybridized for 1–4 hr. The probe is boiled for 10 min to denature it, added to fresh hybridization buffer, and hybridized overnight. To calculate appropriate hybridization temperatures in buffers with or without formamide, and for details of hybridization buffers and washes, see standard protocols.[13] For hybridizations at low stringency to detect signals from fragments of approximately 65–70% overall sequence similarity, we apply 55° for hybridization and washes in 6× SSC without formamide. Similar protocols can be used for colony and plaque hybridizations.

Comments. Figure 3 shows an example of homologous and heterologous hybridization with the group I self-spicing intron of *Azoarcus* sp. as a probe. Cloning and sequencing of the fragment of *Agrobacterium tumefaciens* DNA which hybridized to the intron probe confirmed the occurrence of a related group I intron.[10] The nature of the weak signal in DNA of *Nitrosomonas europeae* has not yet been confirmed by sequencing.

Instead of DNA probes, labeled RNA probes can be used as well, especially for small-sized fragments. If the respective inserts are cloned adjacent to a T3 or T7 promoter, transcripts can be labeled with [α-^{32}P]UTP by *in vitro* transcription (see below). Hybridizations with filter-bound DNA are carried out as described above for GTP-labeled introns at appropriate temperatures. This approach has been used to clone cyanobacterial introns.[6]

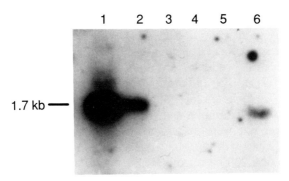

FIG. 3. Southern hybridization of the group I intron of *Azoarcus* sp. BH72 to genomic DNA of selected eubacteria. A cloned fragment derived from PCR amplification of the tRNA^{ILE} gene (see section on amplification of putative intron-containing genes by polymerase chain reaction) was used as a probe. Three micrograms of genomic DNA per lane was digested with *Pst*I. Hybridization with the ^{32}P-labeled DNA probe was carried out at low stringency (55°). Lane 1 serves as positive control for hybridization with homologous DNA on a 1.7-kb fragment. Lane 1, *Azoarcus* sp. BH72; lane 2, *Agrobacterium tumefaciens* A136; lane 3, *Synechocystis* PCC6803; lane 4, *Anabaena* PCC7120; lane 5, *Borrelia burgdorferi;* lane 6, *Nitrosomonas europeae.*

Amplification of Putative Intron-Containing Genes by Polymerase Chain Reaction

If an intron is assumed to be inserted into a known gene, evidence for it can be obtained by amplification of the respective DNA fragment by the polymerase chain reaction (PCR). In the PCR, oligonucleotides flanking the fragment to be amplified serve as primers for DNA polymerase during several cycles of synthesis. For this strategy, it is assumed that the gene in the organism under study shares sequence similarity with homologous genes of related organisms. Consensus sequences are used to design degenerate primers that are likely to anneal to the unknown target sequence. Guidelines for reaction conditions and design of primers can be found in Innis *et al.*[15] The size of the amplification products is determined by agarose or polyacrylamide gel electrophoresis. If an amplified fragment is larger than the expected exon, it can be analyzed for the presence of introns by direct PCR sequencing[15] or by *in vitro* transcription and sequencing after cloning into a suitable vector. This approach has been used to detect introns in genes for leucine tRNAs of cyanobacteria, using primers specific for the exons of sequenced tRNA^{Leu} genes (UAA anticodon) of land plant chloroplasts and *Cyanophora*,[7] which were previously known to have introns in the anticodon sequence.

An example of intron detection via the PCR and some obstacles associated with it is the detection of an intron in a tRNA^{Ile} gene in a gram-negative eubacterium.[10] We anticipated that a putative self-splicing intron detected by the GTP-labeling assay in *Azoarcus* (Fig. 1) might be localized in the same gene as in cyanobacteria. Degenerate primers were designed from consensus sequences of published tRNA^{Leu} (UAA) genes of eubacteria. Amplification products from genomic DNA of *Azoarcus* (lanes 3–6), *Escherichia coli* (lane 2), and the control (water lane 1) are shown in Fig. 4. Bands of the size expected for amplified tRNA genes [~70 base pairs (bp)] were detected in all samples but the control. Additional amplified fragments of larger size were present in the *Azoarcus* lanes, with intensities varying somewhat with the Mg^{2+} concentrations used. Fragments of the size expected of a 200-bp intron (see Fig. 1) inserted into tRNA (arrow in Fig. 4) were obtained but represented only a minor fraction; they were therefore reamplified after agarose gel purification before cloning. In contrast to the situation in chloroplasts and cyanobacteria, this group I intron is inserted into a different gene, for tRNA^{Ile} (CAU anticodon), which had been unexpectedly amplified by the degenerate primers designed for tRNA^{Leu}.[10] Other large fragments were not completely characterized but

[15] M. A. Innis, D. H. Gelfand, J. J. Sninsky, and T. J. White (eds.), "PCR Protocols: A Guide to Methods and Applications." Academic Press, San Diego, 1990.

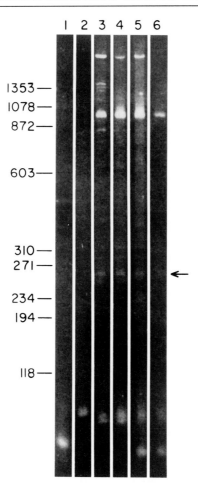

FIG. 4. PCR amplification products with degenerate primers designed for eubacterial tRNA^{Leu} (UAA anticodon). Cycle conditions were as follows: denaturation, 94°, 1.5 min; annealing, 42°, 2 min; extension, 72°, 2 min; 30 cycles; [MgCl$_2$] was 3.0 mM except in lane 3 (1.5 mM), lane 5 (4.5 mM), and lane 6 (6.0 mM). Genomic DNA (100 ng) of *Escherichia coli* JM109 (lane 2) and *Azoarcus* BH72 (lanes 3–6), or sterile distilled water (lane 1), were used as templates. Arrow indicates the amplified tRNA^{Ile} gene harboring a group I intron. Scale at left indicates position of *Hae*III restriction fragments of phage ΦX174 DNA in nucleotides.

may represent clusters of tRNA genes in *Azoarcus*. Therefore, the appearance of amplification products larger than the exon has to be interpreted with care. Having carried out the GTP-labeling assay prior to PCR amplification was a major advantage since it enabled us to predict the correct size of the amplification product.

Confirmation of Self-Splicing by *in Vitro* Transcription of Cloned DNA of Putative Introns

Analysis of products of *in vitro* transcription can give evidence for self-splicing of cloned putative introns. If self-splicing occurs efficiently *in vitro,* splice products of the size of ligated exons and intron are detectable in addition to the primary runoff transcript. The respective DNA fragment has to be cloned into a vector which contains promoters for RNA polymerase adjacent to the cloning site. The pUC-derived vector pBSM13+/− (Stratagene, La Jolla, CA) for example, allows transcription of the insert in both directions under control of promoters of either T3 or T7 phage RNA polymerase. Prematurely terminated or partially degraded transcripts can complicate the analysis of RNA products arising from splicing. In our hands, the template should not exceed 1–2 kilobases (kb) for efficient transcription into a full-length product, best results being obtained with the shortest transcripts possible. Therefore, runoff transcripts should not extend into the vector, but the DNA template should be truncated at the multiple cloning site or within the insert by digestion with suitable restriction enzymes. If a physical map of the insert is available, truncation to different lengths can be used to locate the intron within the insert (Fig. 5), no splice products being detected from transcripts ending within the intron.

A protocol for *in vitro* transcription used in our laboratory is given below. Rapid plasmid minipreparations made by alkaline lysis without RNase treatment are usually sufficient to prepare DNA templates. DNA samples linearized by restriction digestion are extracted with chloroform, and precipitated with ethanol, and 0.2 to 0.5 μg of DNA dissolved in 1 μl of distilled water is used per assay. Contamination with RNases must be avoided during the following steps. The reaction mixture for eight reactions is prepared at room temperature to avoid precipitates.

Reagents

Label: [α-^{32}P]UTP, 20 mCi/ml, 800 Ci/mmol (e.g., Amersham, No. PB.20383)

Reaction mixture: 13 μl sterile distilled water, 8 μl of 5× transcription buffer (200 mM Tris-HCl, pH 8.0, 40 mM MgCl$_2$, 10 mM NaCl), 8 μl of 0.75 M DTT, 1 μl of RNase inhibitor (40 U/μl, e.g., Promega RNasin, No. N2511), 1.6 μl each of 10 mM ATP, CTP, and GTP, 1.6 μl of 0.5 mM UTP, and 1.6 μl label

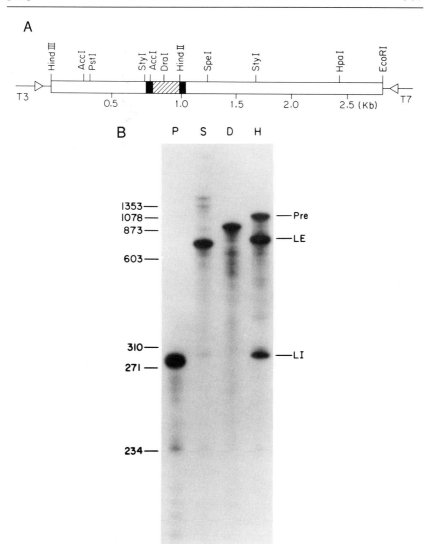

FIG. 5. *In vitro* transcription of a cloned DNA fragment from *Anabaena azollae* containing a self-splicing intron. (A) Restriction map of a 2.7-kb insert in pBSM13−. Arrows indicate direction of transcription from the T3 and T7 promoters in the vector. tRNA exon (solid) and intron (hatched) sequences are indicated. (B) Plasmid DNA truncated by *Pst*I (P), *Sty*I (S), *Dra*I (D), and *Hin*dII (H) (in the 3′ exon), downstream of the T3 promoter. After transcription with T3 RNA polymerase, products were fractionated on a 5% polyacrylamide–8 *M* urea gel; the autoradiogram is shown. Scale at left indicates position of *Hae*III restriction fragments of phage ΦX174 DNA in nucleotides. Labels indicate positions expected for the unspliced run-off transcript (Pre), ligated exons (LE), and linear intron (LI). (From Xu *et al.*[6] Copyright 1990 by the AAAS.)

In Vitro Transcription Procedure. Add 0.4 μl of T3 or T7 RNA polymerase (50 U/μl) just before dividing the reaction mixture into 4-μl aliquots and starting the reaction by adding 1 μl of DNA template per tube. Incubate at 37° for 1 hr. Stop the reaction by adding 8 μl of gel loading solution (see section on detection of self-splicing group I introns by [^{32}P]GTP labeling assay) and denature the samples by heating to 65° for 15 min before applying to a denaturing polyacrylamide gel (4–6%). GTP levels present in the buffer allows self-splicing of group I introns.

Comments. An example of anticipated results is given in Fig. 5. Transcription from the T3 promoter resulted in a full-length run-off transcript when templates were truncated at sites upstream of or within the putative intron (lanes P, S, and D, fig. 5B). Additional RNA bands, consistent with the sizes of ligated exon and intron, were derived from a longer transcript (H), indicating that the complete intron was included in DNA within 1.1 kb of the T3 promoter.[6] When samples are incubated at conditions favorable for splicing (see Detection of self-splicing group I introns by [^{32}P]GTP labeling assay) after *in vitro* transcription, accumulation of splice products, with concomitant disappearance of runoff transcript, can give further evidence for the presence of introns.[10]

Determination of Splice Junctions

When sequences of exons and the intron are known, the splice junctions can often be deduced from sequence comparison with related genes and by applying the rules governing splice-site selection of group I introns. Primer extension sequencing of RNA, using reverse transcriptase, gives experimental evidence for the exact boundaries of intron excision. When primers that anneal to sequences in the 3′ exon are extended toward the intron, RNA preparations that contain both unspliced and spliced transcripts will yield a unique sequence until the position of intron excision. After the splice junction, the intron sequence will be superimposed on the sequence of the 5′ exon. Alternatively, after *in vitro* transcription and splicing, ligated exons can be excised from gels and sequenced by reverse transcription. Comparison with the DNA sequence of the entire cloned fragment will indicate the splice boundaries. A description of the experimental procedure of reverse transcriptase sequencing is given by Hahn *et al.*[16] Examples of both applications of RNA sequencing to determine splice junctions can be found in Shub *et al.*[17]

[16] C. S. Hahn, E. G. Strauss, and J. H. Strauss, this series, Vol 180, p. 121.
[17] D. A. Shub, J. M. Gott, M.-Q. Xu, B. F. Lang, F. Michel, J. Tomaschewski, J. Pedersen-Lane, and M. Belfort, *Proc. Natl. Acad. Sci. U.S.A.* **85,** 1151 (1988).

[36] Site-Directed Mutagenesis in Analysis of Protein–Protein Interactions

By Lauren N. W. Kam-Morgan, Thomas B. Lavoie, Sandra J. Smith-Gill, and Jack F. Kirsch

Introduction

The replacement of specific amino acid residues in proteins using site-directed mutagenesis allows for a rationally designed, systematic, and quantitative analysis of the interactions that govern molecular recognition among proteins. The strategy involves analyzing the interface of a structurally defined protein–protein complex and examining the effects of specific side-chain substitutions on the free energy of the interaction.

The reactions of monoclonal antibodies (MAbs) with their target proteins provide an excellent model system for experimental analysis of molecular complementarity. We shall use as an example monoclonal antibodies to the well-characterized chicken lysozyme (HEL), which was chosen as a model antigen because extensive information on structure, function, and immunological response is available.[1,2] Evolutionary variants of lysozyme have been used to predict critical residues in epitopes recognized by various monoclonal antibodies.[3,4] Genetic engineering provides the capability of making precise mutations in both the antibody and antigen, thus raising the art of epitope mapping and antibody structure and function to high resolution. Three-dimensional structures of complexes between HEL and the Fab fragments of three HEL-specific antibodies, namely, D1.3,[5] HyHEL-5,[6] and HyHEL-10[7] have defined the structural interfaces of these complexes. The goal of this study is to delineate the forces involved in an antibody–antigen complex, utilizing biochemical, serological, and X-ray crystallographic data as well as site-directed mutagenesis techniques.

[1] E. F. Osserman, R. E. Canfield, and S. Beychok (eds.), "Lysozyme." Academic Press, New York and London, 1974.

[2] S. J. Smith-Gill and E. Sercarz (eds.), "The Immune Response to Structurally Defined Proteins: The Lysozyme Model," Adenine Press, Guilderland, New York, 1989.

[3] S. J. Smith-Gill, A. C. Wilson, M. Potter, E. M. Prager, R. J. Feldmann, and C. R. Mainhart, *J. Immunol.* **128,** 314 (1982).

[4] S. J. Smith-Gill, T. B. Lavoie, and C. R. Mainhart, *J. Immunol.* **133,** 384 (1984).

[5] A. G. Amit, R. A. Mariuzza, S. E. V. Phillips, and R. J. Poljak, *Science* **233,** 747 (1986).

[6] S. Sheriff, E. W. Silverton, E. A. Padlan, G. H. Cohen, S. J. Smith-Gill, B. C. Finzel, and D. R. Davies, *Proc. Natl. Acad. Sci. U.S.A.* **84,** 8075 (1987).

[7] E. A. Padlan, E. W. Silverton, S. Sheriff, G. H. Cohen, S. J. Smith-Gill, and D. R. Davies, *Proc. Natl. Acad. Sci. U.S.A.* **86,** 5938 (1989).

METHODS IN ENZYMOLOGY, VOL. 224

Methods

Production of Mutant Lysozymes

Selection of Residues for Mutation. The rationale for designing mutants to date has been (1) to concentrate on positions where naturally occurring (evolutionary) replacements have been shown to affect binding or (2) to examine residues that are predicted to be important for binding based on their charge, side chain volume, and polarity effects at a given position.

Site-Directed Mutagenesis. Mutations at specific residues are introduced into a cloned HEL gene, and mutant proteins are expressed in the yeast *Saccharomyces cerevisiae*.[8] Site-directed mutagenesis is performed by the Kunkel method[9,10] with the following modifications. T7 DNA polymerase is used instead of the T4 enzyme because the former has high levels of single- and double-stranded DNA $3' \rightarrow 5'$-exonuclease activities, which appear to be responsible for the high fidelity of the enzyme.[11]

Transfection and DNA Sequencing of Mutant Lysozymes. Competent cells of kanamycin-resistant *Escherichia coli* strain DH5αF'*lacI*^{Q} are transfected with an aliquot of the mutagenesis mix and grown accordingly.[12] Single-stranded DNA templates are prepared from the phage in the supernatants[12] and sequencing of each template is performed using the reagents supplied in the Sequenase kit (U.S. Biochemical Corporation, Cleveland, OH) and an appropriate oligonucleotide primer. 7-Deaza-dGTP nucleotide mixes and Sequenase reagents (U.S. Biochemical Corporation) are used to sequence the entire gene to verify that no other changes have occurred. The 7-deaza-dGTP nucleotide mixes alleviate the problem of band compression found when conventional Sequenase mixes [deoxynucleoside triphosphates (dNTPs)] are used. After sequencing has confirmed the mutation, double-stranded replicative form (RF) DNA is isolated from the bacterial cell pellets[12] and used for subcloning.

Subcloning into Yeast Expression Vector. Mutant chicken lysozyme sequences located in a 2.5-kilobase (kb) *Bam*HI fragment are subcloned into the *Bam*HI site of the yeast shuttle plasmid pAB24[13] according to the procedure of Gibco BRL Life Technologies, Inc., Grand Island, NY. This

[8] B. A. Malcolm, S. Rosenberg, M. J. Corey, J. S. Allen, A. de Baetselier, and J. F. Kirsch, *Proc. Natl. Acad. Sci. U.S.A.* **86,** 133 (1989).
[9] T. A. Kunkel, *Proc. Natl. Acad. Sci. U.S.A.* **82,** 488 (1985).
[10] T. A. Kunkel, J. D. Roberts, and R. A. Zakour, this series, Vol. 154, p. 367.
[11] R. L. Lechner and C. C. Richardson, *J. Biol. Chem.* **258,** 11185 (1983).
[12] J. Sambrook, E. F. Fritsch, and T. Maniatis, "Molecular Cloning: A Laboratory Manual," 2nd Ed. Cold Spring Harbor Laboratory, Cold Spring Harbor, New York, 1989.
[13] P. J. Barr, H. L. Gibson, V. Enea, D. E. Arnot, M. R. Hollingdale, and V. Nussenzweig, *J. Exp. Med.* **165,** 1160 (1987).

procedure uses polyethylene glycol 8000 (PEG 8000) and optimizes ligation conditions. Ligated DNA is transformed into competent cells of *E. coli* strain DH5α*F'lacI*^{Q}. Bacterial colonies are picked and grown, and the plasmids are purified and analyzed for the *Bam*HI fragment containing the chicken lysozyme cDNA sequence. Plasmid DNA is transformed into *Saccharomyces cerevisiae* strain GRF180 (gift from Chiron Corporation, Emeryville, CA) for protein production according to the procedure of Spencer *et al.*[14] using a yeast lytic enzyme (zymolyase) isolated from *Arthrobacter luteus* (ICN Biomedicals, Inc., Costa Mesa, CA).

Protein Production, Isolation, and Purification. The expression and purification of chicken lysozyme mutant proteins in yeast are performed as described by Malcolm *et al.*[8] with the following modifications. The 50-ml minimal medium second seed yeast culture is used to inoculate a 2.8-liter Fernbach flask containing 500 ml of 1% yeast extract/2% Bacto-peptone/8% glucose (w/v) medium and is then incubated for 7–9 days at 30°. Cells are harvested, washed twice with 60 ml of 0.5 *M* NaCl, and collected by centrifugation. The supernatants are pooled, diluted 5-fold with deionized water, and loaded onto a 20-ml column of CM Sepharose Fast Flow (Pharmacia, Piscataway, NJ) equilibrated with 0.1 *M* potassium phosphate, pH 6.24. The column is washed with the same buffer, and lysozyme is eluted with 0.5 *M* NaCl/0.1 *M* potassium phosphate, pH 6.24. Fractions are assayed by activity (decrease in A_{450} of *Micrococcus lysodeikticus* cell wall suspensions per minute). Fractions containing lysozyme are concentrated in Centricon-10 (Amicon, Danvers, MA) filter units, washed with 0.1 *M* potassium phosphate buffer, pH 6.24, and stored at 4°. The protein concentration is determined from $\epsilon_{280}{}^{1\%} = 26.4$.[15]

Polyacrylamide Gel Electrophoresis. To determine the purity of protein preparations, denaturing gels are run [5% polyacrylamide stacking gel/15% polyacrylamide (w/v) separating gel] with protein samples denatured in Laemmli buffer.[16] Native gel electrophoresis is performed in a 3.6% polyacrylamide stacking gel in 0.1 *M* Tris/0.16% (v/v) sulfate buffer, pH 6.1, and a 15% polyacrylamide separating gel in 0.1 *M* Tris/50 m*M* glycine buffer, pH 8.6. This high-resolution gel system is used to verify specific charge changes in mutant lysozymes. Fixing and Coomassie Brilliant Blue R250 (Pierce, Rockford, IL) staining of denaturing and native gels are performed according to Sambrook *et al.*[12] Acrylamide, bisacrylamide, and

[14] J. F. T. Spencer, D. M. Spencer, and I. J. Bruce, "Yeast Genetics: A Manual of Methods." Springer-Verlag, Berlin and Heidelberg, 1989.
[15] T. Imoto, L. N. Johnson, A. C. T. North, D. C. Phillips, and J. A. Rupley, *in* "The Enzymes" (P. D. Boyer, ed.), Vol. 7, p. 666. Academic Press, New York, 1972.
[16] U. K. Laemmli, *Nature (London)* **227**, 680 (1970).

sodium dodecyl sulfate (SDS) are purchased from Bio-Rad Laboratories (Richmond, CA).

Production of Native and Mutant Antibodies

Cloning and Expression of HyHEL-10. The methods utilized for the cloning, expression, and mutagenesis of antibody genes are described in this series. Expression of wild-type and mutant HyHEL-10 is accomplished through transfection of vectors containing the genomic rearranged heavy and light chain genes in SP2/0 cells, in order to avoid the potential problems associated with refolding antibodies expressed in *E. coli.* The genomic heavy and light chains are isolated from EMBL4 and EMBL3 phage λ libraries.[17] The rearranged heavy chain is cloned into vector pSV2-*gpt,* and the rearranged light chain is cloned into pSV2-*neo.* Cell lines expressing the HyHEL-10 light chain are produced by transfection of the light chain vector by spheroplast fusion. The cells are then transfected by the heavy chain vector as described.[18] Mutagenesis of the V regions is by the method of Zoller and Smith.[19] Whenever possible, the coding changes introduced in the V regions also introduced an unique restriction site, allowing rapid characterization of the mutants. The V regions are subcloned from M13 into the expression vectors and are sequenced using the Sequenase Kit (U.S. Biochemical Corporation) and V-region-specific oligonucleotide primers. Antibody production is assayed by particle concentration fluorescence immunoassay (PCFIA, see below), and antibodies are purified from tissue culture media by protein A affinity chromatography.[20]

Sources of Naturally Occurring Evolutionary Variant Lysozymes

Chicken egg-white lysozyme (HEL) is purchased from Sigma Chemical Company (St. Louis, MO) or from Worthington Chemicals (Freehold, NJ). For immunoassays, HEL is routinely further purified by gel filtration on a BioGel PD-10 column or by high-performance liquid chromatography (HPLC) using a TSK-2000 column. Bobwhite quail, Japanese quail, and turkey lysozymes used for immunoassays are isolated from egg whites by ion exchange and gel filtration.[3] Bobwhite quail, Japanese quail, Montezuma quail, and turkey lysozymes used in the enzyme assay are isolated from egg whites and are a generous gift from Dr. Ellen Prager (University of California, Berkeley).

[17] A. M. Frischauf, N. Murray, and H. Lehrach, this series, Vol. 153, p. 103.
[18] M. Better and A. H. Horwitz, this series, Vol. 178, p. 476.
[19] M. J. Zoller and M. Smith, this series, Vol. 100, p. 468.
[20] Pharmacia technical bulletin, *Separation News,* 13.5 (1986).

Measurements of Antigen–Antibody Interactions

The objective of the experiments is to compare the affinity of wild-type antibody and wild-type immunogen (chicken lysozyme) with that of wild-type antibody and mutant antigen or with that of mutant antibody and wild-type immunogen. All the assays involve incubation of a constant concentration of one reactant with varying amounts of the complementary reactant, along with estimating the concentration of either bound or free reactant by an immunochemical or enzymatic method. In each case, the assumption is made that the measurement step does not disturb the equilibrium between antigen and antibody, and it is important that this assumption be validated experimentally. We summarize below several alternative methods.

Competitive Inhibition Enzyme Assays. Estimates of antibody–lysozyme dissociation constants can be obtained by taking advantage of the fact that most monoclonal antibodies efficiently inhibit enzymatic activity.[3,5] The combining site of HyHEL-10, which is presented as an example, also has been demonstrated by X-ray crystallography to overlap a portion of the catalytic site of lysozyme.[7] A constant concentration of lysozyme is incubated with varying amounts of antibody, and amounts of free (unbound) lysozyme molecules are estimated by the proportion of catalytic activity remaining. The assay assumes that the addition of *Micrococcus lysodeikticus* cell walls and concurrent dilution of the antibody–antigen mixture do not disturb the equilibrium.

Activity is measured by the procedure of Shugar.[21] To 2.9-ml cuvettes (1 cm path length), diluted lysozyme (ranging from 0.1 to 0.5 nM) and antibody (ranging from 0.013 to 50 nM) are added to 66 mM potassium phosphate buffer, pH 6.24, and 0.1% bovine serum albumin (BSA) (w/v) to a volume of 900 μl. The solutions are kept at 25° for 1 hr to allow the lysozyme–antibody complexes to come to equilibrium. The activity assays are initiated by the addition of 100 μl *Micrococcus lysodeikticus* (Sigma Chemical Company) cell walls (2 mg/ml in 66 mM potassium phosphate, pH 6.24) to a final A_{450} of 0.8–1.0. Cuvettes are wrapped with Parafilm to prevent evaporation, inverted several times to mix, and placed in a Perkin-Elmer, Norwalk, CT) Lambda 4B spectrophotometer. Reactions are monitored by the decrease in A_{450} for 70 min with a data point collected every minute.

A typical data set is shown in Fig. 1a. Samples lacking lysozyme and antibody had a slope of 3.0–4.5 $\times$ 10^{-4} A_{450} units/min owing to settling of

---

[21] D. Shugar, *Biochim. Biophys. Acta* **8**, 302 (1952).

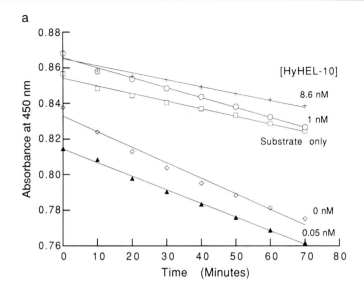

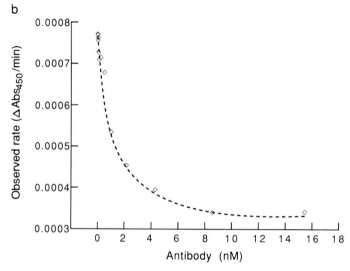

FIG. 1. Effect of HyHEL-10 on *M. lysodeikticus* lysis activity of chicken lysozyme at 0.1 n*M* concentration using the competitive inhibition enzyme assay described in the text. (a) Raw data with varying concentrations of antibody and the substrate-only control. (b) Plot of antibody concentration versus decrease in absorbance at 450 nm per minute at 0.1 n*M* enzyme (uncorrected for substrate-only control).

the particulate *M. lysodeikticus* cell wall suspension. The lowest concentration of lysozyme that can be assayed by the method is 0.1 n*M*. Most assays employed 0.2 to 0.4 n*M* of the mutant lysozymes. Dissociation constants are calculated by nonlinear regression to Eq. (1):

$$V = \frac{V_O - (V_O - V_I)(K_D + [Ab] + [Lz]) - \{(K_D + [Ab] + [Lz])^2 - 4[Ab][Lz]\}^{1/2}}{2[Lz]} \quad (1)$$

where [Ab] is the concentration of antibody, [Lz] the concentration of lysozyme, V the remaining activity in solution, and V_O and V_I the activities of lysozyme in the absence of added antibody and in the presence of excess antibody, respectively. The observed rates are plotted against the concentration of antibody added (Fig. 1b).

Disposable cuvettes are used only once to eliminate the possibility of lysozyme carryover. All microcentrifuge tubes and pipette tips are autoclaved, and buffers and BSA solutions are not used beyond 7 days. The *M. lysodeikticus* suspension is prepared 18–24 hr before use and set at 37° with shaking, to hydrate fully. It was found that substrate prepared just before use settled much faster and interfered with the signal at the low concentrations of enzyme used.

Particle Concentration Fluorescence Immunoassay. The PCFIA is a solid-phase immunoassay in which proteins are attached to polystyrene particles by adsorption or covalent coupling for the solid phase and fluorescent-labeled reagents are utilized for product detection.[22] The general principles of the assay are similar to those of the enzyme-linked immunosorbent assay (ELISA), which has been reviewed extensively elsewhere.[23] PCFIAs are performed in specially designed 96-well format Fluoricon assay plates utilizing an automated "Screen Machine" (Idexx Corporation, Research Product Division, Portland, ME).

Fluorescein isothiocyanate (FITC)-labeled anti-mouse immunoglobulin G (IgG) is purchased commercially (Idexx Corporation, or Southern Biotechnology Associates, Inc., Birmingham, AL). Custom labeling of our own antibodies or proteins is generally carried out according to the general method using carbonate–bicarbonate buffer[24]; we usually attain *F/P* labeling ratios of approximately 4 by using a FITC to protein ratio of

[22] M. E. Jolley, C. J. Wang, S. J. Ekenberg, M. S. Zvelke, and D. M. Kelsoe, *J. Immunol. Methods* **67**, 21 (1984).
[23] E. Engvall, this series, Vol. **70**, p. 419 (1980).
[24] L. Hudson and F. C. Hay, "Practical Immunology." Blackwell, Oxford, 1976.

approximately 1.2 in the reaction mixture. Lysozyme molecules that have more than one molecule of FITC attached are generally insoluble, so a somewhat different approach must be utilized to optimize conjugation and yield. Five microliters FITC, dissolved in dimethyl sulfoxide (DMSO) at a concentration of 50 mg/ml, is added to 1 ml of lysozyme at 5 mg/ml in 100 mM Na$_2$HPO$_4$, pH 10. The mixture is vortexed briefly; the container is wrapped in foil and incubated for approximately 30 min at 37° with occasional mixing. The mixture is spun for 5 min in a microcentrifuge to remove any insoluble lysozyme. It is then applied to a BioGel PD-10 Sephadex G-25 minicolumn equilibrated with phosphate-buffered saline (PBS) to separate the lysozyme from the unreacted FITC. HEL is covalently coupled to 0.9 μm carboxymethylpolystyrene particles (Idexx Corporation) by incubating a closed glass container containing 100 μg/ml HEL, 0.5% (w/v) particles, and 0.5 mg/ml diethyl pyrocarbonate (DEPC) in 100 mM phosphate buffer, pH 7.4, for 1–2 hr at room temperature on a rotary shaker. The reaction is stopped by adding 0.5 ml of 1 M Tris per milliliter of reaction mixture and mixing well, and the beads are then pelleted by centrifugation in a clinical centrifuge. The beads are resuspended in PBS, repelleted, and finally resuspended in PBS containing 0.08% (w/v) sodium azide to a bead concentration of 0.25% (w/v).

Antigen Inhibition Assay. The antigen inhibition assay measures the relative reactivity of a series of antigens to an antibody by comparing their ability to inhibit binding of antibody to antigen that is immobilized on a solid phase. The amount of antibody bound to the solid phase is estimated by a second FITC-labeled anti-IgG. Because the assay mixture is incubated for only a relatively short period of time, the reactants do not come to equilibrium, and therefore the reactant concentration and incubation times must be carefully controlled to assure reproducibility of results. All dilutions of antibodies and reactants are made in 150 mM Tris, 150 mM NaCl, pH 7.4, containing 2% fetal calf serum (FCS) and 2% normal goat serum. Twenty microliters each of 50 ng/ml HyHEL-10 and a dilution of lysozyme are incubated for 2 min in a well of a Fluoricon 96-well microtiter plate (Baxter Healthcare Corp./Pandex Division, Muskegon, MI). Then 20 μl each of 0.25% HEL particles and FITC-labeled goat anti-IgG (Baxter Healthcare Corp./Pandex Division) are added and the mixture incubated an additional 30 min at room temperature. The particle-bound and soluble reactants are then separated by vacuum filtration, the particles are washed twice with 50 μl isotonic buffered saline (Baxter Diagnostics, McGraw Park, IL), and bound fluorescence is read by the Screen Machine. The percent bound *(PCB)* is calculated relative to a control containing trace amounts of lysozyme and a background reading with diluant in the place of antibody. *PCB* values between 10 and 90% are linearized utilizing

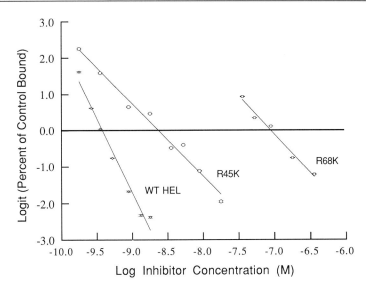

FIG. 2. Log–logit plot of inhibition of HyHEL-5 binding to HEL by mutant lysozymes. The profile of cloned wild-type lysozyme (WT HEL) was not significantly different from that of native wild-type HEL. The reactivity of the R68K mutant was reduced by over 3 log units, similar to that seen previously with bobwhite quail lysozyme [T. B. Lavoie, L. N. W. Kam-Morgan, A. B. Hartman, C. P. Mallett, S. Sheriff, D. A. Saroff, C. R. Mainhart, P. A. Hamel, J. F. Kirsch, A. C. Wilson, and S. J. Smith-Gill, *in* "The Immune Response to Structurally Defined Proteins: The Lysozyme Model" (S. Smith-Gill and E. Sercarz, eds.), p. 151. Adenine Press, Schenectady, New York, 1989], whereas R45K reduced reactivity by approximately 100-fold.

a log–logit transformation, and the I_{50} (concentration of lysozyme giving 50% *PCB*) is determined as the abscissa intercept of a linear regression of the data points. Typical antigen inhibition curves for HyHEL-5 with wild-type and mutant lysozymes are shown in Fig. 2.

Two different sequential saturation assays are modified from the ELISA[25] to determine association constants of antibody with lysozymes. In both assays, a constant antibody concentration is incubated with varying concentrations of lysozyme until equilibrium is reached. The time necessary must be determined empirically for each antibody; for the antibody HyHEL-10, this was determined to be at least 16 hr,[26] and all incubations were performed for 16–24 hr. At the end of the incubation, an aliquot of the mixture is treated with an excess of reagents to sample free (unbound) antibody combining sites by incubating with labeled or solid-phase coupled

[25] B. Friguet, A. F. Chaffotte, L. Djavadi-Ohaniance, and M. E. Goldberg, *J. Immunol. Methods* **77**, 305 (1985).
[26] T. B. Lavoie, W. D. Drohan, and S. J. Smith-Gill, *J. Immunol.* **149**, 3260 (1992).

lysozyme for a very short period. This apparently does not disturb the equilibrium which has been reached. Again, the incubation time must be determined empirically for each antibody. For determining association constants with mutant lysozymes that have a greatly reduced affinity (by at least several orders of magnitude) compared to the unmutated lysozyme, it may be necessary to modify the sampling protocol by reducing the concentration of reagents or shortening the incubation period, since a high-affinity reagent is more likely to dissociate the complex with a low-affinity mutant.

The particle binding affinity assay samples free antibody sites by binding uncomplexed antibody to HEL particles. For determination of antibody affinity for HEL, HyHEL-10 is incubated with serial dilutions of HEL at a final incubation concentration of 83 pM, and 60–80 μl of each sample is incubated with 20 μl of 0.25% HEL particles in a Fluoricon assay plate well for 4 min. For lower affinity mutant lysozymes, such as Japanese quail or Montezuma quail, HyHEL-10 is incubated at a final concentration of 667 pM, and a 60–80 μl sample is incubated with 20 μl of 0.05% HEL particles for 1 min.[27] Following incubation, the particle-bound MAb is separated from free antibody by suction and washing with PBS, and 20 μl of FITC-coupled goat anti-mouse IgG is added and incubated for 10 min, followed by separation, washing, and reading of the epifluorescence.

The antibody capture affinity assay samples the free MAb binding sites by the addition of FITC-labeled antigen to the incubation mixture for 1 min, and the antigen–antibody complexes are immobilized on anti-IgG particles. The plates then are treated as above.

In both assays, the proportion (R) of antibody-binding sites occupied by the soluble antigen is calculated with respect to controls with no inhibitor added ($R = 0$, no binding sites occupied) and with buffer but no antibody added ($R = 1$, all binding sites occupied). The experimentally determined R, which estimates the fraction bound, is plotted versus $R/([Ag] - [Ab]R)$, which is equivalent to [Ab] times B/F, where [Ag] is the total soluble antigen concentration and [Ab] is the total concentration of antibody binding sites. Linear regression analysis is used to calculate the association constant from this modified Scatchard plot. Figure 3 represents an analysis of the particle affinity assay using wild-type and mutant antibody. For the plate binding assay, it is necessary to correct the calculation of R for the fact that bivalent antibody is being trapped on a solid phase, and the assay detects an antibody equivalently with a single or both combining sites free; calculations are performed using a corrected R^*, which is equal to $R^{1/2}$.[28]

---

[27] T. B. Lavoie, Ph.D. Thesis. University of Maryland, 1989.

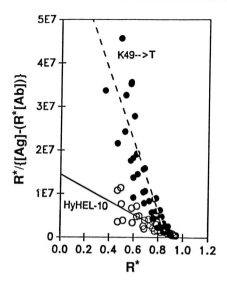

FIG. 3. Affinity analysis of wild-type and mutant $K49_{V_K}T$ of MAb HyHEL-10 sequentially saturated with duck lysozyme in the particle affinity assay. The abscissa (R^*) is the corrected proportion of antibody binding sites filled with soluble lysozyme, and the ordinate is equivalent to [Ab] times B/F, as described in the text. (Reproduced from Lavoie et al.[26])

Results and Discussion

Determination of Effects of Amino Acid Replacements among Evolutionary Variant Lysozymes on Antibody Affinity

The values of the dissociation constants for HyHEL-5 and HyHEL-10 wild-type lysozyme complexes were found to be approximately 0.1 nM using the competitive inhibition activity assay. Previous serological studies using lysozyme variants from various birds showed that Bobwhite quail lysozyme (BWQEL) had 1000-fold reduced affinity for HyHEL-5 and California quail lysozyme (CQEL) bound as tightly to HyHEL-5 as chicken lysozyme.[3] BWQEL and CQEL differ from each other at positions 68 and 121 of their primary amino acid sequences[29,30] (Table I). Because both chicken and CQEL have an arginine at position 68 whereas BWQEL has a lysine, and BWQEL and chicken both have a glutamine at position 121, it was concluded that position 68 is responsible for the reduced affinity.[3] A

[28] F. J. Stevens, Mol. Immunol. 24, 1055 (1987).
[29] I. M. Ibrahimi, E. M. Prager, T. J. White, and A. C. Wilson, Biochemistry 18, 2736 (1979).
[30] E. M. Prager, N. Arnheim, G. A. Mross, and A. C. Wilson, J. Biol. Chem. 247, 2905 (1972).

TABLE I
PRIMARY AMINO ACID SEQUENCES OF EVOLUTIONARY VARIANT LYSOZYMES

```
                         1        10        20        30        40
Chicken      (HEL) ª   KVFGRCELAAAMKRHGLDNYRGYSLGNWVCAAKFESNFNTQATNR
Turkey             b   --Y----------L------------------------H----
Japanese     quailᶜ    --Y-------------K-Q--------------------
Montezuma    quailᵈ    -------------------W-----------------S-----
California   quailª    -----------------------------------S-----
Bobwhite     quailᵉ    ------------------------------------S-----


                         50        60        70        80        90
Chicken      (HEL) ª   NTDGSTDYGILQINSRWWCNDGRTPGSRNLCNIPCSALLSSDITA
Turkey             b   ---------------------------K----------------
Japanese     quailᶜ    --------------------------------------------
Montezuma    quailᵈ    -S-------V----------------------------------
California   quailª    ---------V----------------------------------
Bobwhite     quailᵉ    ---------V-----------K----------------------


                         100       110       120
Chicken      (HEL) ª   SVNCAKKIVSDGNGMNAWVAWRNRCKGTDVQAWIRGCRL
Turkey             b   --------A-G-------------------H--------
Japanese     quailᶜ    -----------VH----------------N--------
Montezuma    quailᵈ    T----------------------------H--------
California   quailª    T----------------------------H--------
Bobwhite     quailᵉ    T-------------------------------------
```

ª Ibrahimi et al.[29]

b J. N. LaRue and J. C. Speck, J. Biol. Chem. 245, 1985 (1970).

ᶜ M. Kaneda, I. Kato, N. Tomzinaga, K. Titani, and K. Narita, J. Biochem. (Tokyo) 66, 747 (1969).

ᵈ Lavoie et al.[31]

ᵉ Prager et al.[30]

ᶠ Dashes in the sequences indicate identity at that position to the chicken sequence.

single mutation, R68K, was made to test this hypothesis, and immunochemical tests showed that it bound HyHEL-5 at least 1000-fold less tightly than HEL.[31,32]

Chicken lysozyme residue Asp-101$_{HEL}$ in the HyHEL-10 epitope[7] is of additional interest because it contributes to the free energy of association of substrate ligands with the enzyme;[33] turkey lysozyme (TEL) has a glycine at position 101.[34] Nine replacements were designed to test the importance

[31] T. B. Lavoie, L. N. W. Kam-Morgan, A. B. Hartman, C. P. Mallett, S. Sheriff, D. A. Saroff, C. R. Mainhart, P. A. Hamel, J. F. Kirsch, A. C. Wilson, and S. J. Smith-Gill, in "The Immune Response to Structurally Defined Proteins: The Lysozyme Model" (S. Smith-Gill and E. Sercarz, eds.), p. 151. Adenine Press, Schenectady, New York, 1989.

[32] T. B. Lavoie, L. N. W. Kam-Morgan, C. P. Mallett, J. W. Schilling, E. M. Prager, A. C. Wilson, and S. J. Smith-Gill, in "The Use of X-Ray Crystallography in the Design of Anti-Viral Agents" (G. W. Laver and G. Air, eds.), p. 213. Academic Press, San Diego, 1990.

[33] D. C. Phillips, Proc. Natl. Acad. Sci. U.S.A. 57, 484 (1967).

[34] J. N. LaRue and J. C. Speck, J. Biol. Chem. 245, 1985 (1970).

of both the size and charge of the side chains at this position. Turkey lysozyme has a 5- to 6-fold reduced affinity for chitotriose compared to chicken lysozyme at pH 4.5–6.0,[35] and it also binds about 7 times less tightly to HyHEL-10. Of the six additional amino acid replacements between chicken and turkey lysozymes, two are contact residues in the HyHEL-10 epitope, namely, Leu-15$_{TEL}$ and Lys-73$_{TEL}$, which may account for the difference in affinities for HyHEL-10 observed between turkey (Gly-101$_{TEL}$) and mutant D101G.

Previous serological epitope mapping investigations of HEL and JQEL identified positions 19, 21, 102, and 103 as critical residues for association with HyHEL-10.[4] R21Q and R21W are of particular interest because two evolutionary variants found in nature, Japanese quail (JQEL) and Montezuma quail (MQEL) lysozymes, have glutamine and tryptophan, respectively, at position 21. Mutations at position 21 were also designed to explore the importance of both size and charge of side chains in addition to determining the effects of the JQEL and MQEL replacements. Both Japanese and Montezuma quail lysozymes have five additional substitutions compared to chicken[32,36] (Table I). In the case of Montezuma quail, the R21W substitution is the only replacement within the epitope. Competitive inhibition assays show that Japanese quail and Montezuma quail have an approximately 30- to 35-fold reduced affinity for HyHEL-10 compared to chicken as do their respective single mutations. Thus it appears that the R21W replacement by itself accounts for the lowered affinity of MQEL for HyHEL-10, since position 21 is the only difference from chicken in the epitope. It also appears that the lower affinity of Japanese quail to HyHEL-10 can be accounted for by position 21, although additional substitutions are present. Japanese quail has two additional replacements in the chicken epitope, positions 19 and 102[36] (Table I). Mutations were made to the JQEL replacements to determine their affect on affinity. Competitive inhibition activity assays showed that both of the mutants, N19K and G102V, bound MAb HyHEL-10 as tightly as the wild type.

Conclusions

In this chapter, we have discussed the methods used in our studies to investigate the application of site-directed mutagenesis techniques in the analysis of the interaction of proteins, specifically between an antibody and an antigen. Ours is a unique system in that we can mutate both the

[35] N. Arnheim, F. Millett, and M. A. Raftery, *Arch. Biochem. Biophys.* **165**, 281 (1974).
[36] M. Kaneda, I. Kato, N. Tomzinaga, K. Titani, and K. Narita, *J. Biochem. (Tokyo)* **66**, 747 (1969).

antibody and the antigen, which increases the power of this strategy. Previous serological analysis and the X-ray crystal structure of the lysozyme–antibody complex has made it possible to predict the importance of specific contact residues at the interface of the two proteins. From there, site-directed mutagenesis methods were applied to cloned sequences of antibody and antigen using general methods that we have improved in order to attain the highest efficiencies. Immunochemical and inhibition assays were used to quantitate the effect of specific mutations on antibody–antigen binding. Thus, we have been able to determine the effect on affinity of specific amino acid residues at specific positions.

A concern frequently raised over the study of proteins produced by site-directed mutagenesis is the possibility of long-range changes in globular protein structure affecting parameters such as thermodynamic stability. A simple amino acid replacement to study enzyme structure and affinity may result in altering several interactions.[37,38] Generally, the stability of a protein is not perturbed on a large scale when replacements are made at surface residues.[37]

In summary, we have shown that site-directed mutagenesis is an important tool which allows precise investigation into the detailed chemistry of protein–protein interaction. HEL has proved to be ideal for these studies. Monoclonal antibodies, X-ray crystal structures of protein complexes, and evolutionary lysozyme variants found in nature provide an excellent system for understanding the molecular basis of protein recognition. Site-directed mutagenesis offers the ability to alter a protein at a specific site. Interpretation of the results within the context of an X-ray crystal structure allows the quantitative assignment of the free energy contributions from individual amino acids to the stability of the complex.

Acknowledgments

We thank the late Dr. Allan C. Wilson for reviewing the manuscript. This work was supported by National Institutes of Health, National Service Research Award AIO7989 from the National Institute of Allergy and Infectious Diseases (to L.N.W.K.-M.) and by National Institutes of Health Grant GM35393 (to J.F.K.).

[37] T. Alber, S. Dao-pin, K. Wilson, J. A. Wozniak, S. P. Cook, and B. W. Matthews, *Nature (London)* **330**, 41 (1987).
[38] A. Fersht, R. J. Leatherbarrow, and T. N. C. Wells, *Biochemistry* **26**, 6030 (1987).

[37] Comparative Studies of Mammalian Y Chromosome

By PRISCILLA K. TUCKER and BARBARA LUNDRIGAN

Introduction

Vertebrate sex chromosomes are typically a heteromorphic pair of chromosomes that carry one or more genes involved in sex determination. Heteromorphic sex chromosomes can take one of two forms, depending on the mode of sex determination. In an XX female/XY male sex chromosome system (e.g., mammals), males produce two kinds of sperm, X- and Y-bearing, whereas females produce only X-bearing eggs. Zygotes receiving two X chromosomes develop into females, and zygotes receiving one X chromosome and one Y chromosome develop into males.[1] In a ZW female/ZZ male sex chromosome system (e.g., birds), females produce two kinds of eggs, Z- and W-bearing, whereas males produce only Z-bearing sperm. Zygotes receiving one Z chromosome and one W chromosome develop into females, and zygotes receiving two Z chromosomes develop into males.[1]

Heteromorphic sex chromosomes were hypothesized to have evolved from a homologous pair of chromosomes by the suppression of crossing-over between the ancestral pair of sex chromosomes and subsequent loss of gene function on the ancestral Y (or W) chromosome.[2] One prediction of this hypothesis is that recombination between the X and Y chromosomes (or between the Z and W chromosomes) will be reduced or nonexistent. This prediction is well supported for eutherian mammals: with rare exception, recombination between the X and Y chromosomes is reduced to a small region referred to as the pseudoautosomal region. As a consequence, most of the mammalian Y chromosome is isolated in the genome and is thus clonally inherited from father to son.

There are two evolutionary questions relevant to the clonal inheritance of the Y chromosome. First, do haploid nuclear sequences that do not recombine in meiosis evolve in a different manner and at a different rate than autosomal or X chromosomal sequences which do recombine? Second, are clonally inherited nuclear sequences useful for phylogeny reconstruction?

Mitochondrial DNA (mtDNA), which is clonally inherited through maternal lineages, has proved to be a powerful genetic marker for phyloge-

---

[1] J. J. Bull, "Evolution of Sex Determining Mechanisms," p. 11. Benjamin/Cummings, London, 1983.

[2] H. J. Muller, *J. Exp. Zool.* **17,** 325 (1914).

netic studies. One reason for this is that mtDNA provides an historical account of molecular evolution that is not confounded by the effects of recombination.[3,4] Analogously, the clonal inheritance of the Y chromosome may make it equally well suited to phylogenetic inference. In addition, phylogenies constructed from sequence variation in paternally inherited Y chromosome DNA will complement phylogenies constructed from sequence variation in maternally inherited mtDNA, as well as phylogenies constructed from variation in Mendelian inherited genes.

To conduct a comparative study of Y chromosome DNA sequences, one must first identify a DNA fragment that is conserved on the Y chromosome in a group of taxa and then develop the experimental protocols necessary to isolate and sequence it from representatives of those taxa. This is not a straightforward problem, as few conserved genes have been identified from the Y chromosome in mammals. For example, although numerous Y chromosome sequences have been isolated and characterized from humans and laboratory mice,[5-21] only three of the sequences are conserved on the Y chromosome in at least five orders of eutherian mammals. These are ZFY (Zfy in laboratory mice), a zinc finger-encoding gene of unknown

[3] C. Moritz, T. E. Dowling, and W. M. Brown, *Annu. Rev. Ecol. Syst.* **18**, 269 (1987).

[4] J. C. Avise, J. Arnold, R. M. Ball, E. Bermingham, T. Lamb, J. E. Neigel, C. A. Reeb, and N. C. Saunders, *Annu. Rev. Ecol. Syst.* **18**, 489 (1987).

[5] L. M. Kunkel and K. D. Smith, *Chromosoma* **86**, 209 (1982).

[6] H. J. Cooke, J. Schmidtke, and J. R. Gosden, *Chromosoma* **87**, 491 (1982).

[7] D. C. Page, M. E. Harper, J. Love, and D. Botstein, *Nature (London)* **311**, 119 (1984).

[8] M. Koenig, J. P. Moisan, R. Heilig, and J. L. Mandel, *Nucleic Acids Res.* **13**, 5485 (1985).

[9] R. P. Erickson, *J. Mol. Evol.* **25**, 300 (1987).

[10] J. Wolfe, R. P. Erickson, P. W. J. Rigby, and P. N. Goodfellow, *Ann. Hum. Genet.* **48**, 253 (1984).

[11] J. Wolfe, R. P. Erickson, P. W. J. Rigby, and P. N. Goodfellow, *EMBO J.* **3**, 1997 (1984).

[12] R. D. Burk, P. Ma, and K. D. Smith, *Mol. Cell. Biol.* **5**, 576 (1985).

[13] D. C. Page, R. Mosher, E. M. Simpson, E. M. C. Fisher, G. Mardon, J. Pollack, B. McGillivray, A. de la Chapelle, and L. G. Brown, *Cell (Cambridge, Mass.)* **51**, 1091 (1987).

[14] A. H. Sinclair, P. Berta, M. S. Palmer, J. R. Hawkins, B. L. Griffiths, M. J. Smith, J. W. Foster, A.-M. Frischauf, R. Lovell-Badge, and P. N. Goodfellow, *Nature (London)* **346**, 240 (1990).

[15] J. Gubbay, J. Collignon, P. Koopman, B. Capel, A. Economou, A. Munsterberg, N. Vivian, P. Goodfellow, and R. Lovell-Badge, *Nature (London)* **346**, 245 (1990).

[16] Y. Nishioka and E. Lamothe, *Genetics* **113**, 417 (1986).

[17] T. H. K. Platt and M. J. Dewey, *J. Mol. Evol.* **25**, 201 (1987).

[18] E. M. Eicher, K. W. Hutchinson, S. J. Phillips, P. K. Tucker, and B. K. Lee, *Genetics* **122**, 181 (1989).

[19] P. K. Tucker, B. K. Lee, and E. M. Eicher, *Genetics* **122**, 169 (1989).

[20] M. J. Mitchell, D. R. Woods, P. K. Tucker, J. S. Opp, and C. E. Bishop, *Nature (London)* **354**, 483 (1991).

[21] P. K. Tucker, K. S. Phillips, and B. Lundrigan, *Mammal. Genome* **3**, 28 (1992).

function found also on the X chromosome,[13] *SRY* (*Sry* in laboratory mice), the male sex-determining locus,[14,15,22] and *Ubely-1,* a candidate spermatogenic gene.[20]

As an example of Y chromosome locus work, we describe the isolation, sequencing, and comparative analysis of a portion of the male sex-determining locus, *Sry,* in the rodent family Muridae, subfamily Murinae. Male sex determination in mammals is effectively equivalent to testis determination, because the primary event in male sex determination is differentiation of the testes from a bipotential gonad.[23,24] The strategy used in isolating the sex-determining locus was based on two assumptions. First, Y-specific sequences involved in male sex determination should map to the region of the Y chromosome known to be critical in the formation of testes, and second, these sequences should be highly conserved on the Y chromosome in mammals.

SRY was first cloned from the human Y chromosome using the following approach: (1) the region on the Y chromosome involved in testes determination was localized using genome analysis of sex-reversed XX males (XX individuals with presumed additions of Y chromatin including the sex-determining locus) and XY females (XY individuals with presumed deletions in the sex-determining locus),[13,14,25,26] and (2) single-copy Y-specific sequences from the "sex-specific" region were subsequently cloned and used as hybridization probes to determine whether they were conserved across mammals.[14] A portion of the male sex-determining locus *(SRY),* a single-copy Y-specific 2.1-kilobase (kb) fragment (pY53.3) conserved in at least six orders of eutherian mammals, was isolated from the human Y chromosome in this manner. The homologous sequence was subsequently isolated from an inbred mouse strain by screening a size-selected inbred mouse library with the human probe, pY53.3.[15] As in humans, the mouse *Sry* sequence was found to be present in the smallest part of the mouse Y chromosome known to be male sex-determining, and it was deleted from a mutant Y chromosome carried by sex-reversed XY female mice.[15] Direct evidence that *SRY/Sry* is necessary for testis formation comes from human clinical studies in which sex-reversed XY females

[22] P. Koopman, J. Gubbay, N. Vivian, P. Goodfellow, and R. Lovell-Badge, *Nature (London)* **351,** 117 (1991).

[23] A. Jost, *Arch. Anat. Microsc. Morphol. Exp.* **36,** 271 (1947).

[24] A. Jost, *Rec. Prog. Horm. Res.* **8,** 379 (1953).

[25] G. Vergnaud, D. C. Page, M.-C. Simmler, L. Brown, F. Rouyer, B. Noel, D. Botstein, A. de la Chapelle, and J. Weissenbach, *Am. J. Hum. Genet.* **38,** 109 (1986).

[26] M. S. Palmer, A. H. Sinclair, P. Berta, N. A. Ellis, P. N. Goodfellow, N. E. Abbas, and M. Fellous, *Nature (London)* **342,** 937 (1989).

carry *de novo* mutations in *SRY*[27,28] and from transgenic studies in which XX mice transgenic for *Sry* develop as males.[22]

The mouse *Sry* gene is contained within a 14-kb DNA fragment and includes a putative translated region of 321 base pairs (bp), with a 240-bp highly conserved apparent DNA-binding domain.[15] The translated region is defined at the 5' end by a stop codon, two putative splice acceptor sites, and an inframe start codon. The 3' end is undefined. We used the polymerase chain reaction (PCR) to amplify enzymatically a 471-bp fragment containing the 321-bp putative exon from the Y chromosome of representatives from eight genera of murine rodents. We sequenced single-stranded PCR products from five of the genera. Individual nucleotides were treated as character data in a subsequent phylogenetic analysis.

Methods

DNA Samples and Preparation

Samples included male and female specimens from the following species in the family Muridae, subfamily Murinae (Old World mice and rats): *Mus musculus, Mus minutoides, Stochomys longicaudatus, Mastomys hildebrandtii, Hylomyscus alleni, Aethomys chrysophilus, Hybomys univitatus, Praomys fumatus, Rattus everetti,* and *Rattus exulans.* In addition, male and female representatives from one species in each of three other murid subfamilies were sampled: Sigmodontinae (New World mice and rats), *Peromyscus maniculatus;* Arvicolinae (voles, lemmings, and muskrats), *Cleithrionomys gapperi;* and Gerbillinae (gerbils, jirds, and sand rats), *Gerbillus gerbillus.* DNA samples were prepared from frozen tissue on permanent loan from the Museum, Texas Tech University; The Field Museum of Natural History; and the National Cancer Institute. High molecular weight genomic DNA was prepared from frozen kidney, spleen, or liver following the methods of Jenkins *et al.*[29]

Isolation of Y-Specific Sequences

Y-Specific sequences were enzymatically amplified from genomic DNA using the PCR.[30] Two oligonucleotide primers, 21 bp in length, were synthesized. One of these matches a portion of the 5' end of the published 471-bp fragment from *Sry.* The second is the reverse complement of a

[27] P. Berta, J. R. Hawkins, A. H. Sinclair, A. Taylor, B. L. Griffiths, P. N. Goodfellow, and M. Fellous, *Nature (London)* **348,** 448 (1990).

[28] R. J. Jager, M. Anvret, K. Hall, and G. Scherer, *Nature (London)* **348,** 452 (1990).

[29] N. A. Jenkins, N. G. Copeland, B. A. Taylor, and B. K. Lee, *J. Virol.* **43,** 26 (1982).

[30] R. F. Saiki, S. Scharf, F. Faloona, K. B. Mullis, G. T. Horn, A. Erlich, and N. Arnheim, *Science* **230,** 1350 (1985).

portion of the 3' end of the sequence. Both primers were purified by high-performance liquid chromatography (HPLC). Next, a double-stranded amplification of the 471-bp sequence was performed. The concentrations of genomic DNA, *Taq* DNA polymerase, reaction buffer, and deoxynucleotide triphosphates (dNTPs) per 100-μl reaction were those recommended by Perkin-Elmer Cetus (Norwalk, CT); we used one-half the recommended concentration of each 10 μM primer. The amplification protocol was as follows: 25 cycles with denaturation at 95° for 1 min, annealing at 50° for 1 min, and elongation at 72° for 1 min, 15 sec. Because PCR amplification is highly sensitive to contamination with foreign DNA, all amplification experiments included a control sample which contained *Taq* DNA polymerase, reaction buffer, and dNTPs but no template DNA. Double-stranded products were separated by size on a 3% NuSieve (FMC, Rockland, ME) gel, stained with ethidium bromide, and visualized under UV light. To verify that our amplified sequences were Y-specific, we conducted preliminary amplification experiments that included male and female genomic DNA from each species under investigation. Amplification of the target size sequence, in this case the 471-bp fragment from *Sry,* in males only was taken as evidence that the sequence was amplified from the Y chromosome.

Double-stranded products were purified using centrifugation dialysis following the protocol described by Allard *et al.*[31] Purified double-stranded products were subsequently dried under vacuum and resuspended in 15 μl of 1X Tris/EDTA buffer (TE). To obtain single-stranded DNA, we used 2 μl of the purified double-stranded PCR product as the template. The concentrations of *Taq* DNA polymerase, reaction buffer, dNTPs, and 10 μM primer were the same as those used in the double-stranded amplification, but only one primer was included. The amplification protocol was the same as that used for the double-stranded amplification, except that the annealing temperature was set to 55°, and only 20 cycles were performed. Single-stranded products were visualized and purified using the procedure described for double-stranded products.

DNA Sequencing

Purified single-stranded products from representatives of six species were sequenced using the Sanger dideoxy-sequencing technique[32] and the Sequenase Version 2.0 sequencing kit (U.S. Biochemical Corp., Cleveland, OH). We followed the protocol that comes with the kit, except that the reactions tubes were heated to 85° (instead of 65°) during the annealing reaction.

[31] M. W. Allard, D. L. Ellsworth, and R. L. Honeycutt, *BioTechniques* **10,** 24 (1991).
[32] F. Sanger, S. Nicklen, and A. R. Coulson, *Proc. Natl. Acad. Sci. U.S.A.* **24,** 5463 (1977).

Sequence Analysis

Sequences were aligned with the published sequence from the laboratory mouse,[15] using the SS2 algorithm[33] available through Eugene, an interface which accesses the Molecular Biology Information Resource software tools,[34] on a UNIX operating system. The aligned sequences were then exported from Eugene, converted to NEXUS format using a Hypercard Stack[35] available for the Macintosh, and transferred to PAUP, Version 3.0q.[36] In our phylogenetic analysis, characters were unordered (Fitch parsimony) and uniformly weighted. We chose uniform weighting for the characters because preliminary examination of the data indicated that sequence divergence is low. We used an exhaustive search, which guarantees the identification of the set of most parsimonious trees.[37]

Results

The target size sequence (471 bp) was successfully amplified from male DNA of each of the ten murine species representing the eight genera investigated in the study. No amplification was apparent from female DNA of these species, in accord with the prediction that the 471-bp fragment from the male sex-determining locus, *Sry,* is Y-specific (Fig. 1).

We could not amplify the 471-bp fragment from female or male DNA of species from the subfamilies Sigmodontinae, Arvicolinae, or Gerbillinae. Presumably, our primers, which were designed using the published *Mus* sequence (subfamily Murinae), were not an adequate match for amplification to take place. We repeated the experiments at lower annealing temperatures but were still unable to amplify *Sry* in representatives of these subfamilies.

We sequenced an average of 384 bp per individual. Our sequence data, along with previously published sequences from rabbit and human,[14,15] suggest that this region of *Sry* is conserved. Percent sequence divergence averaged 6.8% (range 3.6–9.5%) between species from different genera in the subfamily Murinae and 27% between mammalian orders, including our data from Rodentia and previously published sequences from rabbit (Lagomorpha) and human (Primates). The frequency of nucleotide substitutions was relatively low within the 240-bp DNA-binding domain but increased markedly both 5′ and 3′ to the DNA-binding domain.

[32] F. Sanger, S. Nicklen, and A. R. Coulson, *Proc. Natl. Acad. Sci. U.S.A.* **24,** 5463 (1977).
[33] S. F. Altschul and B. W. Ericson, *Bull. Math. Biol.* **48,** 603 (1986).
[34] Baylor College of Medicine, Houston, Texas, 1989.
[35] D. Eernisse, *CABIOS* **8,** 177 (1992).
[36] D. L. Swofford, Illinois Natural History Survey, 607 E. Peabody Drive, Champaign, IL 61820 (1990).
[37] M. J. Sanderson and M. J. Donoghue, *Evolution* **43,** 1781 (1989).

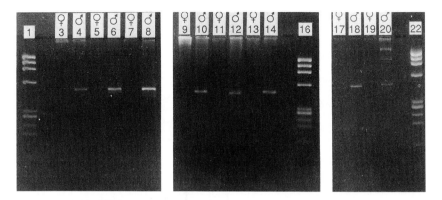

FIG. 1. Ethidium bromide-stained agarose gels demonstrating amplification in males only of the 471-bp fragment from the Y-specific male sex-determining locus, *Sry,* from the following murine species: *Mus musculus* (lanes 3 and 4, *Hylomyscus alleni* (lanes 5 and 6), *Aethomys chrysophilus* (lanes 7 and 8), *Hybomys univitatus* (lanes 9 and 10), *Praomys fumatus* (lanes 11 and 12), *Rattus everetti* (lanes 13 and 14), *Mastomys hildebrandtii* (lanes 17 and 18), and *Stochomys longicaudatus* (lanes 19 and 20). PCR products were separated on 3% N*u*Sieve agarose. The size standard is *Hae*III-digested φX174 RF DNA (lanes 1, 16, and 22). Additional fragments were amplified in male *Stochomys longicaudatus* (lane 20). These spurious fragments are unrelated to the 471-bp *Sry* fragment.

A phylogenetic analysis was performed using the published rabbit and human sequences[14] as the outgroup. These taxa were chosen as outgroups because it was not possible to select a reasonable outgroup from the murine taxa we examined, and human and rabbit were the only nonmurine taxa for which sequences were available. A single most-parsimonious tree was found (Fig. 2). The length of the tree is 212, and the consistency index, excluding uninformative characters, is 0.815. This analysis places *Mus musculus* and *Mus minutoides* as sister taxa, *Hylomyscus* and *Mastomys* as sister taxa, these three genera as the sister group to *Stochomys,* and *Rattus* as an outgroup to the clade formed by these four genera.

Conclusions

Our data suggest that a portion of the male sex-determining locus, *Sry,* is highly conserved across genera within the rodent subfamily Murinae. Nonetheless, Y-chromosome-specific sequence data obtained for the six taxa examined provide phylogenetic information at the generic level. This is in contrast to allozyme studies,[38] which were unable to resolve generic level relationships among murines. However, additional sequence from

[38] J. X. She, F. Bonhomme, P. Boursot, L. Thaler, and F. Catzeflis, *Biol. J. Linn. Soc.* **41,** 83 (1990).

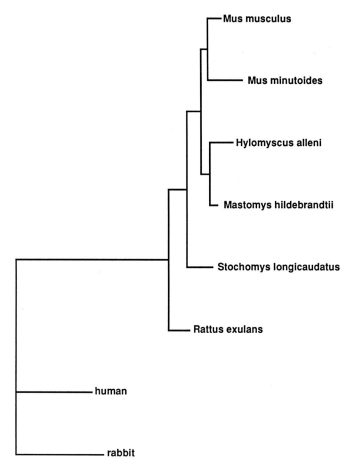

FIG. 2. Phylogenetic tree constructed using sequence from a portion of the Y-chromosome-specific male sex-determining locus, *Sry*. The length of the tree is 212, and the consistency index, excluding uninformative characters, is 0.815.

these and other representatives of the 117 murine genera is needed before well-supported hypotheses of relationship at the generic level in the Murinae can be constructed. In addition, a more appropriate outgroup, ideally the sister group to the subfamily Murinae, should be used.

It may be possible to investigate phylogenetic relationships at lower taxonomic levels using *Sry* by focusing on nontranslated regions. Introns are generally more variable than exons and thus are likely to be phylogenetically informative at lower taxonomic levels. Such variability, however, may make it difficult to construct primers that can amplify the target sequence across a broad range of species. This problem might be avoided by constructing primers from the exons bordering the target intron.

Despite the high conservation of this portion of *Sry,* we were unable to amplify the homologous sequence in the subfamilies Sigmodontinae, Arvicolinae, or Gerbillinae using primers designed from the *Mus* sequence (subfamily Murinae). It may be possible to amplify a portion of the sex-determining gene in distantly related taxa using primers constructed from highly conserved regions, such as the 240-bp DNA-binding domain.

Finally, paternally inherited sequences, whether introns or exons of a gene, provide a distinct new class of characters for use in phylogeny reconstruction, as they represent one of the three modes of inheritance: paternal, maternal, and Mendelian. For this reason, the use of paternally inherited Y chromosome sequences as character data in phylogenetic analyses will add a new dimension to our understanding of the evolutionary history of mammalian taxa.

Acknowledgments

We thank B. Patterson (The Field Museum of Natural History), R. J. Baker (The Museum, Texas Tech University), and M. Potter (National Cancer Institute; Contract NO1-CB-71085) for providing frozen tissues. This research was supported by funds from a University of Michigan Rackham faculty grant, a University of Michigan grant from the Office of the Vice-President for Research, and a grant from the National Science Foundation (BSR-9009806) to P.K.T.

[38] Detection and Quantification of Concerted Evolution and Molecular Drive

By GABRIEL A. DOVER, ANDRÉS RUIZ LINARES, TIMOTHY BOWEN, and JOHN M. HANCOCK

Introduction

Since the initial demonstration of gene duplication by classical genetic experiments in *Drosophila,*[1] a considerable amount of research has taken place on the possible mechanism, function, and evolutionary implications of gene duplication. Early speculations concentrated on gene duplications as an important mechanism for the origin of new genes.[2,3] Experimental

[1] A. H. Sturtevant, *Genetics* **10,** 117 (1925).
[2] J. B. S. Haldane, "The Causes of Evolution." Longmans, Green, New York, 1932.
[3] H. J. Muller, *Genetica* **17,** 237 (1935).

work from the mid-1960s to the mid-1970s showed that DNA repetitions are a widespread characteristic of eukaryotic genomes,[4] and that, surprisingly, such sequences do not evolve independently but in a way that has been termed variously as coincidental, horizontal, or concerted evolution.

Concerted evolution is a distribution pattern of mutations in multiple copy sequences such that there is a greater similarity in sequence between members of a repeated family from within a species than there is between members of the family drawn from different species (see Fig. 1 and below for details of subfamily homogenization, etc.). There are several genomic mechanisms responsible for such patterns which are involved, in their different ways, with the nonreciprocal transfer of information between members of a family. Prominent among these "turnover" mechanisms are unequal crossing-over and gene conversion, but other mechanisms such as DNA transposition, RNA-mediated sequence transfers, and DNA slippage are also involved. The continual, stochastic or biased, gain and loss of sequence variants within individuals can ultimately lead to the concomitant spread (molecular drive) of a variant through a family (homogenization) and through a sexual population (fixation). Molecular drive is a *process,* consequential on the mechanisms of DNA turnover, that can explain observed *patterns* of concerted evolution. It can bring about, like natural selection and genetic drift, a long-term change in the genetic composition of a population with respect to a given family of sequences.[5]

Levels of Detection of Concerted Evolution

The study of concerted evolution seeks to detect homogenized mutations that are diagnostic for members of a repetitive family (or subfamily) of a species, through comparison with other species. This analysis is greatly dependent on the method employed to detect similarity between family members and on the degree of phylogenetic closeness between the species under study. The initial studies describing concerted evolution used protein sequencing of immunoglobulin (Ig) light chains and showed that certain amino acids are shared between the different Ig light chains within a species which differ from those found in the Ig light chains of other species.[6,7] Subsequent studies have approached similarity between loci mainly at the level of the DNA. This similarity can be detected at increasing levels of resolution progressing from DNA hybridization through re-

[4] R. J. Britten and E. H. Davidson, *Q. Rev. Biol.* **46,** 111 (1971).
[5] G. A. Dover, *Nature (London)* **299,** 111 (1982).
[6] L. Hood, K. Eichmann, H. Lackland, R. M. Krause, and J. J. Ohms, *Nature (London)* **228,** 1040 (1970).
[7] C. Milstein and J. R. L. Pink, *Prog. Biophys. Mol. Biol.* **21,** 211 (1970).

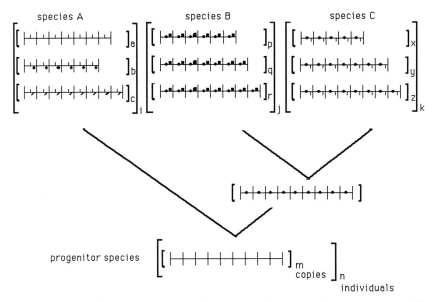

FIG. 1. Schematic pattern of concerted evolution of three related species, A, B, and C. Each repeat contains a tandem array of subrepeats. Homogenized subrepeat arrays, containing different sets of diagnostic mutations, are contained in the smaller brackets. In species A, the family is divided into subfamilies, each with its own diagnostic mutation. In species B and C, all subarrays share one diagnostic mutation (circle) but differ in other homogenized mutations (square, short vertical line). Lengths of subrepeat arrays can be polymorphic within each species, with different lengths present at different copy numbers $(a-c, p-r, x-z)$ in different populations (sizes $i-k$; larger brackets). Diagnostic mutations are symbolized by circles, squares, vertical lines, and angled lines.

striction mapping to DNA sequencing. Hybridization techniques depend on an overall degree of sequence similarity between family members, and concerted evolution can be detected by comparing the stability of hybridization between homologous and heterologous mixtures of equivalent sequences drawn from different species. Restriction mapping looks directly at the nucleotide sequence of a given DNA region, and concerted evolution is identified by the presence of species-specific restriction sites common to family members. Nucleotide sequencing of several repeats of a sequence family from different species is the finest level of analysis of concerted evolution. From an alignment of nucleotide sequences, concerted evolution can be detected as a greater level of nucleotide diversity in interspecific alignments relative to that found in intraspecific alignments (see below). The method used for calculation of nucleotide diversity can produce different estimates, particularly if the sequence alignment contains many gaps and different weights are given to such gaps in the calculations. However, as the definitive criterion of concerted evolution is the identification of species-specific nucleotides at defined positions among repeats, it is clear that a metric such as overall nucleotide diversity could overlook such sites. In the last analysis species-specific nucleotides can only be found by direct comparison of all the sequences available. As shown below, special computer programs are also of use in the detection of concerted evolution using DNA sequence data.

It is important to bear in mind that a higher level of resolution is needed for detecting homogenized species-specific variants among closely related species. In such cases the time since divergence of the species can approach the time required for the homogenization of species-specific characteristics in a given family. This makes concerted evolution increasingly difficult to detect in some instances.

One illustration of the methodology used for the detection of concerted evolution among closely related species concerns the analysis of satellite DNA families among seven sibling species of the *melanogaster* species subgroup of *Drosophila*.[8,9] Two such satellites, the "360" and "500" families (each with several thousand repeats per individual) were initially analyzed using Southern blotting and restriction mapping, which showed concerted evolution among the sequences as evidenced by species-specific levels of hybridization of satellite probes of the *D. melanogaster, D. orena,* and *D. erecta* species. Also, restriction mapping permitted the identification of species-specific restriction sites in the closely related species *D. teissieri* and *D. yakuba.* However, no concerted evolution was detected through restriction analysis between the two most closely related species, *D. mauritiana* and *D. simulans,* suggesting that there might be too little nucleotide difference between the two species to be detected with restriction enzymes. In contrast, nucleotide sequence analysis of the "360" and "500" satellites confirmed the occurrence of concerted evolution among all member species of the *melanogaster* subgroup. Pairwise comparisons, based on percentile sequence divergence of consensus sequences between most species pairs, indicated an interspecific nucleotide diversity generally one order of magnitude higher than the intraspecific diversity. Nevertheless, the same measure of divergence between the two most closely related species (*D. simulans* and *D. mauritiana*) failed to reveal any difference in the level of nucleotide diversity within and between the species.

To detect concerted evolution in such an instance it is necessary to abandon the approach using percentage differences of consensus sequences (which operationally ignores useful sequence information) and to examine

[8] T. Strachan, E. S. Coen, D. A. Webb, and G. A. Dover, *J. Mol. Biol.* **158,** 37 (1982).
[9] T. Strachan, D. Webb, and G. A. Dover, *EMBO J.* **4,** 1701 (1985).

directly all the available sequences drawn at random from among repeats and individuals. When the sequences of the two satellite DNAs of *D. mauritiana* and *D. simulans* were compared directly, it was easy to identify many species-specific nucleotides. Furthermore, direct comparison of the satellite sequences enabled the identification of "transition stages" during the homogenization process and provided both evidence for the underlying mechanisms (unequal crossing-over and gene conversion) and some insight into their rates in the establishment of the concerted evolution patterns.[9]

Concerted Evolution of Length Variants and Nested Levels of Unequal Crossing-Over

So far we have concentrated on concerted evolution detected as within-species similarity in nucleotide sequence between family members. Another pattern of concerted evolution relates to variation in repeat unit lengths and to difference in copy number of length variants. This pattern of concerted evolution is best illustrated by the example of the ribosomal RNA (rRNA) genes in *Drosophila*. The rRNA genes in *D. melanogaster* are organized as tandem repeats of about 250 units on each of the X and Y chromosomes. The tandem genes are separated by spacers (IGS) which themselves contain internal arrays of repeated sequences of 95, 330, and 240 base pairs (bp) in length (Fig. 2). Unequal crossing-over at the level of the whole rDNA unit is responsible for the homogenization of sequence variants throughout the rDNA family, identified by particular restriction sites in the rDNA units of *D. melanogaster* that are absent in those of the other sibling species.[10-12] For example, analysis of the 240-bp subrepeats in *D. melanogaster* has shown that they are also evolving in concert, as indicated by the presence in all of them of a species-specific *AluI* site. Concerted evolution of these repeats is presumably also due to unequal crossing-over but at the level of the subrepeats (Fig. 3).

Thus, in the rDNA family there are two levels of unequal crossing-over. One occurs at the level of the IGS 240-bp subrepeats, leading to the homogenization of their sequences within an rDNA unit, and the other occurs at the level of the entire rDNA unit, leading to the homogenization in sequence of units within an entire rDNA array. Furthermore, as unequal crossing-over generates variation in the number of subrepeats, it leads to heterogeneity in length of this region between rDNA units. The operation

---

[10] E. S. Coen, J. M. Thoday, and G. A. Dover, *Nature (London)* **295**, 564 (1981).
[11] E. S. Coen, T. Strachan, and G. A. Dover, *J. Mol. Biol.* **158**, 17 (1982).
[12] E. S. Coen and G. A. Dover, *Cell (Cambridge, Mass.)* **33**, 849 (1983).

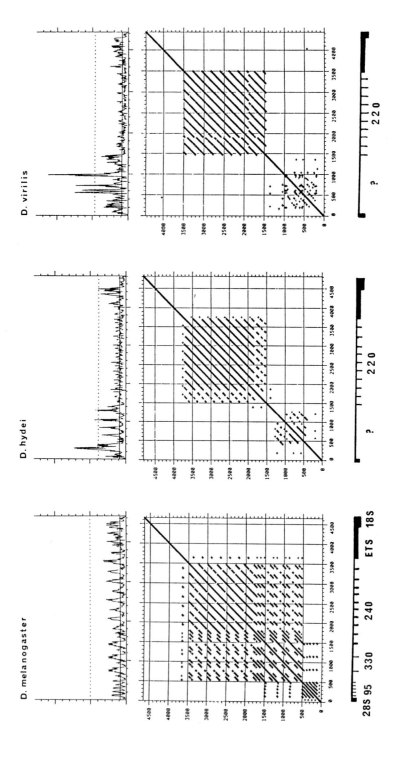

D. melanogaster　　　　　D. hydei　　　　　D. virilis

of unequal crossing-over at the higher level of whole rDNA units then generates variation in the number of particular IGS length variants (Figs. 1 and 3).

Paradox of Partially Homogenized Mutations

So far, we have discussed concerted evolution as if it were a blanket homogeneity of all repeats within and between individuals of a given species, for mutations that are diagnostic for that species. Such an extreme distribution pattern is not achieved overnight, however, and the molecular drive process, like all evolutionary processes, requires considerable time to replace an existing family of repeats in all individuals with a variant repeat. Hence, it is to be expected that partial distribution patterns of variant repeats are to be found which are indicative of variants that are on the way in, or on the way out, of a given family of sequences. The precise patterns of partial distribution and the extent of homogenization through populations of genomes depend critically on (1) the differences in rates of "turnover" within and between the relevant chromosomes on which the repeats can be found, (2) the total number of repeats and individuals, and (3) the time the first mutant repeat was produced (for further discussion, see Refs. 13 and 14). It is important to attempt to quantify all relevant parameters involved with variant repeat homogenization if errors in interpretation are to be avoided. For example, if a variant repeat of rDNA is found to be restricted to, say, the X or Y chromosomal arrays in *D. melanogaster,* it

13 G. A. Dover, *J. Mol. Biol.* **26,** 47 (1987).
14 G. A. Dover, *Genetics* **122,** 249 (1989).

FIG. 2. Patterns of internal repetition and "DNA simplicity" in the IGS region of the repetitive rDNA unit of three species of *Drosophila.* Dot-matrix plots of DNA sequences cover the region from the end of the 28 S rRNA gene throughout the whole spacer until the start of the 18 S rRNA gene. The simplicity profile of each sequence is shown above the dot-matrix plots [see D. Tautz, M. Trick, and G. A. Dover, *Nature (London)* **322,** 652 (1986), for a description of the method used to generate the simplicity profiles]. The solid horizontal line represents the mean overall simplicity factor derived from 10 randomized sequences, each of 10 kb, with the same base composition as the natural sequence. The dotted horizontal line represents the position of the highest peak that any of the 10×10 kb of randomized sequence was able to muster. The complex pattern of internal repeats in *D. melanogaster* is caused by subrepeats (330 bp) which comprise various pieces of two other basic repeats, namely, the 240-bp repeats in the center of the spacer and the 95-bp repeats at the 5' end of the spacer. Slippage-generated simplicity has almost obliterated the 95-bp arrays of repeats in *D. hydei* and *D. virilis* found in the equivalent regions in *D. melanogaster.* Slippage is operating on units that are smaller than the units of unequal crossing-over and at a seemingly faster rate (see text).

Unequal exchange within rDNA units:

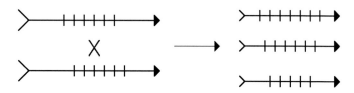

Unequal exchange between rDNA units:

FIG. 3. *(Top)* Unequal exchange at the periodicity of an internal region of repetition within a larger repeating unit generates differences in length of the longer repeat by multiples of the internal repeat length. *(Bottom)* Unequal exchange at the periodicity of the longer repeat unit (e.g., total rDNA unit) generates variation in copy number of the different lengths of the unit generated by the first level of unequal crossing-over.

would be inadvisable to interpret such a distribution pattern as indicative of the separate independent evolution of X- and Y-located rDNA arrays, without proper consideration of the parameters listed above (see, e.g., Ref. 15). Similarly, the finding of restriction sites that are partially, or fully, homogenized to a single chromosome bearing an rDNA array from among the five pairs of such chromosomes in the human species is not necessarily indicative of separate haplotype evolution of human rDNA arrays (see e.g., Ref. 16). In both species it is known that there are other mutations that are indeed shared by all X- and Y-bearing rDNA arrays in *D. melanogaster* (see above) or by all five pairs of human rDNA arrays.[17,18]

How can different distribution patterns within the same species be reconciled? This seeming paradox is resolvable in a variety of ways, all of which are experimentally testable. For example, variants that are homogenized and restricted to specific chromosomes might be relatively young such that insufficient time has elapsed to allow for the transfer of the mutation in question to other relevant chromosomes (homologous or

[15] S. M. Williams, J. A. Kennison, L. G. Robbins, and C. Strobeck, *Genetics* **122**, 617 (1989).
[16] P. Separack, M. Slatkin, and N. Arnheim, *Genetics* **119**, 943 (1988).
[17] M. Krystal, P. D'Eustachio, F. H. Ruddle, and N. Arnheim, *Proc. Natl. Acad. Sci. U.S.A.* **78**, 5744 (1981).
[18] N. Arnheim, *in* "Evolution of Genes and Proteins" (M. Nei and R. K. Koehn, eds.), p. 38. Sinauer, Sunderland, Massachusetts, 1983).

nonhomologous chromosomes).[12,14] It has been calculated that the rate of "turnover" within chromosomes may be as little as one-sixtieth the rate of turnover between chromosomes without overly disturbing the extent of sequence similarity between repeats sampled at random from among all chromosomes.[19,20] Hence, it is to be expected that a young mutation will be either lost or relatively rapidly homogenized within the array of origin before it is transferred to another chromosome by a much rarer exchange between arrays. Once this has happened, however, the appearance of the variant repeat on the homologous (or nonhomologous) chromosome ensures either a further loss or relatively rapid spread down the new array. Alternatively, chromosome-specific homogenized variants might be indicative of physical barriers to interchromosome transfer, possibly localized to the region within a repeat in which the mutation occurs.

It cannot be overstressed that the mechanisms of turnover operate at different rates, with different biases and on different unit lengths of DNA, depending on the chromosomes and the species involved, for reasons that are not well understood. For example, exceptionally fine and detailed studies of mutant distribution patterns among some of the subfamilies that make up the chorion gene superfamily in the silk moth, *Bombyx mori,* reveal that whereas in the so-called late high-cysteine subfamily gene conversion has equitably distributed mutations between all A and B genes of the subfamily (the subfamily consists of approximately 15 paired genes of the A and B type), the early subfamily has mutant distribution patterns which indicate that gene conversion is limited to only the A genes in an A–B pair, leaving the B genes to evolve independently and to accumulate gene-specific mutations.[21,22] In this example, it is intriguing that high intergenic and interchromosomal gene conversion is associated with high numbers of chi-like recombination signals which, in turn, have been generated by a slippage-like mechanism of turnover. A similar close association of such signals at the borders of gene conversion is documented in the human classical class II genes of the major histocompatibility complex (MHC) multigene family.[23] Hence, not only do these examples raise the interesting point that the activities of one turnover mechanism intimately affect the rates and locations of occurrence of another, creating complex patterns of concerted evolution (see below for the joint activities of slippage and unequal crossing-over in rDNA), but they also illustrate the possible

[19] T. Ohta and G. A. Dover, *Proc. Natl. Acad. Sci. U.S.A.* **80,** 4079 (1983).
[20] T. Ohta and G. A. Dover, *Genetics* **108,** 501 (1984).
[21] T. H. Eickbush and W. D. Burke, *J. Mol. Biol.* **190,** 357 (1986).
[22] B. L. Hibner, W. D. Burke, and T. H. Eickbush, *Genetics* **128,** 595 (1991).
[23] U. B. Gyllensten, M. Sundvall, and H. A. Erlich, *Proc. Natl. Acad. Sci. U.S.A.* **88,** 3686 (1991).

molecular basis for the absence of turnover by gene conversion in some parts of a gene family.

There are other examples of supposedly haplotype-restricted homogenized mutations in which the molecular constraints on homogenization throughout all relevant chromosomes are unknown. For example, the limited distribution of variants of the α-satellite DNA to some subset of human chromosomes[24] and the evolution of seeming haplotypes at some loci of human minisatellite DNA[25] cannot be simply explained on the basis of the presence or absence of important signals for the initiation or cessation of turnover. Interestingly, in the former case, other species of higher apes also show variant α repeats that are restricted to the same equivalent subset of chromosomes, indicating that the constraints on concerted evolution in this case are ancient.

All the above examples, from the *Drosophila* rDNA to human repeated sequences, indicate the extent to which a proper detailed analysis of variant repeat distributions that are either in states of transition or are permanently restricted yield important information on the rates, biases, and constraints of the underlying molecular mechanisms, the extent to which they are involved with the activities of one another, and the subtleties of the molecular drive process that underpin concerted evolution. They should discourage both the simplistic view of concerted evolution as an all-or-nothing phenomenon and the naive generalization that the dynamics of genomic turnover operate in the same way, in all species, for all time.[13,14]

A New Mapping Procedure for Monitoring Repeat Sequence Evolution

From the discussion in the preceding section it is clear that a full understanding of the dynamics of concerted evolution requires an experimental procedure that can monitor the evolutionary history of a mutation from its initial occurrence in one repeat, within one chromosomal array in one individual. The development of the MVR–PCR method for mapping variant repeat distribution maps within single chromosomal arrays provides a powerful method that can probe deep into repeat sequence evolution.[26] Using two PCR primers that can distinguish between two variant repeat types (e.g., A and B) within an array (where A and B vary by only one or a few nucleotides in a given region), it is possible to map efficiently

[24] H. F. Willard and J. S. Waye, *Trends Genet.* **3,** 192 (1987).
[25] A. J. Jeffreys, R. Neuman, and V. Wilson, *Cell (Cambridge, Mass.)* **60,** 472 (1990).
[26] A. J. Jeffreys, A. Macleod, K. Tamaki, D. L. Neil, and D. G. Monckton, *Nature,* **354,** 204–209. (1991).

and systematically the interspersion patterns of As with Bs in given arrays. In the cases where these patterns are hypervariable from array to array (e.g., in defined loci of human minisatellite DNAs) it is possible to recognize recombinant exchanges between arrays and to examine whether the exchanges are reciprocal or nonreciprocal, equal or unequal, inter-sister chromatid or inter-non-sister chromatid events.[26] (For a short review, see Dover, 1992.[27]

The successful application of MVR–PCR mapping to the 240 bp subrepeat array within the rDNA IGS of *D. melanogaster,* exploiting a variation of 5 bases in a stretch of 19 between A and B variant repeats, reveals both unexpected interspersion patterns and the seemingly widespread homogenization of such patterns.[28] On a stochastic model of unequal crossing-over at the subrepeat level (see above, and Fig. 3) it is expected that repeats at the array center would be more similar to each other relative to the repeats at the array ends which would tend to be excluded and slowly decay. End repeat exclusion and decay has indeed been observed in several systems including a 220 bp subrepeat array in the IGS of *D. funebris.*[29] In contrast, the 240 bp subrepeat array of *D. melanogaster* is composed of two halves, with a block of A repeats abutting a block of B repeats at the array center. This bipartite structure has been observed in a cloned IGS, in PCR amplifications of IGS, and in whole rDNA analysis of genomic DNA, suggesting that the structure has been considerably homogenized (Fig. 4). Investigations of its distribution between X and Y based arrays, or between spacers that are known to vary considerably in length in wild populations have yet to be completed; nevertheless, the use of MVR–PCR for monitoring the dynamics of evolution of a particular mutant repeat as it is bobbed and buffeted by the mechanisms of repeat family turnover such as unequal crossing-over, gene conversion, and slippage, is open for exploitation.

Computer-Based Methods for Detection and Quantification of Concerted Evolution in Ribosomal DNA Sequences

The essential aspect of computer-based analysis of concerted evolution is the search for similarity between physically separated parts of a given sequence. Two approaches can be used for such searches: graphical comparison of an entire sequence with itself (or with a carefully selected test sequence) or a straightforward listing of regions of sequence similarity

[27] G. A. Dover, *Trends Genet.* **8,** 45–47 (1992).
[28] A. Ruiz Linares, T. Bowen, and G. A. Dover, *J. Mol. Evol.* (submitted).
[29] A. Mian, G. M. Church, and G. A. Dover, *J. Mol. Biol.* (submitted).

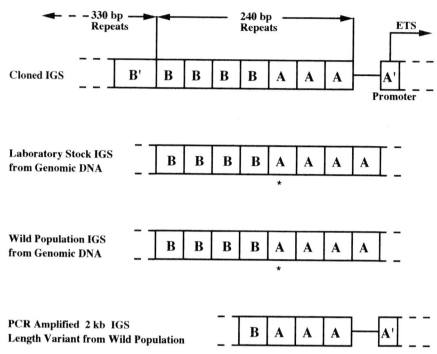

FIG. 4. Distribution of A and B variant repeats in the 240 bp subrepeat array in the IGS of *D. melanogaster* as observed in a cloned IGS, IGSs from genomic DNA of a laboratory stock and a wild population, and PCR amplified products of a short IGS length variant. B' and A' refer to B and A deviants in the flanking 330 repeat and promoter region which contain regions with sequence similarities to the 240 bp repeat. The asterisk indicates a point of ambiguity concerning the A or B status of the repeat at that position. All interspersion patterns were derived from the new and powerful mapping procedure devised by A. J. Jeffreys.[26] The technique provides a tool for probing deep into the evolutionary dynamics of concerted evolution in tandem arrays of repeats.[27,28]

within the sequence. In addition, statistical analysis of the sequence composition of regions undergoing concerted evolution can give an insight into the mechanisms responsible for this process.

The most revealing graphical method for detection of concerted evolution within a sequence is dot-matrix comparison. Many versions of the basic dot-matrix algorithm are available in both commercial and public domain packages. All operate on essentially the same basis: each position within the sequence is compared with each other position, and a graphical representation of matches is generated. A number of refinements of the basic algorithm are commonly built into dot-matrix comparison software, and these can be used conveniently to investigate different levels of concerted evolution. Some parameters of dot-matrix comparisons need to be defined.

Window Size (Span Length). If every identical nucleotide between two sequences were identified in the graphical output, the result would be a dense field of matches within which nothing interesting could be distinguished. To reduce this high background, the unit of comparison, or window size, is invariably greater than 1. The actual window size used in particular comparisons will depend on the natures of the sequences to be compared; acceptable values range at least from 9 to 35, with popular packages settling at around 11.

Match Stringency. To further manipulate the clarity of the graphical output, it is usually possible to select a threshold level of matching within a window that will be considered to be a match. In all cases except those of virtual sequence identity, stringency can usually be set at levels below a perfect match.

Proportional or Identity Matching. Even for a given stringency, matches can be detected in two ways. Take as an example a quoted stringency of 8/10. This can mean one of two things: either 8 consecutive matches in a window of 10 were detected (identity match), or any 8, not necessarily consecutive, bases out of 10 match (proportional match). Perfect matching is a more stringent test at a given nominal stringency, but the use of proportional matching can also be useful.

High levels of stringency in dot-matrix analysis can be used to detect concerted evolution in sequences composed of a number of identical, or near identical, subsequences. An example of this is the *D. melanogaster* rDNA intergenic spacer (IGS) (Fig. 2). On dot-matrix analysis these subrepeats are revealed as blocks of parallel lines, reflecting the near sequence identity of the individual subrepeats.[30] It is interesting to note that the array of 95-bp subrepeats seen in *D. melanogaster* are progressively much less obvious in *D. hydei* and *D. virilis*. An algorithm that detects scrambled short motifs reveals that this region in the two latter species is subject to DNA slippage, probably operating at a higher rate than unequal crossing-over and hence destroying the longer 95-bp subrepeats and their homogeneity patterns (see Fig. 2).

High-stringency dot-matrix analysis of the *D. melanogaster* large ribosomal subunit (LSU) rRNA gene reveals nothing of interest, but at lower stringencies patterns of sequence similarity between LSU rRNA expansion segments begin to appear.[31] In this case, optimal separation of a true pattern of concerted evolution from the general background can be

---

[30] D. Tautz, C. Tautz, D. A. Webb, and G. A. Dover, *J. Mol. Biol.* **195,** 525 (1987).
[31] J. M. Hancock and G. A. Dover, *Mol. Biol. Evol.* **5,** 377 (1988).

achieved by manipulation of stringency parameters; in the case of *D. melanogaster* LSU rRNA expansion segments, the clearest pattern is observed at 19/35 proportional matching. It must be remembered, however, that at such relatively low matching stringencies the base composition of the sequences can make a major contribution to the apparent pattern of sequence evolution. Because of this, it may be worthwhile to include dummy sequences in the analysis. In the case of the human LSU rRNA gene, for example, in which the expansion segments show strong patterns of concerted evolution but also have highly biased base compositions, two dummy control sequences were generated: one with the same base composition as the entire rRNA sequence, to test for accidental patterns of matching, and a second, more stringent control, which mimicked the variation of base composition along the sequence.[31]

Further characterization of concertedly evolving sequences, and insight into the role of slippage in their evolution, can be achieved by analysis of sequence compositions at the level of short sequence motifs. A convenient level of analysis is the trinucleotide motif, as the number (64) of different trinucleotide motifs is manageable and their expected frequencies are high enough within short stretches of sequence for statistical analysis to produce valid results. The distribution of all 64 trinucleotide motifs can be displayed conveniently using a modification of the dot-matrix method. Instead of comparing two DNA sequences, one of the sequence files (preferably the one displayed on the horizontal axis) comprises an array of blocks of trinucleotide motifs, from AAA through to TTT, interspersing contiguous trinucleotides within the artificial sequence with 'padding' characters to present false positives. If the file is compared to a DNA sequence, the resultant plot will show the distribution of all trinucleotides within the sequence. Figure 5 presents such a display for the human 28 S rRNA gene, which shows a characteristic preferential use of a subset of trinucleotide motifs within the expansion segments.[32]

Repetitiveness within a sequence can also be detected on the basis that a dot-matrix self-comparison of the sequence shows more matches at a given level of stringency than would be expected by chance. The DIAGON/SIP program[33] provides such a facility. This approach has the advantage that it allows direct analysis of the statistical significance of patterns observed in dot-matrix analysis. It has been used to detect internal repetition within dipteran LSU rRNA expansion segments in the absence of strong dot-matrix patterns.[34]

Alternatives to graphical methods to reveal patterns of concerted evo-

[32] J. M. Hancock and G. A. Dover, *Nucleic Acids Res.* **18,** 5949 (1990).
[33] R. Staden, *Nucleic Acids Res.* **10,** 4731 (1982).
[34] A. Ruiz Linares, J. M. Hancock, and G. A. Dover, *J. Mol. Biol.* **219,** 381 (1991).

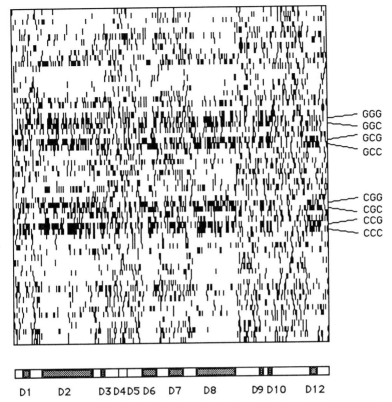

FIG. 5. Distribution of individual trinucleotide motifs within the human 28 S rRNA gene visualized using the DIAGON/SIP program.[33] Locations of individual trinucleotide motifs are indicated by a vertical bar in each of 64 horizontal rows representing motifs from AAA (bottom) to TTT (top) in alphabetical order. Below the display is a diagram of the arrangement of expansion segments (D1–D12) within the gene, showing the correlation between expansion segments and nonrandom occurrence of a subset of motifs.

lution within and between sequences are methods which search for and display textually regions of similarity between sequences, or repetitious motifs within sequences. We have found the LFASTA option of the FASTA program[35] particularly useful for this purpose. The aim of using such techniques is to identify rapidly regions of similarity between and within sequences. Searching for self-similarity within a sequence can reveal not only the location of coevolving regions of the sequence, but also the extent of their similarity. The latter measure is valuable because arrays of sequences, such as subrepeats within rDNA intergenic spacers that are

[35] W. R. Pearson and D. J. Lipman, *Proc. Natl. Acad. Sci. U.S.A.* **85,** 2444 (1988).

homogenized by unequal crossing-over, are expected to show higher degrees of similarity between units at the center of an array (where rates of unequal crossing-over per subrepeat are higher) than at their ends (where rates are lower).[36,37] However, evidence that this is not always the case is given in the preceding section.

Analyses of sequence motif composition can be taken a stage further to consider the statistical probability of the observed pattern of trinucleotide frequency or distribution being due to random chance or to other processes. The χ^2 analysis of observed frequencies of trinucleotide occurrence within a given sequence (ideally more than 322 bp long) compared to expected frequencies derived from base composition can give an indication whether the usage of trinucleotides within a given sequence is random. This can give an indication of the specific amplification of certain motifs within the sequence by processes such as slippage,[32] but it is also likely to be affected by phenomena such as CpG avoidance and biased codon usage within protein-coding regions.

Detecting Absence of Concerted Evolution

Finally, it is important to develop procedures that can accurately assess the absence of concerted evolution. We have pointed out above that high levels of overall divergence between repeats, while suggestive of independent repeat evolution, could miss key single-base changes that are shared between all or some of the repeats. This is particularly problematic in many situations where the domain of gene conversion can be very much shorter than the repeat unit, generating mosaic genes.

Several analytical procedures have been developed for the detection and quantification of micro gene conversion.[38,39] One study[39] analyzed nine actin genes in plants of the family Solanaceae, which, unlike their animal counterparts, can be up to 26% divergent in nucleotide sequence. The use of the "compatibility test"[40] on the genes allowed for the construction of a "maximal clique" in which all sites are distributed among genes to give rise to an internally consistent gene phylogeny. All remaining incompatible sites can be considered, in the first instance, as due to gene conversions between pairs of genes not related by descent. A nonparametric runs test can then be used to examine the physical clustering of incompatible sites along the genes, and the distribution among genes of each site

[36] R. F. Barker, N. P. Harberd, M. G. Jarvis, and R. B. Flavell, *J. Mol. Biol.* **201**, 1 (1988).

[37] A. Mian and G. A. Dover, *Nucleic Acid Res.* **18**, 3795 (1990).

[38] S. Sawyer, *Mol. Biol. Evol.* **6**, 526 (1989).

[39] G. Drouin and G. A. Dover, *J. Mol. Biol.* **31**, 132 (1990).

[40] G. F. Estabrook, *Syst. Bot.* **3**, 146 (1978).

in such clusters can be tested against a maximum-likelihood phylogeny.[41] If all sites within a cluster are shared by the same gene pairs, then a gene conversion can be invoked. The study of actin genes[39] using these and further procedures managed to assess that very few, either long or short, domains of gene conversion had actually taken place in the evolutionary history of these particular genes. The appropriate combination of analytical procedures could be applicable to a wide variety of alternative situations.

[41] M. J. Bishop and A. E. Friday, *Proc. R. Soc. London B.* **226**, 271 (1985).

[39] Assaying Differential Ribosomal RNA Gene Expression with Allele-Specific Probes

By ELDON R. JUPE and ELIZABETH A. ZIMMER

Introduction

Any assay for differential gene expression must be able to detect specific RNA transcripts such that transcripts from one gene can be distinguished from those of another. This requirement is easily met when assaying transcript levels for genes whose coding regions have diverged significantly. In these cases standard blot/hybridization experiments using cloned genes as probes will determine differential transcript levels. These methods cannot, however, be applied to the analysis of differential expression of closely related members of a multigene family such as the ribosomal RNA gene (rDNA) coding regions in which minimal sequence divergence has occurred. For example, in many inbred maize lines rDNA sequences are essentially conserved except for a few restriction enzyme polymorphisms present in some lines in a significant portion (10–50%) of the repeat units of an individual.[1,2] DNase I sensitivity and undermethylation of a fraction of repeat units containing a 26 S coding region *Eco*RI polymorphism suggested that these polymorphic repeats are preferentially active.[2,3]

To further investigate this structural and functional variation in rDNA, we have utilized oligonucleotide probes containing single base pair mismatches to RNA transcripts to determine the levels of specific transcripts in maize hybrids in which both rDNA variants are being expressed to different extents.[3] The experimental basis for these methods is the fact that properly designed oligonucleotides will hybridize specifically to their com-

[1] E. A. Zimmer, E. R. Jupe, and V. Walbot, *Genetics* **120**, 1125 (1988).
[2] E. R. Jupe and E. A. Zimmer, *Plant Mol. Biol.* **14**, 333 (1990).
[3] E. R. Jupe and E. A. Zimmer, *Plant Mol. Biol.* **21**, 805 (1993).

plementary DNA or RNA sequences.[4,5] The utility of oligonucleotides containing single base pair mismatches to different allele types was first used for detection of the sickle cell β^S-globin allele in genomic DNA.[6] This method has since been applied to the detection of point mutations at numerous other loci[7-9] and to identify genomic clones or cDNAs coding for specific members of a multigene family.[10] Oligonucleotide probes containing single base pair or more mismatches to certain alleles have been successfully applied to the analysis of RNAs.[11-13] In this chapter we discuss the rapid identification and characterization of multigene family sequence variants, the design of allele-specific oligonucleotide probes, and procedures for the use of the oligonucleotide probes to analyze differential rRNA transcript levels.

Identification and Confirmation of Sequence Variation

Restriction Enzyme Polymorphisms

Variation in multigene families such as rDNA may be identified by several methods. The most obvious method involves cloning and sequencing of specific variants present in the samples of interest. Even with the advent of rapid cloning and sequencing methods this can be a time-consuming process when several polymorphisms and/or variable samples are of interest. Potential sequence polymorphisms may be most rapidly identified by screening restriction enzyme digests of rDNA by Southern blot/hybridization mapping experiments.[1] These restriction enzyme polymorphisms are described as potential sequence polymorphisms because in many cases in higher plants (and vertebrates) restriction enzyme polymor-

[4] R. B. Wallace, J. Shaffer, R. F. Murphy, J. Bonner, T. Hirose, and K. Itakura, *Nucleic Acids Res.* **6**, 3543 (1979).

[5] R. B. Wallace, M. J. Johnson, T. Hirose, T. Miyake, E. H. Kawashima, and K. Itakura, *Nucleic Acids Res.* **9**, 879 (1981).

[6] B. J. Conner, A. A. Reyes, C. Morin, K. Itakura, R. L. Teplitz, and R. B. Wallace, *Proc. Natl. Acad. Sci. U.S.A.* **80**, 278 (1983).

[7] V. J. Kidd, R. B. Wallace, K. Itakura, and S. L. C. Woo, *Nature (London)* **304**, 230 (1983).

[8] H. H. Kazazian, S. H. Orkin, A. F. Markham, C. R. Chapman, H. Youssoufian, and P. G. Waber, *Nature (London)* **310**, 152 (1984).

[9] D. J. G. Rees, C. R. Rizza, and G. G. Brownlee, *Nature (London)* **316**, 643 (1985).

[10] D. H. Schulze, L. R. Pease, Y. Obata, S. G. Nathenson, A. A. Reyes, S. Ikuta, and R. B. Wallace, *Mol. Cell. Biol.* **3**, 750 (1983).

[11] J. Geliebter, R. A. Zeff, D. H. Schulze, L. R. Pease, E. H. Weiss, A. L. Mellor, R. A. Flavell, and S. G. Nathenson, *Mol. Cell. Biol.* **6**, 645 (1986).

[12] R. A. Zeff, J. Gopas, E. Steinhauer, T. V. Rajan, and S. G. Nathenson, *J. Immunol.* **137**, 897 (1986).

[13] O. P. Das and J. W. Messing, *Mol. Cell. Biol.* **7**, 4490 (1987).

phisms that appear to be site losses are due to methylation of cytosine residues (or possibly adenine in some plants),[14] thus rendering the sequence refractory to cleavage by methylation-sensitive restriction enzymes. We have utilized various methods discussed below to distinguish rapidly genuine site loss sequence polymorphisms from methylation polymorphisms identified by restriction enzyme screens.[3] The only requirement for utilizing these methods is that a sequence for a representative member of the multigene family is available for designing oligonucleotide primers flanking the polymorphic region of interest. When evolutionarily conserved sequences such as rDNA coding regions are being analyzed, the reference sequence does not necessarily have to be from the same species or genus.

Screening Polymerase Chain Reaction Products

After restriction enzyme mapping experiments have identified potential sequence polymorphisms, we have utilized restriction enzyme analysis of DNAs synthesized by the polymerase chain reaction (PCR) to distinguish rapidly genuine sequence polymorphisms from methylation polymorphisms.[3] The details of designing primers and performing the PCR are presented elsewhere in this volume (see [28]). A standard PCR synthesis is designed to produce a target sequence containing the restriction enzyme polymorphism sites of interest. Depending on the distribution of sites of interest, fragments from 50 to 1000 (or more) base pairs (bp) may be synthesized. We have routinely synthesized and analyzed fragments approximately 250 bp in length with the potential polymorphic site near the middle of the fragment. The PCR-synthesized fragment is subjected to digestion with the restriction enzymes that indicated potential polymorphisms. The digestion products are analyzed on the appropriate type of agarose gel. Smaller fragments (50–300 bp) are separated on 4% NuSieve (FMC, Rockland, ME) gels, but larger fragments may be analyzed on standard 1% agarose gels. The gels are stained and photographed to reveal the fragment pattern of the in vitro synthesized products. We also routinely subject the gels to Southern blotting to nylon membranes and hybridization with an end-labeled oligonucleotide primer to determine if any minor digestion products are present. Because the products of a PCR amplification do not contain methylated bases, the PCR products represent demethylated copies of the genomic sequence. Thus, any sites that are not cleaved in genomic DNA because of methylated bases will be cleaved in the PCR products. The in vitro synthesized DNAs containing base se-

[14] Y. Gruenbaum, T. Nahev-Many, H. Cedar, and A. Razin, Nature (London) 292, 860 (1981).

quence polymorphisms will not be cut by the restriction enzyme, just as it was not cut in the genomic DNA. This procedure provides a rapid method for screening many genomic DNA samples to distinguish base sequence polymorphisms from methylation polymorphisms.

Determining Base Changes

Following the confirmation of sequence polymorphisms, the base sequence of the region of interest must be determined in order to design allele-specific probes. This does not require large amounts of sequencing because only a short region has to be sequenced from a polymorphic and a nonpolymorphic sample. The PCR products synthesized for restriction enzyme analysis may be directly sequenced (see this volume [28]) or cloned and sequenced. Alternatively, for rDNA coding regions we have determined the sequence of polymorphic and nonpolymorphic sites by directly sequencing rRNA transcripts. These methods are described in this volume (see [25, 27]) and by others.[15]

Designing Allele-Specific Oligonucleotide Probes

Identification of the base sequence differences between two samples allows the design of allele-specific oligonucleotide probes for use in detecting specific transcripts. Figure 1A,B shows the oligonucleotide probes we have utilized. A probe is designed to match exactly the rRNA from one inbred maize line (Fig. 1A, left). A single base pair mismatch is present when the probe is hybridized to rRNA from a different inbred line (Fig. 1A, right). A second probe is designed with the opposite specificities (Fig. 1B). Probes may be designed to detect sequences having from a single base change up to several bases changed. Commonly, we observe regions containing one or two bases changed with other surrounding sequences being conserved.

The oligonucleotide probes must be designed so that specific stable hybrids can be formed with the rRNA containing exact matches, but the hybrids formed also must be unstable enough that they will disassociate from the rRNA containing a single base mismatch under the appropriate temperature and wash conditions. We have followed guidelines for the length of the probe and position of the mismatch recommended by previous single base pair mismatch experiments in which oligonucleotide probe–DNA hybrids were under investigation.[6,16] The oligonucleotides are at least 19 bases in length to ensure a high probability of hybridization to a

[15] R. K. Hamby, L. Sims, L. Issel, and E. A. Zimmer, *Plant Mol. Biol. Rep.* **6,** 175 (1988).
[16] S. Ikuta, K. Takagi, R. B. Wallace, and K. Itakura, *Nucleic Acids Res.* **15,** 797 (1987).

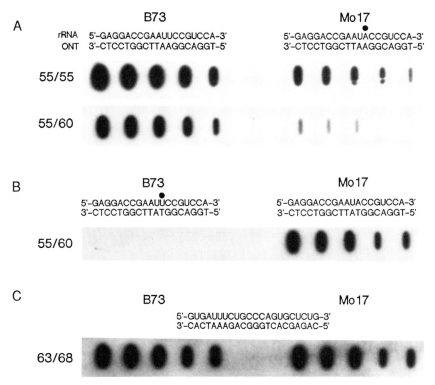

FIG. 1. Testing allele-specific oligonucleotide probes in slot blot/hybridization experiments. The specificity of oligonucleotide probes was tested using a series of replicate slot blots containing dilutions of maize rRNAs. Four replicate slot blots hybridized to various oligonucleotide probes are shown. The set of five slots on the left-hand side shows B73 total RNA applied at 1.0, 0.75, 0.50, 0.25, and 0.10 μg/slot from left to right. The set of five slots on the right-hand side shows the same amounts of Mo17 total RNA. The sequences of the target rRNAs and the oligonucleotide probes hybridized to them are shown above each set of samples. Positions of mismatched bases are indicated by solid circles. The temperatures at which hybridizations/washes were performed are shown at left. (A) Results obtained when the probe specific for the B73 sequence (T_d 60°) was hybridized and washed at 55° *(top)* or hybridized at 55° and washed at 60° *(bottom)*. (B) Mo17-specific probe (T_d 60°) when hybridized at 55° and washed at 60°. (C) Hybridization of the control probe to B73 and Mo17. The sequence detected by this probe is identical in both lines.

unique sequence. This length is short enough that a single base pair mismatch will destabilize hybrids when the appropriate wash and temperature conditions are used. The mismatch base or bases should be located in the center or as near the center as possible in order to maximize thermal instability. In addition, the base composition of the oligonucleotides should not be extremely AT-rich for using the standard hybridization and wash protocols we discuss. The sequence of the region to be spanned by the oligonucleotide will dictate how closely these rules can be followed. We and others have been successful using oligonucleotides in which the mismatch was offset from the center.[3,6]

Finally, in addition to identifying potential oligonucleotide probes in regions with sequence mutations, an oligonucleotide probe from a region that is 100% conserved in all samples under analysis is useful as a control when characterizing allele-specific probes and setting up the assays for quantitating relative levels of transcripts (Fig. 1C). These regions are usually easily identified on the same sequencing gel that was used to determine the mismatch sequence.

Procedures

RNA Isolation

Total RNA may be purified from tissues by any number of methods. We have successfully purified RNA from a number of plant tissues using modifications of either a hot (100°) borate- or a guanidinium isothiocyanate-based method.[15] For the assays described in this chapter we scaled down and slightly modified a hot borate method[17] to purify RNA from individual maize seedlings (0.5 g). Fresh seedling tissue is harvested, rapidly frozen in liquid nitrogen (sample may be stored at this point for several months at −80° in heat-sealed plastic bags), and ground to a fine powder in a prechilled mortar and pestle. The extraction buffer [0.2 M sodium borate, pH 9.0, 30 mM EDTA, 5 mM dithiothreitol (DTT), 1% sodium dodecyl sulfate (SDS)] is heated to 100° in a water bath prior to use. The powdered tissue is added to 10 volumes (5 ml) of the preheated extraction buffer, and the mixture is ground with a Polytron homogenizer twice for 30 sec each time at room temperature. The homogenate is immediately filtered through one layer of Miracloth before proteinase K (100 μl of a 10 mg/ml solution) is added. The sample is mixed briefly and incubated at 37° for 1 hr. If multiple isolations are being performed, the proteinase K incubation should be started before beginning the next sample.

After the incubation, 300 μl of 2 M KCl is added, and the sample is chilled on ice for 10 min. Following centrifugation at 10,000 rpm for 10 min at 4°, the supernatant (containing the RNA) is decanted into a fresh tube, and 300 μl of 10 M LiCl is added. The RNA-containing solution is stored overnight at −80° and the next morning is placed on ice or at

[17] T. C. Hall, Y. Ma, B. U. Buchbinder, J. W. Pyne, S. M. Sun, and F. A. Bliss, *Proc. Natl. Acad. Sci. U.S.A.* **75**, 3196 (1978).

4° for 15–30 min prior to centrifugation. The RNA is pelleted by centrifugation (10,000 rpm, 10 min, 4°), resuspended in 0.6 ml of 2 M LiCl, and centrifuged again as above. The RNA pellet is resuspended in 0.6 ml of 2 M potassium acetate, pH 5.5. The RNA is reprecipitated by adding 2.5 volumes of ice-cold 95% (v/v) ethanol and storing (2 hr or overnight) at $-20°$. The RNA is pelleted by centrifugation (as above), redissolved in 0.6 ml of STE (10 mM Tris-HCl, pH 7.5, 10 mM NaCl, and 1 mM EDTA), and reprecipitated overnight with ethanol. The RNA is pelleted again by centrifugation, and the final pellet is resuspended in 500 μl of TE (1 mM Tris-HCl, pH 7.5, 0.1 mM EDTA). The concentrations of samples are determined spectrophotometrically, and the integrity and purity of the RNA are checked on 1% agarose gels under neutral or denaturing conditions.[15] Routinely, 100–200 μg of total RNA is obtained from 0.5 g of single seedling leaf material.

Oligonucleotide Preparation

Oligonucleotides may be synthesized on any number of commercially available automated synthesizers or ordered from a supplier of custom synthesized oligonucleotides. Generally, oligonucleotides prepared by automated synthesis may be used for most applications (PCR, library screening) with minimal purification. The standard procedures we use for purification of oligonucleotides have been covered in detail elsewhere.[15] The manuals for automated synthesizers often contain detailed purification protocols. Briefly, we recommend that oligonucleotides used as allele-specific probes be purified by electrophoresis on 20% acrylamide/8 M urea gels. The oligonucleotide band is detected by UV shadowing, excised from the gel, and purified using a Sep-Pac C_{18} column (Waters Division of Millipore, Bedford, MA).[15]

End Labeling

Oligonucleotides used as hybridization probes are end labeled by phosphorylation of 5′ termini with [γ-^{32}P]ATP catalyzed by T4 polynucleotide kinase. We routinely label 50–100 ng of oligonucleotide in each reaction. Prior to setting up the labeling reaction, the proper amount of oligonucleotide is diluted to a final volume of 30 μl with distilled water, heated for 5 min at 65°, and chilled on ice. The reaction mixture consists of 10 μl of 5× kinase buffer (0.25 M Tris HCl, pH 7.5, 50 mM MgCl$_2$, 25 mM DTT, 50 μM spermidine, and 50 μM EDTA), 8 μl of [γ-^{32}P]ATP (>7000 mCi/mmol), and 5–10 units of T4 polynucleotide kinase in a final volume of 50 μl. The reaction is incubated for 1 hr at 37° and stopped by adding 10 μl of 0.5 M EDTA. The labeled oligonucleotide is separated from unincorporated label using either a Sephadex G-25 spin column made in a 1-ml syringe or a Sephadex G-50 gravity column made in a glass pipette.

Preparation of RNA Slot Blots

RNAs are denatured by formaldehyde/heat treatment. All of the stock solutions for RNA slot blots are made using sterile diethyl pyrocarbonate (DEPC)-treated water. RNAs are diluted as necessary from stock solutions with water, and 3 volumes of 6.15 M formaldehyde in 10× standard saline citrate (SSC) is added to give a final RNA concentration of 10–100 $\mu g/ml$. The RNA dilutions are heated to 65° for 15 min and quick-chilled on ice. The denatured stock is further diluted with 4.16 M formaldehyde in 7.5× SSC such that the desired concentration of RNA may be applied to each slot in a total volume of 400 μl. The nylon membrane is prewet in water and then soaked in 10× SSC for 20 min. Slot blots are performed using a commercially available apparatus hooked to a vacuum source. After the samples are blotted through, each well is washed with 400 μl of 10× SSC. The membrane is removed from the apparatus and baked in a vacuum oven at 80° for 2 hr.

Hybridization with Oligonucleotide Probes

The temperatures at which hybridizations and washes are carried out is estimated by using the empirically derived formula T_d (in °C) = [2 × (A + T) + [4 × (G + C)] to calculate the approximate dissociation temperature for oligonucleotides.[18] In our experience dissociation temperatures predicted by this formula are quite accurate. However, we have observed situations where optimal results are obtained at temperatures 2–3° different from those predicted. Experimental calibration of the dissociation behavior of each oligonucleotide used should be performed as discussed below.

Filters are prehybridized in 5 ml of a solution containing 25 mM potassium phosphate, pH 6.5, 1% sarkosyl, 5× SSC, 1× Denhardt's solution, and 200 $\mu g/ml$ of denatured salmon sperm DNA for at least 2 hr. The prehybridization fluid is discarded, and a fresh aliquot containing 1–2 × 10^6 counts/min (cpm)/ml of end-labeled oligonucleotide is added. The temperature of prehybridization/hybridization used is 5° below the calculated T_d for an oligonucleotide. Hybridizations are typically incubated overnight in a shaking incubator. After removing the probe solution, the filters are washed twice in 6× SSC, 0.1% SDS for 20 min at room temperature. The initial wash is followed by two high-stringency washes in 6×

---

[18] S. V. Suggs, T. Hirose, T. Miyake, E. H. Kawashima, M. J. Johnson, K. Itakura, and R. B. Wallace, *in* "Developmental Biology Using Purified Genes" (D. D. Brown, ed.), p. 683. Academic Press, New York, 1981.

SSC, 0.1% SDS at the T_d (preheat the buffer to this temperature) for 5–10 min each. Excess moisture is removed by blotting the membranes between filter paper, and the membranes are immediately wrapped in plastic wrap while still damp. It is important not to allow the membrane to dry out if it will be sequentially washed (see below). The membranes are then exposed to preflashed X-ray film using one intensifying screen.

Applications

Characterizing Oligonucleotide Probes

Figure 1 shows examples of experiments done to establish the conditions for using a group of oligonucleotide probes to characterize specific maize rRNAs. The samples shown are loaded in sets of five slots containing a decreasing concentration series from left to right. The set on the left-hand side contains B73 RNA, and the set on the right-hand side contains Mo17 RNA. Figure 1A shows the results obtained when an oligonucleotide probe specific for B73 rRNA is hybridized to either B73 (left) or Mo17 (right) RNAs. The blot at the top of Fig. 1A was hybridized and stringently washed at a temperature 5° lower than the estimated T_d for the allele-specific probe. Considerable cross-reactivity with the mismatched Mo17 rRNA occurs. In the lower portion of Fig. 1A, cross-reactivity with Mo17 is eliminated when the stringent wash is performed at the estimated T_d. Figure 1B shows the results obtained when the probe specific for Mo17 rRNA is hybridized to the same samples at 5° lower than the T_d and washed at the T_d. These experiments determined the hybridization and wash conditions at which the allele-specific probes detect only the exact-match rRNAs. Figure 1C shows an example of a slot blot from which an allele-specific probe was stripped and the filter rehybridized to a control probe containing a sequence that is identical in B73 and Mo17. Hybridization with a control probe ensures that rRNA loading errors did not occur in setting up the blot.

In addition to verifying the specificity of probes, the experiment in Fig. 1 also determines the range of sample concentrations in which a linear response is obtained for the probes. The autoradiograms are scanned with a densitometer equipped with an integrator to determine the peak areas for each slot. The peak areas are then plotted against the amount of RNA loaded per slot. A linear increase in peak areas is observed when between 0.1 and 0.75 μg of RNA is applied per slot. We find that the signal response is nonlinear above 1 μg per slot, suggesting that at this point filter-bound sequences are present in excess of the probe.[19] Thus, quantitation is accu-

[19] M. L. M. Anderson and B. D. Young, in "Nucleic Acid Hybridization: A Practical Approach" (B. D. Hames and S. J. Higgins, eds.), p. 73. IRL Press, Oxford, 1985.

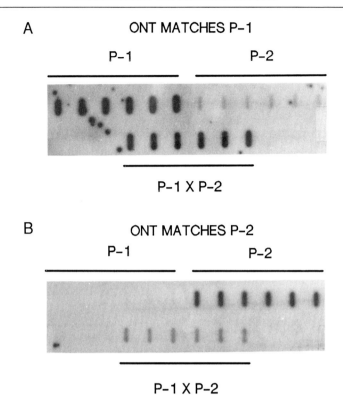

A   ONT MATCHES P-1

P-1   P-2

P-1 X P-2

B   ONT MATCHES P-2

P-1   P-2

P-1 X P-2

FIG. 2. Analysis of parental and hybrid maize rRNAs with allele-specific oligonucleotide probes. In the cross illustrated the maternal parent P-1 is B73 and the paternal parent P-2 is Mo17. The samples were applied at 0.5 μg/slot with six replicates of each. The parental and hybrid samples are identified. (A) The oligonucleotide probe used is the same one shown in Fig. 1A. In (B) the probe is the same as in Fig. 1B.

rate only for slots in which 0.75 μg of total RNA or less is applied. For analysis of multiple hybrid samples, we routinely apply 0.5 μg of total RNA in each slot (see below). If low-copy mRNAs are being detected in a total RNA preparation, a range of 0–10 μg of RNA per slot should be tested in calibration experiments.[19]

Analysis of Ribosomal RNA Levels in Hybrids

Figure 2 shows an example of a quantitation experiment in which five replicate slots containing each of the two parental RNAs and five replicate slots containing hybrid RNAs have been applied to two duplicate blots. To maintain uniformity between samples we prepare dilutions of a particular RNA sample such that 0.5 μg can be applied to each of ten slots. One of the duplicate slot blots is hybridized to a probe specific for one of the parental rRNAs (Figure 2A), and one is hybridized with the probe specific for the other parental rRNA (Fig. 2B). The washes are performed such that the parental specific probe remains hybridized with its complementary rRNA but does not detect the mismatched rRNA. In the example shown in Fig. 2, the hybrid rRNA sample contains more transcripts matching the probe P-1.

A quantitative estimate of the percentage of the hybrid RNA detected by each specific probe may be calculated by determining a value for the total amount of 26 S rRNA present. Initially, hybrid RNAs are prepared as triplicate slot blots, with one hybridized to the P-1 probe, one to the P-2 probe, and the other to the conserved probe shown in Fig. 1C. The areas of the peaks are determined as discussed above. We consistently find that the sum of the areas of peaks obtained for probes P-1 and P-2 hybridized to hybrid rRNA is equivalent to the area of the peak for the conserved 26 S probe hybridized to hybrid rRNA. In other words, the sum of the peak areas observed with the two allele-specific probes hybridized to the hybrid RNA is the same as the peak area obtained for the hybrid RNA analyzed with the conserved probe. Thus, in the analysis of hybrid rRNAs, we routinely consider the sum of the probe P-1 and probe P-2 peak areas as reflecting 100% of the 26 S rRNA. In testing the assay, we also observed that the sum of the peak areas obtained for the hybrid peak areas are essentially equivalent to those observed for a parental RNA hybridized to its homologous probe. This internal consistency in the assay further suggests that all of the 26 S rRNA transcripts present are being detected in the hybrid samples by the two specific probes. In Fig. 2 approximately 80% of the rRNA is homologous to probe P-1, and 20% is homologous to probe P-2.

Comments and Suggestions

The highly radioactive end-labeled oligonucleotides should not be used for more than 1 week after preparation because they are subject to radiochemical degradation. Labeled probes not used immediately are stored at −20°. We routinely prepare several groups of slot blots and perform the hybridization to oligonucleotide probes on the day they are labeled.

When characterizing the allele-specific probes as shown in Fig. 1, we initially performed washes at lower temperatures, then immediately wrapped the filters in plastic wrap and exposed them to film. If signal remains on the mismatch slots, the filter can be rewet in 6× SSC and washed at a higher temperature. The wash, exposure, wash at a higher temperature cycle can be repeated numerous times until the exact temperature at which the probe is removed from mismatched RNAs is determined. This procedure allows one to gain a thorough understanding of the behavior of specific oligonucleotide probes when detecting their target sequence and mismatched sequences. Remember that even exact match oligonucleotide probes are rather easily removed from their target sequences if wash conditions are too stringent.

As an additional characterization, we hybridize the allele-specific probes to rDNA Southern blots to confirm that they hybridize to restriction fragments containing the region from which they were designed. The Southern blots are hybridized and washed using the same conditions described for RNA slot blots.

In the experiments analyzing differential levels of hybrid RNAs, it is essential to maintain uniform conditions when producing the slot blots and labeled probes. Each of the slot blots for parallel analysis with the different probes should be prepared from dilutions of the same RNA samples. The two probes exhibiting reciprocal reactivity (as well as any control probes) are prepared in parallel using the same batch of radioactive label. This routinely leads to the production of probes having comparable specific activities. Following these guidelines we have found this method gives highly reproducible values when determining differential rRNA levels.

Acknowledgments

The authors were in the Department of Biochemistry (E.A.Z., E.R.J.), the Department of Botany (E.A.Z.), and the Louisiana Agricultural Experiment Station (E.A.Z.), Louisiana State University, Baton Rouge, Louisiana, when this research was performed. Support was provided by setup funds from the LSU College of Basic Sciences, a grant from the National Science Foundation (DEB-BSR-8615212) to E.A.Z. and a Sigma Xi Grant-in-Aid of Research to E.R.J. This work was approved for publication by the Director of the Louisiana Agricultural Experiment Station as manuscript number 91-125203.

[40] Assays for Copy Number, Differential Expression, and Recombination in Lysozyme Multigene Family

By David M. Irwin

Introduction

Gene duplication and the diversification of gene families have been essential for the development of the complex genomes of modern vertebrates.[1] Many of the intricate physiological processes found in vertebrates involve cascades of related enzymes or families of similar, but functionally distinct, proteins. The blood coagulation factors[2] provide an example of a cascade of related enzymes, and the visual photoreceptors[3] illustrate a family of related proteins with similar functions but differing specificities (in this case, color). Related proteins can often take part in very different functions in diverse physiological systems, as demonstrated by the serine proteases involved in blood coagulation, digestion, and processing of neuropeptides[4] and by the case of the tyrosine kinases[5] and G proteins[6] used in signal transduction for many types of biological systems. In fact, most genes found in any vertebrate are related, in total or in part by either gene duplication or exon shuffling, to other genes found in the same genome.[7] An understanding of the relationships among the different members of gene families should provide insight not only into the origin of the different genes but also into how the different physiological processes may be related and how these processes may be interconnected.

Unfortunately, most of the duplication events and the diversification of gene families involved in the evolution of the complex physiological processes occurred very early in the evolution of the vertebrates, or in some cases the early metazoans (i.e., more than 600 million years ago). Many of the early events in the evolution of these gene families have been obscured by subsequent events. To try and understand the molecular processes and mechanisms involved in the origin and evolution of new physiological systems, it is more useful to study genes which arose very recently and for which a change of function is clearly understood. This would allow us to study the evolution of an essential gene (the new function) before multiple mutational events can obscure the evolutionary history. The origin and evolution of stomach lysozymes in two groups of mammals meet these criteria and have been extensively explored.[8-13] Methods are presented

[1] T. Ohta, in "Oxford Surveys in Evolutionary Biology" (P. H. Harvey and L. Partridge, eds.), Vol. 5, p. 41. Oxford Univ. Press, Oxford, 1988.

[2] L. Patthy, Cell (Cambridge, Mass.) 41, 657 (1985).

[3] J. Nathans, D. Thomas, and D. S. Hogness, Science 232, 193 (1986).

[4] H. Neurath, Science 224, 350 (1984).

[5] S. K. Hanks, A. M. Quinn, and T. Hunter, Science 241, 42 (1988).

[6] M. I. Simon, M. P. Strathmann, and N. Gautam, Science 252, 802 (1991).

[7] R. F. Doolittle, Trends Biochem. Sci. 10, 233 (1985).

[8] C.-B. Stewart, J. W. Schilling, and A. C. Wilson, Nature (London) 330, 401 (1987).

[9] D. M. Irwin, A. Sidow, R. T. White, and A. C. Wilson, in "The Immune Response to Structurally Defined Proteins: The Lysozyme Model" (S. J. Smith-Gill and E. E. Sercarz, eds.), p. 73. Adenine Press, Schenectady, New York, 1989.

[10] D. M. Irwin and A. C. Wilson, J. Biol. Chem. 264, 11387 (1989).

[11] D. M. Irwin and A. C. Wilson, J. Biol. Chem. 265, 4944 (1990).

which have been used to study the evolution of ruminant lysozyme genes; these include the determination of the number of genes within the gene family, the isolation of cDNAs of expressed members, and phylogenetic analysis of the sequences.

Determination of Gene Number

An initial step in characterizing any gene typically involves determination of its copy number. If multiple copies of a gene are found in one species, it may be possible to determine when this change in gene number occurred by comparisons of gene numbers in other related species. The most common method used in the estimation of gene copy number is Southern blotting.[14] The Southern blot is a technique where DNA fragments are separated by size, and then the fragments containing the sequence of interest are detected by hybridization with a labeled probe. Southern blots are useful if the number of genes in the gene family detected by hybridization is relatively low (e.g., 1 to 20 copies). In many cases only a fraction (a subfamily) of all the members of a gene family will be detected; the size of the subfamily depends on the amount of sequence similarity among members of a gene family. By changing the stringencies of hybridization and washing, genes (or subfamilies of genes) with differing amounts of sequence similarity can be detected. The methods we have used are modified from those of Kan and Dozy[15] and have been suitable for exploring the evolution of the lysozyme gene family in mammals.[9,10] Shown in Fig. 1 are examples of genomic blots hybridized with bovine lysozyme probes that detect multiple genes in ruminants (including cow) and lysozyme genes in species as diverse as mice and humans.

Methods

Southern Transfer

1. Digest approximately 10 μg of genomic DNA with the restriction enzyme of choice under appropriate conditions.

2. Separate the digested DNA by electrophoresis through a 25 × 20 cm 0.8–1% agarose gel overnight. Both the gel and the running buffer are 1×

---

[12] J. Jollès, P. Jollès, B. H. Bowman, E. M. Prager, C.-B. Stewart, and A. C. Wilson, *J. Mol. Evol.* **28,** 528 (1989).

[13] J. Jollès, E. M. Prager, E. S. Alnemri, P. Jollès, I. M. Ibrahimi, and A. C. Wilson, *J. Mol. Evol.* **30,** 370 (1990).

[14] E. M. Southern, *J. Mol. Biol.* **98,** 503 (1975).

[15] Y. W. Kan and A. M. Dozy, *Proc. Natl. Acad. Sci. U.S.A.* **75,** 5631 (1978).

TAE ($1 \times$ TAE is 40 mM Tris–acetate, 2 mM EDTA), with the gel containing 10 μg/ml ethidium bromide to stain the DNA. Addition of approximately 5000 cpm of a ^{32}P-end-labeled *Hind*III digest of λ DNA to an unlabeled digest will provide radioactive markers that correspond to the visible markers.

3. The separated DNA fragments, which are stained with ethidium bromide are visualized by transillumination with UV light and photographed. UV exposure also damages large DNA molecules, allowing more efficient transfer.

4. DNA within the gel is denatured by soaking the gel in a solution containing 0.5 N NaOH and 0.6 M NaCl for 45 min without shaking.

5. The gel is then neutralized by soaking in two changes of neutralization buffer (1 M Tris base, 0.6 M NaCl, adjusted pH to 7.4 with HCl).

6. DNA is transferred from the gel to nitrocellulose membrane by the Southern method in $10 \times$ SSC ($1 \times$ SSC is 0.15 M NaCl, 15 mM sodium citrate, pH 7.2) by capillary diffusion for 40–48 hr.

7. After transfer is completed, the nitrocellulose membrane is washed briefly in $2 \times$ SSC and allowed to air-dry. DNA is bound to the membrane by baking for 2 hr at 80° in a vacuum oven or for 6 hr at 68° in a conventional oven. Both conditions give similar results. The nitrocellulose must always be handled with care (i.e., with clean gloves) to prevent later background problems.

Preparation of Probe. Probes for genomic Southern blots have been made by a variety of methods and must be of high specific activity. A simple method that yields a probe with a very high specific activity utilizes the polymerase chain reaction (PCR).[16] An advantage of PCR over methods based on nick translation or primer extension is that only one strand of DNA is synthesized, so that reassociation of denatured single-strand probe in solution is minimized.

1. The DNA fragment to be labeled can be either a cloned DNA, cDNA, or a PCR fragment. If a PCR fragment from genomic DNA is used, it is advisable to purify through agarose gels and reamplify at least once more to dilute out contaminating genomic DNA (especially repetitive DNA). Cleanest results are from amplifications starting with cloned material, and this also ensures that the single primer added to the polymerase chain reaction will match perfectly.

2. PCR amplification is performed in a 50-μl volume using 200–250 μM of each of three cold nucleotides (dGTP, dATP, and dTTP) with 50 μCi of ^{32}P-labeled dCTP ($\sim$3000 Ci/mmol) in $1 \times$ *Taq* buffer [67 mM

---

[16] K. B. Mullis and F. A. Faloona, this series, Vol. 155, p. 335.

A

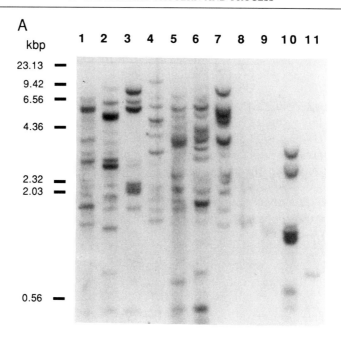

B

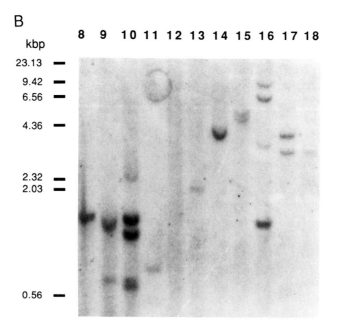

Tris, at pH 8.8, 3 mM MgCl$_2$, 16.6 mM NH$_4$(SO$_4$)$_2$], 0.5 μM PCR primer, and 0.5 units of *Taq* DNA polymerase. The reaction mixture is overlaid with mineral oil, and approximately 25 cycles are performed.

3. Labeled product is purified by extracting the mixture with phenol, followed by precipitation with ethanol to remove excess primers, nucleotides, and unincorporated label.

Hybridization and Washing. Genes are identified by hybridization of a probe to the blot and washing at low stringency followed by autoradiography. Changes in the stringency of hybridization and washing can be used to limit detection to subfamilies of gene families. The conditions described here are of low stringency and have successfully detected lysozyme *c* genes in all mammals tested.[9,10]

1. Blots are prehybridized to block all nonspecific binding (which causes background). To do this, one first wets the nitrocellulose in 3 × SSC, then in 50% formamide, 3× SSC, 0.1% sodium dodecyl sulfate (SDS), 1 mM EDTA (pH 8.0), 10 mM Tris (pH 7.5), 10× Denhardt's solution [1 × Denhardt's is 0.02% Ficoll, 0.02% bovine serum albumin (BSA), and 0.02% polyvinylpyrrolidone 360], 0.05% sodium pyrophosphate, and 100 μg/ml denatured DNA. Denatured DNA is sheared herring or salmon sperm DNA that has been boiled for 10 min before addition to the prehybridization mix. Blots are sealed within plastic bags, placed between two glass plates to give uniform distribution of prehybridization solution over the blot, and incubated in a water bath at 37° for at least 1 hr.

2. Blots are hybridized under the same conditions as for prehybridization with the addition of denatured labeled probe. The labeled probe has a specific activity of at least 0.5 × 10^8 counts/min (cpm)/μg and is present at a concentration of at least 1 × 10^6 cpm/ml. The probe is denatured by treatment with 1/100 volume of 10 N NaOH for 10 min at 68° followed by neutralization with 1/10 volume of untitrated 1.5 M NaH$_2$PO$_4$. The blots are hybridized for 40–48 hr at 37° between glass plates in a water bath.

FIG. 1. Determination of gene number by Southern blotting. Genomic DNAs from a range of mammalian species [(1) cow, (2) sheep, (3) pronghorn, (4) giraffe, (5) axis deer, (6) fallow deer, (7) chevrotain, (8) llama, (9) camel, (10) hippopotamus, (11) peccary, (12) pig, (13) dolphin, (14) zebra, (15) rhinoceros, (16) elephant, (17) mouse, and (18) human] were hybridized to either (A) whole lysozyme cDNA clone from cow stomach or (B) exon 2 from a gene encoding cow stomach lysozyme. The positions and sizes of molecular weight markers appear to the left of the blots. The presence of multiple bands indicates the existence of multiple genes, whereas a single band may indicate a single gene. Smaller probes (e.g., single exons) are better for estimating gene number as they usually detect only a single DNA fragment for each gene.

3. Blots are washed extensively at low stringency to remove as much background as possible. The initial wash is at room temperature in 2 × SSC and 1 × Denhardt's for 60 min. This is followed by two 90-min washes at 50° in 0.1 × SSC, 0.1% SDS. The blots are then rinsed twice at room temperature in 0.1 × SSC, 0.1% SDS, and four times in 0.1 × SSC.

4. Blots are allowed to dry in air and are then exposed to X-ray film with an intensifying screen at −70° for 1 to 7 days. Major bands (including markers) are usually visible after a 24-hr exposure, whereas weak bands, representing very divergent sequences, may take several days to appear. Additional factors influencing the intensity of the band include the number of exons on each fragment and the sizes of hybridizing targets.

Isolation of Lysozyme Complementary DNA Clones

Many methods have been described for the construction of cDNA libraries and isolation of clones from them.[17] In evolutionary studies it is often useful to obtain sequences of genes from several species, and any method which avoids construction of libraries would be useful. The advent of the PCR[16] simplifies the ability to obtain DNA sequences rapidly from multiple samples. PCR methods have also been developed which amplify DNA from RNA using a reverse transcriptase in an initial step.[18] These methods require knowing two primer sequences. The introduction of versatile primers, which amplify DNA segments from a wide range of species,[19] suggested that this could be applied to RNA amplification as well. Because mRNA is polyadenylated, an oligo(dT) primer could be used to amplify any mRNA, reducing the requirements to the knowledge of one primer sequence. A method which has successfully amplified several stomach lysozymes from diverse ruminants[11] is described below.

Methods for the isolation of both total RNA and mRNA[20] and the synthesis of first-strand cDNA[21] are not described here. For this method, cDNA synthesis must be primed with oligo(dT) because this is also used as one of the amplification primers. Examples of products of the amplification of cDNA are shown in Fig. 2.

[17] A. R. Kimmel and S. L. Berger, this series, Vol. 152, p. 307.
[18] E. S. Kawasaki, *in* "PCR Protocols" (M. A. Innis, D. H. Gelfand, J. J. Sninsky, and T. J. White, eds.), p. 21. Academic Press, San Diego, 1990.
[19] T. D. Kocher, W. K. Thomas, A. Meyer, S. V. Edwards, S. Pääbo, F. X. Villablanca, and A. C. Wilson, *Proc. Natl. Acad. Sci. U.S.A.* **86,** 6196 (1989).
[20] S. L. Berger, this series, Vol. 152, p. 215.
[21] M. S. Krug and S. L. Berger, this series, Vol. 152, p. 316.

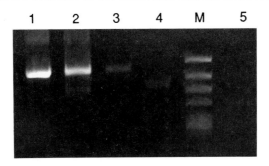

FIG. 2. Electrophoretic demonstration of amplified lysozyme cDNA. The primers used were oligo(dT) and a lysozyme-specific primer. The templates were mRNAs purified from stomach mucosa of (1) cow, (2) sheep, and (3, 4) axis deer; lane, 5 contained no mRNA. Separations were performed in a 1% agarose gel of low melting temperature. The lysozyme cDNA product is approximately 900 bp in all species. In lane 4 the axis deer product was digested with *Pst*I, which cuts at an internal site, giving a visible product of 700 bp. The molecular weight markers (ϕX174 DNA digested with *Hinc*II) were run in lane M.

Methods

1. The products of first-strand cDNA synthesis are treated with RNase A to remove excess mRNA and extracted with phenol to remove enzymes. Excess nucleotides and salts are removed by ultrafiltration in a Centricon 30 tube (Amicon, Danvers, MA).

2. Lysozyme cDNA is amplified in a volume of 25–100 μl under the conditions described above (see Preparation of Probe). The two primers added are (1) the lysozyme-specific primer DI5 (5′ AAGATCTTTGAGA-GATG 3′) and (2) the nonspecific oligo(dT) primer with an *Eco*RI linker (5′ GAATTCTTTTTTTTTTTTTTTTTTTT 3′). Both primers are added to a final concentration of 0.5 μM. Restriction endonuclease sites are introduced into both primers to facilitate cloning. Amplification is for 32–40 cycles, and the annealing step is conducted at 50° owing to the high T content of the general primer.

3. After amplification, the products are cut with appropriate restriction endonucleases, either at sites in the amplification primers or at internal sites. The digestion products are separated electrophoretically in agarose of low melting temperature and purified prior to cloning.

4. Purified amplified cDNA is cloned into M13 vectors,[22] and resulting clones are isolated and sequenced using M13 primers and lysozyme-specific primers.

[22] J. Messing, this series, Vol. 101, p. 20.

Tree Analysis

The acquisition of multiple DNA sequences allows a tree analysis, which can be done using either distance, maximum likelihood, or parsimony methods.[23-26] These three approaches aim to provide an estimate of the true genealogical relationships of the sequences. It is also possible to examine subsets of the sequence data and estimate phylogenetic relationships for each subset. Different types of subsets that can be compared phylogenetically include positions which have been classified as silent or replacement sites, or the subsets may reflect different physical portions of a gene, such as different exons. A prediction of differing evolutionary histories from different subsets of a DNA sequence may suggest that evolutionary forces have acted differently on subdivisions within the sequence. These different predicted evolutionary histories may be due to differences in selection (e.g., convergent evolution[27]) or molecular histories (e.g., recombination).

Recombination between members of a multigene family can result in different portions of a gene having differing evolutionary histories. One of the most common results of recombination within multigene families is concerted evolution,[28-31] which results in the different members of a gene family having more similarity than expected based on time since their origin. Several examples of this phenomenon have been described in mammalian multigene families.[32-35] If concerted evolution is limited to only a portion (or portions) of a gene, then the inferred phylogeny of the regions undergoing concerted evolution should differ from the phylogeny of an adjacent region which is not participating in concerted evolution.

[23] H. Kishino and M. Hasegawa, this series, Vol. 183, p. 550.

[24] N. Saitou, this series, Vol. 183, p. 584.

[25] J. Czelusniak, M. Goodman, N. D. Moncrief, and S. M. Kehoe, this series, Vol. 183, p. 601.

[26] D. L. Swofford, and G. L. Olsen, in "Molecular Systematics" (D. M. Hillis and C. Moritz, eds.), p. 411. Sinauer, Sunderland, Massachusetts, 1990.

[27] K. W. Swanson, D. M. Irwin, and A. C. Wilson, J. Mol. Evol. 33, 418 (1991).

[28] E. A. Zimmer, S. L. Martin, S. M. Beverley, Y. W. Kan, and A. C. Wilson, Proc. Natl. Acad. Sci. U.S.A. 77, 2158 (1980).

[29] G. A. Dover, Nature (London) 299, 111 (1982).

[30] N. Arnheim, in "Evolution of Genes and Proteins" (M. Nei and R. K. Koehn, eds.), p. 38. Sinauer, Sunderland, Massachusetts, 1983.

[31] J. B. Walsh, Genetics 117, 543 (1987).

[32] S. L. Martin, C. F. Voliva, S. C. Hardies, M. H. Edgell, and C. A. Hutchinson III, Mol. Biol. Evol. 2, 127 (1985).

[33] B. S. Chapman, K. A. Vincent, and A. C. Wilson, Genetics 112, 79 (1986).

[34] S. M. McEvoy and N. Maeda, J. Biol. Chem. 263, 15740 (1988).

[35] B. F. Koop, D. Siemieniak, J. L. Slightom, M. Goodman, J. Dunbar, P. C. Wright, and E. L. Simons, J. Biol. Chem. 264, 68 (1989).

Differences in the inferred phylogenetic histories of different portions of a gene have been used as evidence for concerted evolution involving part of a gene.[35] A similar series of events may be occurring in ruminant lysozyme genes.[11]

Different inferred evolutionary relationships among genes do not necessarily indicate (or, even less so, prove) that there were in fact different evolutionary histories. One must use the available data to test the various hypotheses to see if statistically significant differences exist.[36] We provide an example of this in Table I for four cDNA sequences encoding ruminant stomach lysozymes.[11] The coding regions of ruminant stomach lysozyme genes are more similar within each species than between species, suggesting that the genes duplicated independently on each species lineage after divergence of the two species (hypothesis I). In contrast, the 3′ untranslated sequences suggest a phylogenetic relationship in which genes from one species are more closely related to genes in another species than to other genes within a species, suggesting that the duplications occurred prior to

TABLE I

TESTING WHETHER DIFFERENT PARTS OF GENE HAVE DIFFERENT GENEALOGICAL HISTORIES[a]

| | | Number of winning sites[c] | |
Hypothesis	Tree[b]	Coding region	3′ Untranslated region
I	S1 D1 / S2 D2	7	1
II	S1 S2 / D1 D2	0	21

[a] DNA sequences for the analysis are from D. M. Irwin and A. C. Wilson, *J. Biol. Chem.* **265**, 4944 (1990).

[b] Two trees are shown relating sheep lysozyme genes (S1 and S2) to deer lysozyme genes (D1 and D2). S1, Sheep 1b; S2, sheep 2a; D1, deer 1; and D2, deer 2. The two hypotheses being tested are as follows: (I) concerted evolution, that is, genes within a species (e.g., S1 and S2) are more related to each other than to genes in another species (e.g., D1 and D2); and (II) divergent evolution, that is, genes in different species (e.g., S1 and D1) are more related to each other than to other genes within the species (e.g., S2 and D2).

[c] Numbers of informative sites that support alternative hypotheses.

[36] E. M. Prager and A. C. Wilson, *J. Mol. Evol.* **27**, 326 (1988).

the divergence of the cow, sheep, and deer (hypothesis II). In Table I the four-taxon test[36] is used to test the two hypotheses against each other using cDNA sequence data.

With the four-taxon test we calculate the probability of producing (by evolution) the observed sequence differences if a specific hypothesis (tree) is true, and we reject a tree if the probability of this result is less than 5%.[36] For part of the coding region of lysozyme [247 base pairs (bp)], as shown in Table I, there are 7 positions in support of hypothesis I and 0 positions in support of hypothesis II. The probability of observing this result (7 to 0), assuming that hypothesis II is true, using a binomial distribution, is less than 5%; therefore, we reject hypothesis II and conclude that it is unlikely to produce the observed results. With the rejection of hypothesis II, in which the genes are duplicated prior to speciation, we accept hypothesis I, in which the duplications occurred on each species lineage after speciation. When sequences of the untranslated region (512 bp) are used to test the two hypotheses, we find, in contrast, 21 positions in support of hypothesis II and only 1 in support of hypothesis I. Here we reject hypothesis I, as the probability of observing the data at random under a binomial distribution is less than 0.1%. Therefore, two different regions of the lysozyme gene statistically support contradictory evolutionary hypotheses, and we conclude that differing histories have occurred for these regions.

Both genomic blots[9] (see Fig. 1) and 3' untranslated sequences[11] (see Table I) support a model where the lysozyme genes duplicated prior to the radiation of cow, sheep, and deer. The observation that the coding regions of the lysozyme genes are more similar within each species can be explained if concerted evolution has occurred within each species and has been limited to the coding regions. The use of statistical methods has helped in identifying regions of genes that have different evolutionary histories.

Conclusions

This chapter presents methods for the estimation of gene number, for rapid isolation of cDNAs for multiple members of a gene family, and for phylogenetic analysis of sequences. These methods should be broadly applicable to the evolutionary study of multigene families. Often, unexpected sequence similarity is found between a newly characterized gene and previously isolated genes, and this information has been used to predict function or properties of the new sequence based on known properties of the previously characterized genes. Therefore, an understanding of the phylogenetic history of members of a multigene family can provide insight into the molecular properties of products of gene families and suggestions as to how the different gene products may interact with one another.

Acknowledgments

I thank L. Kam-Morgan, J. R. Kornegay, S. J. Mack, I. Matsumura, E. M. Prager, P. Shih, A. Sidow, C.-B. Stewart, and A. C. Wilson for discussions and suggestions. This work was supported by fellowships to the author from the Medical Research Council of Canada and the Izaak Walton Killam Memorial Foundation and by a grant from the National Institutes of Health to Allan C. Wilson, in whose laboratory the work described here was conducted.

[41] Identification of Lens Crystallins: A Model System for Gene Recruitment

By Graeme Wistow

Introduction

This chapter gives an overview of techniques that have been used to screen and identify the major protein components of vertebrate and invertebrate eye lenses, pointing out some of the special insights into evolutionary processes which have arisen from these studies and which may have wider relevance. None of the technologies involved are peculiar to the lens. Instead, it is the particular nature of the tissue itself which makes the lens so amenable to studies of molecular evolution and molecular zoology.

The most outstanding features of the lens are the simplicity of its composition and its apparent evolutionary plasticity. Because it provides part of a direct interface with the outside world, the lens is sensitive to changes in environment and habit, such as the change in refractive index from water to air and the varying requirements for soft, accommodating lenses for active, diurnal vertebrates. The cellular lens of vertebrates and the superficially similar lenses of some invertebrates achieve their optical properties by maintaining high concentrations of a limited repertoire of soluble proteins which, for historical reasons, are known as crystallins.[1-3] Coupled with recent technical advances in automated protein sequencing and in the increasing usefulness of GenBank and other sequence databases,

---

[1] J. J. Harding and M. J. C. Crabbe, *in* "The Eye" (H. Davson, ed.), Vol. 1B, p. 207. Academic Press, New York, 1984.

[2] W. W. de Jong, *in* "Molecular and Cellular Biology of the Eye Lens" (H. Bloemendal, ed.), p. 221. Wiley (Interscience), New York, 1981.

[3] G. Wistow and J. Piatigorsky, *Annu. Rev. Biochem.* **57**, 479 (1988).

these features make the lens a system highly amenable to simple analyses of taxon-specific gene recruitment. The molecular zoology of lenses has revealed an unexpectedly high degree of taxon specificity in lens structural proteins.[3,4] In particular, this has demonstrated multifunctionality in familiar enzymes and has illustrated an alternative evolutionary strategy to that enshrined in the idea that gene duplication is a necessary and prerequisite for the acquisition of new protein functions.[5]

Some crystallins are (as far as we know) ubiquitous in vertebrates. Of these αA-crystallin is the most abundant and has been used in large-scale phylogenetic analyses, using classic techniques of isolation and sequencing.[6] However, it has recently been appreciated that other major crystallins show remarkably taxon-specific patterns of expression (Fig. 1). Furthermore, these taxon-specific crystallins are all identical to enzymes, or else rather recently derived from enzymes (Table I). Some modification in a single functional gene led to the acquisition by the protein product of a dual function as both enzyme and structural lens protein. In this model gene recruitment comes first; duplication may or may not follow.[7]

Preparation of Lens Extracts

The pathology departments of many zoos will save and freeze postmortem tissue samples on request. The eye is usually one of the most obvious features of an animal. The lens can be removed directly, but it is often easier to remove the eye first, cutting nerves and connecting tissues. An incision near the junction of the white sclera and transparent cornea will allow the lens to pop out under gentle pressure. The firm lens should be distinguishable from the gelatinous vitreous humor. Any pigmented material that adheres to the lens equator can be removed by rolling the lens on a piece of dry filter paper.

Lenses can be dissected and homogenized with varying degrees of sophistication; however, for most purposes a simple homogenization with a pestle in 1 or 2 volumes of a buffer like TE (10 mM Tris-HCl, pH 7.4, 1 mM EDTA) is usually satisfactory. Lenses contain protease inhibitors and high concentrations of reduced glutathione, so addition of exogenous inhibitors and reducing agents is normally unnecessary. After homogenization the membrane fraction should be removed by centrifugation in a bench-top microcentrifuge. Lens extracts are stable for long periods at

[4] J. Piatigorsky and G. J. Wistow, Cell (Cambridge, Mass.) 57, 197 (1989).
[5] M. Kimura and T. Ohta, Proc. Natl. Acad. Sci. U.S.A. 71, 2848 (1974).
[6] W. W. de Jong, in "Macromolecular Sequences in Systematic and Evolutionary Biology" (M. Goodman, ed.), p. 75. Plenum, New York, 1982.
[7] J. Piatigorsky and G. Wistow, Science 252, 1078 (1991).

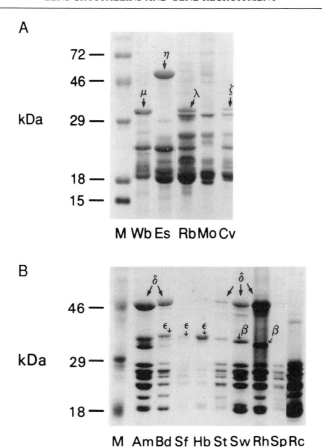

FIG. 1. Taxon specificity in crystallin expression, as revealed by SDS-PAGE analysis of lens extracts from mammals (A) and birds (B). Major enzyme crystallins are indicated. In (B) the variability in βB1 (β) mobility among birds is apparent. [For full species names, see G. Wistow, E. Roquemore, and H. S. Kim, *Curr. Eye Res.* **10,** 313 (1991), and G. Wistow and H. Kim, *J. Mol. Evol.* **32,** 262 (1991), from which this figure is adapted.] (A) M, Marker; Wb, wallaby; Es, elephant shrew; Rb, rabbit; Mo, mouse; Cv, rock cavy. (B) M, Marker; Am, merganser; Bd, black duck; Sf, chimney swift; Hb, hummingbird; St, starling; Sw, barn swallow; Rh, rhea; Sp, house sparrow; Rc, raccoon (a mammal for comparison).

$-20°$, even in frost-free freezers, and survive several episodes of freeze–thawing. Protein concentrations can be estimated by standard methods.[8]

The most convenient way of examining the composition of crystallins is by sodium dodecyl sulfate–polyacrylamide gel electrophoresis (SDS-

---

[8] M. M. Bradford, *Anal. Biochem.* **72,** 248 (1976).

TABLE I
TAXON-SPECIFIC CRYSTALLINS[a]

Type	Source	Enzyme
δ	Birds, reptiles	*Argininosuccinate lyase*
ε	Birds, crocodiles	*Lactate dehydrogenase B*
ζ	Cavies, camels	Alcohol dehydrogenases, enoyl reductases
η	Macroscelids	*Aldehyde dehydrogenase I*
λ	Rabbits, hares	CoA-derivative dehydrogenases
μ	Marsupials	Dehydrogenases?
ρ	Frogs	NADPH-dependent reductases
τ	Many species	*α-Enolase*
S	Cephalopods	Glutathione *S*-transferases
BCP54	Mammalian cornea	*Aldehyde dehydrogenase III*

[a] Major structural proteins related to enzymes by gene recruitment. Italics indicates examples of demonstrated or probable identity between crystallin and enzyme. In other cases the crystallin belongs to an enzyme superfamily but is not identical to a known enzyme. BCP54 is a corneal protein that may have been recruited by the same mechanism as the crystallins. For references, see Ref. 7.

TABLE II
LAEMMLI GELS[a]

Component	10%	11%	12.5%	15%	Upper gel
Lower gel buffer	8.0	8.0	8.0	8.0	—
Upper gel buffer	—	—	—	—	3.75
Acrylamide solution	10.6	12.5	13.4	16.6	2.25
Water	13.4	11.5	10.6	8.0	9.0
10% AP	0.16	0.16	0.16	0.16	45 μl
TEMED	15 μl	15 μl	15 μl	15 μl	15 μl

[a] Recipes for SDS–polyacrylamide gels in the percent acrylamide ranges most useful for resolution of crystallins. All volumes are milliliters except as shown. AP, Ammonium persulfate; TEMED, *N,N,N',N'*- tetramethylethylenediamine. Volumes correspond to those required for a typical mid-sized gel with an upper stacking gel, total dimensions approximately 15 × 15 cm, thickness 1 mm. Thinner and smaller gels are suitable for examination of extracts but are not advised for preparative runs.

PAGE) under reducing conditions,[9] followed by staining with Coomassie blue. This resolves different protein subunits according to size. Crystallins are well resolved by an 11% acrylamide gel following the protocol given below.

[9] U. K. Laemmli, *Nature (London)* **227**, 680 (1970).

Laemmli Gel Electrophoresis

Reagents

Lower gel stock: 1.5 M Tris-HCl, 0.4% SDS, pH 8.8
Upper gel stock: 0.5 M Tris-HCl, 0.4% SDS, pH 6.8
Acrylamide stock: 30.3% acrylamide, 0.78% bisacrylamide
Running buffer: 3 g Tris, 14.4 g glycine, 1 g SDS in 1 liter
Sample buffer (3×): 1.5 ml glycerol, 0.94 ml of 1 M Tris-HCl, pH 6.8, 0.75 ml of 20% SDS, 15 μl of 2-mercaptoethanol, about 1 μg bromphenol blue, and water to a total of 5 ml

Procedure. Follow the proportions in Table II. Mix and pour into a sealed gel former immediately. Leave room for a short upper "stacking" gel. With a fine pipette, layer a small amount of 0.1% SDS solution on top of the unpolymerized gel to give a flat surface and to remove bubbles. As the gel polymerizes the water interface will become indistinct and then will reappear as the gel sets. Pour off the water and add the upper gel. Immediately insert the comb for well formation. Upper stacking gels usually shrink slightly as they polymerize. Mount the gel in a vertical apparatus. Make sure the running buffer contacts both upper and lower surfaces of the gel.

Mix protein samples and sample buffer. A few micrograms of lens extract will usually give a reasonable loading, but this needs to be determined empirically for each sample. In general, for a lens homogenized in about 2 volumes of Tris buffer, 1 or 2 μl of extract is appropriate for a 0.5-cm lane. Samples can be warmed at 65° for 5–10 min before loading. Surprisingly, some crystallins, notably γ-crystallins and cephalopod crystallins, are prone to precipitation and clumping when boiled in SDS-containing buffers, something more usually expected of membrane proteins and which may say something about the peculiar surface properties of these proteins.

A 15 × 15 cm gel should be run at 140 V. Progress of the run is monitored by the dye front or by prestained protein markers, such as BRL (Gaithersburg, MD) high molecular weight protein standards. When the run is finished, the stacking gel can be discarded and the lower gel stained or blotted. For a stained gel, soak in 50% (v/v) methanol–10% (v/v) acetic acid–0.1% (w/v) Coomassie blue for at least 1 hr. Overnight staining often gives improved results. Destain in several changes of 50% methanol, 10% acetic acid. The gel will shrink. If this is a problem, the methanol concentration can be reduced.

Lenses from related species will show clear similarities in banding patterns. An experienced eye can quickly identify the major common subunits and any interesting taxon-specific features (see Fig. 1). An intriguing but untried possibility is that SDS-PAGE patterns themselves could be digitized as phylogenetic bar codes and used as another tool for cladistic analysis.

Various antisera are available for known lens crystallins, and they can be used in Western blot analysis[10] to identify most subunits. Caution should be applied, however. For instance, antisera raised against avian β-crystallins often react with τ-crystallin/α-enolase. This is because column-purified β-crystallin fractions consisting of dimers and higher aggregates have a size range of about 45–200 kDa, a range which includes both monomeric and dimeric τ-crystallin.[11-13] Before τ-crystallin was identified, its presence was not anticipated as a "contaminant" in the β-crystallin fraction. Furthermore, homologous crystallin subunits may exhibit taxon-specific differences in electrophoretic mobility. βB1-Crystallin subunits in birds migrate at a higher apparent size than βB1 subunits in mammals and display variability among birds. This seems to be the result of a specific modification, probably a glycosylation.[14] Protein microsequencing, therefore, gives a less biased view of the true composition and identity of particular protein bands; it is also the most direct way of examining bands of unusual mobility or abundance.

Protein Microsequencing

Microsequencing is now a well-automated procedure. However, it is a major commitment to set up such a facility. Fortunately, microsequencing is available as a service at several sites (including the Microchemistry Facility of the Biological Laboratories at Harvard University, Cambridge, MA). The basic procedure for microsequencing was established by Aebersold et al.[15] This technique has been slightly modified by William Lane and colleagues at the Harvard Microchemistry Facility[16] and by various users.

The effective minimum level for obtaining reliable sequence data is a few picomoles. Densitometry of a stained gel is usually the most practical

[10] H. Towbin, T. Staehelin, and J. Gordon, *Proc. Natl. Acad. Sci. U.S.A.* **76,** 4350 (1979).

[11] G. J. Wistow, T. Lietman, L. A. Williams, S. O. Stapel, W. W. de Jong, J. Horwitz, and J. Piatigorsky, *J. Cell Biol.* **107,** 2729 (1988).

[12] L. A. Williams, L. Ding, J. Horwitz, and J. Piatigorsky, *Exp. Eye Res.* **40,** 741 (1985).

[13] S. O. Stapel and W. W. de Jong, *FEBS Lett.* **162,** 305 (1983).

[14] G. Wistow, E. Roquemore, and H. S. Kim, *Curr. Eye Res.* **10,** 313 (1991).

[15] R. H. Aebersold, J. Leavitt, R. A. Saavedra, L. E. Hood, and S. B. H. Kent, *Proc. Natl. Acad. Sci. U.S.A.* **84,** 6970 (1987).

[16] P. L. Rothenberg, W. S. Lane, A. Karasik, J. Backer, M. White, and C. R. Kahn, *J. Biol. Chem.* **266,** 8302 (1991).

way to estimate fractional abundances. Preparation of samples for analysis is straightforward. Preparative SDS–polyacrylamide gels, unfixed and unstained, are electroblotted onto membranes.[10] We have found it useful to include prestained protein molecular weight markers in the gel as visible indicators of the run and of the transfer. "Dry" blotting using the American Bionetics (Emeryville, CA) Polyblot apparatus is rapid and convenient for highly abundant crystallin samples. However, "wet" electroblotting (12 g Tris, 57.6 g glycine, 1 g SDS, 800 ml methanol per liter) in a tank apparatus for 2–20 hr is usually recommended for best yields. Directions for blotting depend on the apparatus used.

For N-terminal analysis, the protein should be blotted onto a polyvinyl difluoride (PVDF) membrane, which provides a suitable substrate for direct sequencing. However, most crystallins, and many other proteins, are N-terminally blocked, so it is necessary to digest the protein sample and to separate peptides before sequencing. For this purpose the gel should be blotted onto nitrocellulose (see Ref. 10 or the manufacturer's instructions). Blotted proteins are then stained with Ponceau S (0.2% in 1% acetic acid) for about 2 min. Excess dye is removed by brief washing in 1% acetic acid. Ponceau S is a soluble dye that does not interfere with subsequent steps, but it does not give staining equivalent to Coomassie blue; minor and low molecular weight bands may be harder to see than expected. In general the gel should be loaded as heavily as possible to give the maximum concentration of protein in a band, consistent with good resolution. A strong band on a stained blot will have sufficient material for sequencing.

Bands of interest are excised with a razor blade. At this stage the samples are quite stable. They should be kept moist, but not submerged, and should be stored in Eppendorf tubes or sealed in plastic bags. We have successfully shipped bagged samples at ambient temperature in overnight mail, but shipping on dry ice is usually recommended.

Excised protein bands can be digested *in situ* in an Eppendorf tube. In the Aebersold procedure[15] pieces of nitrocellulose are incubated at 37° for 30 min in 0.5% PVP-40 polyvinylpyrolidone, average molecular weight 40,000) dissolved in 100 mM acetic acid, to block absorption of enzyme onto the paper. The PVP-40, which has strong UV absorption, is removed by multiple washings with water. The recommended amount of enzyme used for digestion is 5% of the estimated weight of protein substrate. Trypsin digestion is carried out in 100 mM Tris-HCl, pH 8.2, 5% acetonitrile (by volume) overnight at 37°. Released tryptic peptides are then separated by high-performance liquid chromatography (HPLC) for sequencing. For example, the Harvard Microchemistry Facility uses narrowbore reversed-phase HPLC on a Hewlett-Packard 1090 equipped with a 1040 diode array detector, using a Vydac 2.1 × 150 mm C_{18} column and a gradient modified from Stone *et al.*[17] Buffer A is 0.06% trifluoroacetic acid in water and buffer B is 0.055% trifluoroacetic acid in acetonitrile. A gradient of 5%B at 0 min, 33%B at 63 min, 60%B at 95 min, and 80%B at 105 min with a flow rate of 150 μl/min is used. Fractions are collected by monitoring at 210 nm and are stored at $-20°$. Samples are directly applied to a Polybrene precycled glass filter and subjected to Edman degradation in an ABI (Foster City, CA) Model 477A protein sequencer. The resultant phenylthiohydantoin amino acids are identified manually using an on-line ABI Model 120A HPLC and Shimadzu (Columbia, MD) CR4A integrator.

This procedure is ideal for identification of well-resolved, reasonably abundant proteins and for the design of oligonucleotide probes for subsequent cDNA or genomic analysis. It is not a reliable way of obtaining complete protein sequences, however, since many large and hydrophobic peptides are not efficiently retrieved. For full sequence analysis it is necessary to use the fragmentary protein data as a probe for cDNA cloning.

Analysis of Complementary DNA

Lens RNA turns out to be quite stable, probably because of the presence of endogenous RNase inhibitors. Crystallin mRNAs should be abundant in lens. In one case a duck δ-crystallin clone was identified by sequencing clones picked at random from a lens cDNA library.[18] Lens cDNA libraries can be constructed by standard methods (see Ref. 19) using any of the popular kits. cDNA can be cloned into various vectors, but λ phage insertion vectors, capable of *in vitro* excision of plasmid, have many advantages.[20] We have obtained good results using the reagents and protocols of the Stratagene (La Jolla, CA) λ-zap cloning system.

Complementary DNA clones are usually selected by replica blotting to nitrocellulose or nylon and hybridization with ^{32}P-labeled probes (see Ref. 19). The probes may be oligonucleotides designed from known peptide sequence. As described by Lathe,[21] the best way to design such probes is to make a single "best guess" for each codon, based on codon usage tables, and to exceed 36 nucleotides length. Alternatively a cDNA fragment can be obtained directly from lens total RNA by the polymerase chain reaction

[17] K. L. Stone, M. B. LoPresti, N. D. Williams, J. M. Crawford, R. DeAngelis, and K. R. Williams, *in* "Techniques in Protein Chemistry" (T. Hugli, ed.), p. 377. Academic Press, New York, 1989.
[18] G. Wistow and J. Piatigorsky, *Gene* **96**, 263 (1990).
[19] S. L. Berger and A. R. Kimmel (eds.), this series, Vol. 152.
[20] J. M. Short, J. M. Fernandez, J. A. Sorge, and W. D. Huse, *Nucleic Acids Res.* **16**, 7583 (1988).
[21] R. Lathe, *J. Mol. Biol.* **183**, 1 (1985).

(PCR)[22] using pairs of shorter oligonucleotides designed from peptide sequences. In this case it may be beneficial to introduce some redundancy in oligonucleotide sequence to take into account codon redundancy. For PCR primers the most important nucleotides are those at the immediate 3′ end; these should correspond to invariant first and second positions of codons. PCR products can be sequenced directly[23] or, if restriction sites have been incorporated in the primers, can be subcloned into a sequencing vector. PCR-synthesized DNA can be radioactively labeled by nick translation or random priming[19] and used as a probe to obtain full-sized cDNA clones. Otherwise, if the peptide sequence shows a close relationship with a known enzyme, a previously cloned fragment can be used; clones for duck lens τ-crystallin were selected using a human α-enolase cDNA as a probe.[11] Clones are sequenced by standard methods; chain termination sequencing in double-stranded DNA[19,24] works well for clones in plasmids.

Identifying Sequence Relationships

Once peptide and/or cDNA sequences have been obtained, they are compared with the major sequence databases. Although it may be informative to examine several protein and nucleic acid databases, in practice any important relationships or identities will be found by comparing determined protein sequence with translations of a large nucleic acid database such as GenBank. There has been considerable success at detecting quite distant superfamily relationships using the program SEQFT, part of Kanehisa's IDEAS package,[25] running on a CRAY supercomputer at the Advanced Scientific Computing Laboratory (Frederick, MD). The same functions can be performed by more widely available software, such as TFASTA found in the GCG package.[26]

Several of the taxon-specific crystallins (ε, $\delta 2$, τ, probably η) are identical homologs of known enzymes, so that sequence identification is unambiguous.[27-29] Even for ρ-crystallin there is up to 50% identity with

[22] M. A. Innis, D. H. Gelfand, J. J. Sninsky, and T. J. White (eds.), "PCR Protocols: A Guide to Methods and Applications." Academic Press, New York, 1990.

[23] M. W. Allard, D. L. Ellsworth, and R. L. Honeycutt, *BioTechniques* **10**, 24 (1991).

[24] "Sequenase Version 2.0: Step-by-step Protocols for DNA Sequencing with Sequenase Version 2.0," 5th Ed. United States Biochemical, Cleveland, Ohio, 1990.

[25] M. Kanehisa, "IDEAS 88. User's Manual." Frederick Cancer Research Facility, Frederick, Maryland, 1988.

[26] W. R. Pearson, this series, Vol. 183, p. 63.

[27] G. J. Wistow, J. W. Mulders, and W. W. de Jong, *Nature (London)* **326**, 622 (1987).

[28] G. Wistow and J. Piatigorsky, *Science* **236**, 1554 (1987).

[29] G. Wistow and H. Kim, *J. Mol. Evol.* **32**, 262 (1991).

other members of an NADPH-dependent oxidoreductase superfamily.[30] Others, however, have much weaker similarities. In these cases it is important to have support for proposed relationships that goes beyond a count of scattered similarities. The λ- and ζ-crystallins show only about 20% identity with known enzymes, namely, hydroxyacyl-CoA dehydrogenases and relatives[31] and alcohol and sorbitol dehydrogenase and enoyl reductases.[32] However, all important structural features, such as residues of the pyridine nucleotide-binding domain, are conserved. Similarly, distant relatives of the β- and γ-crystallins were found among microorganism dormancy proteins by considering important features of the internally repeated structural motifs of these proteins.[33] Such relationships can shed light on the functions and origins of proteins. However, weak similarities with no supporting structural or functional insights can be deceptive, as, for example, in the suggestion that β- and γ-crystallins might be related to c-*myc*.[34]

The most important part of the search for sequence relationships is the completeness of the database. Early in the discovery that taxon-specific crystallins were related to enzymes, no relatives of δ-crystallin could be found in GenBank. When an update of GenBank was obtained, a clear relationship between yeast argininosuccinate lyase (ASL) and chicken lens δ-crystallins was seen. In fact, the yeast ASL sequence was published on the same day as a δ-crystallin cDNA but took longer to arrive in the database. In the same week other researchers had already spotted the similarity of δ-crystallin to ASL when entering into a personal database the sequence of human ASL, which was just being published.[28] There is no substitute for complete and up-to-date databases.

Gene Recruitment

Sequence relationships between proteins of different function are common. The distinctive feature of the taxon-specific crystallins is that several are actually identical to enzymes, showing that a new function has been acquired without gene duplication and specialization. Identity is strongly indicated by sequence analyses and by enzyme activity analysis. There are arguments to suggest that crystallins do not require catalytic activity for

[30] D. A. Carper, G. Wistow, C. Nishimura, C. Graham, K. Watanabe, Y. Fujii, H. Hayashi, and O. Hayaishi, *Exp. Eye Res.* **49**, 377 (1989).
[31] J. W. Mulders, W. Hendriks, W. M. Blankesteijn, H. Bloemendal, and W. W. de Jong, *J. Biol. Chem.* **263**, 15462 (1988).
[32] A. Rodokanaki, R. K. Holmes, and T. Borras, *Gene* **78**, 215 (1989).
[33] G. Wistow, *J. Mol. Evol.* **30**, 140 (1990).
[34] M. J. Crabbe, *FEBS Lett.* **181**, 157 (1985).

function; therefore, such activity is preserved by a second, enzymatic role.[28] However the most fundamental proof of identity is to show that the crystallin and enzyme are encoded by the same gene. This has been achieved by Southern blotting[19] using cDNA probes at high stringency.[11,35] One approach is to use cDNA probes for the crystallin and, if available, the homologous enzyme from another species. Thus, the cDNA for both duck lens τ-crystallin and human α-enolase hybridize at high stringency to the same single band in digests of duck genomic DNA.[11] Finally, genomic libraries can be screened and the gene cloned and sequenced.[36]

For molecular biologists the next task is to decipher the mechanism by which the gene is expressed. In the case of the recently recruited enzyme crystallins, this amounts to dissecting the machinery of expression and comparing it with that of homologous genes in closely related species in which the enzyme has not added the dual role of crystallin. Although this analysis is still at an early stage, it is intriguing that in some cases we have already seen that, whereas genes for particular enzymes have typical GC-rich housekeeping promoters, homologous enzyme crystallin genes in other species have TATA and sometimes CCAAT boxes.[36-41] Promoter modification or the acquisition of new lens-preferred promoters may play an important role in gene recruitment.

Wider Implications

The crystallins serve to illustrate three general points that may be worth bearing in mind in other areas of molecular evolution. First, they show that there may be striking differences in the expression of homologous genes even among closely related species. Second, they emphasize the multifunctionality that may be characteristic of many proteins. Third, the enzyme crystallins show that a new function may arise without gene duplication. Some enzymes and stress proteins acquired a structural role in the lens by modifications in gene expression. Normal expression is retained in other

[35] W. Hendriks, J. W. M. Mulders, M. A. Bibby, C. Slingsby, H. Bloemendal, and W. W. de Jong, *Proc. Natl. Acad. Sci. U.S.A.* **85,** 7114 (1988).

[36] R. Kim, T. Lietman, J. Piatigorsky, and G. Wistow, *Gene* **103,** 192 (1991).

[37] J. M. Nickerson, E. F. Wawrousek, T. Borras, J. W. Hawkins, B. L. Norman, D. R. Filpula, J. W. Nagle, A. H. Ally, and J. Piatigorsky, *J. Biol. Chem.* **261,** 552 (1986).

[38] J. M. Nickerson, E. F. Wawrousek, J. W. Hawkins, A. S. Wakil, G. J. Wistow, G. Thomas, B. L. Norman, and J. Piatigorsky, *J. Biol. Chem.* **260,** 9100 (1985).

[39] T. Matsubasa, M. Takiguchi, Y. Amaya, I. Matsuda, and M. Mori, *Proc. Natl. Acad. Sci. U.S.A.* **86,** 592 (1989).

[40] R. D. Abramson, P. Barbosa, K. Kalumuk, and W. E. O'Brien, *Genomics* **10,** 126 (1991).

[41] A. Giallongo, D. Oliva, L. Cali, G. Barba, G. Barbieri, and S. Feo, *Eur. J. Biochem.* **190,** 567 (1990).

tissues while very high concentrations are achieved in the lens. This phenomenon may extend beyond the lens. Work on the soluble proteins of the cornea suggests that similar taxon-specific recruitment of enzymes as abundant proteins has occurred (see Ref. 7 for references).

A protein with a taxon-specific pattern of multifunctionality is likely to exhibit different rates of sequence divergence in different lineages. A second role may accelerate changes in some parts of the sequence while suppressing drift in other regions. This could certainly affect the molecular clock for certain proteins. Indeed, it has already been suggested that the recruitment of lactate dehydrogenase B as ε-crystallin in some birds and crocodiles may have been the cause of problems in a phylogenetic analysis in birds based on this enzyme.[42]

After recruitment, the single gene product is exposed to two sets of selective forces. It seems that there are circumstances in which the dual roles of enzyme and crystallin create an adaptive conflict.[7,43] Sometimes this may be resolved by loss of the newly acquired lens role; otherwise, the solution may be gene duplication, allowing varying degrees of specialization and separation of function. In any case, initial gene duplication is not necessary before a new function is acquired.

Because of its peculiar characteristics, the lens may be showing us recent, accessible examples of a process that may have wider significance. In the case of enzyme crystallins new roles have arisen because of new tissue-specific patterns of expression. However, it is also possible that proteins could acquire additional functions without a modification in gene expression. Several enzymes have mysteriously cropped up in other guises. As just two of many examples, α-enolase, a glycolytic enzyme that also serves some species as τ-crystallin, has been isolated as a plasminogen receptor,[44] and lactate dehydrogenases (related to ε-crystallin) have been implicated as single-stranded DNA binding proteins (see Refs. 7 and 28 for references and other examples). There is no reason why a protein that evolved for a particular function (perhaps as an enzyme) could not become useful in an additional role (perhaps as a receptor or binding protein) as a result of sequence or environmental changes. As in the lens, subsequent adaptive conflict might then provide the selective pressure to drive gene duplication and specialization. Evolution is necessarily an expedient process.

[42] A. C. Wilson, H. Ochman, and E. M. Prager, *Trends Biochem. Sci.* **3**, 241 (1987).

[43] G. Wistow, A. Anderson, and J. Piatigorsky, *Proc. Natl. Acad. Sci. U.S.A.* **87**, 6277 (1990).

[44] L. A. Miles, C. M. Dahlberg, J. Plescia, J. Felez, K. Kato, and E. F. Plow, *Biochemistry* **30**, 1682 (1991).

Other Aspects of Crystallins in Molecular Evolution

Beyond gene recruitment, there are other fascinating stories in the evolutionary history of crystallins. One of the crystallins common to all vertebrate lenses, αA-crystallin, is an important tool in molecular phylogenetic analysis. de Jong's group has made a highly informative study of this protein by classic techniques of peptide analysis. Data for αA-crystallin have been particularly important in indicating the existence of a paenungulate clade, including elephant, hyrax, aardvark, manatee, and, most unexpectedly, elephant shrew.[29,45]

The αA-crystallin gene has its own taxon-specific feature. In some placental and marsupial mammals the αA-crystallin gene contains an extra, alternatively spliced, exon in the first intron.[3,46,47] The sequence of the human αA-crystallin gene revealed the presence of a silent, pseudoexon copy of the sequence.[48] It thus seems that the "insert exon" arose in a distant common ancestor of mammals but has lost expression in many descendant species. A phylogenetic examination of this part of the αA-crystallin gene has begun.[47]

The γ-crystallin gene locus in mammals, a linked cluster of genes subject to repeated gene conversion, is also of great interest. Although six genes are actively expressed in rats, all but two of the genes in humans have lost or reduced expression.[49] γ-Crystallins are associated with hard, high refractive index lenses not suitable for diurnal terrestrial species. Instead of recruiting an enzyme to dilute out the effects of γ-crystallins in the human lens, their expression has been reduced. This is another example of the dynamic nature of molecular evolution in the vertebrate lens.

Acknowledgments

I thank the many colleagues who have helped with advice and protocols, in particular, John Talian of the Laboratory of Molecular and Developmental Biology, National Eye Institute, and William Lane of Harvard Microchemistry and Cambridge Prochem. I also thank Andrea Anderson, Nick Kim, and Michael Shaughnessy for invaluable contributions to our work.

[45] W. W. de Jong, A. Zweers, and M. Goodman, *Nature (London)* **292**, 538 (1981).

[46] W. W. de Jong, J. A. M. Leunissen, W. Hendriks, and H. Bloemendal, in "Molecular Biology of the Eye: Genes, Vision and Ocular Diseases" (J. Piatigorsky, T. Shinohara, and P. S. Zelenka, eds.), p. 149. Alan R. Liss, New York, 1988.

[47] C. J. Jaworski, H. S. Kim, J. Piatigorsky, and G. Wistow, *Invest. Ophthalmol. Visual Sci.* **31**, 507 (1990).

[48] C. J. Jaworski and J. Piatigorsky, *Nature* **337**, 752 (1989).

[49] R. H. Brakenhoff, H. J. Aarts, F. H. Reek, N. H. Lubsen, and J. G. Schoenmakers, *J. Mol. Biol.* **216**, 519 (1990).

[42] Reconstruction and Testing of Ancestral Proteins

By PHOEBE SHIH, BRUCE A. MALCOLM, STEVE ROSENBERG, JACK F. KIRSCH, and ALLAN C. WILSON[†]

Introduction

With tree analysis it is possible to infer sequences of ancestral proteins and probable evolutionary pathways to their modern descendants.[1] The advent of site-directed mutagenesis makes possible the recreation of evolutionary intermediates based on these predictions. One may then compare the properties of reconstructed intermediates with one another and with proteins from contemporary creatures. These comparisons provide a way of testing theories about the mechanism of molecular evolution. For example, this approach has provided a new criterion for distinguishing between neutral and nonneutral events.[2] This chapter describes the use of site-directed mutagenesis to recreate ancestral lysozymes and presents methods of evaluating their properties.

Lysozyme as Model

The bird lysozymes *c*, of which chicken egg white lysozyme (CL) is the most extensively studied example, provide an ideal system to recreate evolutionary intermediates and to study structure–function relationships of reconstructed ancestral proteins. Three major considerations qualify the avian lysozyme system for reconstruction of evolutionary pathways: (1) the biochemistry of the enzymes has been extensively studied and well characterized,[3,4] (2) there are many natural variants available from other birds, and homologous comparisons can be ensured since lysozymes for all game birds are encoded by a single gene,[5] and (3) the three-dimensional structure of CL has been resolved at the atomic level, which allows for structural interpretation of the mutational impact.[2,4] Proteins representing the ancestral, evolutionarily intermediate, and derived states of chicken and related bird lysozymes are made and characterized as described below.

[†] Deceased.

[1] J. Stackhouse, S. R. Presnell, G. M. McGeehan, K. P. Nambiar, and S. A. Benner, *FEBS Lett.* **262**, 104 (1990).

[2] B. A. Malcolm, K. P. Wilson, B. W. Matthews, J. F. Kirsch, and A. C. Wilson, *Nature (London)* **345**, 86 (1990).

[3] P. Jollès and J. Jollès, *Mol. Cell. Biochem.* **63**, 165 (1984).

[4] T. Imoto, L. N. Johnson, A. C. T. North, D. C. Phillips, and J. A. Rupley, "The Enzymes" (P. D. Boyer, ed.), 3rd Ed., Vol. 7, p. 665. Academic Press, New York, 1972.

Molecular Biology and Protein Purification

The CL variants are constructed by site-directed mutagenesis of the cDNA for chicken egg white lysozyme and heterologously expressed in yeast behind a regulated hybrid promoter.[6]

Site-Directed Mutagenesis

A 2.4-kilobase (kb) *Bam*HI cassette containing the alcohol dehydrogenase 2/glyceraldehyde-3-phosphate dehydrogenase (ADH-2/GAPDH) promoter,[7-9] cDNA of CL, and the GAP terminator[8] had been previously cloned into bacteriophage M13mp18 for site-directed mutagenesis.[6] The use of DNA templates containing uracil for mutagenesis directed by synthetic oligonucleotides provides a simple and efficient way of generating desired mutants. The preparation of recombinant CL bacteriophage M13 containing uracil follows the method of Kunkel *et al.*[10] Figure 1 shows the main steps in the procedure. *Escherichia coli* strain CJ236 (*dut⁻ung⁻F'*)[11] is used as the host bacterium.

Mutagenesis with Three Primers

Sequential annealing of three mutagenic primers to the uracil-labeled template is carried out, because CL constructs with single, double, and triple mutations are desired in this case. This procedure allows the desired single mutants as well as combinations of double and triple mutants to be screened simultaneously. Oligonucleotide primers (~22 bases in length) containing desired mutations are synthesized on a DNA synthesizer (Applied Biosystems, Foster City, CA) and purified via a Poly-Pak cartridge (Glen Research, Sterling, VA).[12] The primers are phosphorylated using T4 polynucleotide kinase.[13]

[5] K. M. Helm-Bychowski and A. C. Wilson, *Proc. Natl. Acad. Sci. U.S.A.* **83,** 668 (1986).

[6] B. A. Malcolm, S. Rosenberg, M. J. Corey, J. S. Allen, A. de Baetselier, and J. F. Kirsch, *Proc. Natl. Acad. Sci. U.S.A.* **86,** 133 (1989).

[7] J. Travis, M. Owen, P. George, R. Carrell, S. Rosenberg, R. A. Hallewell, and P. J. Barr, *J. Biol. Chem.* **260,** 4384 (1985).

[8] V. L. Price, W. E. Taylor, W. Clevenger, M. Worthington, and E. T. Young, this series, Vol. 185, p. 308.

[9] S. Rosenberg, D. Coit, and P. Tekamp-Olson, this series, Vol. 185, p. 341.

[10] T. A. Kunkel, J. D. Roberts, and R. A. Zakour, this series, Vol. 154, p. 367.

[11] C. M. Joyce and N. D. F. Grindley, *J. Bacteriol.* **158,** 636 (1984).

[12] K.-M. Lo, S. S. Jones, N. R. Hackett, and H. G. Khorana, *Proc. Natl. Acad. Sci. U.S.A.* **81,** 2285 (1984).

[13] C. C. Richardson, *in* "Procedures in Nucleic Acid Research" (G. L. Cantoni and D. R. Davies, eds.), Vol. 2, p. 815. Harper & Row, New York, 1971.

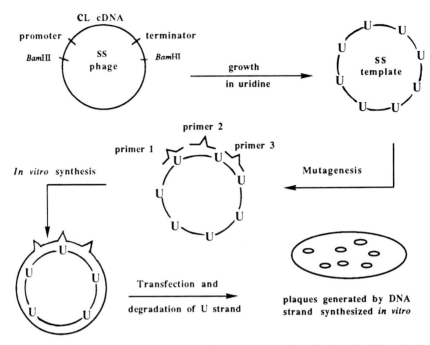

FIG. 1. Oligonucleotide-mediated, site-directed mutagenesis using the Kunkel method. See the text (sections on site-directed mutagenesis and mutagenesis with three primers) for further details. ss, Single-stranded.

1. Two microliters (44 pmol) of each of the previously phosphorylated primers are diluted with 5 μl of deionized water.

2. A mutagenesis mix containing 1 μl (1 μg/μl) of uracil-labeled template, 2.5 μl of deionized water, 0.5 μl of 10× annealing buffer [200 mM Tris, pH 7.5, 100 mM MgCl$_2$, 500 mM NaCl, and 10 mM dithiothreitol (DTT)], and 1 μl of one of the three diluted primers is placed into a 500-μl Eppendorf tube.

3. The tube is heated at 60° for 5 min and allowed to cool slowly to 37°.

4. After the tube is microcentrifuged at 12,000 g for 1 sec, 1 μl of the second diluted primer is added and Step 3 is repeated. This procedure is then repeated for the last of the three diluted primers. The reaction mixture is allowed to cool from 37° to room temperature for 10 min.

5. An enzyme cocktail is prepared by combining 3 μl of a 2 mM deoxynucleotide triphosphate mix (containing dATP, dGTP, dCTP, and dTTP each at a concentration of 0.5 mM), 1 μl of 10 mM ATP, 0.5 μl of T7 DNA polymerase [3 units/μl, United States Biochemical (USB), Cleve-

land, OH], and 1 μl of a solution containing 600 mM potassium phosphate, pH 7.5, and 100 mM MgCl$_2$.

6. One microliter of thoroughly mixed enzyme cocktail from above and 1 μl of T4 DNA ligase (10 units/μl) are added to the annealing mix, which is kept on ice for 5 min, followed by 15 min at room temperature, and finally at 37° for 90 min.

7. The mutagenesis mix is diluted by adding 90 μl of TE (10 mM Tris, 0.1 mM EDTA) at pH 7.5. It may be frozen at this point for later use.

The protocols for preparation of competent cells and for transfection are described elsewhere.[14] *Escherichia coli* strain DH5αF'IQ (Bethesda Research Laboratories, Gaithersburg, MD) is recommended as a good male strain that is deficient in recombination. Ten microliters of diluted mutagenesis mix from Step 7 is used to transfect 200 μl of competent cells, and 200 μl of a fresh overnight culture of the same strain are added to ensure an adequate background lawn. The transfection should yield between 20 and 200 plaques, several of which are picked for single-stranded phage DNA synthesis and sequencing by dideoxy methods. Preparation of single-stranded DNA[15] and DNA sequencing[16] using Sequenase (Version 2.0, USB) are not described here. Once the desired clones are identified, the sequence of the entire gene is determined to ensure that no other mutations occurred during the procedures.

Subcloning

After identification and verification, the replicative forms (RF) of the desired mutants are prepared by alkaline lysis procedures.[17] The 2.4-kb *Bam*HI cassette containing the promoter, mutagenized cDNA, and the terminator is isolated by restriction digestion and gel purification (1% agarose).[18,19] The cassette is inserted into a plasmid (pAB24), termed a shuttle vector,[20] that has been cut with *Bam*HI and dephosphorylated (Fig. 2). This vector contains the *LEU2-D* and *URA3* genes for selection and the entire 2-μm circle for replication as well as pBR322 with its ampicillin gene for selection and replication in *E. coli*.[21] Plasmid pAB24 is propagated in

[14] D. A. Morrison, this series, Vol. 68, p. 326.
[15] J. Vieira and J. Messing, this series, Vol. 153, p. 3.
[16] A. T. Bankier, K. M. Weston, and B. G. Barrell, this series, Vol. 155, p. 51.
[17] A. C. Birnboim, this series, Vol. 100, p. 243.
[18] E. Southern, this series, Vol. 68, p. 152.
[19] R. C.-A. Yang, J. Lis, and R. Wu, this series, Vol. 68, p. 176.
[20] P. J. Barr, H. L. Gibson, V. Enea, D. E. Arnot, M. R. Hollingdale, and V. Nussenzweig, *J. Exp. Med.* **165**, 1160 (1987).
[21] J. D. Beggs, *Nature (London)* **275**, 104 (1978).

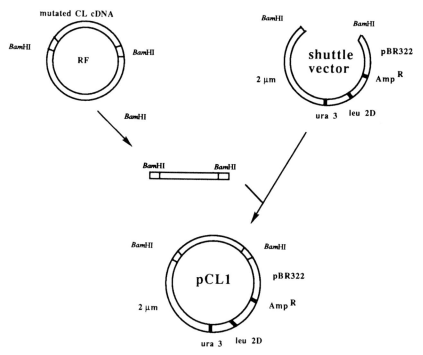

FIG. 2. Subcloning and expression system for chicken egg white lysozyme. See the text (section on subcloning) for further details. RF, Replicative form.

E. coli DH5αF'IQ and prepared using the minipreparation procedure.[22] Molar ratios of vector to cassette of 1:1 or 1:3 are used in the ligation reaction (~500 ng combined in 20 μl for each reaction).

Preparation of *E. coli* DH5αF'IQ competent for transformation follows the protocol of Morrison.[14] To 200-μl aliquots of competent cells in small culture tubes (~8 ml) on ice, 5- and 45-μl aliquots (low and high) of the products of the ligation reaction are added for each mutant. After transformation, the cells are recovered with shaking (100 rpm) for 2 hr in 2 ml of rich medium [2YT: 1.6% tryptone (Difco Labs, Detroit, MI), 1% yeast extract, 0.5% NaCl; or LB: 1% tryptone, 0.5% yeast extract, 1% NaCl (all %, w/v)] to allow bacteria to repair cell walls and express antibiotic resistance. The cells are plated on a rich medium containing 100 μg/ml of ampicillin for selection. Four transformants are picked, and a plasmid minipreparation[22] is carried out for each. Clones are identified by doing a DNA miniscreen with *Bam*HI and *Sal*I restriction endonucleases. A larger scale plasmid preparation is carried out for positive clones.

[22] H. C. Birnboim and J. Doly, *Nucleic Acids Res.* **7**, 1513 (1979).

Expression of Mutant Lysozyme in Yeast

Approximately 1 μg of wild-type or mutant pCLI plasmid DNA is used for transformation of yeast cells. Spheroplasts of *Saccharomyces cerevisiae* strain GRF 180[23,24] are prepared following the protocol of Burgers and Percival[25] with some modification. High copy number transformants are directly selected on minimal medium plates lacking leucine and containing 8% glucose (to repress lysozyme expression) and 1 M sorbitol as an osmotic stabilizer.

Preparation of Competent Yeast Cells

Unless otherwise specified, the following procedure is performed at room temperature. All % of media and solution is expressed as weight by volume.

1. A single colony of GRF 180 is picked from a fresh plate and inoculated into 5 ml of YP [1% yeast extract/2% Bacto-peptone (Difco Labs)] with 2% glucose. The culture is grown at 30° with shaking for 12 to 24 hr. The absorbance at 650 nm (A_{650}) of a 24-hr culture should be around 20.

2. Five microliters of a culture with an A_{650} of 10 is used to inoculate 50 ml of YP containing 2% glucose, and the culture is grown at 30° with shaking until the A_{650} of the culture reaches approximately 0.6 (~ 12 hr for GRF1 80).

3. The cells are harvested by centrifugation in a sterile, 50-ml poly-propylene Corning tube at 500 g for 5 min. The supernatant is removed by carefully pouring it off.

4. The cells are washed successively with 20 ml of sterile water and 20 ml of 1 M sorbitol by resuspension, followed by centrifugation at 500 g for 5 min, and the supernatant is removed with a sterile pipette.

5. The cells are resuspended in 5 ml of SCEM (1 M sorbitol/0.1 M sodium citrate, pH 5.8/10 mM EDTA/30 mM 2-mercaptoethanol).

6. A turbidity measurement of the A_{650} at zero time is made by diluting 100 μl of the cell solution into 1 ml of 0.1% sodium dodecyl sulfate (SDS).

7. One hundred microliters of freshly prepared Zymolyase 20,000 (Seikagaku Kogyo, Tokyo, Japan) in 1 M sorbitol/10 mM Tris-HCl (ST) is added to the cells, and the tube is inverted once to mix.

[23] A. J. Brake, J. P. Merryweather, D. G. Coit, U. A. Heberlein, F. R. Masiarz, G. T. Mullenbach, M. S. Urdea, P. Valenzuela, and P. J. Barr, *Proc. Natl. Acad. Sci. U.S.A.* **81**, 4642 (1984).

[24] E. Erhart and C. P. Hollenberg, *J. Bacteriol.* **156**, 625 (1983).

[25] P. M. J. Burgers and K. J. Percival, *Anal. Biochem.* **163**, 391 (1987).

8. The reaction is incubated at 30° with very gentle shaking (60 rpm). The A_{650} is monitored every 5 min by diluting 100 μl into 1 ml of 0.1% SDS until the A_{650} decreases to 10 to 20% of the turbidity at zero time (about 15–20 min).

9. The tube is centrifuged immediately at 200 g for 3 min. The supernatant is carefully removed with a pipette.

10. The spheroplasts are resuspended very gently in 20 ml of ST and centrifuged as in Step 9.

11. The spheroplasts are resuspended very gently in 5 ml of cold 1 M sorbitol/10 mM Tris-HCl, pH 7.5/10 mM CaCl$_2$ (STC) and centrifuged as in Step 9.

12. The spheroplasts are resuspended in 2 ml of STC by very gentle stirring with a 1-ml sterile pipette tip. The competent cells should be used immediately at best or may be stored for up to 5 hr. The above preparation should be sufficient for 20 transformations.

Transformation of Competent Yeast Cells

1. For each mutant, a sample containing the pCLI plasmid of interest and carrier DNA (e.g., Salmon testes DNA, Sigma) in a total of 5 μg of DNA in less than 10 μl volume is added to 100 μl of competent cells placed in the bottom of a sterile 12-ml Falcon 2051 tube (Becton Dickinson, Lincoln Park, NJ). The contents are swirled gently to mix, and the tube is incubated for 10 min.

2. One milliliter of filter sterilized PEG (10 mM Tris-HCl, pH 7.5/ 10 mM CaCl$_2$/20% polyethylene glycol with average molecular weight of 8000) is added to the tube, and the contents are swirled gently and incubated for 10 min. The tube is centrifuged at 200 g for 3 min, and the supernatant is removed with a pipette.

3. The pellet is resuspended in 150 μl of SOS (1 M sorbitol/6.5 mM CaCl$_2$/8% dextrose/25% YP). The tube is incubated at 30° for 20 to 40 min.

4. After incubation, the cells are plated with 9 ml of soft top agar onto minimal selection plates (the agar and plates both containing 1 M sorbitol, 8% glucose, and amino acid supplement lacking leucine).

5. The plates are incubated at 30°. Yeast transformants typically appear within 4 to 7 days.

Growth and Purification of Recombinant Lysozyme

1. A single colony from the transformation plate is inoculated into 5 ml of minimal medium without leucine containing 8% glucose and incubated at 30° with vigorous shaking for 36 hr. This seed culture is

subcultured into a 250-ml flask containing 50 ml of the same medium for an additional 36 hr at 30° with vigorous shaking.

2. The culture is used to inoculate 500 ml of YPG medium (1% yeast extract, 2% Bacto-peptone with 8% glucose) in a 2.8-liter Fernbach flask. The 500-ml culture is then grown at 30° with vigorous shaking.

3. The culture is harvested by centrifugation at 5000 g and 4° at peak production. Lysozyme activity, which is monitored by the standard turbidity assay of culture supernatant (described in a later section), typically reaches its peak after 5 to 7 days of growth in rich medium. For inactive mutants, the peak of lysozyme secretion is estimated as 12 to 24 hr after the yeast cell density (monitored by A_{650}) reaches its peak.

4. Cell pellets are washed with 100 ml of 0.5 cold M NaCl in 66 mM potassium phosphate buffer, pH 6.2 (KP 6.2), to recover lysozyme that is electrostatically adsorbed to the yeast cell walls.

5. The combined culture supernatant and cell wash are centrifuged at 16,000 g, 4°, for 20 min to remove residual cells and debris. The supernatant is loaded onto a 1.9×14 cm column containing 30 ml of CM Sepharose Fast Flow resin (Pharmacia, Piscataway, NJ) previously equilibrated in KP 6.2 at 4° using a peristaltic pump at a rate of approximately 10 ml/min.

6. The column is washed with 5–10 column volumes of KP 6.2, and lysozyme is eluted with 3 column volumes of 0.5 M NaCl in KP 6.2.

7. Twenty-five 1-ml fractions of the eluent are collected and assayed for lysozyme activity. The fractions containing lysozyme activity (typically 6) are pooled.

8. The protein solution is concentrated and desalted against KP 6.2 in Centricon-10 filter units (Amicon, Danvers, MA) as directed by the manufacturer.

9. Purified lysozymes are stored at 4° at a concentration of approximately 10 mg/ml. Protein yield varies depending on the mutant but averages 4 mg/liter.

Protein Gel Electrophoresis

The purity of the CL mutant protein expressed by yeast is further verified by denaturing polyacrylamide gel electrophoresis with SDS[26] and native gel electrophoresis.[27] SDS stacking (5% top gel and 15% bottom gel) polyacrylamide gel and Laemmli gel buffers are used. Ten percent native

[26] U. K. Laemmli, *Nature (London)* **227**, 680 (1970).
[27] O. Gabriel, this series, Vol. 22, p. 565.

gels at pH 8.9 and pH 4.3 are used.[28,29] Approximately 5 μg per sample is used in the SDS gel, and 10 μg per sample is used in the native gels. Protein samples are visualized with Coomassie Brilliant Blue G250. In addition, lysozyme activity in the native gels can be detected (before Coomassie staining) using *Micrococcus luteus* incorporated into an overlying polyacrylamide gel.[28,29]

Bacteriolytic Assay

The bacteriolytic activity of lysozymes is assayed by a modification of the spectrophotometric method of Shugar.[30] Between 5 and 10 units of lysozyme in 20 μl of KP 6.2 is added to a 2.6-ml suspension of *Micrococcus luteus* (Sigma, St. Louis, MO) in the same buffer prewarmed to 25°. A concentration of 0.2 mg/ml of *Micrococcus luteus* is used, and the absorbance reading at 450 nm (A_{450}) should equal approximately 1. The decrease in A_{450} is followed on a spectrophotometer with kinetic measurement capability (e.g., Shimadzu UV-160, Columbia, MD) at 15-sec intervals for 2 min at 25° against a control suspension without enzyme. One unit is defined as the amount of lysozyme that gives a change in absorbance of 0.001/min under these conditions.

Thermostability Experiments

One global property suitable for evaluation of evolutionary intermediates is protein stability. The unique biological and physical properties of an enzyme are highly dependent on the integrity of its native conformation. The structural stability of a protein can be measured by its resistance to heat, acid, and various chaotropic agents. Measurements of the thermostability of lysozymes provide an easy and an accurate way to assess the protein stability.

Tryptophan Absorbance

The transition midpoints (T_m) of lysozymes can be monitored directly with a spectrophotometer equipped with an automatic temperature-stepping module (e.g., the Gilford UV-Vis apparatus, Oberlin, OH). The decrease in absorbance at 292 nm with rising temperature reflects the shift of the internal tryptophan residues to an aqueous environment,[31] and hence

[28] M. F. Hammer, J. W. Schilling, E. M. Prager, and A. C. Wilson, *J. Mol. Evol.* **24**, 272 (1987).
[29] G. A. Cortopassi and A. C. Wilson, *in* "The Immune Response to Structurally Defined Proteins: The Lysozyme Model" (S. J. Smith-Gill and E. E. Sercarz, eds.), p. 87. Adenine Press, Schenectady, New York, 1989.
[30] D. Shugar, *Biochim. Biophys. Acta* **8**, 302 (1952).
[31] J. G. Foss, *Biochim. Biophys. Acta* **47**, 569 (1961).

unfolding of the enzyme. Data are recorded at 0.5° intervals from 50° to 90°. A typical trace is given in Fig. 3.[32] Pre- and postdenaturational baselines are extended to the region around the transition midpoint, and T_m is assigned as the temperature at the midpoint of the thermal transition. All lysozymes are used at a concentration of approximately 25 μM in a solution of thoroughly degassed KP 6.2. A control that contains buffer only is tested in parallel. Multiple determinations are performed for each lysozyme variant.

Differential Scanning Calorimetry

Although thermostability experiments using tryptophan absorbance as the probe give fairly reproducible T_m values, the thermodynamic parameters such as free energy, enthalpy, entropy, and heat capacity cannot be obtained directly. Calorimetric measurement provides the most complete picture of the thermodynamic process of protein unfolding. This technique, however, requires more material and is more time consuming.

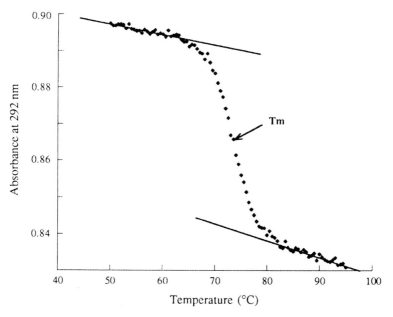

FIG. 3. The transition midpoint (T_m) of thermal denaturation of a synthetic lysozyme is determined spectrophotometrically by monitoring the decrease in the absorbance at 292 nm with rising temperature. The thermally induced transition for CL is cooperative and obeys the two-state model to a good approximation.[32]

[32] J. R. O'Reilly and F. E. Karasz, *Biopolymers* **9**, 1429 (1970).

Figure 4 illustrates the results of a differential scanning calorimetry experiment done on a synthetic lysozyme.[33] Two milliliters of an approximately 2 mg/ml solution of lysozyme, which has been extensively dialyzed against 10 mM potassium phosphate acidified to approximately pH 4, is used per experiment. Lysozyme is very soluble under these conditions, and the aggregation problem encountered at higher temperatures is minimized. Samples are thoroughly degassed before being injected into the cell of a differential scanning calorimeter (e.g., Micro-Cal MC2, North Hampton, MA). The heat capacity of the unfolding process is monitored as a function of temperature. The data are collected and analyzed following the manufacturer's instructions. The transition midpoint of protein unfolding corresponds to the temperature at the peak of the curve, and the thermodynamic parameters ΔH and ΔCp are evaluated by the procedure of Privalov.[33]

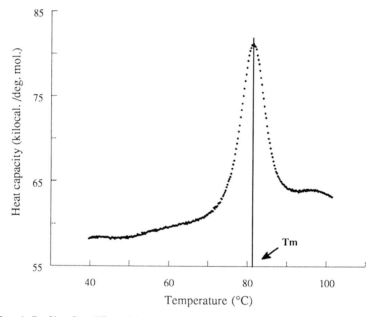

FIG. 4. Profile of a differential scanning calorimetry experiment done on a synthetic lysozyme. The heat capacity (kilocalories per degree per mole) of the unfolding process was monitored as a function of temperature on a Micro-Cal MC2 instrument. The transition midpoint of protein unfolding corresponds to the temperature at the peak of the curve, and the thermodynamic parameters ΔH and ΔCp are evaluated by the procedure of Privalov.[33]

[33] P. L. Privalov and S. A. Potekhin, this series, Vol. 131, p. 4.

Quasi-Physiological Inactivation Studies

Intrinsic stability under physiological conditions is a reflection of many physical interactions, both intramolecular and with the environment (solvent and other macromolecules), which in turn affect the biological viability and utility of the protein. To compare the stability of lysozyme variants under conditions as near to the physiological ones as possible, studies of the irreversible thermal inactivation of enzymatic activity are conducted with a free-thiol trap method.[34] The experimental conditions (50 mM sodium carbonate/bicarbonate buffer, pH 9) simulate those found in egg white during the first week of incubation,[35,36] with the concentrations of lysozyme and ovalbumin found in egg white (0.2 and 1.4 mM, respectively[37]) being used. Ovalbumin (Sigma, grade V), which is the major constituent of bird egg white, acts as a disulfide–sulfhydryl exchange agent and irreversibly traps any accessible sulfhydryl groups of completely or partially unfolded lysozyme.

1. Samples of reaction solution (0.2 mM lysozyme, 1.4 mM ovalbumin, 50 mM sodium carbonate/bicarbonate buffer, pH 9.0) are sealed in microcentrifuge tubes and heated in a water bath previously equilibrated to a specified temperature.

2. Samples are removed and placed on ice to cool at 10 time points over three or more half-lives.

3. Three 5-μl aliquots are removed, and appropriate dilutions into KP 6.2 are made and later assayed (in triplicate) for residual activity using the bacteriolytic assay described above.

4. Approximate rate constants of inactivation (k) are calculated for each lysozyme by fitting the data to a first-order model using nonlinear regression analysis.

5. The experiment is repeated at several different temperatures to generate a wide range of inactivation rate constants.

6. Arrhenius plots (ln k versus $1/T$) are constructed. Temperatures required for 50% inactivation of the enzyme in 15 min for each variant are compared.

The present method reflects one possible mode of degradation of lysozyme and is used because of its relevance to a biophysical property. Other methods that test a relevant attribute of the enzyme, such as protease susceptibility, might also be employed. Emphasis should be placed on *in*

[34] M. F. Hammer and A. C. Wilson, *Genetics* **115**, 521 (1987).
[35] A. Romanoff, "Biochemistry of the Avian Embryo." Wiley, New York, 1967.
[36] L. R. MacDonnell, R. B. Silva, and R. E. Feeney, *Arch. Biochem. Biophys.* **32**, 288 (1951).
[37] F. E. Cunningham and H. Lineweaver, *Poult. Sci.* **46**, 1471 (1967).

vivo significance. For example, when protease sensitivity is used as a probe, preference should be given to proteases that the protein might encounter naturally.

Discussion

This chapter describes a method to reconstruct and examine the properties of ancestral proteins. Because the method is relatively new, few papers have yet been published on its use for testing ideas about the mechanism of protein evolution.[1,2] One of the questions addressed concerns neutral mutations. Many molecular evolutionists think that most of the mutations fixed during protein evolution may be neutral.[38] According to this view, there are two other types of mutations, deleterious and advantageous. Although there may be many deleterious mutations, they are rarely fixed because negative selection normally weeds them out of natural populations. Advantageous mutations, which have a good chance of being fixed by positive selection, are not a major contributor to protein evolution because they occur so rarely, especially in proteins that have kept the same function for billions of years.

There are conditions under which mildly deleterious mutations could contribute to protein evolution. If a second mutation occurs in a protein-coding gene that has experienced a deleterious mutation, it might compensate for the harmful effect of the first. If the doubly mutated gene were functionally equivalent to the wild-type gene, it would be fixed by genetic drift.

The method designed here has been used to search for such cases (Fig. 5). AB and BA represent the two possible intermediate states on the evolutionary pathway linking AA to BB via sequential amino acid substi-

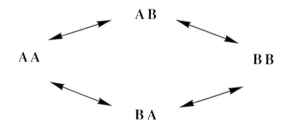

Fig. 5. Scheme showing the two simplest evolutionary pathways leading from the ancestral protein with amino acid sequence AA to the derived protein with sequence BB. AB and BA represent sequences of the intermediates.

[38] M. Kimura, "The Neutral Theory of Molecular Evolution." Cambridge Univ. Press, Cambridge, 1983.

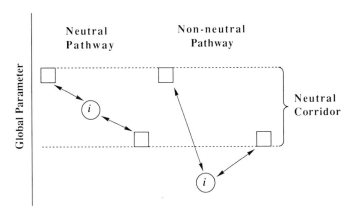

FIG. 6. Scheme illustrating the concept of a neutral corridor. Squares denote wild-type states, and the circle is an evolutionary intermediate; the arrows depict single amino acid replacements. The dashed lines demarcate the limits of the neutral corridor, with the upper line giving the mean plus two standard errors for a global property of one of the wild-type proteins and the lower line giving the mean minus two standard errors for the same property of the other wild-type protein. In the neutral pathway, the intermediate (i) has a parameter value within the corridor, but this is not the case in the nonneutral pathway.

tution. Were the mode of evolution of this protein strictly neutral, then comparable properties should be observed for the four forms.

To reconstruct the evolutionary intermediates and the two end states, the gene encoding the protein is cloned, and site-directed mutagenesis is used to generate the four protein variants. After the protein variants are expressed and purified, the *in vitro* evaluation of some global properties of the variants is undertaken. The global properties of the native state of the two end states are then used to define a corridor of neutrality to which the properties of the intermediates are compared (Fig. 6). The pathways involving intermediates whose properties lie completely within the corridor are regarded as neutral, and those that do not are viewed as nonneutral by this definition.

Two considerations must be kept in mind for the neutrality test. Its relevance relies on the extent to which *in vitro* performance parallels *in vivo* behavior and on how strongly "lying within the neutral corridor" correlates with "actually being neutral." Research on hemoglobin mutants in humans strongly supports the hypothesis that the results of *in vitro* tests on global properties correlate with *in vivo* behavior. The majority of hemoglobin mutants with pathological effects on humans have altered stability or ability to bind oxygen.[39] Most hemoglobin mutants cause no disease and

[39] R. E. Dickerson and I. Geis, "Hemoglobin: Structure, Function, Evolution, and Pathology." Benjamin/Cummings, Menlo Park, California, 1983.

have no detectable difference in properties from the wild-type form. The only cases of hemoglobins that cause disease and yet have normal functional properties under certain conditions are the sickling mutants.[40] Clearly, the correlation, although strong, is not perfect.

The neutral corridor test therefore provides a criterion but not a proof of nonneutrality. If an evolutionary intermediate lies outside the corridor, it becomes a candidate for a nonneutral pathway. The lysozyme mutants described in this chapter have already provided such candidates.[2] The stage is now set for identifying additional cases and for finding out how often evolutionary intermediates lie within the neutral corridor.

Undoubtedly there are often questions about the mechanism of protein evolution that can be addressed by making ancestral proteins and testing their properties, especially the origin of new functions. Lysozyme provides an example for that type of investigation because on two occasions in mammalian evolution it has been recruited for functioning in the harsh conditions of the stomach fluid.[41,42] The sequences of the common ancestor have been inferred by tree analysis, and at least four of the new selection pressures faced by this enzyme are known. Irwin et al.[43] have reviewed this subject and the speculations regarding the possibility that double or triple mutations were required to achieve new functions.

Acknowledgments

We thank D. M. Irwin, J. R. Kornegay, and E. M. Prager for helpful discussion. This research received support from the National Institutes of Health and the Director, Office of Energy Research, Office of Basic Energy Science, Division of Material Sciences and Energy Biosciences, of the U.S. Department of Energy under Contract DE-AC03-76SF00098.

[40] H. F. Bunn, B. G. Forget, and H. M. Ranney, "Human Hemoglobins." Saunders, Philadelphia, Pennsylvania, 1977.
[41] D. E. Dobson, E. M. Prager, and A. C. Wilson. J. Biol. Chem. 259, 11607 (1984).
[42] C.-B. Stewart, J. W. Schilling, and A. C. Wilson, Nature (London) 330, 401 (1987).
[43] D. M. Irwin, E. M. Prager, and A. C. Wilson, Anim. Genet. 23, 193 (1992).

[43] Comparative Method in Study of Protein Structure and Function: Enzyme Specificity as an Example

By CARO-BETH STEWART

Introduction

One of the goals of molecular evolutionary studies is to understand the mechanisms and pathways used by the evolutionary process to produce proteins of altered or new function. These types of issues are best addressed through the use of the comparative method,[1] wherein explicit evolutionary models are used to interpret the biochemical data.

The primary mode of the evolution of molecules is divergent; that is, over evolutionary time homologous (i.e., genealogically related) molecules become both different from their common ancestor and different from each other. Homologous molecules can diverge either because of speciation of the organisms that contain them or as a consequence of gene duplication. Homologous molecules that are related by speciation are termed orthologous, whereas those that are related by gene duplication are termed paralogous.[2] Divergent members of enzyme families usually retain their overall tertiary structure and enzymatic mechanism, yet they may evolve regarding functional characteristics such as substrate specificity.[3]

The purpose of this chapter is to outline how explicit evolutionary models are constructed and used in comparative studies of protein structure and function, taking as example the evolution of substrate specificity in the pancreatic carboxypeptidase family. Amino acid parsimony methods are emphasized for the following reasons. First, more protein sequences are available than are gene sequences for many of the well-studied protein families; in these cases, it is necessary to use the protein sequence information in establishing the phylogeny of the family. Second, good and versatile amino acid parsimony programs are now widely available. Third, parsimony methods deal with the amino acids or nucleotides in sequences as "characters" that can be analyzed cladistically. Thus, specific amino acid replacements often can be assigned along given lineages and correlated with changes in protein function.[4] Computer programs that are useful for

[1] P. H. Harvey and M. D. Pagel, "The Comparative Method in Evolutionary Biology." Oxford Univ. Press, Oxford and New York, 1991.

[2] W. F. Fitch, *Syst. Zool.* **19**, 99 (1970).

[3] T. E. Creighton, "Proteins: Structures and Molecular Principles." Freeman, New York, 1984.

[4] C.-B. Stewart and A. C. Wilson, *Cold Spring Harbor Symp. Quant. Biol.* **52**, 891 (1987).

these purposes are briefly discussed. The approach outlined below can be applied to any molecular family to study the evolution of structure and function.

Functional Characterization

Before the evolution of protein function can be studied, functional differences first must be demonstrated between members of the family. At present, the best comparative biochemical data exist for classically studied protein families, such as the globin family and several families of digestive enzymes.[3]

The vertebrate pancreatic carboxypeptidases are one such family of digestive enzymes. Within the family, two broad substrate preferences are known. Enzymes with carboxypeptidase A (CPA) activity cleave hydrophobic and aromatic residues from the carboxyl terminus of peptides and proteins, whereas those with carboxypeptidase B (CPB) activity prefer substrates with arginine and lysine residues at the C terminus.[3] Several duplicates within the family have CPA-like activity but display a range of substrate preferences.[5,6]

We wish to know which amino acid replacements in the carboxypeptidases were responsible for the changes in substrate specificities during the evolution of the family. Pinpointing specific amino acid replacements (as compared to simple sequence differences) which might be responsible for changes in function requires several steps, including alignment of the protein sequences, reconstruction of the phylogeny of the family, cladistic analysis of the amino acid sequences and functional changes on this phylogeny, and examination of these replacements in light of the three-dimensional structure of the molecules, if available. Site-specific mutagenesis experiments can allow us to test if the pinpointed amino acid replacements can indeed change enzyme function in the manner predicted. These steps are described below, with pancreatic carboxypeptidases as example.

Sequence Alignment

The simplest level of sequence comparison is alignment of the primary structures of homologous molecules. A multiple sequence alignment allows one to look for amino acid differences between family members

---

[5] S. J. Gardell, C. S. Craik, E. Clauser, E. J. Goldsmith, C.-B. Stewart, M. Graf, and W. J. Rutter, *J. Biol. Chem.* **263**, 17828 (1988).

[6] M. Natsuaki, C.-B. Stewart, P. Vanderslice, L. B. Schwartz, M. Natsuaki, B. U. Wintroub, W. J. Rutter, and S. M. Goldstein, *J. Invest. Dermatol* **99**, 138 (1992).

which might be responsible for differing functions. However, such pairwise comparisons do not have the resolving power that can be attained by cladistic analysis of amino acid replacements on a phylogenetic tree, which can allow assignment of specific amino acid replacements to specific lineages.

A prerequisite of phylogenetic tree building is an historically accurate alignment of the sequences of interest. Alignment of multiple sequences can be a difficult task to accomplish by hand, and it is usually implemented with a computer. The problem of multiple sequence alignment is not always trivial but is outside the scope of this chapter (the interested reader is referred to Volume 183 of this series). Despite difficulties sometimes encountered, alignment of protein-coding sequences often is straightforward, especially if the primary structures are highly similar owing to recent common ancestry or sequence conservation. Although there are many computer programs that aim to produce a mathematically optimal alignment, there is no program that can tell if the alignment is historically correct. An excellent program for multiple protein sequence alignment is PIMA (pattern-induced multisequence alignment) by Smith and Smith.[7]

If nucleic acid sequences are available for the family of interest, then this information can be valuable for phylogenetic tree building. The multiple protein sequence alignment should be used as a guide to align the DNA sequences based on codons. If gaps are allowed, a mathematically optimal alignment of nucleotide sequences sometimes can be in error regarding alignment of the protein sequences they encode.

Many researchers feel that an alignment produced solely by a computer is inherently superior to one produced by hand, since the manual alignment is more likely to introduce the biases of the investigator. However, the reader should bear in mind that the assumptions and biases (and perhaps mistakes) of the programmer are present in any computer program. Although minimization of researcher bias is a worthy goal, the computer cannot replace the human brain in evolutionary analysis. The output of any computer program should be visually inspected and corrected, if necessary, and problems reported to the programmer. It also should be borne in mind that any sequence alignment is a hypothesis of relationship and can change as new data accumulates.

Alignment of five completely sequenced members of the pancreatic digestive enzyme carboxypeptidase family[5] is shown in Fig. 1, along with hepatopancreatic carboxypeptidase B from crayfish.[8] Only the amino acid

---

[7] R. F. Smith and T. F. Smith, *Protein Eng.* **5,** 35 (1992). This UNIX-based package is available from the authors, or by anonymous ftp from mbcrr.harvard.edu (in subdirectory MBCRR-Package).

[8] K. Titani, L. H. Ericsson, S. Kumar, F. Jakob, H. Neurath, and R. Zwilling, *Biochemistry* **23,** 1245 (1984).

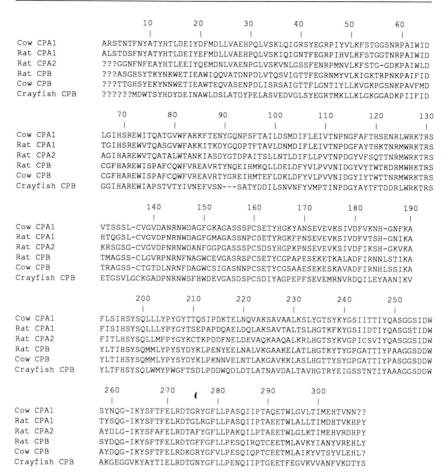

```
                   10        20        30        40        50        60
                    |         |         |         |         |         |
Cow   CPA1   ARSTNTFNYATYHTLDEIYDFMDLLVAEHPQLVSKLQIGRSYEGRPIYVLKFSTGGSNRPAIWID
Rat   CPA1   ALSTDSFNYATYHTLDEIYEFMDLLVAEHPQLVSKIQIGNTFEGRPIHVLKFSTGGTNRPAIWID
Rat   CPA2   ???GGNFNFEAYHTLEEIYQEMDNLVAENPGLVSKVNLGSSFENRPMNVLKFSTG-GDKPAIWLD
Rat   CPB    ???ASGHSYTKYNKWETIEAWIQQVATDNPDLVTQSVIGTTFEGRNMYVLKIGKTRPNKPAIFID
Cow   CPB    ???TTGHSYEKYNNWETIEAWTEQVASENPDLISRSAIGTTFLGNTIYLLKVGKPGSNKPAVFMD
Crayfish CPB ??????MDWTSYHDYDEINAWLDSLATDYPELASVEDVGLSYEGRTMKLLKLGKGGADKPIIFID

                  70        80        90       100       110       120       130
                   |         |         |         |         |         |         |
Cow   CPA1   LGIHSREWITQATGVWFAKKFTENYGQNPSFTAILDSMDIFLEIVTNPNGFAFTHSENRLWRKTRS
Rat   CPA1   TGIHSREWVTQASGVWFAKKITKDYGQDPTFTAVLDNMDIFLEIVTNPDGFAYTHKTNRMWRKTRS
Rat   CPA2   AGIHAREWVTQATALWTANKIASDYGTDPAITSLLNTLDIFLLPVTNPDGYVFSQTTNRMWRKTRS
Rat   CPB    CGFHAREWISPAFCQWFVREAVRTYNQEIHMKQLLDELDFYVLPVVNIDGYVYTWTKDRMWRKTRS
Cow   CPB    CGFHAREWISPAFCQWFVREAVRTYGREIHMTEFLDKLDFYVLPVVNIDGYIYTWTTNRMWRKTRS
Crayfish CPB GGIHAREWIAPSTVTYIVNEFVSN---SATYDDILSNVNFYVMPTINPDGYAYTFTDDRLWRKTRS

                 140       150       160       170       180       190
                  |         |         |         |         |         |
Cow   CPA1   VTSSSL-CVGVDANRNWDAGFGKAGASSSPCSETYHGKYANSEVEVKSIVDFVKNH-GNFKA
Rat   CPA1   HTQGSL-CVGVDPNRNWDAGFGMAGASSNPCSETYRGKFPNSEVEVKSIVDFVTSH-GNIKA
Rat   CPA2   KRSGSG-CVGVDPNRNWDANFGGPGASSSPCSDSYHGKPNSEVEVKSIVDFIKSH-GKVKA
Rat   CPB    TMAGSS-CLGVRPNRNFNAGWCEVGASRSPCSETYCGPAPESEKETKALADFIRNNLSTIKA
Cow   CPB    TRAGSS-CTGTDLNRNFDAGWCSIGASNNPCSETYCGSAAESEKESKAVADFIRNNLSSIKA
Crayfish CPB ETGSVLGCKGADPNRNWSFHWDEVGASDSPCSDIYAGPEPFSEVEMRNVRDQILEYAANIKV

                 200       210       220       230       240       250
                  |         |         |         |         |         |
Cow   CPA1   FLSIHSYSQLLLYPYGYTTQSIPDKTELNQVAKSAVAALKSLYGTSYKYGSIITTIYQASGGSIDW
Rat   CPA1   FISIHSYSQLLLYPYGYTSEPAPDQAELDQLAKSAVTALTSLHGTKFKYGSIIDTIYQASGSTIDW
Rat   CPA2   FITLHSYSQLLMFPYGYKCTKPDDFNELDEVAQKAAQALKRLHGTSYKVGPICSVIYQASGGSIDW
Rat   CPB    YLTIHSYSQMMLYPYSYDYKLPENYEELNALVKGAAKELATLHGTKYTYGPGATTIYPAAGGSDDW
Cow   CPB    YLTIHSYSQMMLYPYSYDYKLPKNNVELNTLAKGAVKKLASLHGTTYSYGPGATTIYPASGGSDDW
Crayfish CPB YLTFHSYSQLWMYPWGFTSDLPDDWQDLDTLATNAVDALTAVHGTRYEIGSSTNTIYAAAGGSDDW

                 260       270       280       290       300
                  |         |         |         |         |
Cow   CPA1   SYNQG-IKYSFTFELRDTGRYGFLLPASQIIPTAQETWLGVLTIMEHTVNN??
Rat   CPA1   TYSQG-IKYSFTFELRDTGLRGFLLPASQIIPTAEETWLALLTIMDHTVKHPY
Rat   CPA2   AYDLG-IKYSFAFELRDTAFYGFLLPAKQILPTAEETWLGLKTIMEHVRDHPY
Rat   CPB    SYDQG-IKYSFTFELRDTGFFGFLLPESQIRQTCEETMLAVKYIANYVREHLY
Cow   CPB    AYDQG-IKYSFTFELRDKGRYGFVLPESQIQPTCEETMLAIKYVTSYVLEHL?
Crayfish CPB AKGEGGVKYAYTIELRDTGNYGFLLPENQIIPTGEETFEGVKVVANFVKDTYS
```

FIG. 1. Aligned sequences of the "pancreatic" carboxypeptidase family. Pancreatic carboxypeptidase sequences are from domestic cow *(Bos taurus)*, Sprague-Dawley rat *(Rattus norvegicus)*, and crayfish *(Astacus fluviatilis)*, as referenced previously [M. Natsuaki, C.-B. Stewart, P. Vanderslice, L. B. Schwartz, M. Natsuaki, B. U. Wintroub, W. J. Rutter, and S. M. Goldstein, *J. Invest. Dermatol.* **99**, 138 (1992)]. The numbering system is according to the primary sequence of cow carboxypeptidase A$_1$. The amino acid sequences are presented in the single-letter code. This alignment was used as the input matrix for PROTPARS in PAUP (D. L. Swofford, "PAUP: Phylogenetic Analysis Using Parsimony, Version 3.0s." Computer program distributed by the Illinois Natural History Survey, Champaign, Illinois, 1991). Gaps are indicated by dashes and were treated as a "new state." Question marks indicate length variation at the N and C termini and were treated as "missing."

sequences are known for some of these enzymes, so phylogenetic analysis of the nucleotide sequences is not possible at this time. This carboxypeptidase family also contains some vertebrate carboxypeptidases that are not expressed in the pancreas, including mast cell carboxypeptidase (which is most closely related to pancreatic carboxypeptidase B)[6] and some bacterial carboxypeptidases.[9,10] For purposes of illustration, the present discussion deals only with the pancreatic enzymes, because they are the best-characterized regarding substrate specificity and tertiary structure.

Evolutionary Tree Building

The goal of evolutionary tree building is to produce an historically accurate phylogeny for the molecular family of interest. Many legitimate methods of tree building exist and have been reviewed by Swofford and Olsen.[11] It is wise to try several tree-building methods on the data; if various methods produce the same tree, this is a reassuring sign that the tree may be correct. Of course, one can never be entirely certain that a given phylogeny is historically correct; therefore, a phylogeny should be viewed as a hypothesis concerning relationships, and alternative phylogenies can be examined if warranted.

Many parsimony programs do not deal adequately with amino acid sequences, in that all amino acid replacements are considered equally likely regardless of the genetic code. A good amino acid parsimony program, PROTPARS, is now widely available in both Joseph Felsenstein's PHYLIP package[12] and David Swofford's Phylogenetic Analysis Using Parsimony (PAUP).[13] Although PROTPARS was originally developed by Felsenstein, the PAUP version has several advantages: PAUP can guarantee to find the most parsimonious tree for at least 10 protein sequences, it currently is available for the Macintosh in a "user-friendly" format, and it has many features that are useful in later stages of analysis. (In practice, PROTPARS is slow compared to nucleotide parsimony programs. It can require several days of computing time on a Macintosh II to find the shortest tree for 9 or more protein sequences.) The multiple sequence alignment program, PIMA,[7] generates a generic sequence alignment file that can be used by PAUP after a little editing.

Establishing the phylogenetic tree for a molecular family takes three

[9] Y. Narahashi, *J. Biochem. (Tokyo)* **107**, 879 (1990).

[10] S. V. Smulevitch, A. L. Osterman, O. V. Galperina, M. V. Matz, O. P. Zagnitko, R. M. Kadyrov, I. A. Tsaplina, N. V. Grishin, G. G. Chestukhina, and V. M. Stepanov, *FEBS Lett.* **291**, 75 (1991).

[11] D. L. Swofford and G. J. Olsen, *in* "Molecular Systematics" (D. M. Hillis and C. Moritz, eds.), p. 411. Sinauer, Sunderland, Massachusetts, 1990.

[12] J. Felsenstein, "PHYLIP (Phylogeny Inference Package), Version 3.5, Manual," University of Washington, Seattle, 1993.

[13] D. L. Swofford, "PAUP: Phylogenetic Analysis Using Parsimony, Version 3.1." Computer program distributed by the Illinois Natural History Survey, Champaign, Illinois, 1993.

steps: (1) constructing an unrooted tree that has the lineages joined in the order that requires the fewest evolutionary events (i.e., amino acid replacements or nucleotide substitutions); (2) determining where to place the most ancient node on the tree (i.e., "rooting" the tree); and (3) assigning the locations of gene duplications, if they exist. These steps are illustrated below, using the pancreatic carboxypeptidases as an example.

Constructing Unrooted Trees

The six aligned sequences in Fig. 1 were used to build amino acid parsimony trees by PROTPARS in PAUP.[13] An exhaustive search of the 105 alternative unrooted trees was performed; the resulting distribution of possible trees was skewed[14] positively, with a long tail (not shown) containing the shortest tree. This shortest unrooted tree (Fig. 2) requires 601 amino acid replacements; the next shortest tree requires 625 events. In this case, the most parsimonious tree appears rather trustworthy.

Rooting Trees

In a bifurcating tree, the ancestral node (or root) could theoretically fall along any of the lineages (see Fig. 2). Parsimony methods for constructing molecular phylogenies do not produce a rooted tree automatically; this is a

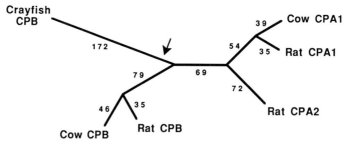

FIG. 2. Unrooted phylogenetic tree of the pancreatic carboxypeptidase family. An exhaustive search of all possible trees was performed using the amino acid parsimony program PROTPARS (J. Felsenstein, "PHYLIP: Phylogeny Inference Package, Version 3.5 Manual." University of Washington, Seattle, 1993) in PAUP (D. L. Swofford, "PAUP: Phylogenetic Analysis Using Parsimony, Version 3.0s." Illinois Natural History Survey, Champaign, Illinois, 1991) for the sequences shown in Fig. 1. The most parsimonious tree (in this case, the one requiring the fewest amino acid replacements, based on the genetic code but allowing silent substitutions to occur freely) is shown. This tree requires 601 amino acid replacements; the next shortest tree requires 625 amino acid replacements. The branches are drawn in proportion to their lengths, as reconstructed by PAUP; the branch lengths (e.g., number of reconstructed amino acid replacements) are indicated. The location of the midpoint root is indicated by the arrow.

[14] D. M. Hillis, in "Phylogenetic Analysis of DNA Sequences" (M. M. Miyamoto and J. Cracraft, eds.), p. 278. Oxford Univ. Press, Oxford and New York, 1991.

separate process that requires decision-making by the investigator. Several approaches to rooting trees exist, each requiring prior knowledge and/or a choice of assumptions. (1) A tree can be rooted by a known outgroup. For example, a given protein sequence from an egg-laying mammal could be used to root the tree for orthologous proteins from the placental mammals. Ideally, an outgroup should be chosen that is closely related to the "ingroup," or sequences of interest. (2) In molecular families having gene duplications, a paralogous gene can be used as an outgroup to root the tree. For example, β-globin can be used to root the α-globin side of the globin family tree. (3) The root can be placed at the midpoint of the longest reconstructed distance found between two sequences in the tree. Midpoint rooting assumes that the molecules are evolving at fairly constant rates along the different lineages, and that each major lineage is represented adequately. (Adding new sequences along a lineage increases the number of reconstructed events along that lineage and therefore can make a well-represented lineage appear longer than an underrepresented one.) These assumptions are not always valid. However, if a trustworthy outgroup is not known, then midpoint rooting is a useful default. It is best to try several approaches, and to consider the implications if different methods produce different rooted trees. Ideally, different methods of rooting will produce the same result.

Such is the case for the pancreatic carboxypeptidases. The rooted tree for this family is shown in Fig. 3. The same root is found either by using the crayfish carboxypeptidase sequence as an outgroup to the vertebrate sequences (justified on the basis that most invertebrates appear to have only one hepatopancreatic carboxypeptidase[15]) or by midpoint (see Fig. 2; the longest distance reconstructed by PAUP is between crayfish CPB and cow CPA1, and the midpoint of this distance falls along the crayfish CPB lineage, as indicated by the arrow). The same topology among the vertebrate enzymes is found[6] when the tree is rooted using a bacterial carboxypeptidase[9] as an outgroup.

Assigning Gene Duplications

Evolutionary tree building programs do not tell which sequences are the products of gene duplications (paralogs) and which are the products of speciation (orthologs). Again, this decision must be made by the investigator. After the phylogenetic tree is constructed and rooted, gene duplications are assigned to internal nodes on the tree in the most parsimonious manner that explains the known species distribution of the family members.

[15] H. J. Vonk and J. R. H. Western, "Comparative Biochemistry and Physiology of Enzymatic Digestion." Academic Press, London, 1984.

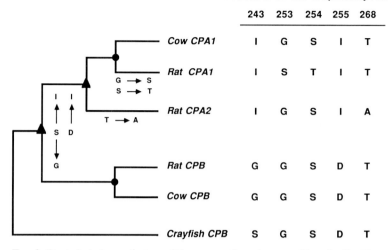

Residue in S1' Site of Specificity Pocket

	243	253	254	255	268
Cow CPA1	I	G	S	I	T
Rat CPA1	I	S	T	I	T
Rat CPA2	I	G	S	I	A
Rat CPB	G	G	S	D	T
Cow CPB	G	G	S	D	T
Crayfish CPB	S	G	S	D	T

Fig. 3. Rooted phylogenetic tree of the pancreatic carboxypeptidase family. The tree in Fig. 2 was rooted using the crayfish carboxypeptidase as outgroup; the same topology is found when the tree is rooted by midpoint (see text). Gene duplications (represented by triangles) are placed in the most parsimonious manner required to explain the modern species distribution of the pancreatic carboxypeptidases. Speciation events are represented by filled circles. Variable amino acids found in the S1' substrate binding pocket (i.e., residues within 4.5 Å. of the C-terminal residue of the substrate) are shown at the end of the lineages in the single-letter code. Amino acid replacements were assigned manually in the most parsimonious manner along the lineages, taking account of the genetic code. Arrows indicate the inferred directions of the amino acid replacements. For example, the most parsimonious explanation for the presence of S^{253} and T^{254} in rat CPA1 is that two amino acid replacements occurred along the rat CPA1 lineage, $G \rightarrow S^{253}$ and $S \rightarrow T^{254}$. Thus, G^{253} and S^{254} are inferred to be ancestral in this family (which we could not state if the root of the tree had fallen along the rat CPA1 lineage). Therefore, S^{253} and T^{254} are derived states in rat CPA1, which suggests that the narrower specificity of this enzyme (see text) may also be derived. Likewise, D^{255} is inferred to be ancestral on this tree, suggesting that CPB-like activity is the ancestral state in the pancreatic carboxypeptidases and that enzymes with CPA-like activity evolved from enzymes with CPB-like activity in the animals.

For example, three distinct paralogous carboxypeptidases have been purified from mammalian pancreases, enzymatically characterized, and sequenced. The laboratory rat contains pancreatic carboxypeptidases A_1, A_2, and B.[5] A phylogenetic tree containing these paralogous carboxypeptidases must account for the gene duplications. In Fig. 3, gene duplications were assigned to the fewest nodes required to explain the presence of these three carboxypeptidases in the rat. This final tree implies that the gene duplications which gave rise to the vertebrate carboxypeptidases occurred after the lineage leading to crayfish diverged from the lineage leading to modern vertebrates.[6] Furthermore, this tree implies that the well-characterized cow CPA is orthologous to rat CPA1 and paralogous to rat CPA2, and that this duplication occurred before the mammalian radiation.[5]

Cladistic Analysis

After the branching order, root, and placement of gene duplications have been decided, the phylogeny for the molecular family is complete. This "known" phylogeny now can be used as a framework for studying the evolution of sequences at a fine-grained level, with the hope of correlating specific amino acid replacements along lineages with known changes in protein structure and function. This is accomplished by assigning amino acid replacements along the lineages in the most parsimonious manner considering the genetic code,[16] then looking to see if given replacements correlate with known differences in structure and function between members of the protein family.

Cladistic analysis of molecular sequence characters differs from that described by Hennig[17] for organismal characters in several important ways. Molecules do not leave fossils, thus there is no "hard" record of which character state (i.e., amino acid residue) is ancestral and which is derived at any position in a sequence. The 20 common amino acids are found in all living forms and therefore have nothing inherently "ancestral" or "derived" about them. Furthermore, amino acid replacements have no intrinsic directionality, even though some replacements are more likely to occur than others. In other words, amino acid replacements are not inherently "polarized." In addition, amino acid replacements and nucleotide substitutions are reversible. For these reasons, the character state(s) of the outgroup molecule(s) cannot be assumed to be ancestral.

For each position in the sequences of interest, the conditions of "ancestral" and "derived" must be inferred logically by cladistic analysis on a known phylogeny, taking the genetic code into account.[15] The computer program MacClade is helpful in character analysis.[18] MacClade interfaces easily with PAUP. Some of the output from the two programs can be helpful in making logical inferences regarding character evolution. Nevertheless, every position of interest should be analyzed by the investigator.

[16] W. F. Fitch, in "Cladistics: Perspectives on the Reconstruction of Evolutionary History" (T. Duncan and T. F. Stuessy, eds.), p. 221. Columbia Univ. Press, New York, 1984.
[17] W. Hennig, Annu. Rev. Entomol. 10, 97 (1965).
[18] W. P. Maddison and D. R. Maddison, "MacClade: Interactive Analysis of Phylogeny and Character Evolution, Version 3.0." Sinauer, Sunderland, Massachusetts, 1992.

Ancestral reconstruction of a long protein sequence can be a tedious task. Fortunately, it may not always be necessary to analyze the entire protein sequence cladistically. Functional characteristics often are thought to be determined by limited regions in a protein molecule; if so, the analysis can be limited to these regions.

For example, the specificity of a given carboxypeptidase is thought to be determined primarily by the amino acid side chains in its binding site for the C-terminal residue of the substrate molecule.[5] Thus, cladistic analysis of amino acid replacements likely to be responsible for changes in substrate specificity initially can be limited to the residues that comprise this binding site. An example of cladistic analysis of the amino acid replacements in the substrate binding pocket of the vertebrate pancreatic carboxypeptidases of known sequence and substrate preference is shown in Fig. 3. Displayed on this phylogeny are all of the amino acid replacements that are in the binding pocket and are therefore most likely to confer the differences in substrate specificities exhibited by the carboxypeptidases.[5] These amino acid replacements can be examined in light of the tertiary structure of carboxypeptidase.

Examination and Alignment of Tertiary Structures

Many computer programs currently exist that allow the viewing and manipulation of protein crystal structures. The more powerful of these programs allow superposition of homologous structures in three-dimensional space and modeling of hypothetical amino acid replacements. Two of the better molecular modeling programs are Insight[19] and MIDAS.[20] These programs require a graphics-capable computer and considerable expertise. Researchers committed to the study of protein evolution should consider investing the time and money needed for these resources; others should seek collaboration.

The carbon backbones of homologous proteins (especially closely related ones, such as orthologs) are often nearly superimposable in three-dimensional space.[3] This type of observation led to the widespread belief that the observed differences in function between homologous enzymes are due primarily to replacements of amino acid side chains, rather than to rearrangements of the carbon backbone. For example, the backbones of the substrate binding pockets of cow CPA1 and CPB are nearly superimposable in three-dimensional space (not shown), suggesting that the differ-

---

[19] "Insight II, Version 2.2.0." Biosym Technologies, San Diego, CA (1993).

[20] T. E. Ferrin, C. C. Huang, L. E. Jarvis, and R. Langridge, *J. Mol. Graph.* **6**, 13 (1988). The program is available by writing to Midas Software Distribution, Computer Graphics Laboratory, University of California, San Francisco, CA 94143-0446.

ences in substrate preference might be due to replacement of amino acid side chains in the pocket. Indeed, the original sequencing[21] and crystallographic[22] studies on these molecules suggested that the change in substrate specificity is conferred by a single amino acid replacement of an isoleucine (in CPA) for an aspartic acid (in CPB) at position 255.

Computer modeling of the amino acid replacements shown in Fig. 3 on the three-dimensional structure of cow CPA1 suggested the following: (1) the two replacements along the rat CPA1 lineage appear to make the substrate specificity pocket smaller; (2) the single replacement along the rat CPA2 lineage appears to make the pocket larger[5]; and (3) the replacement of isoleucine at position 243 by a glycine in CPB may be necessary so that arginine and lysine can fit properly into the pocket and have access to the aspartic acid at position 255. Consistent with these computer modeling studies, rat CPA1 prefers substrates with smaller side chains than does cow CPA1, and rat CPA2 prefers substrates with larger side chains.[5]

The crystal structures of rat CPA2[23] and pig proCPA1 and proCPB[24] have been determined. More detailed analyses of these structures may lead to more refined predictions regarding the relationship between changes in protein sequence and function.

Testing Hypotheses through Site-Directed Mutagenesis

Hypotheses regarding the consequences of specific amino acid replacements on substrate specificity can now be tested experimentally through site-directed mutagenesis of carboxypeptidase.[25] Site-specific mutations can be introduced into cloned DNA by several highly efficient and simple methods. These methods have been summarized[26] and are not detailed here.

The predictions from the above cladistic analysis and modeling are being tested through site-directed mutagenesis of cloned rat CPA1. Preliminary studies[27] suggest that rat CPA1 can be converted to a cow CPA1-like enzyme through the amino acid replacements shown in Fig. 3. Likewise, an enzyme with CPA2-like activity can be created, although it is not as

[21] K. Titani, L. H. Ericsson, K. A. Walsh, and H. Neurath, *Proc. Natl. Acad. Sci. U.S.A.* **72,** 1666 (1975).

[22] M. F. Schmid and J. R. Herriott, *J. Mol. Biol.* **103,** 175 (1976).

[23] Z. Faming, B. Kobe, C.-B. Stewart, W. J. Rutter, and E. J. Goldsmith, *J. Biol. Chem.* **266,** 24606 (1991).

[24] A. Guasch, M. Coll, F. X. Avilés, and R. Huber, *J. Mol. Biol.* **224,** 141 (1992).

[25] M. A. Phillips, R. Fletterick, and W. J. Rutter, *J. Biol. Chem.* **265,** 20692 (1990).

[26] L. Hedstrom, L. Graf, C.-B. Stewart, W. J. Rutter, and M. A. Phillips, this series, Vol. **202,** p. 671.

active as wild-type CPA2. The creation of a CPB-like enzyme from rat CPA1 has not proved to be so easy; these mutant enzymes are unstable, suggesting that further mutations are needed.[27]

Concluding Remarks

The above analysis leads to a better understanding of the evolution of substrate specificity in the pancreatic carboxypeptidase family and presents testable hypotheses (see Fig. 3). The phylogenetic tree for the family provides a rough time scale regarding when these duplications occurred and therefore suggests which species are likely to contain the various gene duplicates. Larger three-dimensional structural differences are expected between products of ancient gene duplications than between products of orthologous genes, such as cow CPA1 and rat CPA1. Therefore, the phylogeny reveals which enzymes may be the best targets for mutagenesis studies. Ideal enzymes for such studies are closely related, yet show interesting changes in function.

The products of ancient gene duplications, such as CPA1, CPA2, and CPB, have been diverging for long evolutionary periods. Ancient duplicates generally show larger changes in backbone structure, and simple amino acid replacements are less likely to confer altered specificities. For example, chymotrypsin and trypsin are homologous endopeptidases with substrate preferences analogous to carboxypeptidases A and B, respectively.[3] Chymotrypsins and trypsins are the products of ancient gene duplications and differ substantially in primary sequence. Mutagenesis of amino acid residues in trypsin that had been identified through crystallography as being responsible for trypsin-like substrate specificity to those predicted to produce chymotrypsin-like activity did not result in highly active enzymes with altered specificity.[28,29] The successful mutagenesis of trypsin to chymotrypsin-like activity required replacement of surface loops outside of the binding pocket.[29] Similar loop replacements may be necessary to change CPA1 so that it has CPB-like activity. Thus, to understand the evolution of paralogous enzymes, features larger than amino acids may need to be examined in a phylogenetic framework.

This analysis was presented as a simple model of how the comparative method[1] can be applied to the study of protein evolution and does not fully illustrate the potential resolving power of cladistic analysis compared to

[27] Caro-Beth Stewart and William J. Rutter, unpublished results (1990).
[28] L. Gráf, A. Jancsó, L. Szilágyi, G. Hegyi, K. Pintér, G. Náray-Szabó, J. Hepp, K. Medzihradszky, and W. J. Rutter, *Proc. Natl. Acad. Sci. U.S.A.* **85,** 4961 (1988).
[29] L. Hedstrom, L. Szilágyi, and W. J. Rutter, *Science* **255,** 1249 (1992).

simple sequence comparisons. Conclusions regarding the evolution of the pancreatic carboxypeptidase family presented here should be considered tentative because of the small number of proteins on which the analysis is based. To increase the reliability of the cladistic reconstruction at any given position, more sequences from diverse species will need to be determined and analyzed in the proper phylogenetic framework. Similarly, these enzymes will need to be characterized kinetically and the changes in specificity correlated with changes in sequence. Information gained from these types of analyses will further our understanding of the structural basis of enzyme specificity.

Designing mutant proteins with altered functions is a goal of numerous research groups worldwide. Many of the successful attempts to modulate function by site-directed mutagenesis have used primary and tertiary structural comparisons in designing mutants, although explicit evolutionary models provided by phylogenetic trees do not appear to be widely used. Explicit evolutionary models will lead to a better understanding of how the evolutionary process modifies protein structure and function. This increased understanding should prove to be helpful in experimental protein design.

[44] Acquisition of New Metabolic Activities by Microbial Populations

By Barry G. Hall and Bernhard Hauer

Introduction

Acquisition of new metabolic activities by microorganisms has been studied deliberately since the early 1960s. Most of those studies, many of which are reviewed in detail in Mortlock,[1] were conducted using well-characterized laboratory strains of *Escherichia coli, Klebsiella,* or *Pseudomonas aeruginosa* and were intended to provide insights into evolutionary processes. In most cases the new activities were catabolic functions simply because the organisms in question were already capable of all required anabolic functions. There is no reason, however, why the methods outlined below should not be applicable to isolation of mutants with novel biosynthetic capabilities, provided that a sufficiently strong selection for those capabilities can be devised. The purpose of this chapter is to provide some

[1] R. P. Mortlock (ed.), "Microorganisms as Model Systems for Studying Evolution." Plenum, New York and London, 1984.

examples of such selection methods in both "laboratory" organisms and in organisms of industrial interest. We shall not concern ourselves with cases in which novel activities are gained by acquisition of genetic material from other organisms (i.e., by plasmid transfer) but shall restrict ourselves to cases in which there has been a change in the genetic material of the host itself.

Major Means by Which Microorganisms Acquire Novel Functions

Microorganisms have three major means by which they can acquire new activities. The first is by regulatory mutations; classical cases are those in which the novel substrate for growth is already a substrate for an existing catabolic pathway but is not an inducer for expression of that pathway. Some well-studied cases include evolution of xylitol utilization by constitutive expression of the ribitol dehydrogenase in *Klebsiella aerogenes,*[2,3] utilization of D-arabinose by constitutive expression of L-fucose isomerase in *E. coli* and in *Klebsiella,*[4,5] and formamide utilization by *Pseudomonas aeruginosa.*[6] The mutations can alter either regulatory sites or regulatory proteins; for example, both *lacI* and *lacO^c* (*lac* operon constitutive) mutations can be isolated by selection for growth on the "novel" substrate phenyl-β-galactoside, which is a noninducing substrate of β-galactosidase.

The second way of acquiring new activities is by mutations that alter properties of "structural gene" products such as enzymes and transport proteins. The best-studied examples of this class include mutations that alter the aliphatic amidase of *Pseudomonas aeruginosa* to permit utilization of butyramide and other amides that are poor substrates for the wild-type *amiE* gene product and mutants of the *ebgA* encoded β-galactosidase of *E. coli* that greatly increase the catalytic efficiency of the enzyme so that the enzyme can hydrolyze a variety of β-galactoside sugars well enough to permit their utilization as sole carbon sources. Both of these systems have been extensively reviewed.[6-9]

[2] R. P. Mortlock, D. D. Fossitt, and W. A. Wood, *Proc. Natl. Acad. Sci. U.S.A.* **54**, 572 (1965).
[3] T. T. Wu, E. C. C. Lin, and S. Tanaka, *J. Bacteriol.* **96**, 447 (1968).
[4] K. P. Camyre and R. P. Mortlock, *J. Bacteriol.* **90**, 1157 (1965).
[5] D. J. LeBlanc and R. P. Mortlock, *J. Bacteriol.* **106**, 90 (1971).
[6] P. H. Clarke, *in* "Microorganisms as Model Systems for Studying Evolution" (R. P. Mortlock, ed.) p. 187. Plenum, New York and London, 1984.
[7] B. G. Hall, *in* "Evolution of Genes and Proteins" (M. Nei and R. Koehn, eds.), p. 234. Sinauer, Sunderland, Massachusetts, 1983.
[8] B. G. Hall, *in* "Microorganisms as Model Systems for Studying Evolution" (R. P. Mortlock, ed.), pp. 165–185. Plenum, New York and London, 1984.
[9] B. G. Hall, *BioEssays* **12**, 551 (1990).

The final mechanism is by mutations that activate "cryptic" or silent operons. Many microorganisms, including *E. coli*,[10,11] *Pseudomonas putida*,[12] and the eukaryote *Saccharomyces cerevisiae*,[13] have been shown to possess silent genes that require a mutation in order to be expressed. The best-studied cases involve the cryptic *bgl* and *cel* operons for utilization of β-glucoside sugars in *Escherichia coli*. Both operons include fully functional genes for the transport and hydrolysis of arbutin and salicin, and the *cel* operon also allows transport and hydrolysis of cellobiose.[14] Neither operon can be expressed in wild-type strains.[11] The strains in which one of the operons has been activated by a spontaneous mutation usually express that operon only when induced by one of the substrates, but some isolates express the operon constitutively.[14-16] Two kinds of mutations have been shown to activate these silent operons: (1) base substitutions in a gene within the operon that encodes a regulatory protein and (2) insertion of one of the mobile genetic elements *IS1, IS2,* or *IS5* in a region that is just upstream of the operon itself.[15,17,18] In neither case is the mechanism by which insertion sequences activate the operon known, although it is known that they do not provide promoters for transcription. Despite the regulatory and functional similarities between the two operons, they exhibit no homology at either the DNA or amino acid sequence levels.[16]

It must be emphasized that the three mechanisms of acquiring new activities are not by any means mutually exclusive. In almost every system that has been studied both regulatory and "structural" gene mutations are required for full expression of the new growth phenotype. In most cases these multiple mutations are isolated in a series of independent steps, either by selecting for improved growth on the original novel substrate[19] or by selecting on a series of related substrates.[20] In the *ebg* system, however, all of the spontaneous lactose-utilizing mutants of a Δ*lacZ* strain had both a regulatory and a structural gene mutation.[21]

Although the foregoing discussion has emphasized the acquisition of

[10] B. G. Hall, S. Yokoyama, and D. Calhoun, *Mol. Biol. Evol.* **1,** 109 (1983).
[11] B. G. Hall and P. W. Betts, *Genetics* **115,** 431 (1987).
[12] J. H. Slater, A. J. Weightman, and B. G. Hall, *Mol. Biol. Evol.* **2,** 557 (1985).
[13] C. E. Paquin and V. M. Williamson, *Mol. Cell Biol.* **6,** 70 (1986).
[14] M. Kricker and B. G. Hall, *Mol. Biol. Evol.* **1,** 171 (1984).
[15] L. L. Parker and B. G. Hall, *Genetics* **124,** 473 (1990).
[16] L. L. Parker and B. G. Hall, *Genetics* **124,** 455 (1990).
[17] A. E. Reynolds, J. Felton, and A. Wright, *Nature (London)* **203,** 625 (1981).
[18] A. E. Reynolds, S. Mahadevan, S. F. J. LeGrice, and A. Wright, *J. Mol. Biol.* **191,** 85 (1984).
[19] S. A. Learner, T. T. Wu, and E. C. C. Lin, *Science* **146,** 1313 (1964).
[20] B. G. Hall, *Genetics* **89,** 453 (1978).
[21] B. G. Hall and N. D. Clarke, *Genetics* **85,** 193 (1977).

novel catabolic activities (because these are the best-studied cases), it should not be thought that new anabolic activities cannot be selected. *Lactobacillus casei* is naturally multiply auxotrophic, but in one case spontaneous mutants were isolated that no longer required 7 of 12 amino acid that were required by the wild-type strain.[22]

Techniques for Isolating Mutants with Novel Capabilities

Although it is counterintuitive, chemical or UV mutagenesis appears to be no more effective than reliance on spontaneous mutations to produce the desired phenotypes. The key elements for success in isolating mutants with novel capabilities, especially those capabilities that require multiple mutations, appear to be (1) large populations, (2) intense selection, and (3) prolonged periods of selection that may last up to several weeks.

Selection can be accomplished either in liquid cultures or on plates. Plates are preferred when it is desirable to isolate many independent mutants. For liquid cultures the simplest approach is to inoculate several cultures of a convenient size in which the medium contains a limiting supply of some resource so that the culture density is limited to about 10–20% of the saturation density. When one is selecting mutants with novel catabolic capabilities, the culture is provided with a limiting concentration of a utilizable carbon or nitrogen source and an excess of the novel resource. For *Escherichia coli* and other members of the family Enterobacteriaceae, 0.02% (w/v) glucose or 0.02% (v/v) glycerol limits cells to about 2×10^8/ml. Thus, even cultures as small as 10 ml provide quite large populations. Selection for anabolic functions is even more straightforward. Cultures are simply grown to an appropriate density with a limiting supply of the required nutrient. The cultures are inspected daily until a sudden increase in turbidity in one or more cultures indicates that the desired mutation has occurred and that the mutant has grown. Inspection is facilitated by including, for comparison, one or two cultures in which the novel resource is not present. The resulting desired mutant is subsequently purified by streaking from the turbid culture onto plates that contain only the novel resource (or that do not contain any of the required nutrient, in the case of anabolic functions).

The time required for mutants to appear (i.e., for the culture to become turbid) usually reflects the time required for the mutation to occur, rather than the time required for the rare mutant to grow. The time also depends on the nature and number of mutations that are required. Mutants of *Aerobacter (Klebsiella) aerogenes* capable of utilizing D-arabinose were

---

[22] T. Morishita, T. Fukada, M. Shirota, and T. Yura, *J. Bacteriol.* **120,** 1078 (1974).

typically detected after 1–2 days of incubation of the cultures, whereas L-xylose-utilizing mutants appeared only after 40–60 days of incubation.[23] In both cases, inoculation of fresh cultures with a few mutant cells resulted in full turbid growth within 2–3 days. Similar results were obtained when a spontaneous citrate-utilizing mutant of *E. coli* K12 was selected after 14 days of incubation in liquid medium. The growth rate of the citrate-utilizing mutant was such that a single mutant cell could saturate the medium in 2.2 days; thus, the mutation could have occurred no earlier than day 11. This was a particularly interesting case, because citrate utilization required mutations in two widely separated genes, *citA* and *citB*.[24]

Selection on plates may be accomplished with solidified media identical to the media described above or on a variety of "indicator" plates such as MacConkey medium (Difco, Detroit, MI), TTC (triphenyltetrazolium chloride), or EMB (eosin methylene blue) medium[25] when the selection is for sugar utilization. Indicator media typically provide amino acids and small peptides as carbon–energy–nitrogen sources, and they support growth of cells whether or not they can use an added sugar. If the sugar is fermented, the colonies are a different color from those colonies that cannot ferment the included sugar, thus facilitating identification of the sugar-utilizing mutants.

Whatever the medium that is used, plates are spread with about 100 cells so that distinct colonies are formed. Growth ceases when the limiting nutrients are exhausted. When a mutation that permits utilization of the novel resource occurs, the mutant cell begins to grow and within 1–2 days forms an outgrowth called a papilla on the surface of the colony. The advantages of this approach are that (1) papillae on different colonies are independent mutants, and thus it is often possible to isolate a dozen or more independent mutants from the same plate; (2) a single plate can support about 10 times as many cells as a single nutrient-limited liquid culture; and (3) it is often more convenient to incubate 100 plates than 100 liquid cultures. Disadvantages include (1) desiccation of plates during prolonged incubation, (2) the need to examine each plate carefully at frequent intervals, (3) contamination resulting from occasional opening of the plates when condensation is present, and (4) the appearance of "false" papillae. Desiccation can be avoided by incubating the plates in tightly closed boxes into which is placed a small beaker of water with a few tissue papers as wicks to provide a humid atmosphere. Contamination can largely be avoided by working under a laminar flow hood or other sterile

[23] E. J. Oliver and R. P. Mortlock, *J. Bacteriol.* **108**, 287 (1971).
[24] B. G. Hall, *J. Bacteriol.* **152**, 269 (1982).
[25] J. H. Miller, "Experiments in Molecular Genetics." Cold Spring Harbor Laboratory, Cold Spring Harbor, New York, 1972.

hood. "False" papillae result from adaptive mutations other than those of interest, and they are false only from the perspective of the investigator. False papillae typically appear as small bumps on colonies, namely, bumps that do not continue to grow after their initial appearance. It is not unusual to observe that after a couple weeks the majority of colonies resemble the surfaces of raspberries. It is usually wise to wait 1–2 days after the initial appearance of papillae, in order to determine if additional growth occurs, before restreaking the putative mutant onto selective medium for purposes of purification and isolation. On balance the advantages of plates often outweigh the disadvantages.

Time Dependence of Mutations

The time element should not be ignored when selecting mutants with novel capabilities. In a number of studies incubation times of 2–4 weeks have been required, following which papillae began to appear such that within a few days the majority of colonies had papillae.[26,27] Time dependence is particularly evident when the selected phenotype requires multiple mutations,[27] but we have observed situations in which single-event mutations required several weeks of incubation (B. G. Hall, unpublished results, 1992). It is becoming increasingly evident that, during prolonged intense selection, time-dependent mutations occur in nondividing cells.[28] The mechanism(s) responsible for those mutations is not yet understood, but the process is so powerful that it can select double mutations (base substitutions in two different genes) at rates that are five to eight orders of magnitude higher than would be predicted by the individual mutation rates under identical conditions.[29]

Organisms of Industrial Interest

Coryneform bacteria are irregular, nonsporulating gram-positive bacteria that are widely distributed in nature. In recent years some strains have been reclassified into new genera.[30] Corynebacteria are widely used for amino acid production.[31] Some have relevance in bioconversions, for example, steroid conversion,[32] terpenoid conversion,[33] and the conversion of

[26] B. G. Hall and D. L. Hartl, *Genetics* **76**, 391 (1974).
[27] B. G. Hall, *Genetics* **120**, 887 (1988).
[28] B. G. Hall, *Genetics* **126**, 5 (1990).
[29] B. G. Hall, *Proc. Natl. Acad. Sci. U.S.A.* **88**, 5882–5886 (1991).
[30] M. P. Starr, H. Stolp, H. G. Truper, A. Balows, and H. G. Schlegel, "The Prokaryotes." Springer-Verlag, Berlin, 1986.
[31] O. Tosaka, H. Enei, and Y. Hirose, *Trends Biotechnol.* **1**, 70 (1983).
[32] A. Constaninides, *Biotechnol. Bioeng.* **22**, 199 (1980).
[33] Y. Yamada, C. W. Seo, and H. Okada, *Appl. Environ. Microbiol.* **49**, 960 (1985).

acrylnitrile to acrylamide.[34] Other applications include degradation of hydrocarbons[35] and the production of emulsifying and surface active agents.[36]

The strains listed in this chapter are useful for L-lysine production *(Corynebacterium glutamicum)*, reduction of 2,5-diketo-D-gluconic acid to the vitamin C precursor 2-keto-L-gulonate *(Nocardia* sp., *Arthrobacter ilicis, Clavibacter rathayi)*, the degradation of papaverine (*Nocardia* sp.), or the hydroxylation of phenylacetic acid (*Nocardia* sp.).[37] Table I shows the results of selection experiments using those strains (B. Hauer, unpublished results). None of the wild-type strains can utilize the listed carbon sources. Dilute suspension of the cultures were spread onto plates containing mineral salts medium plus 50 mg/liter yeast extract. Colonies grew to about 3×10^7 cells (cfu, colony-forming units) per colony, and the number of

TABLE I
CARBON SOURCES THAT CAN BE UTILIZED BY MUTANTS OF CORYNEFORM BACTERIA[a]

Carbon source[b]	Corynebacterium glutamicum	Nocardia sp.	Arthrobacter ilicis	Clavibacter rathayi	Microbacterium laevaniformans
D-Glucose	−	+	+	−	−
D-Galactose	−	−	+	−	−
L-Gulonic	−	−	+	−	−
L-Idonic acid	−	+	−	−	−
5-Keto-D-gluconic acid	−	+	+	+	−
D-Ribose	+	−	+	−	−
D-Arabinose	−	−	+	−	−
L-Arabinose	−	+	+	−	−
D-Xylose	−	+	+	−	−
L-Rhamnose	−	+	+	−	−
D-Arabitol	−	−	−	−	−
L-Arabitol	−	−	+	−	−
Xylitol	−	−	+	−	+
Lactose	−	−	+	−	+
Cellobiose	−	−	+	−	−
Melibiose	−	−	+	−	−
Salicin	−	+	+	−	+
Arbutin	−	−	−	−	−
Starch	−	−	−	−	−
L-Lactic acid	−	−	+	−	−

[a] (+) Papillae appear that can utilize the carbon source; (−) papillae that can utilize the carbon source do not appear.
[b] None of the carbon sources can be utilized by the wild-type organisms.

[34] T. Nagasawa and H. Yamada, *Trends Biotechnol.* **7,** 153 (1989).
[35] D. G. Cooper, J. E. Zajic, and D. E. F. Gracey, *J. Bacteriol.* **135,** 795 (1979).
[36] Z. Duvnjak and N. Kosaric, *Biotechnol. Lett.* **3,** 583 (1981).
[37] B. Hauer, K. Haase-Aschoff, and F. Lingens, *Hoppe-Seyler's Z. Physiol. Chem.* **363,** 499 (1982).

viable cells remained constant for 35 days. After 5 to 60 days, depending on the carbon source and the organism, the first papillae appeared on the colonies, and the number of colonies with papillae increased over the next several days. Papillae continued to grow and eventually overgrew the original tiny colonies, forming full-sized colonies containing about 10^9 cells. In each case, when cells from a papilla were restreaked onto the appropriate selective medium, normal colonies grew up within 2 days.

Arthrobacter ilicis was especially interesting in that it was able to evolve the ability to utilize 16 of the 20 tested carbon sources. The kinetics of papillation on lactose, L-arabitol, and L-gulonic acid were studied by depositing single cells in an equally spaced grid onto plates by using a flow cytometer in combination with a cell sorter. Using this technique 120 cells were placed in a regular order with equal spacing. Figure 1 shows the kinetics of papilla formation on the three substrates.

In exponentially growing liquid cultures, the spontaneous mutation rate to lactose utilization *(lac+)* is high (Table II), which may explain the early appearance of Lac+ papillae. However, neither L-gulonic acid-utilizing nor L-arabitol-utilizing mutants were ever obtained by plating liquid cultures directly onto selective media.

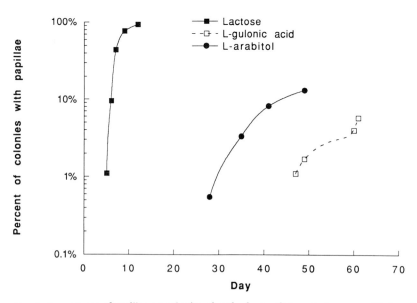

FIG. 1. Appearance of papillae on colonies of *Arthrobacter ilicis* on lactose, L-arabitol, and L-gulonic acid selective plates. No papillae were observed before the first point shown.

TABLE II
MUTATION FREQUENCIES IN EXPONENTIALLY
GROWING CULTURES OF *Arthrobacter ilicis*

Phenotype	Mutation frequency
Streptomycin resistance	2.8×10^{-9}
Lactose utilization	3.2×10^{-6}
L-Gulonic acid utilization	$<1.1 \times 10^{-9}$
L-Arabitol utilization	$<9.8 \times 10^{-10}$

Indirect Selection

When multiple mutations are required for the desired new phenotype, it may not be possible to select that phenotype in one step. A case in point is selection for lactobionate utilization by $\Delta lacZ$ strains of *E. coli*. Lactobionate utilization requires three mutations in the *ebgA* gene to produce an enzyme with sufficient catalytic capability, along with a mutation in *ebgR* to permit constitutive expression of the *ebg* operon.[20] Constitutive (*ebgR⁻*) mutants with a single substitution in *ebgA* were selected in one step by selecting for lactose utilization. A second round of selection, for lactulose utilization, provided the second substitution in *ebgA*, following which selection on lactobionate itself produced the desired mutant. Thus, when selection on a novel substrate proves unsuccessful, it is worthwhile to attempt selection on structural analogs of the novel substrate.

With *Arthrobacter ilicis* the time required for the appearance of L-arabitol-utilizing papillae is dramatically reduced for rhamnose-utilizing or ribitol-utilizing mutants, compared with the wild-type strain (Fig. 2). Indirect selection need not be a multistep process, however. When 110 L-arabitol mutants of *Arthrobacter ilicis* were tested, 42 could also utilize D-ribose, and 18 of the 42 could also utilize L-rhamnose. Similarly, 16 of 60 Lac⁺ mutants could also utilize L-arabinose.

Finally, B. G. Hall (unpublished results, 1985) has found that cellobiose-utilizing mutants of *E. coli* K12 could be isolated much more easily by selection on the related sugar arabinose than by selection on cellobiose itself.

Other Factors That May Affect Selection

The various factors that affect the number, rate, and timing of appearance of papillae are not well understood. Crowding of colonies on plates affects the number of papillae that appear on *Arthrobacter ilicis* colonies:

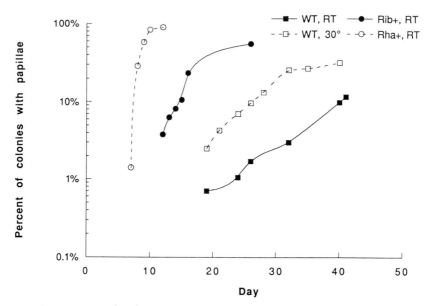

FIG. 2. Appearance of papillae on colonies of wild-type and mutant *Arthrobacter ilicis* on L-arabitol plates. WT, Wild type; Rib+, ribitol-utilizing mutant; Rha+, rhamnose-utilizing mutant; RT, room temperature.

the closer the colonies are, the fewer papillae they produce (B. Hauer, unpublished results, 1987). Both the number and time of appearance of L-arabitol-positive papillae are affected by incubation at 30° instead of at room temperature ($\sim 22°$) (Fig. 2). The rate at which plates dehydrate can also affect the papillation rate in *E. coli* (B. G. Hall, unpublished observations).

Predicting Success of Selection Schemes

Despite several careful and intense studies, we do not yet have much insight into the cellular properties that will predict the chance of success in selecting novel capabilities.[9] Some general observations, however, can be made. (1) If even a very weak activity can be detected *in vitro*, it is very likely that mutations which will increase that activity to a point of functionality can be selected. (2) If activity toward an analog of a novel substrate can be detected, it is likely that the desired mutation can be selected. (3) If a fairly close relative of the organism in question exhibits the desired phenotype, it is likely that the phenotype can be selected in the organism of interest.

Comments

It should be pointed out that the approaches outlined here are certainly not new, especially to microbiologists. In the 1947 book, "The Bacterial Cell"[38] Rene Dubos includes a chapter on the variability of bacteria, in which he points out that "bacteria also exhibit orderly, predictable changes which are related to the age of the culture and which are the expression of a definite growth cycle." It seems likely that he was on the right track; however, it is difficult to reproduce his experiments because the cultures were not deposited or classified.

Acknowledgments

Studies in the laboratory of B. G. Hall have been supported by grants from the Medical Research Council of Canada, the National Institutes of Health, and the National Science Foundation, and by grants from the University of Connecticut and the University of Rochester.

[38] R. J. Dubos, "The Bacterial Cell." Harvard Univ. Press, Cambridge, Massachusetts, 1947.

[45] Chemostats Used for Studying Natural Selection and Adaptive Evolution

By DANIEL E. DYKHUIZEN

Introduction

Comparisons of anatomies, metabolic pathways, and DNA sequences from different species have provided us with our understanding of the history and process of evolution. In contrast, experimental studies of evolution and adaptation have been less informative, primarily owing to the impracticality of working with large populations over many generations. Experimental studies can be done with microorganisms, which are very small and have short generation times. An experiment of 50 generations can be done between Monday morning and Friday afternoon and long-term selection experiments of 5000 generations in one year. Experimental studies are also hampered by heterogeneity of genetic background and a varying and complex environment. Use of isogenic strains, defined media, and continuous culture devices like chemostats can reduce genetic and environmental heterogeneity to a minimum.

Chemostats have been used in two ways to study evolution. These are distinguished as the study of natural selection and the study of adaptive evolution. The first approach attempts to characterize natural selection in terms of how selection coefficients change as either the genetics or the environment is changed.[1] Two strains are mixed, and the initial selective difference between the strains is measured over the first 50 generations. This is similar to studying initial rate kinetics in enzymology. The purpose of the second approach is to follow the course of adaptive evolution of a population in a particular environment.[2-4] A single initial strain is monitored over hundreds to thousands of generations to determine the evolutionary changes. The advantage of the first approach is that the genetic differences are known and the effects of these differences in various environments can be explored. The advantage of the second approach is that the actual evolutionary trajectory of fitness changes is followed. This has provided some surprising results such as nontransitivity[5] and "speciation." [6] The disadvantage is that the precise genetic changes are usually hard to determine and investigate.

Chemostats

Evolutionary experiments can be done in tubes, on petri plates, or in various continuous culture devices. Each type of culture provides different environmental conditions and different advantages and disadvantages.[7] Although there are many different kinds of continuous culturing devices,[8] most evolutionary experiments have been carried out in chemostats. Chemostats are designed to provide a constant, homogeneous environment in which the cells grow at a constant rate.

Chemostats are open systems in which nutrients are continually added at a constant rate and spent medium plus cells removed at the same rate, such that a constant volume (V) is maintained. The rate of addition of medium (f) is such that the dilution rate $(D = f/V)$ is less than the maximal growth rate of the bacteria in the medium used. At equilibrium, the organismal growth rate (μ) equals D, the dilution rate.[9,10] D is estimated by

[1] D. E. Dykhuizen and A. M. Dean, *Trends Ecol. Evol.* **5**, 257 (1990).

[2] H. E. Kubitschek, *Symp. Soc. Gen. Microbiol.* **24**, 105 (1974).

[3] J. C. Francis and P. E. Hansche, *Genetics* **74**, 259 (1973).

[4] D. Dykhuizen and D. Hartl, *Evolution* **35**, 581 (1981).

[5] C. E. Paquin and J. Adams, *Nature (London)* **306**, 368 (1983).

[6] R. B. Helling, C. N. Vargas, and J. Adams, *Genetics* **116**, 349 (1987).

[7] D. E. Dykhuizen, *Annu. Rev. Ecol. Syst.* **21**, 373 (1990).

[8] J. W. T. Wimpenny (ed.), "Handbook of Laboratory Model Systems for Microbial Ecosystems," Vol. 1. CRC Press, Boca Raton, Florida, 1988.

[9] H. E. Kubitschek, "Introduction to Research with Continuous Cultures." Prentice-Hall, Englewood Cliffs, New Jersey, 1970.

[10] D. E. Dykhuizen and D. L. Hartl, *Microbiol. Rev.* **47**, 150 (1983).

measuring both the flow of medium either into or out of the chemostat during the experiment (usually 3 to 4 times to confirm that the rate of pumping has remained constant) and the culture volume after the experiment ends.

The medium is constructed so that one required nutrient, called the limiting nutrient, is added in amounts that limit the number of bacteria. Normally the limiting nutrient is added in a concentration such that the number of bacteria per milliliter is 5- to 50-fold lower than the maximal density for the medium. If the limiting nutrient is truly limiting, addition of more of the nutrient increases cell numbers proportional to amount added.

Chemostat Design

The chemostats used by evolutionary biologists are small and simple compared to those typically used in microbiology and biotechnology. These chemostats have working volumes of 30 to 100 ml and are operated without sensors or regulating devices. The small size and simple design means they are comparatively inexpensive. The complicated parts can be constructed by a local glassblower and the rest from standard laboratory materials. The low price makes it possible to run many chemostats at the same time, allowing an experimenter to perform simultaneously replicated experiments and controls. Replication is critical since evolution is a dynamic process where chance events like mutation are important. Lack of replication makes interpretation of some experiments difficult.[11]

Because the chemostats are not generally purchased, we briefly discuss the various design criteria and some of the solutions so that individuals can assemble their own system. The major design requirement of a chemostat is that the culture remain homogeneous in space and constant in time. Figure 1 is from the design of Kubitschek[9] as modified for use in our laboratory. Figure 2 is from a design from Porton Down[12] and is similar to the design used by New Brunswick Scientific Co. (East Brunswick, NJ) for their Bioflo, a bench-top chemostat. Even the small New Brunswick chemostats are quite expensive when used for evolutionary studies. Figure 3 is from the design of Searcy and Levin[13] as modified by L. Chao (personal communication, 1992) and is the least expensive to construct. Because one can put together parts of various designs, the various options for each component are described together, and we refer to Figs. 1–3 as Models 1–3, respectively.

[11] S.-D. Tsen, *Biochem. Biophys. Res. Commun.* **166,** 1245 (1990).
[12] C. G. T. Evans, D. Herbert, and D. W. Tempest, *in* "Methods in Microbiology" (J. R. Morris and D. W. Ribbons, eds.), p. 2277. Academic Press, London and New York, 1970.
[13] L. Chao, B. R. Levin, and F. M. Stewart, *Ecology* **58,** 369 (1977).

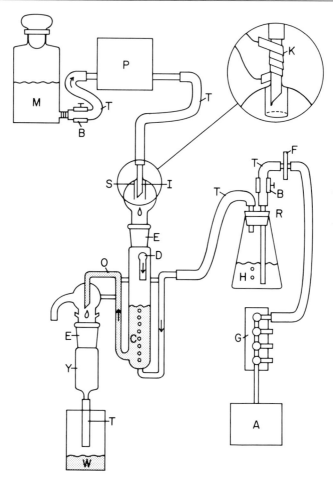

Fig. 1. Kubitschek style chemostat. A, Air pump; B, hosecock clamps (closed during autoclaving); C, growing culture; D, air exit port; E, ground glass joints; F, syringe filter; G, gang valve; H, bottle containing water; I, input area; K, heating tape; M, fresh medium; O, siphon; P, peristaltic pump; R, rubber stopper with holes; S, splash shield; T, silicone rubber tubing; W, waste; Y, waste collection tube. Circles in the liquids represent bubbles. Stippling indicates media with cells.

Media Storage and Pump

The medium used is a buffered defined medium. We use Davis salts [7 g K_2HPO_4, 2 g KH_2PO_4, 1 g $(NH_4)_2SO_4$, 0.5 g sodium citrate, 0.2 g $MgSO_4 \cdot 7H_2O$ in 1 liter of distilled water] and 0.1 g sugar per liter for sugar-limited chemostats. For nitrogen-limited chemostats, the amount of ammonium sulfate is decreased and sugar increased. As Davis salts uses the phosphate for buffering, phosphate-limited chemostats require a completely different medium. Medium is stored in 4-liter glass aspirator bottles, plugged at top with a cotton plug (Figs. 1 and 3). The bottles can be easily refilled by pouring in freshly autoclaved medium. Only if very large

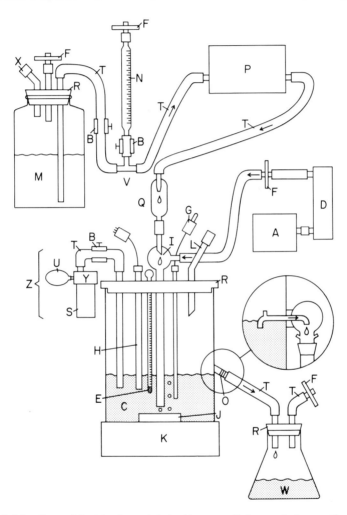

FIG. 2. New Brunswick style chemostat. A, Air pump; B, hosecock clamps; C, growing culture; D, air flowmeter; E, thermometer; F, syringe filter; G, thermistor sensing probe; H, heater; I, input area; J, impeller or stir bar; K, magnetic stirrer; L, inoculation port; M, fresh medium; N, graduated tube; O, output area; P, peristaltic pump; Q, medium break tube; R, rubber stopper with holes; S, screw-top bottle; T, silicone rubber tubing; U, rubber bulb; V, glass T joint; W, waste; X, addition port; Y, autoclavable plastic screw top with two holes; Z, sampling device. Circles in the liquids represent bubbles. Stippling indicates media with cells.

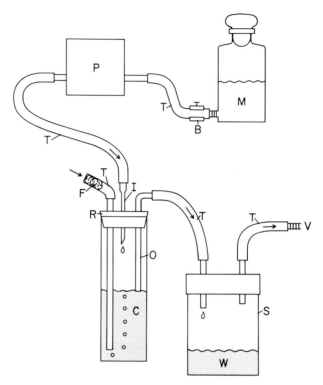

Fɪɢ. 3. Searcy–Levin style chemostat. B, Hosecock clamp (closed during autoclaving); C, growing culture; F, cotton; I, input of fresh medium using an 18-gauge stainless steel needle; M, fresh medium; O, output area; P, peristaltic pump; R, rubber stopper with holes; S, screw-top bottle; T, silicone rubber tubing; V, vacuum; W, waste. Circles in the liquids represent bubbles. Stippling indicates media with cells.

amounts of medium are to be stored (i.e., volumes for which aspirator bottles are prohibitively expensive) is the top entry system (Fig. 2) recommended.

Medium is pumped into the chemostat at a constant rate using a good quality multichannel peristaltic pump with roller heads, for example, the Wiz pump by ISCO (Lincoln, NE). The pumps should use thick-walled silicone rubber tubing as a pumping chamber, tubing with walls of 1.6 mm for a 1.6 mm internal diameter. The thick wall prevents the enlarging of the pumping chamber during the experiment as the wall wears and thus prevents any increase in flow rate. The pumping chambers are connected to the silicone rubber tubing (T in Figs. 1–3) with glass nipples. Heating the glass nipples before inserting them into the ends of the tubing helps seal the joints.

Most modern pumps are microprocessor controlled, allowing accurate resetting of pump speed from experiment to experiment. However, if adjustment of the flow rate at the start of a chemostat run is required, it is done by measuring the time (t, sec/drop) between drops coming into the chemostat. Before the experiment, the chemostat volume (V, ml) and the amount of liquid in each drop (d, ml/drop) from the inflow spout are measured. The estimated generation time, g (hr^{-1}), is

$$g = [(1.69)tV]/3600d$$

This is a less accurate but much quicker way to estimate generation time than measuring the amount of outflow over a few hours (Models 1 and 3). Model 2 has incorporated a pipette in the tubing between the bottle of medium and the pump to measure the flow rate. The pipette is filled by gravity when the stopcock leading to the pipette is opened. When the pipette is full, the stopcock to the medium bottle is closed. The rate of removal of medium from the pipette is measured to obtain the flow rate.

Inflow

The design of the inflow spout should make the drops of incoming medium as small as possible and prevent growth of bacteria in it. The design in Fig. 1, the simple spout, has a beveled end with a point to minimize drop size and a shield around the spout to prevent contamination. This design works well for short experiments (< 1 week), but the spout inevitably becomes contaminated over longer periods. Because only a few cells contaminate the spout, the ratio of strains in the spout is usually quite different than in the chemostat, and the migration of cells from the spout into the culture will cause a misestimation of the selection coefficient. This is prevented by wrapping heating tape (1.1 × 20 cm Thermolyne/Briskheat, Thomas Scientific, Swedesboro, NJ) around the top of the inlet (Fig. 1, inset) and adjusting the temperature using a rheostat. The temperature is set high enough that the medium coming out of the inflow is too hot for the cells to grow but not so hot that cells in the culture are hurt as the drop mixes into the culture. For *Escherichia coli*, the tape is set to about 60° which heats the medium to somewhat more than 45° at the tip of the spout. The design in Fig. 2 adds an air barrier (medium break) so that the contaminating cells cannot travel up the tube. This design feature alone is unsatisfactory for selection experiments. Model 2 also incorporates positive air pressure across the spout to prevent contamination. The air plus fresh medium is driven into the culture by the air pressure, preventing droplets of culture from reaching the spout. Model 3 uses a syringe needle, which can be replaced periodically.

Mixing

The mixing must be sufficient to maintain a homogeneous culture. This does not seem to be much of a problem with small chemostats even when air is used, but it can be a serious problem with large fermenters, where mechanical mixing is required.

Air coming in the bottom provides mixing for most small chemostats, referred to as air-lift fermenters. Spargers should never be used since the small bubbles remove too much CO_2 and the pores tend to become clogged. Otherwise, the chemostat is stirred with a magnetic stir bar or paddles. The stir bar requires that the chemostat have a flat bottom and be set on a magnetic stirrer, whereas the paddles are driven from the top. Anaerobic chemostats are mixed mechanically with either stir bars or paddles.

Outflow

Three main devices are used to remove used culture and cells. The simplest is an overflow, which is a tube set in the side of culture vessel such that excess culture flows out (Model 2). The position of the overflow sets the culture volume. Sometimes a little device is added to the inside (Fig. 2, inset) so that the amount of overflow is not influenced by turbulence on the surface of the liquid. The second system is a siphon (Model 1), which removes culture from the bottom. The bore of the siphon needs to be narrow so cells do not settle during passage in the siphon tube. Siphon systems should not be used for chemostats with long generation times or for large cells like yeast. The narrow bore of the siphon can easily be clogged with small bubbles, preventing flow and causing volume fluctuations. To prevent this, the input air pressure must be low enough that small bubbles are not created. The position of the outside end of the siphon determines level of the culture and consequently culture volume. When chemostats of this type are made by a glassblower, positioning of the end of the siphon is critical if a set of chemostats are to have equal volumes, which is recommended since most pumps can run four or more chemostats. Model 3 uses vacuum to draw the medium, cells, and air out. A tube is stuck down from the top of the chemostat to the level required for a desired volume. As soon as the culture level rises above the end of the tube, liquid is drawn out. Changing the level can be used to set the flow rate.[13] This creates negative pressure in the chemostat so air enters passively, but nevertheless provides sufficient mixing. The negative air pressure pulling in contaminating organisms is not typically a problem. The system will also work with positive pressure, if the rubber stopper is securely sealed down.

Aeration

Aquarium air pumps purchased at local pet shops provide sufficient aeration. Using gang valves, one air pump can aerate a number of chemostats. Pressure gauges (Model 2) are unnecessary to obtain repeatable experiments. In Model 1, the air is sterilized by passing it through sterile 25-mm disposable syringe filters (F in Figs. 1–3) with a 0.2-μm pore size and then hydrated by bubbling through water. This prevents culture volume being lost in evaporation, a problem in Model 3, which also uses outflow to measure the flow rate.

Temperature Control

The preferred solution to maintaining culture temperature is a constant temperature room. We have found that unless a constant temperature room is heavily contaminated with bacteriophage, contamination of the chemostats is not a problem even when the rooms are shared. The second solution is use of water baths with covers having holes for insertion of chemostats. We use plexiglass water baths that are screwed together and sealed with silicone. The temperature is regulated with immersion circulators. The simple overflow (Fig. 2) prevents the chemostat culture from being fully submerged in the water bath. Larger systems can be heated by an internal heater (Fig. 2), which requires that a heater, a thermistor, and a thermometer be inserted into the culture.

Devices to Regulate Media

Some chemostats have systems that regulate physical parameters inside the chemostat like pH and oxygen tension. These systems add considerable cost and reduce the number of chemostats that can be run simultaneously. These devices should be used only if manipulation of the environment is an important part of the experiment (e.g., a study of how selection changes as pH changes).

Chemostat Operation

Measuring Initial Selection Kinetics

Short-term experiments of about 100 hr are designed to measure the initial rate of selection between two strains. As I use chemostats similar to those shown in Fig. 1, the following description of procedures will be in reference to this setup.

For continuous culture, selection coefficients are calculated by

$$\ln[x_1(t)/x_2(t)] = \ln[x_1(0)/x_2(0)] + st$$

where $x_1(t)$ and $x_2(t)$ represent the relevant density or number of the two competing strains at time t, measured in hours, $x_1(0)$ and $x_2(0)$ are the numbers at zero time, and s is the selection coefficient per hour. This equation shows that when the natural logarithm of the ratio of numbers or densities of the two strains is plotted against time, the slope of the generated straight line is the selection/hour.[10] An example is given in Fig. 4. Since there is no significant cell death in chemostats over a range of generation times, the selection measures differences in growth rate. Consequently, it is useful to normalize the selection to a per generation selection coefficient by dividing by D, the dilution rate. Then s is between -1 and $+1$, with these maximum values obtained when one strain does not grow. Frequency-dependent selection can be investigated by mixing the strains in various proportions and density-dependent selection by changing the amount of limiting nutrient in the medium.

Setup and Inoculation of Chemostats

The chemostats are assembled before autoclaving, with the ground glass joints lubricated with Lubriseal from Thomas Scientific (Swedesboro, NJ). Lubriseal grease melts at a temperature higher than that used in autoclaving and is removed at the end of each experiment with toluene before the chemostats are washed. A number is assigned to each chemostat and its

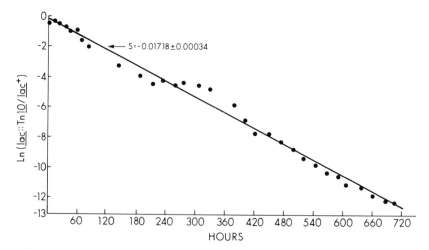

FIG. 4. Selection against the strain of E. coli carrying Tn10 in the lactose operon in a glucose-limited chemostat.

aspirator bottle. When medium is made, the part of the medium containing the limiting nutrient is made in the aspirator bottle, with the common components in flasks. Chemostats are autoclaved as a unit: medium bottle, chemostat, and air line to filter (F in Fig. 1) connected together. Open outlets are wrapped with aluminum foil.

Inoculation procedure is important for selection experiments since strains should be in the same physiological condition at the start of the experiment. Strains are grown separately in Nephlo culture flasks (Bellco Glass, Inc., Vineland, NJ) in medium similar to that in the chemostats until they are growing at the maximum rate given the medium. Nephlo or side-arm flasks have a tube coming from the side so that the optical density of the growing culture can be followed using a Klett–Summerson colorimeter or a spectophotometer without sampling from the culture. The growth rate is determined by converting optical density to cell number by a previously created conversion chart (Fig. 5) and estimating the rate (μ) from the equation

$$\ln(N_t) = \ln(N_0) + \mu t$$

When both strains are growing at their maximum rate and contain about 4×10^8 cells/ml, they are mixed in the desired proportion. At the same time the strains are growing, the chemostats are filled with fresh medium and the flow of fresh medium started. The flow is adjusted and the chemostats equilibrated. For inoculation, 2 ml of freshly mixed culture is

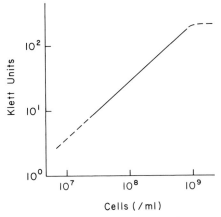

FIG. 5. Converting Klett units to number of cells per milliliter. A blue filter was used in the Klett–Summerson colorimeter, and the cells were *E. coli*. Zero units were set with Nephlo flasks containing the growth medium (0.2% glucose in Davis salts) without cells. The culture was started with an inoculation that gave $5–8 \times 10^6$ cells/ml. The chart is most accurate between 15 and 170 units.

added to the 30 ml of medium in the chemostat. The culture then grows up over the next 4 to 6 hr, and the first sample is taken after the maximal density is reached. This is a step-down inoculation because the cells go from a maximal growth rate to that of the chemostat. The other inoculation procedure, which seems to give the same result in terms of the measured selection coefficient, is a step-up inoculation, that is, the growth rate of the cells go from zero to that of the chemostat. The strains are grown separately, overnight, in the chemostat medium, then mixed in desired proportions to an amount equal to the volume of the chemostat. The mixed culture is put into the chemostat, and only then is the flow of fresh medium started.

Sampling

Samples can be taken either from the overflow if it is accessible and protected from contamination as in Fig. 1 or from within the chemostat by a sampling port. A sampling port can be made of the top half of a screw-cap tube if the cap has a good gasket. We have never found a significant difference in the ratio of the strains between sampling from the overflow and sampling within the chemostat. The sample is then diluted and plated to determine the ratio of the two types. The variance in this ratio is minimized when the diluting and plating are done immediately after sampling.

Determination of the ratio requires easy differentiation of the strains. Although a cell sorter and strain-specific antibodies could be used, traditionally strain differentiation has been done by differential growth on agar plates. Either the gene being studied or a linked neutral marker can be used to discriminate strains. Whichever is used, the marker being scored (the selectable marker) must have 100% penetrance and full viability given the conditions for scoring. For the experiment in Fig. 4, a proper dilution is spread on plates containing a concentration of tetracycline such that all *E. coli* carrying Tn*10*, which carries a gene for resistance to tetracycline, produce colonies and none of the sensitive ones do. This gives the number of cells per milliliter carrying Tn*10*. The culture is also plated on nonselective plates so that all cells produce colonies. This gives the total number of cells per milliliter in the culture. The estimated number of *lac*+ cells per milliliter is the difference of the total and the number of cells carrying Tn*10*. When the percentage of the scored cell type goes above about 60%, the error introduced by the subtraction becomes large. The scored cell type should never be more than about 75% of the culture. Any data points above this limit should be discarded. Usually, the strains are mixed such that the scored strain is 5% of the culture. This particular experiment could also have been followed by plating the cells on lactose indicator plates.[14] Then both types are directly scored from one plate. This gives more accurate estimation around 50%. When selection coefficients are small, this is the preferred method of scoring in that it avoids dilution errors and some of the sampling variance. However, scoring from a single plate becomes less accurate as one or the other genotypes becomes rare. As in Fig. 4, it is best to have the scored cell type as the selectively inferior strain and start it at a percentage of less than 50%.

Normally, one is studying phenotypes that cannot be easily scored, for example two alleles of an enzyme, distinguishable only by protein gel electrophoresis or DNA sequencing. Because bacteria reproduce asexually, the experiment can be set up such that genetic exchange is impossible. Thus, a neutral marker will remain 100% linked with the studied loci. A neutral marker must be selectively equivalent with its paired allele; for example, resistance to phage T5 is selectively equivalent to T5 sensitivity if the limiting nutrient is a sugar (a carbon energy source) but not if the limiting nutrient is nitrogen or phosphorus. Another marker often used is lac^+ versus lac^-. Control experiments should be done to show that the scored marker is selectively neutral on the various media used in the study and is selectively neutral with each of the studied alleles.

The plating can be done either by spreading the cells on the plate with a spreader (a glass rod shaped like a hockey stick) or by mixing the cells in molten soft agar and pouring this over the plate. If the counting is done by hand, either method is good. When the counting is done using an electronic colony counter, where each colony needs to be touched, 75–150 colonies/plate is optimal. Higher numbers become too tiring when counting many plates. Lower numbers require too many plates for adequate statistics. However, if the counting is done by an automatic colony counter (Artek model 880, Dynatech Labs Inc, Chantilly, VA, or similar image analyzing system), pour plating is preferred, with another layer of agar over the top. We use a base of 20 ml [15 g agar/liter in LB (Luria broth, which is 5 g yeast extract, 10 g NaCl, 10 g tryptone per liter water)] for the plates. On top of this is poured 2.5–3 ml of soft agar (8 g agar/liter in LB) with cells and, after this soft agar hardens, another 3 ml of agar (8 g agar/liter in water). This procedure gives small, uniformly sized colonies with sharp edges, and up to 1200 colonies/plate can be counted accurately. For each sample from each chemostat, we make four plates of the nonselective medium to estimate total numbers and four of the selective medium to estimate numbers of the strain carrying the marker. Each plate is counted

[14] J. H. Miller, "Experiments in Molecular Genetics" Cold Spring Harbor Laboratory, Cold Spring Harbor, New York, 1972.

four times in different orientations using the Artek system. The 16 counts for number of colonies on each type of plate are averaged and the percentage of the selectable marker estimated. The percent and time of sample are entered into a computer program for an IBM PC (available from the author on request). The selection coefficient and standard error are calculated, and various statistical tests are done following standard procedures for linear regression statistics.

Controls

Chemostat cultures are highly dynamic systems which can be misleading, causing incorrect conclusions without adequate controls and repetition.

System Controls. Selective and neutral media are defined by competing strains with a wild-type allele and a null mutation.[15] Selective medium demonstrates the occurrence and magnitude of selection against the null, showing that the gene being studied is important for the fitness of the organism under these conditions. This is a positive control and is important for the interpretation of selectively equivalent alleles. A neutral medium eliminates selection against the null. It provides an environment where the gene being studied is not important to fitness. Its purpose is to verify that the observed selection differences are due to the loci under study (see below). It is sometimes difficult to construct a neutral environment that is similar to the selective environments. For example, both fructose and galactose in the proper proportion are required for selective neutrality between *pgi+* (phosphoglucose isomerase) and *pgi−* strains.[16]

Strain Isogenicity. Experiments measuring the initial selection require strict isogenicity. Strains received from other laboratories labeled isogenic are seldom isogenic in the sense required for the experiments. Isogenic strains are constructed using P1 transduction or selecting mutations. With P1 transduction, a single colony of the recipient is picked. Although this can be frozen and used later, we prefer to transduce all the alleles that are going to be used for a particular study into this recipient as soon as possible to prevent accumulation of genetic change. The manipulations of growth on selective plates, streaking out transduced colonies to purify them, and testing of the genotype of transductants should be done as quickly as possible, not leaving the colonies on plates any longer than necessary. The cultures are grown up in a minimal medium similar to the type to be used in the selection experiments and frozen at $-80°$ with 20% (v/v) glycerol. After being frozen, the strains can be stored for 2 or 3 years.

[15] D. Dykhuizen and D. L. Hartl, *Genetics* **96,** 801 (1980).
[16] D. E. Dykhuizen and D. L. Hartl, *Genetics* **105,** 1 (1983).

Even when the isogenic strains are prepared as carefully as possible, background effects are common, as seen in Table I. The four strains of *E. coli* in Table I were constructed the following way. A culture from a single colony of a strain that was *his⁻*, *gnd⁻* was divided and part transduced with P1 phage grown on a strain carrying the *gnd*(K12) allele and part transduced with P1 phage grown on a strain carrying the *gnd*(RM73C) allele, selecting on minimal medium for His⁺ transductants. The transductants were purified by streaking out on the same medium. From the streaks, single colonies were grown up as cultures and then assayed for 6-phosphogluconate dehydrogenase activity. From these, a Gnd⁺ cotransductant was chosen for each allele and the strains immediately frozen. The two alleles of *gnd* were checked using protein electrophoresis. A T5-resistant mutant was selected from each strain by plating the strain in soft agar on an LB plate with excess T5. The resulting colonies were purified by streaking for single colonies. For each allele, a single T5ᴿ mutant was grown up and frozen.

A strain which carries the T5ᴿ marker is mixed with a strain of the other *gnd* allele which is T5 sensitive so that the T5ᴿ strain is 5% of the population. The selection for or against the strain carrying the T5ᴿ marker is measured. Glucose and gluconate are selective environments for *gnd*, whereas ribose plus succinate is a neutral environment. If there are no problems with either the neutral marker or strain isogenicity, then the selection in the neutral environment should not be significantly different from zero and any selection in the selective environments should be of different signs but the same magnitude. It is clear from Table I that the *gnd⁺*(K12) T5ᴿ strain is selected against unexpectedly. The T5ᴿ allele either is interacting with the *gnd⁺*(K12), or there is a mutation in the strain background. When another T5ᴿ mutation was selected from the T5-sensitive strain carrying the *gnd⁺*(K12) allele, there was no selection on either

TABLE I

CONTROL EXPERIMENTS SHOWING PROBLEM WITH BACKGROUND OF ONE STRAIN

Genotype		Selection on limiting nutrients		
T5ˢ	T5ᴿ	Glucose	Gluconate	Ribose + succinate
gnd⁺(K12)	*gnd⁺*(RM73C)	−0.002 ± 0.002	−0.018ᵃ ± 0.002	+0.001 ± 0.001
gnd⁺(RM73C)	*gnd⁺*(K12)	−0.005ᵃ ± 0.001	+0.014ᵃ ± 0.002	−0.006ᵃ ± 0.001

ᵃ Selection is significantly different from zero. A + represents selection for the T5ᴿ type and a − selection for the T5ˢ.

ribose plus succinate or on glucose, showing that the observed selection was caused by a mutation in the background.

Wall Growth

The wall makes the chemostat a heterogeneous environment. Cells stick to the glass wall and, once on the wall, cannot be washed out. Fortunately, for *E. coli* the proportion of the cells that stick to the wall is very small, as can be seen in Fig. 6A. Initially, in this experiment, the *lac⁻* strain is eliminated at the dilution rate in a lactose-limited chemostat. After the ratio has dropped 150-fold, the rate of selection slows. At this point, enough *lac⁻* cells are coming off the wall to counteract the elimination of *lac⁻* cells by selection. The gradual decline implies a gradual depletion of *lac⁻* cells on the wall. If the change in the rate of selection is due to the population on the wall, then the original rate should resume if the entire liquid culture is transferred to a fresh chemostat. For most experiments, where the selection rate is lower, these effects of the wall population are not serious compared to the effects of periodic selection (see next section). Putting silicone on the walls does not prevent cells from sticking to the wall. Sometimes, as in Fig. 4, the wall population can be eliminated by

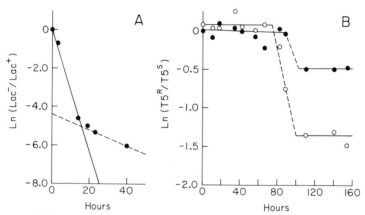

Fig. 6. (A) Effect of cells attaching to the wall on the selection against a *lac⁻* strain of *E. coli* in a lactose-limited chemostat. The solid line is the theoretical wash out rate, and the dotted line shows the effect on the rate of change of the ratio of *lac⁻* cells because of cells sticking to the chemostat wall and being slowly released. (From unpublished data of A. M. Dean. (B) Effect of periodic selection on two experiments: ●, DD1245[*pgi*⁺(RM72C), T5ᴿ] versus DD1194[*pgi*⁺(RM77C), T5ˢ]; and O, DD1245 versus DD1192[*pgi*⁺(BL00), T5ˢ]. Both experiments are conducted in fructose-limited chemostats, and the strains are selectively equivalent. At the position of the dotted lines, periodic selection causes the change in frequency. Note that scales on the *y* axis for (A) and (B) are different.

careful cleaning of new, unscratched chemostats with detergents designed to remove radioactive spills.

Periodic Selection

Periodic selection is another name for adaptive evolution.[7] Mutants occur in the chemostat that are fitter and replace the current population. These are the changes being monitored and studied in experiments studying long-term evolution (next section); however, they confound the experiments designed to measure the fitness difference between two alleles. Because advantageous mutations are rare, there is little chance that they happen equally frequently in both strains, even though occasionally the effects seem to balance out (Fig. 4). For example, Fig. 6B shows the effect of periodic selection on the frequency of two selectively neutral alleles. From about 80 to 100 hr, the population turned over, causing a change in frequency of the two types. If the entire data set were used to estimate the selection coefficient, there would be an artifactually significant selection for one of the alleles. When the two strains are mixed at equal proportions, periodic selection would be expected to favor them randomly. However, since advantageous mutations will also be present in the stored frozen cultures, repeated chemostats from the same frozen cultures could give artifactually similar results.

We find that the first periodic selection happens at about 100 hr, with 100–200 hr between subsequent ones. It can range from as early as 50 hr, if there is a common mutation which is strongly selected such as lactose constitutive mutations in lactose-limited chemostats,[17] to as late as about 120 hr. We sample for about 100 hr and then eliminate deviant final points (those that significantly increase the variance) to eliminate frequency changes caused by periodic selection. It is this limitation of the time over which selection can be measured that limits the resolution of the measurement of the selection coefficient to about 0.002/hr.

Preadaptation to avoid periodic selection is useless. If the strains are preadapted for the first mutation, a second advantageous mutation will replace the population after 100 hr or so. If the strains are preadapted for the first and second, then there is a third, and so on.

Adaptive Evolution or Periodic Selection

Studies of adaptive evolution are long-term experiments, run for hundreds or thousands of hours, whose purpose is to elucidate the adaptive mutations occurring over time. These experiments are usually started with

[17] D. Dykhuizen and M. Davies, *Ecology* **61,** 1213 (1980).

a monoculture so there is little concern about the initial physiological state of the cells. Because selection coefficients are not measured, the experiment can be stopped for a period by freezing the culture and then restarted by inoculating the frozen culture into a fresh chemostat after several weeks.

Contamination. Unlike the short-term experiments where contamination by other *E. coli* strains during the course of the experiment is obvious, *E. coli* contamination is of serious concern for these long-term experiments. Replacement of the culture by the contaminant will appear the same as an adaptive replacement. Strains should be tagged with various genetic markers so they can be distinguished from contaminants. These can be subtle changes such as electrophoretic variants and DNA sequence changes or more obvious ones such as resistance to certain antibiotics. Alternate chemostats should have strains that are marked differently so that cross-contamination can be detected.

For all chemostat experiments, checks for contamination by other species of bacteria and bacteriophage are recommended. Cells from the chemostat are plated as a lawn in soft agar. After incubation, the lawn is studied for plaques and papillae of other species. Other species can be observed more easily when most of the *E. coli* are killed by a species-specific bacteriophage.

Monitoring Adaptive Changes. There are three ways of determining the appearance of an adaptive change or periodic selection event. The first is to mix two strains differentiated by a neutral marker and follow the frequency over time. When there is a change such as in Fig. 6B, strains of both marker types are isolated. These can then be competed against the original strains (the strains that were used to start the chemostats) of the opposite marker to show that they now contain an adaptive change. A second pair of marked strains is made from one of the strains carrying the first adaptive mutation, mixed, and followed until there is a second adaptive mutation. The strains are isolated and tested against the strains which contain the first adaptive mutation. This procedure can be continued for a third and fourth adaptive mutation or as long as the adaptive increase of each mutation is sufficient that it can be measured. This method is often used to obtain a first or second mutation, but not for more.

The second method is to establish monocultures of different neutral markers and remove a sample every 50 or 100 hr. These are then tested for increase in fitness by a competition with a sample of the opposite marker type isolated at the previous time period.[4]

The third and most common method is to follow the increase in frequency of a selectively neutral mutation which can be easily screened, like T5^R. A monoculture is started. This monoculture will linearly accumulate the neutral mutations at the mutation rate. Being asexual, these evolving populations may be considered a series of clones. Because the clones with the neutral mutation are rare, the initial frequency of selectively neutral mutations in the adaptive clones will be zero. The frequencies of the neutral mutation in the newly emerging adaptive clones will be less than those in the preexisting clones, as the former has had less time to accumulate neutral mutations. Therefore, during an adaptive changeover in the population, there will be a transient decrease in the frequency of the neutral mutation. By monitoring the frequency of a selectively neutral mutation, population changeovers because of selection of adaptive mutations can be determined. This works only if there is a single population in the chemostats. If there are coexisting populations, the method breaks down.[6]

Acknowledgments

Support for this research was provided by National Institutes of Health Grant GM30201. I thank A. M. Dean for the data shown in Fig. 6A and P. Silva, A. M. Dean, D. Guttman, and I.-N. Nang for critical comments on the manuscript.

[46] Experimental Determination of Rates of Concerted Evolution

By Sue Jinks-Robertson and Thomas D. Petes

Introduction

Members of a repeated gene family within a given species tend to be very similar in DNA sequence, whereas comparable families of repeated sequences (e.g., ribosomal RNA genes) are not well conserved between closely related species. This observation indicates that organisms possess mechanisms that conserve sequence homogeneity within a family of repeated genes, a phenomenon termed concerted evolution. Two of the mechanisms that contribute to concerted evolution are related to homologous recombination and are the subject of this chapter: namely, unequal sister chromatid exchange and gene conversion.[1] Unequal sister chromatid exchange (SCE) occurs between tandemly repeated genes. As illustrated in Fig. 1, unequal crossing-over within a tandem array of 10 genes containing a single variant can lead to one array without a variant and a second array

---

[1] G. M. Edelman and J. A. Gally, *in* "The Neurosciences: Second Study Program" (F. O. Schmitt, ed.), p. 962. Rockefeller Univ. Press, New York, 1970.

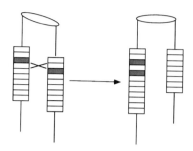

FIG. 1. Unequal sister chromatid exchange. Crossing-over between misaligned sister chromatids is shown. Each rectangle represents one gene of a tandem array of repeated genes. The shaded rectangles represent a variant repeat (or an insertion of a selectable gene) within the tandem array. As a result of the exchange, the variant is deleted from one array and duplicated in the other. In addition, the number of repeats in each array is reciprocally altered.

with two copies of the variant. Repeated cycles of unequal crossing-over will tend to produce either arrays that have lost the variant or ones in which the variant replaces the original sequence.[2]

Whereas unequal SCE events are reciprocal in nature, gene conversion events represent the nonreciprocal transfer of DNA sequences between two homologous genes. Such events were first described in fungi such as yeast, where it is possible to analyze all four products derived from a single meiosis. When a diploid yeast cell that is heterozygous at a given locus (A/a) is induced to undergo meiosis, most tetrads contain two spores with the A allele and two spores with the a allele ($2A:2a$ Mendelian segregation). Gene conversion events occur at a median frequency of about 5% and are signaled by non-Mendelian spore segregation patterns, resulting in either $3A:1a$ or $1A:3a$ tetrads.[3] A variety of genetic experiments indicates that conversion events reflect the transfer of a single DNA strand from one allele to the other, followed by repair of the resulting mismatch.

Gene conversion events involving regions of DNA sequence similarity located at equivalent positions on homologous chromosomes have been called "classical" gene conversion events. All eukaryotes thus far examined, however, contain repeated genes, thus affording the opportunity for other types of recombination events involving homologous sequences at nonidentical genomic locations. Such events are termed "ectopic" recombination events. Both classical recombination and the various types of ectopic interactions are illustrated in Fig. 2. Since the net result of each

[2] G. P. Smith, *Cold Spring Harbor Symp. Quant. Biol.* **38,** 507 (1973).
[3] S. Fogel, J. W. Welch, and E. J. Louis, *Cold Spring Harbor Symp. Quant. Biol.* **49,** 55 (1984).

gene conversion event is a loss of sequence heterogeneity, repeated cycles of ectopic gene conversion, like unequal SCE, would preserve sequence homogeneity within gene families. It should be noted that although ectopic gene conversion and unequal SCE (a type of ectopic crossing-over) appear to be quite distinct events, there is substantial evidence indicating a mechanistic relationship between gene conversion and crossing-over.[4] In meiosis, in yeast and other fungi, conversion events are associated with crossing-over of flanking markers about half of the time.[3]

Most of the information concerning recombination between repeated sequences has been derived from studies in the yeasts *Saccharomyces cerevisiae* and *Schizosaccharomyces pombe*,[5] although some classes of ectopic recombination have also been studied in mammalian cells.[6] The yeast studies indicate that the rates of ectopic recombination are surprisingly high (similar to the rates of classical recombination) and that the rates of these events are much higher in meiosis than in mitosis. Conversion events involving repeats on nonhomologous chromosomes are associated

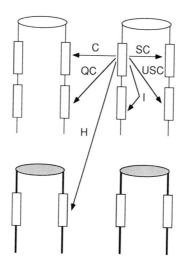

Fig. 2. Classes of recombination events. The rectangles represent repeated genes. Thin lines represent one pair of homologs, and thick lines represent another. Arrows indicate recombinational interactions. The different types of recombination are as follows: C, classical; QC, quasi-classical; H, heterochromosomal; SC, sister chromatid; USC, unequal sister chromatid; and I, intrachromatid. All types of recombination except classical are considered ectopic; sister chromatid, unequal sister chromatid, and intrachromatid recombinations are defined as intrachromosomal events.

[4] T. L. Orr-Weaver and J. W. Szostak, *Microbiol. Rev.* **49**, 33 (1985).
[5] T. D. Petes and C. W. Hill, *Annu. Rev. Genet.* **22**, 147 (1988).
[6] R. J. Bollag, A. S. Waldman, and R. M. Liskay, *Annu. Rev. Genet.* **23**, 199 (1989).

with crossing-over about half of the time, whereas ectopic intrachromosomal gene conversion events are usually not associated with crossing-over. Because the rate of ectopic recombination events is much higher than the rate of mutation, these events are likely to be important as mechanisms of concerted evolution.

In considering the role of ectopic recombination events in concerted evolution, the following questions are relevant. (1) What are the rates of ectopic gene conversion for the various types of recombination shown in Fig. 2? (2) What fraction of the ectopic gene conversion events are associated with ectopic crossing-over? (3) What are the relative rates of meiotic and mitotic ectopic recombination? (4) How is the rate of recombination affected by the size and degree of sequence similarity between the interacting repeated genes? (5) What are the rates of ectopic recombination for naturally occurring families of repeated genes? Below, we describe various methods for studying ectopic recombination between artificially constructed or naturally occurring repeats in *S. cerevisiae.*

Construction of Yeast Strains Containing Artificial Gene Duplications

It is often more convenient to analyze ectopic recombination using artificial duplications generated by yeast transformation procedures rather than to examine recombination between naturally occurring repeats. The most widely used type of construction is one in which a single-copy, selectable gene encoding an enzyme necessary for the biosynthesis of an amino acid or nucleotide is duplicated. The duplicated copies of the gene usually contain different, defined mutations; such different mutant forms of the same gene are referred to as heteroalleles. Mitotic or meiotic recombination between the heteroalleles can be easily detected by selecting for prototrophic segregants on appropriate minimal medium. Below, we discuss the construction of strains with intrachromosomal duplications and strains in which the duplication is located on a nonhomologous chromosome. It should be emphasized that there are many ways of constructing such strains, and we present only a few. For information concerning growth of yeast strains and various types of media, the reader is referred to a recent review by Sherman.[7]

Intrachromosomal Recombination Systems

Intrachromosomal recombination events are defined as those occurring within a single chromatid (intrachromatid) or between sister chromatids. A general method of constructing strains to monitor intrachromosomal re-

---

[7] F. Sherman, this series, Vol. 194 p. 3.

combination is illustrated in Fig. 3. In this example, a haploid yeast strain containing a nonreverting *his3* mutation *(his3-x)* and an additional auxotrophy *(ura3)* is transformed with an integrating plasmid containing a different, nonreverting *his3* allele *(his3-y)* plus a selectable marker (wild-type *URA3* gene). As first shown by Hinnen et al.,[8] plasmids that have no origin of DNA replication (YIp plasmids) undergo homologous recombination with the host genome when introduced into a yeast cell by transformation. Either a lithium acetate[9,10] or spheroplast[8] transformation proce-

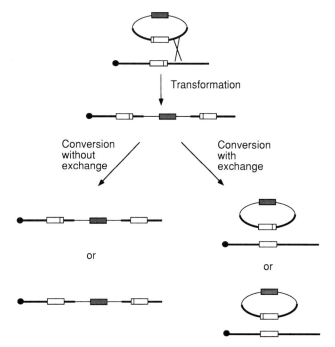

FIG. 3. Assaying intrachromosomal recombination with direct repeats. Open rectangles represent *HIS3* sequences; vertical lines indicate the positions of *his3* mutations. The *URA3* gene is represented by the shaded rectangle. Thick and thin lines correspond to genomic and plasmid sequences, respectively, and circles represent centromeres. A *ura3* haploid strain with the *his3-x* mutation is transformed with a plasmid containing *his3-y* and the selectable *URA3* gene. The resulting strain contains duplicated *his3* heteroalleles separated by the selectable marker and vector sequences. Conversion events between the duplicated sequences that are not associated with crossing-over (left-hand side) will result in a strain with one wild-type allele and one mutant allele. Conversion events associated with crossing-over (right-hand side) will result in a strain with a single wild-type allele. Note that the excised plasmid expected as one product of the crossover would not be retained in the cell, since the plasmid contains no replication origin.

[8] A. Hinnen, J. B. Hicks, and G. R. Fink, *Proc. Natl. Acad. Sci. U.S.A.* **75**, 1929 (1978).
[9] H. Ito, Y. Fukuda, K. Murata, and A. Kimura, *J. Bacteriol.* **153**, 163 (1983).
[10] R. H. Schiestl and R. D. Gietz, *Curr. Genet.* **16**, 339 (1989).

dure can be used to introduce exogenous DNA, and transformants are identified as Ura$^+$ colonies on minimal medium lacking uracil. In general, it is useful to linearize the plasmid by cutting it with a restriction enzyme that has a single recognition sequence within the yeast DNA segment containing the *his3-y* gene. This linearization protocol increases the efficiency of the integration event and targets the plasmid to integrate at the *HIS3* rather than the *URA3* locus.[11] The resulting Ura$^+$ transformant contains two mutant *his3* heteroalleles that are directly repeated and are separated by vector sequences containing the wild-type *URA3* gene. The structure of the artificial duplication can be confirmed by Southern analysis of yeast genomic DNA.

One simple way of generating nonreverting mutations for recombination assays is by "filling-in" a restriction site in a cloned gene.[12,13] The plasmid containing the gene of interest is treated with a restriction enzyme that leaves 5' protruding ends and that cuts only once in the plasmid within the coding sequence of the relevant gene. The enzyme-generated 5' overhangs are then filled in with the Klenow fragment of DNA polymerase. Following ligation, the cloned gene will contain a small insertion [usually 4 base pairs (bp)]. As alternatives to the fill-in reaction, the enzyme-generated overhangs can be enzymatically removed or oligonucleotide linkers can be inserted. If there are no convenient restriction sites, site-directed mutagenesis can be used to create a frameshift mutation. The *in vitro* constructed mutant gene can be inserted into the yeast genome using standard two-step transplacement methods.[14] For example, a *HIS3 ura3* strain could be transformed with an integrating plasmid containing the selectable *URA3* gene and an *in vitro* constructed *his3* allele. Using appropriate restriction sites, the plasmid can be targeted to integrate at the *HIS3* locus, thus creating a duplication of *HIS3* sequences (one wild-type and one mutant sequence) separated by the vector sequences. Excision of the plasmid from the chromosome can be identified by selecting directly for Ura$^-$ segregants using 5-fluoroorotate (5-FOA). Ura$^-$ cells are resistant to 5-FOA, whereas Ura$^+$ cells will not grow on minimal medium containing 5-FOA.[15] Approximately 50% of the plasmid excision events will leave the mutant *his3* allele behind on the chromosome, and these can be identified by screening the Ura$^-$ isolates for those that are also His$^-$.

[11] T. L. Orr-Weaver, J. W. Szostak, and R. J. Rothstein, *Proc. Natl. Acad. Sci. U.S.A.* **78,** 6354 (1981).
[12] H. R. Borts and J. E. Haber, *Science* **237,** 1459 (1987).
[13] L. S. Symington and T. D. Petes, *Mol. Cell. Biol.* **8,** 595 (1988).
[14] R. Rothstein, this series, Vol. 194, p. 281.
[15] J. D. Boeke, J. Trueheart, G. Natsoulis, and G. R. Fink, this series, Vol. 154, p. 164.

Analysis of Intrachromosomal Recombination between Direct Repeats

An intrachromosomal gene conversion event in which the wild-type region of *his3-x* replaces the mutation in *his3-y* (or vice versa) in the example above would lead to one wild-type *HIS3* gene and one mutant gene (Fig. 3). Such wild-type recombinants can be easily identified on minimal medium lacking histidine. Which mutant gene has been converted to wild-type can be deduced using Southern analysis to identify the mutant allele still present. If the conversion event is associated with crossing-over, then the cell would become Ura⁻ at the same time it becomes His⁺. If the conversion event is not associated with crossing-over, the His⁺ derivative will remain Ura⁺. In addition to selecting conversion events involving the *his3* heteroalleles, recombination events that delete the *URA3* gene can be selected directly on minimal medium containing 5-FOA. Although many of the events that delete the *URA3* gene presumably result from crossing-over (either intrachromatid or unequal sister chromatid), some of these events are likely to represent loss of the *URA3* gene by a type of unequal sister-strand conversion.[16-18] As it is not possible to distinguish among these mechanisms with this particular construction, we refer to events that result in loss of the *URA3* gene as "pop-outs" without specifying the mechanism. It should be noted that, since the expression of resistance to 5-FOA has a phenotypic lag, the rate of mitotic pop-outs is underestimated approximately 15-fold by this technique.[19] In summary, intrachromosomal gene conversion events in the strain depicted in Fig. 3 are detected by plating cells on medium lacking histidine, and intrachromosomal pop-outs are detected by plating the cells on medium containing 5-FOA.

The fraction of mitotic gene conversion events associated with crossing-over can be determined by screening independently derived His⁺ derivatives for the ability to grow without exogenously added uracil. His⁺ recombinants are identified by individually patching nonselectively grown colonies onto minimal medium lacking histidine. After 3–5 days, prototrophic papillae appear against the background patches of nongrowing cells. A single His⁺ papilla from each patch is purified and then further analyzed either genetically or physically. Picking only a single papilla from each patch ensures that each prototroph represents an independent recombination event.

Because meiosis is a property of diploid rather than haploid cells,

[16] M. Fasullo and R. W. Davis, *Proc. Natl. Acad. Sci. U.S.A.* **84,** 6215 (1987).
[17] R. Rothstein, C. Helms, and N. Rosenberg, *Mol. Cell. Biol.* **7,** 1198 (1987).
[18] R. H. Schiestl, S. Igarashi, and P. J. Hastings, *Genetics* **119,** 237 (1988).
[19] H. Ronne and R. Rothstein, *Proc. Natl. Acad. Sci. U.S.A.* **85,** 2696 (1988).

intrachromosomal meiotic recombination cannot be studied with the strain described above. There are several ways to approach this problem. One option is to cross the haploid with the *his3-x– URA3–his3-y* duplication to a haploid in which the resident *his3* gene contains both the *his3-x* and *his3-y* mutations. Since this double-mutant gene cannot be used as a donor of wild-type information, any wild-type *HIS3* genes produced by recombination must be the result of intrachromosomal recombination. A second approach is to cross the haploid strain with the duplicated heteroalleles to a haploid from which the *HIS3* coding sequence has been completely deleted.[20] A third approach to the analysis of intrachromosomal meiotic recombination is to construct the *his3-x– URA3–his3-y* duplication in a haploid strain that contains a *spo13* mutation and expresses both types of mating type information. Haploid *spo13* strains that express information derived from both mating types undergo haploid meiosis and produce dyads instead of tetrads.[21] Expression of both *MATa* and *MATα* information can be accomplished by making the haploid strain disomic for chromosome III, by introducing the second *MAT* locus on a plasmid, or by including a *sir* mutation to derepress the silent mating type loci.[22] As in the mitotic studies, minimal medium lacking histidine is used for determining the rate of meiotic intrachromosomal gene conversion, and minimal medium containing 5-FOA is used for determining the rate of intrachromosomal pop-outs. Because of the phenotypic lag in the expression of 5-FOA resistance, the spores should be allowed to germinate for 6 hr in nonselective medium before plating on medium containing 5-FOA.[23]

Other Intrachromosomal Recombination Assay Systems

Although the most common type of intrachromosomal recombination assay system involves direct repeats such as those described above, it is impossible to distinguish intrachromatid from sister chromatid events in such a system. Two other systems that allow one to examine specific types of intrachromosomal events are illustrated in Fig. 4. By modifying the protocol for constructing strains with the heteroalleles directly repeated, one can construct strains with the heteroalleles in inverted orientations with respect to one another.[24] Intrachromatid exchange events associated with prototroph formation invert the segment between the repeats; SCE

[20] M. T. Fasullo and R. M. Davis, *Mol. Cell. Biol.* **8**, 4370 (1988).

[21] J. E. Wagstaff, S. Klapholz, and R. E. Esposito, *Proc. Natl. Acad. Sci. U.S.A.* **79**, 2986 (1982).

[22] R. E. Esposito, M. Dresser, and M. Breitenbach, this series, Vol. 194, p. 110.

[23] M. Kupiec and T. D. Petes, *Genetics* **119**, 549 (1988).

[24] H. L. Klein, *Nature (London)* **310**, 748 (1984).

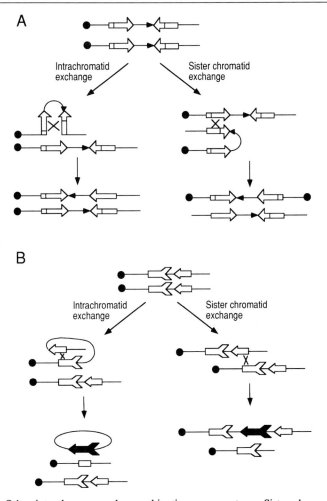

FIG. 4. Other intrachromosomal recombination assay systems. Sister chromatids in a haploid strain are shown with detached centromeres. In (A), open rectangles represent heteroalleles; the direction of transcription is indicated by the arrowhead. The orientation of the region between the duplicated sequences is also indicated by an arrowhead. As shown on the left-hand side, intrachromatid recombination inverts the segment between the inverted repeats. Unequal SCE is illustrated on the right-hand side; these events are not detected since they result in inviable products. A system that detects only SCE is illustrated in (B). Open rectangles represent duplicated sequences truncated at either the 5′ (rectangle with feathers) or 3′ end (rectangle with arrowhead). Intrachromatid exchange is shown on the left-hand side and yields a circular molecule containing a full-length gene. Because the circle contains no origin of replication, it is lost from the cell, however, and such events are not detected. In contrast, unequal SCE (or conversion) yields a chromosomal full-length gene as illustrated on the right-hand side.

events are lethal because they yield a disomic chromosome and an acentric fragment (Fig. 4A). Inversion of the region flanked by inverted repeats can be detected by Southern analysis[24] or by changes in gene expression associated with the inversion.[25] A second system has been designed to examine specifically sister chromatid interactions.[16] In this system, direct duplications of 5′ and 3′ truncated versions of a selectable marker are constructed using transformation procedures (Fig. 4B). If the nontruncated ends of the duplications are juxtaposed, an intrachromatid exchange will lead to excision of the wild-type gene, which will then be lost from the cell as it divides. In contrast, a sister chromatid gene conversion or exchange event will yield a chromosomal wild-type gene.

Construction of Strains with Duplicated Genes on Nonhomologous Chromosomes

The approaches used to construct strains for examining recombination between repeated genes located far apart on the same chromosome or located on nonhomologous chromosomes (heterochromosomal recombination) are similar to those described above. The construction of a heterochromosomal recombination assay using *his3* heteroalleles is illustrated in Fig. 5. In this example, a *ura3 his3-x* haploid strain is transformed with a plasmid-derived linear DNA fragment containing the *his3-y* allele inserted adjacent to the *URA3* gene (*URA3::his3-y*). Replacement of the resident *ura3* allele with *URA3::his3-y* as a result of a one-step transplacement reaction[14] will yield Ura+ transformants that can be selected on minimal medium lacking uracil. These transformants will have two heteroalleles on different chromosomes: *his3-x* on chromosome XV at the *HIS3* locus and *his3-y* on chromosome V adjacent to the *URA3* locus. The presence of the artificial duplication can be confirmed genetically by the production of His+ recombinants or physically by Southern analysis of genomic DNA.

Analysis of Heterochromosomal Recombination

Gene conversion events between the *his3* heteroalleles in Fig. 5 can be identified by selecting for His+ segregants on medium lacking histidine. If the repeats have the same orientation relative to the centromeres of their respective chromosomes, crossing-over will result in reciprocal translocations. As illustrated in Fig. 6A, such translocations can be conveniently diagnosed by Southern analysis, using restriction enzymes that cut asymmetrically in sequences flanking the duplication.[26-28] Alternatively, one can identify translocations by orthogonal-field alternation gel electropho-

[25] K. K. Willis and H. L. Klein, *Genetics* **117,** 633 (1987).

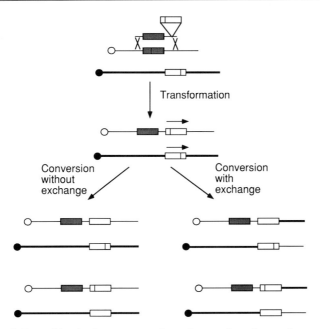

FIG. 5. Recombination between genes located on nonhomologous chromosomes. Symbols are as described in Fig. 3. Arrows indicate the relative orientations of the *his3* heteroalleles. The strain for the analysis of this type of recombination is constructed by transforming a *ura3 his3-x* haploid strain with a linear DNA fragment. The linear DNA fragment contains chromosome V DNA with a *his3-y* allele inserted into sequences flanking the *URA3* gene. Following transformation, Ura⁺ recombinants contain the *his3-x* allele on chromosome XV (thick line and solid centromere) and the *his3-y* allele on chromosome V (thin line and open centromere). As in Fig. 3, conversion without crossing-over results in a strain with one wild-type allele and one mutant allele (left-hand side). An associated exchange (right-hand side) would result in a strain with a reciprocal translocation.

resis (OFAGE) analysis[20] or by the effect of the heterozygous translocation on spore viability.[26] Finally, exchanges can be rapidly identified using polymerase chain reaction (PCR) technology.[29] In PCR analysis, one primer is homologous to chromosome XV sequences upstream of the *his3-x* allele, and the other primer is homologous to chromosome V sequences downstream of the *his3-y* allele (Fig. 6B). A PCR product will be produced only if chromosome V sequences are physically linked to chromosome XV sequences as a result of crossing-over. For the analysis of

[26] M. D. Mikus and T. D. Petes, *Genetics* **101**, 369 (1982).

[27] N. Sugawara and J. W. Szostak, *Proc. Natl. Acad. Sci. U.S.A.* **80**, 5675 (1983).

[28] S. Potier, B. Winsor, and F. LaCroute, *Mol. Cell. Biol.* **2**, 1025 (1982).

[29] T. J. White, N. Arnheim, and H. A. Erlich, *Trends Genet.* **5**, 185 (1989).

A

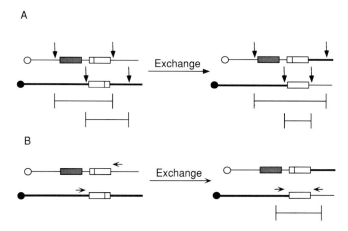

B

FIG. 6. Physical detection of reciprocal translocations. The chromosomes shown are identical to those in Fig. 5. Vertical arrows represent recognition sites for a single restriction enzyme; horizontal arrows correspond to synthetic oligonucleotide primers; and lines below the chromosomes indicate the sizes of restriction or PCR fragments. In (A), the alteration in restriction fragment size as a result of exchange is illustrated. Such alterations can be detected by Southern analysis using the duplicated sequence as a probe. In (B), the production of a PCR product from one of the exchange chromosomes is illustrated. Neither parental chromosome directs synthesis of a PCR product.

meiotic recombination between repeats on different chromosomes, the haploid strain with the *his3* heteroalleles can be mated to a haploid with one copy of the *his3-x* gene at its normal location. Alternatively, haploid meiosis systems similar to those described for intrachromosomal recombination can be used.

Other Heterochromosomal Recombination Assay Systems

In contrast to the system above which yields both gene conversion and exchange events, heteroallelic systems have been designed that specifically detect one or the other type of event. In a system designed to detect only gene conversion events, one heteroallele is full length and contains a centrally located, nonreverting mutation; the other heteroallele contains only an internal fragment of the duplicated gene.[16] Conversion events will generate a wild-type allele that can be selected on minimal medium; exchanges will generate only undetectable mutant alleles truncated at either the 5′ or 3′ end. To detect only exchange events, one constructs a strain with one heteroallele truncated at the 5′ end and the other heteroallele truncated at the 3′ end.[16,28] Only crossing-over will yield a full-length, wild-type gene that can be experimentally detected.

Analysis of Recombination between Naturally Occurring Repeated Yeast Genes

Whereas all of the methods discussed thus far concern ectopic interactions between artificially constructed repeats, there have also been a number of studies designed to examine recombination between naturally occurring repeated sequences. Recombination between both dispersed and tandemly repeated sequences has been examined, and some of the approaches used are described below.

One useful method of examining unequal recombination within a tandem array of repeated genes is to insert a selectable yeast marker into the tandem array and then follow the fate of that marker (see Fig. 1). For example, *leu2* mutant strains have been transformed with a recombinant plasmid containing the wild-type *LEU2* gene and a portion of the yeast rDNA repeat unit.[30,31] Homologous recombination between the plasmid and genomic sequences yields Leu⁺ transformants with insertions of the plasmid within the 140-copy tandem array of rDNA genes. Loss of the insertion is signaled by Leu⁻ segregants and occurs at a high frequency in both mitosis (1% per division) and meiosis (10% per division). Southern analysis can be used to distinguish unequal sister-chromatid from quasi-classical recombination events. In addition to marking a tandem array with a selectable marker, it has been possible to follow meiotic recombination at the amplified *CUP1* locus using nonselective methods.[32]

The two most abundant classes of dispersed repeats in *S. cerevisiae* are the transposable Ty element and the telomeric Y′ element (found in 30–40 and 10–30 copies per haploid genome, respectively). One method of detecting gene conversion events involving these repeats is to insert a wild-type *URA3* gene into one copy of the repeated element in a *ura3* mutant strain.[23,33–35] Gene conversion events in which the marked repeated gene is a recipient of sequence information from an unmarked gene will delete the *URA3* sequences. The resulting Ura⁻ derivatives can be selected with 5-FOA, as described previously. Although most of the Ura⁻ segregants will be the result of gene conversion events between the repeated genes, a variety of other types of genetic events (e.g., a new mutation within the *URA3* gene) can also produce the Ura⁻ phenotype. Southern analysis or genetic methods are useful in determining the types and frequencies of these other classes of events.[23,34]

[30] T. D. Petes, *Cell (Cambridge, Mass.)* **19**, 765 (1980).
[31] J. W. Szostak and R. Wu, *Nature (London)* **284**, 426 (1980).
[32] J. W. Welch, D. H. Maloney, and S. Fogel, *Mol. Gen. Genet.* **222**, 304 (1990).
[33] G. S. Roeder and G. R. Fink, *Proc. Natl. Acad. Sci. U.S.A.* **79**, 5621 (1982).
[34] M. Kupiec and T. D. Petes, *Mol. Cell. Biol.* **8**, 2942 (1988).
[35] E. J. Louis and J. E. Haber, *Genetics* **124**, 547 (1990).

The gene conversion events described above represent events that delete the selectable marker inserted into one of the repeated sequences being examined. Conversion events that duplicate rather than delete a marker can also be selected. Yeast strains containing both the suppressible mutation *ade2-101* and the antisuppressor *ASUx* are phenotypically Ade⁻ if there is one copy of the *SUP11* suppressor but are Ade⁺ if there are two copies of *SUP11*.[35] Thus, by inserting the *SUP11* gene into a repeated sequence, gene conversion events between the repeated genes that result in duplication of *SUP11* can be readily detected.

In the examples given above, a repeated element was marked with an easily assayable gene in order to detect ectopic recombination events. There have been studies, however, where such marking is not necessary. Recombination between delta elements (330-bp direct repeats at the ends of Ty elements) at *SUP4,* for example, results in a loss of suppressor activity that can be selected for in appropriate genetic backgrounds.[17] In addition, recombination between the naturally occurring homeologous genes *SAM1* and *SAM2* has been studied by constructing a mutation in each gene and then selecting for prototrophic recombinants.[36]

Measuring Ectopic Recombination Rates

Because mitotic recombination events can occur at any time during the growth of a culture, the frequency of these events will vary considerably when independent cultures are compared. Early occurring events will produce many recombinant progeny, whereas events occurring late will give rise to relatively few recombinants. The rate of ectopic recombination can be calculated by determining the frequency of these events in a number of independent cultures and then using the experimentally determined median frequency to calculate a rate.[37] One method of performing this analysis is to treat nonselectively grown, individual colonies as independent cultures of cells. Each colony to be analyzed is resuspended in a small volume of water; approximately 20 colonies should be examined. A small aliquot is removed, diluted, and plated nonselectively in order to determine the total number of cells within the colony. The remainder of the volume is plated on the appropriate selective medium to determine the total number of recombinants in the colony. If the total numbers of cells in the original 20 colonies are similar (within a factor of 2), the number of recombinants in each culture is normalized using the average number of

[36] A. M. Bailis and R. Rothstein, *Genetics* **126**, 535 (1990).
[37] D. E. Lea and C. A. Coulson, *J. Genet.* **49**, 264 (1948).

cells per colony. The median number of recombinants (r_o) is ascertained using the normalized values. The median is then used to calculate the mean (m) using the r_o/m values given by Lea and Coulson.[37] The rate of recombination is obtained by dividing the mean by the average number of cells per colony.

Because the rate of meiotic ectopic recombination is identical to the frequency of tetrads containing a recombinant spore, fluctuation analysis is not necessary. For measuring meiotic recombination rates using selective systems, one usually grows the cells in presporulation medium containing acetate instead of glucose as a carbon source; the cells are then transferred to standard sporulation medium. A fraction of the premeiotic culture should be plated on the selective medium in order to determine the frequency of prototrophic, mitotic recombinants in the culture prior to meiosis. In general, the number of mitotic recombinants is negligible relative to the frequency of meiotic recombinants. After the culture has sporulated, spores are prepared[38,39] and plated nonselectively to determine the total number of viable spores, and selectively to determine the frequency of recombinant spores. Often diploid strains are heterozygous for a recessive drug resistance marker such as canavanine. By selecting for canavanine-resistant spores, growth of unsporulated, heterozygous diploids can be eliminated. Recombination in several independent sporulated cultures should be examined to obtain an accurate rate.

In contrast to the rate of mitotic recombination, the rate of meiotic recombination is often high enough to be examined in unselected tetrads. In one study, for example, a haploid strain was constructed by transforming a *leu2* mutant strain with an integrative plasmid containing a wild-type *LEU2* gene.[40] The resulting strain with the duplicated genes was mated to a haploid strain with one wild-type *LEU2* gene. Although most tetrads derived from the diploid strain segregated 4 Leu$^+$:0 Leu$^-$ spores, 4% segregated 3 Leu$^+$:1 Leu$^-$ spores. In half of the aberrant tetrads, the Leu$^-$ spore contained a duplication in which both genes were the mutant *leu2* allele; one Leu$^+$ spore had a duplication with one mutant and one wild-type gene; and two spores had one wild-type gene. Such tetrads represent intrachromosomal gene conversion events. In summary, if one is interested in studying only meiotic recombination events, either selective or nonselective methods will work. For analyzing mitotic recombination events, with certain rare exceptions, only selective methods are appropriate.

[38] L. S. Davidow and B. Byers, *Genetics* **106,** 165 (1984).
[39] B. Rockmill, E. J. Lambie, and G. S. Roeder, this series, Vol. 194, p. 146.
[40] H. L. Klein and T. D. Petes, *Nature (London)* **289,** 144 (1981).

Concluding Remarks

From studies done primarily in *S. cerevisiae,* it is clear that ectopic recombination events occur at frequencies that are high relative to mutation frequencies. These events, therefore, are likely to be important in concerted evolution and perhaps in other types of evolutionarily important processes (e.g., genome rearrangements). The genetic control of ectopic recombination as well as the effects of repeat size and/or sequence heterogeneity on the frequency and types of ectopic recombination are active areas of investigation.

Acknowledgments

We gratefully acknowledge the help of researchers (present and past) in our laboratories. The work described from our own laboratories was supported by National Institutes of Health Grants GM38464 (S.J.-R.) and GM24110 (T.D.P.).

[47] Experimental Testing of Theories of an Early RNA World

By Andrew D. Ellington

Introduction

The discovery of catalytic RNA revolutionized how the origin and evolution of life has been envisioned by deftly resolving the molecular version of the "chicken and egg" paradox: Which came first, informational polynucleotides or functional polypeptides? Although the simple elegance of Watson–Crick base pairing has long made it easy to imagine that the first genomes were composed of nucleic acids, the complexity of the translation apparatus made it difficult to conceive of how protein-associated phenotypes could have existed in primordial organisms, and hence how specific sequences could have been selected for their ability to replicate and adapt. The phosphodiester transferase activity of group 1 introns and RNase P immediately suggested how a self-replicase might have evolved from random sequence information.[1,2] A plethora of models[3-6] have now

---

[1] T. R. Cech, *Proc. Natl. Acad. Sci. U.S.A.* **83,** 4360 (1986).
[2] P. A. Sharp, *Cell (Cambridge, Mass.)* **42,** 397 (1985).
[3] S. A. Benner, A. D. Ellington, and A. Tauer, *Proc. Natl. Acad. Sci. U.S.A.* **86,** 7054 (1989).
[4] A. M. Weiner and N. Maizels, *Proc. Natl. Acad. Sci. U.S.A.* **84,** 7383 (1987).
[5] J. E. Darnell and W. F. Doolittle, *Proc. Natl. Acad. Sci. U.S.A.* **83,** 1271 (1986).
[6] A. I. Lamond and T. J. Gibson, *Trends Genet.* **6,** 145 (1990).

proposed that the catalytic ability of primordial ribozymes may have extended well beyond simple phosphodiester bond transfers, and that ribozymes may have been involved in processes that ranged from self-replication to the biosynthesis of primary metabolites. These theories are all based not only on the appealing notion of self-replication but also, to some extent, on historical evidence. For example, the ubiquity of nucleotide moieties in many modern cofactors can be seen as evidence that these cofactors originated during a period when nucleotides were the primary building blocks of cellular catalysts, and acted as diffusable "modified bases" to augment ribozyme chemistry in reactions ranging from hydride transfers to carbon–carbon bond rearrangements.[7–9] Similarly, the role of ribosomal and transfer RNAs in the translation apparatus has been interpreted as being a vestige of a time when protein synthesis was catalyzed primarily by ribozymes.[10,11]

It is impossible to determine whether a complex metabolism based solely on ribozymes actually existed, because any putative RNA catalysts from this era have long since been supplanted by protein enzymes and, thus, the information necessary for exactly reconstructing a hypothesized "RNA world"[12] has largely been lost. Nevertheless, it has been possible to verify experimentally whether critical features of RNA world models are chemically feasible. In particular, experiments can discern (1) whether the precursors of an RNA world, ribonucleotide monomers and oligomers, can be synthesized under conditions similar to those that existed in the prebiotic environment; (2) whether ribozymes can catalyze template-directed replication; and (3) whether ribozymes can catalyze a range of reactions similar to that of protein enzymes. Until recently, experiments were largely restricted to comparing schemas for the prebiotic synthesis of oligonucleotides and their precursor monomers (point 1 above). However, advances in nucleic acid amplification technologies have made possible the isolation and engineering of new ribozyme activities (points 2 and 3). The bounds of RNA catalysis have already expanded beyond phosphodiester bond rearrangements with the discovery of ribozyme esterase[12a] and peptidyl-

[7] C. M. Visser and R. M. Kellogg, *J. Mol. Evol.* **11**, 171 (1978).

[8] H. B. White, *in* "The Pyridine Nucleotide Coenzymes" (J. Everse, B. Anderson, and K.-S. Yu, eds.). p. 1. Academic Press, New York, 1982.

[9] L. E. Orgel and J. E. Sulston, *in* "Prebiotic and Biochemical Evolution" (A. P. Kimball and J. Oro, eds.), p. 89. North-Holland, Amsterdam, 1971.

[10] L. E. Orgel, *J. Mol. Biol.* **38**, 381 (1968).

[11] F. H. C. Crick, *J. Mol. Biol.* **38**, 367 (1968).

[12] W. Gilbert, *Nature (London)* **319**, 618 (1986).

[12a] J. A. Piccirilli, T. S. McConnell, A. J. Zaug, H. F. Noller, and T. R. Cech, *Science* **256**, 1420 (1992).

transferase[12b] activities, and in the next few years seminal enzymes such as kinases, oxidoreductases, and, most importantly, replicases should be added to this list.

Prebiotic Syntheses of Oligoribonucleotides

Finding an efficient prebiotic route to RNA monomers has long been a virtually intractable problem in organic chemistry.[13-15] The logical conflict between the plausibility of an early RNA-based life form and a possible deficiency of the raw materials necessary for its formation has led several authors to propose alternative scenarios for the origin of living systems. For example, Wächstershäuser has speculated that the self-organization of chemical cycles preceded the biosynthesis of RNA,[16] whereas Cairns-Smith believes that the original templates for heritable information were inorganic minerals rather than organic compounds.[17] There is little experimental evidence to support or contradict these proposals, however, and practical objections to the prebiotic genetic transmission of heritable information have begun to be explored and are in the process of being overcome.

Although there are a number of inefficient steps in most proposed prebiotic syntheses of ribotides, the major objection to RNA as the primogenitor of life has been the relatively small yield of ribose in the formose reaction, a simple condensation of glycoaldehyde. Müller et al.,[18] however, have discovered a variation of the formose reaction that produces a limited mix of pentose diphosphates in which the ribose forms predominate (52:14:23:11, ribose:arabinose:lyxose:xylose). Although many critical chemical roadblocks remain (such as the extremely low yield of pyrimidine nucleosides following the condensation of ribose and free bases), this advance belies the previously held view that products of the formose reaction are necessarily so chemically diverse that they are "the carbohydrate analog of petroleum."[19]

Given a ready supply of ribonucleotides, it is tempting to imagine that

[12b] H. F. Noller, V. Hoffarth, and L. Zimniak, *Science* **256,** 1416 (1992).

[13] G. F. Joyce, *Nature (London)* **338,** 217 (1989).

[14] R. Shapiro, *Origins Life* **18,** 71 (1988).

[15] R. Shapiro, *Origins Life* **14,** 565 (1984).

[16] G. Wächtershäuser, *Microbiol. Rev.* **52,** 452 (1988).

[17] A. G. Cairns-Smith, "Genetic Takeover and the Mineral Origins of Life." Cambridge Univ. Press, New York, 1982.

[18] D. Müller, S. Pitsch, A. Kittaka, E. Wagner, C. E. Wintner, and A. Eschenmoser, *Helv. Chim. Acta* **73,** 1410 (1990).

[19] A. H. Weiss, R. B. LaPierre, and J. Shapira, *J Catal.* **16,** 332 (1970).

the catalysts of an RNA world were selected from pools of random RNA sequences, and that functional molecules were propagated by template-directed polymerization.[20] The first problem is, of course, the generation of a pool of templates. The *de novo* formation of short oligonucleotides (2–6 bases) from ribotide monomers using prebiotic activating agents and mineral catalysts have been reported, but the internucleotide linkages are generally 2′ to 5′, rather than 3′ to 5′.[21–23] Longer oligomers can be constructed by ligating short pieces together: self-complementary dimers have been shown to form chains as long as 30 bases in length.[24]

The difficulties encountered in the untemplated synthesis of "natural" nucleic acids is one factor that has prompted many researchers to propose alternatives to RNA as the initial genetic material. For example, a derivative of 3′-aminoguanosine can form untemplated chains of up to 20 bases in length.[25] In addition, both cyclic and acyclic nucleotide bisphosphate derivatives polymerize without the aid of a template to form nucleic acids with pyrophosphate linkages between the monomers, rather than the more "natural" phosphodiester linkages.[26–29]

Assuming the existence of an RNA template, the polymerization of activated ribonucleotides into polymers containing 3′, 5′-phosphodiester linkages proceeds readily. Using RNA homopolymers as templates, products of up to 50 bases in length have been observed,[30] and when RNA heteropolymers are used as templates the reaction is faithful.[31] These reactions can be catalyzed by a variety of metal ions,[32,33] and seem to occur most easily with ribose-linked nucleotides.[34] Short oligonucleotides can also serve as substrates for template-directed polymerization,[35,36] and they

[20] J. E. Sulston and L. E. Orgel, *Phys. Theor. Biol. C.N.R.S.,* 109 (1971).

[21] J. P. Ferris, *Cold Spring Harbor Symp. Quant. Biol.* **52,** 29 (1987).

[22] H. Sawai, K. Kuroda, and T. Hojo, *in* "Nucleic Acid Research, Symposium Series," No. 19 (H. Hayatsu, ed.), p. 5. IRL Press, Oxford, 1988.

[23] H. Sawai and L. E. Orgel, *J. Am. Chem. Soc.* **97,** 3532 (1975).

[24] W. S. Zielinski and L. E. Orgel, *Nucleic Acids Res* **15,** 1699 (1987).

[25] W. S. Zielinski and L. E. Orgel, *J. Mol. Evol.* **29,** 367 (1989).

[26] A. W. Schwartz, J. Visscher, C. G. Bakker, and J. Niessen, *Origins Life* **17,** 351 (1987).

[27] M. Tohidi and L. E. Orgel, *J. Mol. Evol.* **28,** 367 (1989).

[28] M. Tohidi and L. E. Orgel, *J. Mol. Evol.* **30,** 97 (1990).

[29] A. W. Schwartz and L. E. Orgel, *Science* **228,** 585 (1985).

[30] T. Inoue and L. E. Orgel, *J. Mol. Biol.* **162,** 201 (1982).

[31] T. Inoue, G. F. Joyce, K. Grzeskowiak, L. E. Orgel, J. M. Brown, and C. B. Reese, *J. Mol. Biol.* **178,** 669 (1984).

[32] P. K. Bridson and L. E. Orgel, *J. Mol. Biol.* **144,** 567 (1980).

[33] J. H. G. van Roode and L. E. Orgel, *J. Mol. Biol.* **144,** 579 (1980).

[34] R. Lohrmann and L. E. Orgel, *J. Mol. Biol.* **113,** 193 (1977).

[35] R. Naylor and P. T. Gilham, *Biochemistry* **5,** 2722 (1966).

[36] E. Kanaya and H. Yanagawa, *Biochemistry* **25,** 7423 (1986).

have the advantage that the ligated product can exceed the length of the templating molecule.

The prebiotic relevance of these reactions is open to question, though, since early template-directed polymerizations would have occurred in a heterologous mix of nucleotide isomers and enantiomers.[15] In these circumstances, it is unlikely that an all RNA genetic system (and, hence, an RNA world) could have arisen by the reactions so far examined. Some self-selection of enantiomerically pure polymers may have occurred,[37] but RNA is also known to template efficiently the polymerization of monomers other than ribotides, such as acyclic nucleotide derivatives.[38] Thus, the chemical composition of nascent templates could not have been efficiently reproduced. In addition, the replication of RNA templates is inhibited by the addition of stereochemically "wrong" isomers.[39] This phenomenon, which has been termed enantiomeric cross-inhibition by Joyce, Orgel, and co-workers, is another motivation for the idea that ribotides were preceded by prochiral, acyclic versions of the nucleotides.[40] Although the problem of prebiotic chain termination may have been obviated by some mechanisms, such as the preferential formation of inactive cyclic phosphates in non-ribose-based nucleotides[41] (but see also Ref. 42), nevertheless it is difficult to imagine the template-directed polymerization of monomers into long RNA chains in the early biosphere.

So far, we have constructed an unsatisfying picture of the earliest days of an RNA world: although some prebiotic mechanisms may exist for the untemplated formation of oligonucleotides, these molecules would have been short, would have contained a variety of monomers besides ribotides, and could not have been faithfully copied by the template-directed polymerization of monomers. Given this model, it is difficult to imagine the accumulation of RNA sequences necessary for the Darwinian selection of a multitude of active ribozymes. Nevertheless, these precursors may have been adequate for the first critical step in the formation of life: the formation of an RNA replicase.

[37] H. Schneider-Bernloehr, R. Lohrmann, L. E. Orgel, J. Sulston, and B. J. Weimann, *Science* **208**, 809 (1968).
[38] J. Visscher and A. W. Schwartz, *J. Mol. Evol.* **28**, 3 (1988).
[39] G. F. Joyce, G. M. Visser, C. A. A. van Boeckel, J. H. van Boom, L. E. Orgel, and J. van Westrenen, *Nature (London)* **310**, 602 (1984).
[40] G. F. Joyce, A. Schwartz, S. L. Miller, and L. E. Orgel, *Proc. Natl. Acad. Sci. U.S.A.* **84**, 4398 (1987).
[41] A. R. Hill, L. D. Nord, L. E. Orgel, and R. K. Robins, *J. Mol. Evol.* **28**, 170 (1988).
[42] M. Tohidi and L. E. Orgel, *J. Mol. Evol.* **30**, 97 (1990).

Synthesis of RNA Replicase

The transition from the prebiotic to the biotic world would have involved the formation of organic catalysts that were uniquely associated with a heritable genome. The first such catalyst was likely a self-replicase.[43] In an environment in which some form of undirected replication is already taking place, sequences capable of catalyzing their own synthesis would have enjoyed an immense selective advantage. Thus, in trying to develop a coherent model of early metabolism, it is necessary to focus on what such a replicase would have been like. If a self-replicase composed of RNA can be shown to be experimentally plausible, this will bolster the idea that competition among random RNA sequences led to the origin of living systems.

Although self-replicating systems need not be based on nucleic acid complementarity, as Rebek and co-workers have artfully shown,[44] very simple self-replicating RNAs can be easily devised. Both von Kiedrowski[45] and Zielinski and Orgel[46] have examined systems in which a palindromic oligonucleotide templates the ligation of two shorter, appropriately activated oligonucleotides (Fig. 1). Because the template is palindromic, the products of the desired reaction become templates for further cycles of ligation, and the ligated material accumulates autocatalytically. The selection of individual sequences from a complex mixture may be possible in such systems, since different templates have been found to have very different replication proficiencies.[47,48] However, sequence competition is likely to be based on very simple considerations, such as how the 3'-hydroxyl (or 3'-amine, in the case of some analogs) is aligned on a template relative to the 5'-leaving group of an adjacent molecule.

More efficient self-replicating sequences will presumably utilize more mechanistically complex modes of catalysis, for example, stabilizing the 5'-leaving group (e.g., stabilizing the negative charge that develops as a pyrophosphate is displaced from a nucleotide triphosphate), increasing the nucleophilicity of the attacking 3'-hydroxyl, or binding the bipyramidal geometry of the phosphodiester transition state intermediate. These more involved reaction mechanisms will necessarily be found in molecules that can assume conformations more elaborate than that of a simple double helix. Because the relationship between sequence, structure, and function

[43] J. W. Szostak *in* "Redesigning the Molecules of Life" (S. A. Benner, ed.), p. 87. Springer-Verlag, Berlin, 1988.

[44] T. Tjivikua, P. Ballester, and J. Rebek, *J. Am. Chem. Soc.* **112**, 1249 (1990).

[45] G. von Kiedrowski, *Angew. Chem., Int. Ed. Engl.* **98**, 932 (1986).

[46] W. S. Zielinski and L. E. Orgel, *Nature (London)* **327**, 346 (1987).

[47] T. Inoue and L. E. Orgel, *Science* **219**, 859 (1983).

[48] T. Haertl and L. E. Orgel, *J. Mol. Evol.* **23**, 108 (1986).

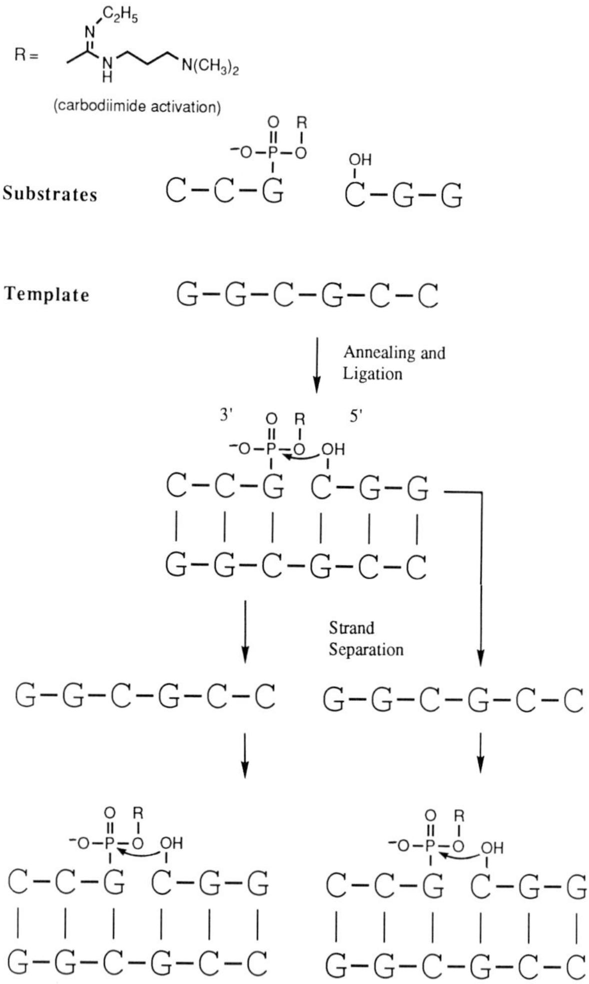

FIG. 1. Autocatalytic replication system examined by von Kiedrowski [G. von Kie-drowski, *Angew. Chem., Int. Ed. Engl.* **98**, 932 (1986)]. The trinucleotide substrates can be aligned on a hexamer template for ligation. The 3'-phosphate of one trinucleotide is activated by treatment with 1-(3-dimethylaminopropyl)-3-ethylcarbodiimide and is attacked by the 5'-hydroxyl of the adjacent trinucleotide. The 3' and 5' ends of the template, and the termini of the trinucleotides that are not adjacent following annealing, are chemically protected to avoid possible side reactions. After ligation, the individual strands of the double-stranded molecule can separate and template new ligation reactions (however, because the hexamer duplex is more stable than the free hexamer, the rate of template accumulation will be proportional to the square root of the template concentration, rather than a higher order exponent).

is nonintuitive, though, it is difficult to envision just how structurally complex the early self-replicases were. Still, it may be possible to develop working models by engineering modern ribozymes.

The Cech and Szostak groups have pioneered the creation of an RNA replicase by using the ligation activity of the *Tetrahymena* group 1 self-splicing intron as an exemplar of ribozyme-directed RNA polymerization. Cech and co-workers have shown that this ribozyme can perform the template-independent elongation of short oligonucleotides using dinucleotide substrates of the form GpN, in which the 5′-terminal guanosine is the leaving group.[49] Bartel *et al.* have further shown that Watson–Crick base pairing to a template can direct the sequential addition of nucleotides from GpN substrates.[50] These efforts demonstrate that monomers can be ribozymatically polymerized, but attempts at template-directed polymerization have so far been limited to only short stretches of template (2–3 bases). In addition, the fidelity of this reaction appears to be limited: at best the correct nucleotide is picked 25 times more readily than a mismatch, while at worst it is preferred by only 2 times.

More substantial progress toward a self-replicase has been made by focusing on the ribozyme-catalyzed ligation of short oligonucleotides. The first step in the splicing cascade, the cleavage of the 5′ intron–exon junction by guanosine, is freely reversible and, hence, can serve as a model for template-directed ligation (Fig. 2a). To allow an engineered replicase to work on templates in trans, a piece of the intron containing the cleavage/ligation junction was separated from the catalytic core (Fig. 2b).[51] Next, the stem–loop containing the cleavage–ligation junction was broken into three pieces: the two oligonucleotides that are joined during the ligation reaction and the template on which they are aligned (Fig. 2c). Finally, it was demonstrated that multiple aligned oligonucleotides could be ligated together into a cRNA product 40 bases in length (Fig. 2d).[52] A potential objection to this scheme as a model for the origin of life might be that the ligation reaction is most efficient with oligonucleotides of from 5 to 9 bases in length, whereas the majority of prebiotically synthesized oligonucleotides may have been shorter. Recently, however, the reaction conditions have been optimized to ligate successfully cRNA pieces of only 3 bases in length.[53] The separation of catalyst, template, and cRNAs should, in theory, allow an engineered replicase to move back and forth between the

[49] M. D. Been and T. R. Cech, *Science* **239**, 1412 (1988).
[50] D. P. Bartel, J. A. Doudna, N. Usman, and J. W. Szostak, *Mol. Cell. Biol.* **11**, 1544 (1991).
[51] J. W. Szostak, *Nature (London)* **322**, 83 (1986).
[52] J. A. Doudna and J. W. Szostak, *Nature (London)* **339**, 519 (1989).
[53] J. A. Doudna, unpublished results (1991).

a

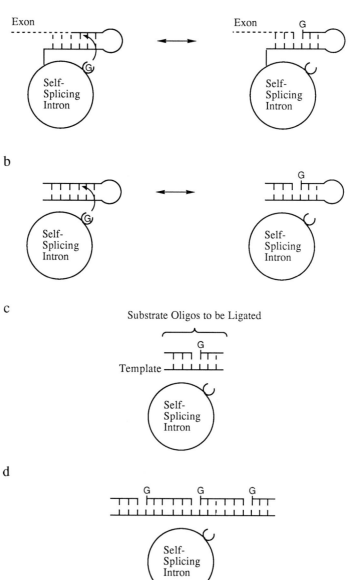

b

c

d

ribozyme phenotype embodied in a "positive" strand and the information carried by cRNA in a "negative" strand.

At this point, it should be noted that there are several possible problems with this system serving as a paradigm for the origin of life. Any replicating system, whether composed of RNA or not, would have faced difficulties if it were free in the primordial soup. The concentration of "food" may have been extremely low; vagaries of the environment, such as metal ions, may have contributed to hydrolysis; and there is no assurance that a new catalyst would function so that its descendents (as opposed to "foreign" molecules) would be preferentially replicated. All these problems can be reduced or eliminated by somehow containing the replicating system, and we must assume that this was an important step in the early history of life. The exact nature of the containment is unknown, but it could have been achieved within vesicles that were conceptually similar to modern cells, as an organic layer on a mineral surface,[16] or even by developing features which favor self-recognition (e.g., Weiner and Maizels have proposed that a stem–loop structure similar to tRNA may have "tagged" early molecules[4]).

Further, the use of oligonucleotides does not eliminate the problems of fidelity that were encountered in the polymerization of monomers. As the length of the substrate oligonucleotide increases, so too does the number of competing oligonucleotides with similar sequences. If faithful copies of the parent ribozyme are to be synthesized, some mechanism for the recognition of correctly base-paired substrate–template complexes must be incorporated into the replicase. This is mechanistically feasible: the *Tetrahymena* ribozyme recognizes the geometry of wobble base pairs at the ligation junction[54]; Herschlag and Cech have demonstrated that the effect of a mismatch on the recognition of a duplex oligonucleotide substrate exceeds the energetic contribution of the mismatch to the formation of the

---

FIG. 2. Engineering a group 1 intron to be an RNA replicase (a) The self-splicing intron binds a free guanosine and cleaves the 5' intron–exon junction. In this reaction, the 3'-hydroxyl of the guanosine displaces the 3'-hydroxyl at the exon terminus. The reaction is freely reversible and can be used as a model for RNA-catalyzed ligations. (b) The stem–loop containing the intron–exon junction is separated from the catalytic core of the ribozyme. The stem–loop can still bind to the intron via tertiary structural interactions, and the reaction is still reversible. By converting an intramolecular reaction to an intermolecular reaction the ribozyme is freed to act on exogenous templates, such as cRNA copies of itself. (c) The detached stem–loop is opened to give two "substrate" oligonucleotides and a "template" oligonucleotide. The catalytic core can now act as a template-directed RNA ligase. (d) Substrate oligonucleotides can be ligated sequentially to give a long cRNA product. The efficiency of the reaction decreases with the length of the cRNA, which may reflect a preference for substrates of the "natural" length, as in (a).

duplex (i.e., some element of the correctly paired structure is recognized by the ribozyme)[55]; and Bartel *et al.* have demonstrated that, for GpN polymerization, mismatches affect not only binding but also the rate of catalysis.[50]

Some authors have also pointed out that an RNA replicase would face inherent difficulties with processivity (owing to intramolecular structure formation) and with separation of a newly synthesized strand from its template (because of the structural stability of double-stranded RNA).[13] However, contemporary sequences, such as Qβ phage[56] and the "X" RNA replicated by T7 RNA polymerase,[57] have managed to solve these problems by using sequences that can gather into stable intramolecular secondary structures, and thus kinetically compete during synthesis with the formation of intermolecular double-stranded RNA. Alternatively, since life may have originated in an environment comparable to a hydrothermal vent,[58,59] a sort of primordial polymerase chain reaction (PCR) might have occurred along a spatial (as opposed to temporal) temperature gradient: cRNA would have denatured from RNA duplexes at high temperatures, and new oligonucleotide substrates would have annealed and polymerized at lower temperatures. The intrinsically high temperature optima of most ribozymes (the *Tetrahymena* ribozyme is active at temperatures in excess of 50°) would have been well-suited to this environment. Finally, the problem of strand separation has been approached in the engineered group 1 RNA ligase by dividing up the ribozyme into three smaller subunits.[60] Each subunit has minimal secondary structure by itself, and thus can be copied readily. Following replication and strand separation, though, the subunits can pair with each other to form the complicated tertiary structure of the active ribozyme. It remains to be seen whether the intersubunit interaction energy is large enough to drive the denaturation of cRNA and template.

There are several reasons to believe that an RNA self-replicase, similar to the RNA ligase described here, could have heralded the evolution of an RNA world from a complicated mix of ribose and nonribose nucleotides.

---

[54] J. A. Doudna, B. P. Cormack, and J. W. Szostak, *Proc. Natl. Acad. Sci. U.S.A.* **86**, 7402 (1989).

[55] D. Herschlag and T. R. Cech, *Biochemistry* **29**, 10172 (1990).

[56] C. Priano, F. R. Kramer, and D. R. Mills, *Cold Spring Harbor Symp. Quant. Biol.* **52**, 321 (1987).

[57] M. M. Konarska and P. A. Sharp, *Cell (Cambridge, Mass.)* **63**, 609 (1990).

[58] N. R. Pace, *Cell (Cambridge, Mass.)* **65**, 531 (1991).

[59] E. G. Nisbet, *Nature (London)* **322**, 206 (1986).

[60] J. A. Doudna, S. Couture, and J. W. Szostak, *Science* **251**, 1605 (1991).

First, given the wide range of prebiotic nucleic acids, ribose-based polymers may be the most eminently suited for catalysis. Eschenmoser has pointed out, for example, that nucleic acids constructed from hexose nucleotides form inflexible ribbon structures,[61] poorly suited for convoluting into the complex shapes that are required for catalysis (e.g., the backbone of the projected tertiary structure of the *Tetrahymena* self-splicing intron folds back on itself a number of times).[62] Conversely, backbones composed of acyclic nucleotides may be too flexible to adopt stable secondary structures (since a great deal of entropy would necessarily be lost on "freezing" into a given conformer). Ribose, on the other hand, has a limited flexibility because of its pseudorotation cycle, and RNA can adopt a variety of helical conformations.

Second, a self-replicase that used oligonucleotides as substrates instead of mononucleotides would partially avoid the problem of enantiomeric cross-inhibition. Nonribose monomers located in the middle of a substrate oligonucleotide would have only an indirect effect on the alignment of the 3′-hydroxyl and 5′-leaving group in a ligation reaction. In fact, some ribose moieties in the "template"[63] and "substrate"[64] strands (see Fig. 2 for description) of the *Tetrahymena* ribozyme can be substituted with deoxyribose without appreciable loss of catalytic activity. Also, since the ligation reaction is readily reversible, cRNA strands whose growth was terminated by oligonucleotides containing nonribose bases at their 3′ ends could be "edited" out and replaced with oligonucleotides that could be productively elongated (of course, if the editing reaction is not specific it could lead to the cleavage of template RNA as well).[1]

Finally, the use of oligonucleotides as substrates may provide a mechanism for preferentially replicating RNA templates. The oligomers in a prebiotic mix that contained a larger proportion or ribose residues should have annealed more readily to an RNA template than those oligonucleotides that contained fewer ribose-linked bases. Duplexes formed solely from RNA are more stable than duplexes in which one of the strands contains even only a minor chemical modification, namely, ribose to deoxyribose.[65] In addition, acyclic residues are found to decrease significantly the melting temperature of DNA duplexes in which they are included.[66]

[61] A. Eschenmoser, unpublished results (1991).

[62] F. Michel and E. Westhof, *J. Mol. Biol.* **216**, 585 (1990).

[63] S. K. Whoriskey, unpublished results (1991).

[64] A. M. Pyle and T. R. Cech, *Nature (London)* **350**, 628 (1991).

[65] P. R. Schimmel and C. R. Cantor, "Biophysical Chemistry, Part 3, The Behavior of Biological Macromolecules," p. 1145. Freeman, San Francisco, 1980.

[66] K. C. Schneider and S. A. Benner, *J. Am. Chem. Soc.* **112**, 453 (1990).

Taken together, the experiments described above suggest a more optimistic model for an early RNA world than that based solely on prebiotic synthesis. Short oligonucleotides would have condensed into longer products via a series of autocatalytic ligation reactions. These self-replicating oligomers can be viewed as extremely primitive living systems that served primarily to amass sequence complexity for further selection. Sequences that could enhance their own syntheses by mechanisms similar to those found in modern ribozymes would have enjoyed the largest selective advantage. The first self-replicating ribozymes would have been mixed polymers but may eventually have evolved to contain only RNA because nucleic acids containing ribose backbones may have catalytic properties superior to nucleic acids containing other enantiomers in their backbones, and because ribose-linked substrates may have annealed most readily to ribose-linked templates. A more advanced replicase would of course be able to exercise stereochemical selectivity for its substrates, just as the *Tetrahymena* ribozyme can today.[67,68] The "crystallization" event that drove the formation of an RNA (read: all ribose) world was not some peculiar attribute of prebiotic chemistry; rather, it was the transition from uncatalyzed polymerization to catalyzed self-replication.

Although engineering a group 1 intron to be an RNA ligase is a practical demonstration that ribozyme-directed RNA polymerization could have existed in the past, it does not prove that the earliest self-replicases were similar to self-splicing introns. Obviously, other phosphodiester transfer reactions could be invoked as potential starting points for RNA self-replicases; for example, the "hammerhead" RNA motif is much smaller than a group 1 self-splicing intron and would have arisen more readily by chance. In addition, the hammerhead ribozyme utilizes nucleoside $2',3'$-cyclic phosphates in ligation reactions, and these can be synthesized by prebiotic routes.[21] Finally, it might be expected that polymerization catalysts that do not have to bind a leaving group would require less chemical sophistication than those that do. On the other hand, two groups have recently demonstrated the existence of homologous group 1 introns in a phylogenetically diverse set of organisms, from cyanobacteria to higher plants such as tobacco.[69,70] The most parsimonious explanation for the occurrence of these introns is an insertion event that would date the group 1 motif to greater than 2.0 billion years ago.

[67] B. L. Bass and T. R. Cech, *Nature (London)* **308,** 820 (1984).
[68] B. Young and T. R. Cech, *J. Mol. Evol.* **29,** 480 (1989).
[69] M.-Q. Xu, S. D. Kathe, H. Goodrich-Blair, S. A. Nierzwicki-Bauer, and D. A. Shub, *Science* **250,** 1566 (1990).
[70] M. G. Kuhsel, R. Strickland, and J. D. Palmer, *Science* **250,** 1570 (1990).

Ancillary Ribozyme Metabolism

It has already been pointed out that a great deal of intracellular biochemistry is based on cofactors, with these cofactors, in turn, often being derived from nucleotides. However, while this indirectly implies the proficiency of ancient RNA catalysts, it does not prove that such catalysts could have existed. Although there are, for example, protein dehydrogenases and esterases, there are no modern ribozymes with similar activities. Just as engineering a ribozyme self-replicase will be an experimental demonstration that life could have arose via RNA, so the production of artificial ribozymes will be a demonstration that a metabolically complex RNA world may once have existed.

Based on chemical considerations alone, ribozymes should be able to catalyze many different types of reactions. Ribozymes can maintain defined secondary and tertiary structures, just as protein enzymes do. Ribozymes can interact with substrates specifically via hydrogen bond networks, just as protein enzymes do. Finally, ribozymes have available to them a chemistry that, while more limited than of proteins, is substantial. RNA contains proton donors and acceptors with pK_a values that cluster at 4 and 9.[71] The critical lack of a good donor/acceptor with a pK_a near 7 can be rectified by any of several simple expedients, such as modification of guanosine to 7-methylguanosine, protonation of triple base-paired cystosine,[72] or inclusion of a proton donor/acceptor in an environment with a different polarity than water (in this respect, it is interesting to note that Dahm and Uhlenbeck have found that the cleavage reaction catalyzed by the hammerhead ribozyme is dependent on some dissociable proton with a pK_a of 8.0).[73]

A broad catalytic repertoire for RNA is not only theoretically possible but experimentally demonstrable. Ribozymes that catalyze reactions other than phosphodiester bond rearrangements have recently been engineered or discovered. Piccirilli *et al.*[12a] have changed the substrate specificity of the *Tetrahymena* self-splicing ribozyme from tetrahedral phosphates to planar esters. Essentially, an ester bond was forced into the active site of a group 1 intron by tethering an amino acid (methionine) to an oligoribonucleotide sequence that would normally position a phosphodiester bond in the active site for cleavage. The ester substrate was hydrolyzed 5-fold better in the enzyme active site than in solution. Similarly, to answer the question of whether ribosomal RNA actively catalyzed protein synthesis, Noller *et al.*[12b] dissected the peptidyltransferase activity of 50 S (large) ribosomal

---

[71] W. Saenger, "Principles of Nucleic Acid Structure." Springer-Verlag, New York, 1984.

[72] R. J. H. Davies and N. Davidson, *Biopolymers* **10**, 1455 (1971).

[73] S. A. Dahm and O. Uhlenbeck, unpublished results (1991).

subunits purified from an obligate thermophile. Deproteination of *Thermus aquaticus* large subunit ribosomal RNA was vigorous and thorough: isolated ribosomes were treated with proteinase K and sodium dodecyl sulfate (SDS) at 60° and subsequently extracted with phenol. These techniques were sufficient to remove up to all but 1% of the protein normally associated with the ribosome, and the remainder appeared to be primarily in the form of small peptide fragments. The deproteinated RNA retained almost full amide bond-forming activity and was found to be inhibited by the same RNA-binding antibiotics that normally inhibit protein synthesis.

The most important criterion for a biological catalyst, however, is whether it can serve as a surface that enhances the reactivity of specific ligands; in other words, can it preferentially bind to a transition state, or provide the correct electrostatic environment for reaction, or juxtapose two substrates in a chemically productive fashion? Amplification techniques such as PCR and transcription-based amplification systems (TAS) have made it possible to select nucleic acid sequences that can present surfaces that are complementary to individual molecular shapes. In these *in vitro* genetic methods, a random or semirandom pool of nucleic acids is constructed, and those variants that can bind to a given ligand are selectively amplified.[74] For example, RNA and DNA sequences that can bind to nucleic acid-binding proteins have been selected from pools containing from 8 to 26 randomized bases.[75-77] RNA molecules that can form surfaces complementary to small organic dye molecules have been selected from pools containing 100 degenerate positions. The interactions with the dye molecules are specific, and particular residues appear to be involved in dye binding.[78]

Although no new RNA enzymes have yet been selected, so-called aptamers,[79] that can bind to specific compounds may be engineered to act as ribozymes. For example, it may be possible to engineer a "ribodiaphorase" from an NAD-binding RNA molecule (Fig. 3). FMN can be positioned near the NAD binding site by inclusion of a site-specific oligonucleotide tail,[80] and electron transfer should be facile owing to proximity effects alone.[81] Such a catalyst would demonstrate that early ribozymes

[74] R. Green, A. D. Ellington, D. P. Bartel, and J. W. Szostak, *Methods* **2**, 75 (1991).
[75] A. R. Oliphant, C. J. Brandl, and K. Struhl, *Mol. Cell. Biol.* **9**, 2944 (1989).
[76] R. Pollock and R. Treisman, *Nucleic Acids. Res.* **18**, 6197 (1990).
[77] C. Tuerk and L. Gold, *Science* **249**, 505 (1990).
[78] A. D. Ellington and J. W. Szostak, *Nature (London)* **346**, 818 (1990).
[79] From the Latin "aptus," meaning "to fit."
[80] T. E. England, R. I. Gumport, and O. C. Uhlenbeck, *Proc. Natl. Acad. Sci. U.S.A.* **74**, 4839 (1977).
[81] W. P. Jencks, "Catalysis in Chemistry and Enzymology," p. 460. McGraw-Hill, New York, 1969.

could have relied on base pairing to position substrates in their active sites. Oligonucleotide tails could have acted as convenient "handles," a role which is even now ascribed to the otherwise superfluous nucleotide moieties of coenzymes.

Selecting or engineering new ribozymes will not only demonstrate that an RNA world was possible, but it will also give us an idea of how probable it might have been. For example, the rough probability that a given RNA molecule will bind to an organic dye is 1:10.[10] This implies that even 1 μg of material (of ~100 bases in length) would contain 1000 functional sequences. In addition, although the original pool contained 100 degenerate positions, the binding sites in the selected molecules were only approximately 20 to 30 bases in length; this is within the range of lengths that can be produced by even untemplated condensations. Most importantly, *in vitro* selection experiments have demonstrated that multiple independent "solutions" to the same ligand binding "problem" may exist. For example, when RNA molecules that could bind to T4 DNA polymerase were selected, a sequence that differed significantly from the wild-type was also found.[77] Similarly, when RNA molecules that could bind to organic dyes were cloned, the aptamers revealed little or no sequence similarity.[78] If it is generally true that there are large numbers of independent sequences with similar functions in a random pool, then it would not have been necessary to search sequence space exhaustively for a particular molecule during the course of early evolution. Even relatively small numbers of random oligonucleotides may have contained phenotypes that could be selected.

Further Experiments

The RNA world hypothesis provides a convenient focus for new experiments that will not only test its validity but will fundamentally advance our knowledge of nucleotide chemistry and RNA catalysis. Several critical experiments should more fully establish (or eliminate) the idea that the forebears of modern life were replicating RNA molecules.

First, the question of where nucleotide monomers may have come from is critical. Given that the formose reaction is the most likely candidate for the synthesis of prebiotic ribose, but yields very little pure material, the role of stereoselective catalysts (clays, amino acids, or lipid aggregates) in directing the reaction should be fully explored. In this respect, Wächtershäuser[16] has advanced a scheme for nucleotide synthesis based on pyrite catalysis than can be readily tested.

Second, the "crystallization" event that led to the selection of RNA as the principal prebiotic biopolymer can be more thoroughly explored. If nucleotide analogs preceded nucleotides as monomeric building blocks,

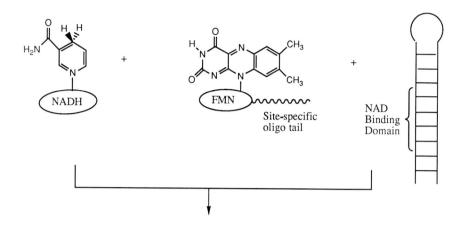

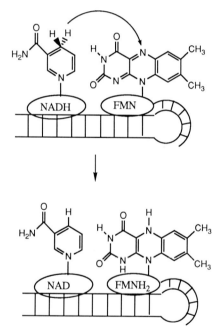

then it should be possible to selectively polymerize activated ribonucleosides on "unnatural" templates. Very little is known about the production of hybrid helices involving both RNA and "other" strands, yet such structures must be postulated as transitory intermediates on the way to contemporary genetic material. In addition, the self-selection of particular RNA oligonucleotides (Fig. 1) can be attempted from random sequence mixtures or from mixtures in which some of the oligonucleotides have been "poisoned" by inclusion of enantiomerically impure monomers (e.g., mixed ribose–arabinose backbones). Ligation may avoid the problems of enantiomeric cross-inhibition described for polymerization.

Third, the search for a self-replicating RNA catalyst can be continued either by deconstructing known catalytic activities or by recapitulating events that may have occurred in the primordial soup. The template-directed polymerization of short oligonucleotides may soon lead to the replication of small ribozyme subsegments and to the creation of replicating RNAs from the bottom up. An alternate top-down approach, though, would be to determine how a whole functional ribozyme could be divided into replicable subsegments. For example, the *Tetrahymena* self-splicing intron could be split into two (or more) portions. Successful template-directed ligation of these portions would lead to an increase in the amount of catalyst available to perform more ligations. The template could be similarly divided, and mixtures of oligonucleotides could undergo amplification cycles in which first catalyst (ribozyme) and then template molecules were reconstructed. By determining where and how scissions could be introduced into the original ribozyme and template molecules, a self-replicating system based on short oligonucleotides could be derived. Finally, a somewhat impractical experiment that mimics both the origin of life and computer simulations dubbed cellular automatons would be to mix random sequence oligonucleotides with nucleotide monomers and determine what sequences could "take over" the population by replicating themselves. Given the probable complexity (and, hence, relative scarcity in a random sequence population) of a ribozyme polymerase, and the problems

---

FIG. 3. A hypothetical ribozyme that can catalyze electron transfer. Aptamers than can bind NAD^+ (and, hence, NADH) are selected, and the binding domain is mapped. An oligonucleotide tail that can bind to an unpaired region near the NAD-binding domain is attached to FMN. The bound FMN–oligonucleotide will be adjacent to NADH when it is bound in the "active site" of the ribozyme. Electron transfer should occur owing to the proximity of the two substrates. The rate of the reaction can be controlled by varying the length of the oligonucleotide tail to vary the distance between NADH and FMN substrate. Although this catalyst is extremely simple (and employs the same principles of "catalysis" found in nonenzymatic template-directed ligation reactions), it would nevertheless demonstrate the ability of RNA to catalyze reactions other than phosphodiester bond transfers.

of self-recognition and hydrolytic side reactions, this experiment is unlikely to work. However, a more modest task could be set to a random sequence RNA pool, such as catalyzing a single ligation reaction. The sequence complexity of catalysts such as the hairpin ribozyme[82] makes it likely that molecules can be found that will enhance the ligation of oligonucleotides over background rates in solution. Such catalysts could be selected by amplification methods similar to those found in Green *et al.*[74] or Robertson and Joyce.[83]

Finally, *in vitro* selection experiments can be used to generate new ribozymes. Catalytic antibodies have been selected by immunizing animals with transition state analogs (TSAs; molecules that mimic high-energy reaction intermediates), and the same approach should work using random sequence pools of nucleic acids. Affinity columns with TSA ligands could be used to purify nucleic acids from random sequence pools. Multiple cycles of selection and amplification should result in the isolation of sequences that can bind the TSA tightly and specifically. These sequences can then be tested for catalytic activity against substrates for the reaction that the TSA mimics. The replicability of nucleic acids *in vitro* suggests other selection schemes as well: a primer for reverse transcription or PCR amplification could be derivatized with a blocking group, and only those sequences in a pool that could hydrolyze the blocking group could be replicated and amplified as well. Similarly, very short oligonucleotide primers (4–6 bases) could be derivatized with a ligand, and only those sequences in a pool that could direct primer binding via the ligand would be replicated and amplified. Although these approaches probably do not resemble events in primordial molecular evolution, they nevertheless can give us a feel for how difficult it is to find nucleic acid catalysts *de novo,* and what such nucleic acid catalysts are capable of doing.

Acknowledgments

I acknowledge Jennifer Doudna, Michael Famulok, and Jack Szostak for helpful discussions during the preparation of the manuscript, and David Bartel for insights into the importance of fidelity in template-directed polymerizations.

[82] A. Berzal-Herranz, J. Simpson, and J. M. Burke, *Genes Dev.* **6,** 129 (1992).
[83] D. L. Robertson and G. F. Joyce, *Nature (London)* **344,** 467.

Author Index

Numbers in parentheses are footnote reference numbers and indicate that an author's work is referred to although the name is not cited in the text.

A

Aarts, H. J., 575
Abbas, N. E., 519
Abel, W. O., 157, 159(27)
Abelson, P. H., 129
Able, K. P., 39
Abrams, E. S., 423
Abramson, R. D., 573
Adamkewicz, L., 300, 308
Adams, D. A., 353, 492
Adams, J., 614, 631(6)
Adams, M., 81, 82(10), 91(10), 98, 104(3)
Adams, R. P., 26–27, 33(15–16), 160
Adey, N. J., 252
Adoutte, A., 354
Aebersold, R. H., 568, 569(15)
Agee, O. F., 408
Ahearn, M. E., 165
Ahlquist, J. E., 232, 237, 238(17), 239(17), 241(17)
Alber, T., 516
Albert, J., 429
Albert, V. A., 309, 445
Alberty, H., 378
Albright, D. G., 10–11, 294, 307
Alcorn, S. M., 70
Aldrich, J. K., 169, 172(15)
Alexopoulos, C. J., 69
Alizon, M., 319
Allard, M. W., 148(19), 150, 165, 201(15), 203, 222, 309, 456, 521, 571
Allard, R. W., 32, 153, 157(10)
Allen, J. S., 504, 505(8), 577
Allner, K., 68
Ally, A. H., 573
Al Moudallal, Z., 140
Alnemri, E. S., 553(13), 554
Altschuh, D., 135, 140
Altschul, S. F., 522

Amato, G. D., 152
Amaya, Y., 573
Ambler, R. P., 142
Ambrose, S., 122
Amici, A., 27, 70
Amit, A. G., 503, 507(5)
Ammirati, J. F., 244
Amos, B., 278, 279(6), 293(6)
Amthauer, R., 178, 184(17), 185
Anastasiou, J., 73
Anderson, A., 574
Anderson, D. M., 237
Anderson, D. T., 129
Anderson, D. W., 39
Anderson, M. E., 413
Anderson, M.L.M., 549, 550(19)
Anderson, S., 178, 180, 184(18–19), 185, 203
Andre, C. P., 163, 164(66–68)
Andrews, P., 251
Angerer, R. C., 213
Antoine, N., 181, 186(16)
Anvret, M., 520
Aono, M., 321
Aota, S., 163(75), 164(75), 165
Apel, I. J., 165
Aplin, K., 236, 237(14), 241(14)
Appels, R., 204
Aquadro, C. F., 11, 103, 113, 184(12), 185, 285, 308
Araya, A., 178, 184(17), 185
Archer, M., 236, 237(14), 241(14)
Archie, J. W., 464
Arctander, P., 199(2), 203
Armour, J.A.L., 279
Arnason, U., 149, 178, 184(23–24), 185
Arnheim, N., 196, 392, 407, 429, 513, 514(30), 515, 520, 532, 560, 641
Arnold, J., 176, 370, 439, 518
Arnold, M., 299, 371

Q

Subject Index

A

Abelmoschus, see Okra
Acetate buffer formulations, 92, 111
Acid phosphatase, staining protocol for plants, 95
Acomys, see Spiny mouse
Aconitase, staining protocol for plants, 97
Aconitate hydratase, staining protocol for plants, 97
Acrylamide
  gel, preparation, 86, 101
  monomer, toxicity, 83–84
  –starch combination gel, preparation, 86–87
Actinomycetes species, definition, 347
Adiantum capillusveneris, see Fern
Aethomys chrysophilus, Y chromosome, comparative studies, 520–524
Agrobacterium tumefaciens, DNA, heterologous hybridization with intron probes, 497
Albumin
  from ancient bone, 124–127, 129
  staining protocol for vertebrates, 111
Alcohol dehydrogenase, staining protocol
  for plants, 93
  for vertebrates, 106
Algae
  cell covering, 169, 172
  cultures
    collection, 66–70
    identification, 68–69
    lyophilization, 73
    ploidy, 69
    storage, 72–73
  DNA, isolation, 168–176
  evolutionary analysis, problems in, 65–66
  field collection, 71–72
  fractionation, 169
  fresh specimens, contamination, 71–72
  genes, sequence analysis, applications, 169

herbarium specimens, 70
identification, 71–72
molecular studies, taxonomy in, 74–75
morphology, 169, 172
multicellular thick-walled, DNA isolation, 175–176
nucleic acids, analysis, applications, 168–169
phylogenetic analysis, 66
shipping, 68
taxonomy, 168–170
type cultures, 69–70
unicellular
  thick-walled, DNA isolation, 174–175
  wall-less, thinly walled, or scale embellished forms, DNA isolation, 172–173
voucher specimens, 71
Alleles, *see* Heteroalleles
Allele-specific oligonucleotide assay
  mitochondrial DNA, DNA isolation for, 182–183
  ribosomal RNA differential gene expression, 541–552
Amylase, staining protocol
  for plants, 95
  for vertebrates, 106
Antibodies
  biotinylated, preparation, 137
  monoclonal, *see* Monoclonal antibodies
Antigen–antibody complex, analysis, 503–516
Antigen inhibition assay, of antigen–antibody interaction, 510–511
Aptamers, engineering to act as ribozymes, 660–662
Aquilegia, DNA isolation, 159
Arabidopsis, ribosomal RNA
  primer extension analysis, 365
  18 S
    secondary structure, 366–368
    sequence, 363–364

ISBN 0-12-182125-0

90018